Pacific Book Review

helping authors succeed!

www.pacificbookreview.com

Book Title

Title: Dealing With The Real Climate Change
Author: Herbert A. Hutchinson
Publisher: Verity Publications
ISBN: 979-8893951011
Pages: 528
Genre: Engineering Research / Environmental Studies / Climatology / Reference / Research
Reviewer: Christina Avina

Pacific Book Review Notable Book

Pacific Book Review

One of the biggest debates and questions for mankind in this current age is what the realities surrounding Climate Change really are. The debate over whether climate change is real or not, and if it is real what the true causes of climate change are, has been waged both in scientific communities and in politics for decades now, and yet the debate still rages on, with the division amongst these communities widening with each passing year.

It is in that divide that author Herbert A. Hutchinson hopes to find real solutions using existing and growing technology to combat both the true causes of climate change and the negative environmental health hazards mankind has caused in his book *Dealing With The Real Climate Change*. The book is a sequel to the previous book in the author's bibliography, *The Facts and Fallacies of Climate Change*, and builds upon the findings of that book to use a mathematical prioritization technique to evaluate emerging and existing technologies which can be used to address the concerns the author has laid out for climate change, and to address these concerns before they do become a crisis.

What stood out immediately was the way the author presented the case for the truth behind climate change - not from a fanatical or conspiratorial way, but a calm and scientific based method, poring over data and using mathematical equations to lay out his arguments and present his findings. The amount of research and data the author included for readers was straightforward and presented with honesty and thoughtfulness and backed it up with great source material and study results, including respected institutions such as the Smithsonian, was brilliantly presented.

Readers who enjoy scientific driven non-fiction readings, especially those that deal with climate change and environmental science as a whole, will be enthralled with this book and the author's work overall. The emphasis the author puts on the impact regional climate changes can evolve and grow over time rather than global changes was thought provoking, and the detail in the presentation of mathematical equations for each point the author presented will speak to the more math and scientific minded readers out there as well, adding a more interactive and educational component to the book. The ideas and presentations for utilizing growing technology were also fascinating, especially with the debate regarding the use of AI and how AI may impact our environment around the world becoming so prevalent over the years. *Dealing With The Real Climate Change* challenges common assumptions, separates myth from reality, and offers a practical roadmap for individuals and communities to take meaningful action. A must-read for anyone who wants truth over hype, and solutions over fear. A memorable and engaging book, be sure to grab your copy of this thoughtful book today!

Why this book?

The climate is changing but is factually proven in the "big picture" to be caused by astrophysical, geophysical, and natural regional causes and little if any by Man-created "greenhouse gas" emissions. Environmental scientists' use of a one-size-fits-all "average global temperature" (land and oceans surfaces) for the planet is factually disproven in this book as a valid indicator of the climate affecting all regions on Earth.

This book has factually shown that the "average global temperature" and "one-size-fits-all" goals of the environmental science groups (especially those in the United Nations Intergovernmental Panel on Climate Change (IPCC)) to limit the change in average global temperature to not exceed two degrees Centigrade by 2035 and achieve net zero "greenhouse gas" emissions by 2050 are unrealistic. Even if fully achieved, the present "combat" of climate change and reduction of "greenhouse gas emissions" will bring little if any change in the surface temperature or climate in most regions. The goal should be to protect human safety, health, comfort and safe transportation, not to preserve the environment at any cost to the human population

A candidate model (LEO for latitude, elevation, and ocean effect) has been developed to assist the climate scientists in more accurately estimating the average annual temperature for any region at any latitude and elevation. The LEO model is far superior to the fallacious "average global temperature" model used for the past several decades.

Humans must survive seasonal extremes of hot and cold temperature in a region, not a benign annual average global temperature. For the United States region, a temperature envelope of the extreme hot and cold temperatures for any latitude within the United States boundaries has been developed. The LEO model, combined with the extreme temperature envelope and with the highly respected Köppen Geiger Climatic model for regions on all continents, are invaluable planning tools for ecologists, agronomists and population center planners challenged by the projected exponential population growth.

An assessment using the weighted importance factor method (proven in credibility in the system engineering of large military applied advanced technology programs) for 25 candidate technologies was made to assist the US population with mitigating of and adaption to the effects of the inevitable climate change on our planet. The methodology and the resulting worksheets are provided in the Appendix to this book.

This book was also written to allay the fears of gullible young people who have been misled to believe their elders and industries are carelessly destroying their future.

There is no "greenhouse gas" emissions caused global warming or climate change crisis on our planet – but there <u>is</u> an urgent need to reduce the volume and types of Man-made pollutants (toxins, carcinogens, and especially plastics) of the land, water, and atmosphere. As the population grows, so does the pollution problem.

Dealing With the Real Climate Change

Identification and assessments of the many technologies that are in various stages of development and prioritizing the candidate technologies by order of usefulness and importance in mitigating the effects of the astrophysical, geophysical, natural, and human factors that dynamically affect regional surface temperatures, climatic anomalies, and rising ocean levels on our planet home

by

Herbert A. Hutchinson

Scientist, System Safety Engineer, Genealogist, Inventor

Durham, North Carolina
2025

VERITY
PUBLICATIONS

Published in the United States of America

ISBN 979-8-89395-101-1 (SC)
ISBN 979-8-89395-102-8 (HC)
ISBN 979-8-89395-103-5 (Ebook)

Library of Congress Control Number: 2025910210

Verity Publications
145 W Broadway
Long Beach, CA 90802, USA
www.stellarliterary.com

Order Information and Rights Permission:

Quantity sales. Special discounts might be available on quantity purchases by corporations, associations, and others. For details, contact the publisher at the address above.

For Book Rights Adaptation and other Rights Permission.
Call us at toll-free 1-888-945-8513 or send us an email at admin@stellarliterary.com.

Disclaimer

The science profession has earned a well-deserved reputation for advancing the knowledge of the natural and physical structure of the solar system and of the environment in which the human inhabitants of the planet Earth conduct their lives. The title of this book and factual contents herein illustrates the scope and complexity of the problems resulting from human/natural environmental interactions. When any scientific observation is made, there is always a possibility of varying interpretation as to the importance and completeness of the variables in the observation. The terminology "fallacious" is not intended to point a finger of blame for a fallacious observation. Instead, constructive criticism is used to show that an original fallacious interpretation of a database of information that is continually being updated (e.g., with advanced technology sensors on satellites) is not an error but requires more complete and more accurate measurements to be made defining the outdated knowns and previously unknowns in that database.

Examples of how the fallacies of the Earth - being a flat circular disc floating on a vast ocean and with the Sun orbiting about the flat disk - were updated by legendary philosophers and mathematicians (and also the first "naysayers) such as Nicolas Copernicus, Galileo de Galilei, and Johannes Kepler in the 1600s are legendary in science history. Updates by more recent scientists, physicists, and mathematicians as Svante Arrhenius, Milutin Milankovic, Charles Keeling, and NASA's Edwin Hubble and James Webb in the twentieth century are equal in significance to the knowledge base of Man.

The learning process in science is a cumulative one that ideally is based on astute observations, common sense logic, accurate and complete measurement of all significant factors, unbiased forensic-like analysis, and with the purest of ethics in interpretation and presentation to peers and the public. However, scientists are also humans, and humans occasionally commit misjudgments in moments of distraction, carelessness, or being unintentionally inconsiderate of the consequences when generating a piece of scientific information that may affect the safety, health, comfort or economics of a segment of the population affected by that information. The interpretation(s) of the myriad informational database(s) in this book – scientific or non-scientific – is/are entirely those of an unbiased qualified and experienced professional scientist.

The Author earned those qualifications and gained the experience to analyze complex databases and to write this book (thanks to those listed in the Dedication section of this book). The Author is also a member (now inactive) of, but not as a representative for, the not-for-profit Mensa International Society. The Author pledges the integrity of the purpose for and the contents of this book in keeping with the Mensa Society Constitution: *"to identify and foster human intelligence for the benefit of humanity."* It should not go unnoticed that the US Constitution also starts with "*We, the People, . . .* ". As the note on the cover of this Author's previous book "The Facts and Fallacies of Climate Change on Our Planet" indicates, *"It is far more about people than about just the environment".* This book "Dealing With the Real Climate Change" updates that previous conclusion to add "*Climate change is a regional non-uniform problem not just an "average" global problem.*"

The book has also been written with the intent to avoid further unrealistic and wasteful expenditure of resources to achieve the then-optimistic – but now outdated - goals (global temperature limiting and net zero carbon emissions) set in the middle of the twentieth century. The more recent assessment is aimed at setting more-achievable goals and realistic priorities of technologies for needed public attention and cooperation in pursuing better (more efficient and with less polluting by-products) methods and technologies for the harnessing and application of natural and geophysical sources of clean and reliable energy.

Further, the Author has taken all measures to assure the information in this book contains no data, information, or less-than-professional interpretation of the environmental database that is believed to have been misleadingly retouched from its published form in reputable scientific journals and reports. Similarly, this book has not "cherry-picked" such that only a portion of the database selected that is forced to agree with a scientific hypothesis that has been incompletely analyzed or presented in a form intended to mislead the public and governmental environmental protection regulators. The seriousness and urgency for resolution of the safety and health "crisis" due to climate change, is much exaggerated by a certain segment of the environmental science profession focused only on protecting the environment. The urgency for corrective actions is clarified herein and presented as a motivation rather than as a threat of an apocalyptic nature to the human population if the unrealistic goals are not met.

The majority of photographs, diagrams, and database information related to climate change, global warming, and ocean levels rising were obtained by the Author from the databases of information in the public domain from the several government organizations identified in the Acknowledgment section of this book. Inasmuch as those organizations were and are being funded by Government grants (whose source is taxpayers' tax monies) and the Author being a taxpayer, the Author is using this book to disseminate factually updated information items – as a public service – that government organizations have already disseminated to the public. Just as that was not done by any government-funded organization to make a profit, the Author also is disseminating that information already in the public domain not for profit but also to add a different interpretation or perspective that results in a more factually based alternate interpretation. Where known, the Author assures that credit for the photograph, diagram, tabulated database, etc. has been given below the item to its creator. Any small infraction on a copyright restriction that may be contained anywhere in this book is purely unintended and for which the Author pre-apologizes to the creator who "used the sweat of his or her brow" to create that item of information (photograph, diagram, analysis document, or statement).

The Author has striven to avoid and dispel any criticism that the Author (long-retired) has published this book in collusion with any person or organization labeled as "nay-sayers" of the environmentalists' claim that Man burning fossil fuels is creating crisis-level emissions having carbon-based content that ominously traps heat in the atmosphere near Earth's surface. This book expressly avoids placing unwarranted blame on humans for the observed and measured temperatures on planet Earth becoming elevated and rising since the start of the "Industrial

Revolution" of the late 19th century. The Author commends the science and engineering professions that have brought the developed countries of the World into a safer, healthier, more comfortable, and more prosperous state of existence for the human population. More needs to be done by the more developed countries to humanely assist those in lesser-developed countries who cannot do that in their present natural environment and status of economy.

Just as done by the Author with his previous book "The Facts and Fallacies of Climate Change on Our Planet", the Author chooses not to accept any form of personal financial gain from publication of this book which was written as a public service. Any and all royalties or profits gained from the sale of this book "Dealing With the Real Climate Change" to better inform and enlist the cooperation of the public (domestic and elsewhere) will be donated to the Salvation Army International and ear-marked to specifically go to their Emergency Services sector. That organization has provided emergency aid to the surviving victims of major natural catastrophes around the world and seek no front-page news coverage for having done that. In the same vein that this book has followed, it is "for the benefit of humanity."

Dedication

This book is quinarily dedicated to:

- This Author's father Karl – who taught his son two basic rules of life: 1) things do not just happen, they are caused to happen; and 2) some things can be prevented from happening, and some things cannot.
- This Author's mother Mildred – who taught her son that the more important things in life are not things but are the people in every person's life who serve as role models, teachers, advisors – and friends.
- The Brooklyn Technical High School – where this Author was taught early in life how to think about scientific problems that need to be solved with a logical thought process and the use of the basic tools of scientific knowledge (especially physics and mathematics).
- The Georgia Institute of Technology – where this Author was taught the basics of science, technology, engineering, and mathematics (STEM) and the systems engineering processes for applying that knowledge to the development of new technologies and systems for the benefit of Mankind.
- USAF Lieutenant General James T. Stewart (Commander, Aeronautical Systems Division) – who enabled this Author (as a Chief System Engineer in Federal Civil Service) to apply everything the above four dedicatees taught this Author in the integration of a set of advanced technologies into a complex system that became, and is still in use (since 1975) in the military aircraft inventory of 28 nations for the defense of the free-World. The most meaningful reward given to this Author is described in an extract of a letter of commendation (October 28, 1975) from Lt. Gen. Stewart:

> Mr. Herbert A. Hutchinson
> Deputy for Engineering
> Aeronautical Systems Division
> Wright-Patterson AFB OH 45433
>
> Dear Mr. Hutchinson
>
> It is a pleasure to award you the Air Force Systems Command Certificate of Merit. Your performance as Chief Systems Engineer, Deputy for the Air Combat Fighter/F-16 System Program Office from July 1974 to July 1975 is deserving of this recognition.
>
> As a result of your dedicated efforts, the Air Combat Fighter made a successful transition from the prototype program, through source selection, into a rigorous full scale development program. The F-16 Air Combat Fighter has introduced many unique technological innovations into weapon systems development.
>
> I commend you for this outstanding performance and wish you continued success.

(This Author subsequently was merit-promoted to Federal Civil Service Grade GS-15 and selected to be the Chief of the USAF Flight Technology Division, 1979 to 1982.)

Acknowledgement

There is a large and constantly growing database of information for the topics of global warming, climate change, and ocean levels rising describing their causes and effects on the land, water, and atmospheric resources and the human population of planet Earth. The total database is compiled from three major scientific observation and measurement activities and a major mathematical analysis to understand and define the factors that are the "causative" variables and those that are the "effect" variables affected by the multiple, complex factors associated with global warming, climate change, and the ocean levels rising. There are many organizations within the government and independent scientific, technology, engineering, and mathematical (STEM) research academies that deserve acknowledgment for the enormous applications of human analytical efforts, administrative labor, and funds to observe, photograph, measure, document, compile, and interpret the large complex databases. Listing the myriad of participants would be analogous to this book appearing to be a telephone directory.

The four primary types of databases and the dominant contributing and originating organizations and their staffs given special acknowledgment for this book are:

1) The Astrophysical and Geophysical factors database (1900-Present)
Special acknowledgement is given to:
The Milutin Milankovic Calculation Team, the original Russian and subsequent International (Russian, United States, British, Scandinavian, and French) Ice Core Teams at the Vostok Research Station (Antarctica), the US Geological Survey, the Smithsonian Institute for Science, the National Geographic Society, the National Science Foundation, the Encyclopedia Brittanica, and the International Union of Geological Sciences.

2) The Natural factors database (1880-Present)
Special acknowledgement is given to:
The National Oceanic and Atmospheric Administration (including the National Weather Service, the National Hurricane Center, the National Tornado Center, the National Snow and Ice Data Center, and the Keeling Atmospheric Observatory at Mauna Loa), and the US Geological Survey Agency (Earthquake Hazards Reduction Program; Volcano Hazards Program).

3) The Weather and Climatic factors database (1900-Present)
Special acknowledgement is given to:
The National Weather Service (of NOAA), and the World Meteorological Organization (WMO).

4) The Earth Research factors database from land-based meteorological stations and space-based satellites and International Space Stations (1960-Present)
Special acknowledgement is given to:

The National Oceanic and Atmospheric Administration, the National Aeronautics and Space Administration, the National Academy of Science, the National Science Foundation, the Smithsonian Society, and the National Geographic Society

Special acknowledgment is also given to the United Nations Population Division and Intergovernmental Panel on Climate Change (IPCC), the US Department of Energy (DOE), the US Department of Commerce (DOC), and to the many other scientific organizations and scientific research and analysis groups world-wide (governmental and institutional) supporting the activities to protect the human population and the environment.

Background

An artist's illustration of the solar system provides a visual appreciation for the enormity and complexity (i.e., the "big picture") of the solar system, and shows the number of days or years the eight major planets – four inner "terrestrials" and four outer "galactics" - expend in their unique orbits of the Sun. (Each of the Earth's complete orbits about the Sun expends 365 days (24 hours per day) and is defined as a "year" for describing the periods of the larger and longer orbits of the outer four planets.)

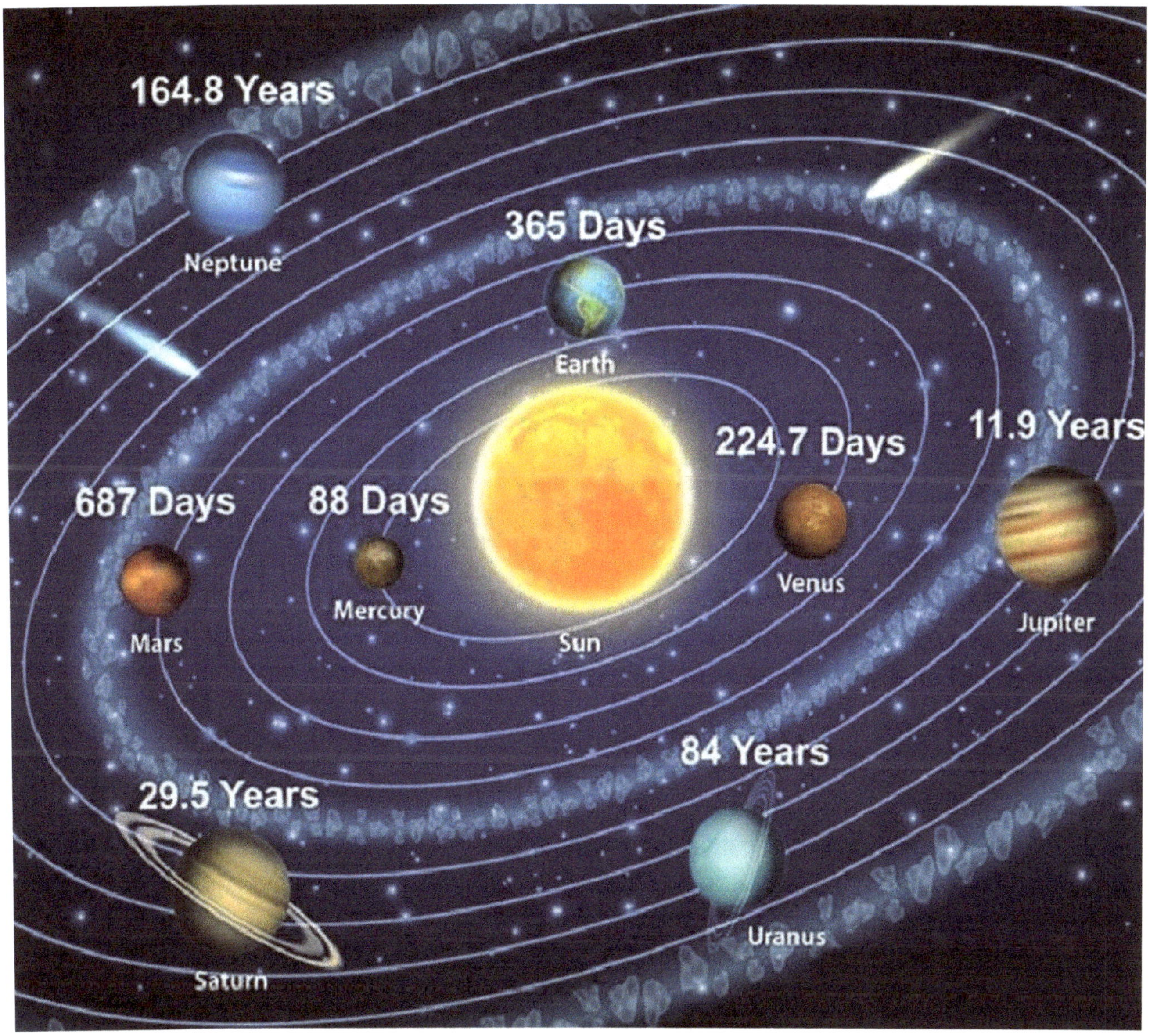

(Credit to unknown graphic artist)

What Readers must also be aware of is that the planet Earth (the "blue planet" near the top center of the diagram above) is not only travelling at a translational speed averaging about 114,000 miles per hour along its orbital path but the Earth is rotating about its unique axis of rotation (one rotation per 24 hour day) which exposes the planet's surface to almost half of the day in darkness and half of the day exposed to the Sun's incoming solar radiation. The axis of rotation is tilted presently at about 23.4 degrees but cycles between about 22.1- and 24.5- degrees tilt with a cyclical period of about 40,000 years. To compound the complexity of Earth's motion, the tilted axis is also wobbling or "precessing" in scientific terminology. Earth's orbit itself is not a regular ellipse but has a characteristic change in elliptical shape known as "eccentricity". It will be explained better later in this book how all these tilt angle changes and Earth's elliptically shaped orbits cause the temperature on Earth's surface to change several degrees. The variations in surface temperature due to "seasonal effects" are affected directly by whether humans and/or industry are burning fossil fuels – but there are small interactions between all the dynamic variables some of which may be affected by atmospheric gases. The largest "gas" having a significant effect on Earth's surface temperature is water vapor – which is largely ignored by scientifically unethical environmental scientists since it detracts from their fallacious hypothesis that the burning of fossil fuels is the dominant cause for global warming and climate change. (Author's Note: There were many moments during the research and analysis for the causes of global warming and climate change where the logic was being "warped" to suit the need and seemed as futile as "a dog chasing its tail").

In the book "Facts and Fallacies of Climate Change on Our Planet", published earlier in 2025 by the Author, the vitally important "big picture" perspective of the factual scientific causes of global warming and climate change was provided. (Author's Note: There was a practical consideration for having two books on the same topic. As Leo Tolstoy found in the creation of his 1869 classic novel "War and Peace", the first publication of "War and Peace" – all 1,652 pages – created a book that was as heavy as "Webster's Encyclopedic Unabridged Dictionary of the English Language". No human can hold a book of that size and weight in one hand and eat an apple with the other hand. To avoid the manual labor inherent in the reading of War and Peace, Tolstoy allowed his "War and Peace" magnum opus to subsequently be published in two volumes – the War portion of the narrative in Book 1 and the Peace portion of the narrative in Book 2.

That same logic was used in the printing of Book 1 and Book 2 for the problem of describing the facts and fallacies in global warming and climate change in Book 1, and Book 2 to provide insight into what needs to be done by humans to mitigate the effects of global warming and climate change.

Book 2 is entitled "Dealing with the Real Climate Change" by the same author and relates to Book 1 as a second part of the question: "What is the cause of the global warming problem and climate change, and what can we humans do to live with the problem?" The two books purposefully overlap – Book 1 concludes with the information pertinent to the first half of the problem and Book 2 starts about ¾ of the way from the end of Book 1 to permit a smooth transition – not a jump - into showing the "stuff and things" that the human population needs to do to "deal with the real climate change", as the book's title relates.

This book begins with a numerical indication of the enormity – the same "big picture" - of the solar system to give a relative size and dimensionality of the inner planets relative to planet Earth. A numerical summary of that "big picture" appears in the following table:

Sizes and Distances of the Inner Four Terrestrial Planets

			Distances from Sun		
	Diameter	Surface Area	Maximum (Aphelion)	Average	Minimum (Perihelion)
Planet	Miles Kilometers	Square Miles Square Kilometers	Million Miles Million Km	Million Miles Million Km	Million Miles Million Km
Mercury	3,032 Mi 4,880 Km	Surface Area n/a No water	43 x 10^9 Mi 70 x 10^9 Km	35 x 10^9 Mi 57 x 10^9 Km	29 x 10^9 Mi 46 x 10^9 Km
Venus	7,521 Mi 12,104 Km	Surface Area n/a No water	68 x 10^9 Mi 109 x 10^9 Km	67 x 10^9 Mi 108 x 10^9 Km	66 x 10^9 Mi 107 x 10^9 Km
Earth	7,926 Mi 12,756 Km	197 x 10^9 Sq Mi 510 x 10^9 Sq Km 71% water; 29% land	94 x 10^9 Mi 152 x 10^9 Km	93 x 10^9 Mi 150 x 10^9 Km	91 x 10^9 Mi 147 x 10^9 Km
Mars	4,221 Mi 6,792 Km	Surface Area n/a No water	155 x 10^9 Mi 249 x 10^9 Km	142 x 10^9 Mi 228 x 10^9 Km	127 x 10^9 Mi 205 x 10^9 Km

(Source: NASA database with tabulation in format chosen by Author)

The author – independently, without any affiliation to any interested organization in the commercial or governmental factions in the public domain, and without any bias - confirmed (from in-depth forensic analyses of several databases of scientific information) that the surface temperatures of the land and oceans and the climate on planet Earth factually are changing. However, the surface temperature of the planet and its associated climate were rapidly discovered to be changing regionally but not globally, and nowhere is the surface temperature and climate changing in an "average global" manner.

A problem encountered throughout the analysis of the complex set of factors influencing climate science for the earlier "Facts and Fallacies..." book was in the understanding of how climate is defined – found to be different in the environmental science databases than how the public perceives climate in the real-world environment. In fact, the root of the problem encountered by this Author (a professional multivariable data analyst) was mostly in the absence of information that environmental scientists were expected to have considered. For identification with a short identifier, that book is referred to hereafter as "Book 1". (The author has published other books on the system engineering aspects of other technical topics – aeronautical engineering - but they do not add any highly pertinent information about global warming or climate change for incorporation in Book 1. The chief value of the Author's other books is in showing how the system engineering processes can be applied to the analyses of most scientific problems.

The books of genealogical history written by the Author provide him with the background understanding of the events of nature that affect the daily and long-term lives of humans on the planet. Historical accounts of natural events – recorded in the island of Barbados Archives which centralized the hundreds of parish church registers dating from the early 1600s to the present – describe the numerous hurricanes that devastated the Author's ancestry on the Caribbean Island of Barbados, at latitude 13° north in the West Indies in the seventeenth and in all centuries since. Those many unexaggerated descriptions indicate the hurricanes are no greater or no less as devastating today as they were in the past four centuries. The emissions from the Industrial Revolution cannot be identified as having any effect on the frequency or severity of the Atlantic tropical storms and hurricanes.

The NASA database indicates the other three planets of the inner four Terrestrials (Mercury, Venus, and Mars) – have very much different surface temperature levels and rates of change than that of planet Earth. Using the logic that if there are little or no "greenhouse gases" on those other planets, as there is on planet Earth, then a good starting place of a temperature analysis would be to determine if the dominant cause for Earth's temperature to be warming is caused by some factor that is changing significantly or if the "greenhouse gas effect" is factually the singularly dominant cause for the temperature increase as environmental scientists' hypothesize. The following diagram was generated (using the NASA database on solar insolation and Johannes Kepler's laws of astrophysics) that clearly confirms

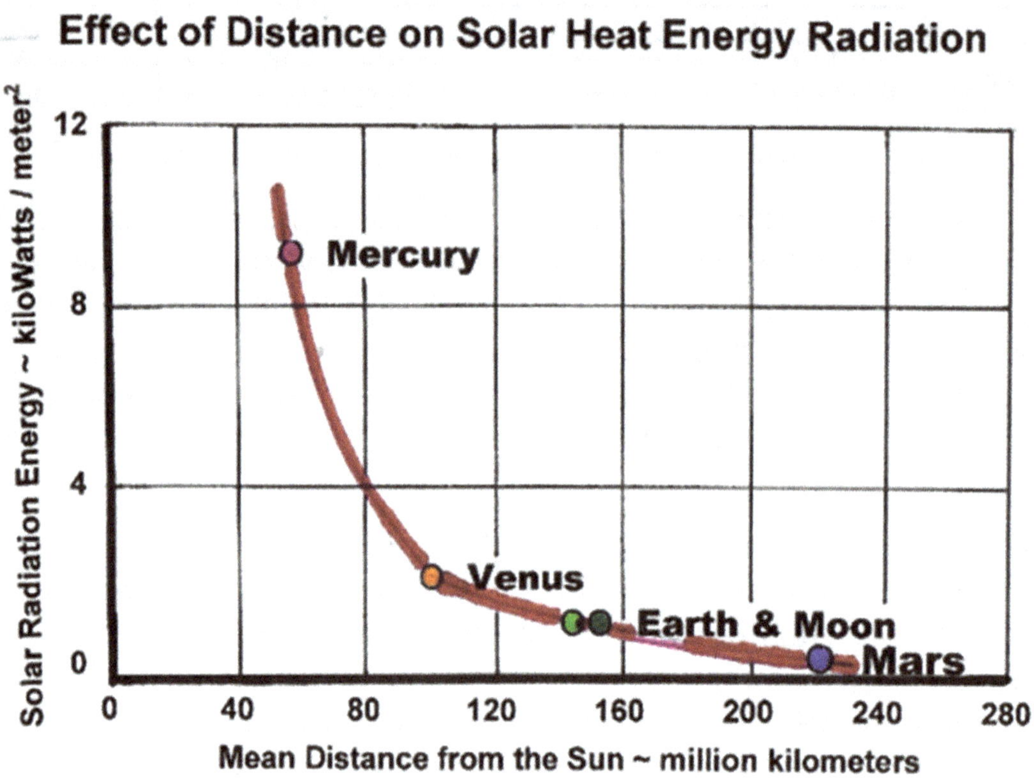

(Source: Physics Textbook (Kepler's Law) with Author Graphics)

A caution note is given that distance alone was found to be a strong factor but not the only factor determining the temperature of the planets - that there is an atmosphere surrounding the Earth through which the Sun's solar radiation must pass before impinging on Earth's surface. It must be noted also that the terminology "mean distance from the Sun" is used for the distance variable. Each of the four planets has a unique orbit about the Sun and the distance from the Sun has a different corresponding solar radiation depending on the maximum and the minimum distance the planet is situated in its unique orbit.

An analysis was done on the statistics NASA has measured (tabulated as seen below) for the interactive effect of the atmospheres for the four inner planets – Mercury, Venus, the Earth and its Moon, and Mars – and their surface temperatures.

Distance,Temperature, & Atmosphere of Inner Planets

Planet	Mean Dist. from Sun		Mean Temperature			Percent CO_2 in and Thickness of Atmosphere
	Miles x 10^6	Km x 10^6	°C (H / L)	°F (H / L)	K (H / L)	
Mercury	36	58	167 (430 -180)	333 (800 -290)	440 (703 93)	0% None
Venus	67.2	108	464	867	737	96% Thin
Earth	93	150	15	59	288	0.04% Thick (with 3% Water Vapor)
Moon	93	150	-5 (129 -130)	21 (250 -208)	270 (400 140)	0% None
Mars	141.6	228	-65 (-1 -125)	-85 (30 -193)	208 (272 148)	95% Thin

Note: There was no reliable data found for the average global hot (H) and average global cold (L) temperature for Venus and Earth.

(Source: NASA, with tabulation by Author)

The following chart reveals a "big picture" evaluation of the four inner planets that was done by the Author – using NASA's measurement database and NASA's measurement of each planet's mean distance from the Sun. To that analysis was added the Kepler Laws of the temperature of the planets being inversely proportional to distance from the Sun:

Effect of Atmosphere on Inner Planets' Temperatures

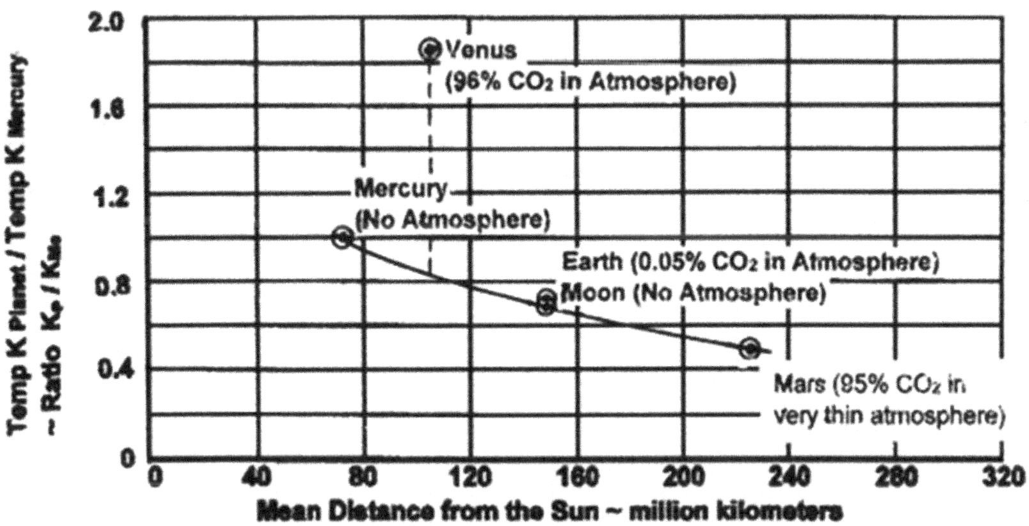

(Source: NASA database with Author's analysis)

The above data can be interpreted in a strict scientific manner to indicate that when the carbon dioxide concentration is very high in the atmosphere of a planet – such as Venus with a 96% concentration of CO_2 in its thick atmosphere – its relative temperature (in a ratio of the Kelvin units for temperature of the planet relative to that of Mercury) is very much higher than the temperature estimated by the Kepler distance law of astrophysical thermodynamics. The other three inner planets having either 1) a thin atmosphere where the heat-trapping capability of CO_2 (at any concentration) does not significantly affect the planet's surface temperature; or 2) the planet such as Earth which has a thick atmosphere but with the concentration of carbon dioxide in Earth's thick atmosphere being less than 0.05% is seen to have almost no effect on Earth's surface temperature.

The above graph was not altered in any way to place planet Earth closer to the Kepler distance line. Earth's Moon had/has no atmosphere and correlated with the Kepler line at the same distance as Earth (with a thick atmosphere but low concentration of CO_2). Of the less than 0.05% concentration of CO_2 in Earth's atmosphere (a recent measurement at the NOAA Mauna Loa Atmospheric research Laboratory places the total to be 0.0435%), less than 0. 0125% can be attributed to human and industrial gases, the remainder of the 0.05% concentration being the result of several other factors of geophysical and natural origin. The small deviation of Earth from the Kepler line is undoubtedly due mostly to the 2-3% of the heat-trapping water vapor presence. That is the more likely explanation for the deviation of planet Earth being higher from the Kepler line than most of the planets having a thick atmosphere but having a low concentration of CO_2. Earth is the only planet with water vapor (1-3%) in its atmosphere and causing water vapor to be over 70 times more effective as a heat-trapper than 0.0125% of CO_2.

The atmosphere of Earth also has random and prevailing wind currents – with differing amounts of vorticity and different levels of velocity. A NASA measurement of the winds in the atmosphere shows – in the following illustration - that any emissions of gases from the lower atmosphere into the upper atmosphere (where CO_2 is said by some environmental scientists to perform the heat-trapping greenhouse effect like a convex lens).

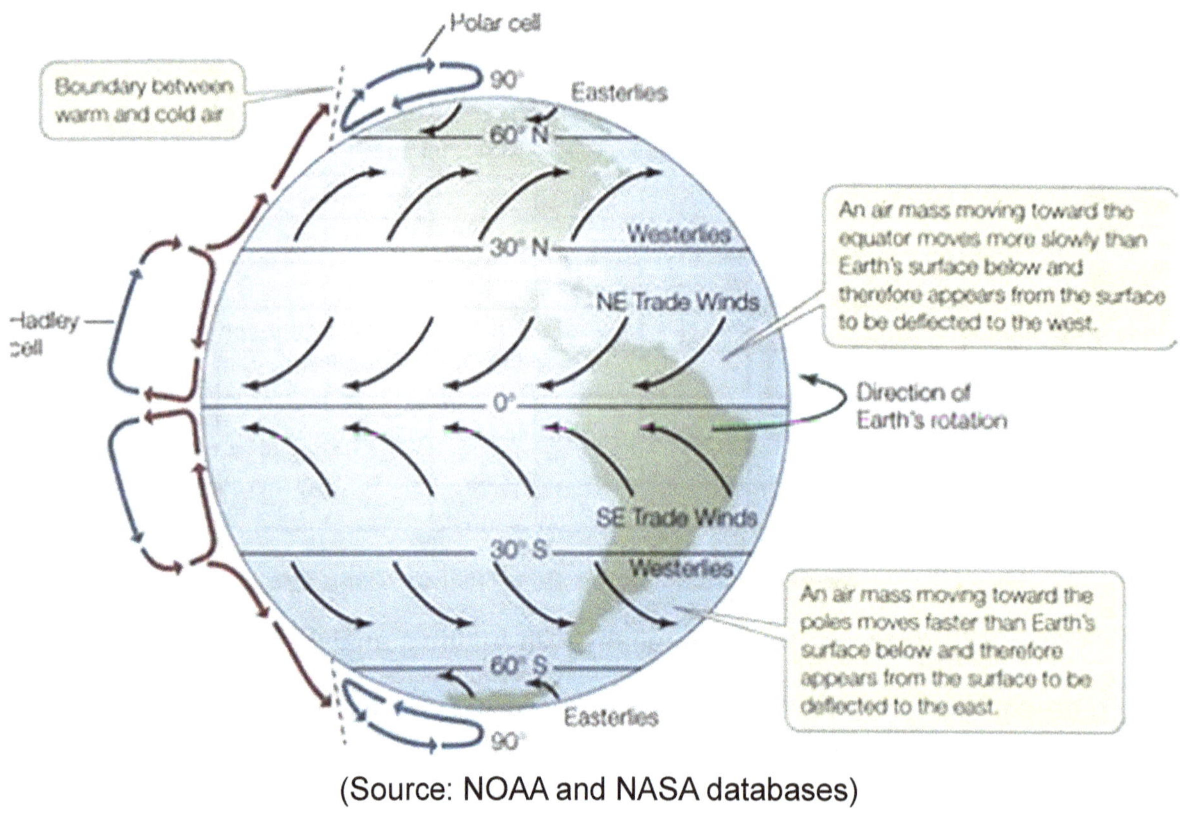

(Source: NOAA and NASA databases)

Creating Book 1 was not channeled toward either 1) confirming an existing set of hypothetical causes for the climate to be changing or 2) only identifying other causes refuting the existing environmental scientists' hypothesis. Because of taking a "big picture" holistic approach, a third possibility and high probability, later proven factual, was that it might be a combination of both. The research into the causes of why the surface temperatures and the climate are changing followed the approach of "turning over a stone to see what is under it and finding more stones to be turned over". In the parlance of science, that is named as being a heuristic branch of a holistic approach. Finding the causes for a problem is the first step toward resolving the problem, and that creates follow-on activity to determine what must be done to physically and satisfactorily resolve the problem. The steps to defining the latter gave rise to the need for this second book, referred to herein as "Book 2".

The population of European countries were very grateful to the use of coal technology for their comfort of warmth but were appalled when their white limestone buildings (cathedrals, government buildings, churches, libraries, etc.) were being stained with the black soot from the coal burning emissions in their atmosphere. The source of that soot was not hidden from public sight very well:

(Source: Encyclopedia Britanica)

(A spoof on the need for controlling the buildup of soot inside the multiple chimneys by chimney "sweeps" of olde London town is seen by the humorous antics of the character Bert – a well-stained chimney sweep in the musical "Mary Poppins".)

An early observation of the carbon dioxide emissions of the coal burning process being a scientific environmental problem is seen in a news article reporting the findings of Dr. Svante Arrhenius in the 1890s in Stockholm, Sweden. Dr. Arrhenius was an atmospheric chemist who experimented with carbon dioxide gas having a heat-trapping property. His experiments were at the scale of a liter-size flask of near 100% concentration of carbon dioxide absorbing the heat energy radiated from a heat lamp. His experimental findings of what he termed as a “hothouse effect” in the atmosphere earned him the Nobel Award in Chemistry in 1903. Those findings were in the near-term and the far-term effects on the environment and were applied to the cooler regions of Scandinavian countries. In the near term, Dr. Arrhenius hypothesized that the warming effect of carbon dioxide in the atmosphere would have a beneficial effect of warming land areas in regions of northern Scandinavia to increase farmlands growing food crops for the exponentially growing population of Europe. In the far term, Dr. Arrhenius cautioned that a high concentration of carbon dioxide in the atmosphere could lead to extremely hot atmospheres endangering the ecology and all life forms on the planet.

During the last decade of the nineteenth century, Arrhenius conducted his research on the effects of carbon dioxide on climate, and in 1903 published a two-volume technical book titled *Lehrbuch der kosmichen Physik* (Textbook on Cosmic Physics). This book was not widely read as it was a textbook for a scientific discipline that did not yet exist. A few years later, Arrhenius published Worlds in the Making, a non-technical book that reached a greater audience. In the latter book, Arrhenius described the “hot-house theory” of the atmosphere stating that the Earth’s temperature was about 30 degrees warmer than it should be due to the “heat-protection action of gases contained in the atmosphere”, a theory based on ideas developed previously by other scientists and mathematicians (Fourier, Pouillet, and Tyndall). Arrhenius’ calculations demonstrated that if the atmosphere had no carbon dioxide, the surface temperature of the Earth would fall about 21 degrees Celsius. At this cooler temperature, the atmosphere would contain less water vapor, resulting in an additional temperature decrease of approximately 10 o Celsius. It is important to note that Arrhenius was not very concerned with rising carbon dioxide levels at the time but rather was attempting to find an explanation for high latitude (above 55 degrees north) temperature changes that could be attributed to the onset of the ice ages and interglacial periods.

By 1904, Arrhenius became concerned with rapid increases in anthropogenic carbon emissions and recognized that “the slight percentage of carbonic acid in the atmosphere may, by the advances of industry, be changed to a noticeable degree in the course of a few centuries.” Arrhenius eventually suggested an increase in atmospheric carbon dioxide due to the burning of fossil fuels could be beneficial, making the Earth’s climates “more equable,” stimulating plant growth, and providing more food for a larger population.

The news media “cherry-picked” the scientific database generated by Dr. Arrhenius and published a small clip in the newspaper in Selma, Alabama, in October 1902 alerting the public to the “crisis” that Dr. Arrhenius had concluded was inevitable.

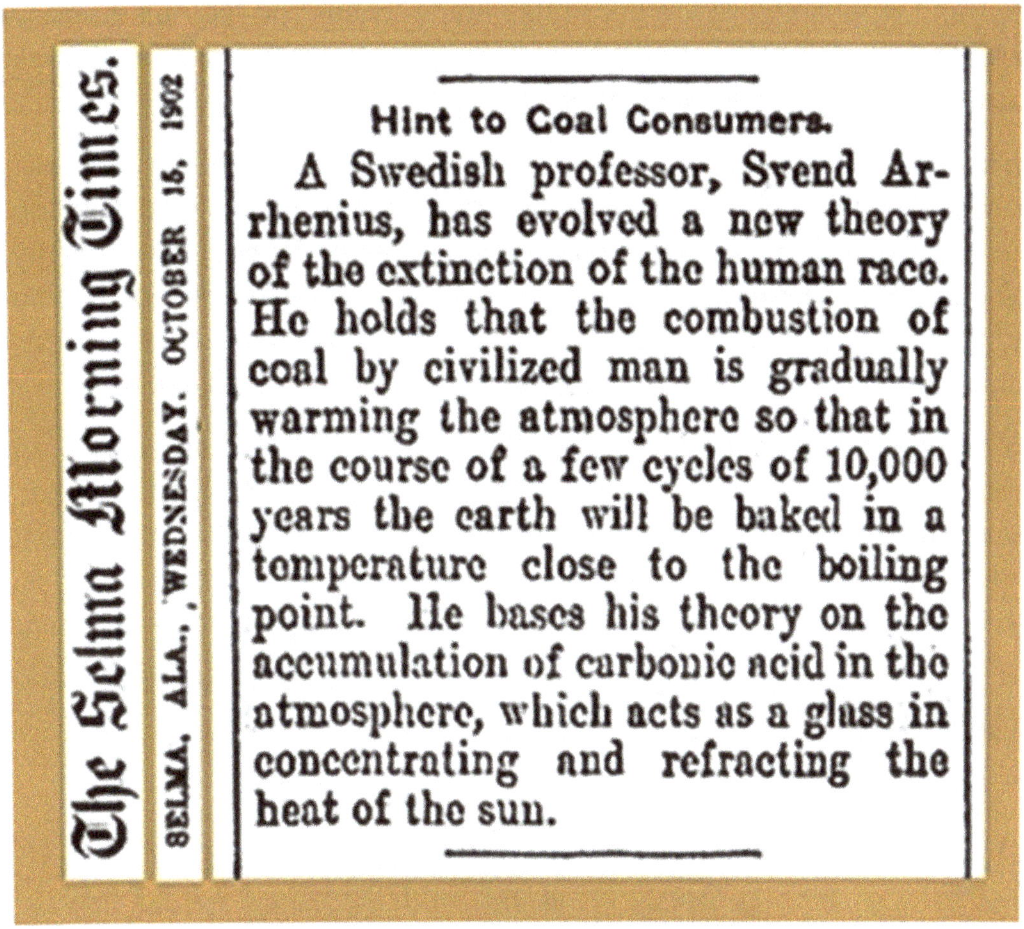

The Selma Morning Times.

SELMA, ALA., WEDNESDAY, OCTOBER 15, 1902

Hint to Coal Consumers.

A Swedish professor, Svend Arrhenius, has evolved a new theory of the extinction of the human race. He holds that the combustion of coal by civilized man is gradually warming the atmosphere so that in the course of a few cycles of 10,000 years the earth will be baked in a temperature close to the boiling point. He bases his theory on the accumulation of carbonic acid in the atmosphere, which acts as a glass in concentrating and refracting the heat of the sun.

(Source: US National Archives)

The alarm sounded in the sensationalistic news media, about humans destroying the atmosphere that could lead to their extinction several tens of thousands of years in the future, was overshadowed by the pre-World War I political/militaristic events in Europe and drew in the United States into World War I which erupted in the 1916 time period.

The environmental science world however, in the United States and in the Environmental Protection Agency, gained stature in the 1930s after the infamous "Dust Bowl" crisis in the mid-west United States in turn causing or exacerbating the economic period known as the "Great Depression". The EPA seized the opportunity to exercise their power over controlling the "greenhouse gas emissions" caused by the Industrial Revolution as their efforts to control pollution from plastics lost momentum. (As a precursor to the conclusions of this book, the recommendation is made that the EPA shift its attention from trying to control the climate and to better control of human, industrial, and agricultural pollution "for the betterment of mankind".

The public has accepted the claims of the environmental science profession that the climate is changing – and factually it is changing. However, the climate is changing for causes other than those which are more factually correct than those expounded by Dr. Arrhenius but which the public apparently cannot or does not want to understand. The clichés of old seem to last for a lifetime. The belief that "Whatever the scientists say must be factual or they would not be saying it" persists today. If the environmental scientists state that the climate is in a crisis state of degradation and is approaching a global catastrophe', the public gullibly accepts that status. The public lives for the "here and now" and in recent polls the public has many "crises" on their plate- gun violence, drug abuse, homelessness, starvation, droughts and floods, mudslides and earthquakes, terrorism, political opposition to the principles of government, etc.

In a children's popular fable, Chicken Little told Henny Penny "The sky is falling". And in a philosophical sense a portion of the sky – scientifically better identified as precipitation – was indeed falling. If there is not enough precipitation falling often enough, the oceans would dry up – in more modern terminology that is identified as "drought" - and the planet would revert to being the way it was 4.67 billion years ago before the oceans were formed. Water is essential to all forms of life, and a chapter of this book is dedicated to the importance of water on the planet.

The planet is not the same planet it was 4.67 billion years ago, and its surface topography and temperature and climate have changed greatly from what they were long ago to what they are now. From all indications, the planet's surface topography and atmospheric environment will continue changing and undoubtedly will be different 500,000 years from now. There is no accurate scientifically based model to predict what the planet's climate will be a decade from now, much less a millennium or more from now.

Many school systems teaching Earth science use an illustration similar to the following to identify what "climate" means in everyday life as it is presently known to be:

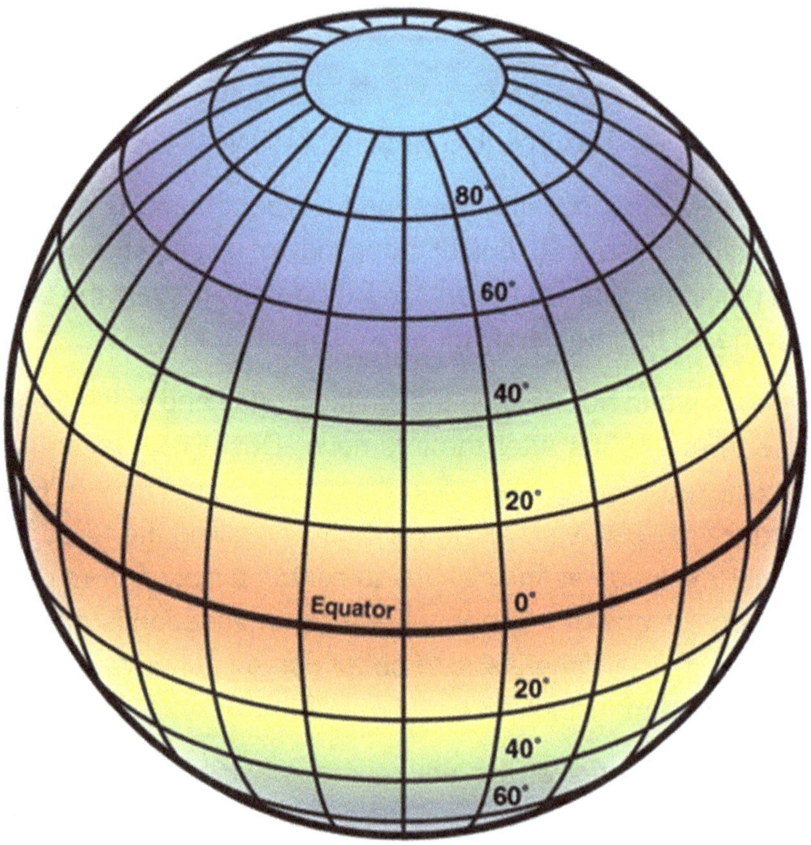

(Source: Encyclopedia Britanica)

There are no numerical definitions of climate taught – the public recognizes climate only by the sensation of comfort when exposed to the temperature (hot – red color, warm - yellow, comfortable - green, cool – dark blue, and cold – light blue) and how "climate change" varies with season.

(Author's note: the environmental scientists have caused the public to confuse climate with seasonal changes attributed to climate change. Seasonal changes do affect the climate but factually there is no "feedback" effect of the climate on seasonal changes.)

Seasonal changes in temperature are the effect of geophysical movement of the rotating planet about a tilted, wobbling axis of rotation in an irregular elliptical orbit each annual trip around the Sun. The 1930s unrest in Europe over geopolitical and social issues revived the interests of scientists, physicists, and mathematicians in gaining a better understanding of the causes for and implications of climate change that Dr. Arrhenius had identified in Sweden around the turn of the century three decades earlier. An unusual set of circumstances (explained more fully in Book 1) generated a military interest in the warming of the cool latitudes in Europe above 50 degrees North – that of northern Scandinavia (Sweden, Norway, and Denmark). A Serbo-Hungarian astrophysicist and mathematician Milutin Milankovic (1879-1958) dedicated his career to developing a mathematical theory of climate based on the seasonal and latitudinal variations of solar radiation received by the Earth. Now known as the Milankovitch Theory, it states that as the Earth travels through space around the sun, cyclical variations in three elements of Earth-sun geometry combine to produce variations in the amount of solar energy that reaches Earth:

1. Variations in the Earth's orbital eccentricity—the shape of the orbit around the sun.
2. Changes in obliquity—changes in the angle that Earth's axis makes with the plane of Earth's orbit and causing the Sun's radiative energy impinging on Earth's surface to decrease as the latitude increases.
3. Precession—the change in the direction of the Earth's axis of rotation, i.e., the axis of rotation behaves like the spin axis of a top that is winding down; hence it traces a circle on the celestial sphere over a period of thousands of years.

Together, the periods of these orbital motions have become known as the Milankovitch Cycles. This database is one of the most vital databases for this book as it explains the "squiggles" (varying waveforms) of the scientific observations and measurements – which scientists go to great lengths to linearize and average out those important "squiggles" in attempts to model them. The Milankovic Cycles are not a predictor of the extent of change in any variable but provide insight into why the astrophysical variables are changing in a cyclical manner without defining the magnitude of the changes. A known shortcoming of the Milankovic Cycles is that they do not explain climate changes in short periods of time less than 50 years. They are not modified by any atmospheric or meteorological facors.

A team of mathematicians headed by Dr. Milankovic laboriously calculated, for the latitudes of 50, 55, and 60 degrees north, the effects of eccentricity, obliquity, and precession of planet Earth on solar insolation (the short term for incoming solar radiation). The time frame covered in the calculations was from the period one million years prior to the 20th century and to about 1900. Graphical illustrations of the team's astrophysical calculations were published in scientific journals in the late 1930s in different formats, but one understandable – less scientifically detailed - format is summarized as follows:

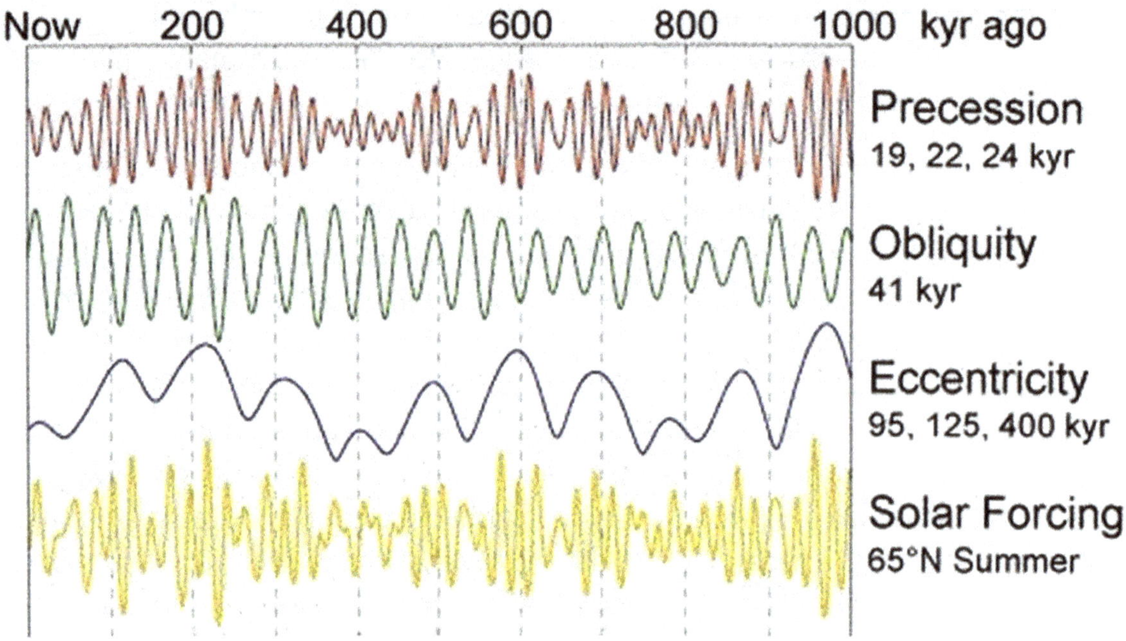

(Source: National Academy of Science)

The scalar magnitudes of the variations in the four variables shown were debated – and still are - for their accuracy, but almost the entire earth science profession considers the cyclic variations having different periodicity defined by the set of calculations to be most valuable in understanding the astrophysical causes for climate change and the rates of change of solar forcing at the latitude 65° North and at the summer equinox position in the orbit relative to the Sun.

It will be shown later in this book a comparison of the calculated (i.e., theoretical) Milankovic Cycles with the measured temperature and gas contents of the Vostok Ice Core Research database. The correlation of the periodicity of the two different – one theoretical and one measured – is later shown to be excellent. Because the Milankovic calculations did not consider the atmosphere of the planet that solar radiation must pass through before impinging on Earth's surface, the magnitude of the solar insolation factually impinging on the planet's surface may justifiably not be of high interest to climatologists using a "grass roots" type of approach in studying the global warming and climate change problems.

Conversely, this Author had committed to a "big picture" approach and quickly realized identified many other factors than carbon dioxide in Earth's atmosphere and summarized the other factors than astrophysical that were contributing to Earth's atmosphere that were of geophysical, natural or anthropogenic origin. The following diagram almost caused the Author to abandon the study of the complex problem of significant interactions between the many factors from different sources that affect the state of the planet's temperature, weather, and its climate.

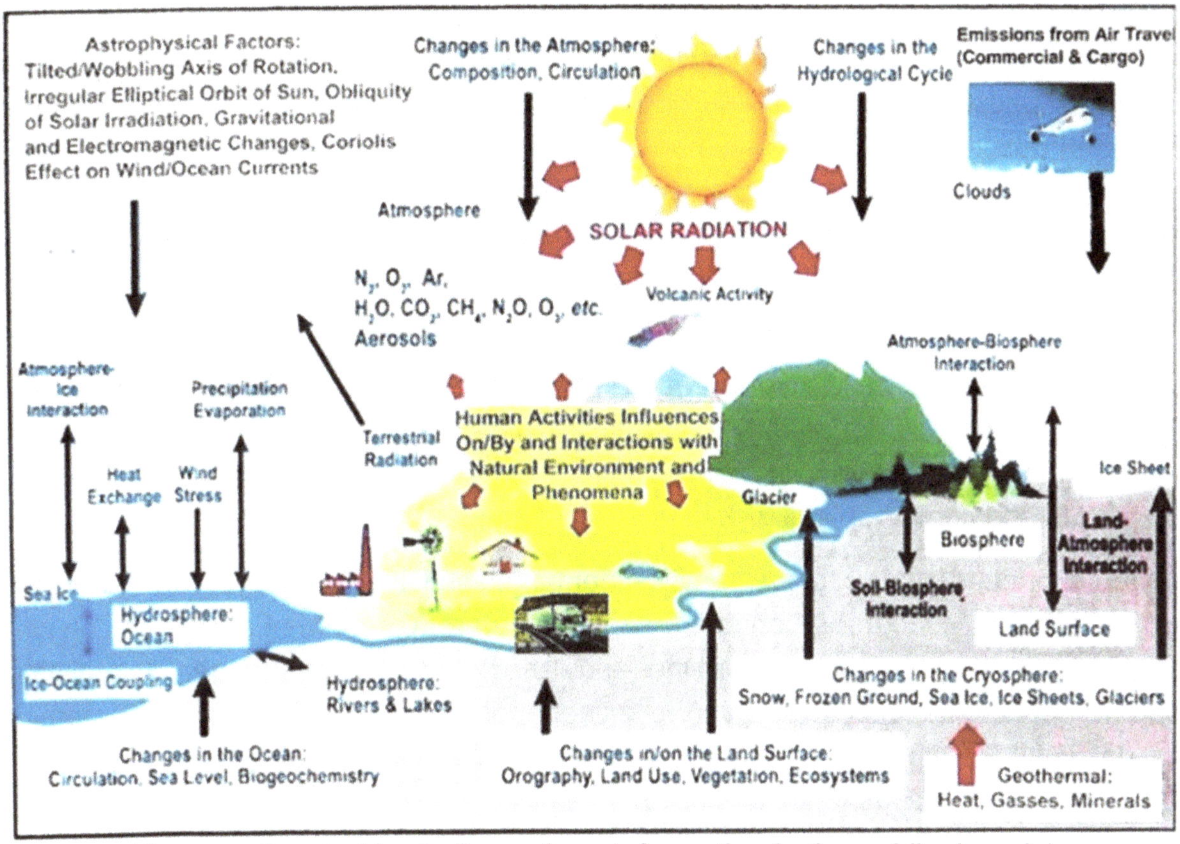

(Source: Created by Author – from information in the public domain)

What is not reflected in the above two-dimensional static diagram is that those many factors are not constant – they vary three-dimensionally with the latitude on the planet's surface they vary being in daylight or darkness, they vary from cold extreme to hot extreme in an annual seasonal manner and vary in a long-term climate change manner from longer cyclical astrophysical and geophysical effects. Natural large scale-events – earthquakes, tsunamis, volcanic eruptions, and lightning-ignited wildfires occur from unpredictable tectonic plate and/or weather-related causes. Hurricanes and tornadoes occur on a somewhat seasonal manner in only generally predictable paths but can/do occur out of season and on unexpected/unpredictable paths in or away from "alleys" in some regions. The toothpaste tube principle is a good analogy with the way the planet functions – a change in pressure applied in one place causes the "toothpaste" to appear in a different place. Two major differences are that the toothpaste tube is static in hand, and its surface can be deformed by human action.

The planet is in an almost unpredictable set of motions – rotating, translating, wobbling, changing shape, and making those movements in sensed but unseen fields (temperature, pressure, electromagnetism), and unseen and unsensed accelerations (gravity, momentum, and inertial effects). The planet's surface is composed of both a hard crust but with several oceans and seas – squeezing on the planet's surface in a small region does not cause the hard surface to deform. Everything on the planet is in one of the three principle states of matter – solid, liquid, or gas. Anything that can possibly happen on the planet probably will, the largest unknown being what, where, when, how much, and at what rate. Each significant abnormal event or factor differing from "normal" has its own schedule and agenda.

Those many possibilities are regarded scientifically as being variables – dependent and independent – and some of the variables vary interactively with one another, all of which causes challenges to be analyzed for cause and effect. These books can be seen to describe and explain in the least scientific (often described as "mumbo-jumbo") language. The Author has attempted to write these climate change books in a conversational style and level of factual content that neither "talks down" to Readers or "talks up" to scientists regardless of their educational background and experience in one of the STEM branches – science, technology, engineering, or mathematics. (A colleague of the Author suggested the title for these books to be "Climate Change for Dummies, Part 1 and 2".)

Everything in any life-form on the planet – in the water, in the air, or on the ground – is along for the magnificent ride. This book hopes only to introduce some of the most important parts of all that to the human population and somewhat prepare them for what they should do and not do to make their odyssey as safe and comfortable as possible.

In the twentieth century and since, environmental scientists have used an "average global temperature" of about 16 degrees Centigrade (approximately 60 degrees Fahrenheit). on which they base the extent and trends of "global warming" and associated "climate change". Scientists involved in agronomy, forestry, and dendrology (the science of tree growth) consider climate to be a combination of temperature and the amount of precipitation which plants and trees require for growth. The news media adds to the confusion by advising the public that whenever a heat wave or a cold wave is shown in a weather report (on television or in the local newspaper), that unusual event is caused primarily by climate change brought about by humans and industry emitting "greenhouse gases" into the atmosphere by burning fossil fuels.

For this book and the previous book on the facts and fallacies of climate change on our planet, the common variable in any of those perceptions of "climate change" is the variable of temperature. As a "conversation starter", the first book addressed what are the causes for the surface temperature of the planet to be changing and which is recognized as being "global warming". This book overlaps the first book so that addresses the measures that can be taken to mitigate the effects of the temperature changes under the umbrella of "global warming".

Most importantly, the changes in the planet's surface temperature were conclusively shown in the previous book to be caused by astrophysical, geometrical, and natural factors which are beyond Man's ability to be controlled. However, the changes in Earth's surface temperature were shown to be regional not in an "average global temperature" manner of increase. Underlying the causes for the "global warming" is the obvious fact (shown both in the illustration on the front cover of the facts and fallacies book and on the diagram on the preceding page is the variation of temperature with latitude.

The illustration reflects the compounded accumulation of knowledge passed down from the countless number of humans in many different walks of life – explorers, sailing ships' crews, scientists and philosophers, pioneers of technology, etc. It is only representative of the "big picture" perspective of the planet as a static huge ball and shown in two-dimensions on two-dimensional paper pages in this book. Later in this book is added many of the important variables that show the planet as a spherical (three-dimensional), rotating about an axis, and in an enormous orbit around the Sun as one celestial body in the solar system and in the Milky Way Universe.

The planet is an almost spherical body of revolution having an average diameter of 7,918 miles (12,742 kilometers) and is rotating – one rotation about every 24 hours – and having an axis of rotation with end points generally located near the North Pole and South Pole of the planet. According to the laws of motion of the bodies in space (defined by the brilliant scientists and mathematician Johanes Kepler), the temperature of a point on Earth's surface is inversely proportional to the distance Earth is from the Sun. This causes the temperature to appear as bands of color and with the same color generally defined by lines visualized on the spherical surface as being circles. The lines defining vertical circles are lines of latitude. The lines defining the horizontal circles are defined as lines of longitude. Readers having cell phones or laptop computers can query those devices for Global Positioning System coordinates of latitude and longitude to locate precisely where the person is on the planet, thanks to the technological developments of science.

Because the axis of the Earth is tilted and "wobbling" (defined scientifically as precession) – changing its angle relative to the Sun - the orbital path and the changing angle of the surface of the Earth relative to the Sun causes the temperature of a point on the surface to vary in a manner named as seasonal. Astronomers have long known of the cause for seasonal temperature variations on the planet and summarize that phenomenon graphically as follows:

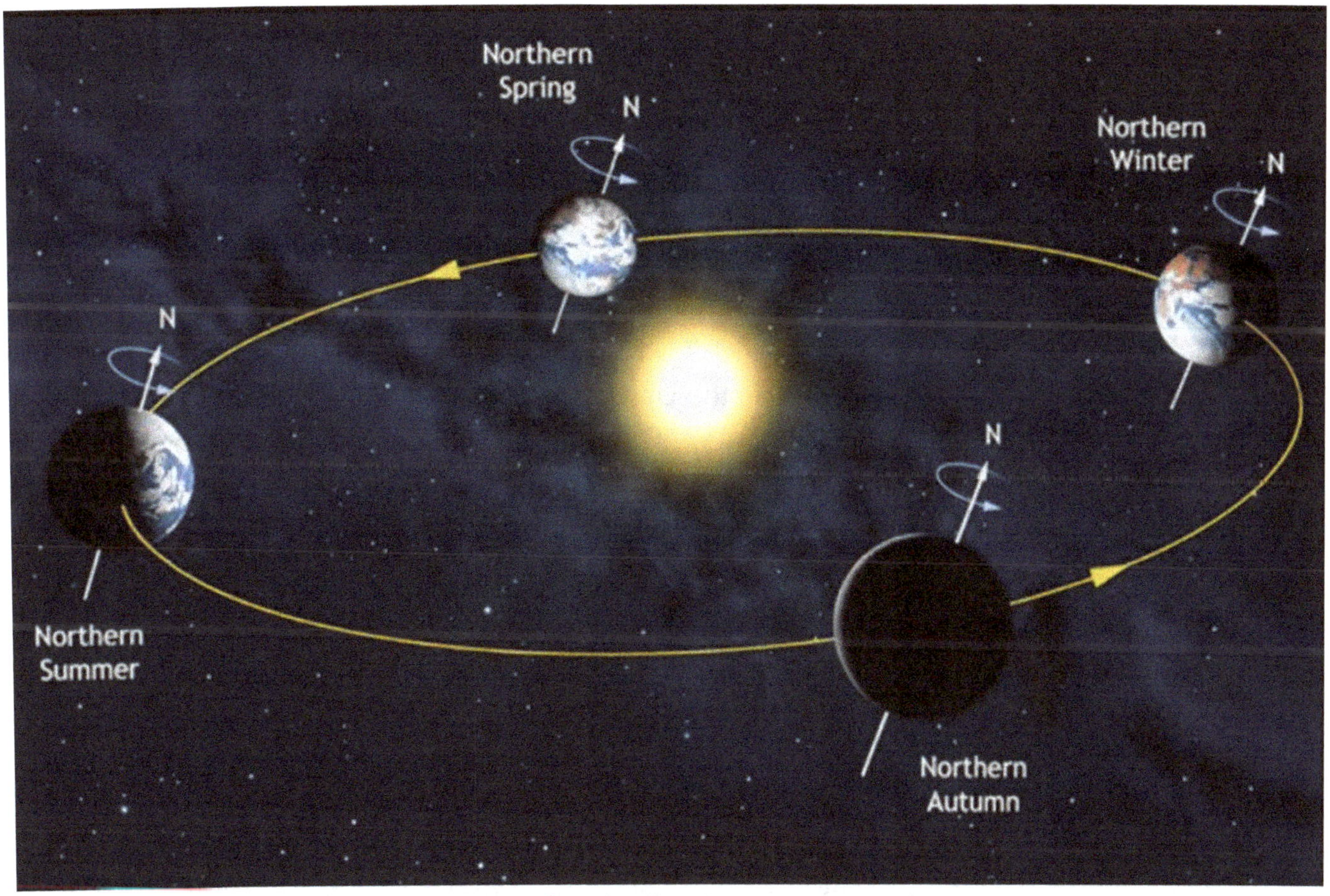

(Source: Any textbook on astronomy)

A graphical illustration of how much the temperature changes seasonally on Earth's surface – at an average latitude for the United States (about 35° north of the Equator), due to the latitude and astrophysical effects on the temperature of a region - is typified as follows:

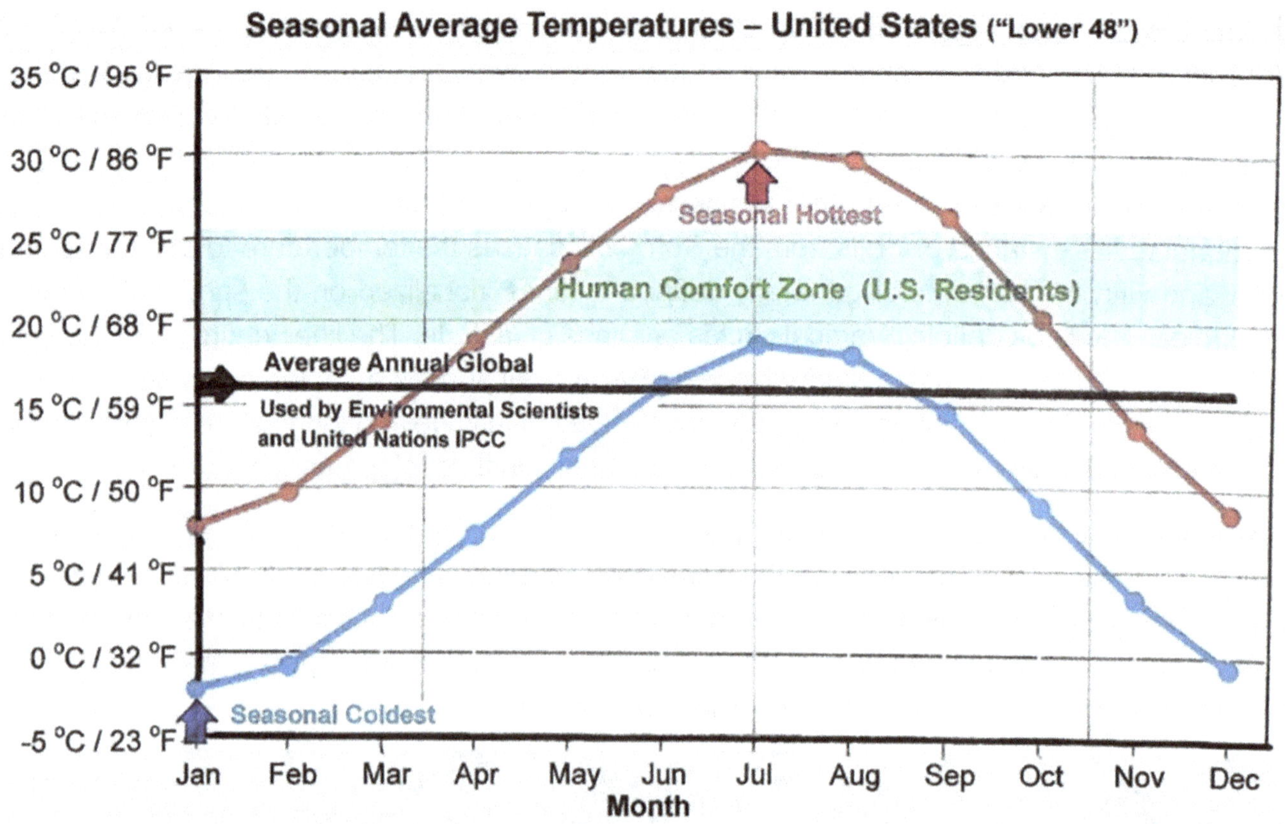

(Source: World Meteorological Organization WMO database, Author Graphics)

The red line is the average hot temperature (the hottest temperature occurring generally in July), and the blue line is the average cold temperature (the coldest temperature occurring generally in January). As an approximation, the average difference each month between the seasonal average hottest and the seasonal average coldest temperatures variation (i.e., between the red and blue "mountain" like curves) is seen to be about 10 °C (18 °F). More importantly, the difference between the seasonal average hottest temperature in the United States (about 35 °C (95 °F) in July) and the seasonal average coldest temperature (about -2 °C (28.4 °F) in January) represents a difference of 32 °C (57.6 °F).

A reality that humans have a "comfort zone" on temperature of the air surrounding them has given rise to the entire industry of heating and cooling (and circulation of the air) industries – manufacturing heating units and cooling units with fan systems for circulating the warm and the cool air in residences and business facilities and commercial buildings. Servicing and maintaining those HVAC systems is another service industry in parallel with the plumbing industry to provide water to all the facilities for human needs – regardless of the temperature outside the residence or building. Adding to the reality, humans do not remain in residences and facilities that are well heated or well cooled and must venture outside into whatever the air temperature is – and whether the wind is blowing or if it is raining – which requires humans to have a personal wardrobe of clothing "for all weather and seasons". Homeless people do not have that luxury of a personal wardrobe, and it is they who suffer the extremes of temperature for the region where they survive the hazards of nature. To every man and woman, their home is their shelter, be it a cave, an apartment, a resort mansion such as Mar-a-Lago – or a cardboard box. Every person in the United States does what he or she needs to do to survive the environment they live in – ideal or not. Airplanes fly within a flight envelope of speed and altitude – it is the same with humans and everything else in ecology. The following is the temperature envelope for the US population – regardless of who or what caused the temperatures:

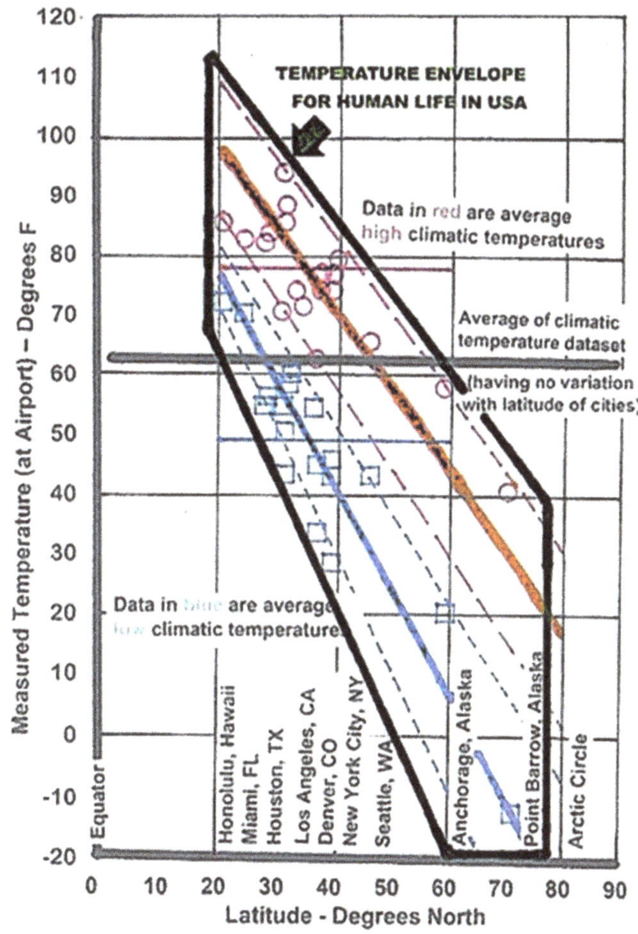

(Source: NOAA National Weather Service database; graphics by Author)

If environmental scientists choose to use this temperature envelope technique for regional temperature modeling purposes, it appears that the upper (extreme hot) boundary is at a lower slope than is the lower (extreme cold) boundary. This difference suggests that the difference between the extreme boundaries is linearly proportional to the latitude. By constructing a median line between the upper and lower extremes, a mathematical relationship could be generated that the mathematical difference of plus or minus the increment of half the difference of the extremes could be applied to the median temperature line at the latitude of any region to establish the extremes of the temperature envelope at the latitude of the region. The LEO method for estimating the average annual temperature of a region should be investigated to determine if the LEO average annual temperature coincides with the median line of the above temperature envelope.

The difference in surviving whatever temperature a human is exposed to within the temperature envelope of the region is apparently not a priority to the environmental scientists and Environmental Protection Agency in the United States (and in many other “developed” countries). The one-size-fits-all “Average Annual Global’ temperature, used by environmental scientists and by the United Nations Intergovernmental Panel on Climate Change as the “line in the sand” reference in defining global warming and climate change, is shown on the above diagram (the solid horizontal black line on the above chart).

(Author's Note: Environmental scientists can engage in all the "battles with the climate" they wish to do - and use taxpayers' money to do it – but preventing a 2 °C (3.6 °F) rise in "average global temperature" (the horizontal bold black line) and averting a fictitious "crisis" will not change the temperature envelope (shown above) within which the US population must survive to continue in existence. Establishing "environmental justice" laws and regulations and penalizing anyone or any organization who burns fossil fuels or who exceeds a "net zero carbon" limit as a war criminal is totally fallacious if expectation is that it will reduce global warming or stop the climate from changing. Politics and "green deals" with carbon credits and carbon trades will not cause the climate to change. Those efforts and those resources should be applied to anyone or any organization who pollutes the environment – land, water, or atmosphere - with toxins and/or carcinogens.)

In the research done for the global warming and climate change "problem" and "crisis" proclaimed by the environmental science group self-titled as "settled science", the Author has read over a hundred news media articles and a greater number of science journal articles showing that a goal for limiting the "average annual global temperature" to not exceed an increase of 2 degrees Centigrade by 2050 is the "line in the sand" temperature beyond which the planet will exceed a "tipping point" and suffer great "catastrophes".

As best as can be determined, the "average annual global temperature" is currently set as being 16 °C (60 °F), and a 2 °C (3.6 °F) rise in that temperature would establish that 18 °C (64.4 °F) is the "tipping point" temperature, beyond which something dire is going to happen to someone or something that is not defined in the news media or scientific journals. One science journal article in the Climate section did publish an article which is in "the sky is falling" – or "the average sea surface temperature (a rise of less than one degree in temperature to 21 °C (69.8 °F) in the past 40 years) will soon be boiling" category:

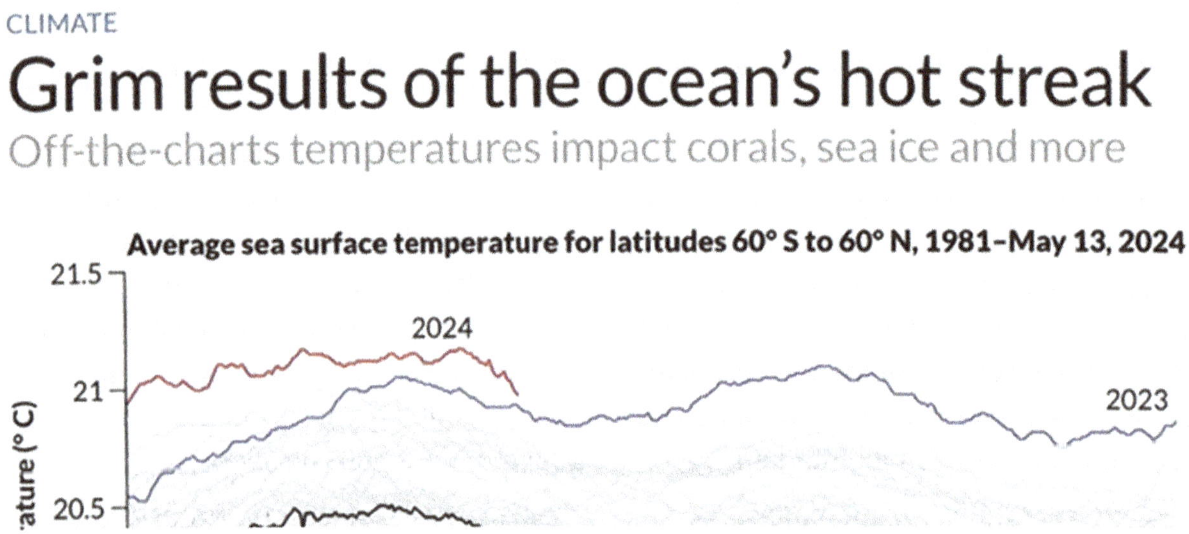
CLIMATE

Grim results of the ocean's hot streak

Off-the-charts temperatures impact corals, sea ice and more

(Source: NOAA Climate Reanalyzer Program)

Another most important factor is that about 70% of Earth's surface is covered with water (saline and fresh). That water standing on Earth's surface is at different latitudes and exhibits the same vaporization of water at the surface into water vapor into the atmosphere. There is a small difference in the vaporization of saline water into water vapor from fresh water, but the principle is the same that the evaporation of water in liquid state into water vapor in gaseous state follows the scientific principle of saturation of an atmosphere with water vapor depends on the temperature of the atmosphere above the water, illustrated as follows:

(Author's Note: Readers are encouraged to gain a full understanding of the existence and vital importance of water vapor in the atmosphere, as this is a most major flaw in the environmental scientists' hypothesis that "by eliminating everything else, it is settled science that greenhouse gas emissions in the atmosphere are the dominant cause of global warming and climate change." Water vapor in the air behaves entirely on the physics of the water temperature and the capability of the surrounding atmosphere to contain the water vapor up to a maximum saturation concentration shown in the following diagram.)

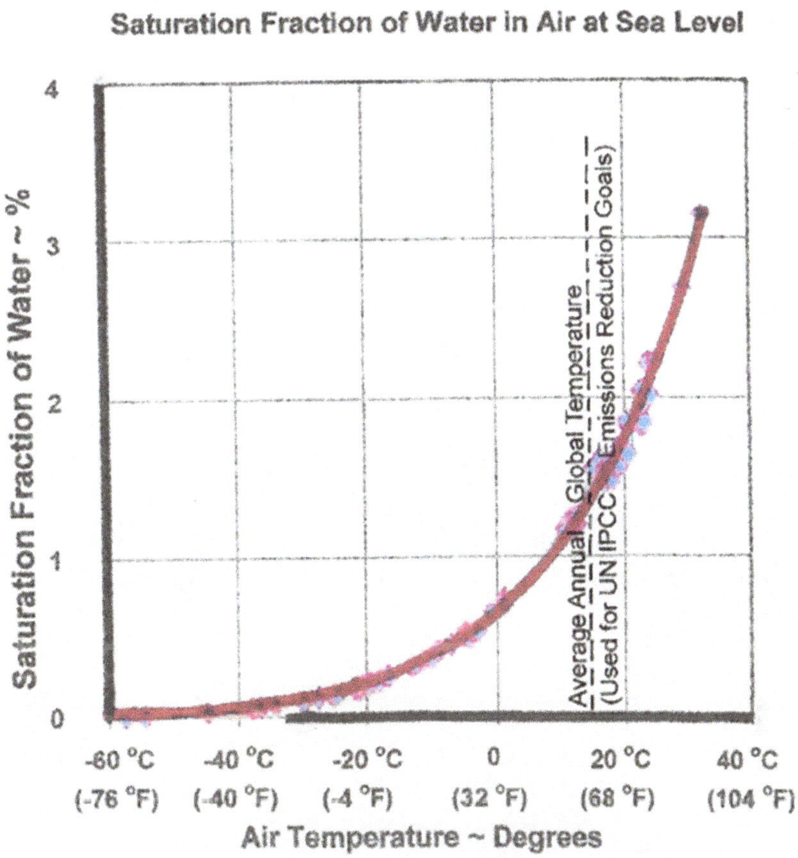

(Source: NOAA database – regenerated by Author and with notations)

The evaporation rate of water into the atmosphere is dependent also on the saline content of the water as well as the temperature of the water and surrounding atmosphere. The amount of water vapor in the atmosphere - being dependent on several variables including the ocean currents transferring water in different patterns of distribution on the planet's surface – was and is measured frequently by NASA satellites in assisting the meteorologists predicting weather for the human population.

The following illustration shows the water vapor content (using the color-coded scale for temperature at the bottom) of the atmosphere above regions at all latitudes of the planet. Just as there has been proven to be a latitude effect on the average annual temperature of land regions, the same latitude effect causes differences in water temperature at different latitudes. The contention that the burning of fossil fuels - to meet human and industrial needs for heat energy - is causing a rise in the ocean surface levels and temperature of the ocean water is another half-truth being fed to the public masked as being the dominant cause. At most, it is of questionable significance to scientists (excepting those in "settled science") – identified as "naysayers" on the opposite side of the global warming and climate change fence.

To be noted of high importance is that the highest saturation concentration (approximately 3.1%) of water vapor (in red color) in the atmosphere is above the warmest water temperature (approximately 30 °C or 86 °F) – found in the Atlantic Gulf Stream and Equatorial zone ocean water. It is not coincidental that high water vapor concentration level is the region where the tropical storms and hurricanes are spawned and travel. Cooler waters (in green color) contain a lower concentration of water vapor, and cold waters (in blue) contain the lowest concentration of water vapor – where there may be only rare tropical storm occurrences.

Although the atmosphere above the land masses is shown to be in black color, there is still water vapor above the land masses as the prevailing winds carry the atmospheric water vapor over the land masses. Without water vapor in the atmosphere there would be no precipitation (rain, snow, hail, sleet, graupel, or simply dense fog). Environmental scientists in the "settled science" segment of their profession become "naysayers" for their hypothesis that carbon dioxide gas in the atmosphere is the dominant cause for global warming and climate change when in scientific fact it is water vapor in the atmosphere that is the dominant cause of global warming and climate change.

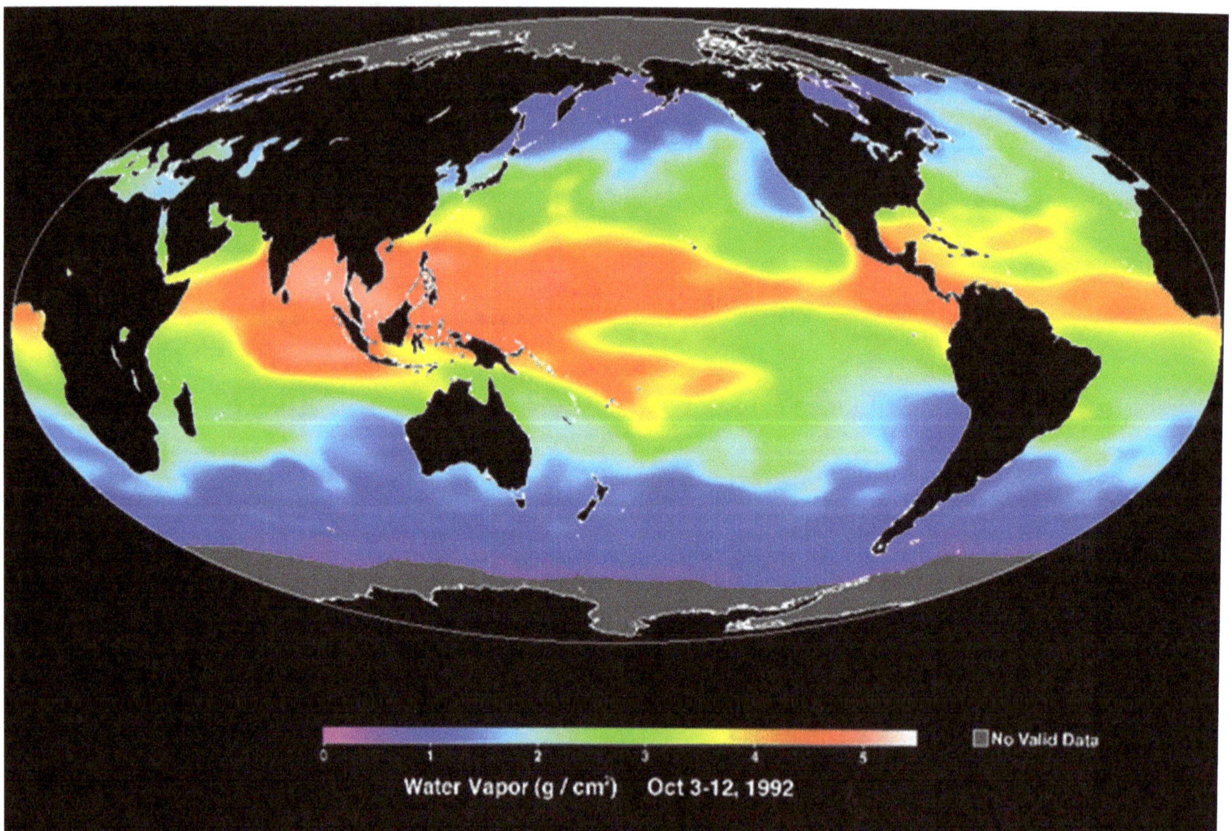

(Source: NASA)

There is valid scientific reasoning that water vapor in the atmosphere above ocean water should be the highest for regions where the water temperature is high and decreases in saturation amount in open water regions where the water temperature is in a cold temperature state. A good correlation is made between the water vapor distribution shown in the NASA satellite illustration above and with the tracking of tropical storms paths and severities shown as follows:

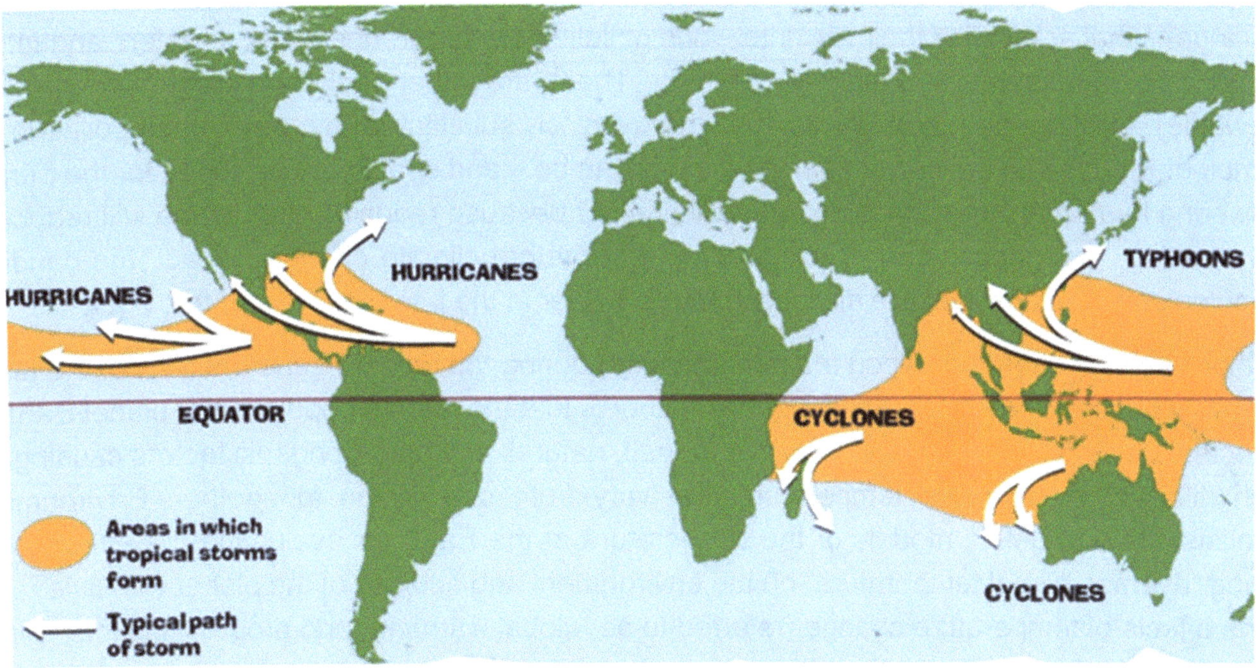

(Source: World Meteorological Organization WMO)

The Coriolis effect is clearly seen in the diagram to cause tropical storms to have a clockwise component of their paths for latitudes north of the Equator. The same Coriolis effect causes tropical storms south of the Equator to have a counterclockwise component of their paths.

The information researched and analyzed for the preceding Book 1 and this Book 2 leads to the proven conclusion that astrophysical, geophysical, and many natural factors combine to be the far more dominant cause of global warming and climate change than all combinations of anthropogenic causes. The environmental scientists' hypothesis that humans and industries are emitting greenhouse gases into the atmosphere and are the dominant cause of global warming and climate change are completely fallacious at best and are extremely devious at worst. The environmentalists issuing the claim are aided and abetted by the sensationalist news media – with their campaign of "combat climate change" is an embarrassment to those scientists who generate and interpret valid scientific databases. The public of all age groups have been misled into accepting the fallacious hypothesis and the headlines of an increasing number of news media releases, similar to the following newspaper editorial title, reflect cult-like loyalty to the belief that has been repeatedly reported to them as being at a "crisis" rate:

Coal is the global climate scourge that just won't go away

If it were not for coal, the United States and almost all technologically developed nations would remain in existence in an underdeveloped state. It is hastened to be added that the technology of coal has the con of a polluting ash residue after it is combusted. Coal ash contains toxins and carcinogens that are finding their way into – and polluting - underground water aquafers and into the soil used for agriculture and animal ranching. There are other fuel technologies (natural and renewable) existing and needing to be improved or substituted by other less-polluting fuel technologies such that the use of coal technology can be – and should be - reduced for the purpose of lowering human and industrial caused pollution not because reducing coal usage will reduce the heat-trapping “greenhouse gas effect” and therefore retard climate change. In fact, the candidate technologies to replace coal are discussed much further in the later chapters of this book

Another major point also surfaced (no pun intended) during the research and analysis of the factors addressed in both Book 1 and Book 2 of this Author is that the human population of planet Earth are subjected to astrophysical, geophysical (seasonal), natural, and anthropogenic factors causing very much different extremes of temperature that vary from one region to another. Environmental scientists strive to define models of the temperature of the Earth for evaluating the effects of the varying “average global temperature” on the environment and ecology of the planet with little regard for the effects of temperature change (referred to as “global warming” and progressing into “climate change”) on the human population. Both Books 1 and 2 point out that the scientific “averaging” and “linearizing” technique used by environmental scientists has misled the public to believe that those factors (temperature and climate) are the result of humans and industry generating enormous amounts of heat-trapping “greenhouse gases”. The process of averaging and linearizing masks the physically essential changes of seasonal effects caused by astrophysical, geophysical and natural phenomena defined by other scientists not associated with environmental science.

A principle scientific factor causing the temperature variation with latitude is the angle (geometrically termed as “obliquity”) at which the column of solar radiation subtends when it reaches the Earth’s surface. Students in geometry and trigonometry will recognize that obliquity is numerically quantified as a cosine variation of the angle subtended. When the Sun’s radiation waves approach Earth in a columnar manner, the insolation (abbreviation of incoming solar radiation) waves are virtually perpendicular to Earth’s surface and the highest amount of solar radiation will be emanated onto the Earth’s surface – in the latitudinal region humans have defined as the Equatorial zone – the surface temperature is theoretically the hottest on Earth’s spherical surface. The over-exaggeration of the greenhouse gas emissions causing a near-term crisis is not warranted when considering geophysics and laws of thermodynamics and the laws of atmospheric chemistry considering the variable of concentration.

As latitude is increased (i.e., moving away from the Equator), the obliquity angle increases, and the solar radiation begins to be deflected from the Earth’s surface. If the solar radiation is deflected, it does not get “absorbed” into the surface, and the annual average temperature at each latitude follows that set of trigonometric events. The numerical significance of that changing obliquity follows very similar to the cosine function of trigonometry. At 0 angle, the cosine of a 0 angle is 1.0. As the obliquity increases above 0 degrees, the cosine decreases trigonometrically. Because the Earth’s surface is that of a sphere, as the latitude increases the curvature of the exterior of that sphere curves away from the incoming solar radiation. In short, the curvature causes the obliquity angle to continually increase such that at the North Pole and South Poles (where the latitude is 90 degrees), the solar radiation is the lowest (i.e., the cosine of the angle approaching 0 degrees is approaching 0.) The lower the solar radiation, the lower the temperature of the surface – hence, the reason that the coldest temperatures on Earth are at the North and South Poles, and the hottest temperatures are in the Equatorial region. There are several factors that influence how intense the oncoming columnar solar radiation waves will impinge on the Earth's surface, but there is no factor that causes more effect on Earth's surface temperature than the "latitude effect" – determined by trigonometry and the geophysics of the solar system.

Humans have no influence or effect on or control of those two contributing factors – water vapor in the atmosphere and/or the geophysical latitude effect.

The familiar Global Positioning System refers to those circles imaginarily drawn on Earth's surface as latitude and longitude. Graphically, the latitude and longitude – developed by mariners centuries ago to aid in navigation on the vast oceans with no visible landmarks to navigate with – is shown as follows:

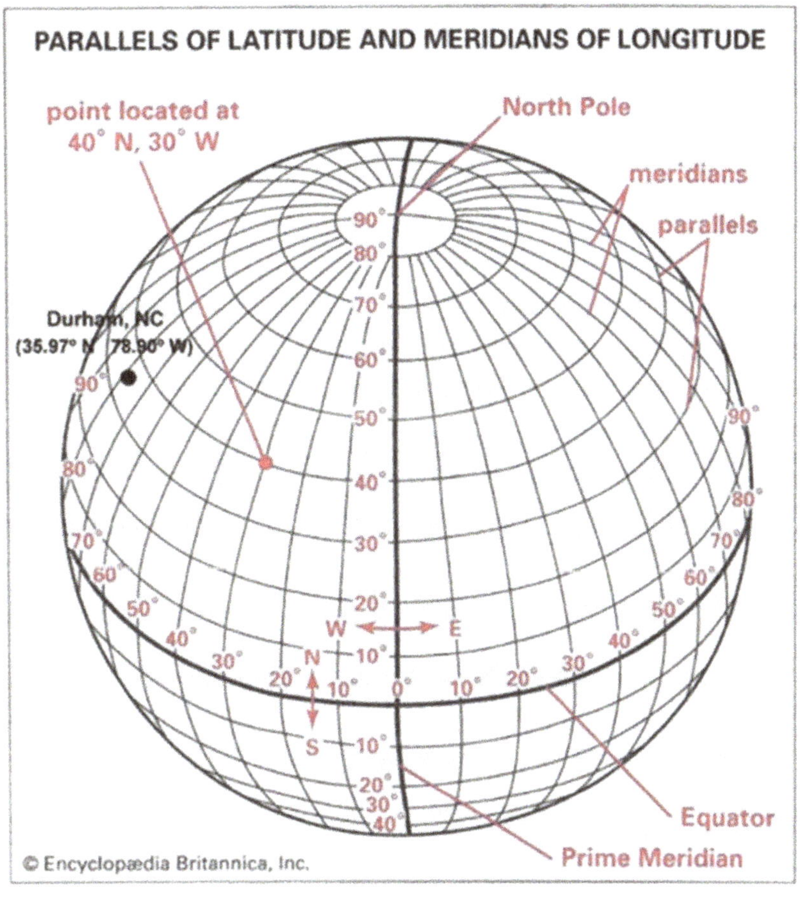

(Source: Encyclopedia Britannica, Inc.)

Jumping ahead for brevity (seemingly *non sequitur*) in this book, the average annual temperature is seen in most of the illustrations in the Facts and Fallacies book to vary in temperature level with latitude. The illustration on the front cover of the Facts and Fallacies book shows from the very start that latitude had/has a very powerful effect on the planet's surface temperature:

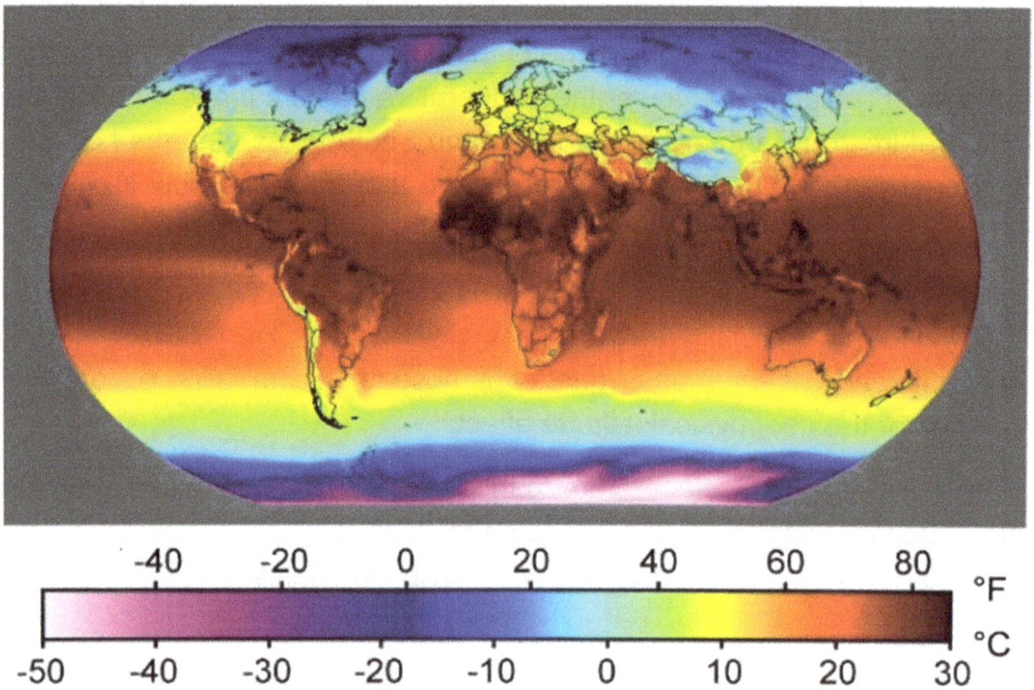

Annual Mean Temperature

(Source: NASA)

An immediate realization came into view that environmental scientists were oversimplifying the global warming perception by using an "average global temperature" – what can be considered "average" on a spherical surface that is rotating and exposing the surface of the planet to the different angles of obliquity relative to the oncoming column of radiation of heat energy from the Sun? There can be an average temperature associated with the circumference of each latitudinal "circle" but everything – animate and inanimate – is located at different latitudes and longitudes and therefore are exposed to different (non-average) temperatures resulting from the Sun's radiated heat energy. The environmental science use of an average global temperature may make it easier for scientists to model the infinite spherical variations in temperature with a one-size-fits-all "average global temperature" but that averages out the several different factual changes in surface temperature and resulting in a fallacious estimation of long-term climate.

What cannot be averaged out are the effects of latitude and seasonal effects on the temperature on the planet. That effect is clearly seen by the color bands on the planet (shown as recorded by NASA satellites in the "big picture") representing the different temperatures of cold, cool, "comfortable = temperate", warm, and hot at different latitudes.

(Author's note: The environmental scientists treat climate change in the simplified manner of a temperature on a two-dimensional flat surface where there are no winds, no atmospheric water vapor, no clouds, no temperature inversions, no albedo effect of obliquity angle, no ocean currents at different temperatures – but have convinced the public and gullible government leaders that humans and industry generating "greenhouse gas emissions" are the "dominant" cause for global warming and climate change. That fallacious conclusion has proliferated for such a long time that young people accept that fallacy as a factual conclusion that past generations burning coal have destroyed their prospects for a safe and happy future on Earth.)

To not mislead the public, the Author simply dove into the research of what is known about global warming and climate change from history by overturning the "first stone to see what was under it". A good first step in researching any formidable problem is to understand why the problem is a problem. Not surprisingly, there appears many articles in the news media and the journals of some scientific groups about humans and industrial gaseous emissions; yet there is a conspicuous absence of articles amidst the information being released by environmental scientists to the public as to what has been learned from the ice core database at the Russian Vostok Ice Research Station in Antarctica.

One of the most important findings before completing the research for Book 1 was the finding of the database resulting from scientists researching the history of the planet as "written" in the ice sheet covering the Antarctic Continent. Russian scientists after World War 2 began drilling through the ice sheet – which is over 2 miles thick in places – and revealed to the scientific world an amazingly good understanding of the many cycles that the planet has undergone during the past five million years in its 4.69 billion years of known existence.

Many forms of life were brought into existence on the only planet that has free-standing water on its surface. Hot peaks caused life-taking droughts. Cold peaks caused glaciation and destroyed the life forms – which were the food of animal and aquatic forms of life. Dinosaurs existed – and dinosaurs became extinct. Such are the realities of life on the planet. All forms of life on the planet must adapt to the temperature conditions and the food supply of their environment or those forms of life that cannot adapt will become extinct. Examination of the fossilized skeletal remains of those extinct life forms found beneath the surface of the planet is the function performed by scientists known as paleontologists.

Soon after World War II ended (in September 1945), the Soviet Union sent a small team of scientists to Antarctica to conduct research as to whatever was contained in the ice sheet covering the southernmost continent of the planet - which was spared the ravages World War II - made on the planet's surface environment. The ice research facilities were named to be the Vostok (translated into English meaning "Southern") Ice Research Center shown as follows:

(Source: Smithsonian Institute for Science)

The housing facilities are the small group of buildings in the back left, the larger buildings being the laboratory facilities where the ice core samples were stored and analyzed, and the orange-colored structures at the left and right foreground were the drilling sites to obtain the core samples. An example of one of the ice core samples (obtained as long cylinders which are/were sliced horizontally in half for scientific analysis – is illustrated as follows:

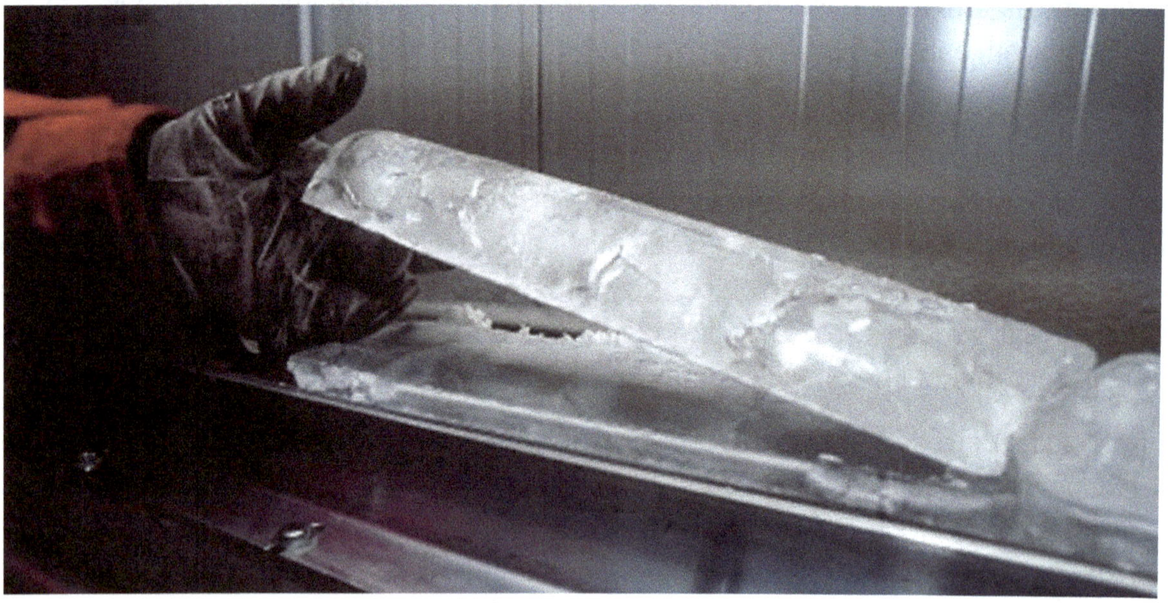

(Source: Smithsonian Institute for Science)

The slicing was done with a mechanical band-saw fine-tooth blade rather than a hot wire (as is used when slicing Styrofoam) as the hot wire would skew the temperature of the inner core being measured with miniscule changes in temperature. The "big picture" research done by the ice research scientists into the history of life on the planet contained within the ice sheet covering the Antarctic Continent captures five million years of history of the continually and dynamically changing planet, gleaned from the ice core samples, summarized (in about ten square inches of area on a two-dimensional page of paper) as follows

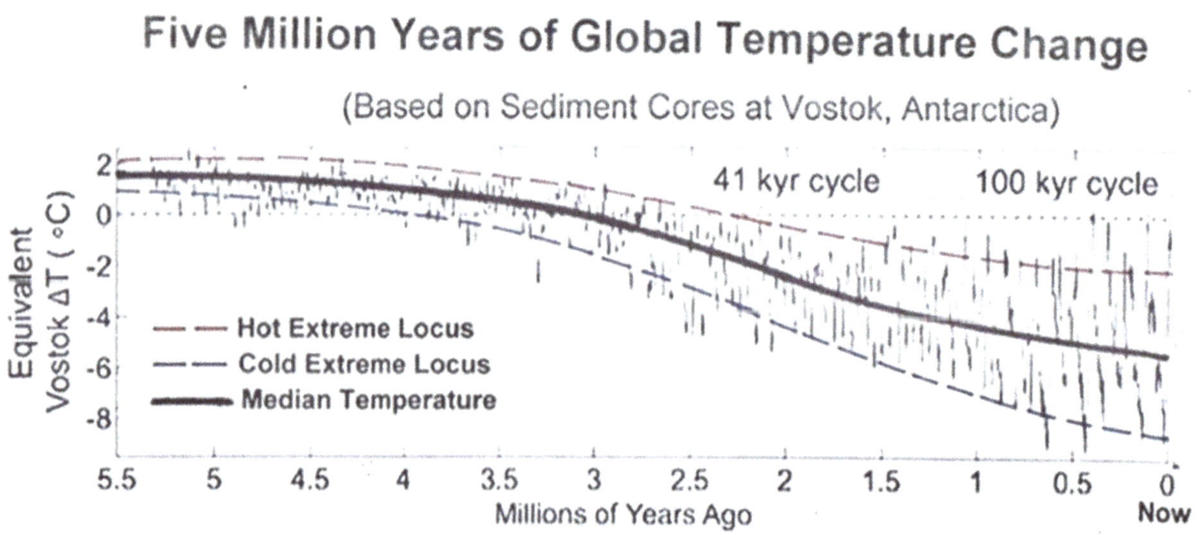

(Source: Smithsonian Institute for Science with Author analytical notations)

(Author's Note: When showing the above diagram to a scientific associate, he commented: "That looks just like a slender cornucopia of food except the temperature substituting for food".) The averaging and use of locus of peaks technique mask the known violent history of the planet – and where the climate changed every 20 years. In five million (5,000,000) years of climate history there were over 250,000 periods of climate change. The Earth is undergoing climate change in the same manner it always has and likely always will – with only miniscule relative changes caused by humans and industry burning fossil fuels. There is valid reason to debate if the same conclusions can be reached if only the most recent 420,000 years of ice core samples are analyzed more closely. That will be shown later in this Background.

The periods of temperature variations from maximum hot to maximum cold are seen to vary – not smoothly but in a closely-timed (several cycles per 100 centuries at the scale of 1.5 million years per inch) increasing and decreasing between peak hot and peak cold. The dashed red line is the locus line through the many peak hot temperature anomalies and the dashed blue line is the corresponding locus line through the many peak cold temperature anomalies. When constructing a median line between the two smoothly varying locus lines, a median line (the solid black line midway between the red and the blue line) is obtained. The valid estimation can be expressed that "based on the data", the planet has been cooling at a variable rate – but if only the ending points of the median line are considered – 1.5 °C hotter than "normal" 5.5 million years ago and 5 °C cooler than normal at time = 0 (now) – a partially valid conclusion can be stated that the planet is cooling – not warming – at a rate of 6.5 °C in 5.5 million years. That rate is 1.2 °C cooler every 1 million years – almost all of which had no presence of humans burning fossil fuels and therefore the possibility of how much "greenhouse gases" affected the warming of the temperatures was of no significance. Is 1.2 °C every 1 million years the cause for alarm to humans who have an average life-span of perhaps 80 years?

The preceding graph reflects only how the temperature as the dominant variable varied with time. The ice core database also includes scientifically important data on the types and amounts of gases that were in the atmosphere and which varied with time. The author's analysis of that same ice core database selecting the most recent 420,000 years is meaningful because it covers 300,000 years without the presence of humans, and then the increasing presence of humans for the most recent 120,000 years. When the most recent 420,000 years of that database is considered and includes the other variables of gas content that are of natural origin when there was no presence of humans, and then when there was the presence of humans adding to the gases in the atmosphere, a different – more realistic - set of conclusions is arrived at.

The following graph was generated for Book 1 and is repeated in this Book 2 because of its profound importance in better understanding of the global warming and climate change phenomena without the presence of any human or industrial activities.

The information in color was the data reported in the original ice core database released circa 1990 from the Vostok Ice Research facility in Antarctica. The Author has added the "smoothing" curves over the jagged-looking peaks and valleys to serve as trend lines in keeping with the "big picture" analytical perspective taken for this book (Book 2) and for Book 1.

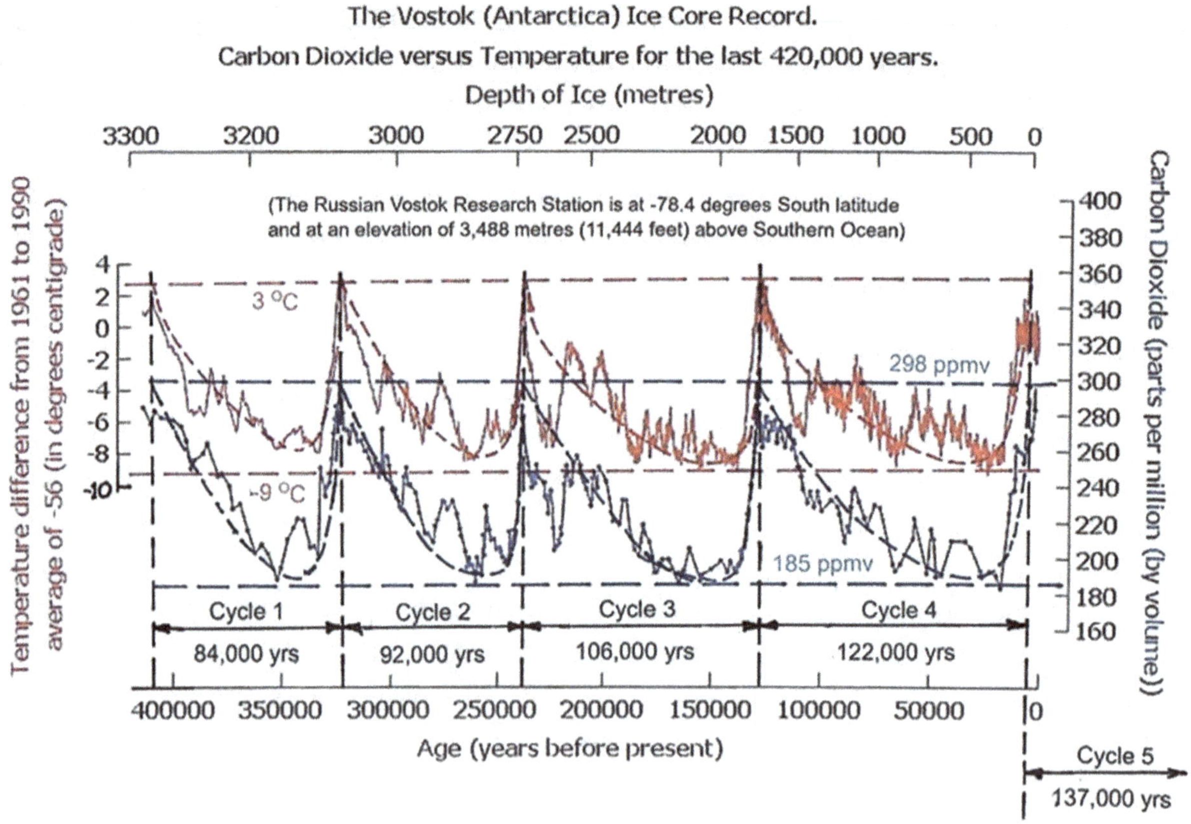

(Source: Smithsonian Institute for Science with Author analytical notations)

Note: There was no known presence of humans to burn fossil fuels during the entire span of 400,000 years shown – to the end of Cycle 4 above – denoted with the vertical dashed line.

When a magnified segment of the ice core database is analyzed, i.e., the most recent 420,000 years instead of the most recent five million years, and when another variable is included in the database – that of carbon dioxide concentration in the atmosphere, a better understanding of how the temperature of the planet is changing becomes apparent. The many peaks and valleys in the temperature are still present but they are seen to have a "big picture" repetition in a cyclic manner. It also is obvious that the carbon dioxide (in blue and measured with the scale at the right) in the atmosphere and the temperature anomaly (difference from normal, in red and at the scale on the left) variations are extremely similar in pattern. The Author added "smoothing" lines (broken and in red and blue color) to illustrate the cyclical patterns better for public understanding. There are seen to be four of the cycles from a peak hot decreasing to a peak cold temperature followed by a quicker increase than was the decrease rate returning to another peak hot. Using the same "locus line" mathematical method of statistical analysis, the five peak hot points in red define the horizontal red locus line, and the similar red horizontal line defines the four peak (valley) minimums of the temperature anomaly. The red peak hot locus line is very close to being 3 °C (5.4 °F) above the normal temperature of Antarctica and has five recorded instances of that same peak anomaly in temperature during the 420,000 years of the Antarctic ice core samples database being analyzed. The peak cold locus line in red is seen to factually not be as horizontal but the two most recent peak cold anomalies are seen to repeat four times with a temperature anomaly of about 9.5 °C (17.1 °F) cooler than normal.

The five peak hot temperature anomalies, each at close to 3 °C (5.4 °F), are not equidistant in number of years between the peaks. Using a quadratic mathematical expression of the estimated years between the five hot peaks, a mathematical estimate indicates the peaks are repeating but in progressively longer periods between those peaks. The quadratic estimate for the periods of the peak temperature cycles evaluates the interval time between peaks to have been:

Cycle 1 84,000 years
Cycle 2 92,000 years
Cycle 3 106,000 years
Cycle 4 122,000 years

Using that same quadratic, a fifth cycle (Cycle 5) could realistically be expected to repeat in the same general shape and have the fifth characteristic end-of-cycle peak hot temperature about 137,000 years from now.

Before environmental scientists declare another "crisis" in temperature 137,000 years from now, the temperature "crisis" declared by environmental scientists is already being experienced (just as predicted by the ice core database) to be currently upon the planet. From the ice care database, the hot peaks were found to have an estimated duration of about 2,000 or 3,000 years (with a temperature anomaly of about 3 °C (5.4 °F). Based on the ice core history, Earth can expect to be experiencing the environmental scientists' hot "crisis" period for several more centuries, even if the "greenhouse gas" problem does not accentuate the crisis. The prospect for a decrease in temperature to soon occur has some perhaps more dire implications – the Earth will likely be headed toward another "little ice age" with a peak low temperature anomaly of 9.5 °C (17.1 °F) cooler than the average global temperature estimated to exist perhaps 100,000 years from now. That should be an even worse "crisis" facing all life forms on the planet and raises the question: "Should the coal mining operations and coal-burning facilities presently being eliminated, abandoned, or converted to burn with oil or natural gas only be "mothballed" such that they can be used again 100,000 years from now to intentionally create greenhouse gases that will <u>warm</u> the planet?".

A scientifically important value of the Vostok Ice Core Research database is that it provides the database for carbon dioxide in the atmosphere measured in the global scale as opposed to that measured at small scale but at a fallaciously high concentration in a controlled laboratory scale and magnified many scale factors to the enormous full scale and non-homogenous "real world" natural atmospheric conditions. Readers should note that the same four cycles of carbon dioxide varied in a cyclical manner and with the same periodicity as the four cycles of the temperature anomaly. There was no presence of humans burning fossil fuels to affect either the temperature data or the carbon dioxide data. The carbon dioxide data is of all-natural source. Repeating the creation of the mathematical locus extremes of the high and low peaks and valleys of the carbon dioxide, the peak high (296 ppmv) and the peak low (185 ppmv) of carbon dioxide in the atmosphere are seen to occur at the same general time as the four cyclic variations of the temperature anomaly.

Earlier in this Background section, mention was made of a correlation of the Milankovic Cycles (mathematically calculated – and without the aid of any digital processor technology) with the Vostok Ice Core Research database. The correlation for the period 250,000 years ago to now is as follows:

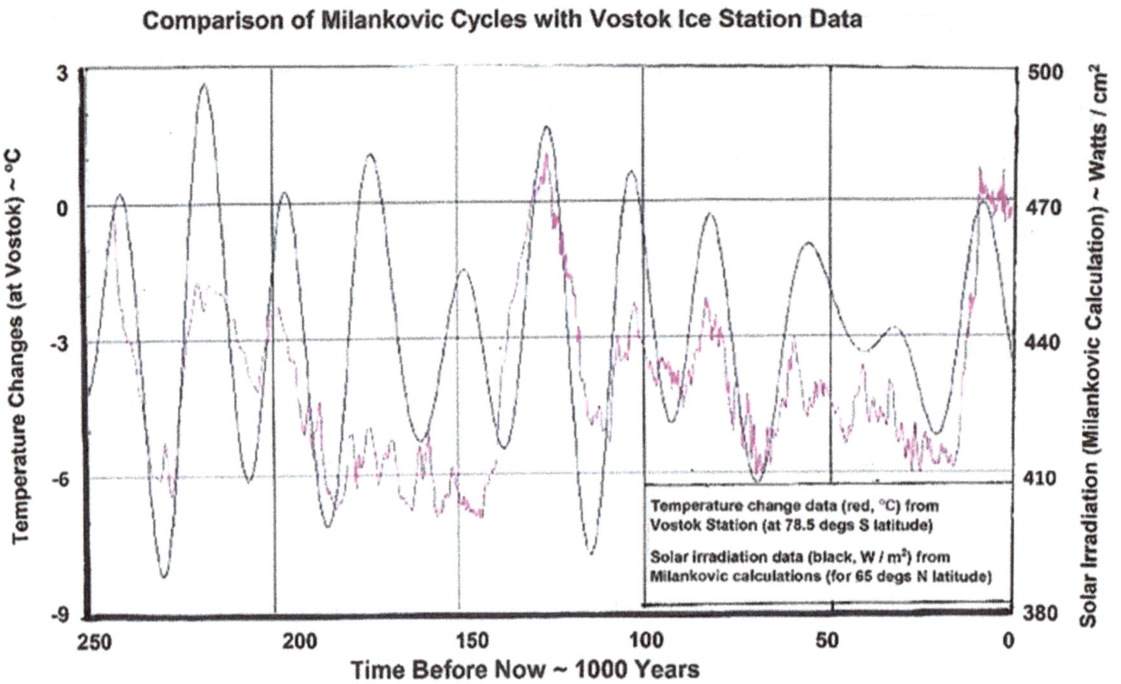

(Source: Smithsonian Institute for Science)

Because the Milankovic Cycle calculations were made for latitude 65° North and the Ice Core measurements and analysis were conducted at the Vostok Ice Research Center at 75° South, the small periodicity of seasonal and Earth's axis of rotation tilt angle variations do not appear at the large time scale shown above. The temperature variation from the ice core analysis having repeated ten times (approximately two times in each 50,000 years) in the 250,000 years before now) correlates well with the Milankovic team calculations (for precession variation cycles of 22, 23, and 24 thousand years) in periodicity but not in amplitude (magnitude), probably due to that large difference in latitude.

NASA Earth Resources personnel have also analyzed the Milankovic Cycles database and conclude that while the Milankovic Cycles provide scientifically valid long-term information about climate change the cycles do not explain the climate changes that occur in short term (50-100 years) periods. That perspective is also very valid, as climate is widely accepted to have resulted from the cumulative effects of daily, seasonal, and short period (less than 25 years) cyclic variations. Within one 23,000 Milankovic cycle, there are undoubtedly a few hundred century period cycles such as that which is presently occurring. The purpose of this book is to point out that it is those near term cycles that are less than 100 years period that people alive now must adapt to - the weather, natural catastrophes (storms, floods, fires that are lightning ignited, blizzards, droughts, tornadoes, etc., any anthropogenic influence (which appears to be minimal on climate change, clearly not dominant). Those unpredictable factors are real and occur differently in different regions.

Coincident with the ice core drilling activities in Antarctica, the National Oceanic and Atmospheric Administration developed a scientific laboratory facility atop the Mauna Loa extinct volcano in Hawaii to conduct scientific measurements directly of the gases in the atmosphere that were being popularly referred to as "greenhouse gases". These are the same gases as the gases referred to by Dr. Arrhenius in Sweden in the 1900s as "hothouse gases". Dr. Charles David Keeling (1928-2005) of the USA was hired by NOAA to be the Director of the Mauna Loa Observatory making the measurements of the atmospheric gas content with emphasis on the carbon dioxide, methane, and sulfur dioxide "greenhouse gases". Dr. Keeling harbored little personal interest in the issue of whether or not humans burning fossil fuels were the dominant cause of global warming – he was truly a dedicated scientist who conducted the measurements with professional care and the best scientific measurement accuracy possible. It was this "Keeling Curve" which environmental scientists in NOAA and the EPA distributed widely throughout the climate science and the public to prove their hypothesis that humans burning fossil fuels were the dominant cause of global warming.

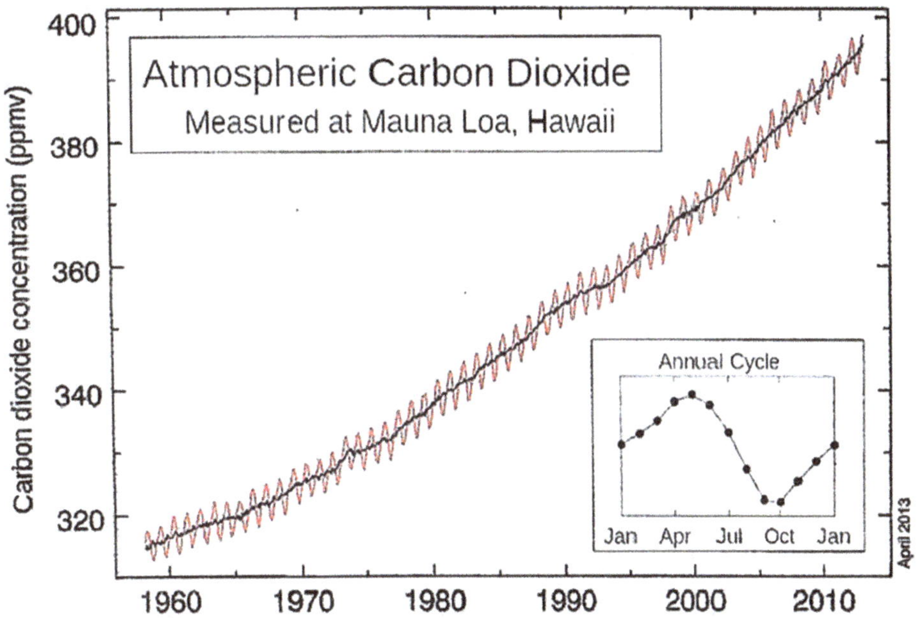

(Source: NOAA)

Because the Keeling Curve carbon dioxide measurements were initiated circa 1958, there is no direct correlation that can be made with the Vostok Ice Core database which produced somewhat unreliable data in the first few feet of the drilling that started in the 1960s and revealed scientifically valid data only prior to 1960.

The "Keeling Curve" was widely touted by environmental scientists and the sensationalistic news media as illustrating the steep increase in the "greenhouse gas" carbon dioxide in the atmosphere ostensibly caused by emissions from the Industrial Revolution practice of burning huge quantities of fossil fuels, especially coal, and those emissions hypothetically being the dominant cause for the increase in "average annual global temperature". The claim of dominance due to anthropogenic causes is a clear indicator that the astrophysical and geophysical causes for variations in temperature and carbon dioxide gas variations due to natural and seasonal causes had not been fully considered. The United Nations Intergovernmental Panel on Climate Change has a membership comprised of only those environmental scientists who are loyal to the fallacious hypothesis. There are close analogies with the "reward for loyalty" politics that are happening since the US Presidential election in 2024 being published by other authors and are listed in the References section near the end of both this book and Book 1.

The vision of every smokestack emitting "greenhouse gases" into the atmosphere were the images shown in newspapers:

(Credit to Unknown photographers)

The photo on the left was taken near sundown at a coal-burning electric generation facility and clearly gives the visual impression that the dark smoke "obviously" contains huge amounts of carbon dioxide from the burning of coal. The photo on the right was taken showing white smoke being emitted at a Pennsylvania steel mill. Why the "obvious" difference in color? Carbon dioxide is a colorless, odorless gas, and not detectable by human vision. The huge clouds of carbon dioxide are in fact water vapor in the state of steam (characteristically white) and are emitted after passage of the emissions from coal combustion water-filled scrubbers (mandated by EPA pollution control regulations) through which the hot gases from the coal combustion must be passed to remove particulate matter. The white steam in both photos is seen to dissipate rapidly as non-visible water vapor a short distance from the facilities. The photo on the left is on the back side of the plumes of smoke where there is a shadow effect – the darkness is not caused by carbon particles in the effluent.

Before being accused of "cherry-picking" of photographs, a photograph must be shown of a true polluter of the atmosphere – that of a lead-acid battery disposal facility which has justifiably been closed:

(Credit to unknown photographer)

There can be no justification for the factual atmospheric pollution of the type shown above which probably contains toxins and carcinogens as well as particulates of carbon, mercury, sulphur, and lead. There is a Lesson Learned however that the above photograph teaches. Batteries of any composition of elements and minerals will continue to be produced for the burgeoning transportation industry – the more people there are in the world, the more vehicles there will be in the transportation system of the world. There is also no "one-type-fits-all" battery for the world transportation system. The developers of batteries for use in public and private transportation vehicles are aware that batteries have a service life and eventually need to be replaced. The disposal of those "dead" batteries cannot be conducted in the environmental polluting manner shown above.

Scientists are advancing the technological chemistry of new batteries for all-electric vehicles. The exponentially growing population will cause the growth in number of electric vehicles on the highways of the planet to also increase in proportion to human growth. Technology developers must develop advanced technologies, but it has become incumbent that the technology of safe disposal of the developed item(s) that wear-out, are damaged beyond repair, or simply have outlived their design life and need to be disposed of must also be developed. The "womb to tomb" concept exists in military aviation technologies (the Author's career field) and that must be considered with the development of all-electric vehicle technology. That factor will be discussed further in the transportation technology section of this book.

(Author's note: The technology of coal burning is justifiably being replaced by cleaner burning alternative sources for heat energy mandated by the EPA with the goal of reducing "greenhouse gas emissions" that ostensibly are causing global warming. Being the son of a man who was employed at the Brooklyn Edison Company in New York from 1921 to 1971 and responsible for converting the coal-burning facilities at the Consolidated Edison Hudson Avenue facility from coal burning to oil burning in the early 1940s, this Author can state with factual confidence that the conversion was NOT done to reduce "greenhouse gas emissions" but WAS DONE to eliminate the requirement of disposal of the ash residue from burning coal. Coal ash contains toxins and carcinogens very harmful to human and animal life, and when the ash residue is dumped into landfills and trash disposal sites, precipitation leaches the hazardous toxins and carcinogens into the aquafers beneath the landfills and trash disposal sites – thereby polluting both the water and the land in the vicinity of large population centers. The EPA emission control features were also installed but cannot be cited as the need to convert the coal burning process to oil burning. The small distance between the Manhattan business district and the coal-burning Brooklyn Edison Hudson Avenue Station (seen in the following photograph (with the four stacks at the left center) located less than a mile from Manhattan across the East River from the Hudson Avenue station):

(Credit to unknown photographer)

The oil storage tanks for the Brooklyn Edison Hudson Avenue Station are seen to the immediate right of the four smokestacks that formerly emitted emissions from the coal-burning process. A perspective of the vitally important Brooklyn Navy Yard (where many military ships for World War II were launched and battle-damage repaired) is seen in the bottom half of the photo.

For his efforts to protect the millions of business workers in Manhattan from those emissions, the Author's father was named Honorary Captain of the oil barge "City of Light":

(Source: Author's Personal Photography File; Consolidated Edison photograph)

There is no visible smoke from the smokestacks as there are much-reduced emissions from the burning of oil. Seen at the right center of the photo are the hoppers and coal storage building that was obsoleted by the conversion of the Hudson Avenue Station from coal to oil. Seen at the left center of the photo is a crane at the Brooklyn Navy Yard adjacent to the Hudson Avenue Station and where many warships during World War II underwent battle damage repair with electric power from the Hudson Avenue facility of the Consolidated Edison.

It should also be noted that the excess steam after passing through the electric generator turbines at the Hudson Avenue Station on the Brooklyn side of the East River is passed through tunnels constructed beneath the East River and that steam is used for the heating of most of the business district of the island of Manhattan. A photo of the Author's father (first on the left) and co-workers inside the tunnel being constructed to transfer steam from the Brooklyn Edison to heat the buildings in the Business District of Manhattan (better known as New York City) in October 1963 is as follows:

Source: Author's Personal Photography File; Consolidated Edison photograph)

(Author's Note: The Author finds the drilling of ice cores by the Russians at the Vostok Ice Core Research to be most fascinating as it parallels the drilling of the granite cores through the riverbed of the East River to guide the excavation bore to move at a speed depending on the content of the earth in the planned path of the steam tunnel. Those granite cores often took a detour by way of the Author's residence where they were discussed at length with Karl Hutchinson who supervised the tunnel construction. Those discussions and "hands-on" experiences with the granite core samples almost persuaded the Author to become a geologist rather than a aeronautical system safety engineer.)

This very "green" activity at the Consolidated Edison facilities was voluntarily done long before the Keeling Curve identified the "greenhouse gas emissions from the Industrial Revolution. (An entire book could be written about the Author's father Karl D. Hutchinson who started employment at the Brookly Edison as a 16-year-old young man from the island of Barbados in the British West Indies and emigrated to New York in the middle of the Great Depression in the early 1920s. The only employment available to him as an unskilled worker was as a maintainer of the coal burners at the Brooklyn Edison facility of the Consolidated Edison Corporation. The shoveling of toxic coal ash residue from beneath the boilers (converting water from the East River into steam to power the electric generators) destroyed the skin on both of his hands and was an affliction he endured for the remainder of his life. His stories told at the dinner table every night inspired his two sons to become high achievers – but that story is left untold, as are so many of that Greatest Generation. This book is co-dedicated to both Author's parents – as described in the Dedication pages at the start of this book.

It was seldom if ever mentioned that of the concentration of total carbon dioxide in the atmosphere having risen from about 320 ppmv in 1960 to about 390 ppmv in 2010 – "an increase of nearly 22% in just 50 years" – 296 ppmv should be attributed to the naturally caused CO2 in the atmosphere as shown by the Vostok Ice Core Database (which is also seldom mentioned). Doing simple conversion to percentage form, the natural-caused pre-Industrial Revolution carbon dioxide at 296 ppmv is 0.0296%. The total carbon dioxide of 320 ppmv is 0.0320% in 1960, and 390 ppmv in 2010 is 0.0320%. The rise in the Industrial Revolution generated carbon dioxide concentration in the atmosphere is factually a rise of 0.007% in 50 years:

[(0.0390% – 0.0296%) – (0.0320% – 0296%)] or [0.0094% - 0.0024%] = 0.007%

The logic for declaring a "crisis" because a gas (one which is not in the neurotoxin category) rises in concentration 0.007% in 50 years – 0.0004% per year – defies scientific reasoning. There is no issue with the scientific validity of the Keeling atmospheric measurement database – only in its interpretation and lack of caveats about being <u>total</u> carbon dioxide in the atmosphere and of that total only about one-third is human and Industrial Revolution generated carbon dioxide.

There persists the inference by environmental scientists that the rise in average global temperature is primarily driven by the "greenhouse gas emissions" from human and industrial burning of fossil fuels. Neglecting to consider the Vostok Ice Core database, environmental scientists use the statistical database from their meteorological database and use the "temperature departure" from an "average global temperature" shown as follows:

Observed Global Surface Warming, 1900-1996

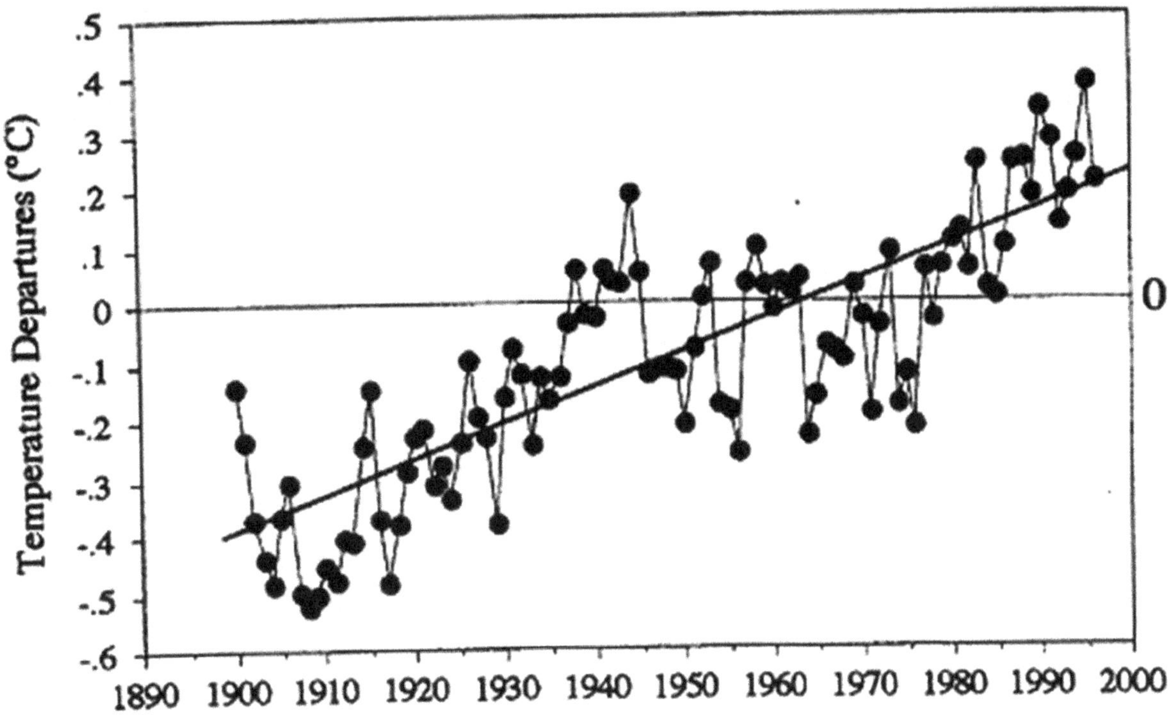

(Source: National Oceanic and Atmospheric Administration, NOAA)

At first glance, the trend of the data points and the linearized trend line through the data points portray a steeply increasing temperature departure with time and the visual perception is that the steep rise in temperature illustrates an ominous forecast. On closer inspection of the scale used in the graphical presentation it is to be noted that there is a decimal point preceding the numbers in the scale – the scale is not in whole degrees it is in tenths of a degree. The approved convention when using a number in decimal form is to have at least one digit ahead of the decimal point even if that is only a 0 as it should have been inserted in the scale for the above figure. With addition of the "leading 0" the rise in temperature is realized properly to be from a departure temperature of -0.4 degrees in 1900 to 0.2 degrees in 2000. The rate of change of the temperature departure was 0.6 degrees net increase in 100 years.

There again, as shown with the amount of human and industrial carbon dioxide in the Keeling Curve rising 0.007% concentration in the atmosphere in 50 years, the manner of presenting the data can be scientifically correct but can be easily misinterpreted by a non-scientist. The environmental scientists, eager to alarm the gullible public, manipulated the scale of the dependent variable (the numbers on the left vertical axis) to be in decimal fractions of a degree (Centigrade) and which greatly overexaggerates the temperature change. The temperature sensors in human skin exposed to the surrounding air hardly distinguishes a change in temperature of less than two degrees much less a change of one-tenth of a degree. As can be seen in the years between 1900 to 1940, the temperature departure from an already averaged global temperature was at most 0.5 °C. From 1940 to 1970, the temperature departure declined about 0.2 °C. From 1970 to 2000 the temperature departure reversed from a decreasing trend to an increasing trend and rose an increment of 0.5 degree.

(Author's note: There were/are many illustrations presented by environmental scientists in communications where the data is intentionally presented at a magnified scale to be easily interpreted for scientific significance. That magnification for scientific purposes often translates to a misinterpretation of significance (sometimes unintentional but more often intentional) by the scientifically challenged public. The quote of Sir Walter Scott in 1808: "Oh, what a tangled web we humans weave when we practice to deceive" poetically describes the unethical practice of a large group of present-day environmental scientists, who describe the intended misinterpretation to represent being "settled science".)

Humans instinctively form a first impression about a visual scene by its optical appearance – size, shape, color, facial expression, "body language" – before interpreting the meaning of its content. In the scientific illustration above, apperception of there being a very steep rise in temperature is seen by the high slope of the line "linearizing and averaging" the overall trend of the temperature. The fact that one tenth of a degree temperature increase occurred over a one-year span has little meaning to a public observer who typically experiences a variation in temperature of more than 20 degrees during the course of each day. The scientific temperature change (measured with high technology instruments) of one half of a degree in a year is of little significance to the average person who endures a cyclical change in temperature due to seasonal excursions in temperature of near freezing in the early seasonally cold months of the year, getting warmer through Spring season, becoming uncomfortably hot and humid in mid-summer, then in the fall becoming cooler and causing the leaves on deciduous trees to turn spectacular shades of color before falling. The season is officially termed Autumn, but the public regards it as Fall because of the leaves on trees falling to the ground – caused to fall by the unseen scientific accelerative effect of gravity on a mass.

In the same analogy, scientists predict and measure precipitation - caused by gravity to fall as rain or snow – but the public "sees" rain as drops of water or snow as flakes of frozen water. regardless of how they move because of unseen forces (gravity and wind). Those in the US public who were alive at the time replay the music in their mind of a popular singer crooning "I'm Dreaming of a White Christmas". There is no angst about 20 degrees changes in a day – why should there be angst over an increase of one-quarter of a degree in the coming decade? The average person simply chooses the layers of clothing from his/her wardrobe to be comfortable throughout the day despite the temperature changing from a state of cooler than comfortable as he/she departs the residence and adjusts the layers of clothing during the day as the temperature transpires from uncomfortably cool to uncomfortably hot and then reverses to uncomfortably cool once again. There is no "crisis".to the human population. Perhaps oceanographers are concerned about the ocean waters rising an average of three quarters of a degree in a year causing the anemones on a coral reef to bleach white, but realistically that is of little concern to the 8 billion people of the world who will never see a coral reef.

Another example is that of scientists choosing the metrics used in presenting data to be easily misinterpreted – and overexaggerated in meaning. The US Department of Energy issued a report in 1990 when the NOAA and EPA and the UN IPCC environmental scientists were "on a roll" with the global warming and climate change "crisis" ostensibly being caused by "greenhouse gases" emanated by humans and industrial sources. Readers are asked to form a first impression of this tabular presentation (by the US Department of Energy in 2000) of the extent of "greenhouse gases in the atmosphere" before the Author can describe its scientifically valid meaning:

TABLE 1.
The Important Greenhouse Gases (except water vapor)
U.S. Department of Energy. (October. 2000) (1)

(all concentrations expressed in parts per billion)	Pre-industrial baseline	Natural additions	Man-made additions	Total (ppb) Concentration	Percent of Total
Carbon Dioxide (CO2)	288,000	68,520	11,880 (2)	368,400	99.438%
Methane (CH4)	848	577	320	1,745	0.471%
Nitrous Oxide (N2O)	285	12	15	312	0.084%
Misc. gases (CFC's, etc.)	25	0	2	27	0.007%
Total	289,158	69,109	12,217	370,484	100.00%

(Source: US Department of Energy)

The DOE apparently intended to illustrate that there are other gases considered to be important "greenhouse gases" in addition to the main greenhouse gas of carbon dioxide. Readers may not have noticed, but the use of the previously presented concentration being in parts per million by volume (ppmv) has been redefined to be parts per billion by volume (ppb). That magnifying of the scale accounts for the trace amounts of methane, nitrous oxide, and other gases that have miniscule heat-trapping properties, to become of exaggerated importance. A concentration of Man-made methane at 320 parts per billion is given far more significance than is the same concentration expressed as 0.320 parts per million. The public should recognize that the concentration of total methane in the atmosphere at 1,745 ppb is the same as 1.745 ppm and is the same as 0.01745%. "No matter how thin "boloney" (bologna) is sliced to get more slices per pound, it remains as being mere boloney".

That change in convention from ppmv to ppb should also be noted to be in printer's font size approximately half the size of the font used in the remainder of the text and numbers in the table. It must be admitted, a total concentration of "greenhouse gases" in the atmosphere of 370,484 ppb is a huge number, and the impression intended is that that much "greenhouse gas" in the atmosphere must be a crisis. The last column uses the more familiar percentage to emphasize that carbon dioxide is 99.438% of the "greenhouse gases" in the atmosphere. Barely squeezed into the middle column of the Table 1 above with the heading "Man-made additions" is the anthropogenic carbon dioxide addition at a concentration of 11,880 ppb. When expressed in percentage, the numbers in the table become very unimpressive. The Man-made addition of carbon dioxide in the atmosphere shown as being 11,880 ppb is factually a much less ominous 0.01188% concentration.

Not to be overlooked, the title of the Table 1 is "The Important Greenhouse Gases (except water vapor)". Environmental scientists typically ignore the water vapor as being a "greenhouse gas", but the fact is that water vapor is equally if not more a heat trapper than is carbon dioxide and there is on average 2-3% of water vapor in the atmosphere. Using parts per billion (ppb) metrics, the atmosphere may contain as much as 30,000,000 ppbv of water vapor. If environmental scientists consider 368,400 ppb of total carbon dioxide in the atmosphere to be an environmental crisis, then why is 20,000,000 ppb of water vapor in the atmosphere not considered to be even more of an environmental crisis? If there is a true crisis over "greenhouse gas" effects, does it matter whether the source of the greenhouse gas is natural or human-made or some combination of both – and does it matter whether only the environment is affected, or only human safety/health is affected? Why is it not some combination of both?

Dr. S. Fred Singer (1924-2020), renowned physicist and Professor Emeritus in Environmental Science at the University of Virginia, was most critical of his peer environmental scientists for ignoring the effects of water vapor on global warming and climate change. For many years, Dr. Singer was very outspoken of the practice of environmental scientists eliminating water vapor so as to exaggerate the effect of carbon dioxide being the dominant "greenhouse gas" in the atmosphere. Dr. Singer presented two charts (from his "A Closer Look at the Numbers" lectures) that are shown below in juxtaposition illustrating the drastic difference in the "greenhouse gas effect" on global warming when water vapor is considered and when it is not:

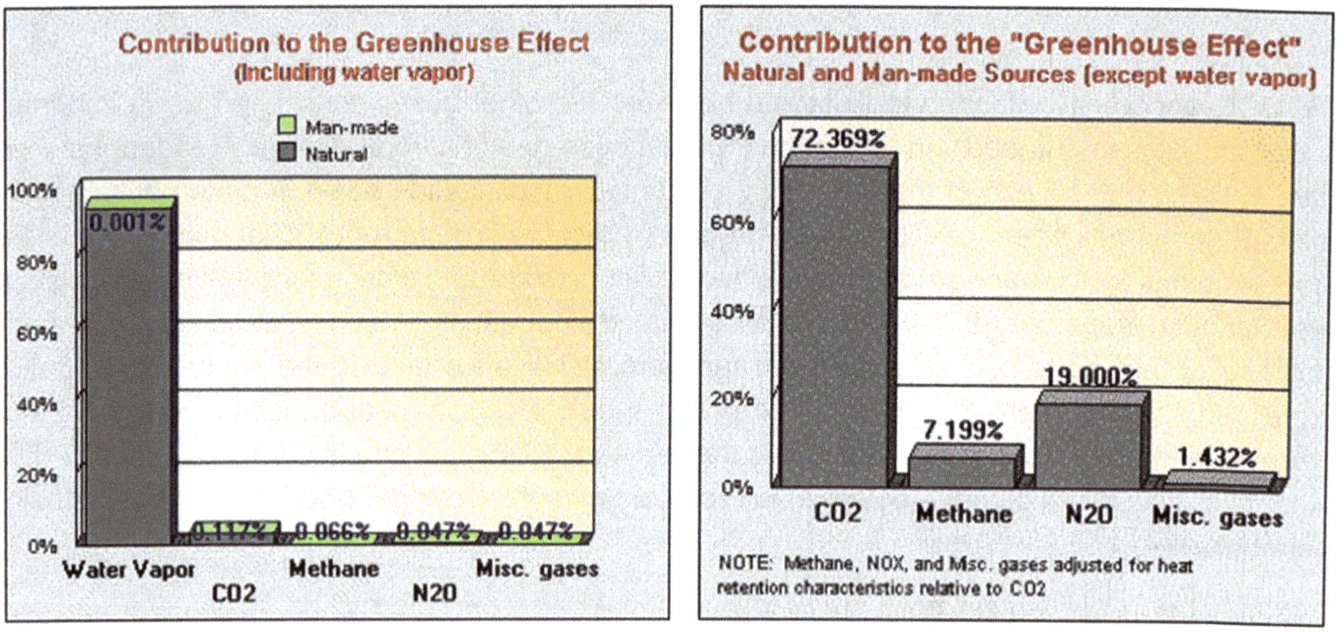

(Source: Dr. S. Fred Singer, University of Virginia)

The environmental science group simply declared that Dr. Singer was just another "naysayer" and continues to display data from which water vapor has been "eliminated" – as shown in the illustration on the above right. At the start of this Background section of this book it was shown that describing a scientist by the term "naysayer" was intended at the time (circa 1600) as a derogatory descriptor. Galileo is recognized in the history of science as the greatest "naysayer" of the scientific hypothesis that the Sun revolved around the Earth. Yet for the next three centuries there were those in the Flat Earth Society that disagreed with Galileo. Sad to say, the "naysayers" of the twenty-first century hypothesis that Man burning fossil fuels is the dominant cause of global warming are labeled also as "naysayers" or as the slightly-less derogatory "contrary scientists". Dr. S. Fred Singer and Dr. Judith Curry are identified as "contrary scientists" by the environmental science cult.

When the Soviet Union economy collapsed circa 1985, the Soviet (Russian) scientists accepted the offer for environmental scientists from France to participate in further drilling and continue the ice research and analysis of the ice cores at the Vostok Ice Core Research Station in Antarctica. The earliest presentation of the results of the Vostok Ice Core Research measurements and analyses appeared as below and it should be noted how the scales for the gases contained in the ice cores analyzed were exaggerated to give the appearance that the "greenhouse gas" effect – even though there is no "Man-made additions" contained therein – was extremely large compared to the temperature departure variation with time.

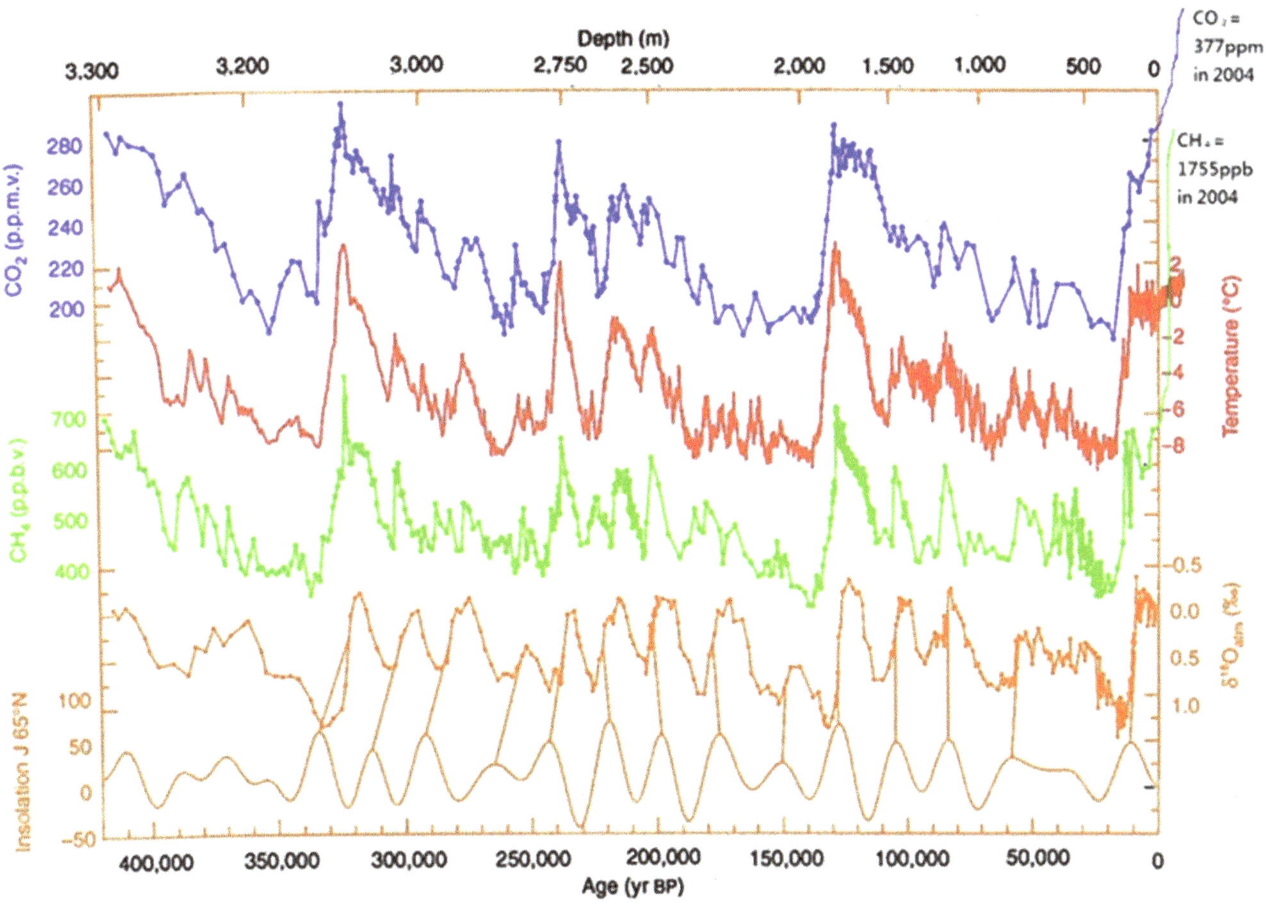

(Source: NOAA, with addition of the CO_2 and CH_4 measurement by NOAA for 2004 from the Mauna Loa Observatory)

The top three lines of data are direct measurements (albeit with a mix of ppmv and ppbv scales for the gases) to imply the relative importance of the temperature data (in red color), the carbon dioxide concentration in ppmv (in blue color), and of the methane concentration (in green color). NOAA environmental scientists added the blue (CO_2) data extension and the green (methane) data extension and continued the red (temperature) information as recorded at Dr. Keeling's Mauna Loa Observatory for Atmospheric Research. The added temperature departure at the original scale is seen to be changing at a very slow rate per recent hundreds of years, whereas - at their exaggerated scales - the fallacious impression is given that the carbon dioxide and methane are "soaring" in magnitude.

Very soon after the original Vostok Ice Core Research Center measurements were introduced into the science records (circa 1995), the French scientist Dr. Nicolas Caillon (and a few associates from the Institut des Geosciences de l'Environnement) forensically analyzed the Vostok ice cores on a closer time scale and concluded the following relationship between the temperature of the atmosphere and the carbon dioxide content of the atmosphere:

CO_2 lags temperature rises in ice core data

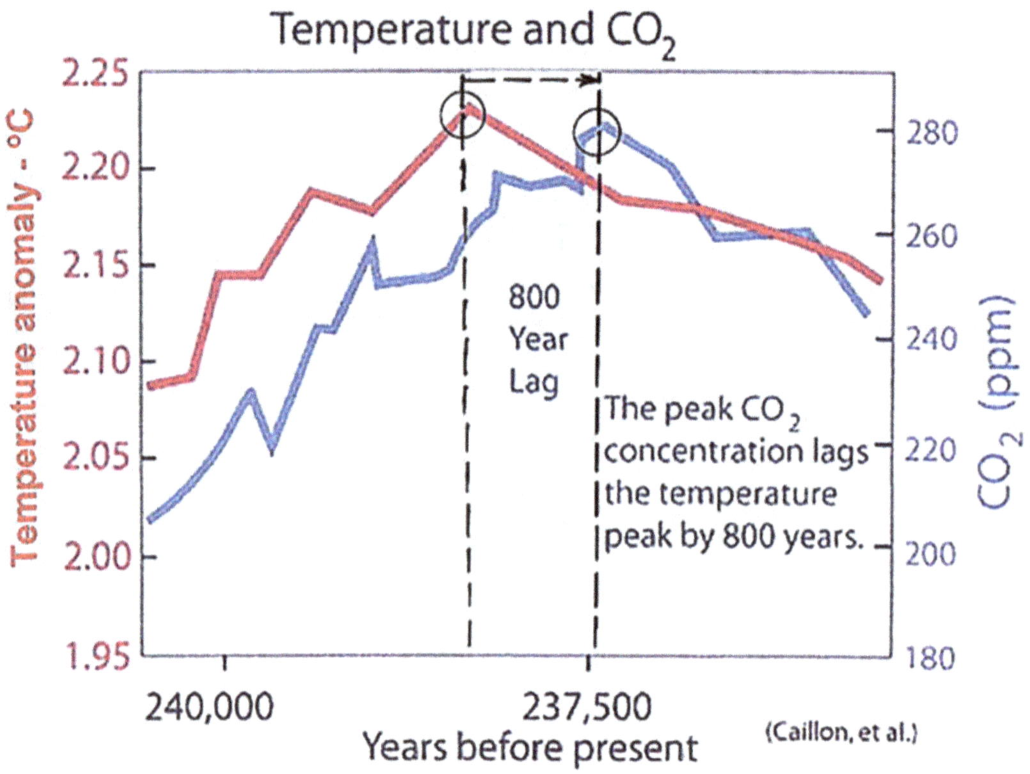

(Source: Smithsonian Institute for Science)

With a closer analysis of the ice cores in the period about 237,500 years before the present (which is the time period at the end of the Cycle 2 and start of the Cycle 3 cyclical variations in the Author's analysis reflected in the previous graph of the full 420,000 years before present Vostok ice core database), the important conclusion was determined that the peak of the all-natural carbon dioxide gas in the Antarctic atmosphere lagged about 800 years after the peak of the temperature anomaly. The significance of that conclusion is that it is not the carbon dioxide concentration level (peaking at 296 ppmv) that drives the temperature anomaly (peaking at about 3 °C (5.4 °F). The inverse of that importantly scientifically valid conclusion is that it is temperature of the atmosphere that drives the amount of natural carbon dioxide in the atmosphere.

A logical extension of that conclusion is obtained by inferential logic that if natural carbon dioxide at a concentration of 296 ppmv does not drive the temperature of the atmosphere, then the addition of 124 ppmv of human and Industrial Revolution carbon dioxide in the atmosphere is neither a dominant nor highly significant driver of the temperature of the atmosphere in 2025. The NASA satellite sensed temperature anomaly measurements of the land and water surface temperatures reflect a very regional distribution of temperature anomalies – the variation in 2023 is provided as follows:

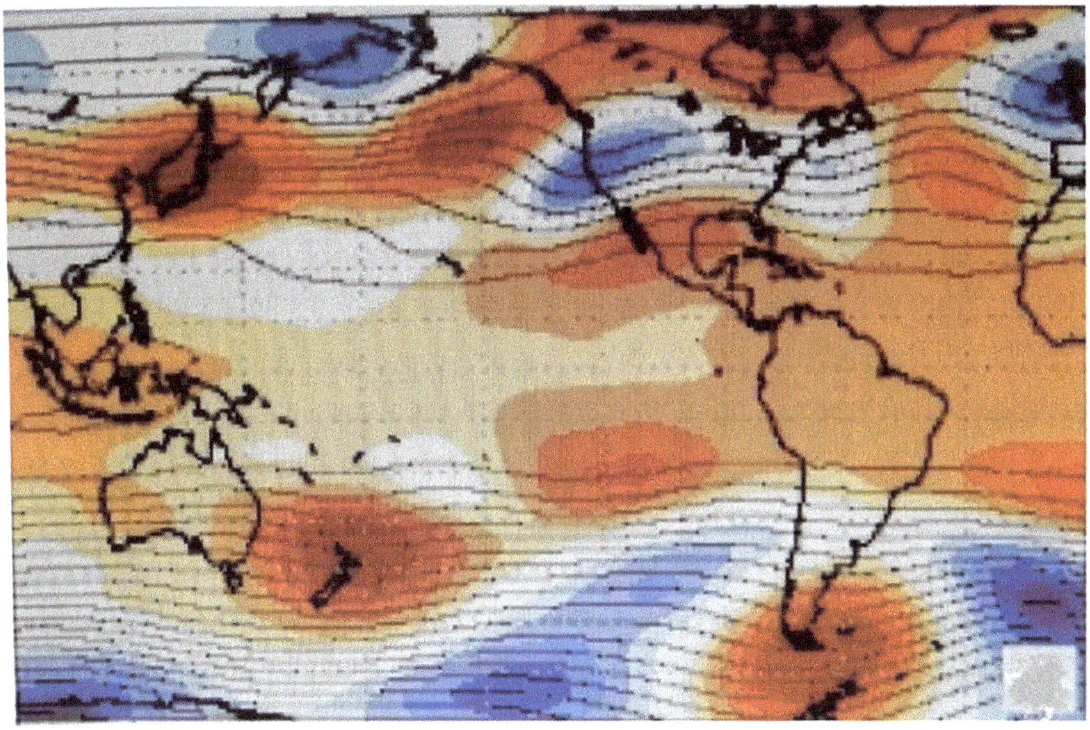

(Source: NASA)

When a map of the oceans' and wind currents of the world is considered (shown as follows), it is readily seen that the "average sea surface temperature" is on a par of unimportance to global warming as is "average global temperature" to climate change on land surfaces.

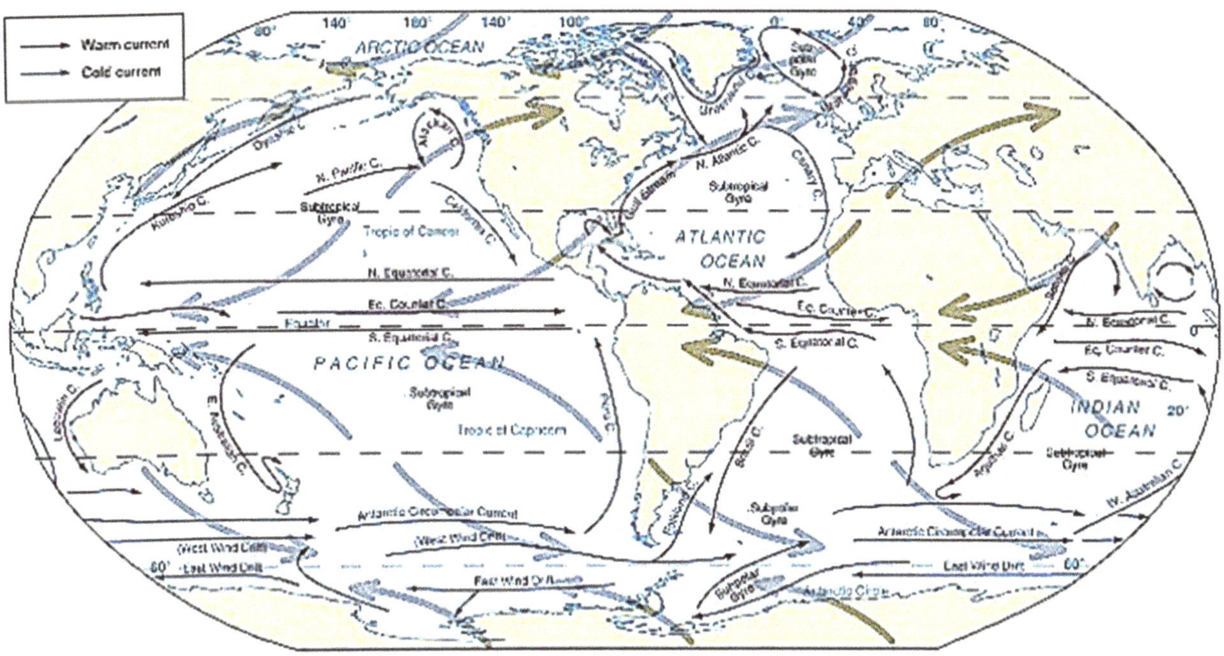

(Source: World Meteorological Organization (WMO)

The temperature of the Gulf Stream is at about 26 °C (80 °F) on the East coast of the United States while (as shown in the world map of ocean currents above) at the same latitude on the West coast the Pacific/Aleutian current is at a temperature of about 16 °C (60 °F). Neither US coast is in a temperature "crisis", and most certainly not from the "average sea surface temperature" of about 21 °C (70 °F).

It should be mentioned in passing that a correlation was attempted of the Milankovic Cycles calculations also with ocean bottom sediment cores (collected by Dr. Hays et al.), similar to that which was done with the Vostok ice core samples. The Author did not analyze the results of that correlation to as much extent as he did the ice core samples because the ocean currents and transfer of temperature in those currents could pollute the scientific interpretation of the ocean floor sediments carried from one region to another.

The specter of the ocean levels rising at an alarming rate is also seen on the following illustration:

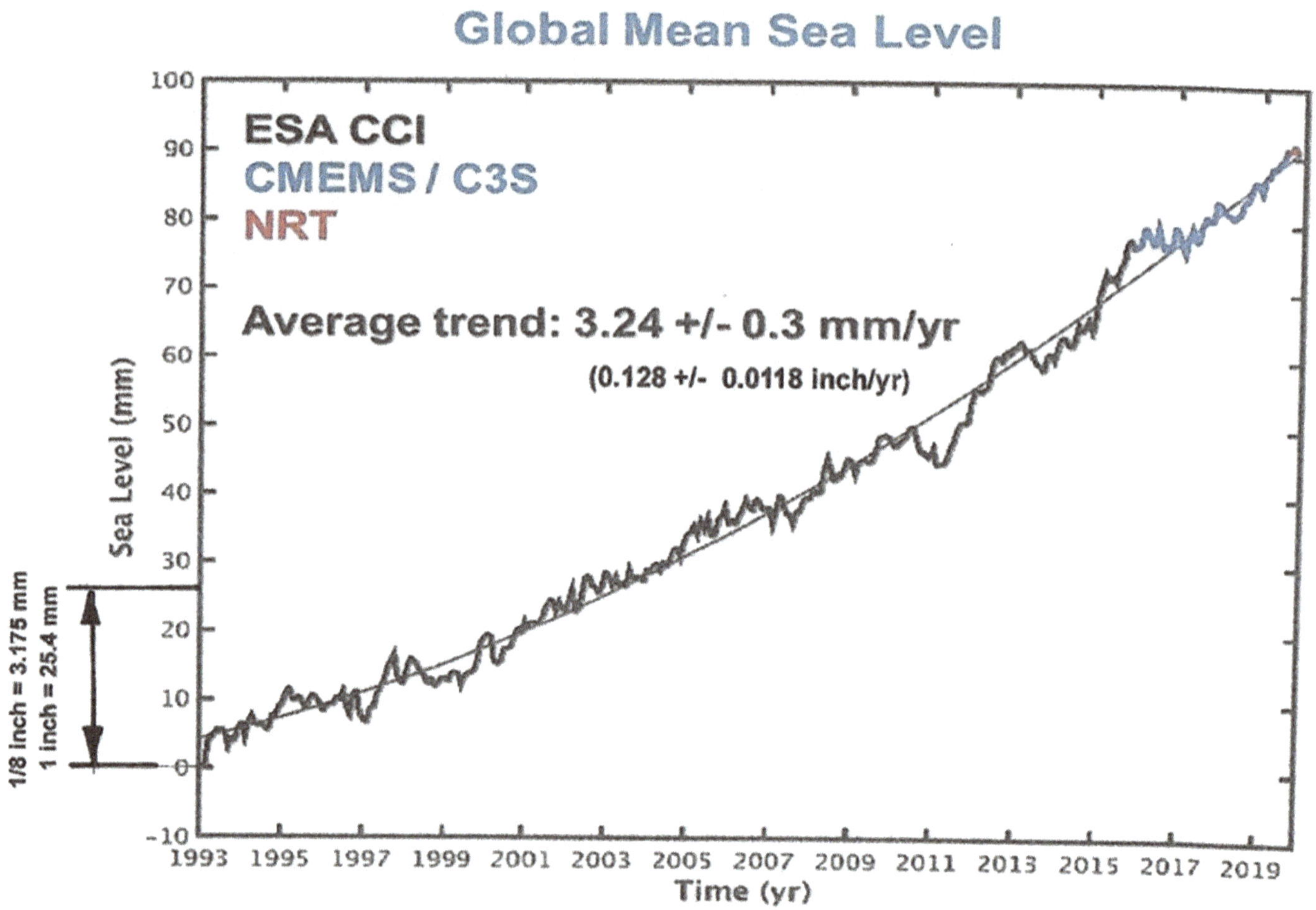

(Source: World Meteorology Organization with Author notations)

The measured rate of increase of average global mean sea level of approximately 1/8th inch per year will result in a rise of about one foot in 96 years and two feet in about 200 years. The useful life of a moderately well-constructed wooden frame house is under 96 years on land not in a flood zone; however, wooden frame homes constructed near a shoreline of an ocean have an expectancy of under 75 years (much less in "hurricane alley or tornado alley"). Logic suggests that oceanside homes and small businesses less than two feet above current sea level when rebuilt should be located further inland at higher elevation than 5 feet above sea level.

A database that the Author found perplexing was that which apparently had been "hidden in the bottom of the filing cabinet" by environmental scientists. Shown in the following diagram is a global "big picture" of the use of coal to meet the world population need and demand for electricity:

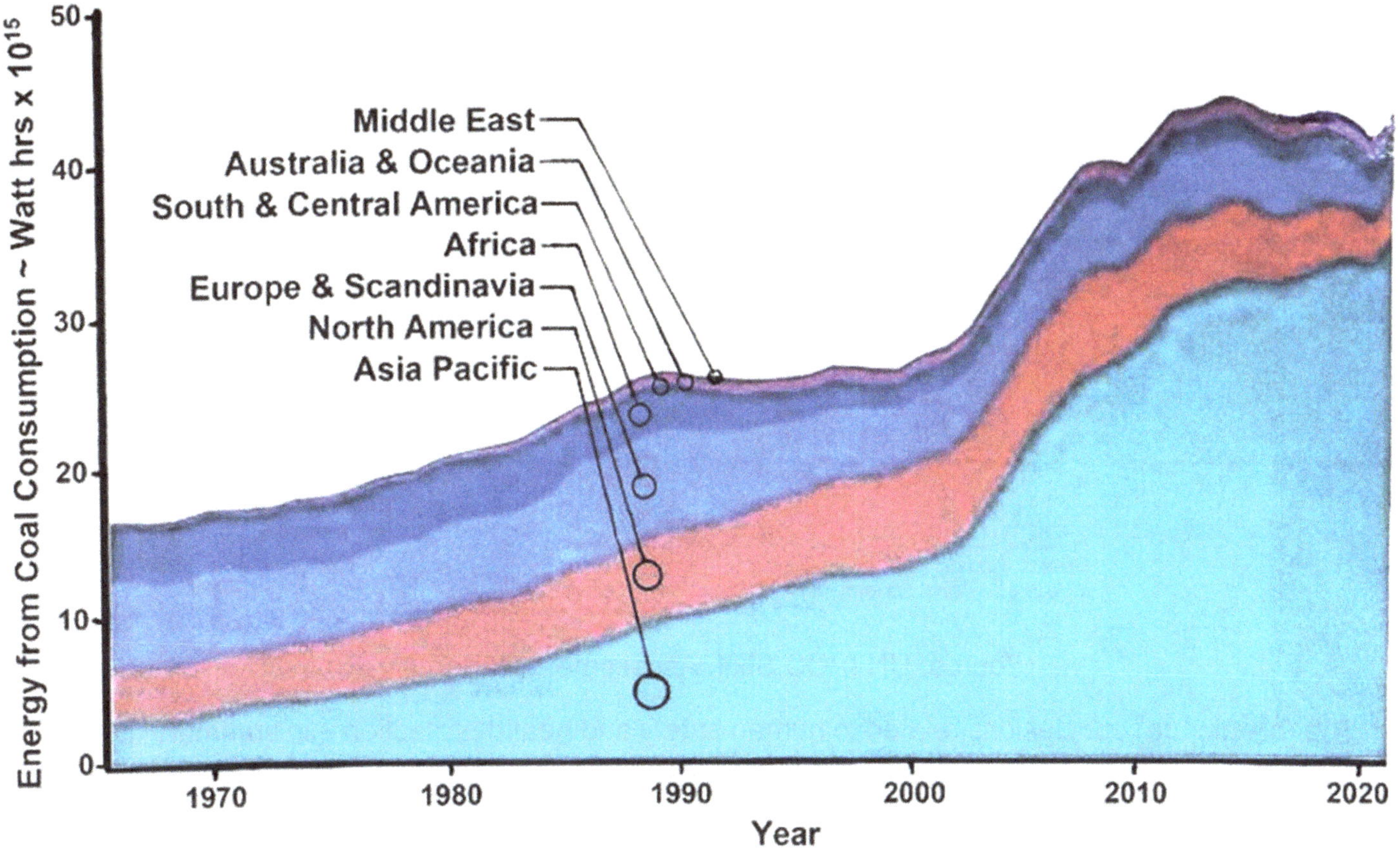

(Source: British Petroleum Company p.l.c.)

As is easily seen, from about 1970 to 1990, the seven major regions (listed at the top of the diagram) of the planet were increasing the burning of coal for meeting the planet's population demands for electrical energy. The Asia Pacific region grew at the fastest rate. When the environmental science and Environmental Protection Agency of the United States launched the "combat against climate change and developed their hypothesis that humans and industry were the dominant causes for increases in "average annual global temperature" and "climate change", the six regions exclusive of the Asia Pacific region can be seen to have reduced their combustion of coal gradually after about 1992 to about 2000. From 2000 to 2020, the Asia Pacific region more than doubled its already highest user of coal for generating electricity.

There is a high probability (at over a level of 90% in this Author's estimation) that Asia Pacific analysts of the same databases found and analyzed by the Author came to the same factual conclusions as did the Author from his "big picture" analysis of those astrophysical, geophysical, and natural databases. The United Nations Intergovernmental Panel on Climate Change (IPCC) corroborates the British Petroleum numbers showing that there are regional (not simply global average) climate differences found shown in the above diagram, but there was found no recognition of the "greenhouse gas emissions" from human and industrial sources being the dominant cause for the global warming and climate changes shown as follows:

While changes in climatic impact-drivers are projected everywhere, there is a specific combination of changes each region would experience

(a) World regions grouped into five clusters, each one based on a combination of changes in climatic impact-drivers

Assessed future changes: Changes refer to a 20–30 year period centred around 2050 and/or consistent with 2°C global warming compared to a similar period within 1960–2014 or 1850–1900.

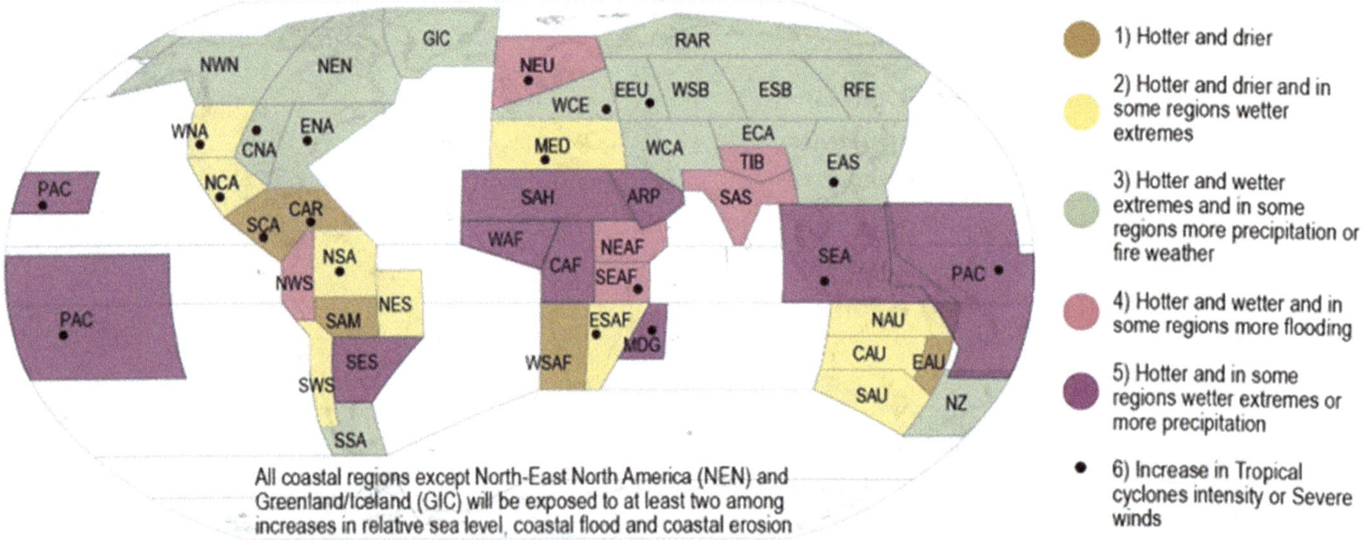

(Source: UN IPCC Sixth Assessment Report (AR-6)

The "big picture" conclusions related to human safety and health protection – in summary from Book 1 and from this Background in the foregoing pages of this book – are found by the Author to be:

1) There has been more than ample data analyzed that confirms that climate changes are factually occurring on the planet.
2) The databases analyzed reveal the climate changes are however dominantly caused by other factors (astrophysical, geophysical, and natural) than are caused by humans and/or industrial activities generating "greenhouse gases" that are being emitted into the atmosphere.
3) The technology of burning coal to be used for industrial energy needs should be discontinued in the United States because the ash residue from coal burning is now known to be toxic and carcinogenic, not solely because it causes "greenhouse gases" in the atmosphere leading to global warming and climate change. Disposal of hazardous coal ash residue in landfills and trash dumps is polluting the water aquafers and nearby land used for growing food products.
4) The discontinuance of the fossil fuel coal does not require the discontinuance of other forms of fossil fuel (oil and/or natural gas) to reduce "greenhouse gas" emissions – in the United States or any other region – as neither oil nor natural gas generates significant toxic or carcinogenic ash residue causing disposal pollution problems.
5) Pollution (especially from plastics) is a far greater hazard to humans and other forms of animate life than is the "greenhouse gas" emissions hazard affecting the climate.
6) The databases analyzed indicate that the magnitude of the changes in temperature of the land, of the ocean waters, and of the atmosphere are not increasing equally everywhere and are much slower in rate of change than the public has been led to believe and has been coerced into acceptance as factual.

7) The changes in the climate are regional in nature and not of an "average global" nature. Prediction models applicable to different regions (using the Koppen Geiger method and the LEO (Latitude, Elevation, and Ocean proximity) model developed herein) should replace the "average global temperature" model. (A Patent Application has been filed for the LEO Method by the Author; however, the patent holder encourages any person, organization, or entity wishing to use the LEO model to feel free in using the model without fear of patent infringement.)
8) The fallacious hypothesis of environmental scientists which has caused the unnecessary angst in the public that there is a "doomsday crisis" existing, and/or if a "tipping point" in "average global temperature" is exceeded is scientifically proven in these two books on the factual causes and effects of global warming and climate change to be unwarranted.
9) The technologies that were developed to reduce "greenhouse gas" emissions – e.g., solar, wind, alternative fuels, carbon removal from the atmosphere, increased use of all-electric vehicles, etc. - are useful but have been found inadequate to meet the increasing need for reliable, clean, affordable electrical energy in the applications that have mushroomed in recent decades.
10) The prospect of the world population growing exponentially creates technological needs in many different categories of technology associated with human and industrial activities (alternative fuels, electric and heat energy storage batteries, desalination of ocean water, non-polluting fertilizers in agriculture, transportation, infrastructure, etc.) The technologies most useful in a region are dependent on the unique resources of that region and on the stage of technological development in that region).
11) The setting of unrealistic emissions goals and "average global temperature" and "net zero" goals is a futile attempt to control global warming and climate change. The legislation and judicial actions that penalize any entity for failure to meet those emissions and global temperature goals are not for the benefit of man-kind. They are veiled attempts for organizations and government departments to exercise authority in the guise of controlling pollution, and they need to be terminated.
12) Educators at all levels of education need to instruct students in the factual scientific causes for global warming, climate change, and ocean levels rising, and for the need of the public to be more considerate in reducing or eliminating their out-of-control activities causing pollution problems and/or wasteful expenditures of energy that are not "for the benefit of mankind and to the environment.

The foregoing background information and resulting conclusions are presented as the "past" portion of the approach stated earlier for this book - "Study the past. What is past is prologue." There is no crystal ball to accurately predict how the climate of the planet will change in the near future but with the now better-known factors that are potentially important for meeting and overcoming the challenges of climate change. It has been repeatedly stated that challenges bring opportunities. Scientists, the public, and government regulators and legislators must form an alliance to plan for and efficiently mitigate the effects of the inevitable climate change. The forestated conclusions and suggestions of this Background, and those in the Introduction and the Chapters forthcoming in this book may be thought of and used as a "strawman plan" for how that can be achieved.

The background information also serves as the transition from the "Facts and Fallacies of Climate Change on Our Planet" book, recently published, to the contents of this book "Dealing with the Real Climate Change" which identifies, evaluates the need, and prioritizes the technologies to assist the human population in the challenges of the changes in the daily and weekly weather, the astrophysical, geophysical, and annual seasonal changes, and the inevitable longer term climate changes (as predicted from the Vostok Ice Core Samples and the Milankovic Cycles databases).

It was only a century ago that the United States was in the midst of "Dust Bowl" environmental and "Great Depression" human and economic catastrophes and recently successfully overcame the medically disastrous effects of a global Covid pandemic. It was shown that when the public and government leaders at all levels performed as a team working toward a realistic common goal, the result will be an improvement for the betterment of the planet and its passengers on the fast trip through life in the solar system. A perhaps "corny" poster that was seen everywhere as World War 2 raged was helpful then – and may be helpful now with the very realistic challenges *du jour* – for the betterment of mankind:

Preface

As a transition from the phase of identifying the factual causes of global warming (which is more of a problem to some regions and not to others) and climate change (which also is a regional not global problem), this book offers suggestions for the actions that the world population should take to mitigate the effects (regardless of cause) of that inevitable global warming and inevitable climate change. It should be apparent that environmental scientists developed a hypothesis that may have been correct for the observation that was made – and was unscientifically interpreted – for the time it was made (early twentieth century). After six or seven decades with more informed observations, that original hypothesis that the burning of fossil fuels was the dominant cause for global warming and climate change can only be regarded as a first step in understanding global warming and climate change.

Sir Arthur Conan Doyle (1859-1930, creator of the legendary but mythical master Scotland Yard detective Sherlock Holmes) stated in 1905 – when asked about the "hothouse gas" conclusions of atmospheric chemist Dr. Svante Arrhenius in Stockholm, Sweden, after Arrhenius had just received the Nobel Prize for Chemistry - *"It is a capital mistake to theorize before one has data. Insensibly, one begins to twist facts to suit theories instead of theories to suit facts."*.

This Author has substituted the descriptor "fallacious" for the descriptor "mistake" as the science profession is beginning to realize environmental scientists derived a partially correct conclusion appropriate eight or nine decades ago but had not arrived at an outright mistaken conclusion. "Greenhouse gases" do trap heat in the atmosphere, but it is the largest greenhouse gas – naturally-generated water vapor (at a concentration of about 2-3% of the atmosphere by volume) – which traps about 100 times more heat than does human/industrial generated carbon dioxide (at a concentration of only 0.02% of the atmosphere by volume). The error in the environmental scientists' original fallacious hypothesis was that water vapor was intentionally disregarded as being a "greenhouse gas" because it was not human made.

Data (sensed in 1992) from satellites reflects the powerful heat-trapping effect that natural water vapor in the atmosphere has over the warmer oceans in the lower latitudes:

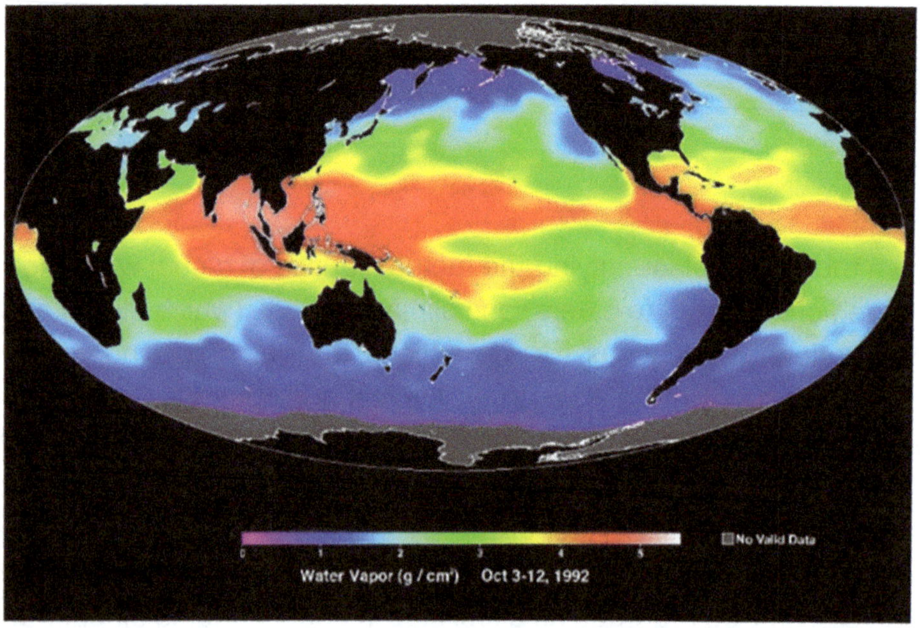

(Source: NASA)

The scale in color and with numerical correspondence of the mass amount of water held in the water vapor ranges from 0 to 5 grams per square centimeter. The amount of water vapor is dependent on the temperature of the atmosphere according to the following principle (graphically shown) of atmospheric physics:

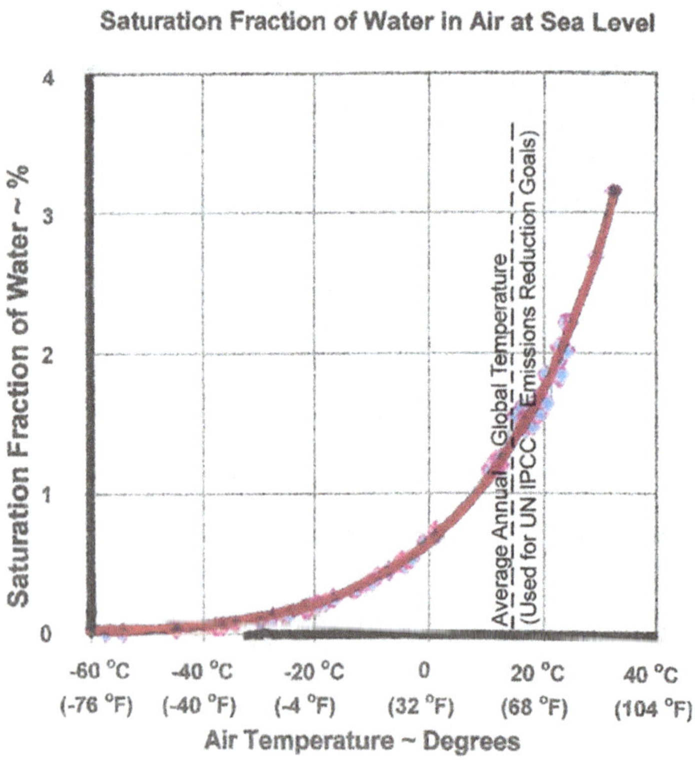

(Source: World Meteorological Organization database, graphic improved by Author)

The dashed vertical black line at 16 °C (61 °F) is the "average global temperature used by environmental scientists in the UN Intergovernmental Panel on Climate Change (UN IPCC) as the reference "line in the sand" for setting the goals on surface temperature limit of global warming (the global temperature goals set in the UN Paris Agreement). At an air temperature at sea level of 16

°C (about 61 °F), the atmosphere could contain about 1.4% water vapor – as it does globally very frequently and up to 3%. However, in the environmental scientists' crusade of eliminating the 0.0128% of human and industrial-made greenhouse gas carbon dioxide emissions the 1.4% of water vapor is disregarded. When Dr. S. Fred Singer (Professor Emeritus in Climate Science at the University of Virginia) brought that omission to their attention, he was branded as being a "naysayer" – much like Galileo was. Nevertheless, Dr. Singer was correct – water vapor is the dominant "greenhouse gas" as he showed in his many lectures on the subject:

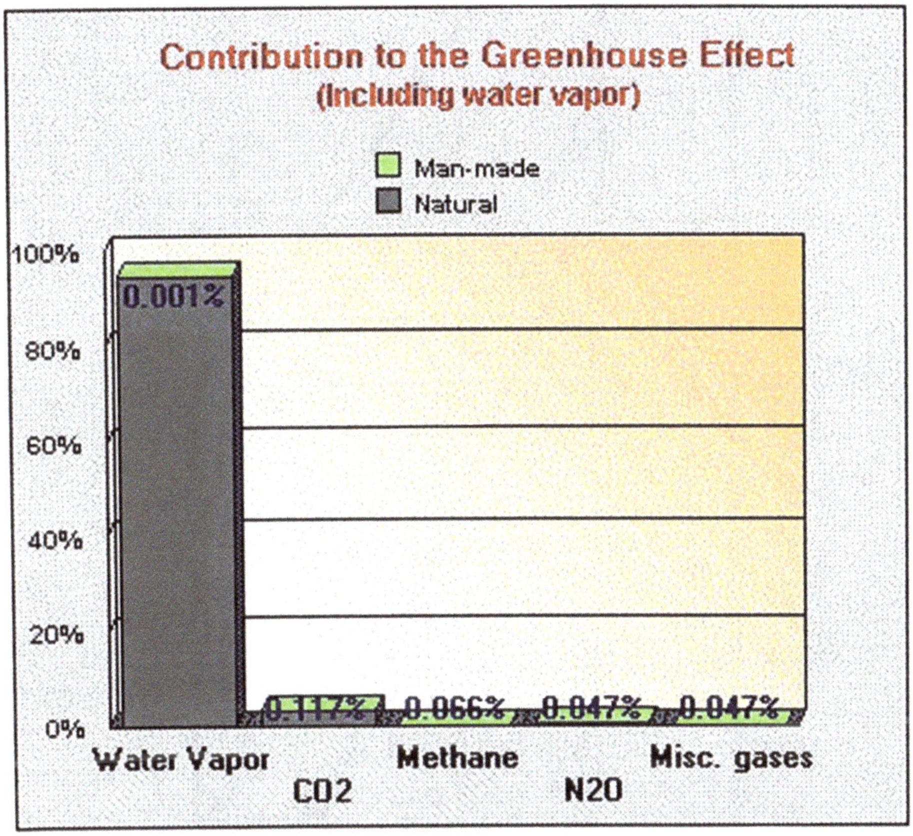

(Source: Dr. S. Fred Singer, Professor Emeritus, University of Virginia)

The Author and Dr. Singer discussed this "oversight" made by environmental scientists and concurred that it was as serious an error in judgment as the error in judgment of the use of an "average annual global temperature" to represent the several complex causes for global warming. The use of any "average" for variables that are affecting the environment, and the human population is meaningless because almost nothing in nature or in human life behaves in an "average" manner. The environment and the population must endure the extremes to which they are and will be exposed. Dr. Singer and the Author concurred - simplifying a complex observation of a natural phenomenon in order for any scientist to be able to derive a "model" for the phenomenon has no place in the ethical world of science. The misleading of the public and government regulators to believe the environmental scientists' misjudgment only compounds the problem of global warming and climate change.

Having mentioned one of the most important factors of the global warming and climate change challenges presented to the population, and stating that the humans and environment must survive the extremes of temperature not merely the average, a preview of what those extremes are like is shown as follows: (It should be borne in mind that this set of temperature extremes is for the latitudes that encompass the United States region. Each different region must develop the temperature extremes from meteorological historical databases for that region.)

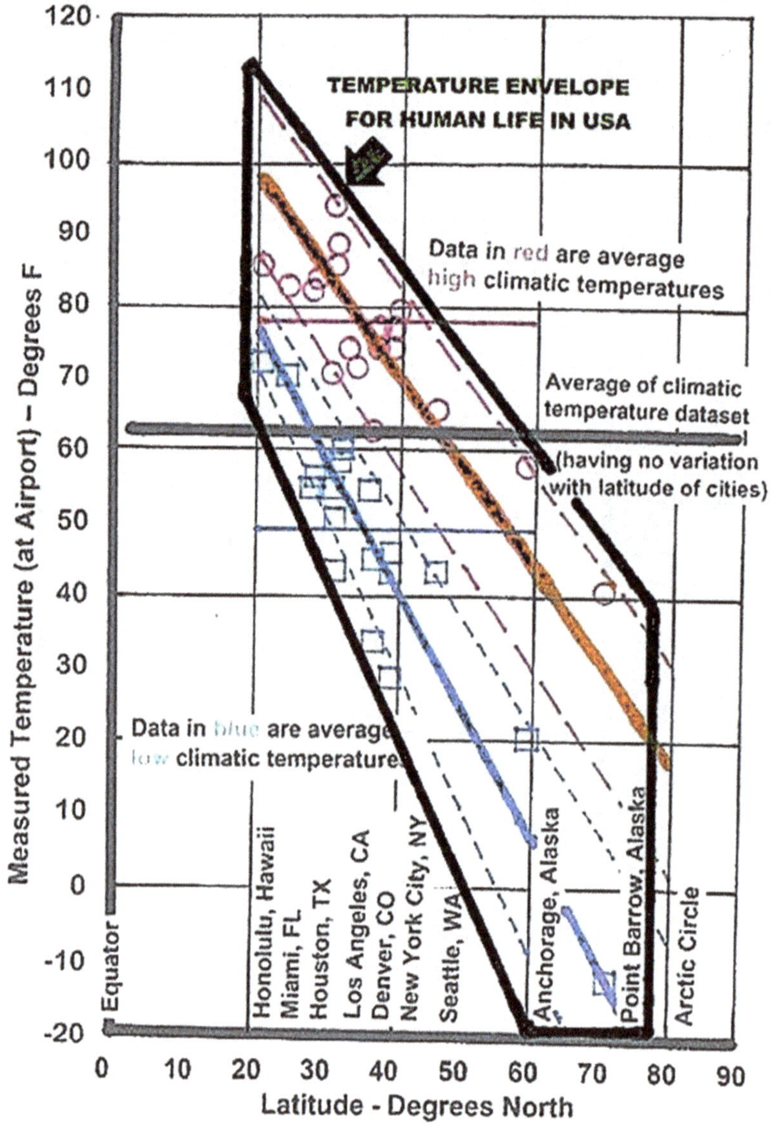

(Source: National Weather Service (NOAA) database, graphed by Author)

For Readers interest, the environmental scientists' "average global temperature" (the horizontal black line at temperature 61 °F / 20 °C) is the temperature for which global warming and climate changes are being referenced in the United Nations IPCC temperature goals. Citing the need to limit that rise to be no greater than 2 °C by 2050 is a contrived "cry wolf". It should be evident that raising that black line 2 °C (3.6 °F) is of relative insignificance to the larger envelope of extreme temperatures to which the US population must adapt. If that 2 °C goal is not attained by reduction of human and industrial generated carbon dioxide emissions, there is no "crisis" or "doomsday catastrophe" lurking in the future of the planet. Of greater significance, the temperature envelope above also defines the temperature range at each latitude that establish the energy needs of heating and air conditioning systems in residences, commercial buildings and workplaces to maintain temperatures into the "human comfort and safety level" of the occupants. The topic of this book "Dealing with the Real Climate Change" was derived from that need to expend energy to permit control of the full range of temperatures generated by nature in its near-term and near future global warming and climate changes. Controlling "greenhouse gas emissions (those that are human and industrial generated) to achieve a comfortable and safe "average annual global temperature" is not a realistic approach. "The tail does not wag the dog."

This book describes in greater detail the six primary factors that discount the environmental scientists' hypothesis that anthropogenic burning of fossil fuels is the dominant cause of global warming. Those factors in order of causative effect on planet Earth's regional temperature variations are:

1) astrophysics (supported by the databases of Galileo and Kepler)

2) geophysics (supported by the databases of Milankovic and several geophysicists identifying the cyclical variations of the planet's dynamic variations in space (its orbital path and tilted, wobbling, angle of rotation causing seasonal effects, the obliquity angle effects causing latitude and albedo to be primary causative variables)

3) natural (supported by the databases of the Vostok Ice Core Research team and of Caillon et al – showing the planet to undergo swings between hot peaks and cold peaks (ice ages) on an approximate 420,000 years cycle with Earth at the hot peak at the end of the fourth cycle; and Caillon showing that it is temperature that drives natural carbon dioxide levels in an 800 years difference between drive and lag) For Readers convenience, the illustration shown in other locations in these two books is repeated – immediately following - because of its extreme significance. (Temperature drives the existence of natural carbon dioxide in the atmosphere, not the reverse of carbon dioxide driving the temperature of the atmosphere as claimed by environmental scientists.)

CO_2 lags temperature rises in ice core data

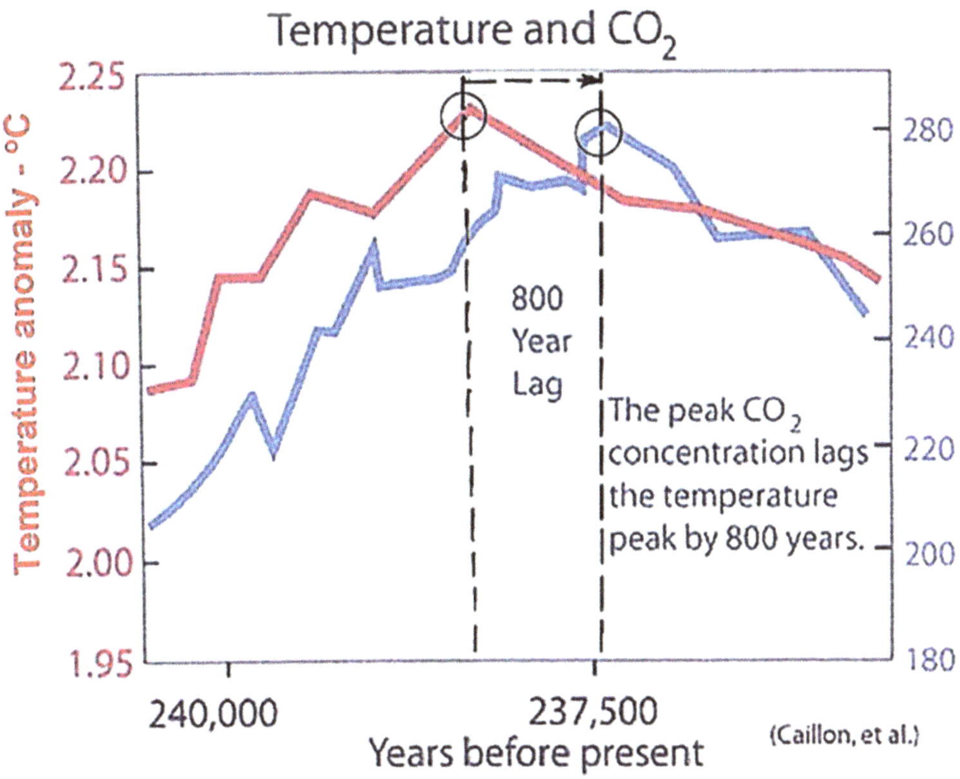

(Source: Smithsonian Institute for Science)

There is an open question how the human and industrial generated carbon dioxide may alter the above illustration, but all indications are that the additional carbon dioxide will cause the magnitude of the carbon dioxide peak to shift upwards and not laterally to alter its lag of about 800 years.

4) water vapor (the discussion preceding and continuing herein showing that the environmental scientists eliminate (by the process of "cherry-picking") variables and factors that refute the hypothesis that the burning of fossil fuels is not the dominant cause of global warming and climate change

5) the NASA satellites "big picture" databases showing that the effects of global warming and climate change are regional and cannot be controlled as "average annual global temperatures or as average sea surface temperatures

6) the anthropogenic effects on the environment may be significant (as collateral damage) but the pollution of the planet by humans and industrial processes far overshadows the effects of the burning of fossil fuels may have on the world population. These effects will grow worse as the population grows at an exponential rate of growth.

The heat-trapping water vapor over the Equatorial region seasonally spawns tropical storms that may become hurricanes, typhoons, cyclones, tornadoes, and increases the frequency of lightning strikes and lightning-ignited wildfires. Environmental scientists' claim that increased "greenhouse gases" increases the severity and frequency but there was no database found that confirms or denies that possibility. What dispels that contention of environmental scientists is that all that "greenhouse gas" carbon dioxide is measured to be at a concentration of 427 parts per million by volume (in February 2025). 427 is a large number; however, of that total 427 ppmv, 285 ppmv is natural carbon dioxide and 142 ppmv is of anthropogenic (human and industrial) causes. At several places in book 1 and in this book, it is explained that the scientific measurement of gases in a laboratory environment is done at small scale where a million molecules of a gas is contained in perhaps a liter size experimental flask. When 427 molecules of carbon dioxide gas is mixed into 9,000,573 molecules of the atmosphere, it is not a large percentage of the atmosphere. 427 ppmv is the equivalent of 0.0427% - a very low concentration. The amount of anthropogenic carbon dioxide is 0.0142% and about one third the concentration of the total carbon dioxide measured to be in the atmosphere. Carbon dioxide is a colorless, odorless, and non-visible gas, but nevertheless it traps certain wavelengths of solar radiation within the carbon dioxide.

Holding that perception of there being 0.0142% of the atmosphere having heat trapping carbon dioxide molecules, the following photograph was taken by satellite of a "big picture" of the real (i.e., not in a well-controlled laboratory) atmosphere of the planet:

(Source: NASA)

The white streaks are well-recognized as cloud formations. Clouds are formed from water vapor in a warm atmosphere ascending to an elevation where the temperature has decreased several tens of degrees lower (colder) than what it was at sea level. The water vapor in the warm atmosphere is condensed into clouds when the warm water vapor is rapidly cooled by the cold ambient air at high elevation. It is the same principle when a person exhales warm air (containing a higher concentration of carbon dioxide gas and water vapor) into the surrounding air where the temperature is at or below freezing. The small cloud is formed in the same manner as clouds are formed by natural conditions of temperature and water vapor. The purpose of showing this photograph is to provide Readers with the visual perspective of the relative amount of water vapor in the real atmosphere comparatively to that of a concentration of 0.0142% carbon dioxide "greenhouse gas" generated by humans and industrial activities.

Environmental scientists do not advertise the amount of water vapor in the atmosphere to the public as that detracts from their hypothesis of humans and industries burning fossil fuels being the dominant cause of global warming and climate change. Any scientific model of the atmosphere that does not account for the water vapor in the atmosphere is misleading to the public.

Another manner in which the public is being misinformed is in the lack of observance of the laws of physics and the associated theorem that mass can only be transformed in state (among the states of solid, liquid, and gas). Mass cannot be created or destroyed – only changed in state. The environmental scientists and the news media are hasty to proclaim that one mass unit of coal results in 14 mass units of carbon dioxide in the atmosphere. Perhaps environmental scientists are not aware that it requires 13 mass units of oxygen "borrowed" from the atmosphere to combust one mass unit of coal to produce heat energy. By the law of conservation of mass, the atmosphere is receiving only the one mass unit of coal, changed in state from solid to gas, and receives back the 13 mass units of oxygen (with which the one mass unit of coal transformed from solid state to gaseous state creates the 14 mass units of carbon dioxide gas into the atmosphere).The net change in the atmosphere is only the one mass unit of coal that has been transformed from a solid state to a gaseous state. If there is measured to be a 0.128% concentration of human and industrial generated carbon dioxide in the atmosphere, only 1/14th of that 0.0128% by volume of carbon dioxide contains the original mass of coal transformed from solid to gaseous state. That may still be an enormous amount of solid coal, but it remains that the coal *per se* is only 0.0009% of the atmosphere by volume.

The previous paragraph highlights the difference between a scientific observation at small numbers in a closely controlled laboratory experiment contrasted with the "big picture" system engineering form of problem analysis and resolution as the following analysis reflects. The small exponents of 10 in the estimates reflect the number of zeroes after the lead digit 1. The seemingly small expression 10^{12} if written in expanded form is 1,000,000,000,000. With 144 square inches per square foot, with 27.9×10^6 square feet per square mile, with Earth's surface area of 196.9×10^6 square miles, with the atmospheric pressure being 14.7 pounds per square inch, and with 2×10^3 pounds in a ton, the mass of the atmosphere is estimated to be 23.26×10^{21} tons (2.326×10^{12} gigatons). 0.0009% of the atmosphere measured to be coal transformed to gas is 2.09×10^{17} tons (2.09×10^5 gigatons) of coal. True, that is a lot of coal, but even much bigger – by approximately 10^7 times - is the 2.326×10^{12} gigatons of atmosphere.

Environmental scientists may use numbers that appear to be large (400 plus parts per million of carbon dioxide in the atmosphere) to mislead the public, but in their scientific analysis process ignore the much larger effect that water vapor in the atmosphere has on global warming. The water vapor has been shown to be up to 3% of the atmosphere, but– on average – in the atmosphere to be 1.4%. In the "big picture", 1.4×10^{-2} water vapor x 2.2326×10^{12} gigatons of atmosphere creates an estimate of 3.13×10^{10} gigatons of total water vapor in the entire atmosphere. How can environmental scientists maintain as their hypothesis that 2.09×10^5 gigatons of coal be the dominant cause of global warming but the 3.13×10^{10} gigatons of water vapor can be ignored?

The public, being misinformed about the factual causes for global warming and climate change, is given the fact that 196 nations of the world have signed an agreement – known as the Paris Agreement and signed in 2015 – to reduce those non-visible heat-trapping carbon emissions (at 0.0142% concentration) in the atmosphere. Since the signing of the UN-sponsored Paris Agreement, which was driven by US environmental scientists and the Environmental Protection Agency membership on an "Intergovernmental Panel on Climate Change", the United Nations IPCC launched a no-holds barred conflict which has been identified as the "combat of climate change". The following news article reflects the stand the United Nations took at the UN IPCC COP25 Conference in Madrid in 2019 (three years after the Paris Agreement in that combat on climate change:

UN chief urges countries not to surrender on climate fight

BY ARITZ PARRA AND FRANK JORDANS
Associated Press

U.N. Secretary-General Antonio Guterres urged countries Monday not to lose hope in the fight against climate change, as representatives from nearly 200 countries gathered in Madrid for a two-week meeting on tackling global warming.

In his opening speech to delegates, Guterres cited recent scientific data showing that levels of heat-trapping gases have hit a record high, reaching levels not seen for at least 3 million years when sea levels were 33-66 feet higher than today.

Unless emissions of carbon dioxide and other greenhouse gases are sharply cut, temperatures could rise to twice the threshold set in the 2015 Paris accord by the end of the century, he warned.

"Do we really want to be remembered as the generation that buried its head in the sand, that fiddled while the planet burned?" Guterres asked.

His appeal came after Chile's environment minister, Carolina Schmidt, said the Dec. 2-13 meeting needs to lay the groundwork for moving toward carbon-neutral economies while being sensitive to the poorest and those most vulnerable to rising temperatures - something that policymakers have termed "just transition."

"Those who don't want to see it will be on the wrong side of history," said Schmidt, who is chairing the meeting. She called on governments to make more ambitious pledges to reduce emissions of greenhouse gases ahead of a deadline to do so next year.

The summit, which moved to the Spanish capital after Chile had to pull out amid anti-government protests, aims to put the finishing touches to the rules governing the 2015 Paris accord.

That involves creating a functioning international emissions-trading system and compensating poor countries for losses they suffer from rising sea levels and other consequences of climate change.

"We have a common challenge but with differentiated needs and urgencies, which we can only overcome if we work together," said Schmidt.

Countries agreed in Paris four years ago to limit global warming to well below 3.6 Fahrenheit, ideally 2.7F by the end of the century compared with pre-industrial times. Already, average temperatures have increased by about 1C, leaving little room for the more ambitious target to be met.

Guterres called out big greenhouse gas emitters that are still building coal-fired power plants, saying that unless the world stops burning coal "all our efforts to tackle climate change will be doomed."

He noted that had countries started cutting their emissions drastically a decade ago, reaching the Paris goal would have been much easier.

"Ten years ago, if countries had acted on the science, they would have needed to reduce emissions by 3.3% each year," he said. "Today, we need to reduce emissions by 7.6% each year."

(Source: Associated Press, Los Angeles Times, Dec 13, 2019)

Five years later, the reality that perhaps the Paris Agreement goals on average global temperature were not going to be met, the UN Secretary-General urged the countries to continue the on-going battle with climate change and to do more to help the less-developed countries being impacted so severely by the effects of global warming and climate change.

UN agency says national climate change plans fall way short

BY PAUL GODFREY
World News - UPI.com

Countries' national plans to tackle climate change are nowhere near what is required to prevent the warming of the Earth from ruining every economy and destroying the lives and livelihoods of their citizens, the United Nations warned in a report Monday.

The so-called Nationally Determined Contributions Synthesis Report, looking at what each country says it is doing to curb the greenhouse gas emissions responsible for global heating, shows combined net emissions of 51.5 gigatons of CO2 equivalent in 2030, just 2.6% below 2019 levels, compared with the 43% cut required to limit warming to 1.5 degrees Celsius, U.N. Climate Change said in a statement.

"The report's findings are stark but not surprising -- current national climate plans fall miles short of what's needed to stop global heating from crippling every economy, and wrecking billions of lives and livelihoods across every country," said U.N. Climate Change Executive Secretary Simon Stiell.

"By contrast, much bolder new national climate plans can not only avert climate chaos - done well, they can be transformational for people and prosperity in every nation.

"Bolder new climate plans are vital to drive stronger investment, economic growth and opportunity, more jobs, less pollution, better health and lower costs, more secure and affordable clean energy, among many other benefits," Stiell said.

However, he noted that while the fractional advances from last year's findings were not unexpected given countries were in the process of preparing new NDCs due in 2025, those plans had to deliver a "dramatic step up" with ambitious economy-wide emissions cuts that keep 1.5 degrees Celsius within reach; are broken down by sector/greenhouse gas type; and have regulatory, statutory and budgetary backing.

With climate disasters worsening at an ever-increasing pace, "every fraction of a degree" mattered, said Stiell, but he warned the greenhouse gas pollution levels the NDC report showed are coming down the track will "guarantee a human and economic trainwreck for every country, without exception."

The gravity of what was at stake placed the new national climate plans "among the most important" policy documents of the 21st century, he added.

The report calls for a 10-year time horizon for the new NDCs along with much tougher targets for 2030 to deliver the drastic emission cuts by the end of the 2020s required to get a head start on reaching the 60% reduction in emissions the Intergovernmental Panel on Climate Change says are needed by 2035 to keep to a 1.5C rise.

U.N. Climate Change said it was working closely with U.N. agencies across the board to provide a range of practical support that took into account "severe capacity-constraints and other headwinds" confronting many countries, especially vulnerable and developing countries.

The agency said it would hold a deliver a series of events next year to that end and initiate a global public conversation on how to deliver on the new NDCs.

Details are set to be announced in November at the COP29 climate conference in Azerbaijan.

(Source: World News – UPI in the October 29, 2024 News Observer, Durham, N. Carolina)

Several takeaways were produced from the COP29 climate conference which took place in Azerbaijan from November 11 – 22, 2024. The two largest were that it was confirmed by the IPCC environmental scientists that carbon dioxide emissions had been reduced by only 2.6% below 2019 levels leaving a 46% reduction to be achieved if the Paris Agreement goals on average global temperature rise are to be met in 2030. The likelihood of that 46% reduction being achieved is considered by the UN IPCC to be highly unlikely. Based on the findings in the "Fact and Fallacies of Climate Change on Our Planet" book, no amount of reduction in human or industry-generated greenhouse gas emissions will limit the average global temperature to any significant degree because it has been proven that astrophysical, geophysical, and natural factors – not carbon emissions - will continue to cause the surface (land and water) temperature to rise and the climate to change regionally.

The largest takeaway from COP 29 was the reaching of a sum of climate finance - $300 billion a year - that developed nations will pay to lesser-developed nations for the destructive consequences of climate change on those lesser-developed nations. The receiving nations immediately began complaining that $300 billion a year is "a paltry sum". Nevertheless, there was a consensus that the level of finance increase from $100 billion per year to $300 billion a year was a welcome outcome from the COP29 conference of participants. India's delegate Chandi Raina told other delegates: "It's a paltry sum. This document (COP 29) is little more than an optical illusion. This, in our opinion, will not address the enormity of the challenge we all face." The Author totally concurs with Ms. Raina that COP29 – and the 28 COPs preceding – are built upon an illusion, although not so "optical" as their primary intent. The illusion is built on a misjudgment of the importance that 0.0128% concentration of human and industrial generated carbon dioxide can have on the enormous real atmosphere of the planet.

It may shed a perspective on the plans the nations have for reducing the coal burning each nation is using and plans to use for developing electrical energy to meet the nation's energy demand to disclose what those plans include:

Coal-Burning Electric Generation Facilities

	Number of Facilities		
Country	**Existing**	**Planned**	**Total**
China (Mainland) + other Asian countries	2,363	1,171	3,534
India	589	446	1,035
United States	225	Reduction	150-
United Kingdom	120	Reduction	120-
Germany	63	9	72
Other EU	240	Reduction	240-
Japan	90	45	135
Indonesia	87	Unk Inc.	87+
South Africa	79	24	103
Russia	71	Unk Inc.	71+
South Korea	58	13	71
Poland	44	7	51
Turkey	33	8	41
Australia/New Zealand and Oceana	32	Unk Red.	32-
Vietnam	26	Unk Inc.	26+
Phillippines	25	Unk Inc.	25+
Czechia	24	Unk Inc.	24+

(Source: US Department of Energy, DOE)

It seems incongruent that India is expecting other nations to pay reparations from burning coal and creating climate havoc in India, yet India (and China) are not only the top two coal users in the world and have plans to increase their coal-burning facilities by over 1,500 new facilities (China by 1,171 and India by 446). All other developed countries are reducing the use of coal to meet each regional population's unique electrical energy needs (for humans and industry) and best utilize each country's unique geological resources. It is hastened to be added that China and India have initiated research and development activities in the development of nuclear facilities to replace some or all of their coal-burning facilities. Although the technology of using uranium as the heat energy source in nuclear-powered electricity generating facilities is well known and is in widespread practice on the planet, the emergence of thorium as a lower cost and safer element to substitute for uranium appears to be a technology advancement of great potential for the exponentially growing population. In both China and India, thorium is more abundantly found and is far safer to use in applications near heavily populated regions of those two large land masses. China and India have apparently recognized that using technology improvements and new developments for economic and safety considerations than is the need to reduce "greenhouse gas emissions" to avert a fictitious "climate crisis".

It is those considerations of regional uniqueness that cause these two books on dealing with global warming and climate change to highly recommend the abandonment of "combating global climate change" and allowing each major region to develop the technologies that are most appropriate for each region's needs and resources. There is much room for protecting both the human population and the environment by making logical compromises in technological development. Each technology has its own unique set of "pros and cons" and the same technology in one region may have a different (better or worse) set of "pros and cons" relative to the regions using that technology. A tongue-in-cheek comment made by an associate to the Author was "It isn't too useful to give a pig a new wristwatch for its birthday."

At the COP 25 through COP 29 meetings, the lesser developed nations pointed the finger of guilt at the developed countries for creating the "greenhouse gases" that are hypothetically caused by humans and industry burning fossil fuels. After all, that is what the settled science environmental scientists and the Environmental Protection Agency has been drilling into the world populations perception as to the cause of climate change. The old adage from the poem "Marmion" by Sir Walter Scott in 1808, is more appropriate now than it was at the start of the research for the causes of global warming and climate change: *"Oh what a tangled web we weave, when first we practice to deceive." The* environmental scientists declared developed countries were the dominant cause of global warming and climate change, and now the damaged countries are expecting to be given the reparations funding given by victors of the war to the losers of the combat. The environmental scientists shot themselves in the foot – or poisoned their lungs - with their own hypothesis about "greenhouse gases".

This book is highly likely to be a further disappointment to UN Secretary-General Guterres but hopefully will give the UN Intergovernmental Panel on Climate Change (UN IPCC) the factual scientific evidence of why the temperature goals were so ill-conceived before the Paris Agreement was signed. The current administration in the United States (President Trump as the newly elected President) who extracted the United States from the poorly conceived Paris Agreement during his first administration and plans to again extract the USA again from the still poorly conceived Paris Agreement in this his second term of office as President). More importantly, this book provides all the information to help the United Nations recognize that climate change is not caused by human and industrial activities creating only a small fraction of the total carbon dioxide in the atmosphere. It is hoped the support of developed nations will continue to be given to those less fortunate in lesser developed nations but as a humanitarian act of governmental aid – similar to that given after natural catastrophes strike a region.

There is no "naysaying" among the science profession that the climate on planet Earth is changing, but there is much naysaying against the causes for the global warming and climate change. The factual causes for climate change are revealed in several databases which virtually absolve humans and industry burning fossil fuels (hypothesized by environmental scientists to be the dominant cause) - were disclosed in the book "The Facts and Fallacies of Climate Change on Our Planet" previously published by this Author (a highly experienced technology developer, major crash accident with multiple fatalities investigator, and system safety engineer). More details about the Author's education and professional experience can be found near the end of this book). Using basic STEM research tools for disclosing the scientifically factual causes of global warming, climate change, and ocean levels rising, an interacting set of factual astrophysical, geophysical, and natural causes of inevitable climate change were vital in arriving at that conclusion.

The brilliant scientist Marie Curie (1867-1934) stated over a century ago: *"Nothing in life is to be feared, it is only to be understood. Now is the time to understand more, so that we may fear less."*

The factual multiple causes shown in several databases logically contradict the fallacious hypothetical singular cause, alleged by environmental scientists, self-described as "settled science" that "Man burning fossil fuels generating heat-trapping emissions, principally that of carbon dioxide, is the <u>dominant</u> cause of climate change". That unproven overexaggeration has been repeatedly expressed by environmental activists and by the sensationalistic news media to create a cry-wolf "crisis" level of concern deeply instilled in the human population.

The environmental science overexaggeration of the "greenhouse gas effect" (from Man-made carbon dioxide, measured to be at a low level of concentration of less than 0.02% of the planet's atmosphere) is realistically assessed as being a nearly negligible cause. Scientifically factual databases essential to understanding climate change (e.g., the Vostok ice-core samples) were intentionally not considered or were unethically "cherry-picked" by environmental protection proponents with the intent of misleading the public to fear there is a real climate crisis. Climate changes may imperil some life-forms (such as coral reef anemones) in the environment but do not endanger the human population which has adapted (in the manner described by Charles Darwin in "Origin of the Species") to regional seasonal temperature extremes and natural climate anomalies. The professional quality research described in the author's previous book leads to the logical need to mitigate those changes – in temperature, climate, and oceans levels - for the betterment of mankind.

This book "Dealing with the Real Climate Change" (written in common-word text and with numerous illustrative images by the same author) extends the wisdom of Marie Curie's vision - gaining of understanding to overcome fear - as an independent public service venture to allay the fears over climate change fallaciously ingrained in the public. A "big picture" forensic analysis confirms climate is changing – more in some regions than others as depicted in the mottled diagram on the front cover – but is predominantly caused by astrophysical, geophysical, and natural factors dynamically varying daily, seasonally, and long-period cyclically. It is futile to combat climate change by the sophomoric activities of eliminating the use of all fossil fuels as the means of changing the global climate. Climate changes beyond Man's control require both Man's adaptation of his lifestyle and by Man's mitigation achieved by redirection and improvement of existing, emerging and/or newly created technologies.

Many (over 25) technologies in various stages of development to achieve meaningful mitigation of climate change effects are assessed herein - for their net value after considering the "pros and cons" each has - and weighing their importance in meeting the human energy needs and availability of the energy stored in the natural resources of different regions. The assessment in this book is applicable to the unique climate changes and energy needs of the United States but can be tailored to each region of the planet for its unique state of resources (natural and financial), its technological development status, its unique climate (present, near-term, and future), and its anti-pollution practices for the safety, health, comfort, transportation, and energy needs of its indigenous human population – not merely that of the environment.

Human pollution of the land, water, and air is a far greater problem and crisis than is climate change. When considering a new technology or improving an existing technology – exploiting its "pros" and minimizing its "cons" – the con of pollution of the environment must be eliminated if possible or at least minimized to being an inconvenience but not a hazard.

A major portion of this book addresses each of 25 technologies – not to reduce climate change but to improve human safety, health, and energy efficiency – by considering the pros and cons of each candidate technology in evaluating the net gain "for the benefit of mankind" by advancing those technologies applicable to regional needs. This book has only assessed those technologies best suited for the United States region. Other regions are encouraged to do the same using the methods and the logic that have been applied for the United States region.

Table of Contents

Introduction

Two statues "Past" (on left) and "Future" (on right), in front of the US National Archives building in Washington, D.C., have the philosophies of *"Study the past"* and "*What is past is prologue*" engraved on their base pedestals.

(Source: Author's Photographs at US National Archives, Washington, DC)

In the age of the philosophers and astronomers many centuries ago those words of wisdom were very sagely advice. Based on his education and experience in development of new technologies for airplanes, which entailed crash investigations when things did not happen quite as planned, the Author added "and present" to the "past" of the National Archives engravings philosophies. That set of information perhaps explains why a system safety engineer has analyzed the scientific topics of global warming and climate change. In the solving of any complex problem, it is also sagely advice to get "a second opinion", or in science and engineering "listen and learn" from another "perspective". Mountain climbers gain one perspective if they are at the base of a mountain looking up, different from at the top of the same mountain looking down. The mountain hasn't changed – only the perspective.

Galileo di Vincenzo Bonaluti de' Gallelei (1564 – 1642) was a gifted Italian astronomer, physicist, engineer and polymath, from the Pisa region, then part of the Duchy of Florence, Italy. Scientists in all branches of physical science regard Galileo as being the father of observational astronomy, modern physics, the scientific method, and modern science. Galileo was the first person labeled as being a "naysayer", although Saint Thomas of the Christian Bible was much-earlier referred to as "doubting Thomas". The impression perhaps created in some readers after reading previous books addressing the topic of global warming and climate change may be that the Author is merely another "naysayer" – as many previous authors have been also named. Anyone who dares question the environmental scientists' fallacious hypothesis - that humans' creating "greenhouse gases" are the dominant cause of global warming and climate change - is labeled as being a "naysayer". Peers are referred to as "contrary scientists".

The term "naysayer" was created by the Roman church centuries ago when accusing Galileo of heresy for scientifically proving that the Earth orbited the Sun and not vice versa as Ptolemy had concluded (but without the more advanced telescopes developed by Galileo and others). Per Ptolemy, the solar system was composed as follows:

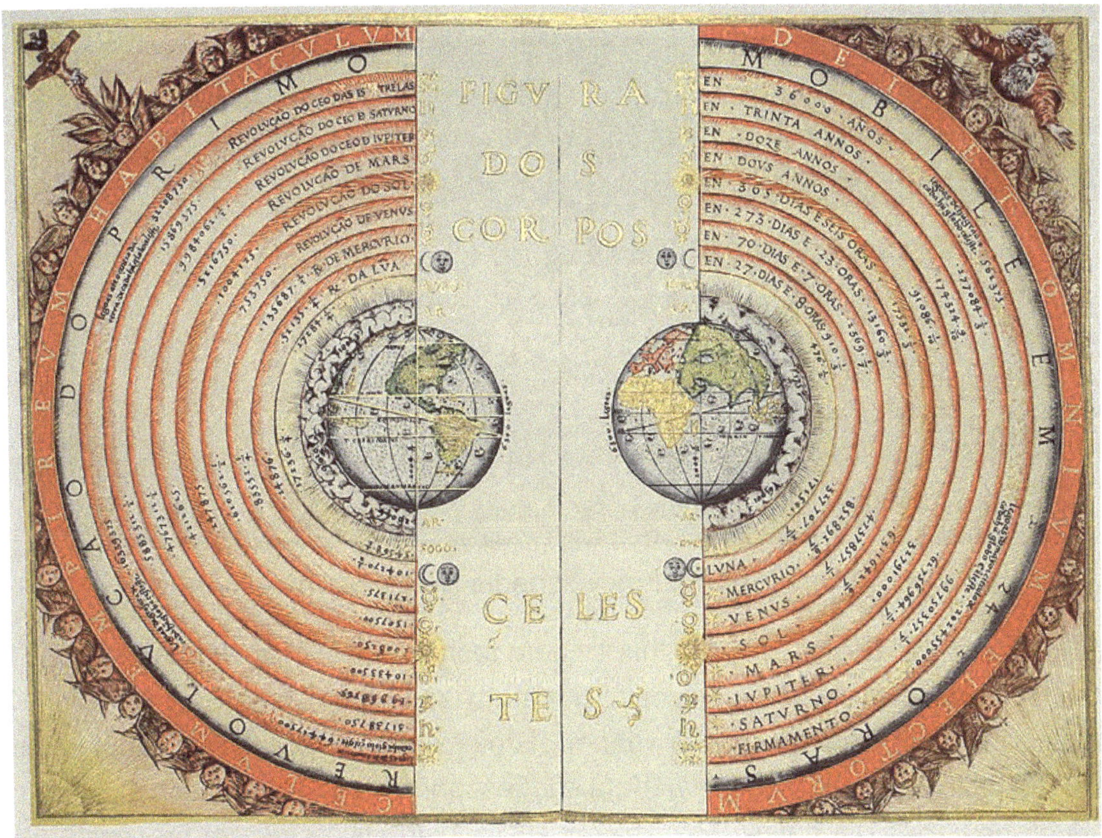

(Source: Encyclopedia Britanica) The solar system by Ptolemy

Galileo's philosophical outlook on everything going on around him, and his many brilliant "naysaying" contributions to the database of scientific knowledge has provided historians and scientists with much guidance. There are at least 25 quotes Galileo issued in that age of science addressed toward his colleagues, one of his more famous was very appropriate at the time - but clearly updated with the many technological advancements in science since then - is *"Those truths which we know are very few in comparison with those which we do not know."*

Not all scientists are rigorous mathematicians – some regard two half-truths (e.g., observations in a science laboratory) to add up to one full truth. Human psychology has many facets – one facet being that humans believe that which they want to believe. A corollary is that scientists simply "eliminate" that which they fear is contrary to what they want to believe. When a group of scientists consolidate an unusual partially factual observation into a full belief that an important discovery has been made, it is most difficult to convince that group otherwise. Galileo encountered that situation with other philosophers who were convinced that the Earth was flat, and that the Earth – not the Sun - was at the center of the Universe: The scientific truth is that neither the Earth nor the Sun were at the center of the Universe. The Universe is composed of several galactic group of bodies in the space known as the Universe, and within the Milky Way galaxy, the planet Earth is among the eight planets which are in different shape and size orbits around the Sun. The Sun is the center of our solar system. In a similar manner to the technological advancements in astronomy by Galileo, today's NASA satellite images of the spherical Earth in its factual position in the solar system totally refute some misinformed scientists' contention that the planet Earth is a flat disk floating on an immense sea. Galileo made a quote regarding those scientists who were adherents of the "flat Earth" cult.

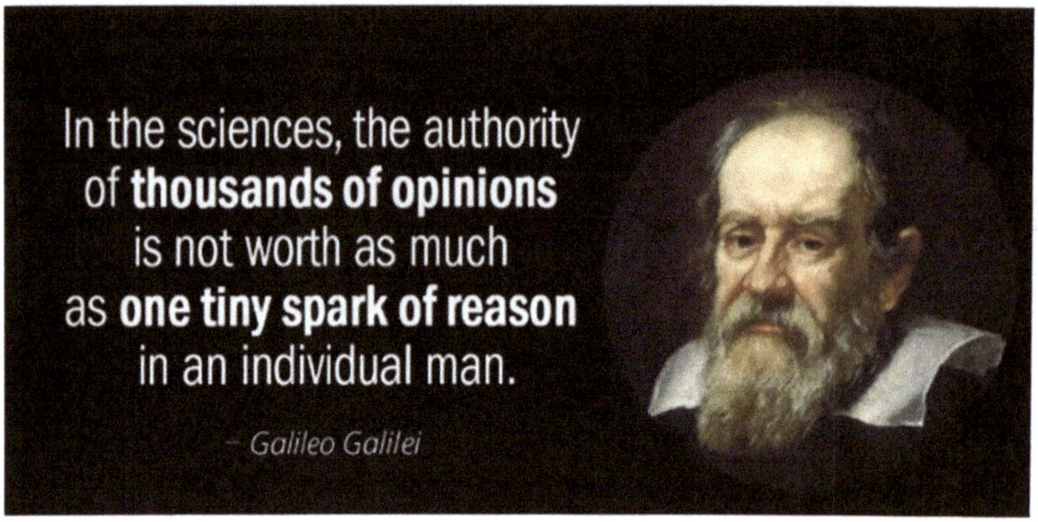

(Source: Encyclopedia Britanica)

Many unknowns – referred to as "unks" in this modern technological age - were converted to knowns by the author with his analysis of the scientific climate change databases researched in "The Facts and Fallacies of Climate Change on Our Planet". Professional ethics dictates that when a flaw is found in the scientific review of a peer's work, a suggestion(s) – i.e., "constructive criticism" - should also be offered how to "fix" the flaw. That review and constructive criticism process is expected in any scientific endeavor and is usually achieved in a "paper" submitted to a host scientific organization for review by other qualified members of that scientific organization. An exception was made in that professional process by a group of environmental scientists in the first half of the twentieth century, claiming to represent "settled science" and who issued the scientifically baseless hypothesis:

"By eliminating everything else, the 'settled science' conclusion is that humans and industry burning fossil fuels are the <u>dominant</u> causes for average global surface temperature rise and climate changes."

The act of "eliminating everything else" is known in statistical analysis circles as "cherry-picking" the data – i.e., accomplished by the removal of any data from an otherwise valid dataset to avoid considering those data which are contradictory to or do not support a preferred theory. (This Author chooses not to identify the source by name of that fallacious hypothesis to avoid professionally causing embarrassment to that scientist and her organization for having eliminated (intentionally or unintentionally) the factual causes for the planet's surface temperatures to be rising and its climate to be changing.

For the benefit of Readers who, for any reason, may not have read the book published earlier by this Author "Facts and Fallacies of Climate Change on Our Planet" there were twelve interim conclusions (IC-1 to IC-12) generated from the research of the many databases of astrophysical, geophysical, geological, and natural information and data. The set of those twelve Interim Conclusions (in abstract form) are repeated for scientific emphasis and for Readers' convenience:

IC-1) Carbon dioxide in the atmosphere is increasing, but the total carbon dioxide measured (at the NOAA Mauna Loa Observatory on Mauna Loa volcano in Hawaii to be 435 parts per million) contains both natural-caused carbon dioxide and Man-made carbon dioxide in an approximate ratio of 2:1 (295 ppmv natural to 140 ppmv Man-made).

IC-2) Environmental protection scientists have overexaggerated and oversimplified an outdated but otherwise valid database of scientific measurements (principally warping the interpretation of the scientifically valid "Keeling Carbon Dioxide Curve"), and eliminating/ignoring other pertinent databases that also have not been fully updated with known more scientifically accurate and more "big picture" data collection techniques and scientific observations and interpretations.

IC-3) The combined effects of astrophysics, geophysics (tectonic plate activity), and natural factors (long-term cyclical, seasonal and random) on the destruction of human-made structures (for housing needs, food supplies, and the far-worse large numbers of human fatalities, are shown to be collectively far greater than are the man-made adverse effects on atmospheric factors).

IC-4) The scientific facts contained in the databases analyses reflect that it is water vapor in the atmosphere - not carbon dioxide - that is the largest greenhouse gas "driver" of warming surface and oceans temperatures.

IC-5) The need for mitigating and adapting to the astrophysical, geophysical, natural and other non-Man caused factors that are uncontrollable must be better balanced with increased building code regulations and penalties for failure to adhere to the essential improved codes.

IC-6) The combined effects of global warming, climate change, and the oceans levels rising are manifested differently in different regions of the planet as shown by satellite imagery having greatly improved sensors measuring a far larger area simultaneously adding to the "big picture" form of analysis

IC-7) The arbitrary setting of unrealistic goals for controlling global temperatures rising, climate changing, and ocean levels rising are based on misleading "averaging", "linearization", "overexaggerations", and "selective considerations" using databases that have known important updates but which have been ignored "for scientific convenience in modeling of trends". Humans must survive the extremes of their surrounding regional environment not just the "average global temperature".

C-8) The presumption that the use of all fossil fuels (coal, oil, and natural gas) must be ceased by 2050 if the climate is to be stopped from reaching an undefined "tipping point" is not supported by the scientific databases that have been correctly analyzed and interpreted with a professionally credible "scientific process".

IC-9) The goal to reduce and eliminate the use of oil and natural gas is not appropriate in many countries, and that "destroy all the eggs in the basket to get rid of one bad egg" decision by devious US environmental protection scientists is deservedly meeting opposition (as shown at the COP28 Conference of Parties in Dubai in late 2023).

IC-10) The pollution aspect of toxic and carcinogenic ash residue (residual from burning coal) causes the need for transitioning away from the use of coal entirely, and to convert coal-burning facilities to oil or natural gas burning <u>in the United States</u> as soon as feasible. That necessity and urgency may not apply to other regions of the planet.

IC-11) The decisions being mandated to shift quickly to the use of all-electric forms of propulsion for commercial and public transportation has proven to be premature as the infrastructure apparently cannot realistically support such an idealistic rapid transition. The more useful transition from fossil fuel-powered vehicles to all-electric vehicles should include a hybrid (electric/fuel) powered vehicle for public use.

IC-12) The primary "one-size-fits-all" goal of reducing Man-made "greenhouse gas emissions" for controlling the climate has been proven to be a wasteful, unrealistic plan with the incorrect goal, and that fact is beginning to be realized. Pursuing the development of alternate technologies to replace the proven but outdated technology of burning coal to satisfy global heat energy demand is timely.

In keeping with the "Study the past; the past is prologue" philosophy being observed in this book; the foregoing 12 interim conclusions are the imaginary door for a person researching the history and future prospects for global warming, climate change, and oceans levels rising, to move from the past to the future. The last sentence of IC-12, suggesting the pursuit of alternate technologies to replace the proven but outdated technology of burning coal, is a good transition from outdated and/or inefficient and unreliable technologies to future technological challenges and opportunities.

It is acknowledged that there were several technologies that had started over six decades ago in most developed countries of the world. However, several of those technologies were being developed primarily for the purpose of reducing emissions from the combustion of fossil fuels (coal, oil, and natural gas) with the aim to constrict the "average annual global temperature rise" as a means of "combatting climate change" and on a "crisis" basis of need. The interim conclusions (IC-1 through IC-12) identified that most of those technologies generated to reduce emissions were not useful for their primary purpose to reduce emissions that affect climate change, but it is hastened to be added that development of alternate technologies – solar and wind - were not a total waste of time or effort (or funding resources). The effort to reduce the use of coal has resulted in the reduction of a serious pollution source – the ash residue from the combustion of coal contains very hazardous impurities (toxins and carcinogens) that are finding their way from the landfills and trash dump sites into the aquafers of potable water located beneath the ground surface level. When there is rainfall on those landfills and trash dumps, the water leaches those toxins and carcinogens into the aquafers servicing human and industrial needs. Those aquafers also serve to provide potable and arable water to agricultural and ranchlands and the animals located thereon.

As the front cover of this book shows (in a mottled temperature image, repeated below and with a notation of the significance of the different patches of color), climate change as of July 2023 and for the near term is a regional concern and not an "average" global problem. There is no "average" population distribution, no average" resources" distribution", no "average" global climate.

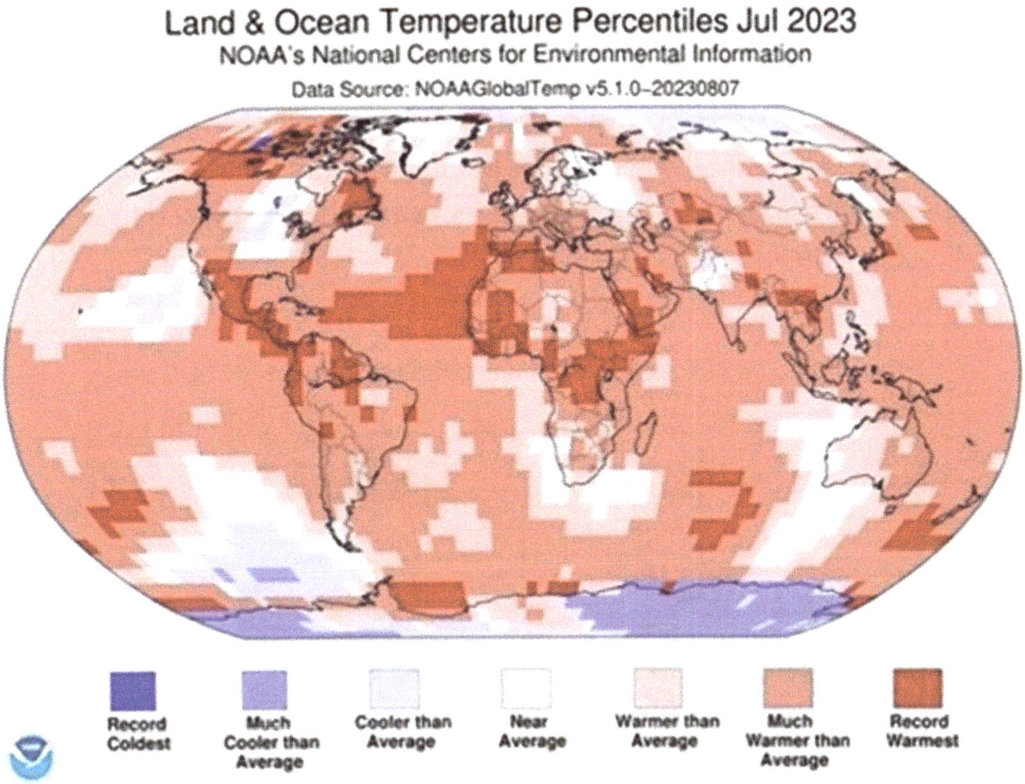

(Source: The National Oceanic and Atmospheric Administration, NOAA)

Based on a visual evaluation of the amount of a color shade representing the climate status, most (peach red color) of the regions – including the East, Gulf, and West coasts of the USA - are much warmer (over 1.5 °C) than average; the next largest group (white color) of the regions – including the many central states of the USA - are near average; next group of regions (pink color) are warmer (over 0.5 °C) than average; next group of regions are those (dark red color) in record warmth (about 2 °C) – but not at "crisis" level to humans; and the next smallest groups of regions (in light blue, medium blue, or dark blue) are those in the Southern Ocean, Arctic and Antarctic polar regions where temperatures are cooler and much cooler than average. The largest area in dark red color ("Record Warmest") are the waters of the South Atlantic Ocean in which there obviously is very low human population.

Many unknowns – referred to as "unks" in this modern technological age - were converted to knowns by the author with his analysis of the scientific climate change databases researched in "The Facts and Fallacies of Climate Change on Our Planet". Professional ethics dictates that when a flaw is found in the scientific review of a peer's work, a suggestion(s) – i.e., "constructive criticism" - should also be offered how to "fix" the flaw. That review and constructive criticism process is expected in any scientific endeavor and is usually achieved in a "paper" submitted to a host scientific organization for review by other qualified members of that scientific organization. An exception was made in that professional process by a group of environmental scientists in the first half of the twentieth century, claiming to represent "settled science" and who issued the scientifically baseless hypothesis.

Meteorological records have been compiled for major cities in both hemispheres of the planet for a few centuries. When listed in tabular form, those temperatures appear to be in very random manner if the tabulation is in alphabetical order for the cities (much like those in a telephone directory). However, as shown in the following illustration, when the latitude of twenty cities in the northern hemisphere (six of which are in the United States) is used (as the independent variable) and the associated average annual temperature (the dependent variable) are placed in symbol form on the illustration, a most useful relationship is observed:

Effect of Latitude on Average Annual Temperature (20 World and US Cities)

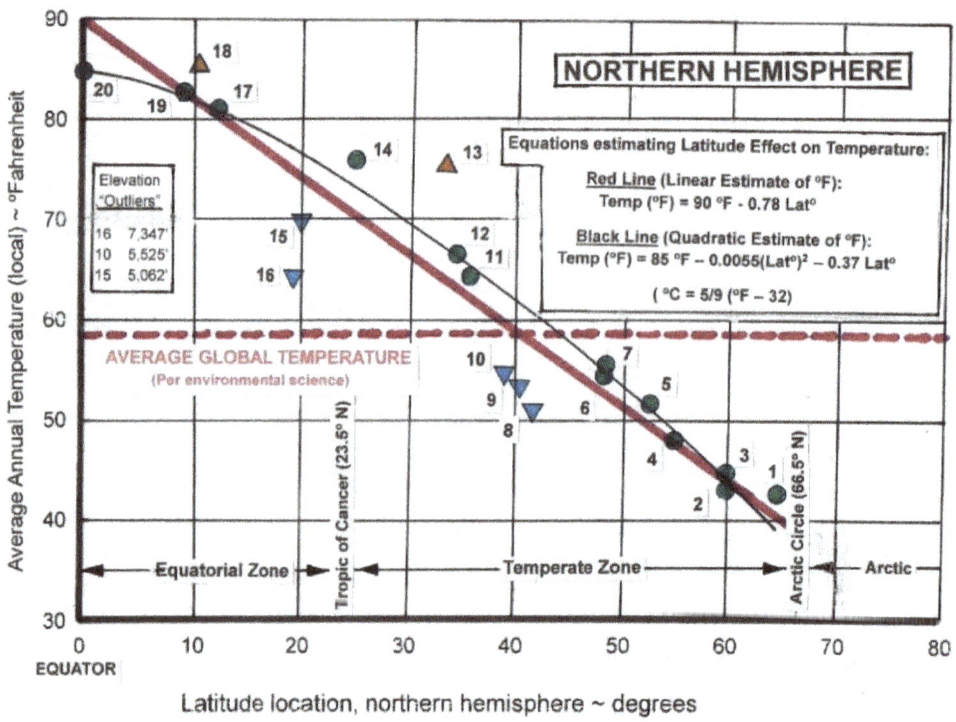

(Sources: Database from World Meteorological Organization (WMO) and NOAA)

A similar pattern of the latitude effect on annual average temperature for twenty cities in the southern hemisphere was shown in the facts and fallacies book and which reflects almost the same effect of latitude as in the northern hemisphere. Students of trigonometry understanding the geometrical shape of the planet and its dynamic motion relative to solar insolation will see that the thin black curved line in the graph above is a first order approximation of the trigonometric cosine variation to a line that "fits" the average annual temperature data points varying with latitude. The placement of the linear (thicker red) line – defined as a point-slope - was guided more to "fit" the trigonometric variation (thin black line) than to "fit" the data points.

A correction factor for elevation also was selected from the Author's frequent use (as an Aeronautical Engineer with the USAF) of a "temperature lapse rate" (ambient temperature of the atmosphere decreasing with altitude increases – from the International Air Transport Association (IATA) analysis of aircraft performance mathematics) – -3.44 °F per 1,000 feet altitude. For this LEO Method, the correction for regional elevation varies (decreases) with elevation up to 3,000 feet. For regions above 3,000 feet elevation, a constant correction was selected as a "cap" – of -10 °F.

More will be described later in this book about the regional effect of latitude and elevation on average annual temperature, and development of the LEO Method), but the primary purpose in this Introduction section is to illustrate that the use of a sophomoric "one-size-fits-all" average global temperature (shown above as the red horizontal dashed line which environmental scientists use and persist in using) as the reference base (i.e., a model) for global warming and climate changes is needlessly misleading. The fallacious claim that a "crisis" looms if the average global temperature rises two degrees Centigrade by 2050 – ostensibly caused by humans and industry burning fossil fuels – is obviously and factually not a scientifically valid interpretation of the factual causes for, or the effects of, global warming and climate change.

The illustration on the front cover of this book "Dealing with the Real Climate Change on the Planet" shows a mottled appearance in color of the regional changes in temperature with some regions warming, some regions not changing, and other regions cooling. An important general observation made during the research for these books is that the land-based temperature and climate change databases used by the National Oceanic and Atmospheric Administration (NOAA) and the Environmental Protection Agency (EPA) are incomplete and/or outdated when compared to the similar satellite-based databases of the National Aeronautical and Space Administration (NASA). The difference in conclusions interpreted from different perspectives is a major issue in these books. Is the myopic laboratory observation perspective requiring enormous scaling-up or is it preferable to take the "big picture" perspective requiring much scaling-down? Statistically, it is beneficial to include more variables in an analysis than too few variables in the analysis.

When a mountain-climber views his objective from a distance, the mountain does not seem to be a difficult climb. However, when the base of the mountain is reached and the sheer vertical face of the mountain is seen close-up (no pun intended), the prospect of an almost vertical climb becomes a daunting challenge. There is no question that the land-based weather-tracking facilities and meteorology databases of both NOAA and NASA are vital to the United States population near-term safety, health, and comfort, but the NASA satellite-based databases clearly are better for estimating the long-term effects of regional changes in surface temperature, climate, and ocean currents differences in the major oceans and seas of the planet.

It was shown in the previous book (which includes a portion of the research conducted by Dr. S. Fred Singer of the University of Virginia in his "A Closer Look at the Numbers" analysis) that water vapor in the atmosphere was a far greater heat-trapping gas than that of a total carbon dioxide at a concentration of less than 0.05% (500 parts per million by volume (ppmv). Others have used a more understandable pie chart format showing the relative contributions of water vapor, natural factors, and human additions to "the green hpuse effect" as follows

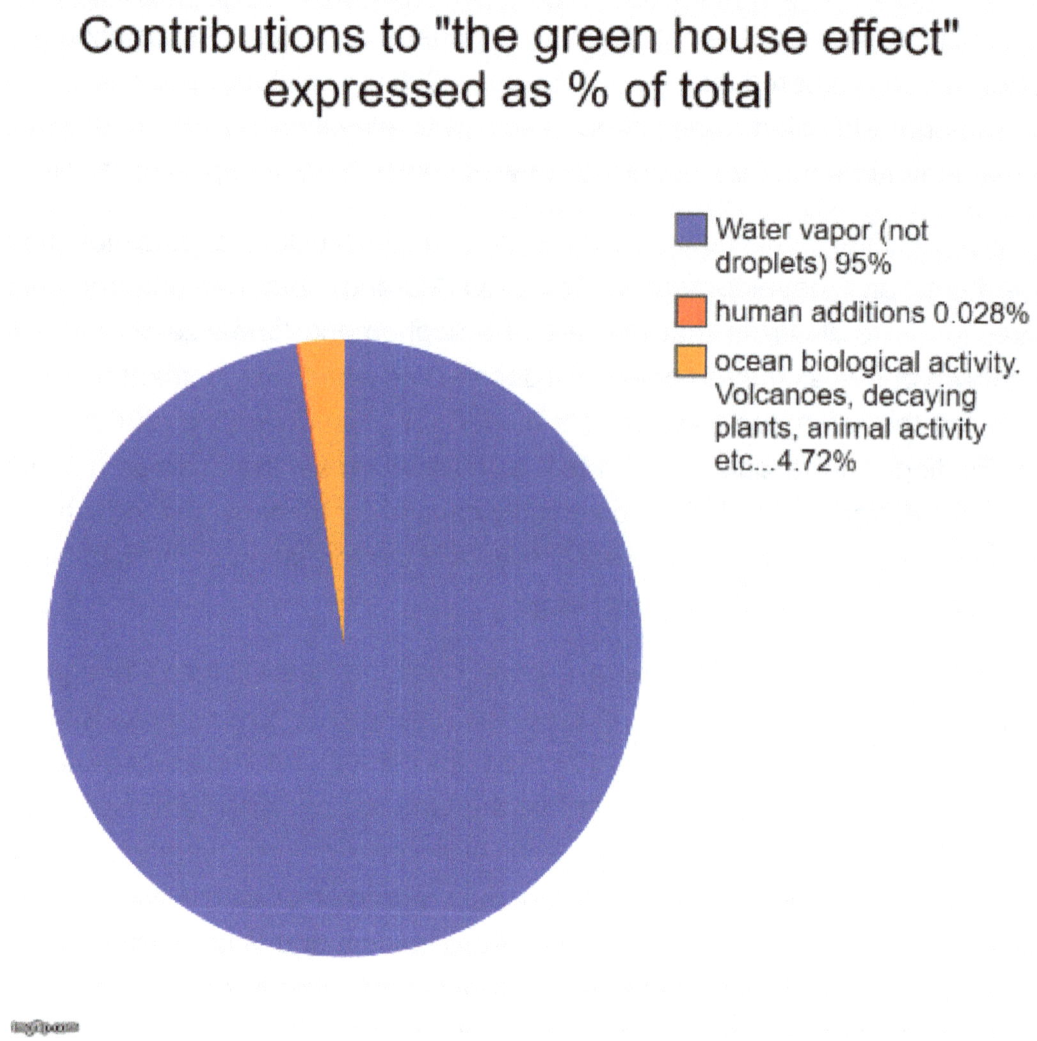

(Source: Internet website – in Public Domain)

With a magnifying glass, one can detect the 0.0278% sliver of the "pie" (in red) sandwiched in between the 95% water vapor (blue) and the 4.72% natural factors (orange).

This Author created a chart comparing the concentrations of water vapor and carbon dioxide in the atmosphere with both presented in a more scientific numerical percentage form that illustrates the same overpowering presence of water vapor:

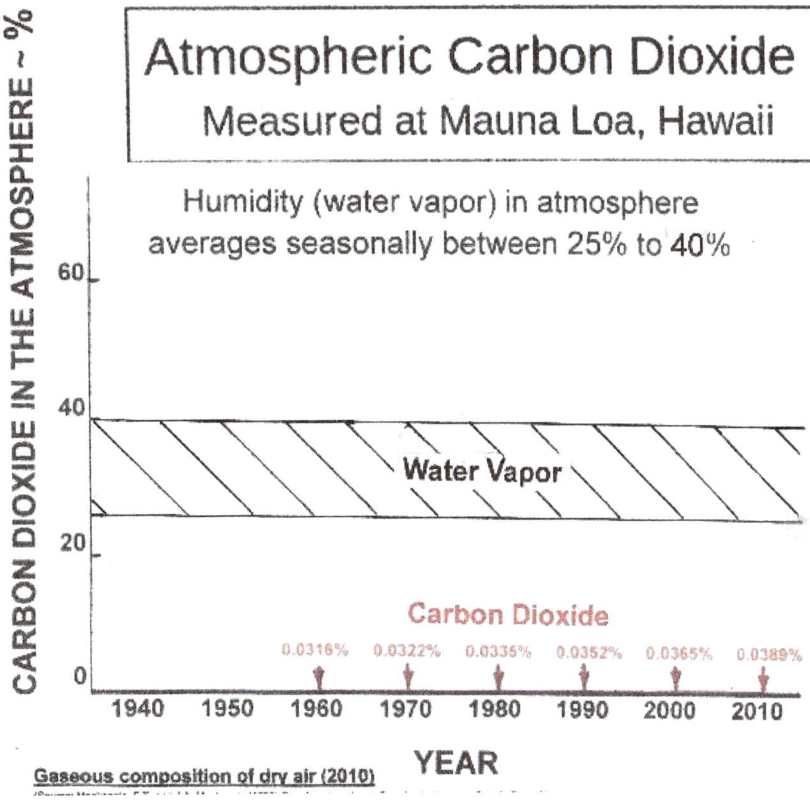

(Sources: Databases of NOAA and the World Meteorological Organization)

NASA satellites with sensors for detecting and measuring water vapor also show that the higher concentration of water vapor is present in the atmosphere but is more concentrated in the equatorial region (at latitudes between about 15° north and south of the Equator) where the predominant spawning grounds for tropical storms and hurricanes are located. The typical flooding as the aftermath from tropical storms is a result of the water vapor in the atmosphere condensing from vapor state into liquid state.

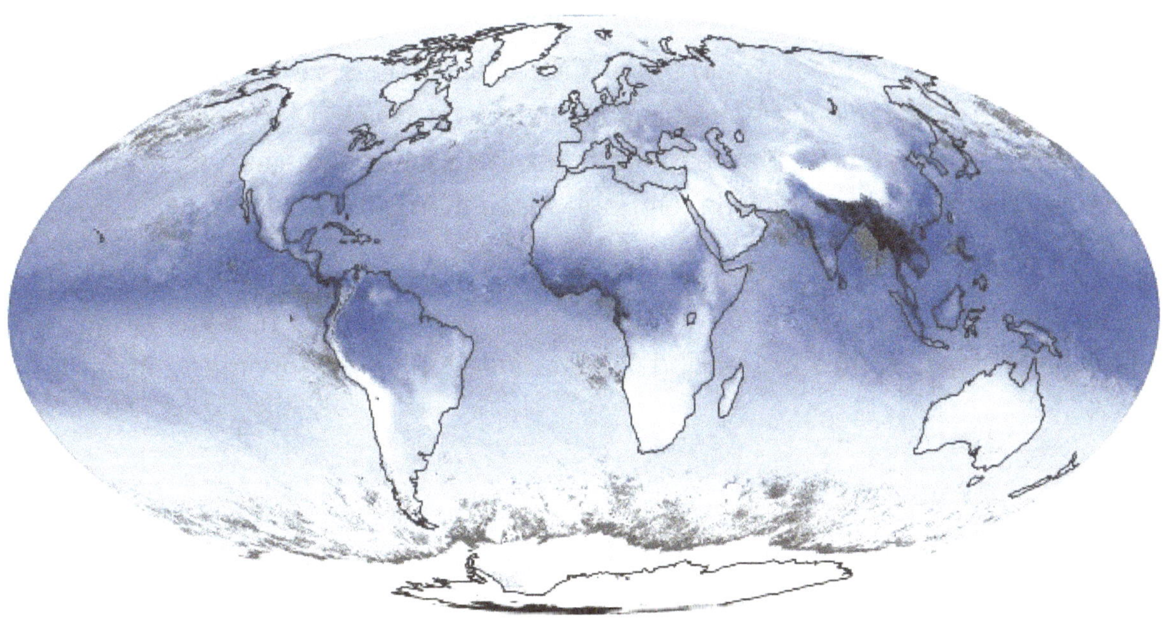

(Source: NASA)

The darker shade of blue indicates where the highest concentration of water vapor is detected, and the lighter the shade of blue becomes, the lower the concentration of water vapor. At the bottom of the illustration is seen the Antarctic continent (in white color) where the atmosphere is extremely cold (average annual temperature of about -46 °F (-43.3 °C) such that there can be no water vapor in the atmosphere. Technically, climatologists categorize Antarctica as having a desert-like climate because of its aridity.

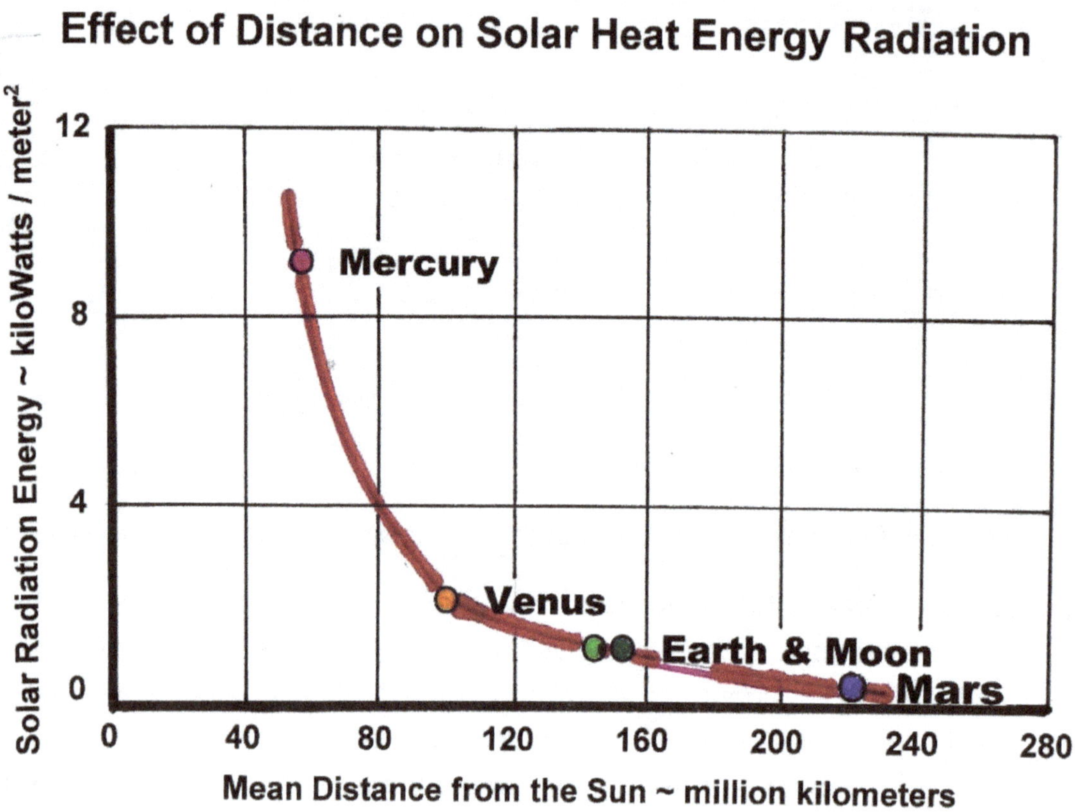

(Source: NASA database; graphics by Author)

It is hastened to be added that realistically those solar radiation levels must pass through the unique atmospheres of the planet. Venus has an atmosphere which is 96% carbon dioxide and has a measured surface temperature higher than the other three inner planets – especially Earth which has less than 0.05% total carbon dioxide. Of Earth's less than 0.05% CO_2, only 0.02% CO_2 is from human and industrial sources.

An important illustration – shown in the previous book and repeated herein is that which was compiled by the National Hurricane Tracking Center of the NOAA parent organization.

Tracks and Intensity of All Tropical Storms

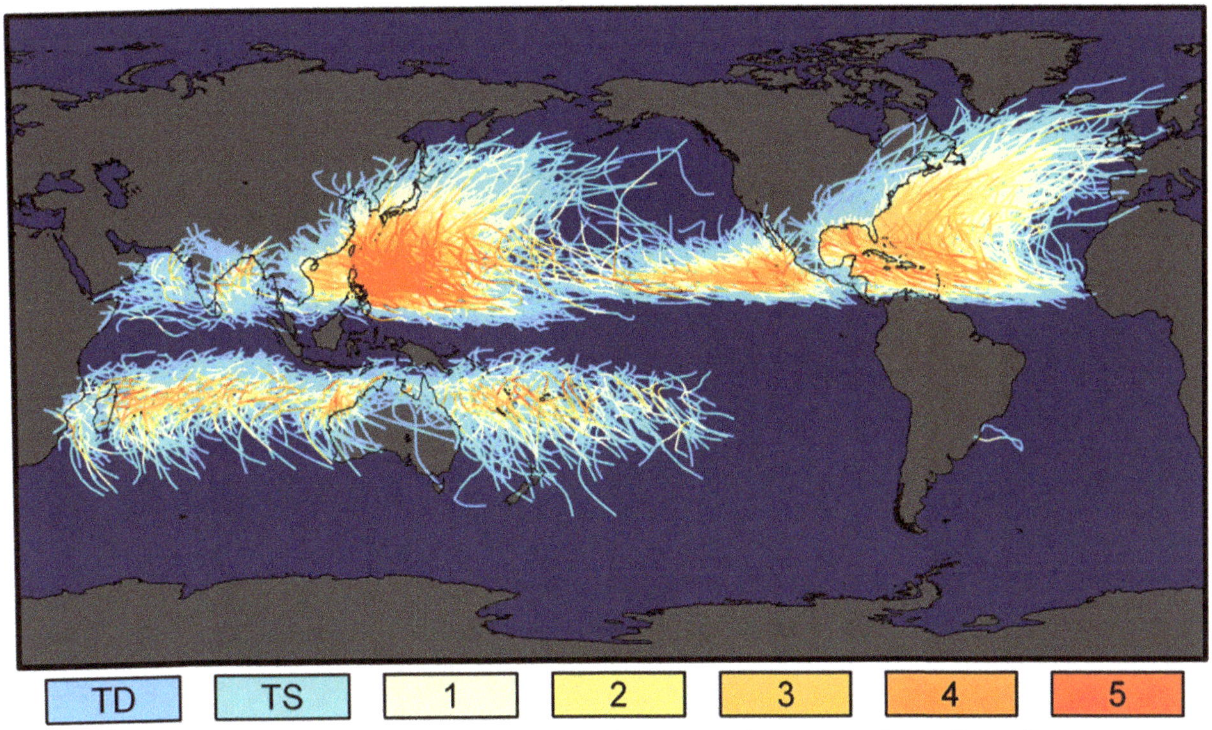

Saffir-Simpson Hurricane Intensity Scale

(Source: National Hurricane Tracking Center, NOAA)

Readers are asked to compare the existence of water vapor in the atmosphere in the "blues" chart above for water vapor" with the "reds" coloration of the paths taken by tropical storms and their intensity in the tropical storms composite of their tracks. It is not coincidental that where there is a combination of warm water, there is a high amount of water vapor (from the evaporation of the warm water) in the atmosphere. The prevailing wind currents enter into the phenomenon of the tropical storm being spawned and the body of the tropical storm being carried on the wind currents along the paths shown.

Coincidentally, the satellite measurements of carbon dioxide in the atmosphere show the higher concentrations of carbon dioxide to be in the Arctic Region (as former Vice-President Al Gore indicates in his "An Inconvenient Truth" video – where (above 65° North latitude) there is very sparse human population. There is seen a lesser concentration in "fingers" over Europe and Asia, and still less over the United States.

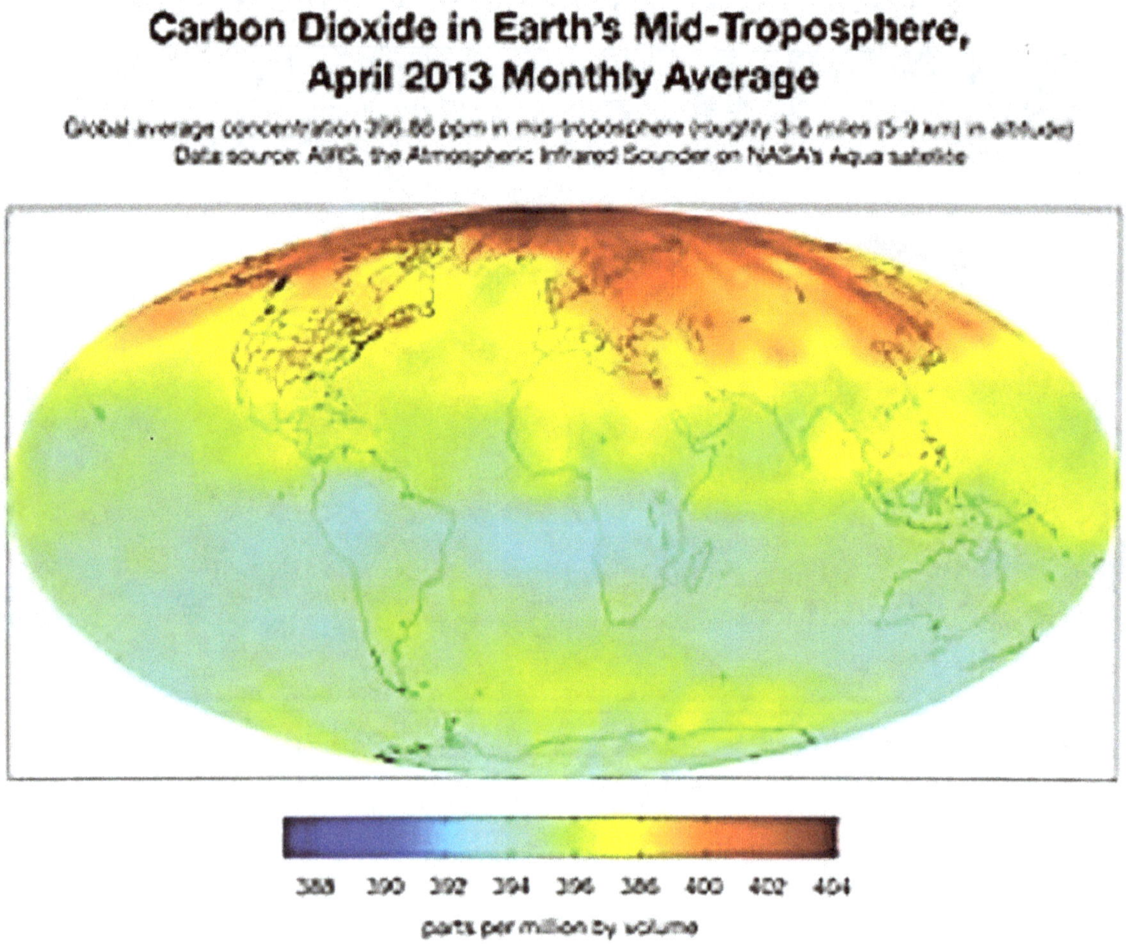

(Source: NASA)

If the environmental scientists' hypothesis that human and industrial activities are the dominant causes of carbon dioxide in the atmosphere causing a "global warming crisis", is that "crisis" created by the human population affecting mostly the region shown in red above? There is seen to be almost no dark red (high concentration of carbon dioxide) in the mid-troposphere over the United States in April 2013. More importantly, if there is 285 ppmv (0.0285%) due to pre-Industrial Revolution causes, then the yellow concentration of the USA at 398 ppmv (0.0398%) must contain 103 ppmv (0.0103%) due to anthropogenic sources. Is not a carbon dioxide concentration of 285 ppmv (0.0285%) from pre-Industrial Revolution sources not more "dominant" than the carbon dioxide concentration of 103 ppmv (0.0103%) from anthropogenic sources? Is that the "new math" students are being taught in science? Basic logic does matter.

Neither the low concentration of total carbon dioxide in the atmosphere nor the concentration of water vapor in the atmosphere above the United States region reflects a basis for a "crisis" by 2050 if the human population and industries do not eliminate the use of all fossil fuels as a source of heat energy to reduce heat-trapping "greenhouse gases". No scientifically valid database was discovered that convincingly shows the elimination of the use of fossil fuels will reduce global warming or significantly alter climate change. Several databases indicate that other factors are far more dominant than humans and industry.

Another most important conclusion drawn from in-depth analysis of the Vostok (Antarctica) Ice Core Sample database (which environmental scientists "cherry-picked" and eliminated when scientists "settled" their fallacious hypothesis) was that the planet's surface temperature was the driver of the amount of natural carbon dioxide in Earth's atmosphere.

Inasmuch as there was/is no water vapor in the Antarctica atmosphere and there were no humans burning fossil fuels 237,500 years before the present, there is no scientific evidence that links the presence of any greenhouse gas effect – either human-caused (by greenhouse gas emissions from fossil fuels) or naturally caused (by water vapor) – for the surface temperature and climate of the planet to be affected in a significant manner by any mechanism other than astrophysical and geophysical causes. A question that persists is if there was/is no vegetation (trees or plants) on the Antarctica continent, what then was/is the source of the carbon dioxide that is being measured in Antarctica with the ice core samples? A parallel question is if there is no water vapor in the Antarctica atmosphere, what is the source of the two-miles thick ice sheet on the continent of Antarctica? This book leaves those two "unk-unks" to the environmental scientists to clarify to the public and to consider in altering their fallacious hypothesis.

After analyzing the meteorological databases for countries on the planet, it was determined that in addition to the seasonal effects on the average temperature there were also the effects of latitude of the region. An illustration was provided earlier for the effect of latitude on regions of the northern hemisphere. The following table lists (by the identification number 1-20) twenty cities south of the Equator and shows their latitude, elevation and average annual temperature (from the World Meteorological Organization – WMO - database).

World Cities Showing Latitude Effect on Temperature

(Southern Hemisphere)

	City	Elevation, E	Latitude, L		Avg. Ann. Temp.
1	Equator	0 ft (0 m)	0° 0'	0°	85.0 °F (29.6 °C)
2	Belem, Brazil	30 ft (10 m)	1° 27'	1.45°	87.4 °F (30.8 °C)
3	Guayaquil, Ecuador	16 ft (5m)	2° 11'	2.18°	78.1 °F (25.6 °C)
4	Papua, New Guinea	2,188 ft (667 m)	5° 30'	5.5°	81.5 °F (27.5 °C)
5	Jakarta, Indonesia	26 ft (8 m)	6° 10'	6.17°	82.5 °F (28.0 °C)
6	Lagos, Nigeria	39 ft (12 m)	6° 27'	6.45°	84.8 °F (29.4 °C)
7	Lima, Peru	7,956 ft (2,426 m)	12° 04'	12.07°	63.5 °F (20.5 °C)
8	Darwin, Australia	26 ft (8 m)	12° 26'	12.43°	81.3 °F (27.4 °C)
9	La Paz, Bolivia	11,975 ft (3,650 m)	16° 30'	16.5°	70.0 °F (20.9 °C)
10	Rio de Janiero, Brazil	1,047 ft (319 m)	22° 54'	22.9°	74.5 °F (23.6 °C)
11	Sao Paulo, Brazil	2,621 ft (799 m)	23° 33'	23.55°	65.0 °F (18.0 °C)
12	Johannesburg, S. Africa	5,297 ft (1,615 m)	26° 12'	26.2°	60.6 °F (15.9 °C)
13	Durban, S. Africa	272 ft (83 m)	29° 53'	29.88°	77.2 °F (25.1 °C)
14	Santiago, Chile	1,887 ft (575 m)	33° 26'	33.43°	61.7 °F (16.5 °C)
15	Sydney, Australia	190 ft (53 m)	33° 52'	33.87°	64.5 °F (18.0 °C)
16	Cape Town, S. Africa	174 ft (53 m)	33° 55'	33.92°	61.0 °F (15.9 °C)
17	Buenos Aires, Argentina	82 ft (25 m)	34° 36'	34.6°	60.0 °F (16.0 °C)
18	Auckland, New Zealand	105 ft (32 m)	36° 51'	36.85°	64.7 °F (18.2 °C)
19	Melbourne, Australia	607 ft (185 m)	37° 49'	37.82°	59.5 °F (15.3 °C)
	"Avg. Global Temperature"(pre-IR)	0° 0'	0 ft (0 m)		59 °F (15 °C)
20	Punta Arenas, Chile	112 ft (34 m)	53° 10'	53.17°	44.5 °F (6.9 °C)

Note: The "average global temperature" calculated by WMO in 1990 became 16 °C (60.8 °F)

(Source: World Meteorological Organization WMO database)

Readers are asked to take note that cities 7, 9, 11, 12, and 14 have high elevations. A city with an elevation of over 5,000 has a noticeably lower temperature than a city at the same latitude but at a lower level (seen tabulated for the cities of Johannesburg and Durban in South Africa at latitudes + / - 2 degrees from 28 degrees). Johannesburg (city number 12) has an average annual temperature of 60.6 degrees Fahrenheit (15.9 degrees Centigrade) whereas Durban (city number 13) has an average annual temperature of 77.2 degrees Fahrenheit (25.1 degrees Centigrade).

The following illustration visually shows the effect of latitude for the set of 20 cities in the southern hemisphere:

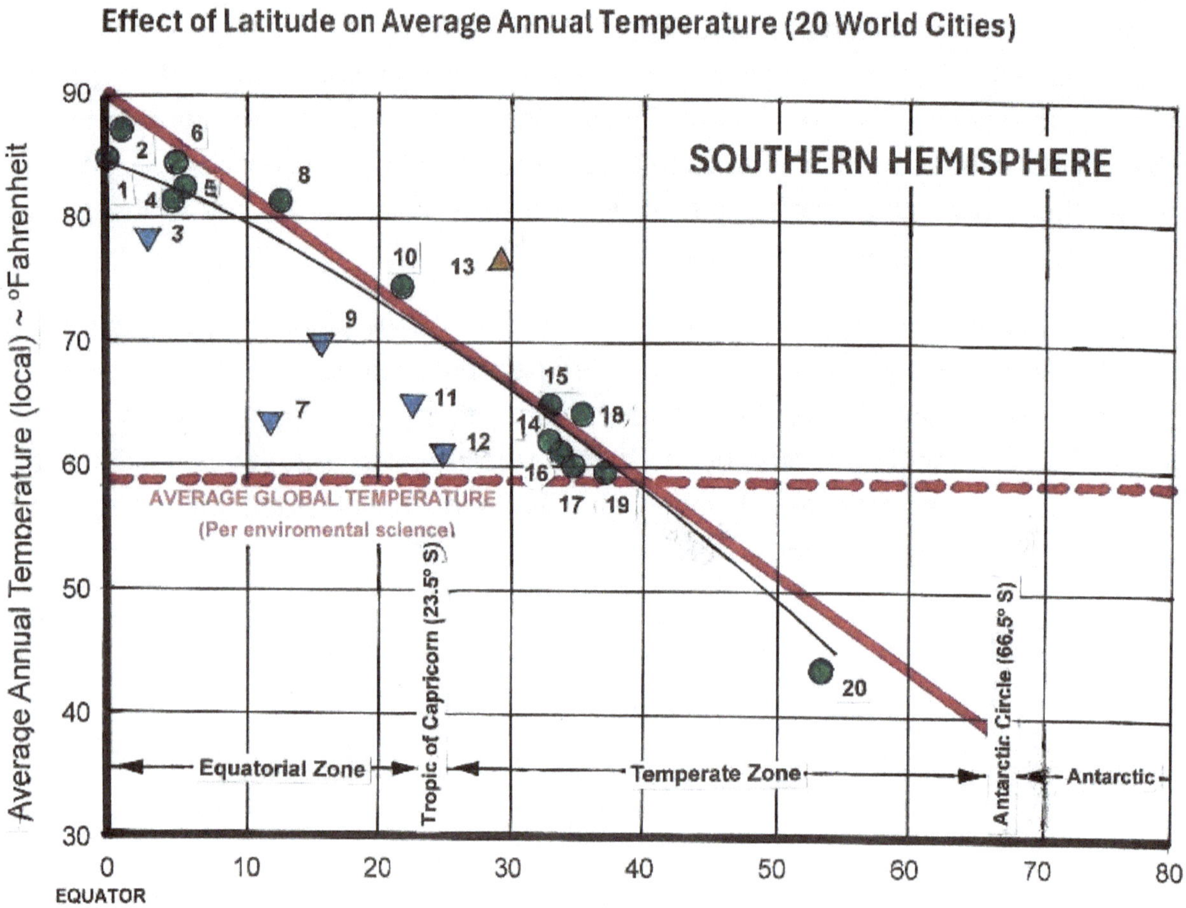

(Source: World Meteorological Organization, WMO)

The actual meteorological measurements of the annual average temperature of each of the cities (numbered 1-20 on the above graph) embody all of the several factors (astrophysical, geophysical, natural, and anthropogenic) that affect the average annual temperature of each city. The thick solid red line is a linearized approximation of the effects of those several factors being used as the first estimate for a "model" for estimating the annual average temperature for any location on the planet's surface. The main point is to show that the effect of latitude is far more dominant on the annual average temperature of a location on the planet's surface than is the sophomoric one-size-fits-all average global temperature (the dashed red horizontal line) used by environmental scientists in attempting to dissuade the public into accepting the fallacious hypothesis that humans and industry burning fossil fuels is the dominant factor causing a two degrees Centigrade (3.6 degrees

Fahrenheit) rise in average annual temperature of the planet's surface. Ceasing the use of fossil fuels will not provide the relief from the heat more urgently needed by those in the lower latitudes – between 10° latitude north and south of the Equator - where fossil fuels are already not in high use.

The average global temperature model (horizontal dashed red line) and the model of the solid sloped red line are seen to intersect at about latitude 40°. It should be obvious though that 19 of the 20 cities that were selected for further analysis are located in latitudes lower than 40°. That observation provides an interesting factor of humanity; most humans prefer to reside where they can be much warmer than average rather than being much cooler than average. City number 20 (Punta Arenas, Chile, at the Southern tip of South America, latitude 53° South) shows that there are a smaller number of humans who have the opposite preference - their preference is to be much cooler than average. From that latitude (about 52°, there are fewer than 2,000 year-round inhabitants.

The estimate of average annual temperature is seen to be grossly inaccurate the further the latitude deviates from 40° latitude. In the real world, those populated regions below 10° latitude (i.e., in the Equatorial zone) are the victims that suffer the greatest effect of global warming if even by only two degrees Centigrade. The human population residing in latitudes between 30° and 45° can do little to assist fellow humans in latitudes below 10° in adapting to the heat extremes in the Equatorial region but there is much help to those unfortunates that can be provided in the form of humanitarian aid. This book certainly endorses those populations in highly developed countries and regions to share their blessings that come with development improvements with those in lesser devloped countries and regions – but that is not a suggestion that Communism is a preferred form of government than is Democracy.

When the "average global temperature" method being used by environmental scientists and that the temperature change effect could more accurately be estimated by accounting for the regional latitude effect, it followed that temperature change from a regional "normal" requires an improved model for estimating that "normal" average temperature for the region. The details for how a replacement method for "average global temperature" are described in the facts and fallacies book (Book 1), but when this book (Book 2) was being written, a number of refinements were revealed in what has been named as the "LEO Model" (from the effect of latitude, elevation, and ocean currents' temperature at the regional location). A visual illustration of the LEO model is shown extracted from "The Facts and Fallacies of Climate Change on our Planet" book:

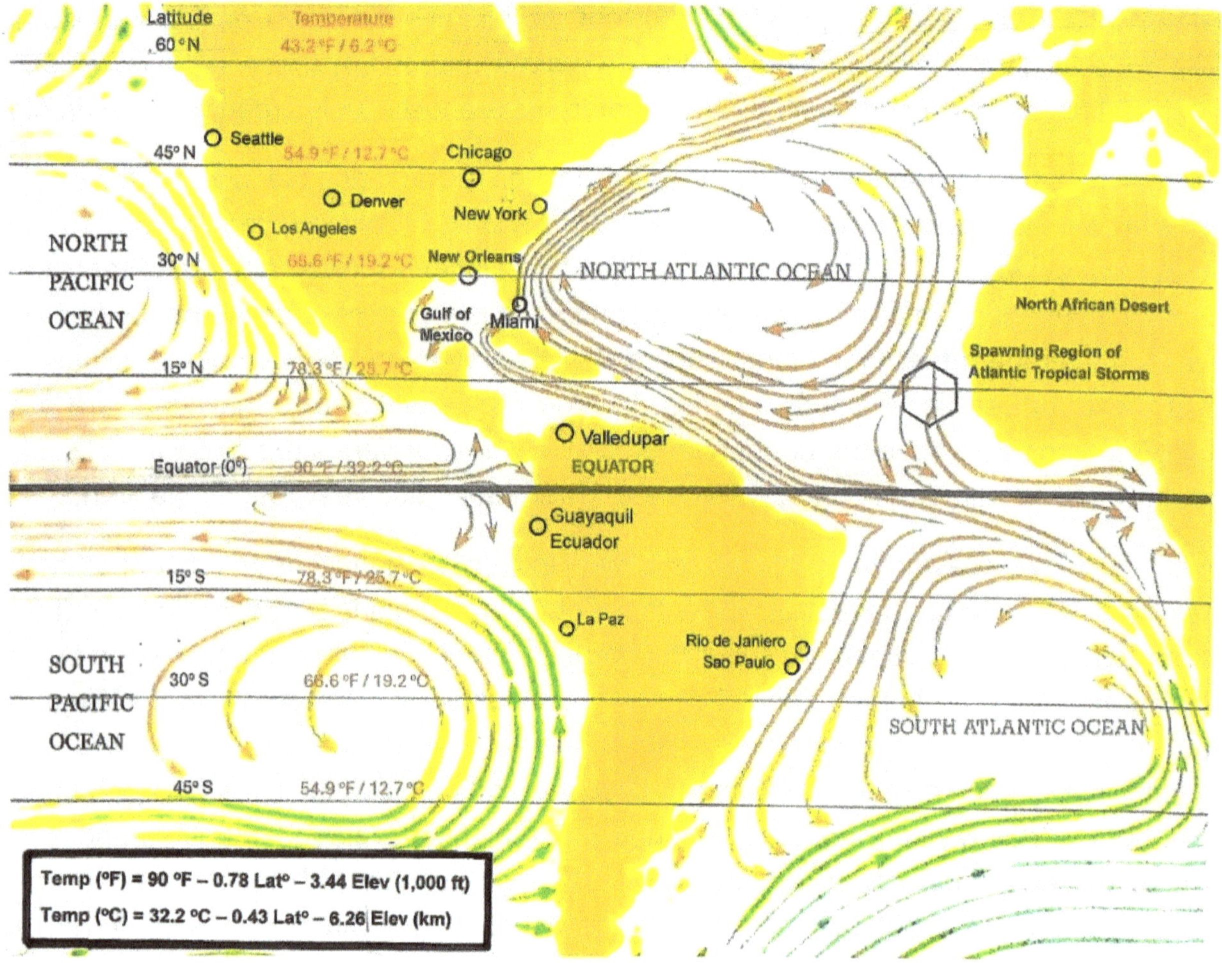

(Source: Ocean Current temperature and circulation map from Rand-McNally Atlas of the World)

The LEO Model (in the bold outlined box in lower left corner) and other notations on the diagram were generated by the Author to show Readers the complexity of global warming, climate change, and the oceans levels rising. It is not an insurmountable set of problems, but the mitigation of their adverse complex interactive effects is more important than expending any major further effort on changing them to suit the population's desires – or that of environmental scientists.

The latitude effect and the elevation effect of a region can be reasonably accurately predicted mathematically with that earlier version of what has become known as the LEO Model (from the effects of latitude, elevation, and the ocean proximity (if any). The earlier LEO model above that was applied to both the northern and southern hemispheres was found to be over-estimating the annual average temperatures for regions between latitudes 10° north and 10° south, and the adjustment for the elevation effect was also detected to be over-estimating the decreasing effect of elevation on the annual average temperature for the higher elevation cities (such as La Paz, Bolivia, at 11,975 feet elevation; Lima, Peru, at 7,956 feet elevation; and Johannesburg, South Africa, at 5,297 feet elevation).

For scientific accuracy, the Author has revised the single linear model for latitude and elevation effect to be a compound linear segment model. The reasoning for the refinements of latitude effect developed after analyzing the geophysical database and realizing the tilt angle of Earth's axis of rotation causes both the seasonal effect and the obliquity effect (angle at which the solar radiation wavelengths impinge on the planet's surface significantly different near the North Pole and the South Pole. The annual average temperature at the North Pole is about 0 °F while at the South Pole the annual average temperature is about -46 °F.

The curvature of Earths surface causes the tilt angle obliquity to vary in a parabolic manner, but to avoid the use of exponentials which are more difficult to model than linear models, the compromise was to approximate the parabolic shape of the estimating function to be a five-linear segmented one and which is slightly different in the two different hemispheres due to the tilt angle. The elevation adjustment was segmented into two linear segments for elevations below 3,000 feet (using 3.33 as the constant for the lapse rate vice the IATA 3.44) and a constant -10 °F for those above 3,000 feet). The LEO model, updated for latitude effect and elevation effect, became as follows – graphical for both hemispheres on the right, and mathematical as shown on the left (slightly different for the northern hemisphere and the southern hemisphere):

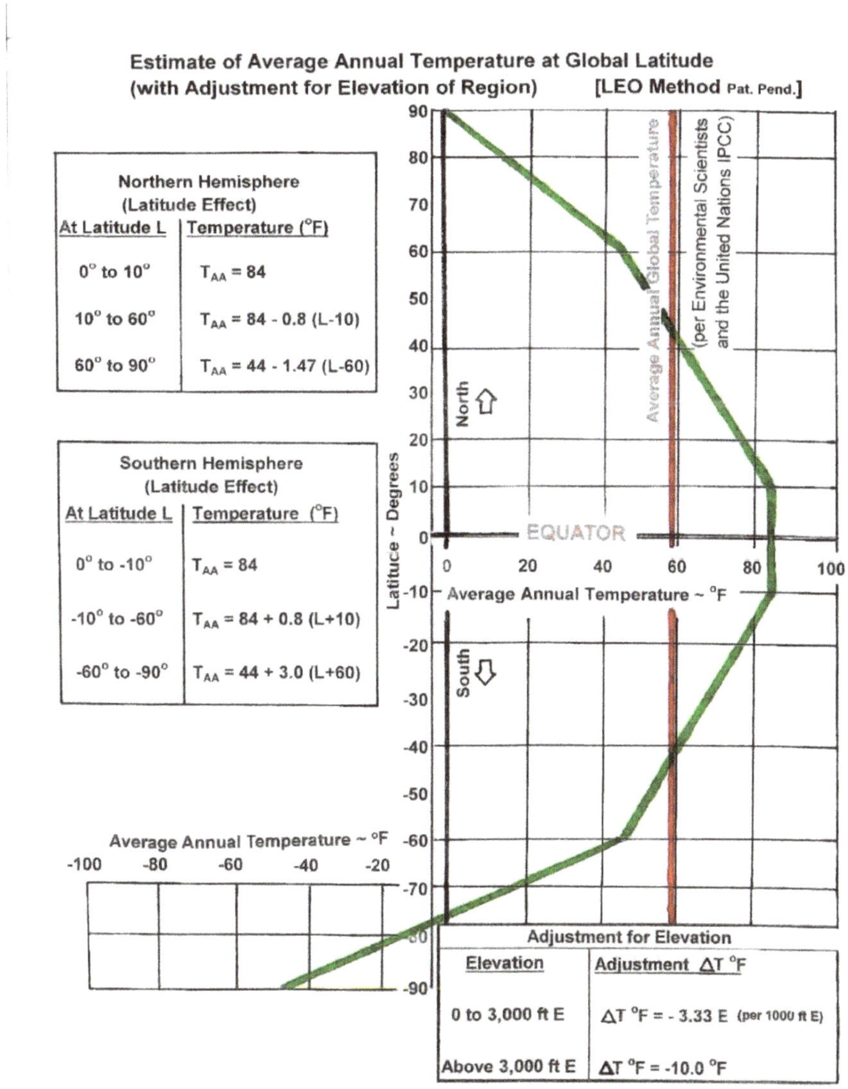

(The LEO Method for Estimating the Temperature of a Region on The Planet)

As shown graphically, the center portion of the green line between 60 degrees latitude north and south could be approximated better than the segmented lines with a parabolic formula (latitude and temperature as the primary variables and with an exponential); however, the planet being tilted causes the polar regions below 20 degrees latitude to have a different slope of its temperature variation with latitude). That difference is attributed to the axis of the planet being tilted and precessing ("wobbling") – presently at 23.5 degrees – with the obliquity angle increase causing the south polar region to be about -50 °F colder due to reduction of the solar insolation than that of the north polar region. Those mathematical improvements on temperature estimation accuracy could be programmed into a computer model. For the purposes of this book, the multiple segmented lines of the LEO model approximate the exponential irregular parabola and suffice. With the LEO Model, the temperature accuracy (after corrections for latitude, elevation, and ocean proximity are made) is within 2 °F estimation accuracy for the any latitude on the entire planet "from pole to pole".

The red vertical line represents the one-size-fits-all "average annual global temperature" method used by environmental scientists with their persistence to convince (misleadingly) the public and governmental regulators that carbon dioxide emissions from humans and industries burning fossil fuels is the dominant cause for global warming and climate change. When the red line is compared to the scientifically-based green line (representing the LEO method for estimating the latitude effect on the average annual temperature of any latitude between the polar caps), it should be obvious when comparing the temperatures between the variable green line and the constant red line that the latitude effect on the planet's surface temperature far overpowers any small effect (including up to a 2 °C or 3.6 °F) rise in "average annual global temperature" constituting a "crisis") that is caused by the "greenhouse gas effect" hypothesized by environmental scientists from the measurements of total carbon dioxide in the atmosphere (i.e., from the Mauna Loa Atmospheric Laboratory. A rise of 3.6 °F (- shifting the sophomoric average annual global temperature (red) vertical line 3.6 degrees Fahrenheit to the right - is seen to not cause a "crisis" except perhaps to those humans residing between 10° of latitude north and south of the Equator).

However, estimating the increment in average annual temperature (for the LEO Method) from the cooling or warming effect of the adjacent ocean current on a shoreline region depends on the proximity of the region to an ocean current (and its temperature) flowing nearby the region and whether there are "on-shore" or "off-shore" winds remains as a matter of judgment pertaining to the unique relationship each region has with a neighboring ocean.

The next two illustrations – the ocean currents of the Pacific Ocean (warmer in dark red and cooler in blue green) and regions those waters affect is shown first and then the ocean currents of the Atlantic and Indian Oceans and regions those waters affect is shown in the second illustration. To be especially noted in the following illustrations is the effect that latitude has on the temperature of the water in the ocean currents. The warmest waters are nearest the Equator and the coolest waters are nearest the polar regional latitudes.

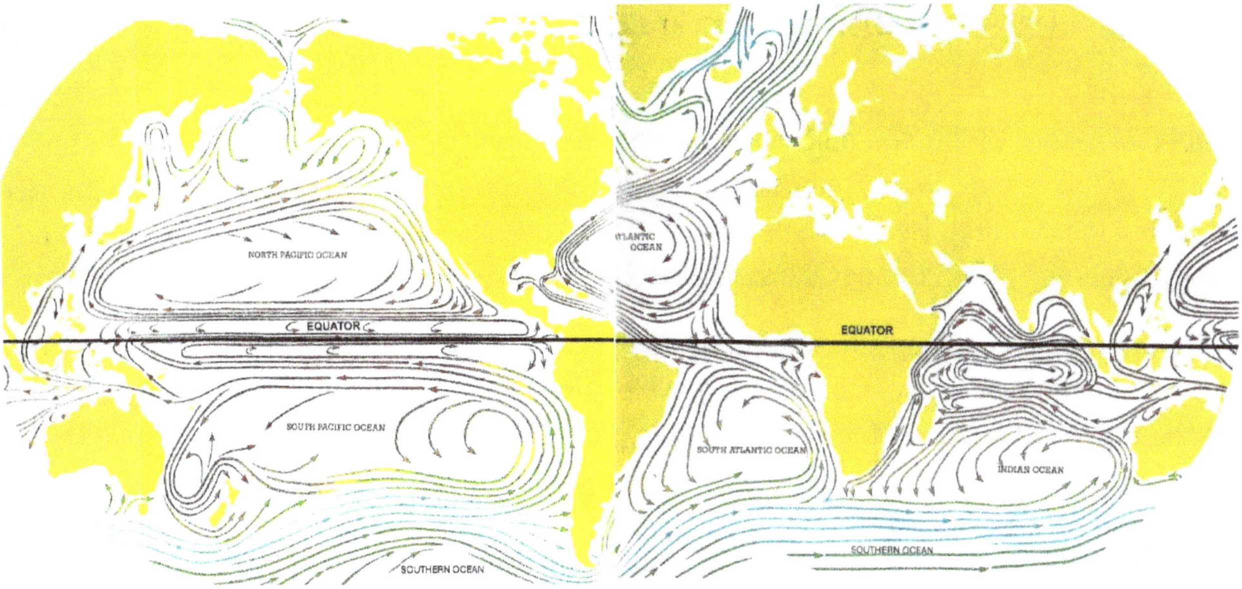

(Source: Atlas of the World, Rand McNally with Author notations)

Although the circulation of the water is not circular by any means (due to the irregular shape of the coastal lands that form a boundary of the water – forming an enormous "basin" - there is an effect of the rotation of the Earth interacting with a principle of physics known as the Coriolis Effect.

Gaspard de Coriolis (1792-1843) was a French mathematician, scientist, and engineer, who basically transformed the equations of motion (which incorporate the acceleration of gravity) developed earlier by Sir Isaac Newton applicable in a fixed reference coordinate system, and applied them to a rotating reference system, such as that for planet Earth. Although the masses for which Coriolis calculated their trajectories were originally artillery shells, the principles of motion known as the Coriolis Effect apply to any rotating mass and have a rotating reference coordinate system. In essence, the Coriolis effect was observed to cause artillery shells to impact to the right of the target aim point when fired from an artillery piece in the northern hemisphere. When fired in the southern hemisphere the shells landed to the left of the target aim point.

Meteorologists applied the Coriolis effect on mass motion to be that of water vapor and clouds – and hurricanes and monsoons – and have proven that the Coriolis effect affects any rotating mass regardless of its state of mater – solid, liquid, or gas. The explanation presented indicates (and observed) that the circulation of large bodies of water north of the Equator have a clockwise circulation. The waters south of the equator have a general counterclockwise circulation pattern from the Coriolis Effect. Likewise hurricanes, tropical storms and monsoons that have an "eye" differ in rotation depending on being north or south of the Equator.

The oceanographers use the "average sea surface temperature" method for global warming tracking of the oceans, analogous to the environmental scientists' use of the "average annual global surface temperature" for global warming tracking of populated land regions. (Author's Note: From personal experience, the Author (during his entire 'teen years) enjoyed many hours on the beaches of Long Island in New York and can attest the water temperature (from swimming in it) of the north-flowing Gulf Stream current was at or above 82 °F (27.8 °C) all year round. The Author also resided in Southern California for 38 years and the water temperature of the Pacific current of the coastal regions was about 59 °F (15 °C) from fishing in it (too cold for swimming or surfing – "hanging ten" - without a wet suit).

In a graphical record of the sea surface temperatures, oceanographers use the same sophomoric "averaging" of the oceans water temperature as the environmental scientists use with the average annual global temperature for the planet's populated surface. They also use an exaggerated scale to magnify the small rise in temperature obtained with the averaging technique.

While the impressive passion the originator has for proliferating the environmentalists' fallacious hypothesis imparting the impression on the gullible public that humans and industry are also causing the oceans to warm because of the "greenhouse gas effect", the originator – envisioned as being a young, impassioned oceanographer – may inadvertently have revealed the major "victim" of the global warming "crisis". But as stated earlier, it is not the human population but is instead the life on the ocean floor in tropical regions and the sea ice on its surface in the Arctic region:

CLIMATE

Grim results of the ocean's hot streak

Off-the-charts temperatures impact corals, sea ice and more

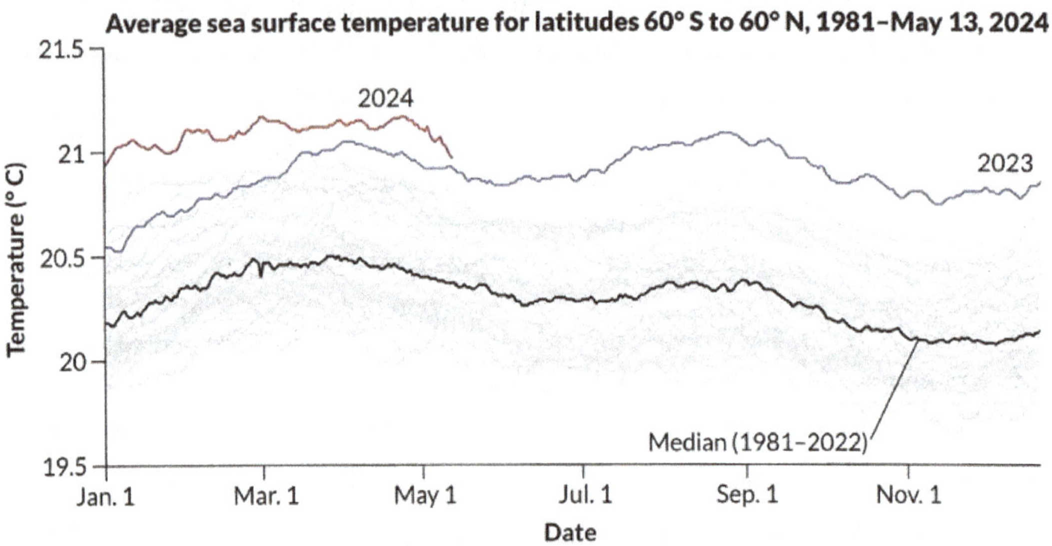

Soaring sea temps Since March 2023, the average sea surface temperature of the extra-polar ocean, at latitudes from 60° S to 60° N, has broken the daily record every single day, NOAA data show. And recent daily temperatures have shattered records that were set last spring.

SOURCE: CLIMATE REANALYZER/UNIV. OF MAINE; NOAA

When presented in a calendar year format, the seasonal and other factors effects are masked by the averaging of the ocean water temperatures.

A striking database that contrasts with the oceanographers' use of "average sea surface temperature" is that of the water temperatures of the oceans recorded in the "big picture" by NASA in 2022:

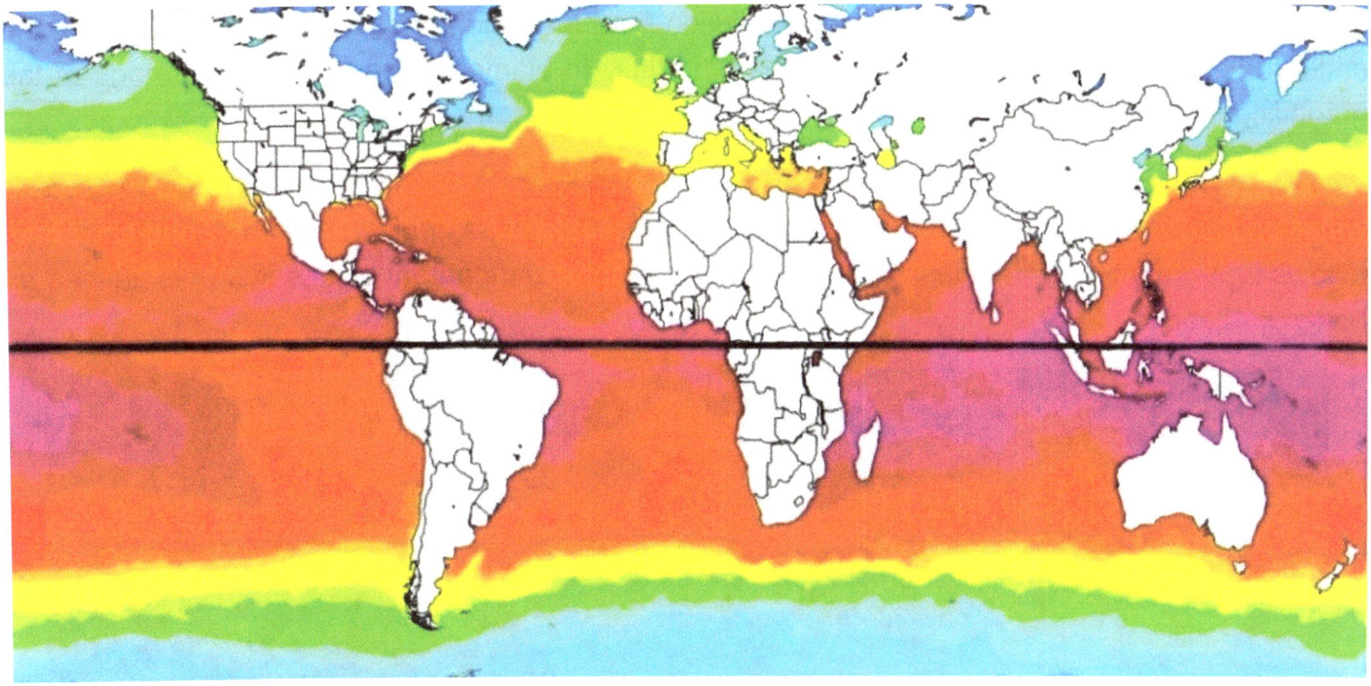

(Source: NASA)

When presented in the "big picture" global format, the latitude effect (presented in color gradient format) becomes a dominant feature.

Although there seems to be small deviations in water temperature from solely latitude effect probably caused by local land masses obstructing and being affected by local ocean currents, the same latitude effect seen in the LEO method "pole to pole" latitudinal format applies rather well compared to the oceanographers' "average sea surface temperature" – showing the sea surface to be a one-size-fits-all 21 °C (70 °F). With the NASA image, the temperature of the Coral Sea off the northeast coast of Australia is "cooking" the anemones on the Great Barrier Reef. Once again, the latitude effect is the dominant cause of global warming temperatures to be increasing, not the "greenhouse gas effect" that environmental scientists' place blame on the human population and on industry burning all forms of fossil fuels.

Virtually nothing is cited in the public news media acknowledging that the westward flowing Equatorial current between the North Pacific and the South Pacific flows unblocked by any land mass until the warm water enters the Coral Sea at Indonesia and northeastern Australia. In that Coral Sea region, the Equatorial current essential divides to flow partially northeasterly and partially southeasterly – dividing at the Great Barrier Reef, as shown in the ocean currents map above for the South Pacific region. There is no flow of the atmosphere with its ostensibly "high concentration of human- and industrial-made carbon dioxide" gases causing the warm waters and "cooking" the coral reef animals 30 or 40 feet beneath the Coral Sea surface.

Coral life flourishes in above average temperature warm water, and the warm waters in that region north of Australia gave mariners the inspiration to name the Coral Sea for that proliferation of coral formations beneath its surface. Environmental scientists remain adamant that it is the greenhouse gases that are cooking the anemones, not the naturally warming waters of the Coral Sea. Almost no mention is made in world-wide news media that residents of the region (northeastern Australia near the port city of Cairns) are concerned that pollution from the runoff of contaminated water from the cattle ranches of eastern Australia are at least partially to blame for the coral reef damage of the magnificent Great Barrier Reef.

Being ignored largely is that the ocean floor has constantly been changing with eons of time since water originally accumulated on Earth's surface. Any geologist can point to photographs shown as follows to explain those changes in ocean floor topography and content of its sediments defined as being its "floor":

(Source: National Geographic Society)

The "party line" of environmental science eliminates any factor that does not support their hypothesis that humans burning fossil fuels are to blame for everything that is trending to be "out of normalcy". The ocean floor at each different level of sediment shown above was the "normal" of its time. Eons in the future, the Great Barrier Reef may appear as a sedimental layer of light (from pulverized "bleached white" anemones?) color in a photograph such as that above.

Another event happening on the ocean floor is the presence of steam vents from within the Earth. Steam vents exist almost everywhere on the ocean floor and are found in highest frequency near the edges of tectonic plates. The region of Japan is on the western edge of the Pacific "Rim of Fire" defining the massive Pacific Plate which should explain why there are such high frequency and large magnitude earthquakes and volcanic eruptions in Japan. The following photograph illustrates the innocuous appearing steam vents which are introducing hot steam into the floor of the Sea of Japan – and thousands of other regions – 24/7/365: These ocean floor thermal vents bear absolutely no relationship – neither cause nor effect - to global warming being attributed to humans and industry emitting "greenhouse gases" into the atmosphere.

(Source: NOAA; Thermal Vents in the floor of the Sea of Japan)

As for the warm waters "cooking" the Great Barrier Reef", some corals exist because of the thermal vents, not despite them:

(Source: National Geographic – credit to unknown photographer)

Another good example of the ocean and wind effect is that observed in Southern California when there is either a "Santa Ana wind" condition (hot air from the inland desert region of California blowing toward the cool Pacific Ocean southerly current) or an "On-shore wind" (cool air blowing from the Pacific Ocean into the hot inland portions of Southern California). When an "El Nino" condition (warm water and easterly wind current) forms in the region of the Hawaiian Island group (now the State of Hawaii), the "mischievous little boy – El Nino" - conveys much water vapor in the atmosphere and precipitates heavy rainfall on the Southern California region. The rain (measured in inches per hour) falls faster than the ground can absorb the excess water and the entire Southern California region endures floods, mudslides, and earth slides. Perhaps the most "beneficial" effect of the "Pineapple Express" is that the heavy rainfall extinguishes wildfires that are ignited by lightning strikes preceding the arrival of the unwanted El Nino phenomenon.

Later in this book is contained a short discussion regarding the State of California lacking the infrastructure (poor planning?) to collect the floodwater into catch-basins or conducted by flood control channels to lakes and watershed areas to be used for overcoming drought conditions that occasionally befall the region. Drought conditions have a detrimental side effect of lowering areas of small lakes that are used as the supply of water for fire-fighting water carrier aircraft (such as the Canadian "Super-Scoopers") when fires are started by lightning strike in heavily forested regions. The recent wildfires in the coastal region of Malibu and Santa Monica went largely uncontrolled because of a lack of water to fight the fire. Many homes and businesses were lost due primarily to the high wind conditions and the lack of water at high enough pressure to extinguish the fires. (The Author's favorite seafood restaurant "Gladstones" – on Santa Monic Bay - was among the restaurants lost to the firestorm).

This book invites the environmental scientists to provide the details to the public of how an exceedance of that "tipping point" temperature will manifest a catastrophic impact on the millions of humans in cities across the United States who already are surviving the extreme temperatures of hot and cold (as shown as follows created for Book 1).

Average Temperatures In United States Cities (8)
(US National Weather Service, NOAA)

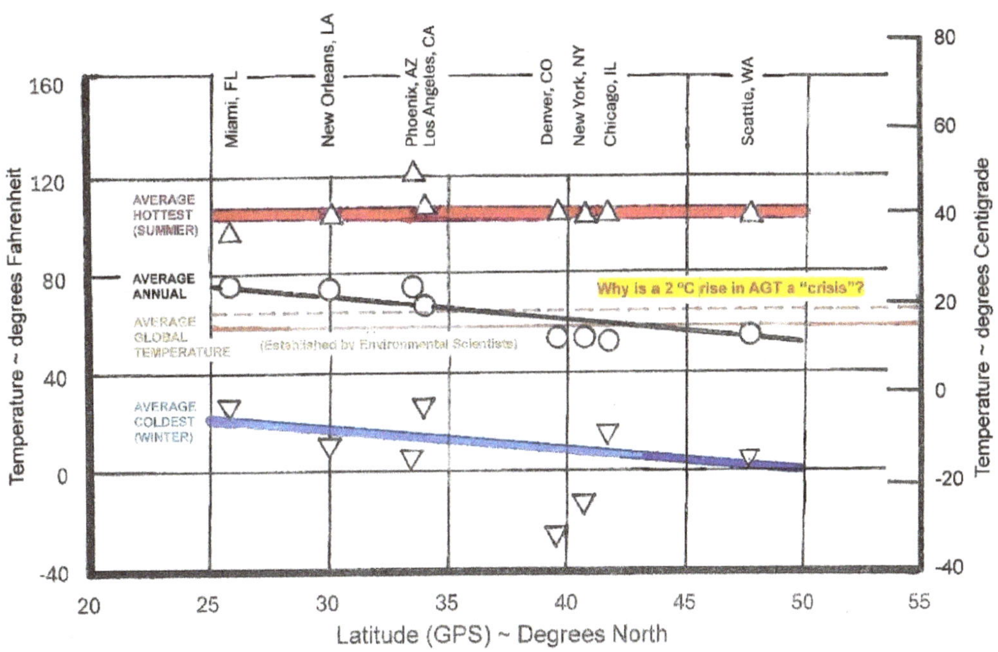

(Source: US Meteorological Service, NOAA; generated by Author)

An observation to be made is that the average hottest temperatures are seen to not vary by latitude and have a maximum of about 115 degrees Fahrenheit. The city of Phoenix, Arizona is the exception but is caused by Phoenix being located in a desert surrounding environment whereas the other cities are not in a desert environment. The average coldest temperatures do have a general trend with latitude. Also shown are the averages of the annual temperature between latitudes 25 and 50 degrees North. Environmental scientists' average global temperature is shown with a horizontal thin red line at about 60 degrees Fahrenheit, and it is repeated for emphasis that their one-size-fits-all annual average global temperature bears little relationship to the extreme average hottest and coldest temperatures to which residents within the USA latitudinal extremes must survive. The average hottest temperature - constant at about 115 degrees Fahrenheit – is about 55 degrees Fahrenheit hotter than the average global temperature. That fact begs the question <u>"Why would two degrees Centigrade (3.6 degrees Fahrenheit) rise in the average global temperature create the environmental scientists' "doomsday crisis" if the 55 degrees Fahrenheit extreme average hottest temperature excess above the average annual global temperature has not already caused an even worse crisis?</u>

In a "big picture" perspective, scientific curiosity causes scientists to understand why real world datasets of information differ from sets of scientific measured data in controlled laboratory small-scale experiments being scaled up and applied to the enormous scale of one planet in the still more enormous solar system. Having shown that there is a powerful effect of latitude on the surface temperatures of the planet, and having shown that there are other effects not related to anthropogenic causes, the question arises as to what are other important factors affecting the temperatures to which humans must accept as existing and need to adapt to in the near future for protection of their health safety.

Combining the effects of season and latitude on the extremes of the temperatures to which humans and other forms of life adapt or die – neither of which are affected to any degree of significance by any activity of humans or industrial source – a temperature envelope was compiled wherein all forms of life and all types of mechanical operating machinery must be designed to survive and function at any latitude and at any time of year:

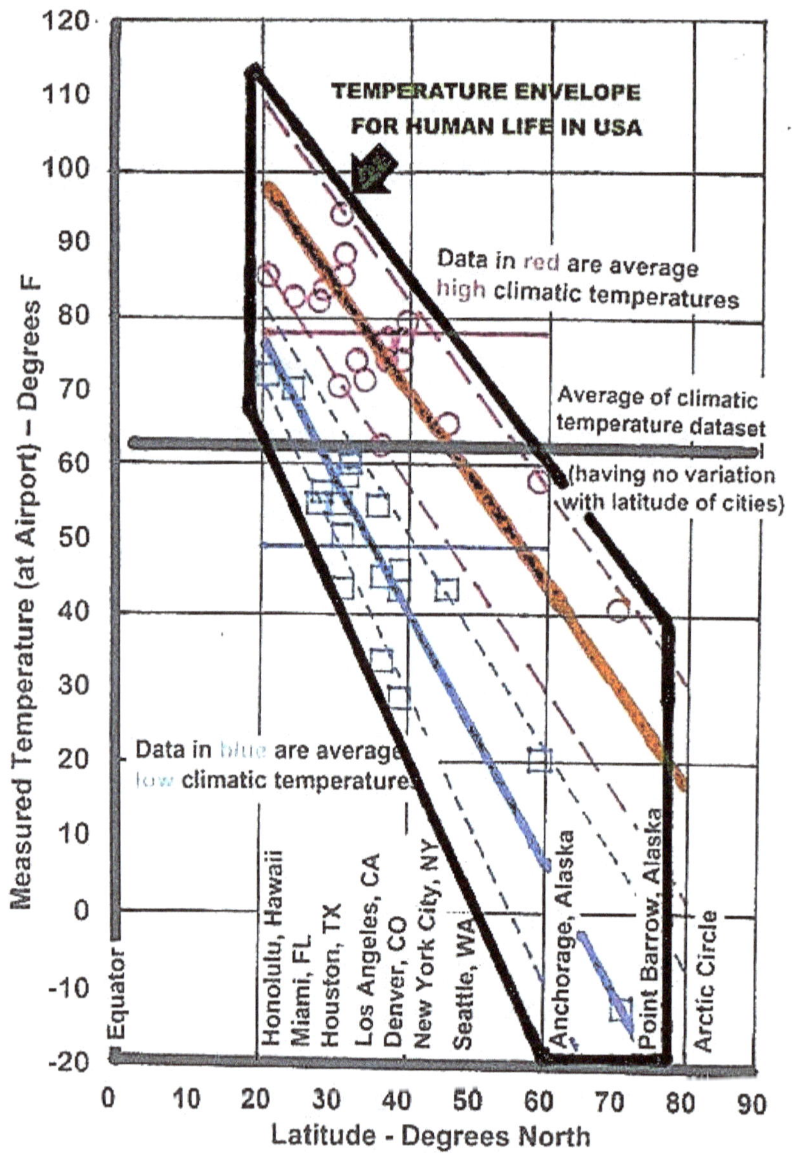

(Source: Author generated with National Meteorological

It should be noted that a horizontal black line labeled as "Average of climatic temperature dataset" is drawn on the illustration at a constant temperature of 61 degrees Fahrenheit (16 degrees Centigrade). If the environmental scientists of the United States are claiming that a rise in that average temperature is not limited by 2 degrees Centigrade (3.6 degrees Fahrenheit) by 2050, the planet will exceed a "tipping point" and a "crisis" will ensue.

It is known that the seasonal effects on temperature have existed on the planet before the presence of Man on the planet. Meteorological databases for the United States have shown the repeatability of the season effect and confirmed by the calculations of mathematicians (including those of Serbo-Hungarian mathematician Milutin Milankovic) provide the seasonal effect on both the hot temperature and the cold temperature extremes of the United States as they vary with time of year:

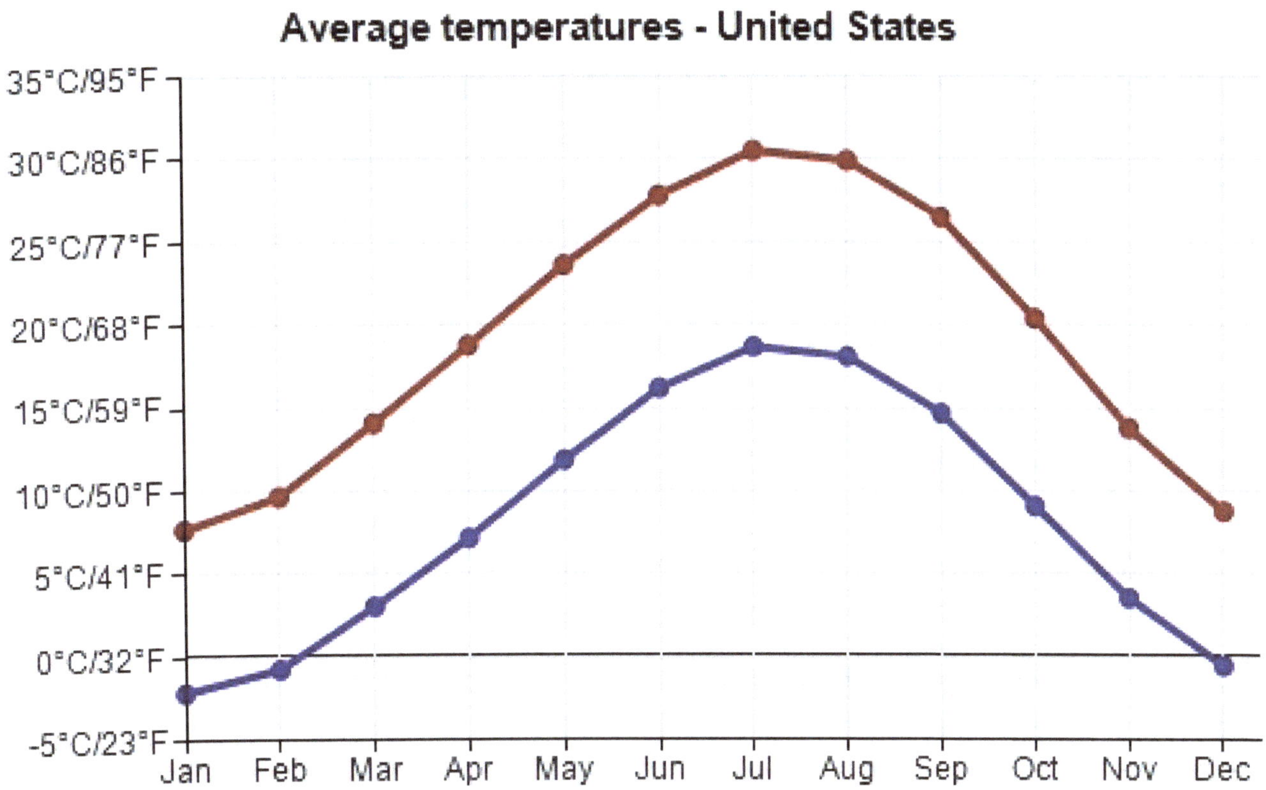

(Source, National Weather Service, NOAA)

Seasonal effects have existed on the planet from its astrophysical origin 4.67 billion years ago. Anthropologists found that homo sapiens dwelled inside natural caves for shelter during the dark hours for protection from carnivorous animals and from the daily and seasonal variations in the outside temperature. Their clothing was from animal skins indigenous to that latitude. In effect, that act of adapting to the regional temperature "right outside the door" for the latitude by early caveman is the message of this book on climate change. He probably did not even know what the process of "averaging" meant – he was far too involved in just surviving each day as it came. His "job jar" had but three notes in it: 1) Find food and water for himself and his cave-mates; 2) Defend against large carnivorous animals outside the cave who had the same instinctive motivation to find their food; 3) Wear an animal skin to stay "comfortable" – within a tolerable range of temperature (as it varied daily, seasonably, and in between volcanic eruptions).

The previous book "Facts and Fallacies of Climate Change on Our Planet" was published as a public service to inform the public of the factual causes for the surface temperatures rising and the climate to be changing. This second book "Dealing with the Effects of Climate Change" echoes the more important factual causes for the surface temperatures to be rising and the climate to be changing –

as they both are unquestionably changing. Neither book found any compelling reason for continuing the "combat" of climate change.

The Author is a veteran of the Korean War and suggests the environmental scientists who want to do "combat" with anything – especially the climate – to rethink that strategy. The "winner" of a combat action is the side that loses the least. Losses can be economic, loss of human lives, loss of expensive weapon systems, or loss of reputation/respect of leadership. This book includes a description of the hazards of nature that are taking human lives which cannot be avoided from happening, but the cost and the losses can be minimized. The best strategy is for humans to take necessary actions to minimize their net losses. Those actions include the development of a set of technologies to mitigate the effects of climate change which are beyond human capability to control. It must include changes that humans must make to reduce those actions and hazardous activities that humans have caused or contribute to as acts of lack of consideration or of careless habits.

The research done for the first book revealed several causes for surface temperature rise and climate change but also revealed the existence and accessibility of many resources that exist or could be transformed to become of use to meeting the needs of the current population on the planet as well as to prepare for the exponentially growing growth of that population. The Author bears no association with any scientific, educational, or industrial group, company, organization, or federal or state entity having a biased interest in the technologies that are herein suggested being advanced or changed in approach (including the reduction of the use of coal in the United States region).

It is highly important that the public realize that each country and continent has its unique set of natural resources and unique needs for energy depending on its state of development. How the government leadership chooses to develop and provide the natural and economic resources to its residents and for protection of its environment also varies greatly from region to region. These two books place emphasis on what is best for the United States, but the logic of choice used in formulating the suggestions can be tailored to the needs of each other country. The present approach of "eliminate the use of all fossil fuels to avert a world "crisis" (which is not revealed in any of the environmental science documentation – except for one paper entitled "Grim results of the ocean's hot streak" discovered in a science journal) to inform the public as to who, where, when, how the fallacious 'crisis" will affect the planet. It was not found during the scientific research that "the sky is falling".

This book does not adequately address political ramifications of global warming and climate change – those ramifications are left to government leaders to negotiate "for the benefit on mankind". Resolving the scientific aspects of global warming, climate change, and oceans levels rising is "one small step for man, but a giant leap for mankind".

There are 25 technologies that are assessed in the latter portion of this book for their pros and cons toward mitigating the adverse effects of temperature rises and climate changes on the people of the United States. The climate science profession apparently has lost sight of the fact that the US Constitution starts with the following:

The technologies chosen for assessment of their value to the US population are evaluated for their overall value and importance in achievement of the following factors:

1. Status of Technology Development (risk, schedule, cost, potential value)
2. Reliability and Sufficiency (time of day/year, weather dependent, climate dependent)
3. Regional Economy and Resources (mineral, water, labor force to operate/maintain)
4. Cost of Resources (domestic and imported, mode of harvesting / distribution)
5. Importance to Industry, Government, and Human Safety, Health, Comfort
6. Importance and Urgency to Economy and Employment
7. Pros and Cons of By-Products (emissions, pollutants, safety/health hazards)
   a. Effect on Human Population (regional in USA) – Strength Codes
   b. Effect on Human Population (regional in USA) – Food Supply
   c. Effect on Human Population (regional in USA) – Safety, Health, Comfort
8. a. Effect on Infrastructure (regional in USA)
   b. Effect on Transportation (regional in USA)
   c. Effect on International Trade (to/from/with USA)
9. a. Effect on Environment (regional in USA) – Atmosphere
   b. Effect on Environment (regional in USA) – Land
   c. Effect on Environment (regional in USA) – Water
10. a. Effect on Animal Life (regional in USA) – Land
   b. Effect on Animal Life (regional in USA) – Aquatic
   c. Effect on Animal Life (regional in USA) – Avian
11. Effect on Forestry and/or Agriculture (regional in USA)

A professional statistical analysis technique employed in system engineering as a "weighted score of importance" is used to score the value of the "pros and cons" in each of 25 technology areas in achieving some degree of mitigating effect on global warming, climate change, and ocean surface levels rising. The worksheets for the scoring are included in an Appendix at the rear of this book as well as a one-page composite showing the scores enumerated against the 11 factor areas shown above. (Other countries can tailor the factors of importance and the weighting of the scores for their unique resources and state of development circumstances.)

The Author hastens to add that research of the fossil fuel databases reveals that a more hazardous aspect of burning coal than the amount of carbon being transformed from solid state (in lumps) to gaseous state in the atmosphere is in the ash residue generated from the coal combustion process. The ash residue contains a high percentage of toxic and/or carcinogenic matter depending on the type of coal (lignite, anthracite, bituminous) and the percent of impurities (unwashed coal having the highest content of impurities). Disposal of coal ash in landfills or trash dumps creates pollution (toxic and carcinogenic) into the land and any water aquafers below the landfilled area. Natural gas and oil are fossil fuels that burn with far less pollution than does coal, and therefore any region – such as the United States – which has an abundance of both coal and natural gas (and oil) should – and has – begun the process of substituting other technologies that are cleaner and more reliable than that of coal.

The efforts of the environmental scientists and the Environmental Protection Agency to rid the use of all fossil fuels in order to abate climate change created the further development of solar and wind power technologies to replace all fossil fuels has been shown to be a good solution but to the wrong problematic need. The pros and cons of coal technology are very much different from the pros and cons of the other two fossil fuels that were targeted for reduction/elimination, and since oil is the heart of the vehicle transportation system of the planet oil technology must continue. The cons of oil technology are mainly with the effect that spillages, pipeline leakages, and transportation accidents have had on the environment and those cons must definitely be reduced to absolute minimum if not zero. The continued use of oil will have virtually no effect on climate change.

More importantly, the energy development and production of natural gas eliminates the two major cons of fossil fuels. Natural gas has become the leading technology for the replacement of coal technology for generating electric power (especially in sight of the population growth demanding more electrical power) as it has none of the cons of coal or oil. That "for the benefit of mankind" advancement is shown in a most revealing summary from the Department of Energy on the trends in energy technologies being used to substitute for the coal technology (as reported in the New York Times in early 2025 but with improved graphical presentation – colorized - by the Author):

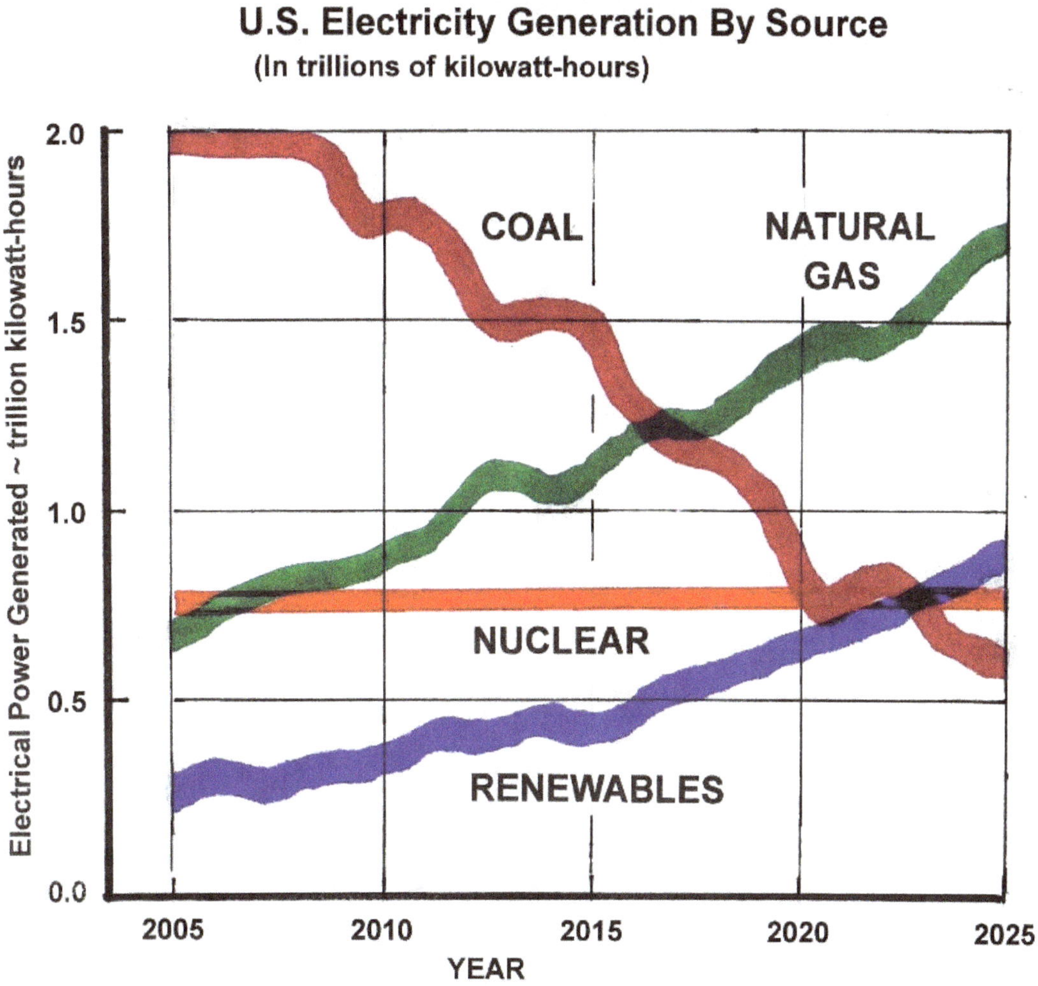

(Source: DOE database; The New York Times; circa January 2025)

The trend in oil technology usage is discussed later in the development of the all-electric vehicle technology which also is a good technology but not for the purpose of improving the climate. The continued use of oil – but with more realistic reduction of its cons – is a major conclusion and recommendation of this book. There can be no doubt that the use of natural gas to replace coal technology is the better choice than the use of solar and wind power, which are too unreliable to be used for meeting the electric power demands of the present and future population growing exponentially with time. There is also little doubt that there will be an increased use of nuclear power (shown as "flat-lined" on the above illustration), and later in this book is shown the Author's evaluation of the nuclear technologies (continued use of uranium-based energy source and the more practical technology of thorium as the energy source). The continued use of renewable technologies of solar and power is certainly recommended but more for residential, commercial, and agricultural uses – less use for generating electric power in large quantity with high 24/7 reliability.

The point in time of any technology when the "cons" become more hazardous than the "pros" are beneficial is the time to use an alternative technology to achieve the same technological function but with a less adverse con. Environmental scientists are correct in stating that the human population must reduce coal burning, but the true need is to reduce/eliminate the polluting aspects of coal burning, not to reduce its effect of being considered a heat-trapping greenhouse gas.

It should be obvious to Readers that if the environmental scientists' goal is to eliminate the use of all fossil fuels by 2050, that would include eliminating coal, oil, and natural gas. Resorting to wind, solar, hydrodynamic, and nuclear power in the United States to replace all fossil fuels by 2050 will be a disaster of global proportion considering that the population of the USA and the world is exponentially growing:

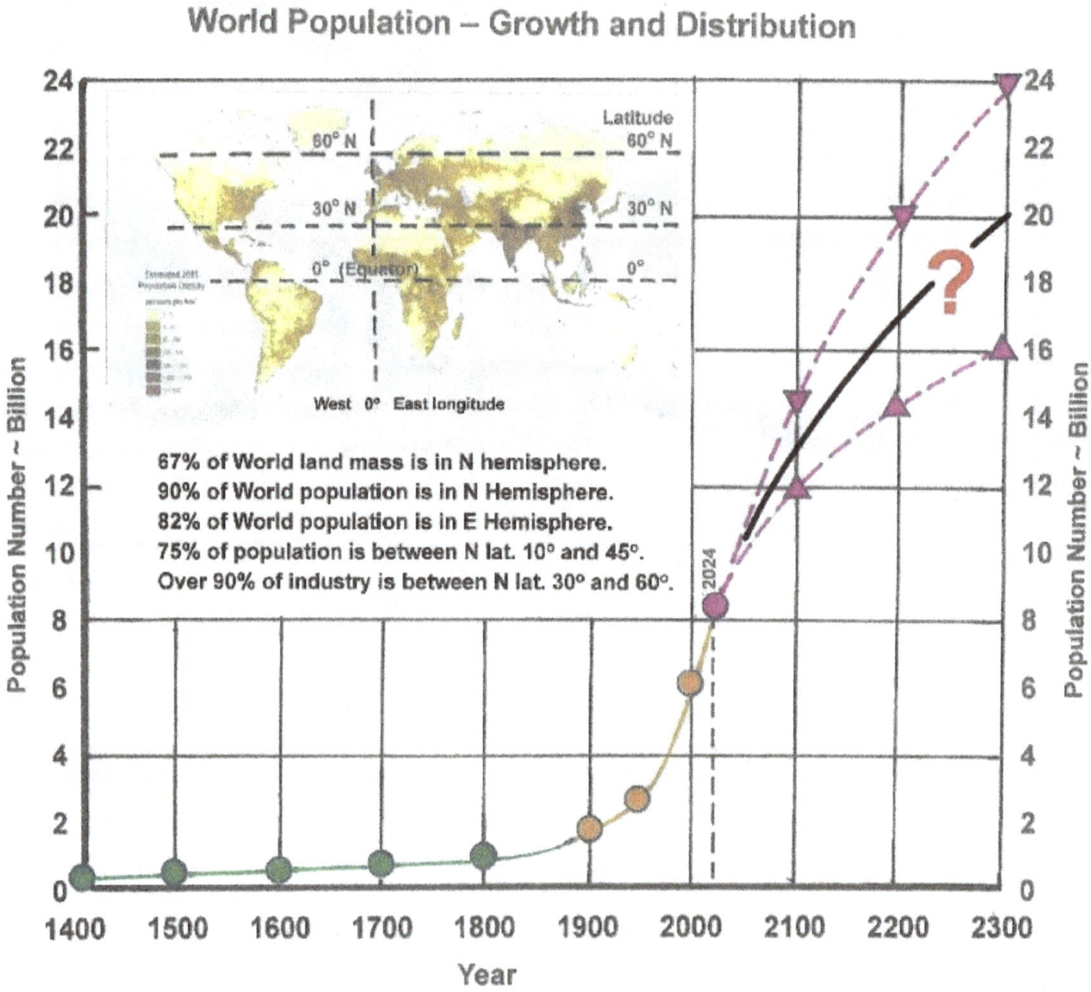

(Source: The UN World Population Division with Author annotations)

It is important to fit together the many pieces of the climate change jigsaw puzzle without "forcing" the pieces to fit. The information presented to the public in these two books provides the pieces of the climate change jigsaw puzzle, but the fitting can best be done by the different regions of the planet to consider the "resources" that are unique to each region. An earlier illustration shows that while the USA is reducing dependency on coal, there are other regions who are not reducing their dependency on all fossil fuels as the US environmental scientists and the USA have fallaciously attempted to do.

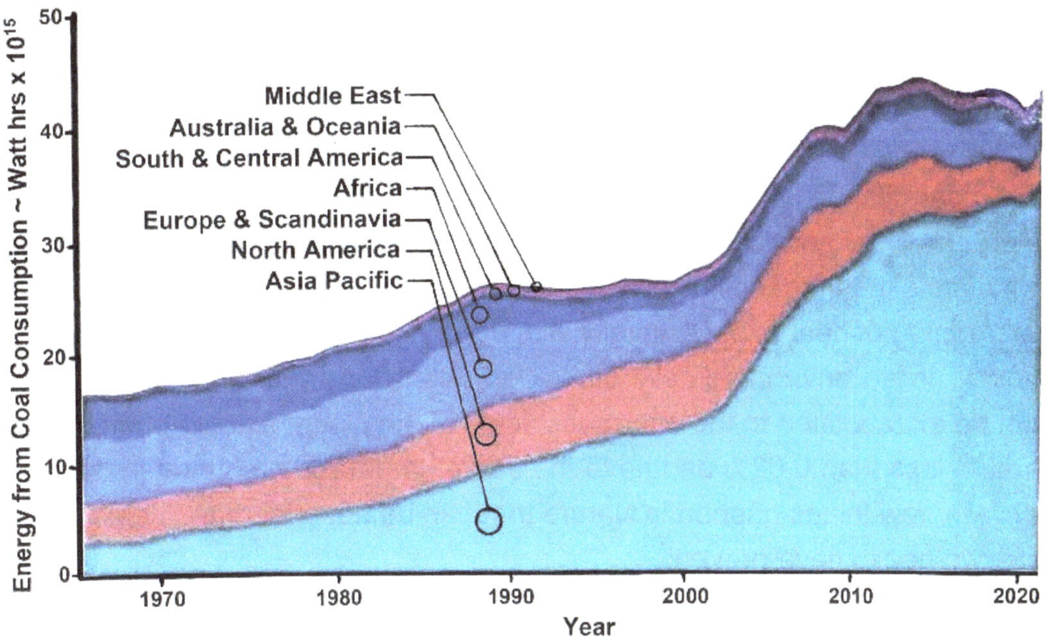

The latitude of the region determines how much solar insolation must be countered. The population of the regions are different and are in differing densities within the region, somewhat dependent on the topography of the region. The population in the Himalayan Mountains is very small compared to the population living in coastal port cities. The state of technology development of each region is different, and not all regions can utilize the technology that has been developed for a different climate and topography and latitude. There was shown a temperature envelope for the United States within which all life forms, machinery, and technologies must observe the temperature extremes and maintaining functionality. That temperature envelope for the USA is not the same as it must be for Upper Scandinavia and for the Equatorial zone regions. The original environmental scientists' "average annual global temperature on which to base fallacious unrealistic emissions control goals to avert a "crisis".

The Vostok ice cores database findings refute the hypothesis created in early decades of the twentieth century by environmental scientists fallaciously claiming: "By eliminating everything else, settled science considers Man burning fossil fuels (and creating heat-trapping "greenhouse gases") to be the dominant cause for "global warming". The expression "when eliminating everything else" apparently included the databases that this Author analyzed which proves that the "greenhouse gas effect" is nearly insignificant and does not pose a "crisis" level of endangerment to the safety or health of the human public.

There is little doubt that the populated regions of the planet that are able to do so should wean away from the use of coal technology as a source of heat energy. It was proven in his first book by the data in several databases that the factors of astrophysics, geophysics, natural seasonal meteorological changes, and the long-term effects of natural anomalistic events (lightning-ignited wildfires, floods and mudslides, volcanic eruptions, hurricanes, and tornadoes, etc.) are of far more significance on the life-forms on the planet than are the nearly insignificant effects of small increases in "average global temperature" due to "Man-made greenhouse gas emissions." The environmental protection scientists' objectives and goals, for eliminating the use of fossil fuels (coal, oil, and natural gas) to improve the changes to the climate, were scientifically proven to be unjustified.

Although there is no harm done to the environment by the reduction of any emissions, the reality exists that reducing Man-made carbon dioxide emissions will have little or no effect on improving the effects of climate change. Climate change is in fact very real, but it cannot be altered by human intervention (especially by environmental protection scientists and/or the Environmental Protection Agency) and it affects the planet differently in different regions Man. What environmental scientists observed (the trapping of heat experimentally with a high concentration (above 96%) of a gas such as carbon dioxide in an environmentally controlled laboratory and at small scale (in a one liter beaker) cannot be extrapolated to the enormous scale of the planet's atmosphere where it is diluted to a concentration less than 0.02% and have the same heat-trapping characteristic as observed in the laboratory. Very few things happen in Nature that can be described as "ideal" or "scalable", as it is hoped these two books have proven.

To derive a "model" of the climate change, scientists resorted to linearization and averaging of trends that suffice for scientific purposes but not for purposes of climate change on the entire planet. As a transitory comment, the first book described the factual causes of the surface temperatures increasing in many regions of the planet, for the climate to be changing accordingly, and the oceans' surface levels to be warming and increasing in height from the thermal expansion. Ensuing chapters in this second book, "Dealing with the Real Climate Change", uses conclusions compiled in the first book to define what must be done to mitigate ("deal with") the effects of the inevitable climate change (caused by several other factors in complex interactions and combinations that defy modeling by climate scientists. Extensive use of the LEO Model for estimating the annual average temperature for any region is only one of the corrective steps needed to be taken. (A patent application has been filed with the US Patent Office for the LEO Model, but that was done not for the purpose of any profit gain from others adopting use of the LEO Model. The Patent if/when approved, assures any world-wide user that it is not merely an interesting idea but in fact is a valuable tool in realistically evaluating the effects of global warming and climate change.

Developing technologies aimed at reducing "greenhouse gas emissions" to avert the fallacious "Man-made climate crisis" is a much lower need than two more factual needs of 1) developing technologies to permit human populations to live and function efficiently and safely within the safe temperature envelope and variable climatic conditions of their surrounding environment for the region's latitude and climate on the planet; and 2) to reduce Man's activities that are polluting the land, the water, and the air with toxic and/or carcinogenic by-products of industrial processes, transportation activities, or unsanitary practices. This book has been scoped to only address the technology assessment for achieving part 1) of mitigating the climate change problem. The pollution of Man identified as part 2) of the major problems facing the human problem is a responsibility of the Environmental Protection Agency (EPA in the United States) scientists. With the projected exponential growth of the human population on the planet, the pollution problem will grow proportionately. Relieving the EPA for any responsibility to pursue technologies for alternative energy sources (as in part 1) for minimizing Man's impact on nature by the use of cleaner technologies will allow environmental scientists presently engaged in part 1) activities to focus on the increasing labor-intensive activities of part 2) for the exponential increasing need for pollution control.

It is also hastened to be added that although this book may seem to be absolving man and industry from being the cause of global warming and climate change, it by no means should be interpreted to mean that man and industry from polluting the atmosphere, land, and water with true chemical pollutants. The very purpose for reducing and eventually eliminating the use of coal is to reduce the amount of coal ash residue that is being produced to generate electricity. The carbon dioxide emissions from coal burning acting as a "greenhouse gas" is of minor concern when a "Closer Look at the Numbers" are considered – as Dr. Fred Singer at the University of Virginia had done. It is the toxins and carcinogens in coal ash residue leaching from disposal sites and landfills into underground aquafers that threatens the health and safety of humans and farm animals.

The combustion of oil and natural gas do not produce any toxins and/or carcinogens of any significant magnitude to hazard human and animal health and safety. Nothing was found in the database research that confirms that all fossil fuels must be eliminated to avert a climate catastrophe. Technologies to mitigate the inevitable climate change in regions will each have a set of attributes and detriments (pros and cons), and the cons must have the minimum possible adverse effects (including harmful pollution). Accidental leakages and spills of polluting materials will occur, but the "owner" of the polluting technology should be held responsible for clean-up of those leakages and spills. Carbon dioxide is not a pollutant, as the courts have ruled, since there is twice as much natural carbon dioxide in the atmosphere necessary for the growth of plants and trees than there is anthropogenic generated carbon dioxide.

Chapter 1
Geophysical Factors Affecting the Planet; The LEO Temperature Model

The category of geophysical factors affecting global warming, climate change, and/or ocean surface levels rising – due to any causal effect – is a major element of this book. This chapter is lengthy due to the many geophysical factors that have a significant causal effect on the planet's surface temperature and/or climate.

Readers were given an introduction in the preceding section of this book to the astrophysical factors as the spherical planet Earth travels in its irregular elliptical orbit of the Sun – one orbit every 365.24 days - and rotates once every 24 hours about its tilted, wobbling axis as it travels. The planet can be regarded as the vehicle on which each species of life travels through time and space during the time span of life expectancy for that species. Passengers on the planet must survive whatever environments – solid, liquid, or gas - and forces and accelerations they will be subjected to during the planet's dynamic and hazardous travel on its "programmed" never-ending travel route. The largest geophysical factors that must be considered for the effects on global warming, climate change, and oceans levels rising are those of the oceans and other large bodies of water covering over 70% of the Earth's surface, the location of humans being near those bodies of water – which serve as heating and cooling sources, the latitude of the cities relative to the Equator, and the elevation of the cities (if above 1,000 feet in average elevation).

In the Introduction of this book, it was shown that the astrophysics (described by Milankovic) and geophysical effects of latitude, elevation, and proximity to a large flowing body of water on cities (both in the "lower 48" of the USA and in other locations) in the Northern hemisphere had a very pronounced effect on the annual average temperature of cities in that hemisphere ("top half" of the planet). The Northern hemisphere was chosen first because over 60% of the human population reside in the Northern hemisphere and it has by far the largest industrial activities on the planet in the northern hemisphere serving the population needs

Effect of Latitude on Average Annual Temperature (20 World and US Cities)

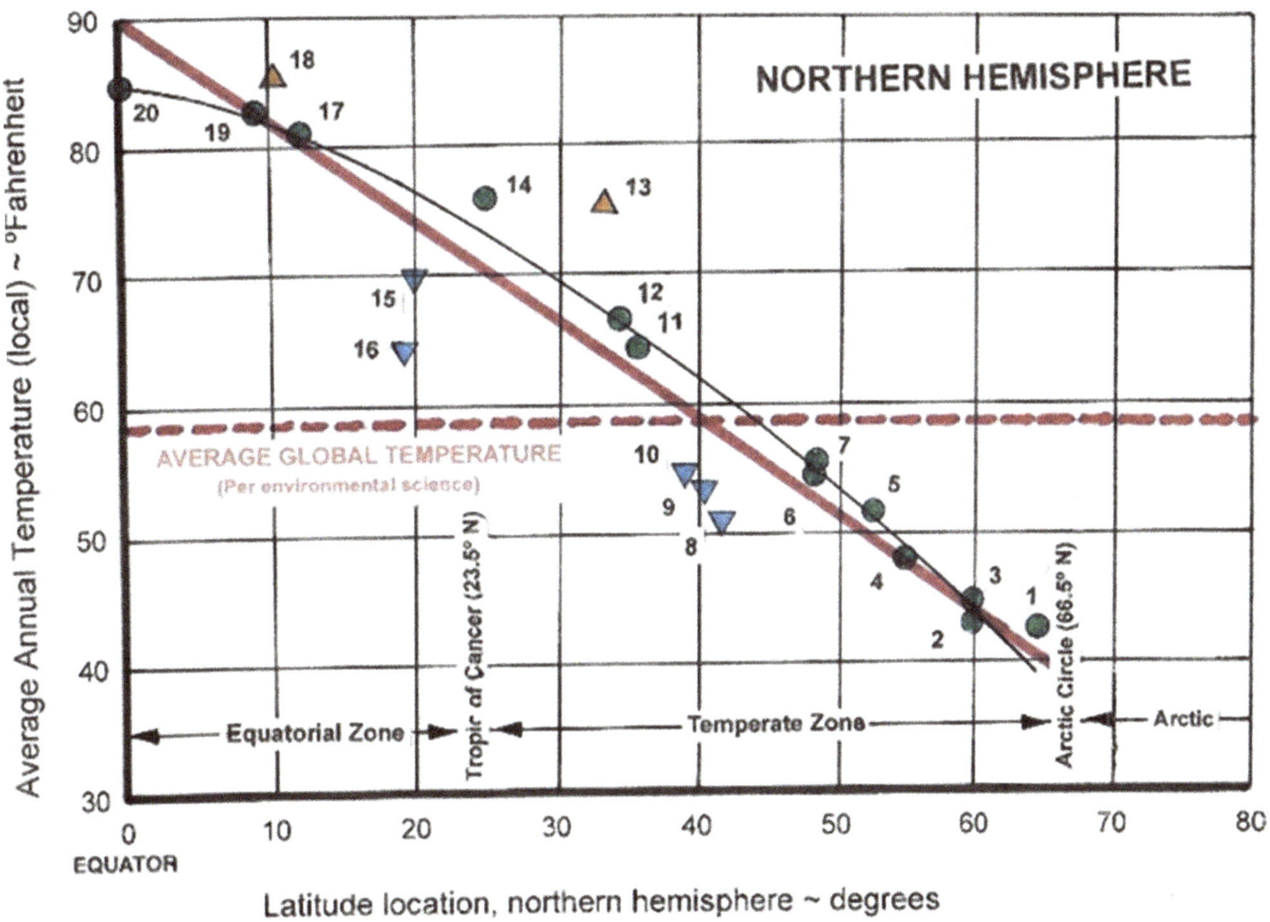

(Source: World Meteorological Organization database, chart by Author)

The Equator is numbered 1 as the reference baseline latitude (as well as being at zero geophysical latitude and zero elevation) and the other 19 cities are numbered 2-20. As seen by the deviations from the solid red line on the graph, elevation of the city and its proximity to an ocean current – being warm or cold - are important geophysical factors that cause the data points to be either above or below the solid red line at their known latitude. A closer statistical "fit" to the data points is seen on the chart as a thin curvilinear black line; however, it is shown later in this book that the solid red line can be modeled as a simple point slop line whereas a quadratic expression is needed to describe the black line as a model.

There is no valid scientific reasoning to not believe that the latitude effect on temperature has the same effect on cities in their unique latitudes distributed over the entire planet. To verify the effect of latitude in the Southern Hemisphere is the same as seen for 20 major cities in the Northern Hemisphere and north of the Equator, a set of 20 cities were selected for the similar analysis in the Southern Hemisphere below (south of) the Equator. The next illustration in tabular form identifies the cities selected in the Southern Hemisphere for a similar analysis of their temperature variation with latitude:

World Cities Showing Latitude Effect on Temperature

(Southern Hemisphere)

	City	Elevation	Latitude	Avg. Ann. Temp.
1	Equator	0 ft (0 m)	0° 0'	85.0 °F (29.6 °C)
2	Belem, Brazil	30 ft (10 m)	1° 27'	87.4 °F (30.8 °C)
3	Guayaquil, Ecuador	16 ft (5m)	2° 11'	78.1 °F (25.6 °C)
4	Papua, New Guinea	2,188 ft (667 m)	5° 30'	81.5 °F (27.5 °C)
5	Jakarta, Indonesia	26 ft (8 m)	6° 10'	82.5 °F (28.0 °C)
6	Lagos, Nigeria	39 ft (12 m)	6° 27'	84.8 °F (29.4 °C)
7	Lima, Peru	7,956 ft (2,426 m)	12 ° 04'	63.5 °F (20.5 °C)
8	Darwin, Australia	26 ft (8 m)	12° 26'	81.3 °F (27.4 °C)
9	La Paz, Bolivia	11,975 ft (3,650 m)	16° 30'	70.0 °F (20.9 °C)
10	Rio de Janiero, Brazil	1,047 ft (319 m)	22° 54'	74.5 °F (23.6 °C)
11	Sao Paulo, Brazil	2,621 ft (799 m)	23° 33'	65.0 °F (18.0 °C)
12	Johannesburg, S. Africa	5,297 ft (1,615 m)	26° 12'	60.6 °F (15.9 °C)
13	Durban, S. Africa	272 ft (83 m)	29° 53'	77.2 °F (25.1 °C)
14	Santiago, Chile	1,887 ft (575 m)	33° 26'	61.7 °F (16.5 °C)
15	Sydney, Australia	190 ft (53 m)	33° 52'	64.5 °F (18.0 °C)
16	Cape Town, S. Africa	174 ft (53 m)	33° 55'	61.0 °F (15.9 °C)
17	Buenos Aires, Argentina	82 ft (25 m)	34° 36'	60.0 °F (16.0 °C)
18	Auckland, New Zealand	105 ft (32 m)	36° 51'	64.7 °F (18.2 °C)
19	Melbourne, Australia	607 ft (185 m)	37° 49'	59.5 °F (15.3 °C)
	"Average Global Temperature"(pre-IR)	0 ft (0 m)	Everywhere	59 °F (15 °C)
20	Punta Arenas, Chile	112 ft (34 m)	53° 10'	44.5 °F (6.9 °C)

Note: The "average global temperature" calculated by WMO in 1990 became 16 °C (60.8 °F)

(Source: World and US Geodetic Organizations; World and US Meteorological Organizations. Compilation and analysis by Author)

As done for the cities in the Northern Hemisphere, the tabulation for the cities in the Southern Hemisphere is transformed into the easier to visually interpret graphical form shown as follows:

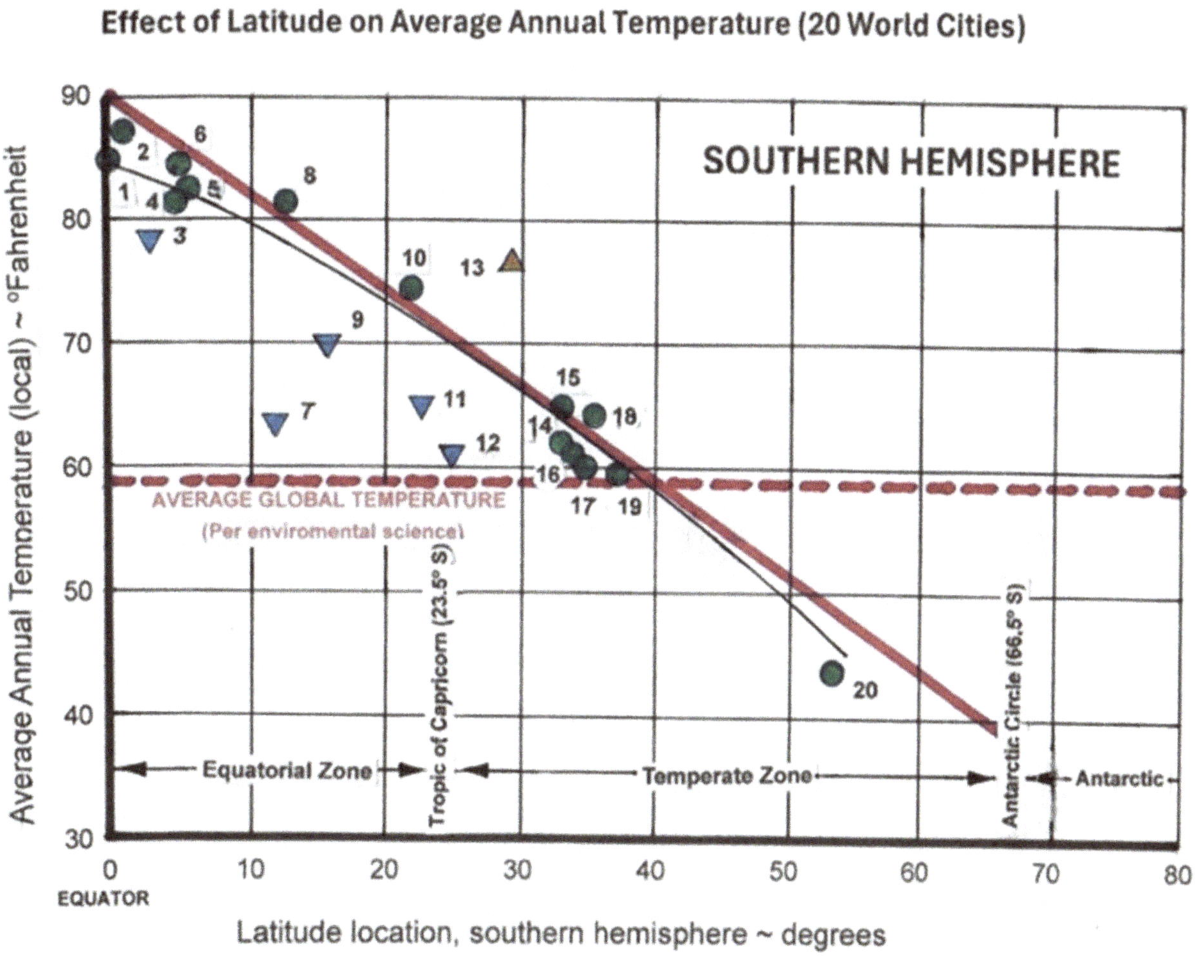

(Source: Table of Latitude Effect represented graphically, by Author)

It can easily be seen by comparing the illustrations of latitudinal effect on temperature of both the Northern Hemisphere and the Southern Hemisphere to have almost the same curvilinear variation (the thin black line) and to also have the same average linear variation (the solid red line) of temperature with latitude variation. For mathematicians or for environmental scientists wishing to model the effect of latitude on average annual temperature, the solid red line can be expressed in a simple mathematical equation for a sloped line as:

Temp (° F) = 90 °F – 0.78 L (°)

providing an estimate of the annual average temperature for cities at different latitudes in either hemisphere (North or South). In metric convention of units, the red line "model" of latitude effect on temperature is expressed as

Temp (°C) = 32.2 °C – 0.43 L (°).

Once again, the environmental scientists using the "average global temperature" approach (the horizontal dashed red line) shows no variation with latitude and is considered to be a scientific descriptor having a sophomoric level of scientific importance in describing the complexity of climatology. Mathematically, the environmental scientists' "model" of the one-size-fits-all average global surface temperature of the planet is expressed as Temp (°C) = 16 °C or, in Imperial units, Temp (°F) = 61 °F. The public and much of government has been misled (primarily via the sensationalistic news media) to believe that a small rise of 2 °C (3.6 °F) above the average global temperature (the dashed red horizontal line) will cause a "crisis" to the planet being destroyed by an inferred apocalyptic set of consequences. The fallacy of that misinformation is seen graphically above – raising the horizontal dashed red line vertically upward an increment of 2 °C might change the 19 data points representing cities below latitude 40°.to be 2 °C (3.6 °F) hotter than present. If that measure of the environment being raised by 2 °C translates to the planet being thrown into hyperthermic chaos, why are those cities below 30° latitude not already in a temperature crisis state of chaos? In a later chapter, describing the importance of water on the planet, the article appearing in the Climate section of a scientific magazine was published without its emotional presentation descriptors being "fuzzed" out – and inadvertently revealed what the environmental scientists are concerned about:

CLIMATE

Grim results of the ocean's hot streak

Off-the-charts temperatures impact corals, sea ice and more

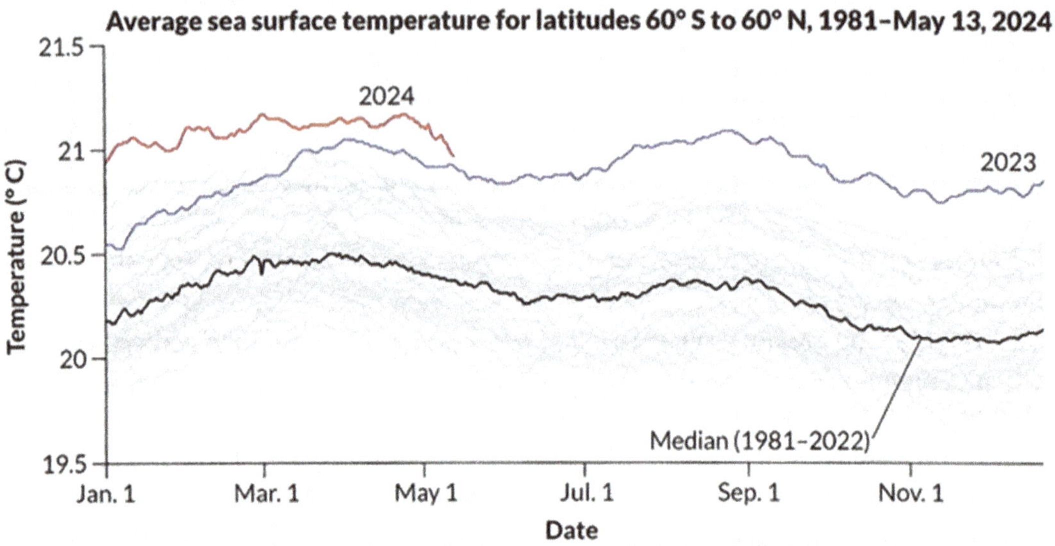

Soaring sea temps Since March 2023, the average sea surface temperature of the extra-polar ocean, at latitudes from 60° S to 60° N, has broken the daily record every single day, NOAA data show. And recent daily temperatures have shattered records that were set last spring.

SOURCE: CLIMATE REANALYZER/UNIV. OF MAINE; NOAA

(Source: Climate Reanalyzer, University of Maine, NOAA)

Readers should note several comments made to arouse public fear over Man-made "greenhouse gases" trapping heat and which is causing the "average sea surface temperatures between latitudes 60° S to 60 o N to soar". That "soar" is measured on the temperature scale (left vertical) to be about 1o C from 1981 to 2024 and is "shattering records". The largest "slip-of-the-tongue" comment is in the second line below the "Grim results of the ocean's hot streak" which reveals the concern: *"Off-the-charts temperatures impact corals, sea ice and more"*. This Author wonders if the "*and more"* includes only other aquatic biological matter or perhaps includes humans? A minor wonder is exactly where are the "off the charts" data shown to be "*soaring off the chart*" – it all appears to be there?

It is also what is not on the chart that causes a wonderment. How does NOAA explain the "El Nino" effect that occurs on a greater than seasonal cyclical period and causes damaging flooding of the

US Pacific Coast states - primarily Southern California? The "average sea surface temperature" method of analysis averages out the important anomalies that causes human fatalities and property damage that do not "soar off the chart". Southern Californians refer to that huge flow of water from the Hawaiian Islands region (where the massively warm "El Nino" Pacific Ocean current – a mischievous "little boy" in Spanish originates) and causes flooding in Southern California. In the "big picture", the Southern Pacific jet stream above the warm El Nino waters accelerates in a generally Easterly direction as shown in the next chart

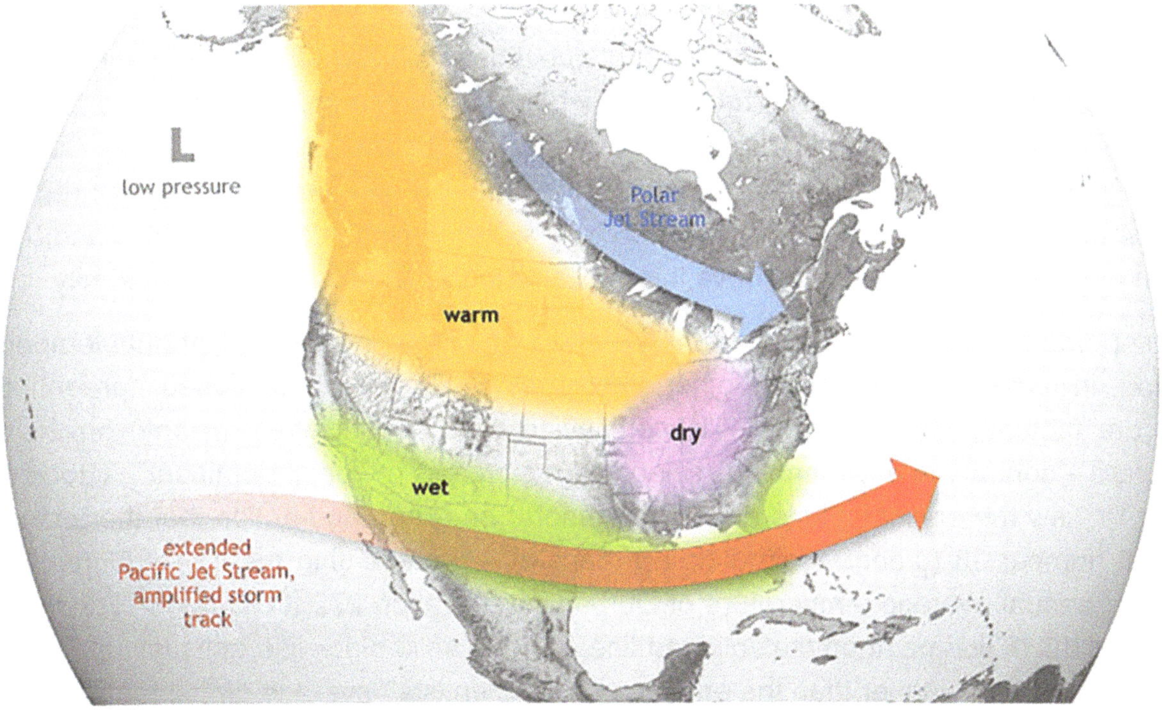

(Source: NOAA)

Californians are forewarned (by the vital National Weather Service of NOAA) of the formation of El Nino events – which occur on an irregular 2- to 7-year cycle - and refer to the wet, fast-flowing storm (in a swath between latitudes about 18° North and 34° North) as "The Pineapple Express". For reference purposes, the latitude of Hawaii is about 21° North, San Diego, California, is at 32° North, Mexico City is at 19° North, and Jacksonville, Florida, is at 30° North. The green areas on the diagram above get much rainfall in a very short period and flooding occurs because the massive precipitation (water) cannot seep into the ground rapidly enough and snow in the higher mountain elevations accumulates rapidly.

Another point of special interest to be noted is that the city number 20 (Punta Arenas, Chile) is the last data point - at latitude 53° 10' S on the earlier graph of latitude effect on temperature - has an average annual temperature of 44.5 °F (6.9 °C). The geophysical significance of that latitude is that Punta Arenas is at the very southern tip of South America and is the populated area in the Southern Hemisphere closest to Antarctica. Punta Arenas has an annual average temperature that is about 16 °C (29 °F) below the horizontal dashed red line representing the environmental scientists' "crisis" reference line. Environmental scientists claim that if/when the average global temperature exceeds the environmental science threshold "crisis" temperature by about 2 °C (3.6 °F), the environment and the human population will face disaster.

The residents of Punta Arenas are not known to be wringing their hands in despair that the temperature of their environment will rise by 2 °C or more. There is no apparent cause to declare an environmental "crisis" in Punta Arenas over Man burning fossil fuels – the 128,000 residents of Punta Arenas have learned to wear a long-sleeve cotton shirt instead of a woolen sweater to stay comfortably warm in Punta Arenas. A much-appreciated humanitarian action would be for the Punta Arenas residents to send their no-longer needed sweaters to the 140,000 residents in Reykjavik, Iceland, or to the 5,000 residents in Point Barrow, Alaska, who then can wear the sweaters from Punta Arenas instead of their present heavy parkas).

To a professional statistician, the more important question is what are the other factors not seen that are causing the data points to not be more closely "spotted" on the solid red line – i.e., latitude is an important effect but what else other than latitude is affecting the average annual temperature of each specific city being analyzed? Two geophysical factors possibly affecting the temperatures are noted to be:1) some of the cities are near a large body of water (which is a "heat sink"); and that 2) the elevation of the city places the city above the heat sink water to which it lies in proximity.

A most important point is that it is not latitude alone that affects the average annual temperature of a location; in fact, there are several other factors and, as already mentioned, "greenhouse gas effects" are factually very low on the list for having a significant effect (if any) on the planet's temperature – contrary to environmental scientists' hypothesis of being "dominant". A good example of that is to view the earlier latitude effect on temperature graph and notice that there is a "rogue" data point number 13 (orange, point side up triangle) – representing the city of Durban in South Africa. The annual average temperature of Durban (at elevation 272 ft (83 m) is 77.2 °F (25.1 °C) and is about 10 °F warmer than the solid red line and thin solid black line at its latitude of 29° 53'S. It is also about 18 °F warmer than the environmental scientists' "average global temperature" of 59 °F (15 °C). If the environmental scientists' goal is to hold the annual global temperature rise to no more than 1.5 °C (2.7 °F) by 2050 or a "crisis"/"tipping point" will destroy human life, how is it possible that the residents of Durban, South Africa, are already 18 °F (10 °C) above that "crisis"/"tipping point" temperature and yet are still alive?

An important factor of influence is that the location of the city to a nearby ocean, with a current traversing the coast at that location, can have its climate affected by that ocean current – varying between very warm or very cold. That ocean current temperature effect on coastal cities is vividly shown in the ocean currents at the southern region of South Africa, and how those currents can cause a most significant difference in temperature for two cities apart by only 4° of latitude – Durban (at 29° 53' S latitude) and Cape Town (at 33° 55' S latitude). The straight-line distance between Cape Town and Durban is about 840 miles (1,352 kilometers), and they are influenced by two different currents as seen on the next map of the oceans and land masses surrounding South Africa:

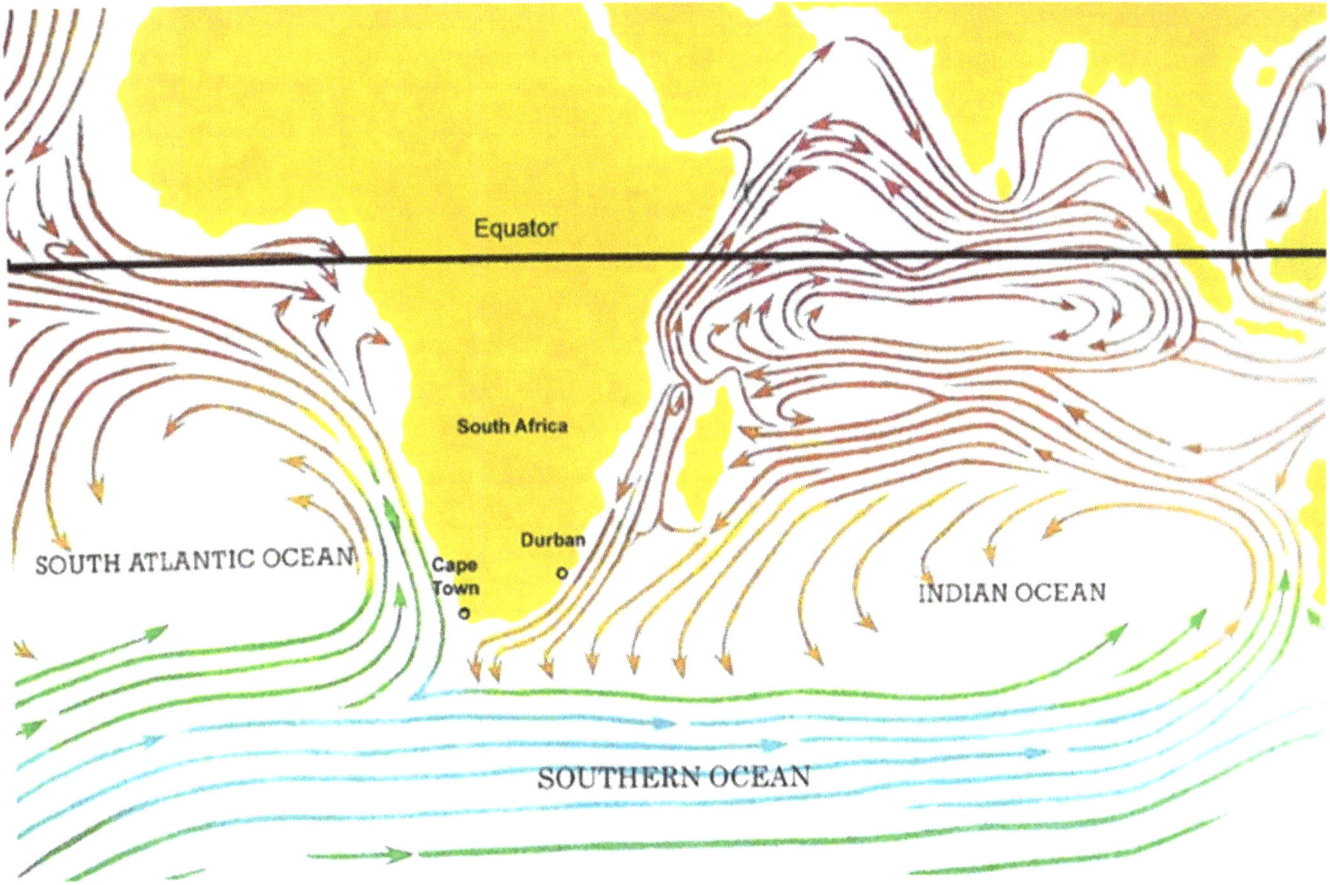

(Source: Basic map from Rand McNally Atlas of the World, with Author notations)

The average annual temperature of Durban (city number 13 on the previous Southern Hemisphere illustration) - warmed by the southwesterly current of the southern Indian Ocean - is 77.2 °F (25.1 °C). In contrast, the city of Cape Town (city number 16 on the previous Southern Hemisphere illustration) is influenced by the northeasterly branch of the current from the easterly-flowing cold Southern Ocean current and has an average annual temperature of 61.0 °F (15.9 °C). There is only a 4° latitude difference in the cities of Durban and Cape Town, but a 16 °F (9 °C) difference in their average annual temperatures. As can be seen on the Southern Hemisphere illustration, it is Durban (number 13) that is much warmer (about 10 °F) than is the solid red line estimate on the earlier chart showing latitude effect on average annual temperature for that latitude.

The elevation of a city high above the ocean surface level also can have an effect on its annual average temperature, but the elevation difference between that of Durban (at an average elevation of 170 meters (558 feet)) and that of Cape Town at sea level (0 meters and 0 feet) is not that large a difference in elevation to cause a difference in their respective annual average temperatures. A geophysical conclusion derived therefore, is that with no significant difference in latitude or elevation, the primary factor causing the 16 °F (9 °C) annual average temperature difference between Durban and Cape Town is attributed to their being exposed to the different temperatures of the two different ocean currents that flow past their coastal locations. Scientifically, there should be a dependency of every coastal city (as many thousands of them are coastal on the planet) on the temperature of the water in the large body of water flowing on their coastal boundary and the prevailing winds being a mixture of both on-shore or off-shore. Anyone familiar with the weather and climate of Southern

California well understands the on-shore and off-shore wind phenomenon on the average annual temperature of the San Diego, Santa Monica, Malibu, and Zuma Beaches.

That leaves the question of the effect of elevation on the data points on the latitude effect figure for the Southern hemisphere shown on the figure on earlier pages of this chapter. It can be seen on the southern hemisphere latitude effect figure that the data points below the solid red line vary in distance (i.e., are at a colder temperature than the solid red line estimates for the latitude effect) from that line. The data numbered 4, 7, 9, 11, 12, and 14 identify the six cities among that group that have substantial elevation above the ocean surface level of the nearest ocean. For readers' convenience, the small tabular listing shown next provides the "raw" numbers that have obscure visual meaning when considered only in tabular form.

			Elevation		Average Temperature	
Point Number	**City Name and Location**	**Latitude**	**Feet**	**Meters**	**°F**	**°C**
4	**Papua, New Guinea**	**5.5° S**	**2,188**	**667**	**81.5**	**27.5**
7	**Lima, Peru**	**12.0° S**	**7,956**	**2,426**	**63.5**	**17.5**
9	**La Paz, Bolivia**	**16.5° S**	**11,975**	**3,650**	**44**	**6.7**
11	**Sao Paulo, Brazil**	**23.5° S**	**2,621**	**799**	**65**	**18.3**
12	**Johannesburg, S Africa**	**26.2° S**	**5,297**	**1,615**	**60.6**	**15.9**
14	**Santiago, Chile**	**33.5° S**	**1,887**	**575**	**61.7**	**16.5**

(Source: World Meteorological Organization)

The effect of elevation on the average annual temperature of a city is approximated by an altitude / Outside Air Temperature / ambient Pressure relationship used in the estimation of aircraft flying at different altitudes in the atmosphere. That relationship appears in Aeronautical Engineering textbooks and was established by the International Civil Aircraft Organization shown below as:

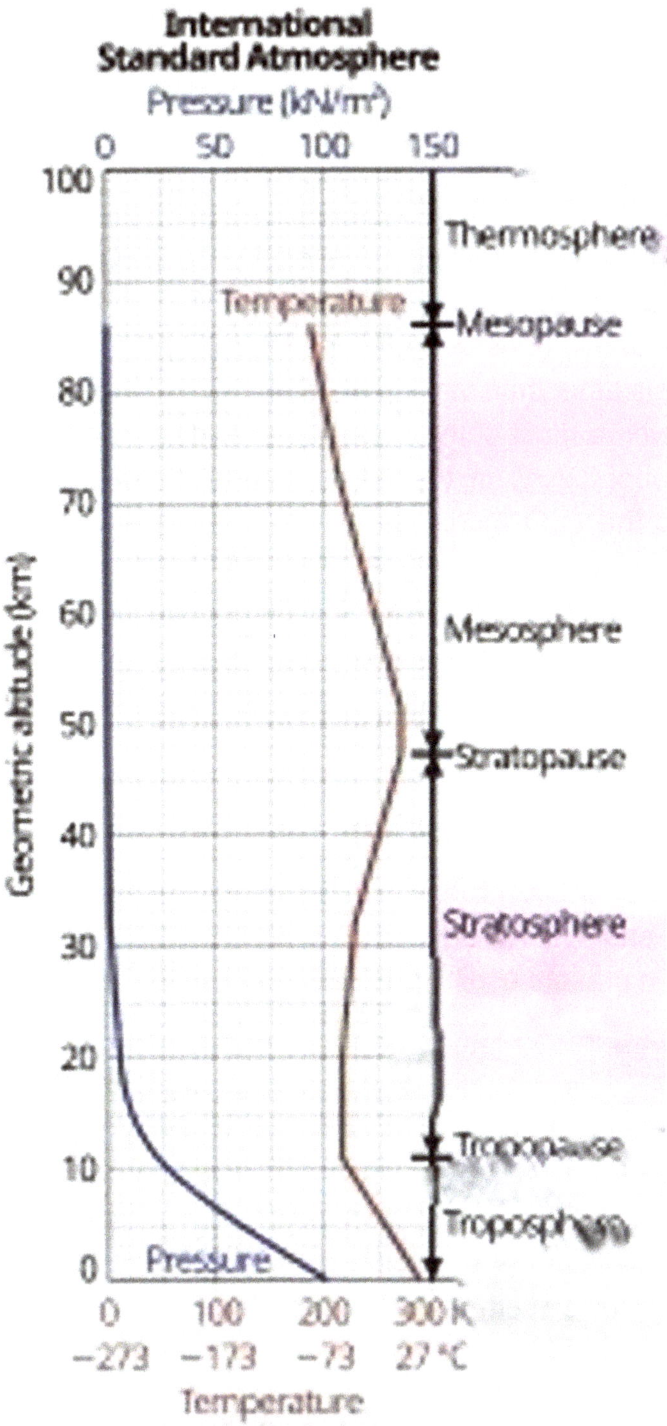

(Source: International Civil Aircraft Organization, ICAO)

The decrease in temperature for the tropopause portion of the atmosphere is referred to as lapse rate and varies linearly from 59 °F (15 °C) at Sea Level (elevation 0) to -65 °F (-54 °C) at 36,089 feet (11 kilometers) altitude/elevation. The linear red line (for temperature) is mathematically defined by the point-slope method to be:

Temp (°F) = 59 °F – 3.44 Elev (1,000 ft) in Imperial units, and

Temp (°C) =15 °C – 6.26 Elev (kilometers) in Metric units.

The "lapse rate" (i.e., the decrease in temperature as elevation increases) is the last half of the equations: -3.44 Elev (1,000 ft) Imperial and -6.26 Elev (kilometers) Metric.

Using the two lapse rates (for Imperial and for Metric) as corrections for the effect of elevation to the solid red line estimates for the effect of latitude on temperature, the correction (i.e., to bring a closer correlation with the solid red line for the six cities tabulated earlier) becomes:

(Imperial) Temp (°F) = 90 °F – 0.78 Lat° – 3.44 Elev (1,000 ft); and
(Metric) Temp (°C) = 32.2 °C – 0.43 Lat° – 6.26 Elev (km)

There was found no mathematical modeling for the effect of ocean proximity and the prevailing wind effect on the estimated temperature on a regional basis, but those can be treated as being corrections (deltas) to the above general equations modeling the combined effects of latitude and elevation. Without any basis other than making an informed estimate (observing that the oceans are warming in the latitudes where most of the population resides – and allowing that the atmosphere has in fact warmed by about 2 °C in the past century), adding an arbitrary 2.8 °C and 6 °F temperature increment to the LEO model as a first order effect of ocean temperature rise and proximity is considered sufficient in making LEO a very good model for estimating the surface temperature of the planet at any latitude, elevation, and proximity of the effects of the atmosphere and the oceans becoming warmer. Adding that temperature increment for the estimated effect of oceans' warming and their proximity to populated coastal cities, the improved LEO model becomes:

(Imperial) **Temp (°F) = 96 °F – 0.78 Lat° – 3.44 Elev (1,000 ft);** and
(Metric) **Temp (°C) = 35 °C – 0.43 Lat° – 6.26 Elev (km)**

The above models are not perfect but are infinitely more reliable for estimating the geophysical effects on the land, water, and atmosphere of the planet than is the "one-size-fits-all" average global temperature (a constant) in present use by environmental scientists.

The Author requests that the LEO method be used in studies and analyses being conducted by scientists, technologists, engineers, and mathematicians (i.e., the STEM community) in the United States - and in other regions of the world - when conducting their activities in developing (or improving) those technologies that may be useful in mitigating the inevitable effects of climate change. An application for patent of the LEO Method for Estimating Average Annual Temperature has been filed with the USPTO for the primary purpose only of indicating that the LEO Method method has significant utility as an item of intellectual property useful to the scientific segment of the population.

If the WMO (World Meteorological Organization) database is ever destroyed (unlikely, but even if only remotely possible), the Latitude/Elevation/Ocean proximity (LEO) model can be of assistance in partially describing the estimated average annual temperature for any city at any latitude on the planet and with reasonable accuracy compared to what is recorded in the WMO database. A comparison was made for the six cities used in deriving the elevation effect comparing the LEO estimated temperatures with the WMO recorded temperatures to indicate the comparative accuracy of the LEO model:

Point Number	City Name and Location	Latitude	Elevation Feet	Elevation Meters	Average Temperature (WMO) (LEO) °F	°C
4	Papua, New Guinea	5.5° S	2,188	667	81.5	27.5
		Estimate by use of LEO:			84.2	28.5
				(Difference)	(+2.7)	(+1.0)
7	Lima, Peru	12.0° S	7,956	2,426	63.5	17.5
		Estimate by use of LEO:			59.3	14.7
				(Difference)	(-4.2)	(-2.8)
9	La Paz, Bolivia	16.5° S	11,975	3,650	44.0	6.7
		Estimate by use of LEO:			41.9	5.1
				(Difference)	(-2.1)	(-1.6)
11	Sao Paulo, Brazil	23.5° S	2,621	799	65.0	18.3
		Estimate by use of LEO:			68.7	19.9
				(Difference)	(+3.7)	(+1.6)
12	Johannesburg, S Africa	26.2° S	5,297	1,615	60.6	15.9
		Estimate by use of LEO:			57.3	13.6
				(Difference)	(-3.3)	(-2.3)
14	Santiago, Chile	33.5° S	1,887	575	61.7	16.5
		Estimate by use of LEO:			63.4	17.0
				(Difference)	(+1.7)	(+0.5)
		Average Net Difference:			-0.3 °F	-0.6 °C
		Maximum + Difference:			+3.7 °F	+1.6 °C
		Maximum - Difference:			-4.2 °F	-2.8 °C

(Source: Meteorological Data from World Meteorological Organization (WMO) with use of Author Generated Latitude / Elevation / Ocean (LEO) Model)

A logical question which might be asked by Readers in the United States is "Can the LEO model be of use in understanding the factors that occur in a smaller region than that of hemispherical proportion? To pursue that question, and to satisfy the Author's affliction with the Curse of Completeness, six large cities were selected in the United States to do a similar comparison for accuracy of the LEO model in the smaller region of the "lower 48" (ConUS) states of the United States. The six cities were shown earlier as being among the 20 in the Northern Hemisphere (and are numbered from that set of 20 cities): 14 – Miami, Florida; 13 – New Orleans, Louisiana; 11 – Los Angeles, California; 10 – Denver, Colorado; 9 – New York, New York; and 7 – Seattle, Washington. The comparison of the World Meteorological Organization (WMO) database and the Latitude / Elevation / Ocean (LEO) Model estimate for the temperature in the six selected cities (selected as having latitude differences, elevation differences, and proximity to a large body of water (ocean, gulf, or Great Lake) is shown as follows:

Point Number	City Name and Location	Latitude	Elevation Feet	Elevation Meters	Average Temperature (WMO) (LEO) °F	°C
14	Miami, Florida	25.8° N	6	2	76.0	24.4
		Estimate by use of LEO:			73.9	23.9
				(Difference)	-2.1	-0.5
13	New Orleans, Louisiana	30.0° N	-7	-2	75.0	23.9
		Estimate by use of LEO:			72.6	22.1
				(Difference)	-1.3	-1.8
11	Los Angeles, California	34.0° N	791	241	67.0	19.4
		Estimate by use of LEO:			66.0	18.9
				(Difference)	-1.0	-0.5
10	Denver, Colorado	39.7° N	5,525	1,684	54.0	12.2
		Estimate by use of LEO:			46.0	7.4
				(Difference)	-8.0	-4.8
9	New York, New York	40.7° N	62	19	53.0	11.7
		Estimate by use of LEO:			64.0	17.4
				(Difference)	11.0	5.7
7	Seattle, Washington	47.8° N	131	40	55.0	12.8
		Estimate by use of LEO:			58.3	14.2
				(Difference)	3.3	1.4
		Average Net Difference:			-0.3 °F	-0.6 °C
		Maximum + Difference:			+11.0 °F	+5.7 °C
		Maximum - Difference:			-8.0 °F	-4.8 °C

(Source: Meteorological Data from World Meteorological Organization (WMO) with use of Author Generated Latitude / Elevation / Ocean (LEO) Model)

There is seen to be good estimation (better than the Author anticipated) for four of the six US cities of being less than three degrees Fahrenheit of difference for those four cities (Miami, New Orleans, Los Angeles and Seattle). The largest difference between the WMO records and the LEO estimate (warmer estimate of 11 °F / 5.7 °C) is seen in the data for New York (where the Author grew up in the first 23 years of his life). The average annual temperature method of averaging the seasonal high temperature in July with the seasonal low temperature in February masks the fact (well known to New Yorkers) that the "lake effect" of Great Lake Erie west of New York during the winter months extends the cold weather months (the four "normal" colder "seasonal" months of November, December, January, and February) by starting in mid-October and ebbing in early March. The residents of Buffalo, New York will confirm the existence of the "lake effect" as that city records record breaking snow depths every year.

The "extra" month of the lake effect (the Great Lakes serving much like a refrigerant fluid modulating the heat energy from the surrounding and downwind areas to the East) in October and retaining that cold temperature (the "lake effect") until well into March. Much if not all of the 11 °F (5.7 °C) warmer over-estimate by the LEO model is explained as being a local anomaly. The underestimate of the Denver data point (8.0 °F (4.8 °C) – the WMO showing Denver to be warmer than estimated by LEO – is most likely due to the opposite of the "lake effect" and being more a "desert effect". Denver is at an elevation of 5,525 feet (1,684 meters) near the ridgeline of the Rocky Mountains that stretch north to south through the desert southwest states on both flanks of the Rocky Mountains. The sands of the desert retain the heat of solar radiation by day but that heat flows upward on the sides of the mountain chain during the evening and thereby raising the temperature in the "Mile High City" on average almost the entire length of the year except for a very brutal winter months season.

The LEO model of the surface temperature on the planet cannot be considered a "giant leap by Man" to environmental science; however, compared to the environmental scientists' one-size-fits-all "average global temperature" method (as shown earlier and which considers Man-made "greenhouse gas emissions" to be the dominant cause for global warming), the LEO model is a far more credible scientific tool in better understanding the "big picture' of climate change and the factually important factors that are causing the surface temperature of regions of the planet to be warming – and cooling - differently in different regions. Averaging is too "fuzzy" a tool for the variable of temperature on which so much of importance to human and ecological lifeforms depends and to which humans must adapt.

A common thread that will be seen throughout this book is that it is not the effect that Man is having on the environment but of far greater importance is the environment – no matter how "inconvenient" – as is the driving factor determining what Man must do to survive within the limits of that environment. The Author had a most successful career developing technologies for advancing knowledge of flight vehicles moving in the atmosphere. The technologies that are useful are those that are reliable, safe to operate, and are affordable to have. The last part of this book is an evaluation of what those technologies are for Man to use within the limits of the environment of the planet. This first part is to identify what are the factors that the environment presents to humans and form the limits within which Man must survive. For an aircraft that is termed as "the flight envelope" – for humans in the environment it is the extremes of the weather and the climate that form the "envelope" within which every human conducts his/her life.

The Author placed a statement on the front cover of this book that sets the tone for the book: *"It's far more about humans than about just the environment."* The development of the LEO model describing the geophysical features of the "temperature envelope" on the preceding pages is of scientific modeling purposes for the average annual temperature of any point on the surface of the planet. The extremes of the temperature are not estimated with the LEO model but are well-recorded in the WMO database. The next figure portrays the extremes of hot and cold for cities in the USA and those extremes form the temperature "envelope" within which Man (in the physical limits of the USA) must function:

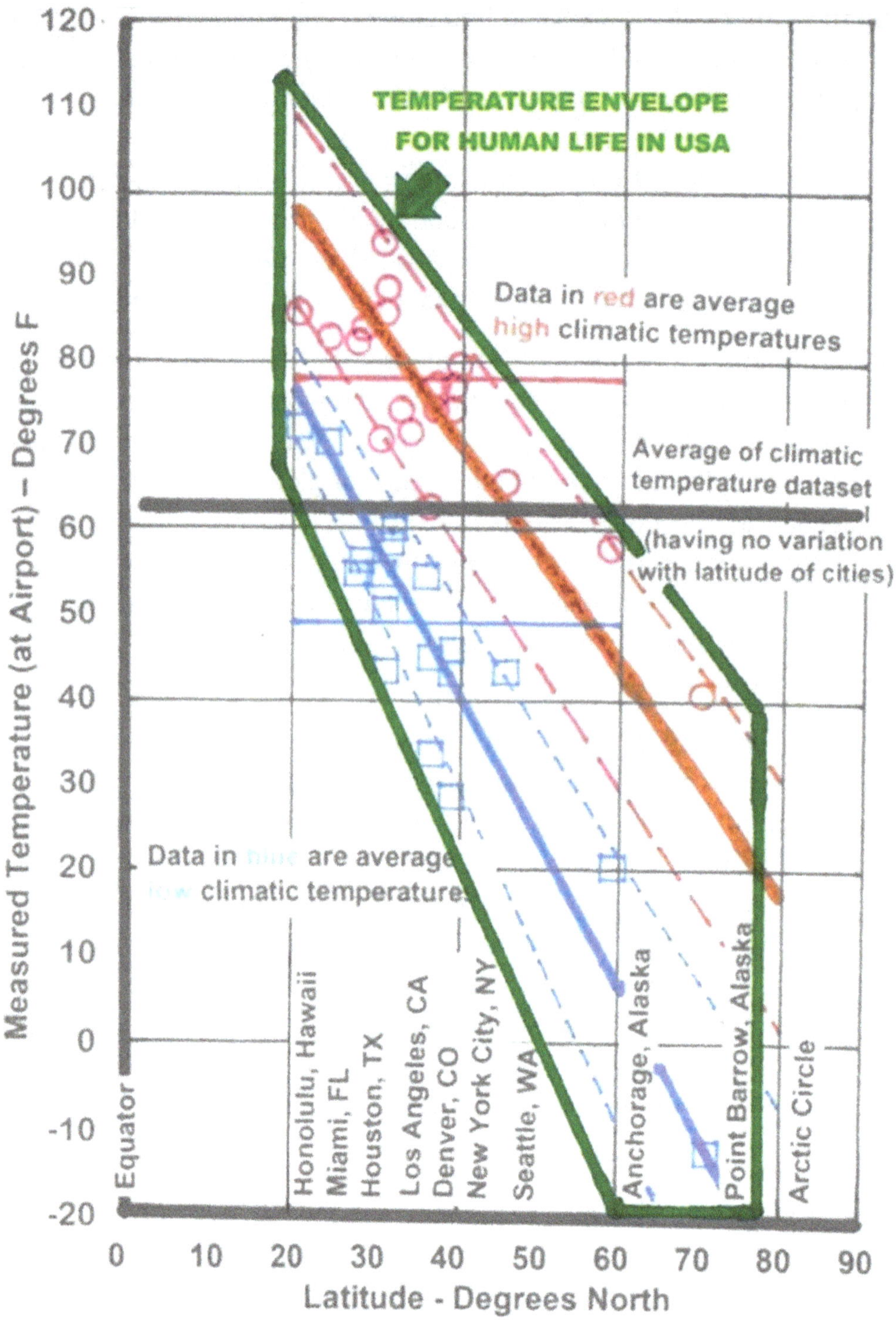

(Source: National Weather Service, NOAA) with Author Notation)

Humans within the latitudes shown above must wear the appropriate clothing and outerwear and/or control the temperature within the structure they inhabit to safely survive long term exposure to the temperatures within the temperature/latitude envelope (solid green line). Additional precautions must be taken if wind chill or heat index factors exist to reduce the limits of the temperature envelope.

Temperature is but one of the two dominant factors that affects the climate of a region – the other major variable being precipitation. The precipitation is a climatic variable and not a geophysical parameter. The precipitation aspect of climate change will be discussed in later chapters when addressing weather and climate change being more in the category of naturally-caused weather events – benign and catastrophic – and resulting in drought or flooding/mudslide/landslide events in a region. Environmental scientists have been misleading the public as to the "cause and effect" principles of science and climatology. Climate change is an "effect" not a "cause". The average annual temperature of a region is changing because of the complex combination of astrophysical, geophysical, and natural factors that are changing with time. Those changes are affecting the changes in the climate – temperature and precipitation.

Environmental scientists, eager to convince the public that Man's polluting activities (and using the ploy of *"eliminating everything else"*) are affecting the climate and creating a climate change "crisis", have perpetuated the fallacious inference that *"Man is the dominant cause of climate change"*. Man is guilty of much pollution – that does need much closer remedial attention by the entire STEM community – but Man is not guilty of changing the climate. Climate change is affecting both the environment and the human population. The scientific approach to mitigating the effects of climate change on the environment are technologically much different than what the approach should be to mitigating the effects of climate change on Man.

This chapter continues with the geophysics of the planet changing the temperature and precipitation on our land and water surface (the surface of the planet creates the extremely important interface with the atmosphere). That interface scientifically is the location where matter in solid and liquid state comprising the planet are transformed into gaseous state. Humans "see" only the visible surface of the planet and only a very limited section of that surface. An estimate is given that a six-foot tall human can see a distance to the horizon (given that the surface has no hills or structures obscuring line-of-sight to the horizon) of about 24 miles (about 39 kilometers). With those 24 miles as a radius, the area perceived by "flat-Earthers" is a maximum of about 1,810 square miles (about 4,778 square kilometers). The area of the interface between the entire planet and the atmosphere surrounding the planet is about 197 million square miles (equivalent to about 510 million square kilometers).

Thc ratio of the interface area between the atmosphere and the planet and the area a human can "see" is 197 million to 1.81 thousand (or in ratio form) 109,000 to 1. In a "big picture" sense, if a group of humans are concerned daily about what the weather will be like in their "What you see is what you get" in their 109,000th portion of the geophysical planet, each regional population is not overly concerned with what might happen in the other 108,999 regions of the planet in 2035. The advent of satellite weather monitoring, tropical storm tracking, and hurricane tracking technologies add to the "big picture" of climate change to be small but very real and on a regional change rather than a global average change. A very enlightening – although apparently ignored by the environmental scientists who claim that global climate change is a "settled science" – but very convincing database was presented by NASA Earth Science that conflicts with the environmental "settled science" proclamation. Using the "normal temperature" calculated over the 30-year baseline period 1951-1980, a series of illustrations for the temperature anomalies (an anomaly is the difference from "normal") was compiled from meteorological (land-based – World Meteorological Organization) and satellite (NASA space-based) records. The magnitude of temperature anomaly (between 2 °C / 3.6 °F warmer – dark red - and 2 °C / 3.6 °F colder – light blue) for the period 1932 to 2022 is indicated by the color barcode shown in the upper left corner of each image arranged as a book is read, top left to top right and proceeding from top to bottom:

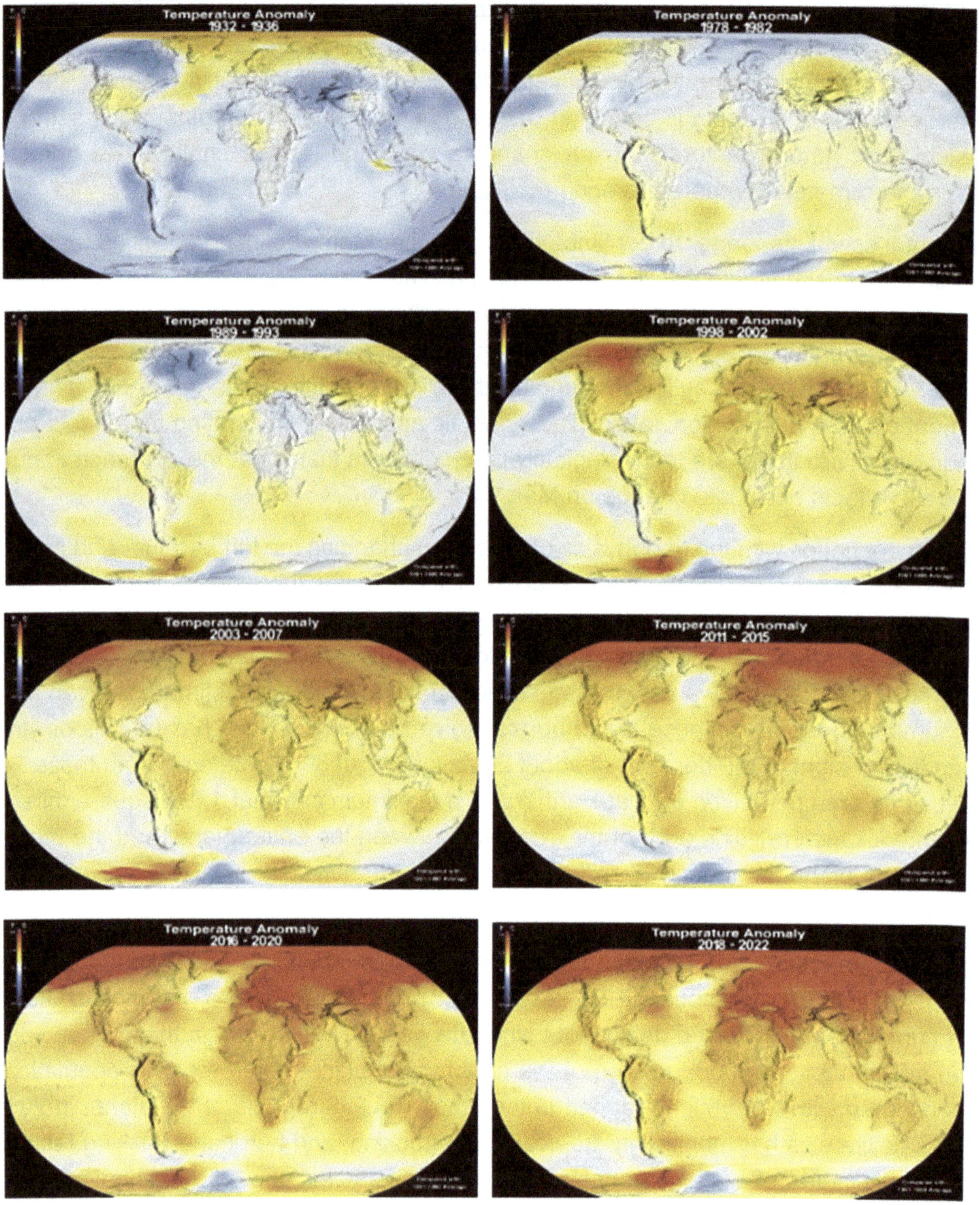

(Source: NASA Earth Resources Center and World Meteorological Organization)

There can be no doubt that in the "big picture" the planet has become warmer – but in regions not globally - since 1932 (based on the amount of red color visible in the montage of temperature anomaly illustrations). A combined "big picture" composite of land and oceans temperatures changes by regions (showing small cells within larger area regions) for 2023 indicates very obviously that some regions of the planet are warming, some regions are cooling, and some show little change independent from whatever efforts Man has been attempting to control "average global temperature".

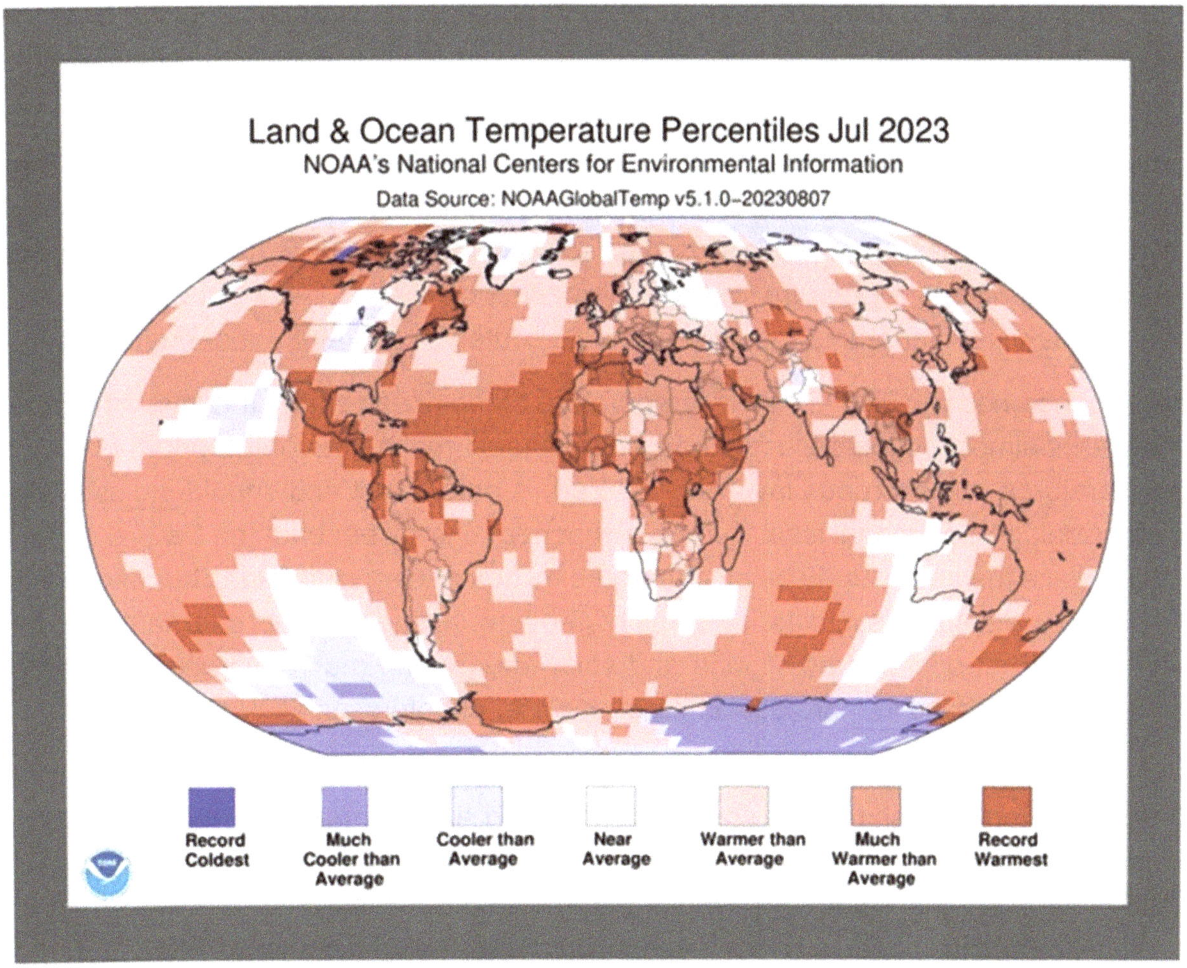

(Source: National Oceanic and Atmospheric Administration, NOAA)

For readers in the USA, the 2023 data indicates the central states have temperatures near normal, the eastern and western states have temperatures warmer than average, and only the desert southwest states (Arizona and southern Nevada) have record-setting hot temperatures. If "greenhouse gases" homogenously distributed in the atmosphere were the dominant cause for global warming on the planets land and ocean surfaces there would not be such "patchy" regional variations – the temperature changes would be much less subtle.

The 2023 data is the most recent and most comprehensive illustration regarding the regional effect of factors other than that of Man burning fossil fuels causing regional temperature changes – and for that reason the illustration on the back cover of this book was chosen for its lasting visual level of importance for Readers to remember.

From the montage set of data, there can also be no doubt that the largest extent of regional warming has been in the lesser population densities of northern Europe, northern Asia, Central America, and northern Canada. The regions of the planet between the Tropic of Capricorn and the Tropic of Cancer exhibit a temperature anomaly of less than 1.5 °C / 5.4 °F with most of the overall warming being in the Eastern hemisphere rather than the Western hemisphere. From the "big picture" (with the planct as a system of many subsystem components) perspective, there appears to be no valid reason for the environmental scientists of the United States to be declaring a "crisis" to the warming of the environment in the United States region. From that same perspective, the population distribution of the planet is such that over 60 percent of the world population is in the northeastern quadrant of the planet and the temperature anomaly montage above graphically shows the predominance of red color to be in that same northeastern quadrant.

Environmental scientists have created climate change to be in a “crisis” but as will be seen later, that “crisis” is not because humans are endangered by the small incremental change in temperature in some regions – the imaginary “crisis” is the “soaring rate” (about 0.75 °C per decade) in the average oceans surface temperature and resulting effect on marine life. Most of the public has no conception of the inner planet and the enormity of its content – and amount of heat.

While being concerned with a 2 °C rise in temperature of the ocean water, the level of temperatures within the planet almost boggles the mind. Readers should make note of the enormous magnitude of the temperatures in the various layers below Earth’s outer crust and should inquisitively form a wonder of: “Wouldn’t it be magnificent if some of that heat energy beneath the Earth’s surface could be used to replace the heat energy being generated by humans and industries burning fossil fuels?” This “big picture” concept falls into the category of geothermal energy as an alternate form of energy that rivals the current approaches of wind and solar energy and much hydroelectric energy. That question will be addressed in more detail nearer the later chapters in this book.

As one of the many “fragments of the “Big Bang” entered into the infinite areas of what is simply termed “space”, Earth is estimated to have been formed approximately 4.6 billion years ago and was formed initially as a ball of molten mass (principally iron in molten and plastic state). As layers of physical matter (in solid, liquid, or gaseous state) were deposited onto the planet with time, the Earth became formed in spherical layers much the same as that of a common onion. Each spherical layer acts somewhat like an insulation layer and such that the layers have a hot boundary (with the hotter layer beneath it) and a cooler boundary (with the cooler layer above it).

The next two diagrams show the spherical layers of the planet from its crust (continental land masses and ocean waters) down to its core, and the temperature increasing as the distance from the from the center of the planet decreases:

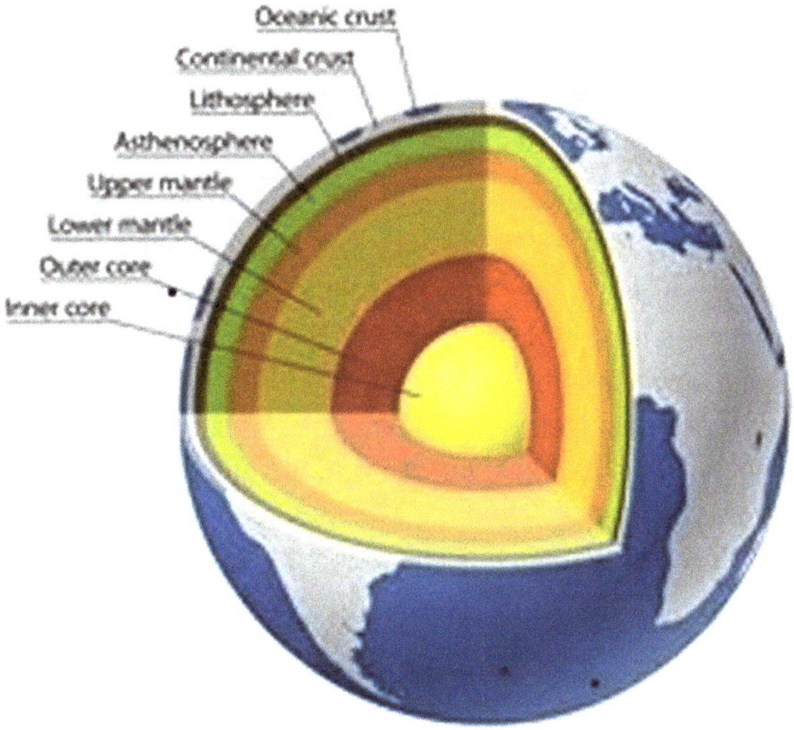

(Source: International Union of Geological Sciences)

The temperature gradients in the several "layers of the onion" are estimated to be as shown in the next diagram:

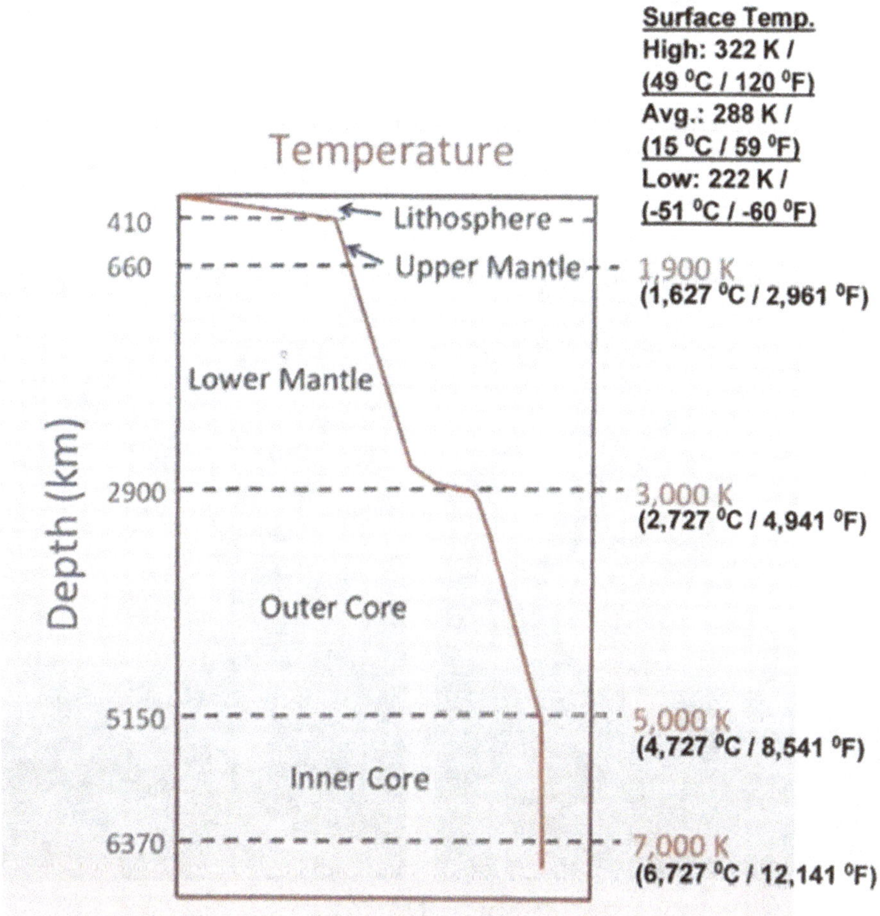

(Source: US Geological Survey with Author modifications)

The above diagrams may be a bit misleading as to the enormity of size of the planet as well as to the uniformity of its composition materials, but as each of its primary geophysical features are described there will be approximate dimensions presented in either/both tabular form, diagrammatic form, or photographic form.

The lithosphere is the most complex layer in composition and that is the layer from the surface of the Earth down (approximately 660 kilometers – 410 miles) to the upper mantle. It is the upper portion of the lithosphere (which contains the several tectonic plates in themselves of very non-uniform composition) and the land and water which are the two mediums supporting the many forms of life on the planet. The diagram has been amended to add the Kelvin temperatures shown in the more common forms of degrees Celsius (^{0}C) and degrees Fahrenheit (^{0}F). It is the surface temperature of the planet which is the least known variable in the environment to which the human population and all other life forms are subjected. The changing surface temperature and its causes and effects are the primary areas of the associated climate change being analyzed in this book.

The above diagrams of the planet and its inner temperature profile should not be interpreted as being truly representative of the inner structure of the planet as being so regularly layered in spherical geometry, because very little in the entire solar system is formed in very regular geometric patterns and move with well-ordered travel paths. Scientists have determined the surface of the planet was a much different surface – far hotter, drier, and more hazardous to volcanic activity, than what it is now. As the surface of the planet cycled between extremely hot and extremely cold conditions, and continued changing, the geophysical and environmental features on the surface of the planet also changed.

With the technological advances made in satellite sensors and imagery able to “see” within the depths of the planet, a somewhat different “big picture” illustrates how unevenly distributed the heat energy is immediately under Earth’s outer crust and having an unevenly distributed thermal effect on the Earth’s surface temperature.

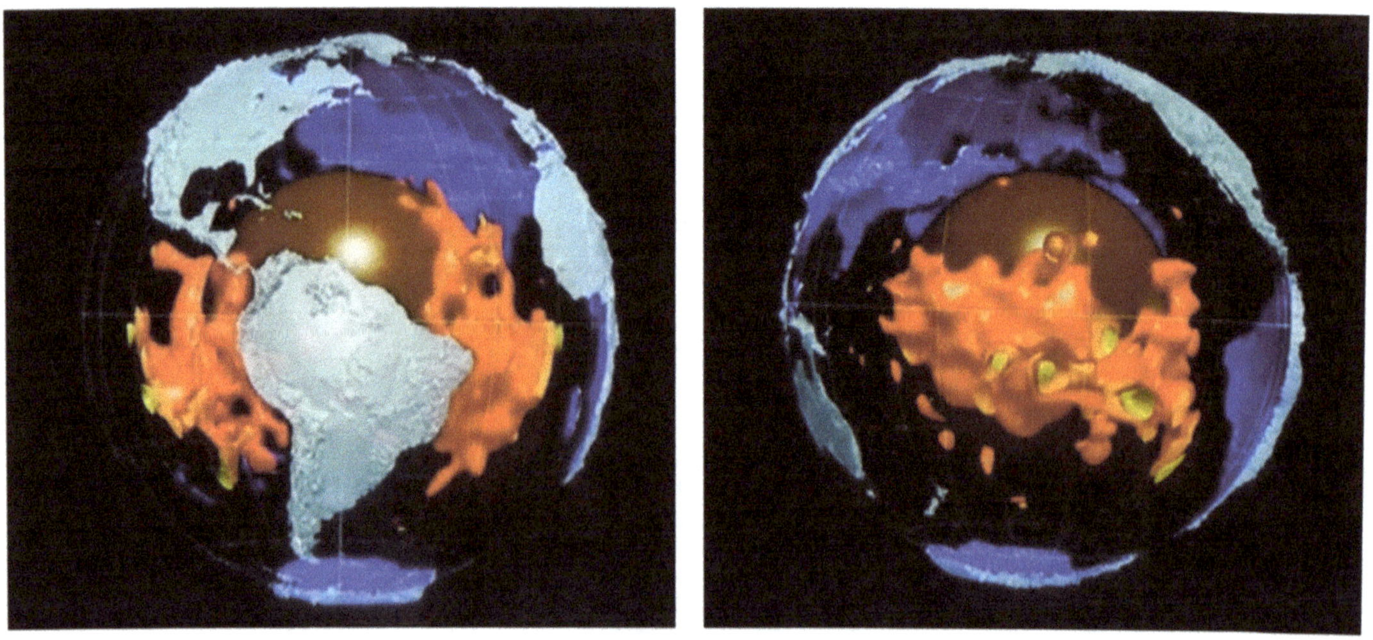

(Source: NASA)

When observing the left image above, there can be seen an immense heat source located under the eastern North Atlantic (where seasonal hurricanes are notorious for forming) and being transferred with the prevailing wind and ocean currents toward the west. Once formed, hurricanes generally take predictable westerly paths into the Caribbean Sea and Islands (Lesser Antilles and then Greater Antilles), then turning northeasterly to impact the southeastern and eastern coastal United States. Some hurricanes make the turn before the state of Florida, but some continue into the northern Gulf of Mexico before making the turn northeasterly. The US Hurricane Tracking Service (part of the National Oceanic and Atmospheric Administration, NOAA) is doing a great job, assisted by satellite imagery, in tracking the hurricanes path and intensity to give forewarning to communities that may be impacted by the approaching hurricane. There have also been scientific models developed to predict the path the hurricanes will follow, and those models are also helpful in identifying which regions should evacuate their inhabitants or to protect their property if they choose to remain.

The question is posed: "Why then are environmental scientists ignoring the above historical and visual databases and persist in distributing the fear-mongering misinformation to the public that it is the Man-made "greenhouse gas emissions" effect that is the dominant cause for the "soaring water temperatures" and the increasing frequency and severity of hurricanes and tornadoes – which are seasonal to start with?" That question is attempted to be answered with this book.

A similar question arises when observing the location of the enormous pools of heat energy under the ocean floor of the central and eastern Pacific Ocean. The islands in the state of Hawaii were formed by the movement of the tectonic Pacific Plate over the "hot spot" in the floor of that plate, allowing the large pool of magma shown in the image to erupt and form each of the major islands above the current Pacific Ocean surface level.

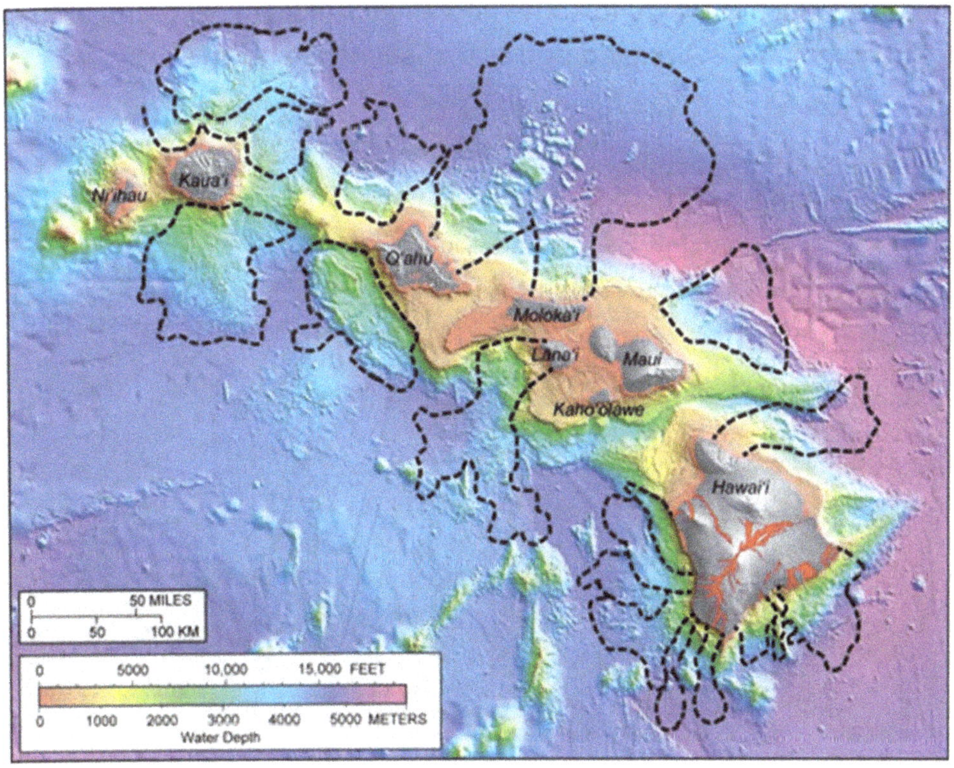

(Source: US National Geological Survey)

A simplified diagram shows the means of how the Hawaiian chain of islands were formed long ago – and are still forming:

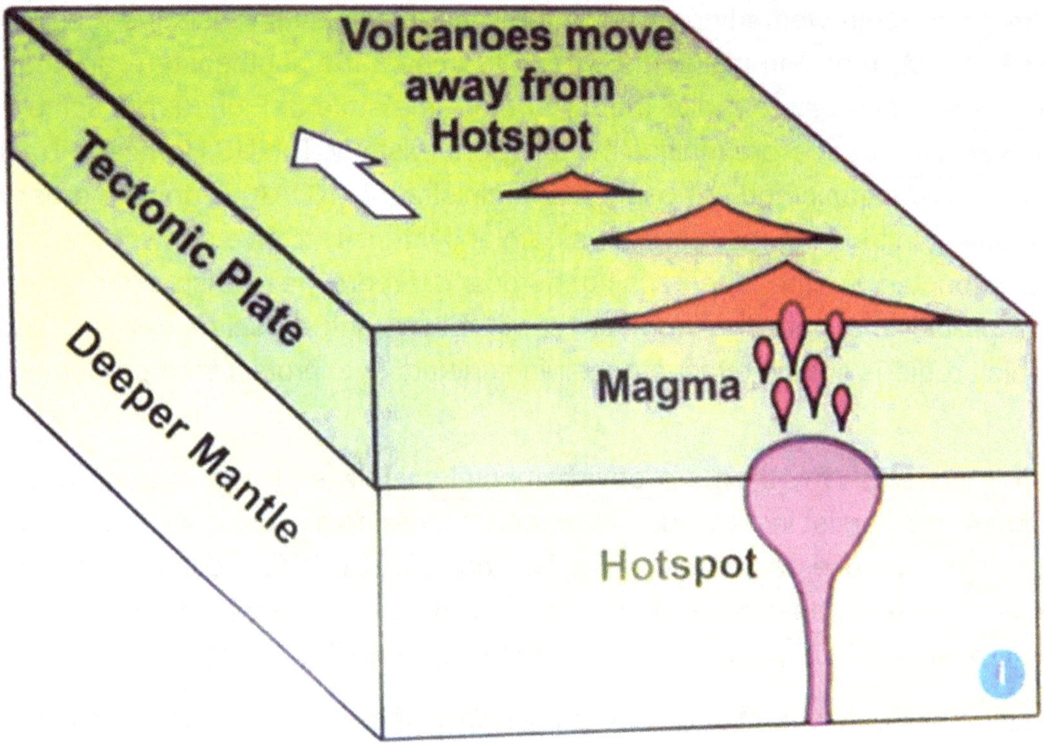

(Source: Internet Educational Website)

The volcano Kilauea on the "Big Island" of Hawaii has been erupting sporadically during the past few centuries, and did recently (June 3, 2024) have a characteristic massive lava flow – but not explosive - that caused much damage to some homes in the lava flow path (which is seen to have spared most of that community). A photograph of that scene is shown below:

(Source: US National Geological Survey)

The known locations (Central Pacific and southern portion of the North Atlantic Ocean) are also exactly where El Nino and La Nina phenomena changing the water temperature are spawned that again travel with the prevailing winds and ocean currents toward the western coasts of the United States and Mexico – the coasts of which are coincident with the Pacific "Ring of Fire" of known volcanoes. The oft referred to "Pineapple Express" is that weather phenomenon (associated with an El Nino ocean temperature and atmospheric weather pattern) when an atmospheric river brings heavy rainfall from the region of the Hawaiian Islands causing flooding in the west coast states (California, Oregon, and Washington).

The persistence of the environmental scientists is admirable – and most of the environmental science has "circled the wagons" as they continue to defend their fallacious hypothesis that the Man-made "greenhouse gas effect" in the atmosphere is the dominant cause for the planet warming, the climate changing, and the ocean levels rising. Their mantra is bult on the premise that *"By eliminating everything else, settled science is proving that "Man-made greenhouse gases is the dominant cause for global warming, etc."*) The sagely old philosophers of the Age of Discovery (who burned fossil fuels to remain safely warm) saw the same effect in their contemporary scientists with the lamentations: *"There are none so blind as those who will not see."* and *"There are none so deaf as those that will not listen."*

Anyone familiar with the basic laws of physics that are taught in high school and college, should be aware of the dynamic motions and forces generated when a large spherical object is in rotation about its axis and translating in a curved trajectory. (Basketball enthusiasts well understand the laws of motion in the "March Madness basketball season). Specifically, a mass not physically fixed in position on the circumference of a circular arc will tend to be moved – by a normal acceleration dependent on the rate of rotation and by the distance from the center of the sphere to the object mass - from a point nearer to the center of the rotating sphere to a location more distant on that sphere's outer surface.

The scientific name for this phenomenon is centrifugal acceleration – in layman's terminology, the capability of "slinging" a mass in an intended direction, as with a stone from a "slingshot" (*a la* David and Goliath). With the assistance of gravity and inertia and pressure forces, molten lava within the Earth's inner layers is "slung" through the irregular cavities and tunnels within earth's inner spherical layers toward the tectonic plates that constitute Earth's surface. In the photos above, the irregular pools of molten material are seen to be beneath the floor of the oceans (Pacific and Atlantic) or seas (Caribbean or Gulf of Mexico). Volcanic eruptions occur by the ejecta being "slung" – assisted by high pressures and high temperatures that exist in the upper mantle of Earth - through a path along the edges of a tectonic plate or through a perforation ("hot spot") in the inner portion of a tectonic plate.

Many environmental scientists have the mind-set to "prove" (rather than to explore "if") it is Man burning fossil fuels that is exclusively causing the temperatures of Earth's surface to rise totally. Not all environmental scientists are scientifically myopic, but those who do, discount – most unprofessionally and ignoring basic principles of science – the proximity of geothermal heating from below rather than surface heating from the atmosphere above Earth's surface. It is that same group/organization that ignores the contribution of volcanic eruptions introducing worse hazards of particulate matter into the atmosphere that adversely affect human health and safety as opposed to the Man-made carbon dioxide in trace concentration which has little comparative effect on human health and safety.

In system engineering, the heat transfer across the interface between two subsystems (for this book, the four major subsystems are the planet, the humans occupying the surface of the planet, the ecology, and the atmosphere) such that one subsystem's functional performance might be adversely affected by an adjacent subsystem functional performance, is mandatory to be considered in a professional analysis – not "*eliminated*" from consideration simply because it is not understood.

Scientists at NOAA have long been measuring how the oceans' water temperature is varying at the depths of the ocean floors (and affecting sea floor organisms) and are theorizing the cause to be the resultant effect of "greenhouse gases" in the atmosphere causing the calculated rise in "average global temperature" and "average ocean water temperature". Life forms on the planet exist because they have adapted to the extremes of temperature in the region where they are located to which they have been and are being exposed. It is a greater probability that the enormous geothermal heat energy immediately below the oceans and seas is the true cause for the surface temperature extremes to be increasing regionally and irregularly, not in an "average global temperature" singular manner.

Scientists in some branches are able to use well-defined mathematics and various independent variables in the analysis of the dependent variables in their branch of science (astronomy, chemistry, electricity, thermodynamics, acoustics, etc.), but in the branches that are not so well defined – tectonics, meteorology ("weather"), climate, volcanism, earthquakes, hurricanes, tornadoes, and most of the environmental disturbances and anomalies – the variables are of complex combinations of factors that are difficult if not impossible to estimate if/when/where they will occur. Those unpredictable independent variables are also nearly impossible to be controlled to a different level more suitable to the safety, health, and/or comfort of Man. Scientists and inventors should be focusing and prioritizing their efforts on those developments of technologies that make the planet a safer, healthier, more comfortable place for humanity to enjoy and thrive, not simply to improve (by "reducing Man's emissions to net zero") the ecological environment.

Millions of years ago, the exterior crust portion of the planet was physically configured much differently from how it is known to currently exist (thanks to the branch of science of cartography – map making). Scientists and geologists have developed the cartographic database (much like a genealogy report for the humanity) describing the changes in Earth's tectonic plate configuration. Many millions of years ago, there was only one tectonic plate named *Pangaea.* A very old French diagram in two parts estimates the "before" (*avant la separation* in French) ad the "after" (*après la separation* in French) configuration of Pangaea separated as shown as follows:

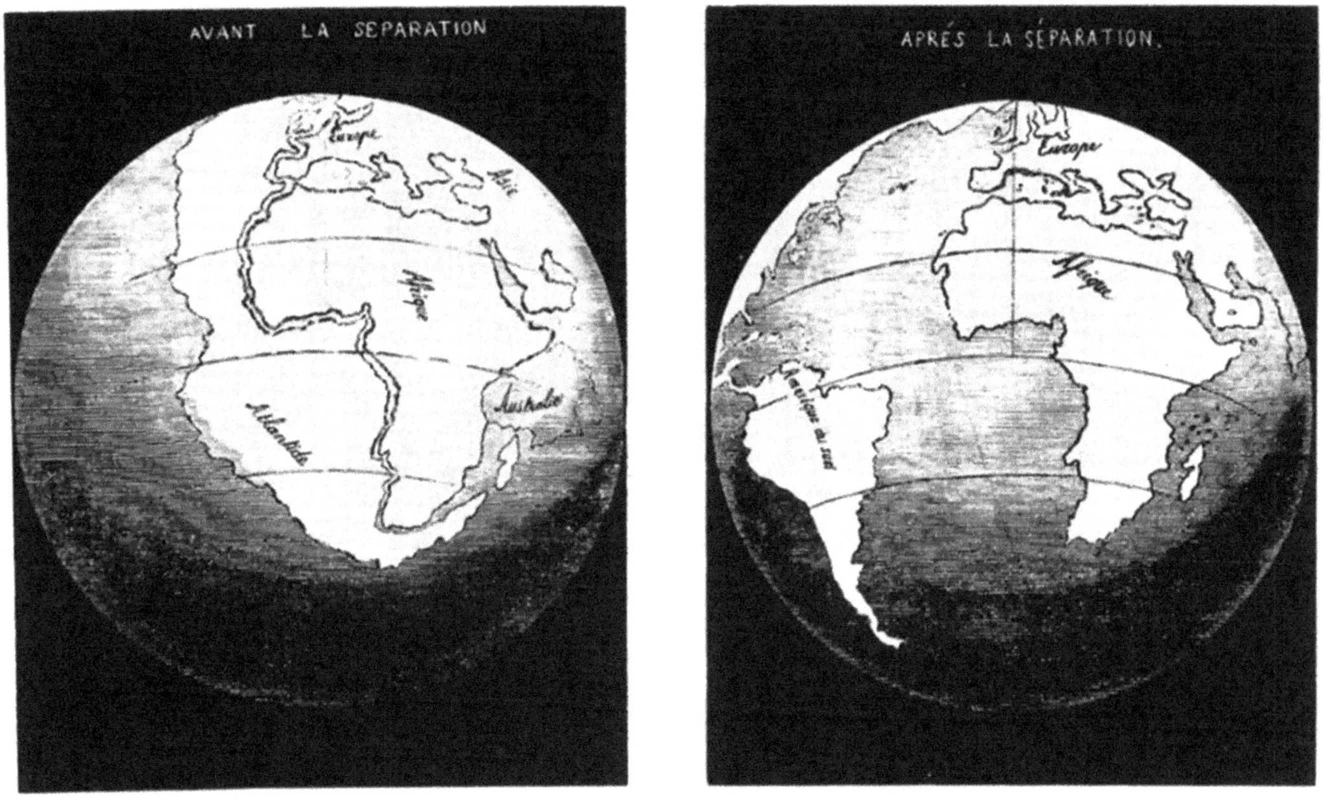

(Source: Encyclopedia Brittanica)

Geologists have developed (as “settled science”) that Pangaea fractured into smaller tectonic plates that have moved apart to eventually form the different tectonic plates as we now know them to exist. (The primary fracture boundary is shown in the left image and is seen to correlate very well with the current coastal regions of the continents in the Western hemisphere and in the Eastern hemisphere as they exist in the current database for tectonics.) Below the continents shown in the before and after diagrams above are the Earths outer crust which is fractured into tectonic plates. The time it took for the tectonic plates to move from their configuration when in the Pangaea configuration (left) to the current configuration (right) is several million years in length.

The continents as we know them today are the visible portions of the tectonic plates hidden from visibility beneath the ocean waters. If one could see the hidden full configuration of the tectonic plates, they would appear as shown next:

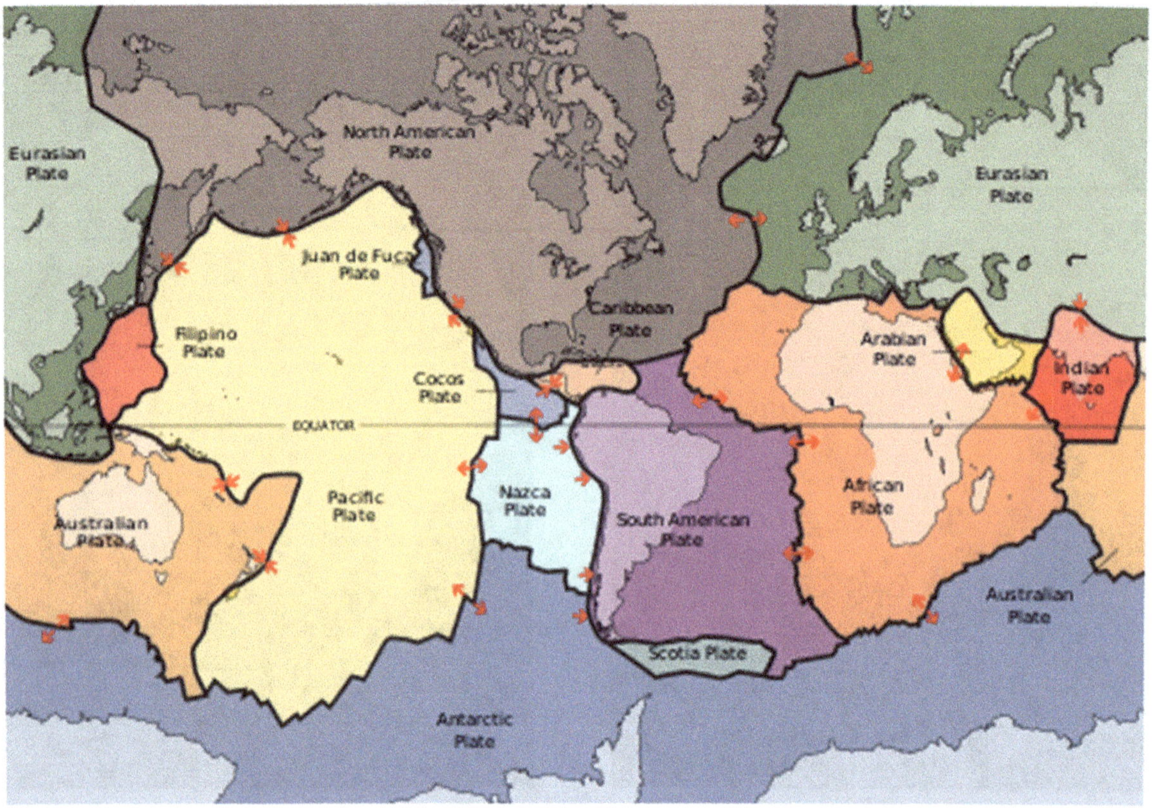

(Source: Encyclopedia Brittanica)

The plates are geologically portions of Earth's outer crust, and on the diagram of the tectonic plates configuration there are red arrows showing how the plates – and the continents (shown above in different colors) atop them – with all of them in continual motion relative to each other.

In the Norse Age (circa 800-1100), the seafarers mapped the waters in the Scandinavia (northernmost Europe) region and gave birth to the science of cartography – the making of maps.

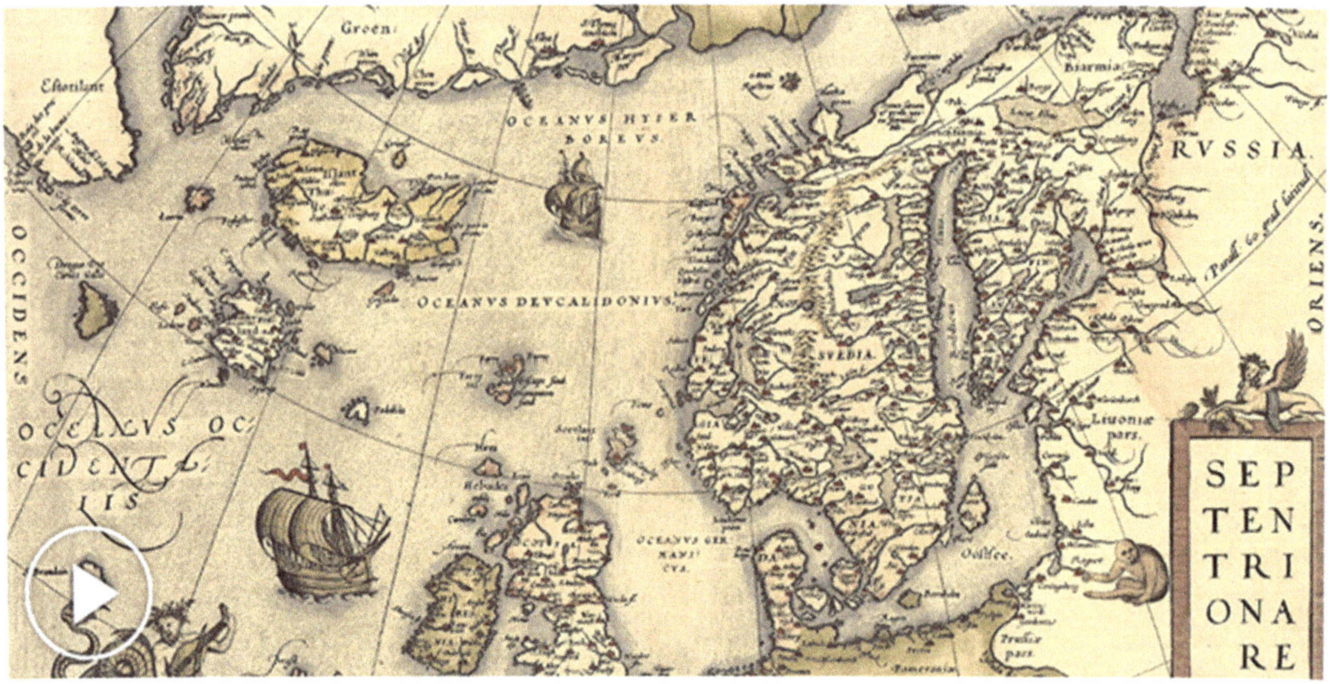

(Source: Encyclopedia Britannica)

These maps were of immense importance to the safety of the Norsemen sailing the treacherous waters of the North Sea but were of lesser importance to the population – or lack of it – in that cold region. By the 1500s and 1600s period, the mapping (cartography) of the entire planet Earth's surface by the astronomers' and mariners' database defined the physical features (land masses and oceans) on the exposed spherical surface of the planet, as shown in the next modern map of the planet showing lines (demarking orthogonal longitude E-W and latitude N-S) in a well-defined grid pattern:

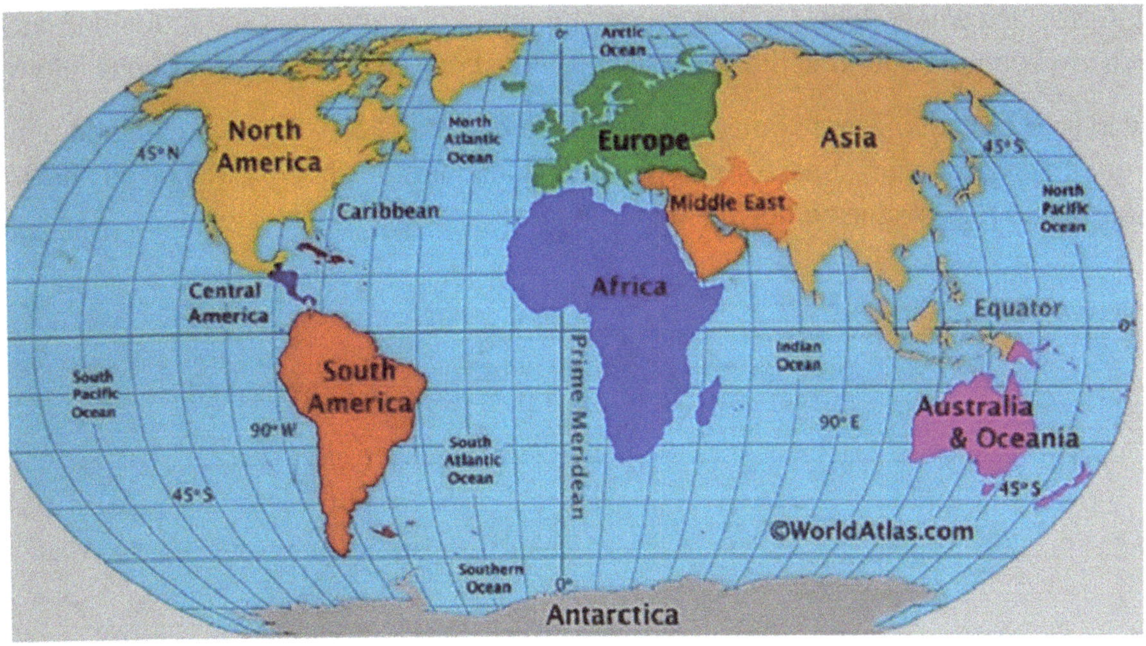

(Source: The International Union of Geological Sciences)

Various forms of aquatic life (vertebrates and invertebrates) and other forms of life formed and evolved over many eons, and it is believed that life on land (above water) evolved from different forms of aquatic species (under water) with the evolvement of lungs that extracted oxygen from the atmosphere. Vegetation and other forms of botanical growth evolved with the photosynthesis process and began to cover the Earth's surface. There was much debate over if/how Man may have evolved from some form of aquatic life (possibly from the octopus' family) and which developed lungs to breath oxygen from the air instead of through gills from the water. Other more credible beliefs are that Man (*homo erectus*) evolved from the Great Ape family of animals. Some religious theologians maintain that Man was created in the image of God. That phase of Man's presence on Earth is left to be debated by others rather than by this Author. It is a formidable challenge to the Author just to describe what "footprint" Man has left on Earth's surface long ago and is leaving now and for the predictable future.

For this book, the baseline in time of Earth's surface was chosen by the Author to be about 400,000 years before now (2024). That selection was not chosen without a basis but is a point in time where the Russian Antarctica Ice Research Station ice core sample data scientifically defines Earth's surface temperature and carbon dioxide and methane in Earth's atmosphere. Those three important factors in the total database are known to have been determined accurately and scientifically by initially Russian scientists and later by scientists from the international environmental science profession using modern-day carbon-dating and spectrographic techniques. It is to be noted that there was no presence of Man burning fossil fuels on the planet 400,000 years ago and so that all other changes that affect global surface temperature, climate, and ocean surface levels can be baselined as the reference in the "before" and "after" presence of Man on the planet.

It is important to note that the "after" in the ice core samples includes the effect of early humans but also includes continuation of the same effects – astrophysical, geophysical, and natural - that existed in the "before" period leading to the baseline. The planet continues to rotate and orbit the Sun and will be continuing that orbital movement regardless of how much fuel of any type humans may combust. Assessing what effect Man has caused by emissions from his (and Industrial) activities on the Earth's environment added to the inspiration for this book to be written for public information.

At a point in time (400,000 years ago), Earth's surface had stabilized considerably, and a map of the planet's geophysical surface had a tectonic plate configuration much like that in the illustration shown earlier and repeated in somewhat different format next:

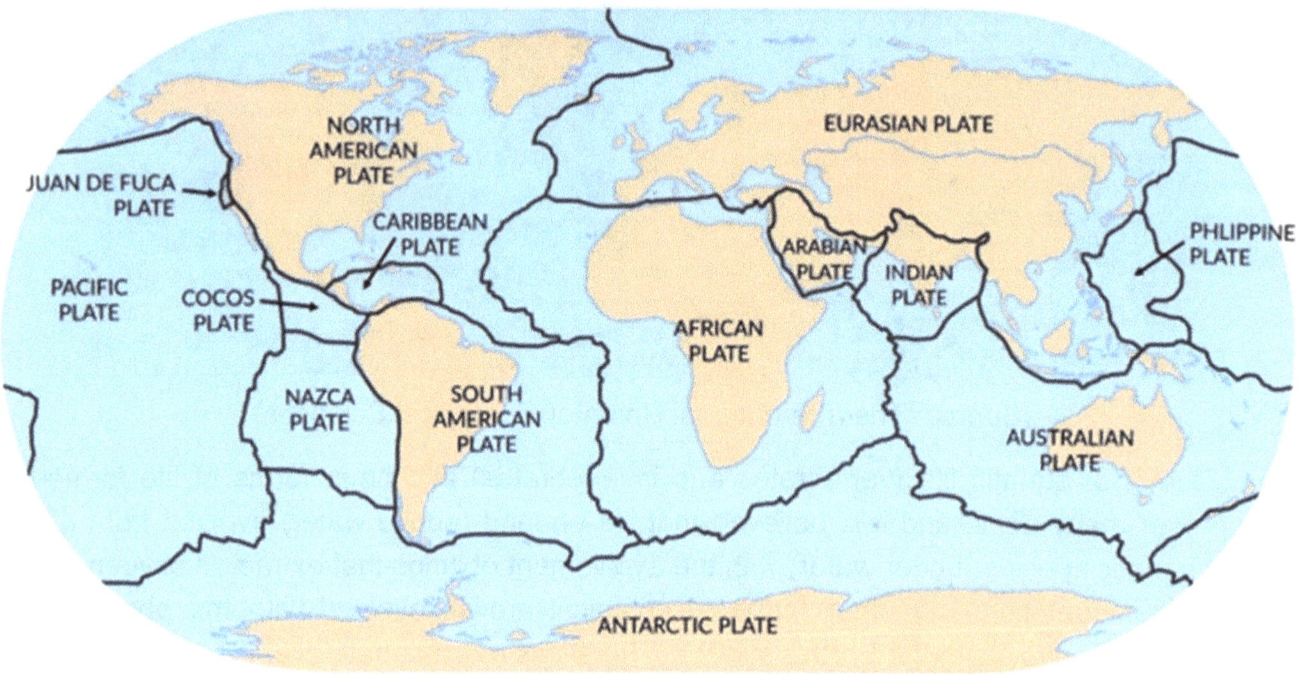

(Source: Brittanica Encyclopedia)

(Readers are asked to envision the oval shaped two-dimensional diagram above to be – in three-dimensional format – the sphere representing the planet Earth. Having a desk globe of Earth would be very useful to Readers of this book. Desk globes are typically mounted on the tilted (about 24 degrees) axis about which the planet rotates and helps to "tie together" the astrophysics and the geophysics of the planet. A larger desk globe, (preferably one that has a topographically shaped surface) is mandatory in a science classroom for students to "see" the scaled down "big picture" of the surface temperature changing daily and the longer-term climate change problem being of regional nature).

On some of the named tectonic plates (in light blue) there is an outline (in a contrasting pink color) of the "continents" as they are known in modern times world geography – North America, South America, Africa, Europe/Asia, Arabia (Mid-East), India, Australia, and Antarctica. The land areas in the pink shade colors (i.e., the continents) are presently above the current surface levels of the world's oceans, and that is the "baseline" topography used in the remainder of this book. In the oval shaped map above, there is 70-75% of the total oval area in blue designating that surface area to be water covered, and the pink colored land area (*terra firma*) is 25% of that area of the oval that is above water.

Where the edge of one tectonic plate meets the edge of an adjacent plate, there is a law of physics stating essentially "Two solid masses cannot occupy the same volume in space at the same time". Accordingly, there are three types of relative motion the plates contesting for that common space can take: 1) Plate A can slide under or over Plate B in a process known as "convergence"; 2) Plate A can move away Plate B in a process known as "divergence"; or 3) Plate A can move sideways relative to plate B in a process known as "transformance". A simplified illustration is shown next to describe what happens to the surface topography of a continent when one of those three events take place (over which Man has no control) and often results in an earthquake (and possible tsunami), a volcanic eruption, a major separation in the surface, or has created mountain ranges above a land mass or below the ocean in a long period of time, etc.

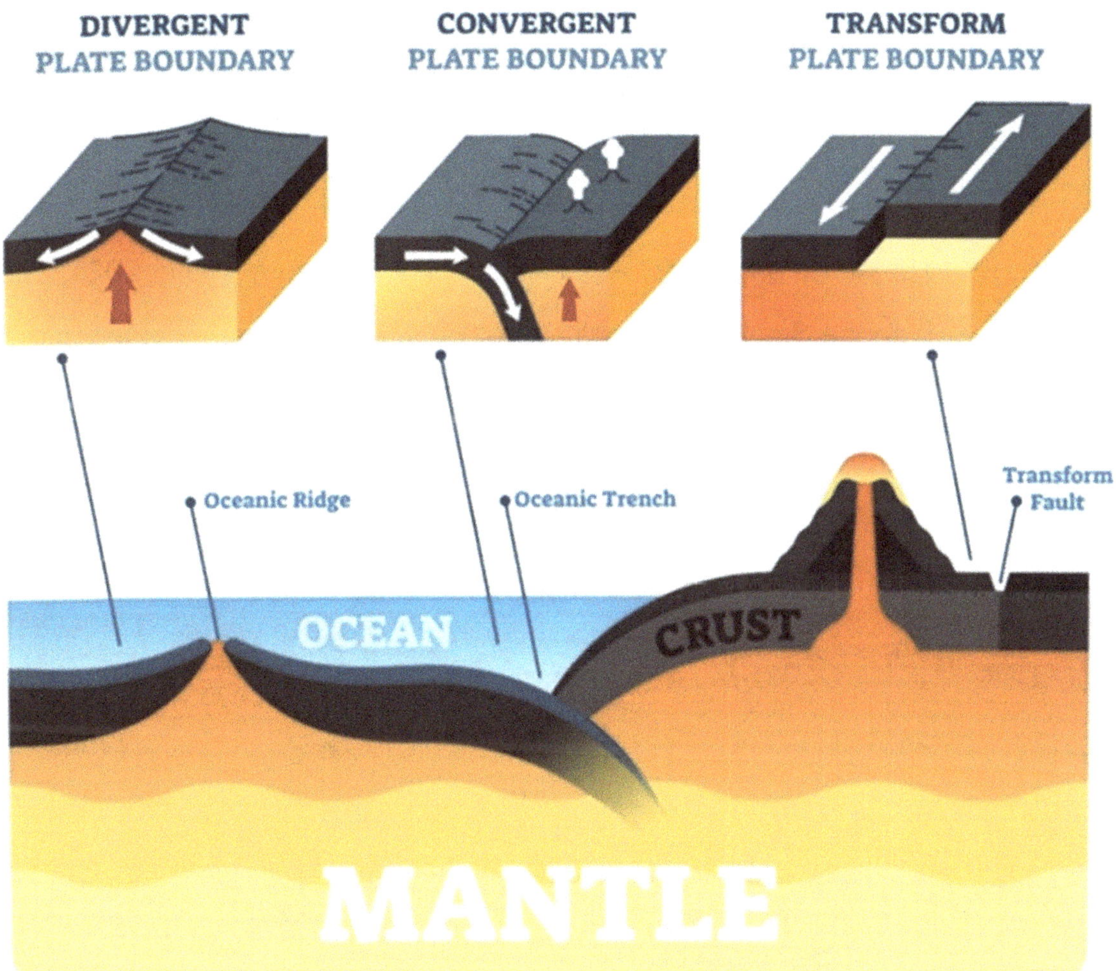

(Source: National Geological Survey)

Another phenomenon can occur along fault lines created by the relative motion of two plates. The diagram above shows the formation of oceanic ridges below the ocean floor and the classic conical shape volcanic formations on land, but there are locations where the lava emerges along a fault line in a very linear fashion – one of the rare instances where Nature kas a linear geometry. Seen below is an aerial view of such a linear eruption on Iceland:

(Credit to unknown photographer for Air & Space Quarterly Winter 2025)

Another perspective (from the Smithsonian Institute for Science) illustrates that the collision of the two plates also produces the magnificent mountain formations of the world:

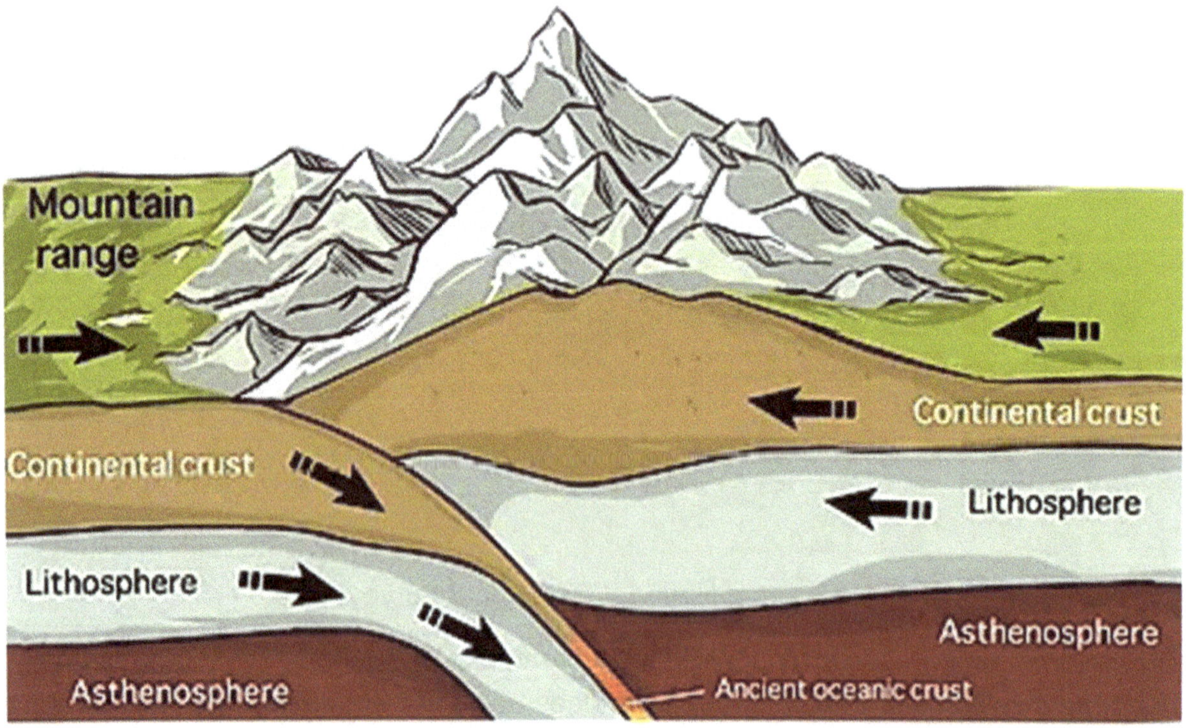

(Source: The Smithsonian Institute for Science)

The movements of the tectonic plates are imperceptible when their motion is at a slow steady rate, but when one considers how much energy it takes to move the incredibly large mass of each plate even a small fraction of an inch in any of the three axes of motion – up, down, or laterally it gives a better appreciation for how earth's surface has been changing for the past 4.65 billion years since the planet was known to have been formed. That formation is often described as being like the layers of an onion, but that is where the similarity ends.

When the plates are in divergent mode (first small inset above), the edges of the plates will cause a ridge to be formed beneath an ocean's surface or a mountain (or mountain chain) to emerge on the planet's surface. This phenomenon occurs slowly over many thousands of years and poses no urgent threat to humans but does cause some forms of the ecosystem to adapt and evolve as the mountain elevates. The "tree line" will remain at an altitude even though the mountain is thrust upward to higher elevations. In the research phase of the problem area being clarified, the statistics on the elevation of the mountains around the world and in the USA became available. The Author has tabulated an impressive database on the elevation and the number of the named mountains as shown next:

Elevation and Number of Mountains in World and in USA

Elevation at Peak			No.Named Mountains	
			In World	**In USA**
Over 8,000 m (26,247 ft)			14 (All in Asia)	None
7,000 m (22,966 ft)	to	8,000 m (26247 ft)	32 (In China, India, Pakistan)	None
6,000 m (19,685 ft)	to	7,000 m (22,966 ft)	117	1 (Denali)
5,000 m (16,404 ft)	to	6,000 m (19,685 ft)	76	3
4,000 m (13,123 ft)	to	5,000 m (16,404 ft)	260	98
3,000 m (9,843 ft)	to	4,000 m (13,123 ft)	342	50+
2,000 m (6,560 ft)	to	3,000 m (9,843 ft)	305	75+
1,000 m (3,280 ft)	to	2,000 m (6,560 ft)	229	150+

Note: The elevation of Denali (formerly Mt. McKinley) in Alaska is 6,194 m (20,316 ft)
The longest mountain range is the Andes in South America at 4,500 miles.

(Source: Tabulation by Author of National Geological Survey databases)

The tallest mountain on the planet is Mount Everest in the Himalayan Mountain Range of Nepal and presently its peak is at 8,849 meters (29,032 feet) in elevation. A photograph of Mount Everest and surrounding terrain is not only visually impressive to non-scientists, but when scientists assess the mass of that amount of the planet's surface being elevated to such extreme elevations, the energy estimated to move that much mass through that distance – albeit it is at a slow rate of movement (speed much slower than the proverbial snail) - the energy from within the outer and inner mantle and outer core (molten state) had to have been mind-boggling in pounds of force and over such an expansive area.

When viewing the table above, Readers are invited to consider the number of mountains and their elevation and imagine the energy from within the planet that was expended to lift the mass of each of those mountains to their elevation indicated. It puts into perspective how small by comparison the amount of energy trapped below a Man-made "greenhouse gas" emission blanket can cause such massive energy effects when compared to that stored within the planet below its outer crust. The later sections of this book suggest tapping into that geothermal energy as a clean alternative form of energy which has thus far largely been neglected by environmental scientists. (The geothermal energy would need to be accessed with long distance drilling into the Earths outer crust and certainly at elevations much closer to sea level than are the mountain peaks tabulated above.

(Credit to Unknown Photographer)

About 800 people attempt to climb to the peak of Mount Everest every year, and as of January 2019, 5,294 have successfully reached the peak. Over 330 have died trying. The view from the peak "at the top of the world" is magnificent by most accounts, but many have reached the peak only to have a magnificent view of cloud tops (that can be seen when in a commercial airline flight at 30,000 feet cruise altitude at far lower cost and with far greater safety).

(Credit to Unknown Photographer)

There can be no doubt that the environment, weather, climate, and human habitation of a region surrounding a tall mountain is affected by the mountain and its proximity to other major features such as adjacent tall mountains or by an ocean. The wind currents, the cloud formations, and the water vapor in the atmosphere on one side of the mountain being different from the other side of the mountain a mere few miles apart.

The eight tallest mountain ranges are all in Asia (China, Pakistan, and India) whereas the United States has the Rocky Mountain/Sierra Nevada ranges near its West coast and the Appalachian range near its East coast (2,500 miles of flat prairie land between). It should be apparent that considering global warming, global climate change, and the global effects of ocean levels rising bears little common patterns within the regions of the 48 contiguous United States, and that in China, India, and Pakistan. Setting a "global emissions goal to control emissions with a "one-size-fits-=all goal of one temperature – regardless of any amount of temperature change – from that one "average global temperature" defies scientific logic. The existence of majestic mountain ranges on the planet's surface is tempered by the fact that humans choose not to live on the sides of mountains as those are prone to have landslides and rockslides far more often than risking their probability of occurrence for daily views of the placid valleys below. Beautiful orchestral music cannot match the sounds of cattle-bells on the necks of a herd of dairy cows grazing in the meadows on an alpine hillside in Switzerland. Each cow has a bell with a unique tone and the cow-herder can tell which of his herd has strayed from the main flock by the absence of that tone in the "cattle-bell symphony".

Seismologists become suspicious when the daily vibrations on a fault-line go silent as that raises the probability of a "big one" occurring soon thereafter when the tectonic plates suddenly overcome the "static friction" force retarding their transition from being at rest into constant movement.

Physicists and engineers often simplify the terminology "static friction" as "stiction", and this variable in the dynamics and forces acting to initiate a mass relative to another mass offers a partial but very basic explanation why earthquakes occur so sporadically and so "spikey" ("jerky") in motion when they do occur. A basic experiment conducted in physics laboratories in colleges placing emphasis on STEM (science, technology, engineering, and mathematics) courses illustrates static friction or stiction. The experiment invariably consists of a mass (usually a large block of wood on a metal table (a plane) that is varied in angle and to an angle where the acceleration of gravity acting on the mass gradually overcomes the static friction between the mass and the plane. Readers can conduct this simple experiment with this book being read as indicated in the diagram next:

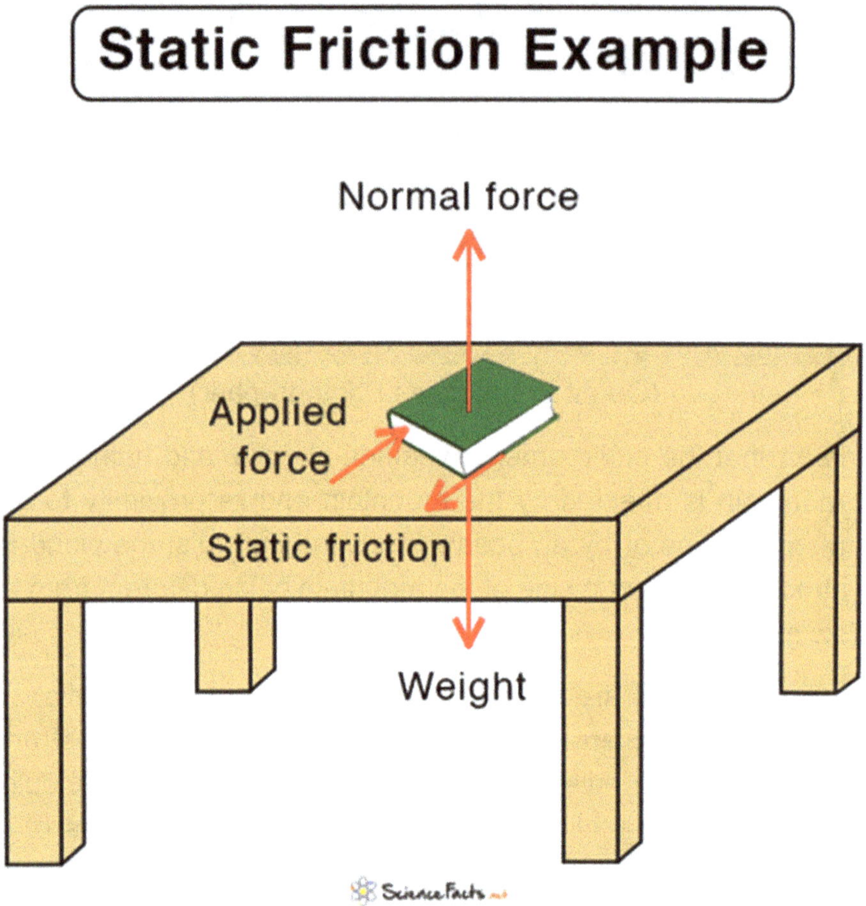

(Source: Any College Physics Textbook)

The applied force can be felt in the fingers gradually building up (to the static friction force) until the applied force is equal to the "stiction" force resisting motion of the mass. When the forces are equal, the book will start ("jump") into a sliding motion relative to the tabletop and the book will remain in motion with a smaller force being applied to the motion resisting frictional force between the two non-smooth surfaces. The weight of the book (mass) and the frictional coefficients (the smoothness of the surfaces in contact - of the bottom cover of the book and that of the tabletop - are quite different and are the variables that are unknown in nature between the two tectonic plates. That "jump" in the seismic waves is explained simplistically in the next diagram:

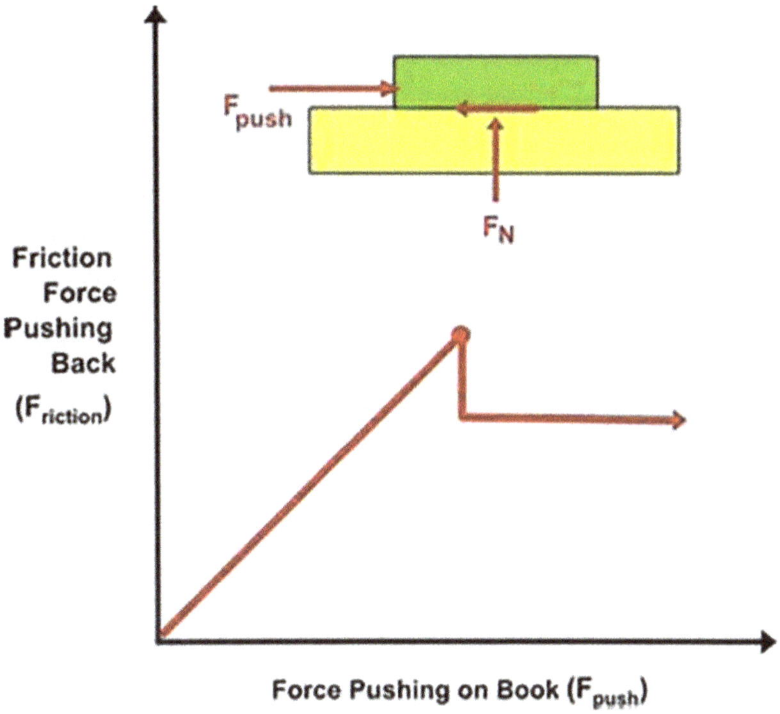

(Source: Any College Physics Textbook)

This is the same principle involved when two tectonic plates abut each other and suddenly start to move relative to each other – and the friction forces transforming into the seismic waves measured by scientists. Because the edges of the tectonic plates are not smooth and are of different materials, that static friction force causing an earthquake at the fault line is quite unpredictable until the seismic waves begin to "jump" up in a spikey pattern as the edges of the tectonic plates grind against each other. For those who may have the question in mind, the atmospheric conditions surrounding the experiment have no significant effect (other than the possibility of the tabletop being damp and the fluid acting as a lubricant to reduce the static friction force). The concentration of Man-made or natural carbon dioxide in the atmosphere is not a factor causing earthquakes – in frequency of occurrence or in severity/magnitude when occurring.

Many books have been written on tectonics and the Author will describe only the two most important aspects that relate to the topic area of this book. Those two topics are volcanoes and earthquakes, both related to the unpredictable dynamics of the tectonic plates. It can be appreciated how frustrating it must be that scientists cannot predict when a volcano will erupt, or an earthquake will occur so that warnings can be issued to nearby residents. The fatalities caused to humans and the property damage done by these two phenomena combined dwarf any estimates that are made by environmental scientists resulting from the factual causes for and effects of global warming, climate change, and ocean levels rising – in the next millennium or two. Knowing however, that most earthquakes and volcanic eruptions occur above or very close to the edges of tectonic plates, there are measurement devices and sensors that have been placed to give the earliest possible warning to the public residing in a known area where an earthquake or a volcanic eruption is most likely to occur. Only the "when" and "where" on which tectonic plate(s) will an earthquake occur, or a volcano erupt is unpredictable in the long term.

It is typical that the buildup of magma (molten rock) below a volcano produces earthquake-like movements deep within the earth. There is usually at least a three- or four-day warning period (not unlike a rattlesnake shaking its tail to issue audible rattling (hence earning its name as rattlesnake) warnings to a human getting near its territory) before a major eruption. Volcanoes also may emit an increasing amount of steam or other heated gases into the atmosphere prior to an eruption and those humans who do not heed those early warning indicators are taking unnecessary risk of property damage and major injury or loss of lives (human and/or animal).

Volcanoes have been erupting on Earth for many millions of years, but their frequency has clearly decreased in recent millennia. Between 1500 and 1700 AD, there were about 600 volcanoes that were recorded (mostly in church records) to have erupted world-wide. The record since 1950 is believed to be almost complete with 2,208 eruptions recorded from 347 volcanoes. The average number of eruptions ongoing per year since 1950 is 63, with a minimum of 46 and a maximum of 85 eruptions recorded per year. In 2023, a major volcanic eruption is occurring in the Iceland region but residents near that volcano have been evacuated such that there will be few if any losses of life. Loss of property and damage to the ecosystem is not yet totaled. Conversely, Iceland is located very near the Artic Circle (at a latitude of 64 degrees north for the capital city of Reykjavik. Later in this book will be discussed the potential for using geothermal energy if there is a source for that form of clean, reliable heat energy in a region and Reykjavik is a role model city for doing that. Almost half of that city's heat energy demand is obtained from nearby geothermal activity at surface level – no drilling required.

An "eye-opener" for many Readers will be a map showing where volcanoes exist on the surface of the planet. It is hastened to be added that most are inactive and have been for a long time. But there are many that are considered active with a probability of erupting from low to high. That map is shown next:

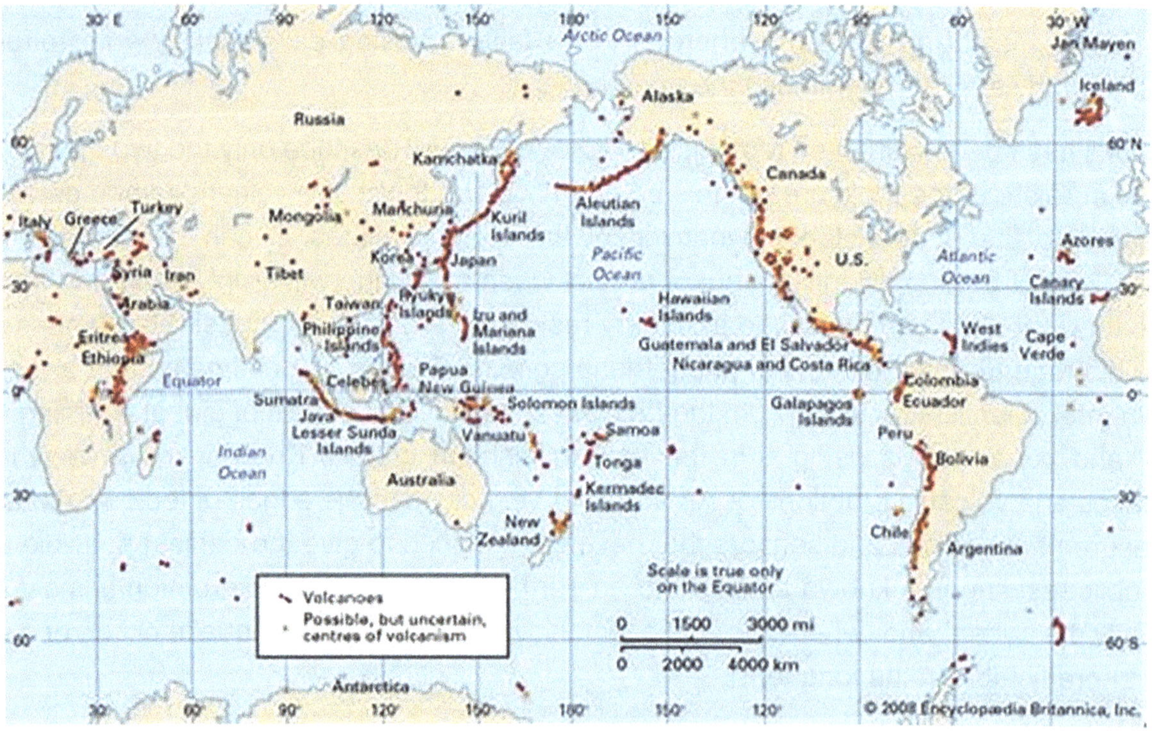

(Source: Smithsonian Institute for Science)

There is a predominance of volcanoes (each shown by a red dot in the map above) around the edges of the Pacific tectonic plate (each shown as a black dot in the lower map) and that has been dubbed as the "Ring of Fire".

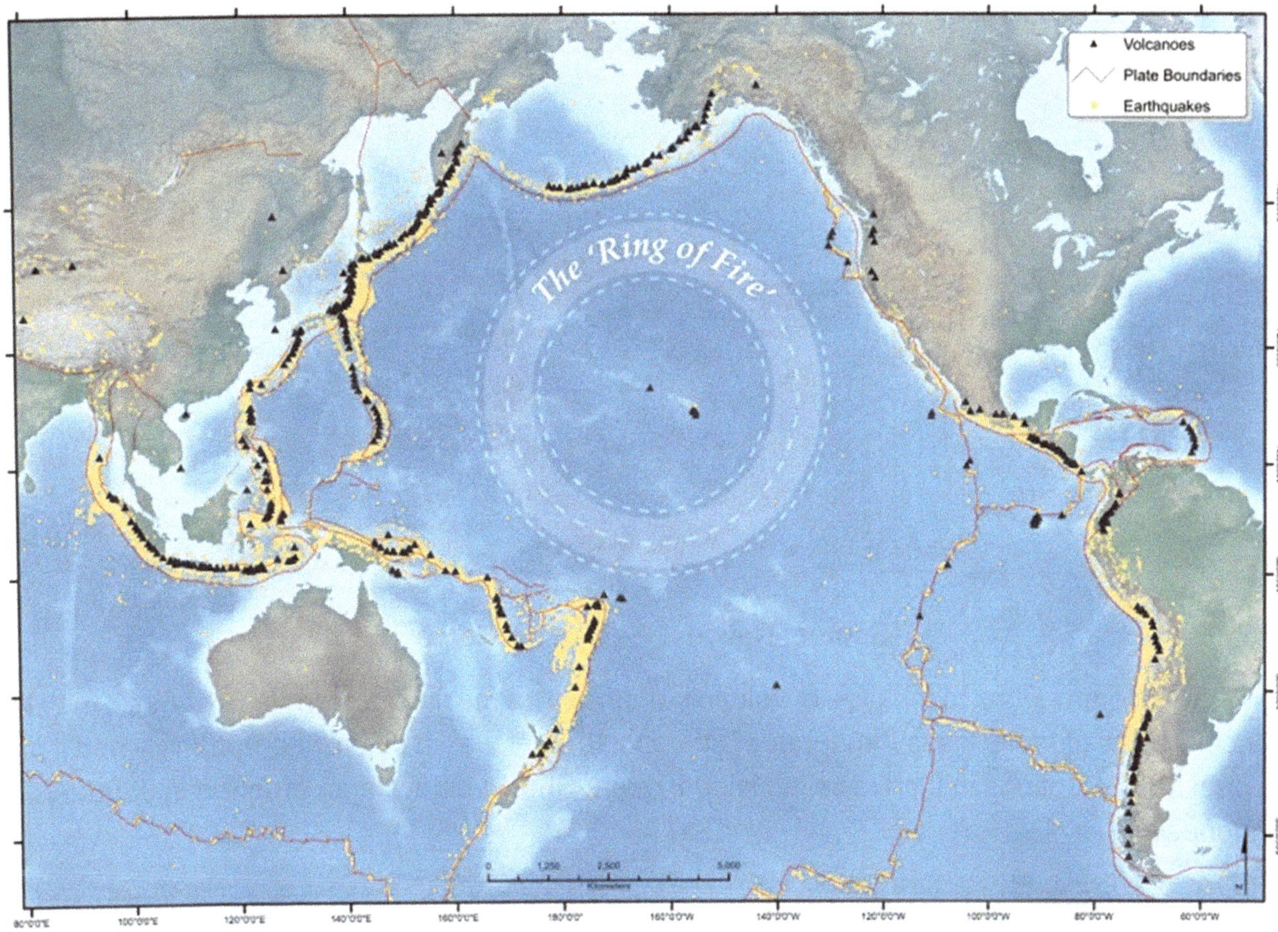

(Source: Smithsonian Institution)

What is of importance to be noticed however, is that the Hawaiian Islands chain is in the center of the Pacific Ocean and well away from the edges of the Pacific Plate. Volcanologists explain that as being a "hot spot" in the outer crust below Hawaii and each of the Hawaiian Islands is the result of a volcano formed upward from the floor of the Pacific Ocean as the tectonic Pacific Plate moved northeasterly the west coast of North and Suth America. Mount St. Helens in the US Pacific Northwest (Washington) which erupted in May 1980 (with 57 fatalities), the eruption of Mt. Pinatubo in the Philippines in 1991 (with 227 fatalities), the eruption of Mauna Loa (Hawaii) in September 2022 (with no fatalities), the eruption of the Puyehue-Cordon Caulle volcano in Chile on June 4, 2011, and currently (December 2023) in Iceland individually may not have contributed much to change the global temperature on Earth's surface.

Another major cluster of volcanoes on Earth's surface is in the Indonesian area in the Pacific Asian region. That formation is not in a ring manner but is in a more linear geometry – suggesting it be distinguished from the "Ring of Fire" and being named the "Line of Fire". That group of volcanoes is illustrated next:

(Source: National Geological Survey)

As a general comment as to the type of volcanic eruptions in the Indonesian "Line of Fire", they have historically been of more "gassy" manner of effluence than "lava" in different eruptions – and it is just the opposite in eruptions in the Hawaiian island chain. Shown below is an eruption in January 2024 on Indonesia:

(Credit to unknown photographer)

The resemblance to a nuclear bomb explosion (such as that in Japan to end World War II) is coincidental. A third historically troublesome region of volcanoes is of those that formed the West Indies in the Caribbean region of the Atlantic Ocean. Shown next is a diagram of the Caribbean Sea region and the chain of islands exposed above the current sea level formed where the Atlantic Plate adjoins the Caribbean Plate:

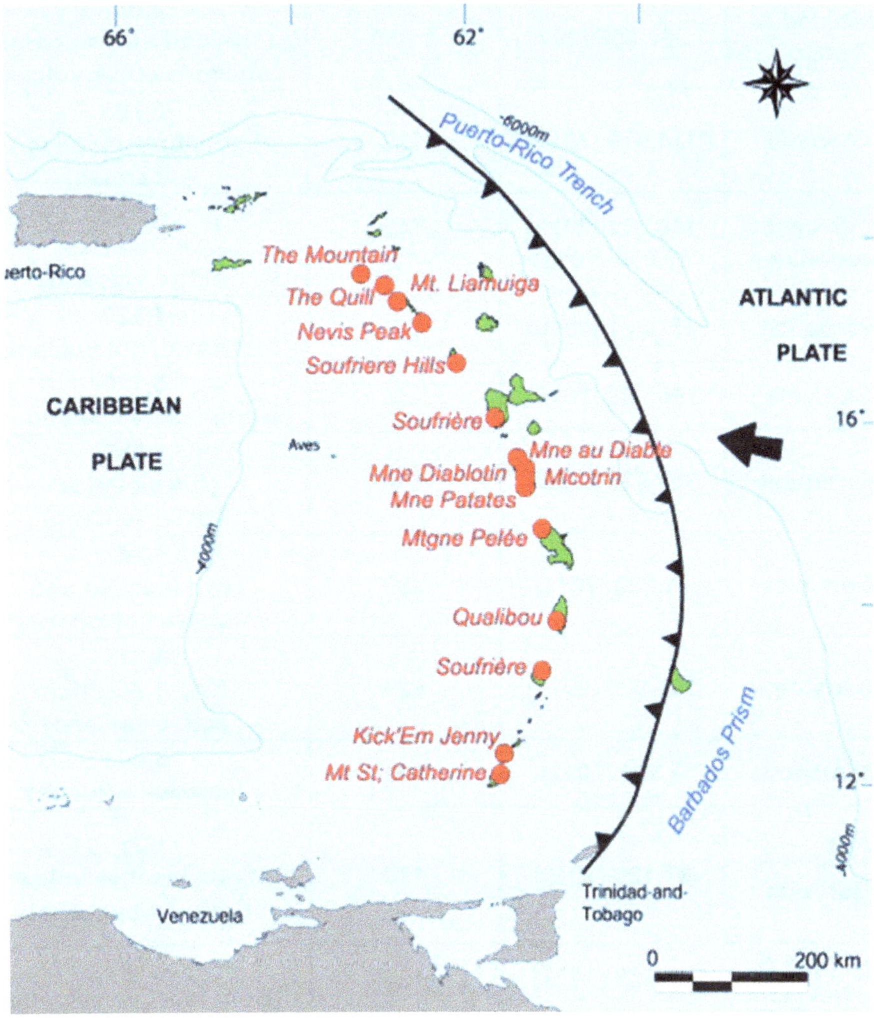

(Source: US National Geological Survey)

An interesting table of geographic and human interest regarding the islands in the Caribbean Sea is shown next, portraying the potential disaster to the over 6 million humans inhabiting the Caribbean islands and which are among the most visited places in the Northern Hemisphere seeking a relief when it is Winter in the Northern Latitudes especially those in the northeastern United States and the southern half of Canada. (The residents of the Caribbean islands refer to those millions of tourists per year as the "Snowbirds".) When volcanic activity threatens humans in larger countries there is typically an advance warning and humans evacuating to distant locations sufficiently far from where seismologists warn of an impending eruption. It can be seen in the next table that there is no distance large enough that residents on each of the volcanic Caribbean Islands can flee. Conversely, if the ocean levels do increase significantly the residents of these islands have the option of relocating further up on the side of the volcano (not a recommended action) or relocating to a larger country such as the North American or South American neighbors to their west (the better of the two options).

Caribbean Islands Most Affected by Ocean Levels Rising
(Coastal beaches and population at low height above Sea Level)

Island	Population (Census Year)	Area of Exposed Land (sq miles)	Highest Point of Island(s)(feet); (Type geologic origin)
Trinidad & Tobago	1,367,558 (2021)	1,980	3,080 (Trinidad non-volcanic; Tobago inactive volcano)
Grenada	124,610 (2021)	135	2,755 (multiple inactive volcanoes)
St. Vincent & Grenadines	110,872 (2022) 10,900 (2012)	142 17	4,049 (La Soufriere active volcano)
Barbados	277,821 (2010)	169	1,120 (upheaval, not volcanic)
St. Lucia	180,000 (2018)	238	3,120 (semi-active volcanoes)
Martinique	349,925 (2024)	436	4,583 (Mount Pelee' active volcano)
Dominica	71,293 (2011)	290	4,747 (two inactive and 3 active volcanoes)
Guadaloupe	395,925 (2024)	629	4,813 (Grand Soufriere active volcano)
Montserrat	4,390 (2022)	102	3,440 (inactive volcano)
Antigua & Barbuda	97,120 (2019)	170	1,319 (Antigua inactive volcanic; Barbuda upheaval)
St. Kitts & Nevis	54,338 (2023)	101	3,793 (Both inactive volcanic)
Anguilla	15,753 (2021)	35	240 (Of upheaval limestone)
US Virgin Islands	108,612 (2000)	134	1,555 (Mix of inactive volcanic and upheaval)
Puerto Rico (US)	3,285,874 (2020)	3,424	4,390 (inactive volcano)
Total:	6,454,991	8,002	

(Source: "Genealogy of the Hutchinson and Edghill Families of Barbados")

The highest public awareness of volcanic eruptions is of those that erupted recently in Hawaii and in Iceland. When former Vice President Al Gore produced his "An Inconvenient Truth" epic video about the loss of sea ice in the Arctic region and placing the cause to be the result of "greenhouse gas emissions" from Man burning fossil fuels, conspicuous by their absence was the volcanic activity on Iceland and the geothermal heat energy under the Greenland Ice Sheet that were not included in that "inconvenient truth".

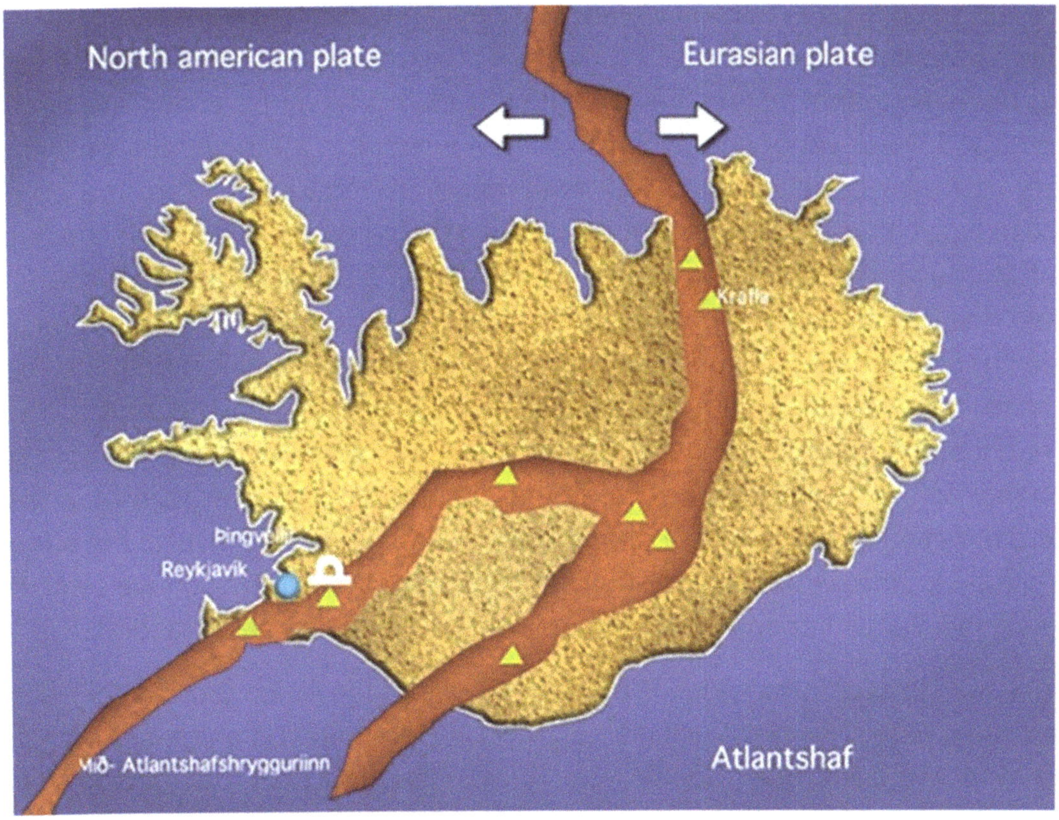

(Source: International Union of Geological Sciences)

While the volcanoes in Iceland are well-shaped in the classic conical shape, the most troublesome activity has been with the eruption of lava from the lateral subsurface tubes connecting the various volcanoes. Evidence of this is shown in the emergence of lava flowing out of a 4 miles long almost linear separation where the North American Plate and the Eurasian Plate are moving apart.

(Source: International Union of Geological Sciences)

Because the nature of the lava flow on Iceland not being a typical cataclysmic eruption but more of an oozing and horizontal flow, nearby residents were able to evacuate without the loss of any human life. It should be noted that the gases being emitted from this type of eruption are largely of natural carbon dioxide and water vapor composition, both of which are the powerful "greenhouse gases" heat trappers and bear no association with Man-made carbon dioxide from the burning of fossil fuels. It is these types of volcanic activity that produced the carbon dioxide in the atmosphere before any effect of Man even appearing on Earth's surface. The Vostok Antarctic Research Station ice core samples "documents" that on at least four cycles in the past 420,000 years the natural carbon dioxide in the atmosphere was at a peak level of about 295 parts per million (0.0295%) and the peak hot temperature in that same cycle had peaked about 800 years prior to the peak carbon dioxide. Reputable scientists interpret that to mean that as a result of whatever natural processes were and are in action (astrophysical, geophysical, and other natural processes without human influence), it is the temperature of the planet's surface that drives the carbon dioxide level in the atmosphere, and not the reverse – i.e., the carbon dioxide in the atmosphere driving the surface temperature of the planet (as "settled science" reports in NOAA publications and in the news media).

To not confuse the claim of former Vice President Al Gore of the Arctic Sea ice melting due to Man-made "greenhouse gases", and the volcanic activity on Iceland being a primary contributor of heat causing the "Inconvenient Truth" ice in the Arctic Ocean, Iceland is geographically located immediately south of the Arctic Circle. The hyperbolic inference of the "Inconvenient Truth" is that it is the atmospheric "greenhouse gas" carbon dioxide (combination of both natural and Man-made) that is collaterally melting the massive ice sheet on Greenland – which is largely north of the Arctic Circle (but not into the northern polar cap). Before the capability of measuring the temperature profile through the ice sheet on Greenland and of the terra firma of the surface on which the Greenland ice sheet resides became available in the latter portion of the twentieth century, the "Inconvenient Truth" pointed the finger of cause to be the Man-made "greenhouse gases", and unless halted, the ice sheets of Greenland and Antarctica will melt and sea levels would rise by estimates of up to 230 feet. It is "convenient" to not mention that of that 230 feet rise, 23 feet of it would be due to the melt of the entire Greenland ice sheet and the remainder (207 feet – a ratio of about 9 to 1) would be due to the melt of the entire Antarctic ice sheet.

Mr. Gore narrated the Academy Award-winning documentary "*An Inconvenient Truth*", in 2006, and which was credited with raising international awareness about the global climate crisis. The gullible public formed many activist groups demanding the cessation of Man burning fossil fuels and governments of many countries instituted legislation that supported that reduction of Man burning fossil fuels. Mr. Gore and the United Nations Intergovernmental Panel on Climate Change (IPCC) was awarded the Nobel Peace Prize for his "Inconvenient Truth" revelation in October 2007. The Nobel Commission citation partially stated:

"On October 12, 2007, former U.S. Vice President Al Gore and the United Nation's Intergovernmental Panel on Climate Change are awarded the Nobel Peace Prize for their efforts to increase public knowledge about man-made climate change."

The former senator from Tennessee who served as President Bill Clinton's vice president from 1993 to 2001, is considered one of the first politicians to recognize (and regrettably over-exaggerate) the adverse aspects of carbon dioxide emissions, postulated as a major cause of human-induced global warming. Gore became interested in the topic of global warming during a college course he took at Harvard University. As a US Congressman, he held hearings on climate change in the late 1970s, a time when most Americans had little or no knowledge of the issue. After losing the 2000 presidential election to George W. Bush, Gore embarked on a new campaign—the fight against man-made climate change—and gave slide-show presentations around the world in an effort to "educate" the public. *An Inconvenient Truth* chronicled Gore's efforts to educate audiences with his "traveling global warming road show" – minus only the sale of the snake oil customarily sold at Wild West medicine-man shows. In the film, he details the facts and fallacies surrounding this "planetary emergency" and describes the events in his own life that led him to become an environmental crusader. It must be wondered if Mr. Gore was aware of the truth – Man burning fossil fuels is causing global warming in perhaps the same ratio of 9 to 1 as is the ratio of the amount of ice in Antarctica as there is in Greenland? Polar bears are magnificently beautiful animals to preserve their existence – but are the many millions of penguins in Antarctica not also worthy of being saved from Man's energy transgressions?

The improvement of the temperature-sensing technology aboard NASA satellites and the International Space Station in the past few decades has revealed (in the image at the top of the next page) that the surface on which the Greenland ice sheet was formed in the past few eons is a very warm surface from geothermal activity (similar to the same geothermal tectonic-geology as exists under Iceland but without the immediate presence of molten matter). In an equally revealing image, the terrain which features a large number of lakes below the ice sheet covering the Antarctica continent (and showing the extensive ice shelves of Western Antarctica that may melt long before the ice sheet above terra firma in Eastern Antarctica) is also shown on the next page:

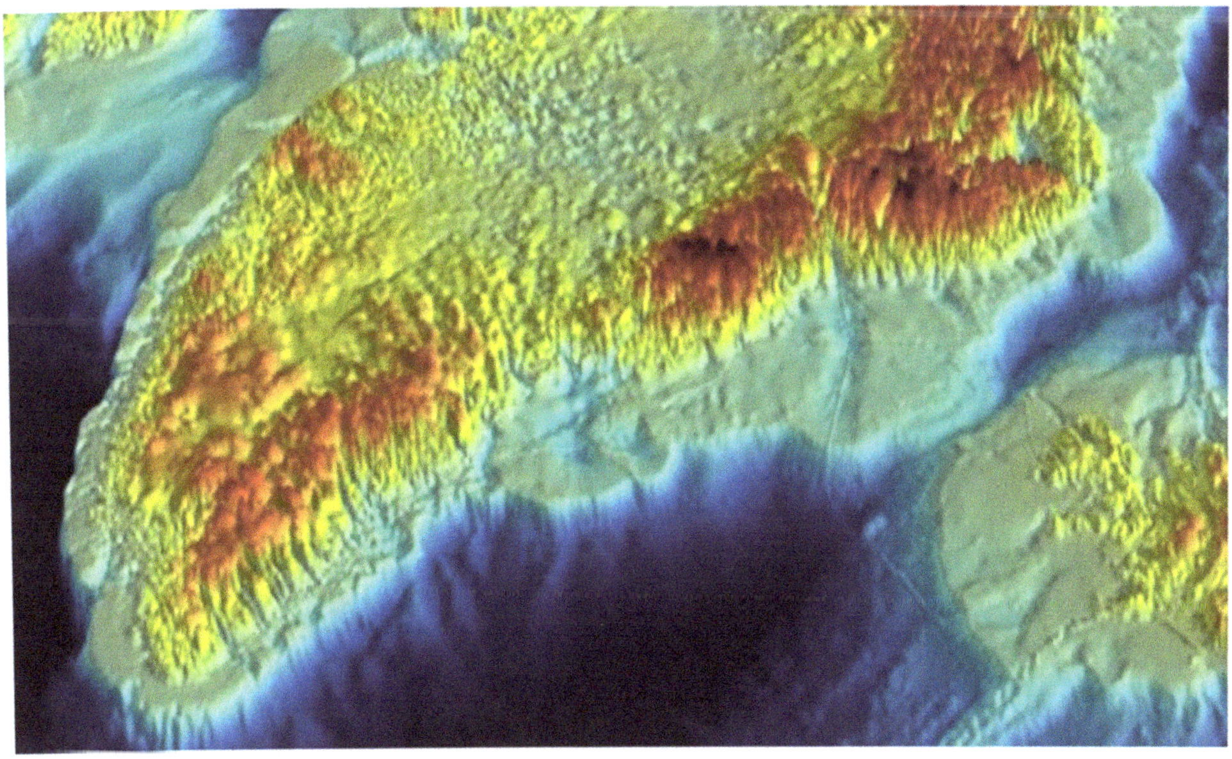

(Source: NASA) Below Greenland Ice Sheet

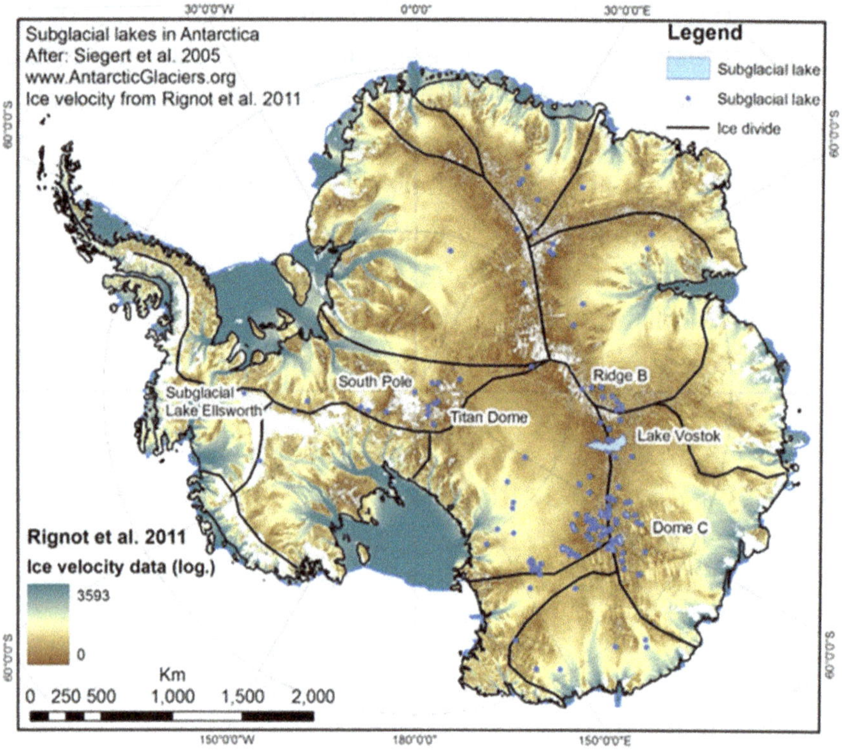

(Source: NASA) Below Antarctica Ice Sheet

A very useful "big picture" Volcanic Explosivity Index (VEI) was developed in 1982 (by Christopher G. Newhall (and an associate scientist Stephen Self) of the US Geological Survey) that reflects, in a qualitative manner since there is no physical way to measure the enormous events of a volcanic eruption, what was the volume of material (solid and/or molten) ejected onto the affected area surrounding the volcano and amount of gaseous and/or particulate material injected into the atmosphere and to what height, and how long did the continuous blast effect (if any) persist. The next chart indicates the definition used for the Volume Explosive Index (VEI ranging from VEI 0 to VEI 8+:

CRITERIA	VEI 0	VEI 1	VEI 2	VEI 3	VEI 4	VEI 5	VEI 6	VEI 7	VEI 8+
Description	non-explosive	small	moderate	moderate-large	large	very large →			
Volume of ejecta (m^3)	$<10^4$	10^4-10^6	10^6-10^7	10^7-10^8	10^8-10^9	10^9-10^{10}	10^{10}-10^{11}	10^{11}-10^{12}	$>10^{12}$
Column height (km)	<0.1	0.1-1	1-5	3-15	10-25	>25 →			
Qualitative description	← "gentle, effusive" →		← "explosive" →		← "cataclysmic, paroxysmal, colossal" →				
				"severe, violent, terrific" →					
Classification		← "Strombolian" →			"Plinian" →				
	← "Hawaiian" →		← "Vulcanian" →			"Ultraplinian" →			
Duration of continuous blast (hours)	← <1 →				← >12 →				
			← 1-6 →						
				← 6-12 →					
Tropospheric injection	negligible	minor	moderate	substantial →					
Stratospheric injection	none	none	none	possible	definite	significant →			

(Source: Newhall and Self; US National Geological Survey)

Of importance to be noted is that the Volume of ejecta (in cubic meters) – to convert a cubic meter to a cubic yard, there are 1.31 cubic yards to one cubic meter – are estimated on the VEI scale to be a range of exponents of ten (e.g., 10^4 cubic meters (10,000 cubic meters) or less is rated (combined with the other factor observations) as being of category VEI 0. The exponent numeric (4, 5, 6, …..12) is the same as the number of zero's after the 1. (As shown in a later chapter, the amount of carbon dioxide gas that environmental scientists claim has been and is being introduced into the atmosphere from Man burning fossil fuels is measured in terms of gigatons. A gigaton is 10^9 tons. Although the frequency of eruptions is only an average of 67 eruptions per year and of a gamut of VEIs, it raised the question in the Author's mind "How much carbon dioxide gas factually results from Man burning fossil fuels (especially coal) in coal-burning facilities and human residential use each year globally and how does that compare to the amount of carbon dioxide gas produced each year by volcanic action?".

To begin to understand that comparison, a statistical analysis was conducted (by the Author) to reconstruct the listed Smithsonian Institution Global Volcanism Program data into the matrix form shown in a table of Volcanic Eruptions (frequency and severity) statistics. The statistical database on volcanic eruptions from 1960 to 2023 compiled by the Smithsonian Institution Global Volcanism Program is a most interesting one – and was considered as a "red flag" needing to be examined closer by the Author.

Volcanic Eruptions Statistics by Decade, Number, and Severity Index

	Total No.	**Number at Average Severity by VEI Level**							
Year Span (No.Yrs)	**Total Eruptions / (Avg/Span)**	**No. at VEI 0-2**	**VEI 0-2 Yr Avg / (Span)**	**No. at VEI 3**	**VEI 3 Yr Avg / (Span)**	**No. at VEI 4**	**VEI 4 Yr Avg / (Span)**	**No. at VEI 5**	**No. at VEI 6**
1960 to 1970 (11)	649 / (59)	305	27.7 / (11)	64	5.8 / (11)	5	0.5 / (11)	1	0
1971 to 1980 (10)	627 / (63)	274	27.4 / (10)	40	5.8 / (10)	4	0.4 / (10)	1	0
1981 to 1990 (10)	626 / (63)	231	23.1 / (10)	45	4.5 / (10)	8	0.8 / (10)	1	0
1991 to 2000 (10)	636 / (64)	283	28.3 / (10)	38	3.8 / (10)	5	0.5 / (10)	1	1
2001 to 2010 (10)	778 / (78)	321	32.1 / (10)	56	5.6 / (10)	10	1.0 / (10)	0	0
2011 to 2020 (10)	705 / (71)	294	29.4 / (10)	52	5.2 / (10)	12	1.2 / (10)	1	0
2021 to 2023 (3)	240 / (80)	63 (2)	31.5 / (2)	2 (2)	1.0 / (2)	1 (2)	0.5 / (2)	1 (2)	0 (2)
(Total Span of 64 Yrs)	Avg 66.5 / Yr.								
Total 1960 - 2023	**4261 / (64)**	**1771 (63)**	**28.1 / (63)**	**297**	**4.7 / (63)**	**45**	**0.7 / (63)**	**6**	**1**

Note: There was only Total Eruptions data available for 2023; no VEI 0-2, VEI 3, VEI 4, VEI 5, or VEI 6 data.

(Source: Table generated by Author from Smithsonian Institute database)

A professional statistical analysis of the 4,261 volcanic eruptions occurring globally in approximate 10-year spans for the 64 years period from 1960 to 2023 reveals some interesting and important trends. (It is not surprising, but neither the environmental scientists nor the news media feel it to be necessary to inform the public about the role volcanic eruptions play in Earth's surface temperature changes, global climate changes, and ocean levels rising.) The information revealed about volcanic eruptions in this analysis and reported in this book hopefully are not among those *'everything's* in the fallacious "settled science" hypothesis of "*when 'everything' else is eliminated, Man burning fossil fuels is considered to be the dominant cause for global warming and climate change.*"

The Smithsonian Institution scientists evaluated each of the 4,000+ eruptions for the entire 63 years and gave each a VEI rating that seemed most appropriate all things considered. That laborious task is most valuable as it not only identified what the frequency of the different eruptions has been but includes a reasonably accurate system for estimating how much ejecta was introduced onto Earth's surface and into its atmosphere, and its oceans. The Smithsonian scientists do a far more scientific analysis than do the environmental scientists who, when they are presented with a database that does not conform with the hypothesis that "Man burning fossil fuels is the dominant cause for global warming and climate change", they simply "eliminate it" from further consideration. When presented in a tabular form (a long list of 64 lines of data), it is difficult to discern any pattern to the frequency of the volcanic eruptions or to their severity (with the VEI assessments assigned to each eruption. By graphically presenting that data, some most important observations are made and conclusions drawn.

To avoid "spotting" over 4,000 data points on the chart (giving it a "measles" appearance), the average number of volcanic eruptions occurring in the time spans, 1960 to 1970 (11 years), from 1971 to 2020 (in 6 spans of 10 years each), and from 2021 to 2023 (3 years) are presented at the mid-point of each span. Because of the large scatter – volcanoes erupt when and where they do, and except for measuring seismic activity in each volcano's vicinity, there is no long-term prediction of any singular volcano erupting.

By constructing a graph of the frequency and VEI index of severity within that frequency of volcanic eruptions (as shown earlier) provides some insight if the severity of the global eruptions is also increasing (from any cause). The graph shown next illustrates that trending of intensity effect for the differing VEI levels of the eruptions:

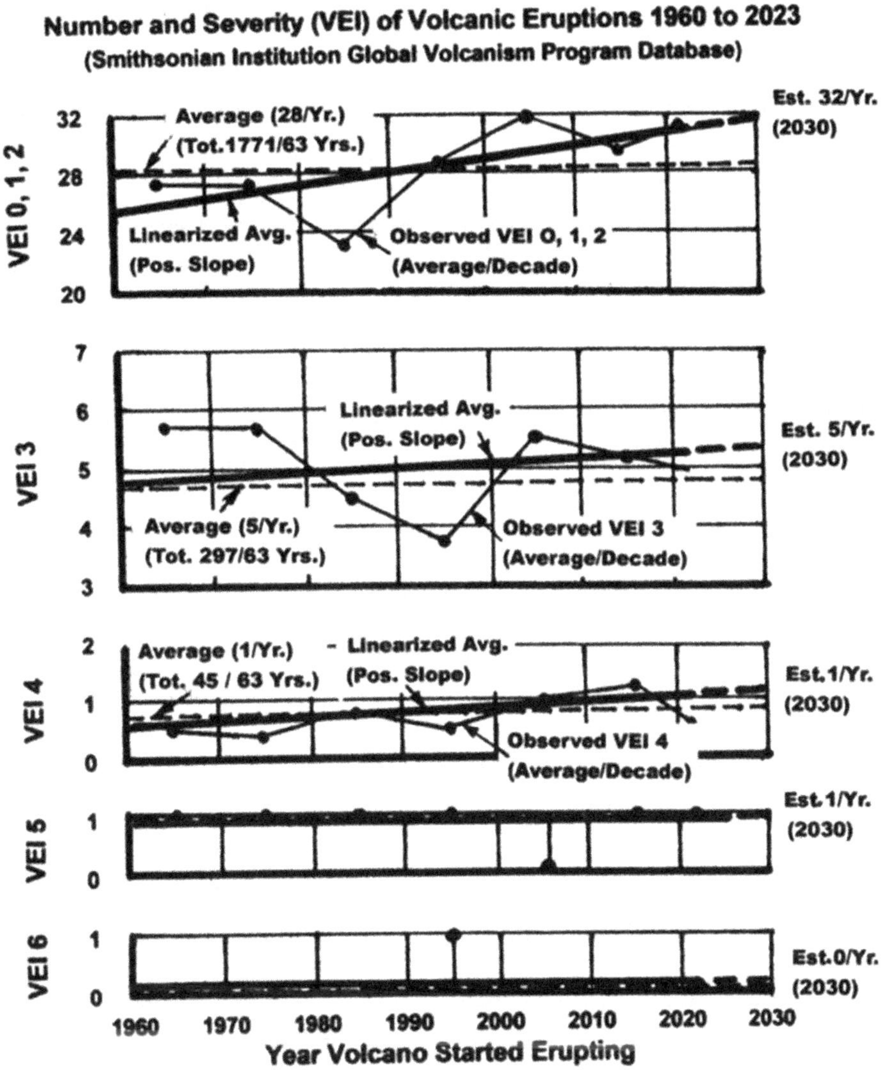

(Source: Author generated graph from Smithsonian Institution database)

The statistical interpretation of the variation in frequency is that there are several more occurrences of volcanic eruptions in the Category VEI 0, 1, and 2 per year, and there is a weak trend toward lesser frequency of the volcanic eruptions as their Category VEI level of severity increases from VEI 3 to VEI 4, 5, and 6. There should be little question whether the changes in both frequency and severity of all VEI levels of volcanic eruptions are due to unpredictable tectonic plate movements, or due to an increase in Man burning fossil fuels.

There can be an average number of all volcanic eruptions per year obtained simply by dividing the total number of all observed eruptions (4,261 below VEI 0 and from VEI 0 to VEI 6) by the 64-year time span between 1960 and 2023, but that is not a very meaningful statistic to a professional database analyst. That average number of total eruptions per year (rounded up to 67) is seen on the graph as a dashed horizontal line. It was stated earlier and restated here (and will be restated later), very little in the real-world behaves in linear form and that is especially true for volcanic activity on the planet's surface.

For the volcanic eruption data from 1960 to about 1990-2000, the predominance of the events (points) are shown to be above the dashed "average" line whereas after about 2000, the predominance of the points was below the "average" line as shown in the graph next:

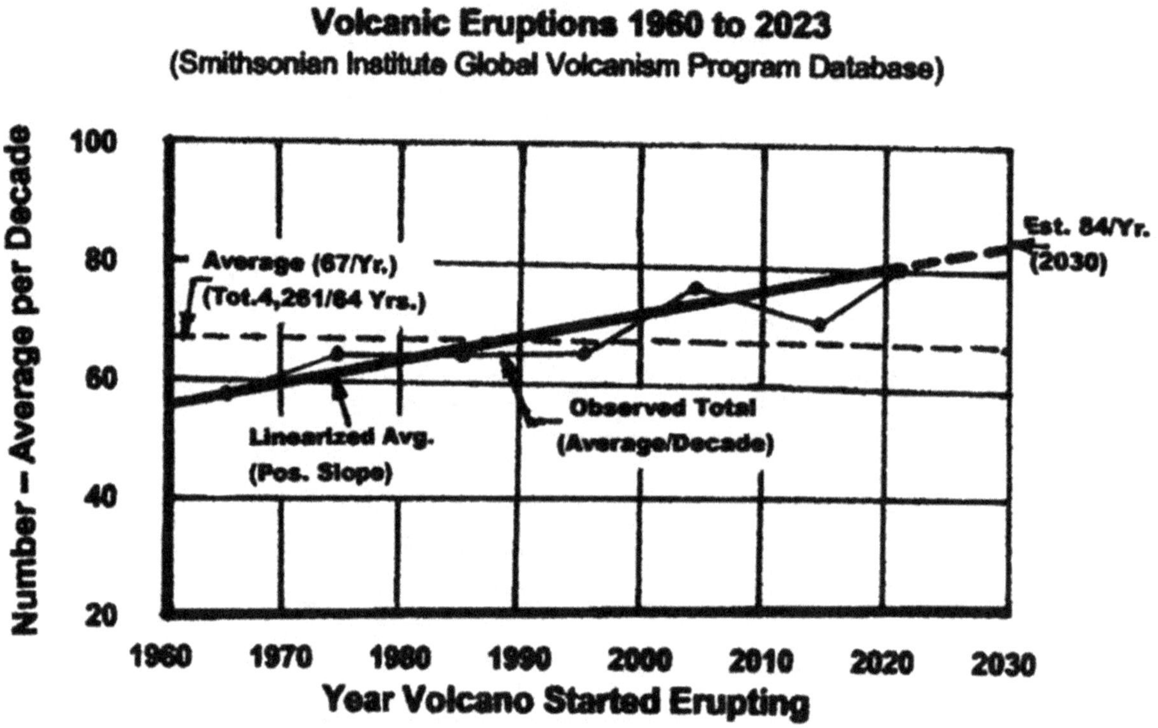

(Source: Author Generated Chart from Smithsonian Institution Database)

The overall "big picture" conclusion is made that the frequency of volcanic eruptions world-wide is increasing in number with time. A near term estimation is that by 2030, the highest probability of the number of eruptions world-wide per year is expected to be about 84 eruptions of all VEI severity ratings.

A lengthy tabulation on a country-by-country arrangement is not feasible in this book, but a summary tabulation of the world-wide eruptions with fatalities and the region identified, during the period 1963 to 2023 is an "eye opener":

Fatalities Caused by Volcanic Eruptions on Earth 1963 to 2023

Year	Yr Total	Country & Number Fatalities
1963	1,584	Indonesia 1,584
1964	139	Indonesia 114; Chile 25
1968	87	Costa Rica 87
1969	3	Philippines 3
1971	30+	Chile 25+ & 5
1973	1	Iceland 1
1977	70	Congo Rep. 70
1979	238	Indonesia 149 & 80; Italy 9
1980	57	USA 57
1982	1,918+	Mexico 1,900+; Indonesia 18
1985	23,000+	Colombia 23,000+
1990	32	Indonesia 32
1991	890	Philippines 847; Japan 43
1992	6	Indonesia 6
1993	79	Philippines 79
1994	99	Indonesia 64 & 30; New Guinea 5
1996	3	Philippines 3
1997	19	Caribbean 19
Total Fatalities 1963-1997 was 28,255+		

Year	Yr Total	Country & Number Fatalities
2002	245	Congo Rep 245
2004	2	Indonesia 2
2005	2	El Salvador 2
2006	1	New Zealand 1
2008	1	Chile 1
2010	356	Indonesia 353; Guatemala 3
2011	35	Eritrea 31; Indonesia 4
2014	23	Indonesia 16 & 7
2017	8	Indonesia 8
2018	3,326+	Guatemala 2,800+; Indonesia 426
2019	23	New Zealand 22; Italy 1
2020	39	Philippines 39
2021	101	Indonesia 69; Congo Rep. 32
2022	1	Mexico 1
2023	10+	Colombia 10+;
Total Fatalities 2002-2023 was 4,173+		

Total Fatalities 1963-2023 was 32,428+
Fatalities 1963-2023 averaged 540+ per year

(Source: International Union of Geological Sciences database, tabulation by Author)

For unknown reason, the fatalities statistics for the global volcanic eruptions in 1998, 1999, 2000, and 2001 were not found. It is possible but not likely that no fatalities due to volcanic activities were incurred for those four years period of time.

If the average number of human fatalities from volcanic eruptions in the period 1963 to 2023 averaged 542 fatalities per year (in the tabulation of fatalities from volcanic eruptions above) and there was an average of 67 volcanic eruptions per year in that same period (1963 to 2023) (in the second table above) then using the "averaging of averages technique" (so freely used by environmental scientists as in "average global temperature"), a conclusion is that Earth loses "only" approximately 8 humans per volcanic eruption. Many of those lost were scientists making measurements and suddenly fatally injured when the volcano unexpectedly erupted nearby.

Again, what occurs in an "average global" manner for several regions may not reflect – either too few or too many – as that common for another region. In the Indonesian region between 1963 and 2023, there were 2,962 human fatalities in 17 eruptions for an average of 174 fatalities per eruption. In the United States ("lower 48") from 1963 to 2023 there was only one eruption (Mt. St. Helens in the state of Washington) and 57 fatalities. Does that substantiate the claim of environmental "settled science" scientists': "When all other factors are eliminated, the conclusion of "settled science" remains that Man burning fossil fuels is the dominant threat to the human population of the World"? How many human fatalities are attributed (with scientific validity) being due to "greenhouse gas emissions" during the period 1963 to 2023? (For volcanic eruptions, that total was over 32,428.) Indonesia clearly must control better where its population resides relative to volcanic eruption hazards and have a better eruption warning system – like that of Hawaii in the United States and other developed countries.

As shown earlier, volcanoes erupt with a wide variety of Volcanic Eruption Indexes and with different "personalities" for the type and amount of ejecta. They also have a unique "kill/damage-mechanism" that varies significantly and with a different effect on the ecosystem. An interesting contrast was seen between the eruption of Mt. Pinatubo in the Philippines (photographs shown earlier of the vast deposit of volcanic ash at Clark Air Base and on local rooftops) compounded by the heavy rainfall from Typhoon Yunya a day after (in 1991), and the earlier eruption of Mt. St. Helens in the US state of Washington in May 1980. The eruption of Mt. St. Helens was the largest eruption in recent U.S. history and one of the best photographed eruptions as it was being well monitored by scientists before, during and after its non-typical form of eruption.

The Mt. Pinatubo eruption in the Philippines in 1991 was a vertical eruption in which there was little lava efflux, but an enormous amount of volcanic ash inserted into the atmosphere and which was carried by winds over the Luzon peninsula. Compounding the heavy ash problem to nearby locations was Typhoon Yunya which passed over the Mt. Pinatubo eruption a day after the eruption. The heavy rainfall from Yunya mixed with the thick layer of ash from Pinatubo on everything on the ground and collapsed many flat rooftops. Nearby Clark Air Base had evacuated its fleet of USAF airplanes based at Clark Air Base after a warning of Typhoon Yunya approaching. A commercial DC-10 which was unable to be flown out because of an on-board equipment malfunction became a victim to the heavy mixture of volcanic ash and typhoon rainfall:

(Credit to unknown photographer at Clark Air Base)

The Mt. St. Helens unusual eruption in the US State of Washington (on the Pacific "Rim of Fire") in early 1980 was a more lateral eruption. Based on seismic measurements being made of Mt. St. Helens which began "rumbling" for weeks before and foretelling its probable eruption, scientists and photographers began observing the bulging of the sidewall of the volcanic cone and very well documented the explosive eruption on May 18, 1980. The Mt. St. Helen eruption was especially well documented photographically as shown next:

(Source: US National Geological Survey; photo by Unknown Photographer)

The photographs of the "during" eruption from ground level (above photo) and from altitude (next photo) shows the extensive amount of mass that was ejected initially horizontally (in the initial "sidewall blow-out") and then the following plume of ejecta from within vertically to an altitude where winds in the atmosphere turned the ejecta cloud again horizontally being carried with the wind after Mt. St. Helen erupted.

(Source: US National Geological Survey; photo by Unknown Photographer)

As the ejecta plume develop and followed its wind-driven path, a GOES satellite photographically tracked (shown next) the path of the plume – providing cautionary warnings to populated areas in its path.

(Source: NASA GOES Satellite)

The Geostationary Operational Environmental Satellites (GOES) is a collaborative NOAA and NASA program providing continuous imagery and data on atmospheric conditions and solar activity (space weather). NASA builds and launches the GOES and NOAA operates them. The GOES database has provided a major portion of the updates to the environmental database for analysis in this book. What the public should "see" in the "big picture" photograph above is not only the long ejecta plume from one volcanic eruption and its shadow on the Earth's surface but notice the extensive cloud cover over the entire portion of the Earth's surface – all of which in combination dwarf the heat-trapping effect of Man-made carbon dioxide from the burning of fossil fuels.

A view several months after the Mt. St. Helen eruption returned to a near-dormant state shows the effect of the eruption of Mt. St. Helen on the surrounding environment (Spirit Lake in the foreground). The explosive blast pressure when Mt. St. Helen erupted in 1980 blew down almost every tree in the large area. The absence of forestry and vegetation with the entire region covered in grey-colored ash should be noted in the photograph. This is not typical for the state of Washington which elsewhere is abundant in green forestry and wild grasses.

(Source: National Park Service)

Acidity of the ejecta, which was blown or flowed into Spirit Lake, destroyed almost all the aquatic life in the lake.

An annotated photograph of the Mt. St. Helens area affected by the 1980 eruption provides some interesting geological statistics regarding the effect that the Mt. St. Helen eruption had on the environment:

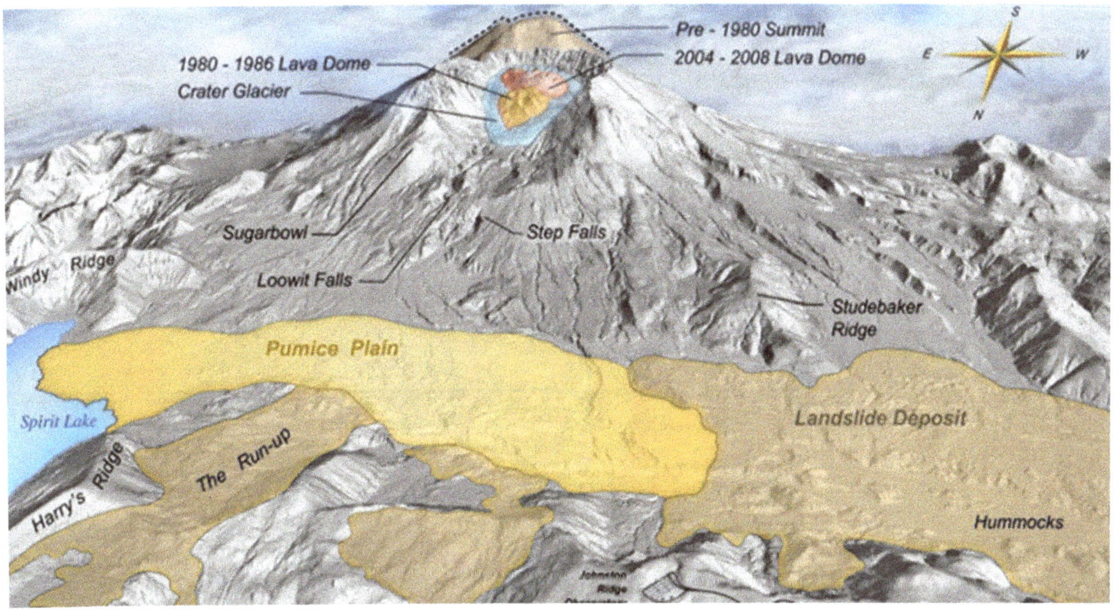

(Source: US National Geological Survey)

Readers should also note that it is natural for assaults on the environment (caused by nature itself or by inconsiderate Man) to trigger other natural forces and processes in the region to return to an equilibrium or "normal" state that existed before the disruption. A photograph of the region most affected by the Mt. St. Helens eruption in 1980 is shown next, illustrating how the region has returned (almost completely) to its natural state in 2023 (over 40 years after the eruption):

(Source: US National Park Service)

With attention turned to the ejecta from global volcanic activity influencing the atmosphere and the surrounding environment, an example of how environmental protection scientists are misinforming (inadvertently or intentionally) the public with over-exaggerated data to prove that a "crisis" looms ostensibly due to Man burning fossil fuels at an unusually high rate is shown next:

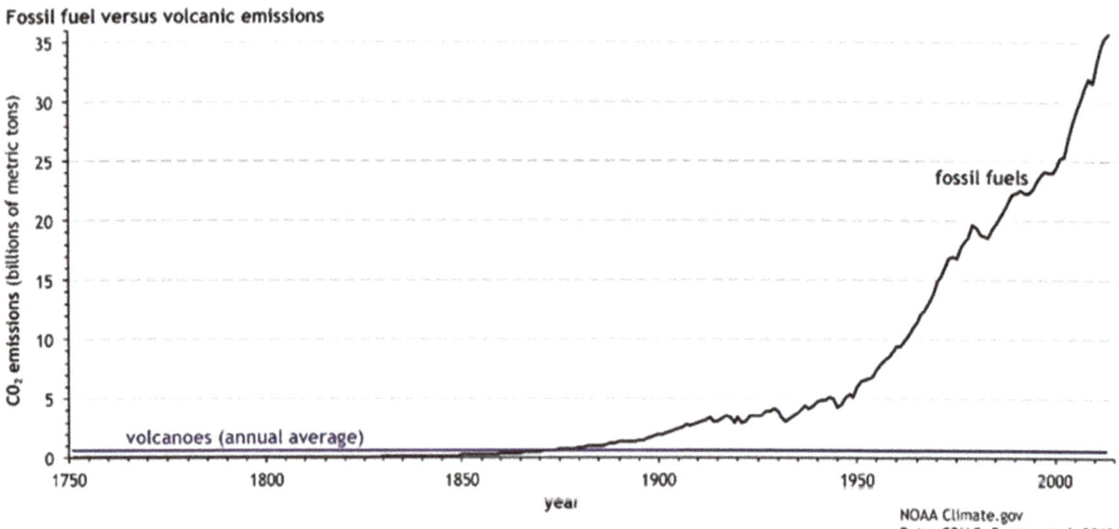

(Source: NOAA)

CAUTION: The above diagram presented to the public has several flaws, none of which may be obvious to the gullible public. Readers are cautioned to observe the intent of the chart, but not the numerical relative magnitudes of its end-effect, to illustrate that Man burning fossil fuels is the dominant cause of the global "greenhouse gas" effect.

1) The above chart graphically reflects the fallacious conclusion drawn by many environmental scientists that the human and industrial use of fossil fuels has gone exponentially out of control, but

indicates comparatively that volcanic eruptions play a negligible part in the global warming and climate change "crisis";

2) The chart shows that the amount of the greenhouse gas carbon dioxide introduced into the atmosphere by World-wide volcanic eruptions is of a suspiciously low magnitude starting in 1750 and remains at that low constant level for the next 250 years (beyond 2000). The linear representation is labeled "Volcanoes (annual average)". There exists no valid database of scientific measurements (by professional volcanologists or environmentalists) of the gaseous content of all volcanic eruptions. Taking measurements of the gases would entail aircraft to fly at those altitudes into the gaseous plume and that would destroy the propulsion systems of aircraft. Therefore, the "average CO_2 in volcanic eruptions" is suspect of gross underestimation to place emphasis on the difference with "fossil fuels CO_2" measurements and overexaggerated estimates by environmental protection scientists.

3) A third question regarding the source of the Fossil Fuels data used for the above comparison and needing clarification is: "How much of the carbon dioxide in the atmosphere (shown on the chart in billions of metric tons) is actually due to Man burning fossil fuels (containing carbon) and how much is produced from natural causes? (The "Keeling Curve", provided in the previous chapter, shows the measurement of total (natural and Man-made) carbon dioxide in the atmosphere (435 parts per million by volume (ppmv) in 2023) but of that 435 ppmv, 295 ppmv of natural carbon dioxide existed in the atmosphere prior to the start of "the Industrial Revolution". Therefore only 140 ppmv of carbon dioxide in the total 435 ppmv is attributable to being Man-made. That begs the question in the above chart as to the source of the "Fossil Fuels" amount of carbon dioxide in that data shown being Man-made (from burning fossil fuels)

When coal (the major offending fossil fuel) is combusted, the combustion process draws air from the atmosphere to complete the combustion process. Air is a collection of gases with varying gas content as altitude increases. At Sea Level, atmospheric physicists use either the composition of a volume of air to be comprised of 77% nitrogen and 21% oxygen with a trace of Sulphur. If considered by mass, air is comprised of 79% nitrogen and 21% oxygen. Depending on the quality of the coal – which does vary considerably by type, (lignite, anthracite, or bituminous) – the "average" coal by mass is 83% carbon, 5% hydrogen, and 2% oxygen. Skipping over the scientific analysis (which can be found in many textbooks of Atmospheric chemistry and physics), by mass it requires 2.66 mass units of oxygen to combust 1 mass unit of coal. However, oxygen is not drawn from air only as oxygen, the mass amount of air (comprised of 79% nitrogen and 21% oxygen) drawn into the combustion process is that which provides the 2.66 mass units of oxygen to combust the 1 mass unit of air. On average, it requires a mass of air 11.28 times as much as 1 mass unit of coal (containing 83% carbon, 5% hydrogen, and 2% of oxygen.) After the thermodynamic process is donc, it takes the oxygen in 11.28 mass units of air to combust 1 mass unit of coal and extract the heat of the combustion process to be used for heat energy purposes. The bookkeeping done by environmental scientists is such that the combustion of 1 mass unit of coal introduces 12.5 mass units of carbon dioxide being introduced into the atmosphere is misleading. It is more factual that only the 1 mass unit of coal transformed from solid to gaseous state that is being added to the atmosphere – the original 11.28 mass units of air (which has the 2.66 mass units of oxygen combining with the 1 mass unit of now gaseous carbon) are also being returned after being "borrowed from the atmosphere" for the combustion process. The ratio of the unit mass of air needed to combust a unit mass of coal is approximately 12 to 1 to produce the desirable heat energy.

A law of physics is that mass can neither be created nor destroyed, regardless of any change in state (solid, liquid, or gas). When considering the Laws of Conservation of Mass, the scientific fact is that 3.66 mass units of undesirable carbon dioxide gas are being introduced into the atmosphere for each mass unit of coal, not the reported 12.5 mass units of carbon dioxide for each mass unit of coal. (Other Authors of the "nay-sayer" books in the Reference and Other Documents section at the end of this book refer to the misleading mathematics being used by the environmental scientists as "cooking the books" on coal combustion.)

Therefore, in the NOAA diagram shown earlier, of the 26 billion metric tons of total carbon dioxide generated in 2000 and introduced into the atmosphere from the combustion of fossil fuels (predominantly coal), only 2 billion metric tons of coal (the fossil fuel cited transformed from solid state to gaseous state) were introduced as the net mass change into the atmosphere – the 24 billion metric tons of air "borrowed" from and then returned" to the atmosphere should not be double-counted.

It is of high importance at this point in the research of identifying the factors adding to the understanding of global warming and climate change to note that the "big picture" approach considers the full set of pros and cons of the technology being researched. The environmental science profession was sent in the direction primarily by the observation in the late 1890s by many scientists (the most notable being Dr. Svante Arrhenius in Sweden who was awarded the Nobel Prize for Chemistry in 1903) pursuing the chemical effects of the emissions of burning coal on the atmospheric environment.

The preceding sections of this chapter address the thermodynamics of the process of combustion of coal and the gaseous and thermal emissions created in the process. The brilliant research done on the heat-trapping characteristic of the carbon dioxide generated by the combustion of the fossil fuel coal – in a scientific laboratory and starting with only one of the products of the combustion process - diverted environmental scientists away from considering the physical residue of the combustion process that produced the carbon dioxide gas being researched in infinite detail for its heat-trapping "greenhouse gas" effect. In the "big picture" system engineering analysis (which considers the total of the pros and cons of the entire system not merely one important characteristic from scientific observation), the ash residue created from the combustion of coal is a con that has been largely overlooked. It is the disposal of that solid state coal ash residue which contains toxins and carcinogens depending on the type and quality of coal being combusted. Coal is categorized in essentially three basic types: anthracite, lignite, and bituminous, each having a different composition of carbon and other elements and impurities.

A later chapter on the fossil fuels – coal, oil, and natural gas – provides further detail on the ash residue being a pollutant of land and water entering the water and food products consumed by humans and livestock animals. In the chapter near the end of the book in which the complete set of pros and cons of 25 technology areas are assessed and prioritized (with the use of weighted trade studies), it concludes that coal technology should be reduced and eventually eliminated in the United States region for the primary reason of its ash residue being a very hazardous (toxic and carcinogenic) polluting material to human and animal health and safety. In comparative ratings, it is the pollution of land and water not that of the carbon dioxide "greenhouse gas" being emitted into the atmosphere (and having a factually small, if any, effect on global warming and climate change) that is of highest significance.

The NOAA comparison chart (Fossil Fuels versus Volcanic Emissions) also shows that the constant average mass units of volcanic eruption generated carbon dioxide at a total rate shown in the chart as less than 1 billion metric tons per year generated in volcanic eruptions and introduced into the atmosphere. That low rate was constant during the period from 1750 to 2000. The oxygen contained in the carbon dioxide ejected from the volcanic underground source (i.e., the oxygen in volcanic ejecta <u>not</u> being drawn from the atmosphere) is already combined with the carbon element both drawn from within the inner regions of the planet – such that 100% of the mass of carbon dioxide ejected during volcanic eruptions is introduced into the atmosphere as a transformation of carbon in solid state form to gaseous state form without any change in its mass during the eruption process.

It was earlier shown graphically that the volcanic eruption database for the different levels of VEI category of World-wide volcanic eruptions during the period 1960 to 2023 (after "averaging") was increasing in frequency – a higher frequency average for the lower VEI eruptions and a singularly frequency average for the highest two VEIs (VEI 5 and 6). The ejecta of each eruption varies exponentially as the VEI index increases (approximately as shown in the earlier tabulation of the Volcanic Explosivity Index developed in 1982.

The Author has made an isolated data extraction of the volume of ejecta associated with the differing levels of Volcanic Explosivity Index (VEI) as shown as follows:

Volume of Ejecta as VEI Increases

(Newhall and Self, 1982)

Volcanic Explosivity Index	Volume of Ejecta (Observable)				
(Newhall & Self)			Other equivalences (approx)		
VEI	meters^3		kilometers^3	miles^3	yards^3
0	$< 10^2$		< 1	< 1/4	$<10^2$
1	10^2-10^4		1 to 10	1 to 3	10^2-10^4
2	10^4-10^6		10^2 - 10^3	10 - 10^3	10^4-10^6
3	10^6-10^8		10^3-10^5	10^3-10^5	10^6-10^7
4	10^8-10^9		10^5-10^6	10^5-10^6	10^7-10^8
5	10^9-10^{10}		10^6-10^7	10^6-10^7	10^8-10^9
6	10^{10}-10^{11}		10^7-10^8	10^7-10^8	10^9-10^{10}
7	10^{11}-10^{12}		10^8-10^9	10^8-10^9	10^{10}-10^{11}

Note: The ejecta in each eruption are a varying amount of solids (from particulate to boulder size), liquid (molten rock (lava, at extremely hot temperature) and various compounds of minerals/elements), and gasses (including water vapor in the form of steam.)

(Source: The National Geological Survey database reformatted by Author)

It should be noted that the small integer exponents of 10 shown in the table above reflect a multiplier of the volume of ejecta – i.e., each volcanic eruption of a VEI 5 category ejects 10^9 (1,000,000,000) to 10^{10} (10,000,000,000) cubic meters of ejecta (possibly containing only a low percent of carbon dioxide) into the atmosphere. 1 billion (10^9) cubic meters of ejecta (solid, liquid, and gas) introduced into the atmosphere by one VEI 5 volcanic eruption is far more hazardous to humans and the environment than the 13 billion tons of carbon dioxide gas (at a concentration of 435 parts per million (0.0435%) from the fossil fuels indicated in the NOAA chart (shown earlier of Fossil Fuels versus Volcanic Emissions).

Readers should take note of the decreasing frequency of volcanic eruptions as the various levels of the Volcanic Explosivity Index (VEI) increases. To be noted also is that the estimate of the volume of ejecta increases exponentially at the different VEI levels. By multiplying the number of volcanic eruption occurrences on average per year of the different levels of VEI severity by the estimated volume of ejecta associated with the different VEI categories, a simple mathematical calculation provides a high probability estimate of the total volume of ejecta that may be produced in world-wide volcanic eruptions for that year. For the year 2030, that volume is estimated with the calculation:
32 eruptions at VEI 0,1,2 at 10^4 m^3 ejecta volume = 320,000 m^3 of subtotal ejecta;
+ 5 eruptions at VEI 3 at 10^7 m^3 ejecta volume = 50,000,000 m^3 of subtotal ejecta;
+ 1 eruption at VEI 4 at 10^8 m^3 ejecta volume = 100,000,000 m^3 of subtotal ejecta;
+ 1 eruption at VEI 5 at 10^9 m^3 ejecta volume = 1,000,000,000 m^3 of subtotal ejecta
The total ejecta calculated for 2030 = 1,150,320,000 cubic meters. To attach some relative size to that large a volume, since a kilometer is 1,000 meters, a cubic kilometer is 1,000,000,000 cubic meters and the total ejecta for 2030 is estimated to be 1.15 cubic kilometers. In those countries (including the USA) where distances are measured in miles rather than kilometers, there is 0.621 miles in every kilometer. In a cubic kilometer, there will be 0.239 cubic miles. The total estimated ejecta for volcanic eruptions producing 1.15 cubic kilometers equates to an ejecta volume of 0.239 cubic miles.

Those 39 eruptions at the various VEI levels produces a high probability (over 75%) of there being a total volume of ejecta of approximately 1,150,320,000 (1.15 x 10 9) cubic meters equivalent to 1.15 billion cubic meters of high temperature, highly harmful to humans particulate, in solid, liquid/molten, and gaseous state (all heat-trapping worse than is the equivalent volume of carbon dioxide) being introduced into the atmosphere of Earth each year from 2024 to 2030. The "one-size-fits-all" human control of "greenhouse gas emissions" to net zero by 2030 is comparatively a futile and unrealistic goal with high probability of not being achieved and does not alter the volcanic eruption of ejecta estimate of 1.15 x 10 9 cubic meters per year of more ominous volcanic ejecta. The significant but not totally off-setting difference is that the ejecta heavier mass from volcanic eruptions will eventually fall to the Earth's surface whereas the gaseous carbon dioxide (natural and Man-made) remains in the atmosphere for several years.

It is those volcanic eruptions world-wide that are and will be affecting the atmosphere with heat-trapping solids/liquids/gases in total (despite their low frequency rate but considering their massive total volume of heat-trappers and blockers ejected) which actually discounts – but does not totally negate - the environmental scientists' hypothesis that Man burning fossil fuels creates greenhouse gas emissions and is the dominant cause for global warming.

It should be obvious that environmental scientists extrapolating what they observed taking place with a beaker of carbon dioxide at high concentration being heated by a heat-source in a laboratory had leapt to the fallacious conclusion that it is Man causing global warming. That is not what is happening in the very real natural world. It would be most valuable to the public if organizations such as NASA, NOAA, USNGS, and the entire environmental science profession were to pool their talents and resources and use all pieces of the many databases to identify the true scientific problem of the multiple causes of global warming, climate change, and ocean levels rising and their respective effects on the environment (atmosphere, ocean, and land) and/or on the human population. This book is a step (but not the last word) toward that objective to better understand the factual causes for global warming and climate change.

There remains another important "flaw" in the volcanic eruption data shown by NOAA, that "flaw" being that the number of average volcanic eruptions of volcanoes on the planet (reported in the Smithsonian Institute Global Volcanism Program Database and graphically shown earlier) was 57 eruptions per year average in 1960 and steadily increasing to 80 eruptions per year average in 2020. Yet, that 41% increase in eruption average rate in the period from 1960 to 2020 reported by the Smithsonian Institute Global Volcanism Program Database is not reflected on the NOAA volcanic carbon dioxide diagram shown above (which shows no increase in carbon dioxide into the atmosphere from volcanic eruptions in the period 1750 to 2000.

The originator of the chart however seems to have disregarded the fact that volcanic eruptions may not be very frequent but when a volcano erupts and expels so much ejecta into the atmosphere in the form of carbon dioxide, there is an enormous volume of other solids/liquids/gases that are introduced into the atmosphere and having far greater heat-blocking and heat-trapping properties than that of only the carbon dioxide generated by Man burning fossil fuels.

Perhaps it is not statistically valid to compare fatalities from "natural catastrophes" to fatalities from "Man-made catastrophes" – as in comparing "apples to oranges". The National Highway Transportation Safety Administration (NHTSA) reports that there are over 40,000 humans fatally injured in Man-caused automobile accidents annually in the United States alone. For the 60 years period 1963 to 2023, there were over approximately 2,400,000 humans fatally injured in Man-caused automobile crashes on the US highway network of roads. There could not be found any comparison data for the number of human fatalities directly attributed to Man-made "greenhouse gas emissions". Deaths reported to be due to heat-stroke (hyperthermia) were found to approach perhaps 1% that of the NHTSA fatality rate. According to the Center for Disease Control (CDC) records of deaths reported due to hypothermia (extreme cold) and hyperthermia (extreme heat) those due to hypothermia (extreme cold) are significantly higher – at a ratio of about nine to one for victims with otherwise normally healthy respiratory and cardiovascular systems.

Is using NHTSA highway data and Department of Energy "greenhouse gas" data a fair "apples to apples" (Man-caused to Man-caused) statistical comparison? If so, then the technology advancement made by environmental scientists to reduce motor vehicle emissions to net zero by replacing fossil-fueled vehicles with all-electric vehicles by 2050 should prove to be of utmost importance to the US and world transportation system and to the human population (minus those

2.4 million humans who lost their lives in vehicle crashes between 1963 and 2023). The oft referred to DOE chart reporting atmospheric carbon dioxide concentration is shown next:

TABLE 1.
The Important Greenhouse Gases (except water vapor)
U.S. Department of Energy. (October. 2000) (1)

(all concentrations expressed in parts per billion)	Pre-industrial baseline	Natural additions	Man-made additions	Total (ppb) Concentration	Percent of Total
Carbon Dioxide (CO2)	288,000	68,520	11,880 (2)	368,400	**99.438%**
Methane (CH4)	848	577	320	1,745	**0.471%**
Nitrous Oxide (N2O)	285	12	15	312	**0.084%**
Misc. gases (CFC's, etc.)	25	0	2	27	**0.007%**
Tota	289.158	69.109	12.217	370.484	100.00%

(Source: US Department of Energy)

In its zeal to "educate" the public about the importance of the "greenhouse gas effect" (especially that of carbon dioxide) on the climate, environmental scientists in the DOE present the concentration of the greenhouse gases in units of parts per billion instead of the conventional NOAA units of parts per million. The public is well familiar with the per cent (in effect parts per hundred) system of units (as in what percentage of their income is taxable), but unfamiliar with the scientific significance of parts per million or parts per billion. The DOE tabulation is misleading (whether intentional or not) as the mix of per cent (showing carbon dioxide as 99.438% of the "greenhouse gas" concentration) and parts per billion (shown as 368,400 ppb for total carbon dioxide in the atmosphere) mask the reality of there being only a concentration of 0.03684% of total carbon dioxide in the atmosphere (accurately measured by NOAA). Of that total of 0.03684%, the Man-made portion of carbon dioxide in the atmosphere is 0.0073% (i.e., excluding the four times larger 0.0295% as the natural-caused (mostly from volcanic eruptions) carbon dioxide that existed in the atmosphere prior to the start of the Industrial Revolution).

Reducing the Man-caused "greenhouse gas emissions" to net zero by 2050 (set as the "one size fits all") goal by the UN Intergovernmental Panel on Climate Change IPCC) does not reduce the much larger portion of natural-caused carbon dioxide in the atmosphere. The recent 2024 level of 435 ppm (0.0435%) concentration of carbon dioxide in the atmosphere, very accurately measured by NOAA can, even if the IPCC goal of "net zero by 2050" is 100% achieved, at best be reduced to the pre-Industrial Revolution level of 295 ppm (0.0295%).

A not-so-humorous question that is overheard when professional database analysts (scientists and engineers) analyze a complex set of data for root cause (by the process as done for accidents which resulted in multiple human fatalities): "Are we choking on the minnows, while the whale is swallowing us?"

Numbers and percentages are difficult to interpret and when trying to explain a composition of anything with four or more "chunks" in its composition, the use of a common "pie chart" is helpful – everyone wants a big slice of Mom's homemade apple pie. Shown next is a pie chart showing how the atmosphere (without water vapor) consists of its major gases:

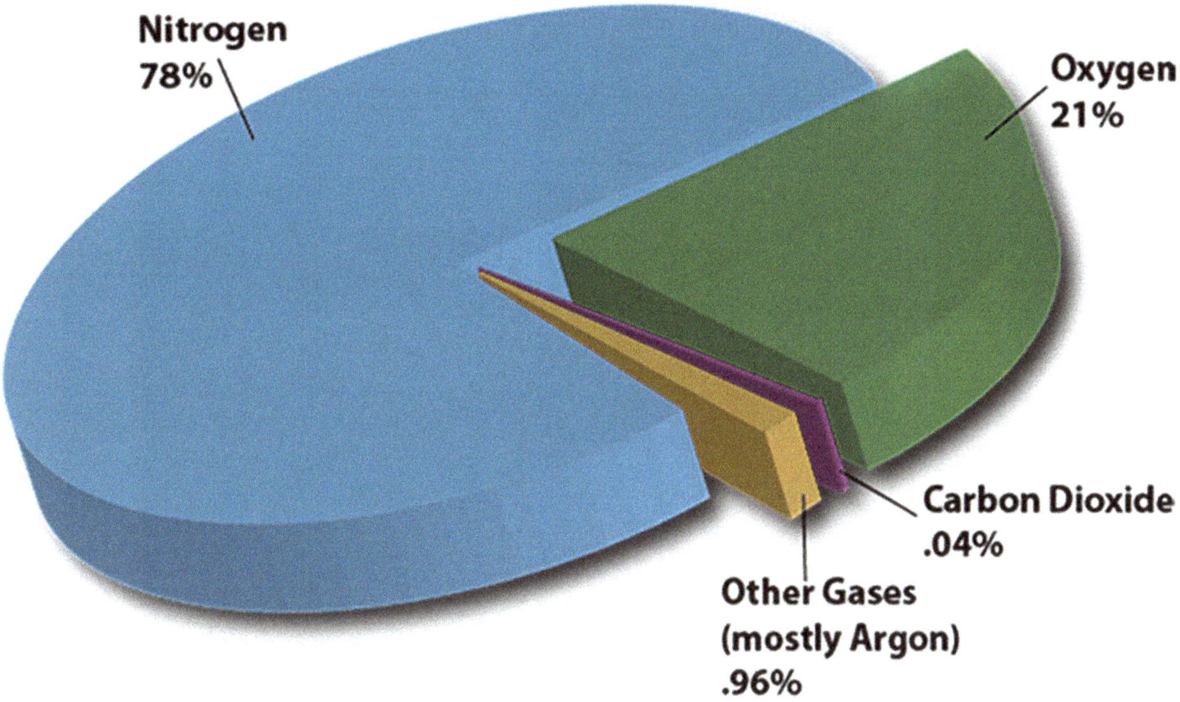

(Source: High School Chemistry Textbooks)

To be noted is the small purple slice of the pie represents the total amount of carbon dioxide in the atmosphere – shown as being 0.04% (equivalent to 400 parts per million or 40,000 parts per billion). Environmental scientists seldom show this form of presentation as it visually shows just how small the amount of carbon dioxide is measured to be as the dominant "greenhouse gas". It is explained throughout this book that of that total 0.04% (the purple sliver of the pie) of carbon dioxide in the atmosphere, there is a natural-caused portion of almost 0.03% (295 parts per million) and very slightly over 0.01% (140 parts per million) which at most is Man-caused. To put that in a better visual perspective, it equates to slicing that small sliver in purple (representing 0.04% of the 100% total pie) in half and then visualizing one of those half-slivers to be the amount of Man-made carbon dioxide in the atmosphere causing a "crisis" level of concern in the environmental science fraternity. It could be acceptable reasoning that the 0.02% of Man-made carbon dioxide was a neurotoxin which is fatal in a concentration of a few drops in a large reservoir. However, carbon dioxide is not a neurotoxin – carbon dioxide is not as effective in trapping heat as is water vapor at an equal concentration. Water vapor in the real atmosphere can be up to 4% (40,000 parts per million or 4 parts per hundred).

Each volcano also has a "personality" that defies being modeled that even the most astute environmental scientist could ever hope to achieve. The volcanic activity in Iceland – a "next door neighbor" of Greenland - tends to be more horizontal than vertical as the lava "oozes" in parallel with the surrounding topography, as opposed to the ejecta being explosively thrown to high altitude providing adequate time for residents in the area to evacuate from "harm's way". Some earthquakes are known "repeaters" (such as the Kilauea volcano on Hawaii) and while it affects the surrounding terrain and ecosystem, the lava flow oozing at a relatively slow speed poses little threat to human life. There may be some steam emitted also but relatively low toxic or climate-affecting gases in the "greenhouse gas" category.

(Source: The US National Geological Survey)

Other volcanic eruptions produce very little, if any, flow of lava, but predominantly emit gas (toxic and non-toxic) and volcanic particulate matter resulting from the eruption. The most recent volcanic event that was a "gasser" was the eruptions of the Eyjafjallajökull Volcano in Iceland which lasted principally from March 20, 2010, to June 23, 2010. Volcanologists assigned a Volcanic Explosive Index of VEI 4 to the main event. Shown next is a photograph (by a land-based photographer) showing the prolific efflux of gaseous and particulate matter without any lava flow:

(Source: National Geographic Society photographic database)

Beginning on April 14, 2010, Eyjafjallajökull created an ash cloud that led to the closure of most of the European IFR (Instrument Flight Rules) airspace which caused about 20 countries to close their airspace to commercial jet traffic, and which affected approximately 10 million air travelers. The commercial jet flights were cancelled for two reasons; that of safety of flight in/through the dense gas/particulate plume, and because the particulate matter when ingested into jet engines damages the turbine blades and could lead to catastrophic results from engine shutdowns or engine damage beyond easy repair.

The Icelandic Meteorological Office (IMO) and the Institute of Earth Sciences (IES) of the University of Iceland estimate the in the ash plume, which reached to about 30,000 feet (9 kilometers) contained 250 million cubic meters (330,000,000 cubic yards) of ejected tephra. Tephra contains fine carborundum grit and glass-like particles which are abrasive to mechanical moving parts but more dangerous to the respiratory system of humans, especially those with already diseased respiratory systems. The extent of the ash plume covering all of Europe and much of the western part of Asia was mapped by the London Volcanic Ash Advisory Centre (VAAC), a part of the United Kingdom Meteorology Office is awesome in its scope and length of hazard presented to the Northern hemisphere as shown on the composite map spanning the period April 14 to 25, 2010:

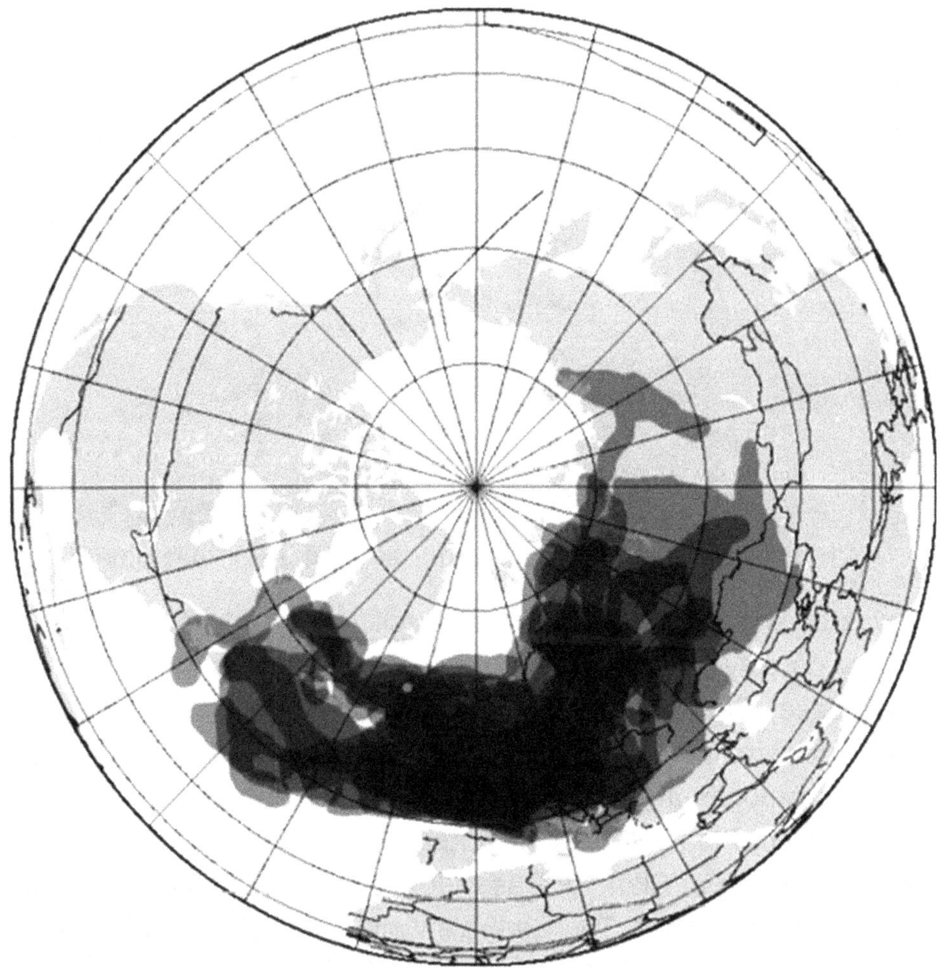

(Credit to the London Volcanic Ash Advisory Centre (VAAC), London Met Office)

Volcanic ash in the atmosphere may be disregarded by the US Environmental Protection Agency as insignificant, but apparently the London Meteorological Office in the United Kingdom has seen it to be more significant and to establish the London Volcanic Ash Advisory Centre (VAAC). The above map supports this Author's observation that weather-related and other natural catastrophic events are important to whatever they have been, are, and will be to the regions where they occur. The London Meteorological Office has the Volcanic Ash Advisory Centre whereas the US National Weather Service (part of NOAA) has the Hurricane Tracking Center and the Tornado Warning Center. They all have some effect on the global surface temperature and therefore on the global climate, yet the United Nations Intergovernmental Panel on Climate Change has representatives only from climate science universities and organizations – and "green" activists' groups.

The most recent major volcanic eruption, causing over 350 human fatalities, was of the Philippines Mount Pinatubo near Luzon in 1991. It was rated at a Volcanic Explosivity Index of 5 which includes an estimate of the ejecta to be in the range of 10^9 to 10^{10} cubic meters. An excellent photograph is shown next indicating the massive pressure that caused the Mount Pinatubo eruption to eject that much particulate (heavy) matter into the atmosphere to an elevation of about 65,000 feet (about 20 kilometers):

(Credit to Unknown Photographer at Clark AB, Philippines)

The above photo was taken from Clark Air Base 15 miles from Mt. Pinatubo and by an unknown photographer possibly on duty with the US Air Force at the vacated base (note that there are no aircraft visible on the base). The severe cumulus cloud formation seen in the photo at the base of the erupting Mt. Pinatubo were in the advancing front of Typhoon *Yunya*. It was a sheer coincidence of nature that a volcanic eruption and a typhoon striking a region would occur within a day of each other. The eruption of Mount Pinatubo deposited heavy ash up to two feet thick on the rooftops of buildings (and roofs of hangars at Clark Air Base) downwind and as far away from the base of Mt. Pinatubo as 20 miles to the northeast. There were initially no reports of damage to any structure or residences caused only by the thick but relatively low-density sponge-like ash deposits from the eruption.

However, compounding the ash deposit problem was the fact that Typhoon *Yunya* (Category 3) inundated the Luzon Peninsula the next day that Mt. Pinatubo erupted (June 15, 1991). The heavy rainfall from Typhoon *Yunya* soaked into the spongy thick ash from the Mt. Pinatubo eruption (much like a dry sponge absorbs water) on many rooftops and collapsed their roofs from the excessive weight of the "mud". 350 people were killed in the 2,013 homes and businesses that were damaged (and mostly under roofs that had collapsed).

[Author's Note: The Author was the US Air Force Chief System Engineer for development of the high technology F-16 Fighting Falcon weapon system. In October 1989 (less than two years before the Mt. Pinatubo eruption) the Author was on an official visit to Clark Air Base in the Philippines located less than 15 miles from the base of Mt. Pinatubo. A squadron of F-16s and a squadron of F-15s were based at Clark Air Base at that time. When the approaching Typhoon *Yunya* storm path initially threatened the Clark Air Base aircraft, the F-16 (and F-15) squadrons were evacuated from Clark Air Base before the typhoon was estimated to arrive and relocated to Kadena Air Base on Okinawa, Japan. The day before *Yunya* arrived, Mt. Pinatubo erupted and Clark Air Base was impacted very heavily by the combination of rain/volcanic ash "mud" that kept gaining weight as the Typhoon *Yunya* slowly passed over. The metal roofs of many buildings (including all the F-16 and F-15 maintenance hangars) eventually collapsed almost simultaneously as shown in the next photograph:

(Source: Author's personal file; credit to unknown USAF photographer)

There are seen no aircraft (either US military operated or Philippine commercial) visible on the base as they had all been pre-evacuated safely when the alert of approaching Typhoon *Yunya* was given 4 days prior to the estimated time of *Yunya's* arrival. The USAF abandoned Clark Air Base, and it remained out-of-use for more than five years until the Philippine Government cleared away all the ash and buildings debris from the Mt. Pinatubo/Typhoon *Yunya* "perfect storm" catastrophe and rebuilt new facilities. The lack of supply of reconstruction materials severely affected the rebuilding supply and caused the US Air Force to cease operations permanently at Clark Air Base. Clark Airfield was reopened (in approximately 2005) and is now used as a commercial cargo aircraft facility serving the Manilla area.]

The amount of volcanic ash from the Mt. Pinatubo eruption on nearby farmlands severely affected the local ecosystem and aggravated the homeless problem with lack of grown food and supply of clean water for years after the 1991 eruption.

The most logical recommendation that can be made for future generations to heed is to not create large communities of the population near the base of a known volcano (active or inactive). A photograph showing the hazards of doing so is shown next:

(Source: Photo by Author on Cruise Ship Approaching Port of Naples, Italy)

The volcano in the background of the above photograph is Mount Vesuvius which is legendary in volcanic eruption history. Mt. Vesuvius, (located six miles from the city of Naples in the Bay of Naples), has had eight major eruptions in the past 17,000 years. The largest Vesuvius eruption on record was in 79 A.D. when it erupted burying Herculaneum (under 66 feet of pyroclastic flow and ash) and Pompeii (under more than 20 feet of ash) – with an estimated 16,000 fatalities. Vesuvius erupted again in 1906 with about 150 casualties, and again in 1944 burying much military equipment in World War II but with no known fatalities. When the Author visited that region (to Pompeii in 2011), there was seen a small amount of wispiness (grey-white smoke) being emitted from the Mt. Vesuvius cone but that is apparently not unusual.

The most catastrophic volcanic eruption (estimated to be a magnitude 8.0) was that in 1556 of the Shaanxi volcanoes in northern China causing death to over 100,000 humans and major crushing injuries to over 730,000 others. The enormous losses and injuries resulted from the impoverished population living in caves of a soft sandstone cliff that collapsed while those killed or injured were asleep in the caves. Those living in a hazardous area do so for a number of differing reasons, but all too often are unaware of what lies beneath the region where they live. The opposite of that was seen by the Author while on that cruise ship which provided the opportunity to take a photograph of the Bay of Naples (shown above). As shown, the residences in the Naples area were built to withstand weather severities which are minimal in that latitude and not being on a major ocean, taking the risk that the volcano looming above them will not erupt during their lifetime. Other communities build on the rim of a volcanic caldera for the spectacular scenery in that caldera. On that same cruise and a few days after departing the Bay of Naples, passengers were treated to a sight that photographs cannot do full attribution to their unusual and awesome scenery. Shown next are three photographs of the region of the Greek islands in the Aegean Sea in the region of the Santorini (also known as Thera) volcano which erupted in circa 1600 BCE. The first is a satellite image showing the immensity of the caldera:

(Source: NASA)

For reference, the distance from Santorini to Therasia is approximately 1 mile. Cruise ships (as well as all other smaller vessels) enter the Santorini caldera at the northwestern channel, and immediately see the spectacular colors (a close second rival to those in the Grand Canyon in Arizona) of the inside walls of the caldera:

(Source: Photograph by Author, from a Princess Cruise Line ship)

To avail themselves and the many annual tourists to the island of Santorini to the entire vista of the large Santorini caldera, the local population have erected residences and commercial buildings on the top of the caldera rim shown as follows:

(Source: Princess Cruise Lines Brochure given to Ship Passengers)

The all-white structures atop the Santorini caldera rim are noted to be whitewashed a gleaming white (with olive oil content to resist the weather elements) but mostly so as to not conflict with the natural varying colors of the caldera. Churches have a classic round dome painted the blue color of Greece and which matches the cobalt blue color of the water inside the caldera. The round domes of the churches and some other structures are shaped other than flat in the event of a small eruption of the Santorini volcano causing accumulation of the ash ejecta to not collapse roofs onto inhabitants.

Closer to home, since 1700 there have been 635 recorded volcanic eruptions with 215,768 major direct human injuries and fatalities and at least 62,600 indirect (from famine and disease) fatalities caused by those eruptions in Alaska (mostly in the Aleutian Islands), Hawaii, and the original "lower 48" United States. Their locations are graphically summarized next, and can be seen to be mostly on the Pacific "Rim of Fire":

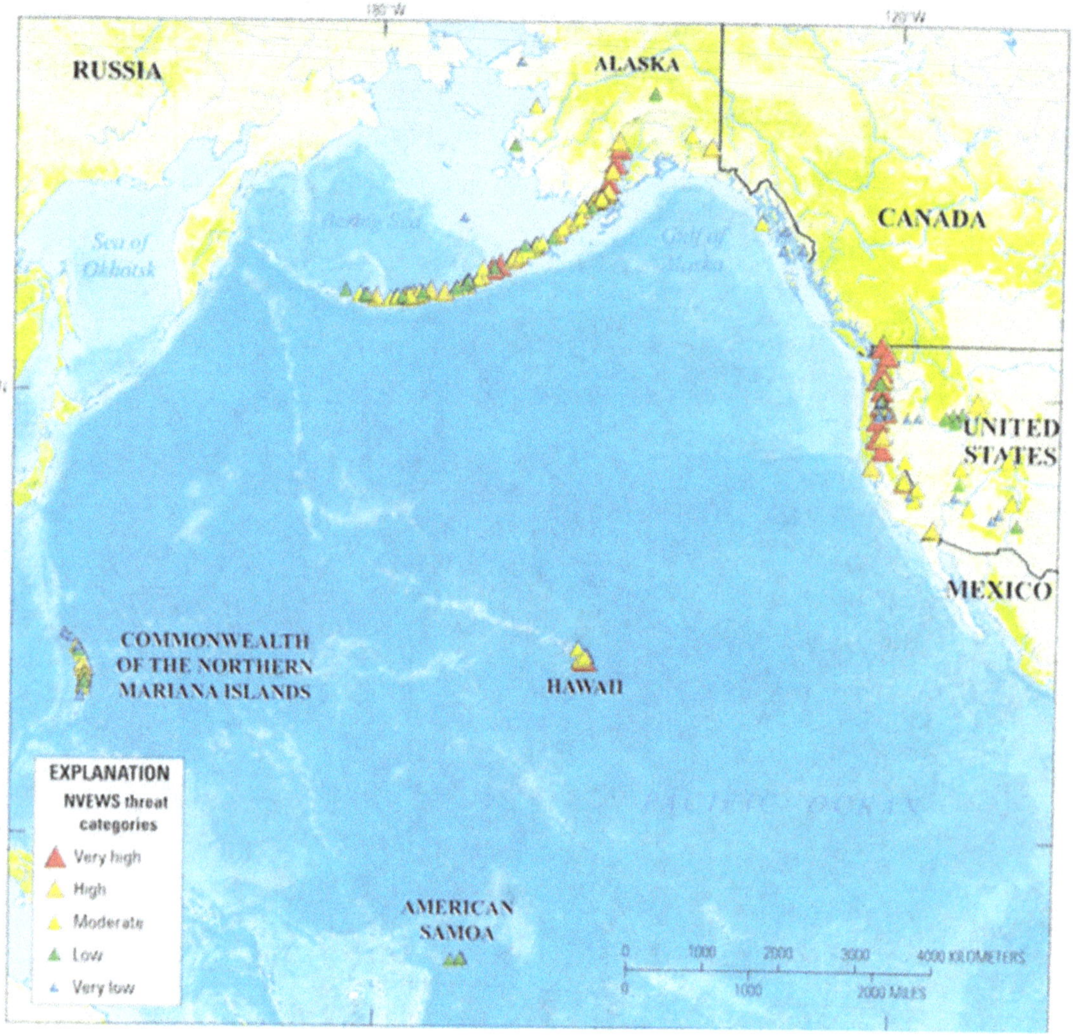

(Source: US National Geological Survey)

There are 205 named volcanoes located around the "Rim of Fire" that are of prime interest to the United States. Those 205 volcanoes are distributed (clockwise in the diagram above) with Alaska (Aleutian Islands) having 101 named volcanoes – 39 erupting since 1700; the 48 United States on the "mainland" having 161 named volcanoes – 6 erupting since 1700 (3 in Washington, 2 in California, and 1 in Oregon); and Hawaii having 25 named volcanoes – 39 erupting since 1700. Of those eruptions, it was shown earlier that the eruption of Mt. St. Helens in 1980 directly caused 57 human fatalities, making that eruption the deadliest in the history of the USA.

The near-term outlook for volcanic eruptions being a major concern (i.e., a crisis) to residents in the populated regions of the United States is concluded to be of a cautionary level only to those residents in the states of Washington, Oregon, California, and Hawaii. The public should be aware though that there is one area in the United States where a long-term outlook for volcanic eruption of cataclysmic proportion to occur cannot, nor should, be dismissed completely. The are a wide range of estimates of the area that would be devastated physically by a major eruption of one or more of the extinct volcanoes below the surface, but the area of the United States would likely have an apocalyptic effect on ecology and population of the Western third of the United States. That large eruption would create its own weather and create havoc to the land area and atmosphere within 600-700 miles (or more depending on winds speed and direction) distance from the center of the erupting super-volcano.

That very active geothermal area is in the region of northwestern Wyoming designated as the Yellowstone National Park. Yellowstone National Park is the first National Park in the US Park System., created in 1872 by then-President Ulysses S. Grant. A diagram of the Yellowstone National Park region, and showing the major calderas contained within Yellowstone Park, is provided next:

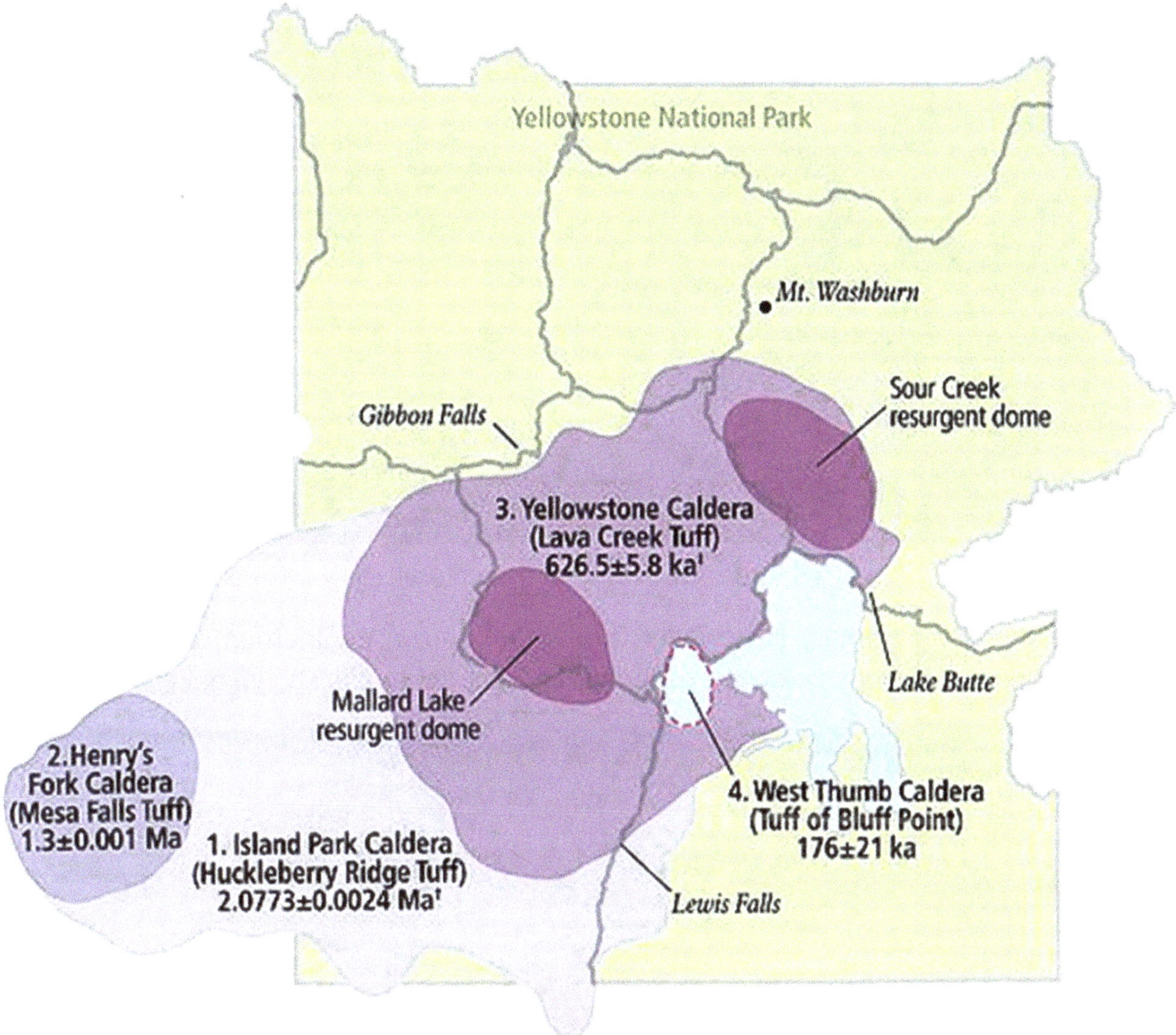

(Source: US National Park Service)

Yellowstone Park is approximately of rectangular shape, about 63 miles north to south, and 54 miles east to west. Its area is approximately 2.2 million acres (3,438 square miles). On the surface, above four major calderas, Yellowstone Park has more natural beauty than most of the US National Parks, one of which is that below showing the Yellowstone River meandering through the Hayden Meadow:

(Source: US National Park Service)

Underneath that natural scenery is however the largest known geothermal area of the world. A diagram indicating the geology below the Yellowstone Park caldera is shown next:

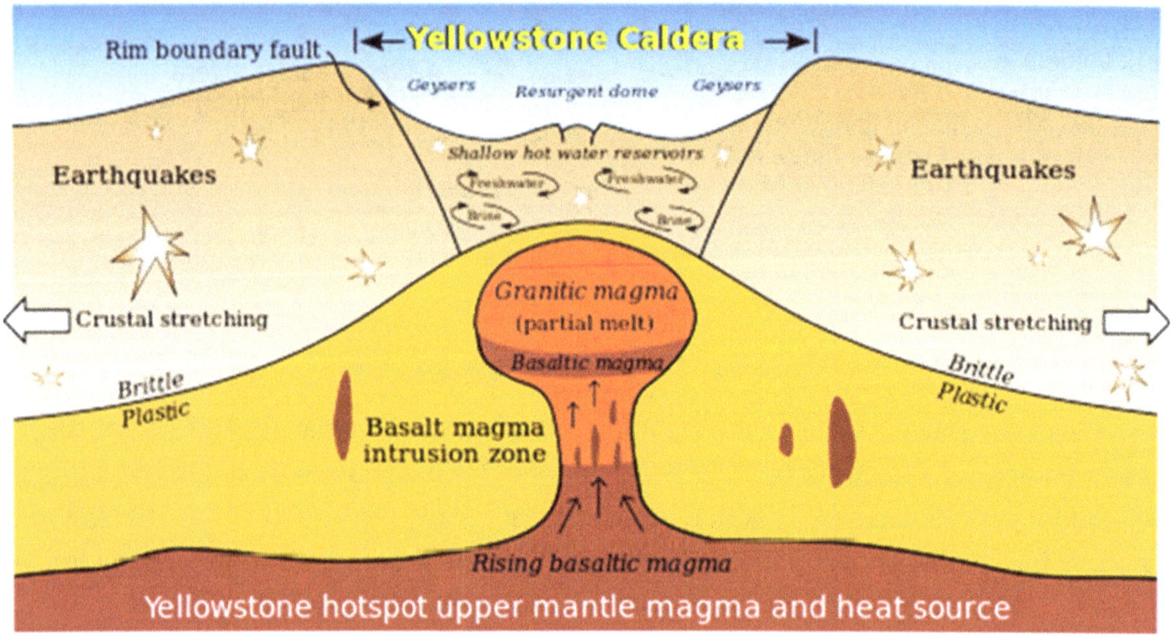

(Source: US National Park Service)

Evidence of the geothermal energy is seen at two of the most visited regions of Yellowstone Park – the first at the wondrous Old Faithful geyser which spouts water and steam into the air very regularly (approximately every 90 minutes between spouts and for a duration of about 90 seconds each spouting) as shown next):

(Source: National Park Service)

The second and by far more colorful scene is that of the Geyser Basins, one of which is shown as follows:

(Source: US National Park Service)

The water temperature in these geothermal features is near boiling or already in the form of steam above boiling (100ºC or 212ºF). As they pose no hazard to the environment (more than they obviously have/do) or to humanity, they possess the geothermal potential for alternative energy sources to the fossil fuels and to the solar and wind power sources which are cleaner than fossil fuels but are not as reliable as geothermal energy has the potential to be. The use of geothermal energy will be discussed in a later chapter dealing with fossil fuels, current natural sources, and other clean, reliable, future forms of alternative energy that should be developed as replacement of coal as the primary energy source in the twentieth century.

It is also very common that where there are volcanic and geothermal geophysical formations, there typically is seismic wave activity preceding volcanic activity. Nearby population areas are alerted to the increased probability of earthquake occurrence doing damage from the earth movement (being at the edges of two tectonic plates which spawn both volcanic eruptions and/or significant earthquakes). That leads to the second major geophysical threat primarily to humans (collapsing their residences on top of them) and with lesser consequences on the environment that results from tectonic plate movement – earthquakes. A diagram indicating the known earthquake-prone locations (along the peripheries of tectonic plates) everywhere in the world is shown next:

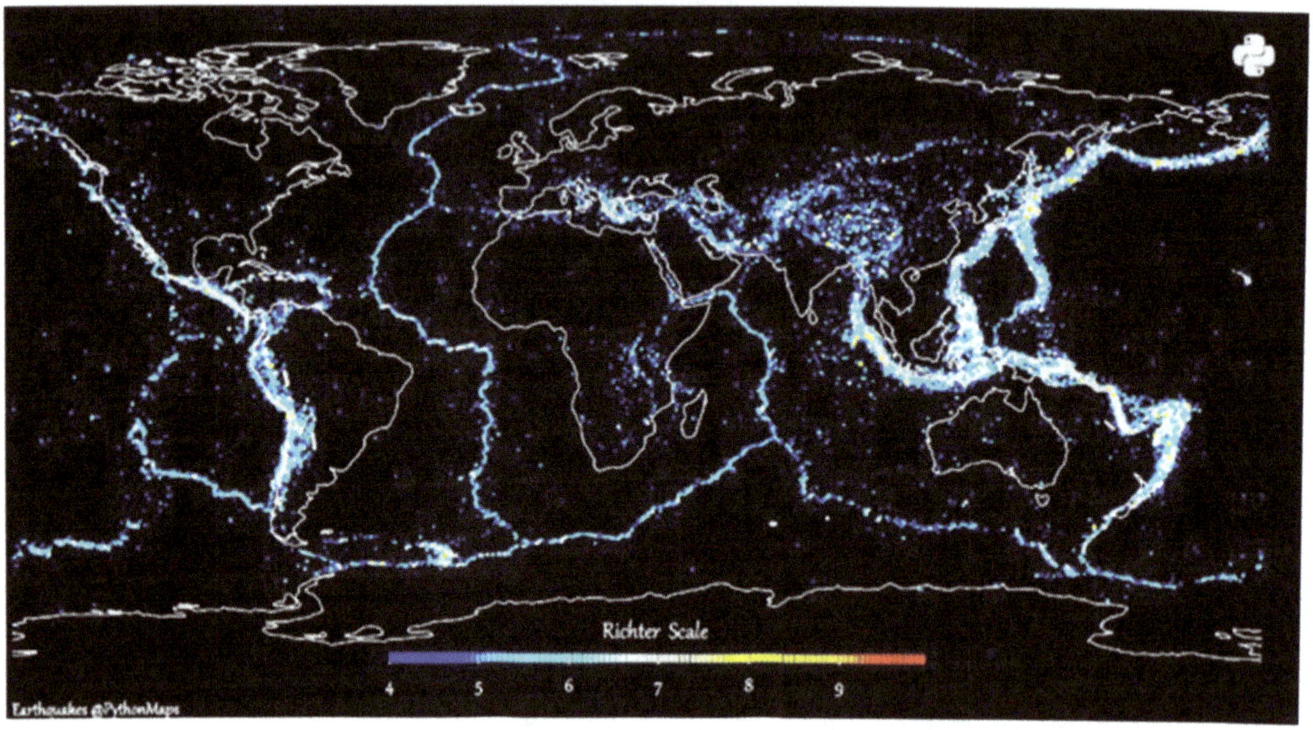

(Source: Python Maps for the US National Geological Survey)

It should be noted the similarity between volcanoes locations and earthquake epicenter locations as they are both most common to occur on the edges of tectonic plates (as shown on earlier charts).

The chart shown next has been created from the US National Geological Survey database (in listed form) of the frequency of earthquakes globally and the average occurrences per year at each of the magnitude levels of interest in this book.

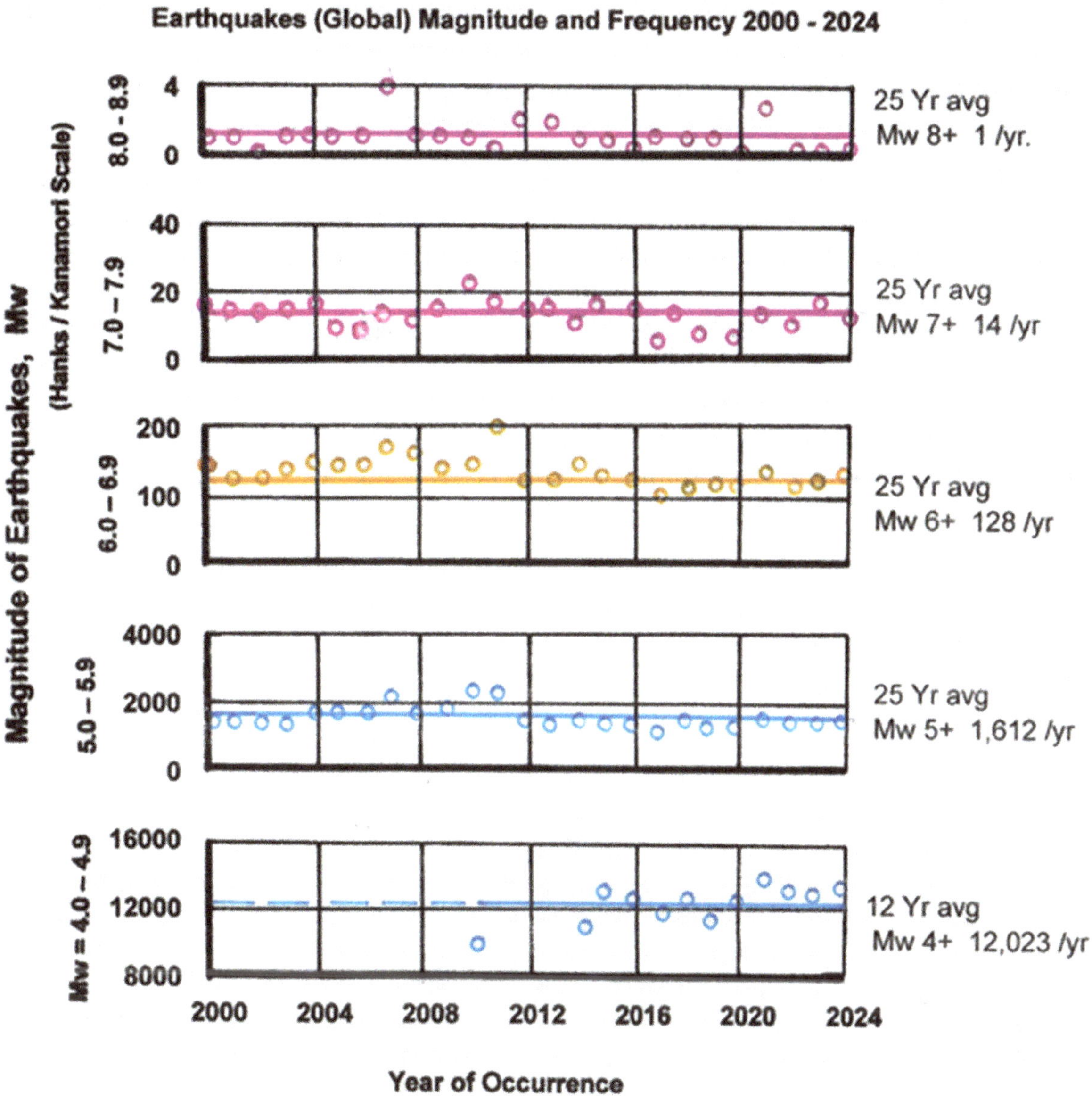

Note: Earthquakes at/above 9 on the Mw scale are too sporadic to be graphically analyzed. There is high probability that a Mw 9+ earthquake will occur every 10 – 12 years.

(Source: Graphical portrayal (by Author) of US NGS tabular database)

It is seen that there is amazingly consistent statistical data for each of the Mw momentum scale (created by Hanks and Kanamori in 1979 to replace the lesser accurate Richter Scale) levels between Mw 4.0 and 8.9. The level Mw 4+ and 5+ are colored blue as there is commonly little damage done and few lives lost at those levels. When levels above Mw 4 in a volcanic existence area are measured (by seismic sensors and recorders) to be rapidly increasing in occurrence, there is high probability that the nearest volcano has the increased potential for erupting. When an earthquake is measured to be in the Mw 6.0 to 6.9 range (colored yellow on the chart), this is considered by many to be the threshold for damage to residential and commercial structures and very dependent on the location of the epicenter relative to populated areas.

It is the Mw categories of 7.0 and above (color coded in red on the chart) that is shown in historical records to be the most damaging and most injurious to humans and animals. Somewhat mercifully, the frequency of the "really bad ones" reduces as the magnitude of Mw increases. As shown on the chart, there is on average 14 per year of the Mw 7.0 to 7.9 category earthquakes occurring globally, and for the category 8.0 to 8.9, there is an average of only 1 per year. The monstrous category Mw 9.0 and above (not shown graphically as the note at the bottom of the chart describes) are sporadic in occurrence and strike only once somewhere on the planet every 10 to 12 years.

The magnitude of an earthquake is only one measure of the damage and effect on virtually everything in its area of influence. Unlike volcanic eruptions which spew lava on the surrounding environment and ejecting enormous amounts of particulate matter and gases into the atmosphere, the mechanism of earthquakes is more physical dynamic and spreads in seismic wavelengths that causes the earth to shake violently. Another important variable is the depth at which the earthquake epicenter occurs and its location relative to a body of water above or nearly above it which creates enormous waves in the body of water named as *tsunamis*. (Tsunami is a combination word from Japanese – "*tsu*" meaning "harbor" and "*nami*" meaning "waves"; prior to the use of the word tsunami, reference was made to the earthquake-generated waves as "tidal waves"). An excellent graphic illustrating the tsunami long-distance side-effect perhaps being worse than the physical effect locally of the earthquake mechanical devastation is shown next (resulting from the largest earthquake in recorded history at Valdivia, Chile, in 1960):

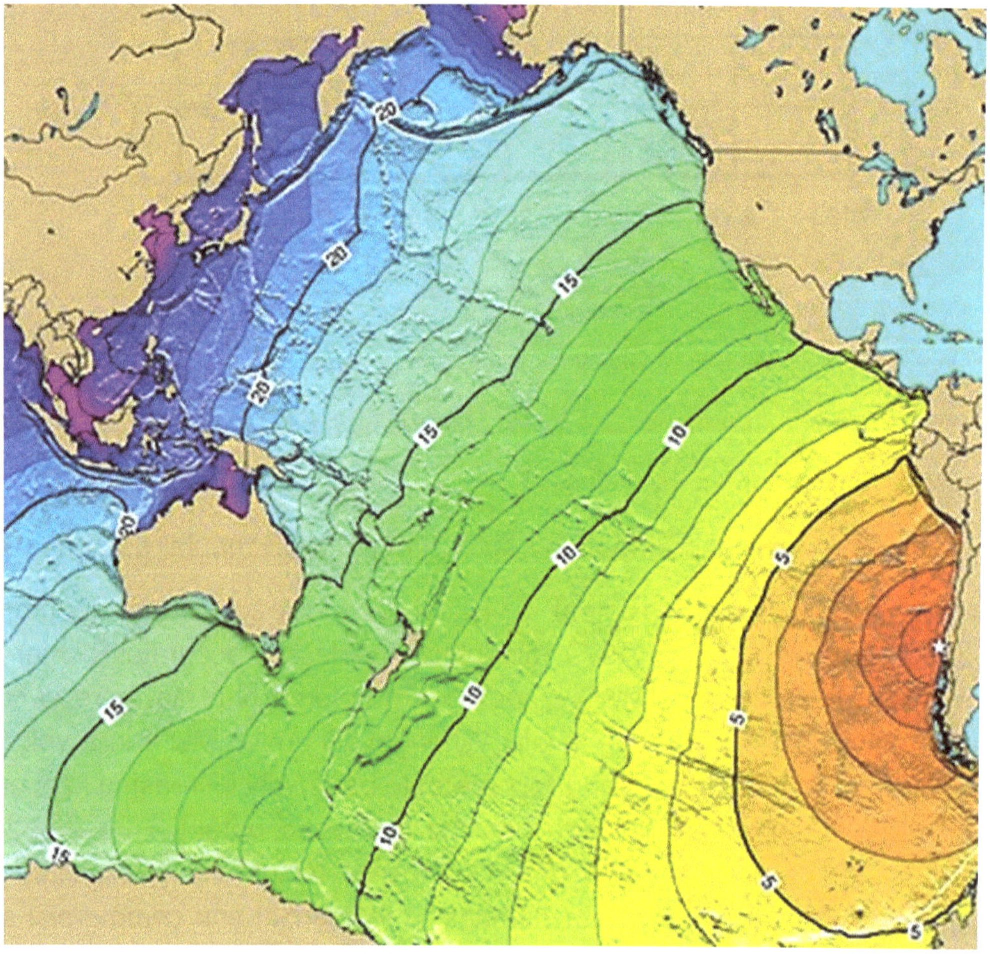

(Source: US National Geological Survey)

During earthquake occurrences globally, there is massive damage and deaths caused from the dynamic shaking and collapsing of inadequately constructed structures (especially residences) as well as by the damage and fatalities resulting on a country far away from the epicenter but impacted by the enormously high tsunami generated by the earthquake. Most importantly, it is the ability of the residents (all affected by the earthquake and tsunami and weather if made homeless) to survive the catastrophic event. A tabulation of the ten most powerful and/or destructive global earthquakes is shown next:

Largest Earthquakes Recorded (Richter or Mw Scale / Fatalities & Damage)

	Place	Date	Magnitude (Richter or Mw Scale)	Fatalities and Major Injury	Damage Cost Estimate	Generated Tsunami?	Ignited Fire?
1	Valdivia, Chile	May 1960	9.5 (Richter)	5,700 in Chile 100+ in Japan 61 in Hawaii	$700 Million	Y (82 ft. Chile; 35 ft. Hawaii; 30 ft. Japan)	N
2	Prince William Sound, Alaska	March 1964	9.2 (Richter)	128	$311 Mllion (Anchorage region)	Y (90 ft.)	N
3	Sumatra Islands, Indonesia	December 2004	9.1 (Mw scale)	300,000 est	Major	Y (100 ft.)	N
4	Tohoku, Japan	March 2011	9.1 (Mw scale)	20,000 est	$309 Billion (Nuclear Facility)	Y (132 ft.)	N
5	Port-Au-Prince, Haiti	January 2010	7.1 (Mw scale)	220,000 est	250,000+ Homes Destroyed	Y (10 ft.)	N
6	Northern Sumatra, Indonesia	April 2012	8.6 (Mw scale)	1,300+	Major	Y (3 ft.)	N
7	Unimak Island, Alaska	April 1946	8.6 (Richter)	5 in Unimak; 159 in Hawaii	Minor in Alaska; $26 M in Hawaii	115 ft. High at Hawaii	N
8	Sanriku-Oki, Japan	March 1933	8.4 (Richter)	500+ in Sanriku; 15,000+ in Honshu	Moderate in Sanriku; Massive in Honshu	100 ft. High at Honshu	N
9	San Francisco, California; USA	April 1906	7.9 (Richter)	3,000+	$524 Million (Collapses and Fires)	N	Y (Fire caused by gas line ruptures)
10	Turkey/Syria (Gazientap)	February 2023	7.8 (Mw scale)	53,537 in Turkey; 7,000+ in Syria	$149 Billion in Turkey; $14.8 Billion in Syria	N	N

Note: The Mw scale (T. Hanks and H. Kanamori, 1979) is more accurate than the Richter Scale for larger earthquakes as it estimates the total energy transformed during large earthquakes. The earlier Richter scale measures the amplitude only on a logarithmic scale.

(Source: Tabulation by Author of US National Geological Survey databases)

The photograph shown next is the damage done to a town (Sanriku) on Honshu, Japan) by the tsunami generated by the Mw 9.5 earthquake at Valdivia, Chile on May 22, 1960. The tsunami was originated at Valdivia and traveled northwesterly over 6,000 miles across the Pacific Ocean before striking Hilo, Hawaii, and then Honshu, Japan. The tsunami height decreased from 82 feet high in Chile, to 35 feet high in Hawaii, and to 30 feet high when it rolled into Sanriku, Honshu, Japan. It is incredible when speculating on how much energy was transferred from the earthquake in Chile and into the Pacific Ocean to cause that height of wave of water and transfer it at a speed of several hundred kilometers per hour, and still have the energy potential to cause the flooding and damage shown in the photo next.

(Source: US National Geological Survey)

The lesser amplitude earthquake (shown at the bottom of the above tabulation as the Gaziantep, Turkey/Syria Mw 7.8 earthquake on February 2, 2023) is a contrast in the number of fatalities, damage cost, and without a tsunami contributing to those deaths/damages compared to the 1960 Valdivia earthquake. The Turkey/Syria earthquakes occurred near the intersection of the African Plate and the Arabian plate, and the initial tremor of Mw 7.8was soon followed by a Mw 7.6 tremor. An amazing statistic however was that there were 25 aftershocks of Mw 4.0 or greater within six hours of the initial tremor. Within the first 24 hours of the initial Mw 7.8 tremor there were over 570 aftershocks, and within three months after the initial tremor over 30,000 aftershocks of less than 4.3 Mw had been recorded. A diagram showing the configuration of the primary tectonic plate fault lines for that region is shown as follows:

(Source: International Union of Geological Sciences, IUGS)

When the 30,000 opening and aftershocks recorded seismologically were superimposed on the tectonic plate diagram shown above, there is an amazingly good correlation matching the geological database for that region.

A comparison of the number of fatalities and damage costs of the Valdivia, Chile largest earthquake ever recorded (Mw 9.5 in May 1960), with the relatively low magnitude (Mw 7.1) earthquake that occurred in Port-Au-Prince, Haiti, in January 2010 but having an unusually high number of fatalities (220,000), and that of the Turkey/Syria earthquake (Mw 7.8 in February 2023) was made. This comparison was done to better understand why there is such a wide disparity between the three earthquakes – the number of fatalities and the cost of damages and repair – not being proportional to the magnitude of the earthquake. Two variables, when considered – the population density of the region affected by the earthquake, and the quality of construction of the structures (residential and commercial) serving the population of the region – brought the correlation into better correlation. Three photographs are shown next to compare the quality of construction of the structures that were destroyed. The first photograph is that of the Valdivia, Chile area, the second photograph is that of the Port-Au-Prince, Haiti area, and the third photograph is that of the Gaziantep region in Turkey with its high damage cost and large number of fatalities.

(Source: International Union of Geological Sciences)

(Source: International Union of Geological Sciences)

(Source: International Union of Geological Sciences)

Before interpreting the statistics of a few of the most significant earthquakes in the United States (the Alaska earthquake in March 1964 and the horrific earthquake and fire in San Francisco in California in April 1906), and the massive number of fatalities and property damage they caused, it must be described that earthquakes are in two distinct modes of motion – "shakers" and "bouncers". Anyone who has lived in southern California for a long time (the Author and his family lived in Ventura County for 34 years and 4 years in San Luis Obispo County in the Central Coast region of California) can attest that it is the "bouncers" that do the most damage. A photograph of the infamous San Andreas Fault under the Southern California area is shown as follows:

(Source: National Geological Survey)

The human injuries/fatalities, damage to property (residential and commercial), and the damage done to the infrastructure (bridges, tunnels, highways, rail lines, multi-level parking structures, etc.) attributed to earthquakes deserves a book dedicated to that branch of earth science and geophysical science. For the USA and California, the epic earthquake in San Francisco in April 1906 is the earthquake that mostly comes to mind of anyone born before 1950. It stands as the largest number of fatalities (over 3,000) and one of the costliest (over $524 million in 1906 dollars - $18.3 billion in 2024 dollars) in developed countries. Two photographs taken after and during the Richter 7.9 magnitude earthquake shown next points to the two factors that dominate the factors causing the high number of fatalities and the high amount of damage:

(Source: US National Geological – formerly Geodetic - Survey)

(Source: US National Geological – formerly Geodetic - Survey)

The upper photograph reflects that it was the initial vertical motion (a "bouncer") of the earthquake that literally bounced the otherwise structurally well-built (with steel reinforced members as a skeleton) buildings causing the non-metallic components attached to the façade and internal steel structure to fail and fall to the ground. It is seen that the skeletons of the buildings remained intact, but their facades and contents had fallen away or down. Had that earthquake of that estimated magnitude of 7.9 been a more horizontal motion (a "slider"), the buildings would have fallen when shifted off their foundation on the hilly area. Of the over 3,000 fatalities, more than 90% (over 2,700) of them were fatalities resulting from crushing type injuries. The major tremor shook the area at about 5:12 in the morning and most of those 2,700 fatalities were of humans in their beds.

The lower photo above shows that there were fires raging in the Mission District which began shortly after the initial tremor. There were again high winds in the region, and the heat from the initial fires created their own hot winds which spread throughout the Mission District. The tremors and falling debris had ruptured the gas lines in most of the residential buildings and in the main distribution pipes in the street serving the Mission District. Coincidentally, the tremor also severed the main water lines serving the Mission District such that there was no water [pressure for fire fighters to extinguish the fires. Attempts were made to explode explosives in the hope of creating firebreaks to halt the fires but those attempts also failed. Unfortunately, the explosives only served to initiate additional fires. Instead of one massive fire-front, there were over 30 fires burning at one point in time. Of the 3,000+ fatalities, the 2,700 early fatalities from crushing injuries were supplemented with over 300 fatalities of persons in the Mission District who were either also crushed instantaneously or were injured but could not be rescued because the fire had engulfed their residences. Many of those 300 succumbed to crushing and/or burn injuries if not completely incinerated by the fast-moving fires.

The fires continued to burn for four days and nights until they were extinguished by hoses from water tanker trucks that had arrived from surrounding cities of the Bay area. Over 25,000 buildings on 490 city blocks had been destroyed. The insurance companies advised the people who had lost their homes that they were not covered for damage done by earth movement but only if they had been damaged by fire. The public saw the "loophole" in that insurance coverage interpretation and some intentionally set fires to the rubble of their destroyed homes to be eligible for insurance coverage of fire damage.

At the start of the catastrophe, San Francisco had a population of over 400,000. Approximately 300,000 were made homeless and left San Francisco to be taken in by family, friends, and co-workers who lived on the other side of the Bay. A map of San Francisco reflects an outlined burn area which destroyed about 80% of the residences and commercial buildings within that outlined area. That map appeared as follows:

(Source: San Francisco Fire Department)

Many “Lessons Learned” were derived from the San Francisco Earthquake and Fires, and a major program was started to improve the safety of buildings (commercial and residential) constructed in Los Angeles County and other surrounding Counties that straddle the many faults associated with the San Andreas Fault that runs the entire West Coast – from California to the state of Washington of the mainland (48) states. Earthquake specialists opine that it is better for the San Andreas Fault to move in several small shocks (below magnitude 7) than to move in one larger jolt of magnitude 9 or more.

In an article written in 1913 by a scientist and geologist John C. Banner made the assessment which clearly applies also to the current problems with global warming, climate change, and ocean levels rising: “The only way we know of to deal successfully with any natural phenomenon is to get acquainted with it, to find out all we can about it, and thus to meet it on its own grounds. That is the way mankind has succeeded thus far, and it is safe to conclude that it is the only way it will ever succeed.”

The Author was in his residence in Thousand Oaks, California, when the Northridge Earthquake (6.7 magnitude) started at 4:30 a.m. PST on January 17, 1994, in the San Fernando Valley about 20 miles north of the City of Los Angeles. He was awakened bythe sound of glass breaking. To his dismay, it was obvious that a bottle of V.S.O.R. (Very Special Old Rum) from the island of Barbados had been shaken off a shelf in the family room, and another bottle of Jim Beam had fallen on top of it with the base of the JB smashing the bottle of VSOR. Some tragedies have no limit to what gets destroyed by natural events. When the Northridge area was identified in the local news broadcasts as the approximate epicenter the number of fatalities began being broadcast. At the end, approximately 72 persons had been fatally injured directly by the damages to buildings in which they were asleep. Another three or four dozen fatalities were considered at least 50% attributable (heart attacks and a few suicides) to the Northridge earthquake. Fortunately, most if not all construction of buildings after about 1908 when building codes were amended to add earthquake-proofing based on the 1906 San Francisco disaster (earthquake and fires). A photograph of an iconic failure of an apartment complex in Northridge is shown as follows:

(Source: The Los Angeles Times, January 18, 1994)

There were no fatalities and the loss of approximately 12 automobiles crushed by the collapse of the building's foundation were covered by automobile insurance policies. There were no reports known to be filed of any gas lines being ruptured and fire breaking out, but there were several reports of plumbing line breakages and leaks in water lines servicing several residences. It is largely due to strict enforcement of the earthquake-proofing provisions of California Building Codes that structural damages and fatality (72) statistics in the Northridge Earthquake of 1994 were not much larger. In comparison, the damages and fatalities in the Turkey/Syria Earthquake were high as over 70% of the structural damage was done to structures for which over 70,000 waivers for construction strength provisions in the Turkish Building Code had been granted waivers (accompanied by much "under the table" payments.)

The prospects for earthquakes in the vicinity of the San Andreas Fault (shown earlier) are summarized well in a USGS report: The U.S. Geological Survey's most recent forecast, known as UCERF3 (Uniform California Earthquake Rupture Forecast 3), released in November 2013, estimated that an earthquake of magnitude 6.7 M or greater (i.e., equal to or greater than the 1994 Northridge earthquake) occurs about once every 6.7 years statewide. The same report also estimated there is a 7% probability that an earthquake of magnitude 8.0 or greater will occur in the next 30 years somewhere along the San Andreas Fault. A different USGS study in 2008 tried to assess the physical, social and economic consequences of a major earthquake in southern California. That study predicted that a magnitude 7.8 earthquake along the southern San Andreas Fault could cause about 1,800 deaths and $213 billion in damage. Meanwhile, in the Los Angeles County region, there are frequent earthquakes (well below a magnitude to cause significant damage or injuries to humans) occurring associated with the San Andreas Fault system.

A total of *234 earthquakes* with a magnitude of four or more have struck within 186 miles (300 km) of Los Angeles, California in the past 10 years. This comes down to a yearly average of 23 earthquakes per year, or 1 per month. On average an earthquake will hit near Los Angeles roughly every 15 days.

A relatively large number of earthquakes occurred near Los Angeles in 2019. A total of 125 earthquakes (mag 4+) were detected within 186 mi of Los Angeles that year. The strongest had a 7.1 magnitude.

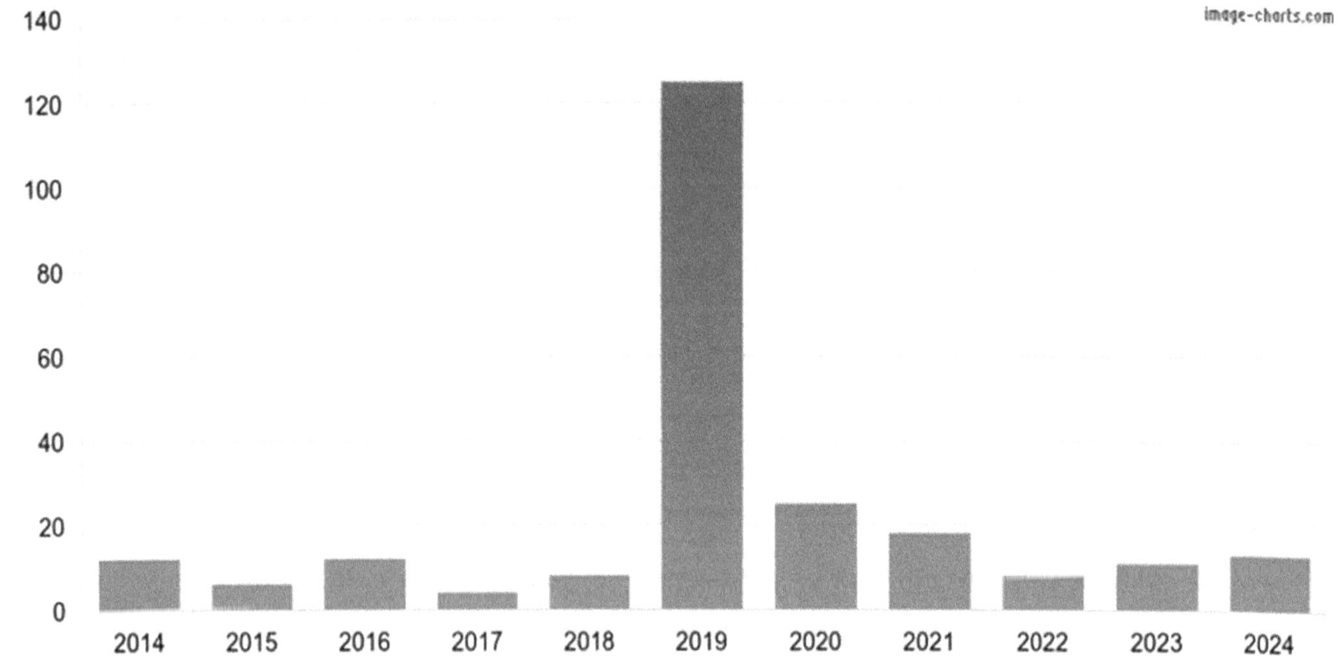

(Source: US National Geological Survey; California Institute of Technology)

The vast majority of earthquakes in the Los Angeles area are below a magnitude 4. A table shown next indicates that most are never reported as even having occurred because residents have adapted to their lifestyles so as to ignore earthquakes having a magnitude of 3 or lower. The statistics for the California region are indicated below:

Average number of earthquakes

Based on data from the past 14 years and our earthquake archive back to 1900, there are about 46,000 quakes on average per year in or near California, USA:

- Mag. 7 or higher: **0.11 quakes** per year (or 1 quake every 9.1 years)
- Mag. 6 or higher: **0.81 quakes** per year (or 1 quake every 1.2 years)
- Mag. 5 or higher: **4.6 quakes** per year
- Mag. 4 or higher: **55 quakes** per year (or 4.6 quakes per month)
- Mag. 3 or higher: **508 quakes** per year (or 42 quakes per month)
- Mag. 2 or higher: **4,200 quakes** per year (or 11.6 quakes per day)
- Mag. 1 or higher: **25,000 quakes** per year (or 68 quakes per day)

California is rated to have a "moderate" level of seismic activity. However, California has had at least 14 quakes above magnitude 7 since 1900, which suggests that larger earthquakes of this size occur infrequently, probably on average approximately every 5 to 10 years.

This chapter has shown the type and extent of damage and cause of human fatalities due to geophysical catastrophes alone and which are totally unrelated to the effect that the small concentration of carbon dioxide in the atmosphere (140 parts per million - 0.014% - of the atmosphere) may have. Environmental scientists have greatly over-exaggerated there being a "crisis" if the Man-made emissions of carbon dioxide are not reduced to net zero by 2050. Is it reasonable to expect that the failure to meet that well-intentioned goal will cause destruction and despair approaching that as shown in the foregoing database of geophysical catastrophes?

The World population – over 8 billion of us – deserve better representation to keep the important balance of control over what the environment is imposing on humans not just what humans are imposing on the environment. Taxpayers' money is well spent to prevent the human population from polluting the environment as that indirectly does protect the safety, health, and comfort of the human population to protect their food supply, the water they drink and cook with, and the air they breathe. However simply using many millions of dollars to protect marshlands and peat-covered areas that are of little value to the population is wasteful. The Environmental Protection Agencies at Federal and State level are prostituting the environmental science profession as a sample grant award of $489 of taxpayers' monies recently awarded shows:

NC lands EPA grant to cut carbon emissions naturally

BY ADAM WAGNER
awagner@newsobserver.com

To cut carbon emissions, North Carolina and three nearby states are going to look to nature.

The U.S. Environmental Protection Agency has awarded the Atlantic Conservation Coalition a $421 million Climate Pollution Reduction Grant. Along with North Carolina, the coalition includes Maryland, South Carolina and Virginia, as well as nonprofit partners like The Nature Conservancy and the N.C. Coastal Federation.

The grant is the largest the EPA has awarded for the use of natural solutions to curb the impacts of climate change.

North Carolina's portion of the award includes $50 million to the N.C. Department of Natural and Cultural Resources to execute "shovel ready" projects, with $30 million passed to the N.C. Coastal Federation to focus on projects like building living shorelines and using dredge spoils to restore threatened salt marshes. The North Carolina chapter of The Nature Conservancy will receive an additional $67.8 million to restore an estimated 33,500 acres of peatlands across North Carolina and Virginia while protecting an additional 10,500 acres.

"Protecting our natural lands for future generations is not only critical to our fight against climate change but also our state's economy and tourism industries," Gov. Roy Cooper, a Democrat, said in a statement.

North Carolina's grant projects are expected to reduce greenhouse gas emissions by more than 22.7 million metric tons by 2050, according to Michele Walker, a N.C. Department of Natural and Cultural Resources spokeswoman.

(Source: The Durham News & Observer, Monday, July 29, 2024)

Inasmuch as the preceding information in this book has proven otherwise, Man-made carbon emissions are not the dominant or even major cause of "Climate Pollution". The most recent measurement of carbon dioxide in the atmosphere, as measured by NOAA (at the NOAA GML Mauna Loa Observatory on July 29, 2024) is measured to be 425.21 ppmv (parts per million). It was shown earlier that natural carbon dioxide in the atmosphere before the start of the Industrial Revolution was 295 ppmv. The environmental scientists attribute the rise from 295 to 425.21 ppmv – a rise of 130 ppmv - of total carbon dioxide in the air is entirely due to Man globally generating "greenhouse gases". The EPA organizations are using the "average global temperature" (which is a false scientific measure) to be held to a rise of not more than 1.5 degrees Celsius by 2050 if "net zero emissions" from Man are achieved (not likely).

The UN IPCC meeting of COP28 in December 2023 in Dubai shows that the goal (to hold "average global temperature" rise to 1.5 ^{0}C by 2050) will not be met. China and India generate over 80 percent of global Man-made emissions – and China and India are building additional fossil fuel generating facilities, not reducing that 80 percent of 130 ppmv (by 104 ppmv). Therefore, if the USA reduces their Man-made "greenhouse gas emissions" to net zero, the most reduction of Man-made CO_2 in the atmosphere would be from 130 ppmv to 104 ppmv. The marshlands and peat areas on the US East coast being protected by the EPA "cutting carbon emissions naturally" would see the total carbon dioxide in the atmosphere reduced from 425.21 ppmv as of July 2024 to not less than 399 ppmv of total CO_2. Any expectation that a realistic reduction of carbon emissions in the global atmosphere by 26 ppmv (which is in percent equivalent to a reduction of 0.0026% of CO_2 concentration in the atmosphere of the US East coast) is ludicrous at best. EPA scientists are due for a remedial course in basic mathematics. The EPA grant of $421 million dollars of taxpayers' monies to protect marshlands and peat area would be better spent building homeless shelter facilities in areas where the homeless now live without air conditioning or heat, no human waste sanitation, and having to scavenge trash cans for food.

The magnificent grandeur that Nature also has created – disguised with the identifier of this chapter as geophysics - in many places on the Earth. One such scene is one of the top favorites for the US public to visit and to appreciate why it is important to protect Man but also to protect the national treasures on its surface. This is the Yosemite Park in Northern California, and showing the Upper Falls and the Lower Falls of the Merced River through the region in its geophysical magnificence:

(Source: US National Park Service)

When viewing this scene for its natural beauty, system engineers note that gravity is the unseen factor of Nature that adds to its beauty – to wit, the magnificent falls (Yosemite Falls – upper and lower) of water. The trees grow upward despite that same gravitational acceleration. The upheaval of that mass of granite stone to such a high elevation requires an energy of the magnitude that boggles the mind.

Chapter 2
Natural Factors Affecting the Planet

The largest factors affecting temperature on the planet Earth's surface were shown earlier to be the astrophysical distance the Earth is from our primary source of heat energy radiation - the Sun, followed by geophysical (especially latitude) factors. There has been found no factual database or scientifically credible factor supporting the environmental scientists' hypothesis that Man burning fossil fuels is the dominant – or even a very significant – cause for global warming, climate change, or ocean levels rising. Disturbingly, there was found much evidence where valid databases were partially or totally ignored if the new data (obtained with "big picture" satellite imagery) disagreed with out-of-date, incomplete ("cherry-picked"), linearized, and averaged, meteorology data. Almost nothing in Earth's astrophysical, geophysical, and natural system functions linearly or in an average manner. This chapter presents the natural factors affecting the planet and all lifeforms on its surface – and to which those lifeforms (human, animal, agricultural, and Forestal) must endure and adapt to survive and thrive.

Water is a most important natural feature in the "big picture" analysis of the planet's system of astrophysics, geophysical, natural, and anthropogenic (Man-made) factors being analyzed in this book for the individual and combined extent of their influence on Earth's surface temperature warming and/or climate change. It is the abundance and dynamics of water on planet Earth that distinguishes Earth from all other celestial bodies in the Universe. Water has some unique properties that cause it to be the most vital factor to all forms of life on the planet. This book prioritizes the safety and health of human life to be of primary value to the public, but in a balance with protection of the ecological system. This chapter is the largest chapter of this book because of the complexities and interactions of the many aspects of nature with the ecology and the human population.

How much water, how it is distributed on the Earth's surface and in its atmosphere, and the dynamics (governed by the laws of physics) associated with water movement on Earth's surface are most important factors needing to factually be established in understanding why and how the Earth's surface temperature and resulting climate (temperature and precipitation) are changing. Water is present on Earth in all three of the principle states of matter – solid (as ice), liquid, and gaseous (as water vapor or clouds). Two images from a "big picture" satellite view are shown next and followed by a detailed tabulation of the amounts of water on the planet and where located:

The Vast Pacific Ocean The Eastern Hemisphere (Source: NASA)

The planet's surface is approximately 70% covered by ocean water. Numerically, the distribution of water on the planet is shown in volume totals and concentrations as water vapor in the atmosphere in the following table (not intended as an eye test chart):

Source of water	Volume of water in km³ (cu mi)	% total water	% salt water	% fresh water	% liquid surface fresh water
Oceans	1,338,000,000 (321,000,000)	96.5	99.0		
Pacific Ocean	669,880,000 (160,710,000)	48.3	49.6		
Atlantic Ocean	310,410,900 (74,471,500)	22.4	23.0		
Indian Ocean	264,000,000 (63,000,000)	19.0	19.5		
Southern Ocean	71,800,000 (17,200,000)	5.18	5.31		
Arctic Ocean	18,750,000 (4,500,000)	1.35	1.39		
Ice and snow	24,364,000 (5,845,000)	1.76		69.6	
Glaciers	24,064,000 (5,773,000)	1.74		68.7	
Antarctic ice sheet	21,600,000 (5,200,000)	1.56		61.7	
Greenland ice sheet	2,340,000 (560,000)	0.17		6.68	
Arctic islands	83,500 (20,000)	0.006		0.24	
Mountain ranges	40,600 (9,700)	0.003		0.12	
Ground ice and permafrost	300,000 (72,000)	0.022		0.86	
Groundwater	23,400,000 (5,600,000)	1.69			
Saline groundwater	12,870,000 (3,090,000)	0.93	0.95		
Fresh groundwater	10,530,000 (2,530,000)	0.76		30.1	
Soil moisture	16,500 (4,000)	0.0012		0.047	
Lakes	176,400 (42,300)	0.013			
Saline lakes	85,400 (20,500)	0.0062	0.0063		
Caspian Sea	78,200 (18,800)	0.0056	0.0058		
Other saline lakes	7,200 (1,700)	0.00052	0.00053		
Fresh water lakes	91,000 (22,000)	0.0066		0.26	87.0
African Great Lakes	30,070 (7,210)	0.0022		0.086	28.8
Lake Baikal	23,615 (5,666)	0.0017		0.067	22.6
North American Great Lakes	22,115 (5,306)	0.0016		0.063	21.1
Other fresh water lakes	15,200 (3,600)	0.0011		0.043	14.5
Atmosphere	12,900 (3,100)	0.00093		0.037	
Swamps	11,470 (2,750)	0.00083		0.033	11.0
Rivers	2,120 (510)	0.00015		0.0061	2.03
Biological water	1,120 (270)	0.000081		0.0032	

(Source: International Union of Geological Sciences database)

Readers are requested not to just visually pass over this large table of numbers but to comprehend the amount of water on Earth on and beneath its surface in the many different bodies of water (saline and fresh) and in its atmosphere.

The water vapor in the atmosphere is in a variable concentration up to 3.1% depending on the temperature of the atmosphere and the water on the surface below. That dependency is graphically presented as follows:

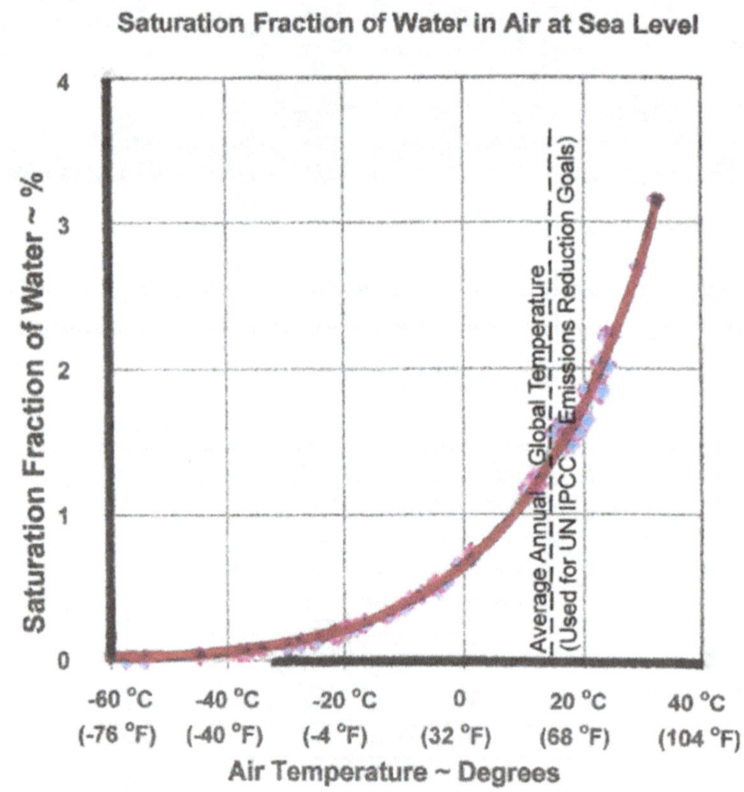

(Source: World Meteorological Organization, with Author Notation)

The point has been made repeatedly that if the environmental science segment persists with the fallacious one-size-fits-all "average annual global temperature" represent the average temperature of the planet, then it must also use an average of 1.4% of water vapor (corresponding to the average annual global temperature of 16 °C (61 °F) in the atmosphere of the planet to be consistent with the water vapor – temperature interaction that exists in nature as shown above. More realistically, environmental scientists should apply the above relationship to the average annual temperature of the regions – not global - being addressed.

When considering the regionality factor rather than average global, over arid (dry) deserts, the concentration of water vapor in the atmosphere is far less and is blown by wind currents into the arid atmosphere from less arid locations. Water vapor is not visible to the human eye but can be sensed by human skin as humidity. When warm water vapor rises to an elevation which is at a cold temperature relative to that containing the water vapor, the water vapor condenses into cloud formations, as are seen in the two photographs above.

It is also of fundamental importance to Readers of this book on climate change to understand that the variables of "global warming", "weather", "climate" and "climate change" are becoming less defined and the databases are being distorted in meaning (i.e., being "averaged", "linearized" so as to be able to be modeled, “misrepresented” or “cherry-picked”) to create fear of a “crisis” hazarding the public rather than as numerically valid analysis tools.

That is important as the environmental scientists fallaciously persist in considering the process of Man and industry combusting fossil fuels and creating ‘greenhouse gas” emissions (in concentration totaling less than 0.02% - 130 parts per million by volume) to be the dominant cause for global warming and climate change. The 0.037% larger volume and more effective heat-trapping water vapor in the atmosphere is disregarded by environmental scientists in forming that fallacious conclusion.)

Global warming is admittedly a somewhat “fuzzy” term when calculated by environmental scientists averaging and linearizing variations in the surface temperature of the entire globe (planet) over a significant number of months and short span of years to prove that an adverse warming trend caused by Man exists. Early attempts by the Environmental Protection Agency to reduce the adverse effects – while disregarding the beneficial aspects - of fossil fuel technology, led environmental activists to induce the public to believe the use of fossil fuels is at a “crisis” level of importance, and needs to be “combatted” in order to bring the climate under control.

What is not being “seen” is the combined effects of unseen other factors (astrophysics, geophysics – gravity, Coriolis effect, natural, and a gamut of “weather” variables changing and having a near-term effect on the planet’s environment and all living forms of life. The most important effect of water temperature is seen in its effect on the coastal cities that are located adjacent to the large bodies of water. An “eye-opener” for the more probable effect of water temperature on the local environment and on the local human population is in figures of the ocean currents that exist – not merely in the NOAA “average sea surface temperature” of 21 °C (69.6 °F).

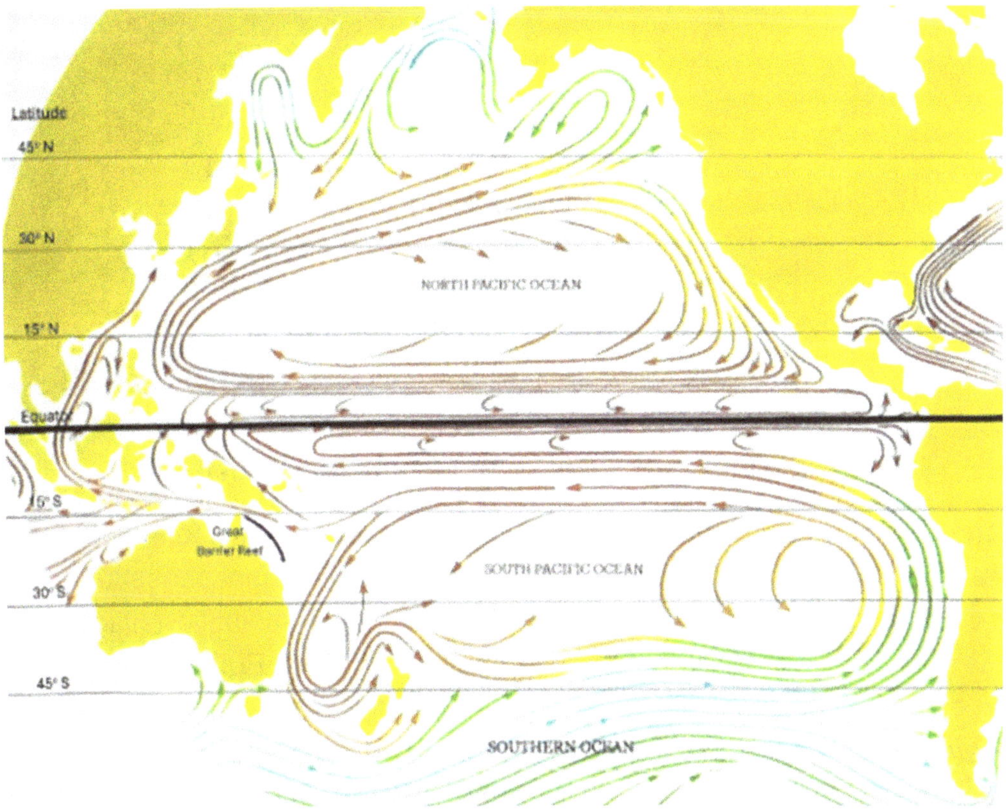

(Source: Rand McNally “Atlas of the World”)

Warm and hot ocean currents (located close to the Equator depicted by the horizontal black line) are depicted with a red coloration and red arrows showing the warm current flow. Cooler and cold ocean currents are depicted with green coloration and green arrows showing the cooler current flow. For illustrative purposes, the Great Barrier Reef off the Northeastern coast of Australia (lower left on the diagram) is affected primarily by the hot Equatorial current of the South Pacific Ocean – not by the NOAA "average sea surface temperature" which is at about 21 °C (69.8 °F). The average temperature variation of the Great Barrier Reef is 24.7 °C (76.5 °F) in Winter, 26.0 °C (78.8 °F) in Spring, 28.9 °C (84.0 °F) in Summer and 27.5 °C (81.5 °F) in Autumn – all much hotter than the NOAA average SST of 21 °C (69.8 °F). Just as humans must survive the extremes of hot and cold that occur seasonally in the atmosphere, the coral animals and all aquatic life in the oceans must survive the extremes of hot and cold that occur in the ocean seasonal pattern annually. The Great Barrier Reef is a naturally beautiful scene (as shown in the next photo) to be protected, but human lives are much more so:

(Credit to unknown photographer (in a helicopter)

The Great Barrier Reef needs to be preserved as a natural wonder – but for environmental scientists to correlate the factor of Man burning fossil fuels as being the dominant cause for any bleaching that is happening on the reef is a grossly unprofessional and non-scientific fallacious attachment of cause. There is no scientific proof that the atmosphere north of the Equator (where over 95% of the Man-made "greenhouse gas emissions" are generated) affects anything south of the Equator – such as the Great Barrier Reef. That is seen (in the "big picture") in the prevailing wind patterns (generated by reputable scientists) near the surface - and in the upper atmosphere - driving the movement of weather cells and ocean currents:

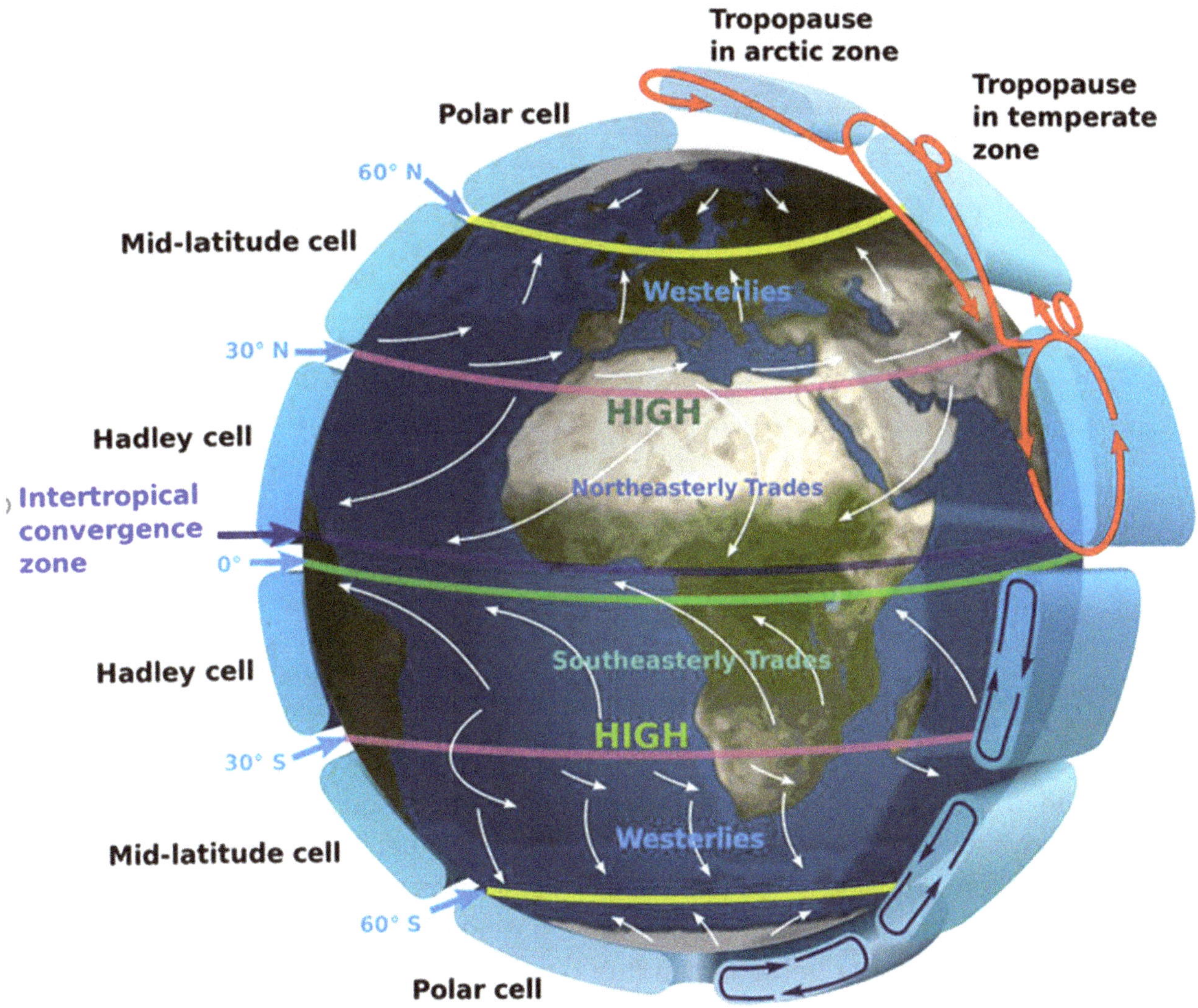

(Source: The Smithsonian Institute for Science)

The illustration of ocean currents also shows (in the upper right portion) the hot ocean currents spawning the tropical storms and seasonal hurricanes in the eastern Atlantic Ocean that devastates coastal regions of land masses interfacing with the Caribbean Sea and the Gulf of Mexico. The sea surface temperature in the Gulf Stream current is also very seasonal, averaging 73 °F in Winter, 77 °F in Spring, 84 °F in Summer, and 81 °F in Autumn. Its extreme temperatures are 70 °F in February and 87 °F in August.

The North American continent would be a much better place to live if the seasonal hurricanes were prevented from being created, but climate science has not fully understood why the tropical storms habitually are formed in the small region of the Cape Verde Islands group off the coast of West Africa. There is undoubtedly a complex interaction of the wind patterns in the atmosphere above the region and a recent satellite photograph reveals the very complex wind patterns in the Cape Verde region:

(Source: NASA and NOAA)

The Author had a career in aeronautical engineering and recognizes those wind patterns to resemble a fluid dynamics phenomenon known as the “Von Karman vortex street” which also has a pattern of vortex features very much like those shown above. The vortex street is seen in wind tunnel testing of cylindrical and near spherical shapes when a wind current flows over the curved surface of the body. A map of the Cape Verde region (the brown spots in the upper right quadrant of the above satellite photo are some of the Cape Verde islands group) does not cause any physical features that could act as vortex generators of the “vortex street flow” and it is surmised that it is the interaction of the colder wind patterns in the atmosphere (shown in the previous wind currents diagram) and the very hot winds flowing from the Sahara Desert region of northwestern Africa:

(Source: Rand McNally “Atlas of the World”)

It would be a most wonderful achievement if climate scientists and aerodynamics engineers could devise a "big picture" technology to control the wind patterns in the atmosphere to prevent the vortex activity that forms initially as the small vortices seen (and with low barometric pressure entering the interaction), but that level of activity is beyond the scope of this book. The system of small vortices in the "spawning ground" coalesce into one larger vortex and that larger vortex moves westerly gaining heat energy from the ocean waters of the Atlantic Ocean and Caribbean Sea. The high temperatures that are associated with the Summer and early Fall seasons at that latitude cause the tropical storms to develop into hurricanes of different severity levels (F1 to F5) to follow a generally predictable track – headed toward the southeastern states of the United States region. The Florida peninsular shape serves as an aerodynamic guide vane for the hurricanes and their typical track westerly turns northerly to follow the Gulf Stream warm current flowing past the east coast of the United States and Nova Scotia.

Climate change is a measure of both the surface temperature of a region <u>and</u> the amount of precipitation (rain, snow, hail, sleet, graupel) on that region and averaged for a longer period of years. For the climate analysis in this book, the period of years for defining the climate of a region – not averaged on a global basis – is considered be no less than twenty-five years but can be an overlap in five-year "sliding" blocks.

How the water being transformed from liquid form into water vapor that rises into the lower atmosphere and is later precipitated back to a liquid form as rain or snow and subsequently into potential other environmental problems such as flooding, mudslides, and bursting through inadequately constructed levees and dams is the topic of other chapters in this book. However, the realization that it is the distribution of massive amounts of water on Earth's surface – too much in some regions causing flooding while too little in other regions causing droughts – that presents challenges to the environment, to the human and animal populations, and to the environment on the planet. Man is powerless to prevent the occurrence (the where, when, and how much) of temperature and precipitation extremes, but instead must develop means to channel and/or divert floodwaters from inundating populated areas and business districts. Ideally, the floodwaters in one region might be channeled into a region suffering drought conditions but each region must adopt the measures for the extreme temperatures and extremes of precipitation that apply in a regional manner not a global "one-size-fits-all" average manner.

Most Insurance companies do not cover damages done to properties by moving water (as in known flood zones), and those that do offer flood insurance charge exorbitant premiums. That should serve notice to humans to build their dream home elsewhere than in a high-risk location, but humans value the aesthetics of where to live and accept the risks of Nature – and higher cost.

(Credit to unknown photographers) "Every man's home is his castle"

Where ocean waves are formed and to what scalar size – on a beach shore or mid-ocean – is caused by the wave energy being transformed into a mechanical form of energy. It is left to ecologists and oceanographers to explain the interactions and disturbances of wind and water affecting the environment and ecology. Informative photos of the interface between winds in the atmosphere and the water on the planet are shown next:

(Credit to Unknown Photographers)

The winds interfacing with the ocean surface can be either a constant velocity (from very low to very high) parallel to the water surface (laminar) or gusty (turbulent) with vorticity in small wind cells. There is no valid fact that Man burning fossil fuels has an adverse effect on the ocean waves. Less water vapor is generated with lower wind action and less of that generated is transformed into the atmosphere over placid waters than is over turbulent water surfaces. The wind, laminar or turbulent, determines whether the motion of the water will be formed into rolling waves that travel on the water surface (as in the left photo above) or become "choppy" with "white caps" in mid-ocean (as in the right photo above). The white in the frothy white caps illustrates that the temperature/warmth of the water and the wind currents play a part in evaporation in generating water vapor, and the generated water vapor being transformed from liquid to gaseous state into the atmosphere. (The explanation of "Where did all the white go?" question of children when Mr. Snow Man melts is left to parents to explain – if they themselves know the answer.)

As in most chemical processes, including that of a liquid evaporating into gaseous form, increasing the temperature and surface area of the water (H_2O) and displacing its molecules with an air current (wind) transforms more of the water molecules into the evaporative process. The energy to cause that evaporative process to occur has been shown in earlier chapters

The presence of water on the planet's surface and water vapor in the lower atmosphere is the singular most important natural resource for several reasons. Without water, there would be no forestry, no agriculture, no botanical growths, and no agricultural foods, and no humans to exist since water is an essential for all forms of life – forestry, agriculture, and animal and human life to exist. Life in the oceans obviously would not exist if the planet became barren as have the other planets in the solar system. The human population that has fish as a main food staple would suffer loss of a main protein source. Animals and poultry (for dairy products and for protein sources) cannot exist without water in their daily diet. Water is also needed for human health, personal hygiene, and for sanitation processes.

It was only until recent years that some environmental scientists acknowledged (most have refused to do so) the fact that water vapor in the atmosphere is a more powerful trapper of heat energy than is the total of all the "greenhouse gasses" (unwanted emissions that trap heat – carbon dioxide, methane, nitrous oxide, etc.) caused by the human population on the planet burning fossil fuels for their safety, health, comfort, transportation, and industrial needs.

The following extract (from an Associated Press article by Seth Borenstein on January 31, 2022) states that acknowledgment based on then-current (2022) information:

'When it comes to measuring global warming, humidity, not just heat, matters in generating dangerous climate extremes, a new study finds. Environmental science researchers say temperature by itself is not the best way to measure climate change's weird weather and down-plays impacts of humidity in the tropics. But factoring in air moisture along with heat shows that climate change since 1980 is nearly twice as bad as previously calculated, according to their study in the January 24, 2022, Proceedings of the National Academy of Sciences."

"The energy generated in extreme weather, such as storms, floods and heavy rainfall for an extended time, is related to the amount of water in the air. A team of scientists in the U.S. and China decided to use an obscure weather measurement called equivalent potential temperature — or theta-e — that reflects the moisture energy of the atmosphere. It is expressed in degrees, like temperature." said study co-author V. "Ram" Ramanathan, a climate scientist at the University of California San Diego's Scripps Institution of Oceanography and Cornell University. *"There are two drivers of climate change: temperature and humidity,"* Ramanathan said. *"And so far, we have measured global warming just in terms of temperature. But by adding the energy from humidity, the extremes — heat waves, rainfall and other measures of extremes — correlate much better. That is because as the world warms, the air holds more moisture, nearly 4% for every degree Fahrenheit (7% for every degree Celsius). When that moisture condenses, it releases heat or energy. That is why when it rains, now it pours,"* Ramanathan said.

"In addition, water vapor is a potent heat-trapping gas in the atmosphere that increases climate change. From 1980 to 2019, the world warmed about 1.42 degrees Fahrenheit (0.79 degrees Celsius). But taking energy from humidity into account, the world has warmed and moistened 2.66

degrees Fahrenheit (1.48 degrees Celsius), the study said. And in the tropics, the warming was as much as 7.2 degrees Fahrenheit 4 degrees Celsius). When judging by temperature alone, it looks like warming is most pronounced in North America, mid-latitudes and especially the poles — and less so in the tropics, But that is not the case, because the high humidity in the tropics "juices up" storm activity, from regular storms to tropical cyclones and monsoons. This increase in latent energy is released in the air which leads to weather extremes: floods, storms and droughts.", Ramanathan said.

University of Illinois climate scientist Donald Wuebbles, who was not part of the study, said: *"It makes sense because water vapor is key in extreme rainfall. Both heat and humidity are important."* Environmental scientist Katharine Mach of the University of Miami, who also was not part of the study, said: *"Humidity is key in shaping the impacts of heat on human health and well-being, at present and into the future."*

Water is the second most dominant factor that controls the surface temperature of the planet and does so by the cycle of free-standing water warming to a temperature and evaporating into the atmosphere, rising (opposing the gravitational acceleration by expending the heat energy in a mass into potential energy of that mass) to an elevation where cooler temperatures exist (due to astrophysical reasons), condensing into cloud formations, and then falling back to Earth (accelerated by gravity) as precipitation (rain, snow, hail, or graupel). That process is done cyclically in accordance with the laws of energy conservation, laws of mass conservation, and in accordance with the laws of motion of a mass. This book stops short of being a textbook in atmospheric chemistry or physics – and only contains the most basic principles of fluid dynamics (i.e., liquids and molten materials flow downhill and those with high viscosity travel slower than those with low viscosity but have more stored heat energy and more kinetic energy).

The small-scale uses of water hydrodynamic properties are illustrated later in this chapter, but its hydrodynamic power used to generate electricity on large scales will also be discussed in a later chapter of the book along with other natural sources of energy.

Logic, and a strategy of applied reasoning, was/is used extensively in this book to analyze the problems of global temperature change, climate change, and ocean levels rising, and without any bias toward selecting only those factors that are considered important by environmental scientists as the main variables "driving" the resulting effect of the factors. Using discipline and correct logic to forensically evaluate a large number of factors and set them into their correct "system" level perspective is the primary difference of this book and the flawed and biased analyses that were done by environmental scientists thirty and forty years ago - which resulted from the use and misuse of incomplete databases and/or the linearization/averaging of information. Few living and natural things perform in a "linearized" or "average" manner.

Many reputable scientists have criticized those environmental scientists and "green" activists for their flawed contention that: *"The science is settled. Humans burning fossil fuels and creating "greenhouse gas" emissions is the dominant cause for global warming and being at a "crisis" level of urgency needing to be reduced or eliminated".*

The most vocal and scientifically valid criticism of environmental scientists dismissing water vapor as the more extensive heat energy "trapper" in the atmosphere compared to the "greenhouse gasses" (carbon dioxide, methane, nitrous oxide, etc.) came from Dr. S. Fred Singer (1924 – 2020). Dr. Singer was a Professor Emeritus and was formerly Chair of the Department of Environmental Science at the University of Virginia. He did much analysis of the causes (natural and/or anthropogenic) for global warming, and climate change and reported the analytical results in a series of papers "A Closer Look at the Numbers". Perhaps the most important illustration of the importance of water vapor compared to that of natural and anthropogenic causes of the "greenhouse gasses" was shown in Dr. Singer's (and in several other reputable environmental scientists') frequent lectures and speeches as shown next:

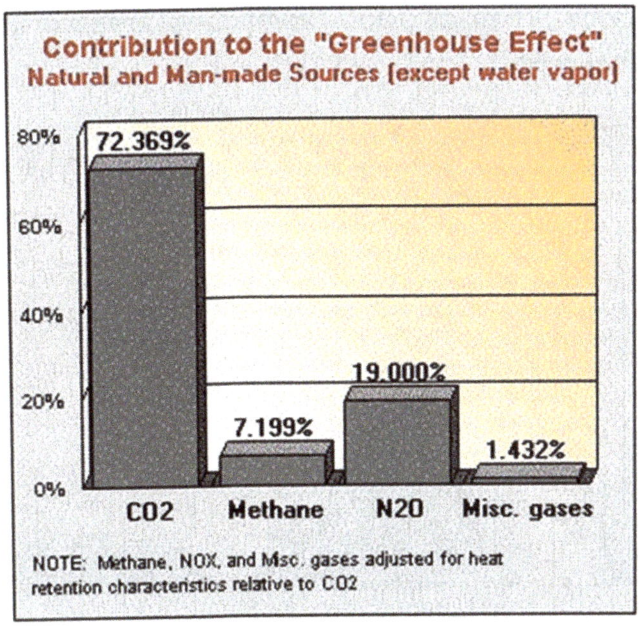

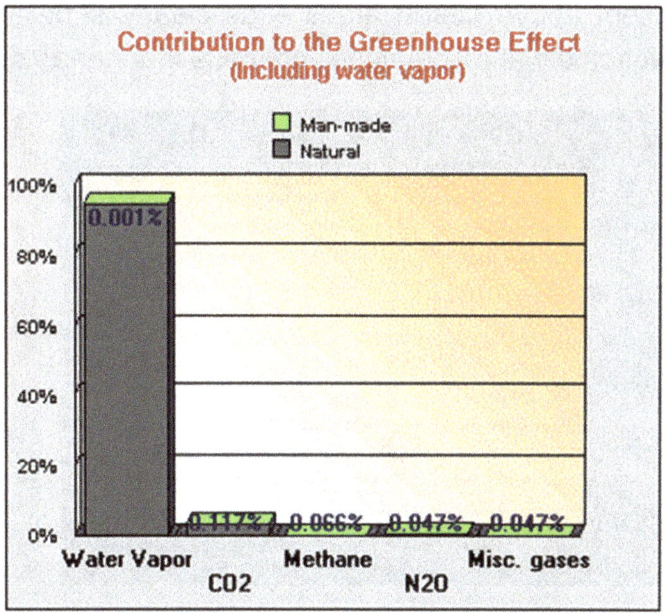

(Source: "A Closer Look at the Numbers", Dr. S. Fred Singer, formerly at U. of Virginia)

The chart on the left is based on the environmental scientists' database on "greenhouse gas effect" and not differentiating how much of the total is natural and how much is Man-made (but as the title indicates in the fine print tag-on "except water vapor"). It is obvious that the environmental scientists are concealing – probably intentionally - that water vapor is a major factor in causing the insulation effect (trapping heat) of the "Greenhouse Effect". That is tantamount to the news media asking President Abraham Lincoln's wife after his assassination: "Other than that Mrs. Lincoln, how did you enjoy the stage play?"

Readers should note that Dr. Singer indicates in the chart at the right how much of the individual gasses and water vapor is from natural cause (in grey color – and over which Man has no control) and how much is of anthropogenic (human) cause (in green color. Charts show the relative effectiveness each factor has on global warming directly. The green-colored Man-made "greenhouse gasses" (principally carbon dioxide) data are small (but not negligible) in comparison to the grey-colored natural-caused amount of the "greenhouse gas" effect; yet Man-made carbon dioxide is the main thrust of environmental scientists to eliminate or to be reduced drastically). The indirect and feedback effects of each (water vapor and Man-made greenhouse gasses) are discussed later in this chapter.

<u>The significance of the environmental scientists claim which disregards the factual dominant effect of water vapor on the "greenhouse gas effect", coupled with the fact that there is about 80% or more</u>

of the measured total carbon dioxide in the atmosphere (0.0417% as of October 2023 compared to the 0.0350% circa 1880 pre-Industrial Revolution) that is naturally-caused, and less than 20% that is Man-caused (human and Industrial), dispels the environmental scientists' fallacious claims that Man burning fossil fuels is the dominant cause for the "greenhouse gas effect" on the planet's surface temperature. The remainder of this chapter is presented with the intent of informing the public and to generate a higher appreciation for the important role water has on their everday life as well as their life-long dependency on water for their safety, health, and comfort.

Clouds are commonly seen above water covered surface areas, and equally common seen to be absent above desert areas – as clearly shown above. They also form over land areas in the awesomely and sometimes ominous manner as shown next:

(Credit to unknown photographers)

A very common form of a cloud formation is that of the altocumulus which is sometimes referred to as a "flock of sheep". Cloud formations – especially cumulus - are even more awesome to see in the "big picture" perspective when flying in an aircraft above them.

(Credit to unknown photographers)

The most ominous cloud formation the Author found was the one shown next:

(Credit to a very talented unknown photographer)

In any cloud formation there is a high concentration of water vapor which could fall as precipitation – rain, snow, hail, graupel, etc. No matter the shape of form or their color, it is the amount of precipitation from cloud formations and their reduction on solar insolation that contributes to the climate of a region that is of importance. Contrary to environmental scientists' misinformation to the public, the climate of the region does not have direct "feedback" causing the cloud formations to exist – but there is a cyclical "interaction" in that clouds (having an "albedo" effect) alter the temperature of the water surface which alters the evaporative rate of the water that is transformed into the atmosphere as water vapor. In common language, "the tail does not wag the dog" – the climate does not change the weather or the clouds and precipitation.

The next table of the water (ocean, seas, bays, gulfs) database numerically illustrates just how much total water in liquid and solid (floating sea ice) state there is (over 80% of Earth's surface area), and its distribution on the planet's surface: The volume of water in the five major oceans is also shown:

	Surface Area of Ocean / Sea (Est. 2021)				Volume of Oceans (Est.)	
Ocean/Sea Name	**Sq. Miles**	**Sq. kilometers**	**% of Planet**	**Rank**	**Cubic Miles**	**Cubic Km**
Pacific Ocean	65,091,238	168,723,000	33.05%	1	160,515,659	669,880,000
Atlantic Ocean	32,843,254	85,133,000	16.67%	2	74,380,203	310,410,900
Indian Ocean	27,221,172	70,560,000	13.82%	3	63,259,291	264,000,000
Southern Ocean	8,471,895	21,960,000	4.30%	4	17,204,610	71,800,000
Arctic Ocean	6,002,083	15,558,000	3.05%	5	4,490,695	18,750,000
Sub-Total Surface Area of All Oceans	**139,629,642**	**361,934,000**	**70.89%**		**319,850,458**	**1,334,840,900**
Philippine Sea	2,197,080	5,695,000	1.12%	1		
Coral Sea	1,848,308	4,791,000	0.94%	2		
Arabian Sea	1,489,912	3,862,000	0.76%	3		
South China Sea	1,185,597	3,500,000	0.69%	4		
Weddell Sea	1,080,205	2,800,000	0.55%	5		
Caribbean Sea	1,062,459	2,754,000	0.54%	6		
Mediterranean Sea	968,327	2,510,000	0.49%	7		
Gulf of Guinea	906,601	2,350,000	0.46%	8		
Tasman Sea	887,311	2,300,000	0.45%	9		
Bay of Bengal	837,931	2,172,000	0.43%	10		
Bering Sea	771,575	2,000,000	0.39%	11		
Sea of Okhotsk	610,702	1,583,000	0.31%	12		
Gulf of Mexico	597,970	1,550,000	0.30%	13		
Gulf of Alaska	591,412	1,533,000	0.30%	14		
Barents Sea	540,103	1,400,000	0.27%	15		
Norwegian Sea	533,544	1,383,000	0.27%	16		
East China Sea	481,849	1,249,000	0.24%	17		
Hudson Bay	474,519	1,230,000	0.24%	18		
Greenland Sea	464,874	1,205,000	0.24%	19		
Somov Sea	443,656	1,150,000	0.23%	20		
Sub-Total Surface Area of Seas/Gulfs/Bays	**17,973,935**	**47,017,000**	**9.22%**			
Total Surface Area of All Waters on Planet	**157,603,577**	**408,951,000**	**80.18%**			

Notes: Total surface area (land and water) of planet is 196,961,745 sq miles and 510,065,665 sq kilometers.

(Source: Tabulation by Author of International Union of Geological Service database)

In a "big picture" of water on the planet, there is seen to be about twice as much water surface area in the southern hemisphere as there is in the northern hemisphere, and over 90% of the human population lives on the land area predominant in the northern hemisphere and east of the Greenwich Meridian:

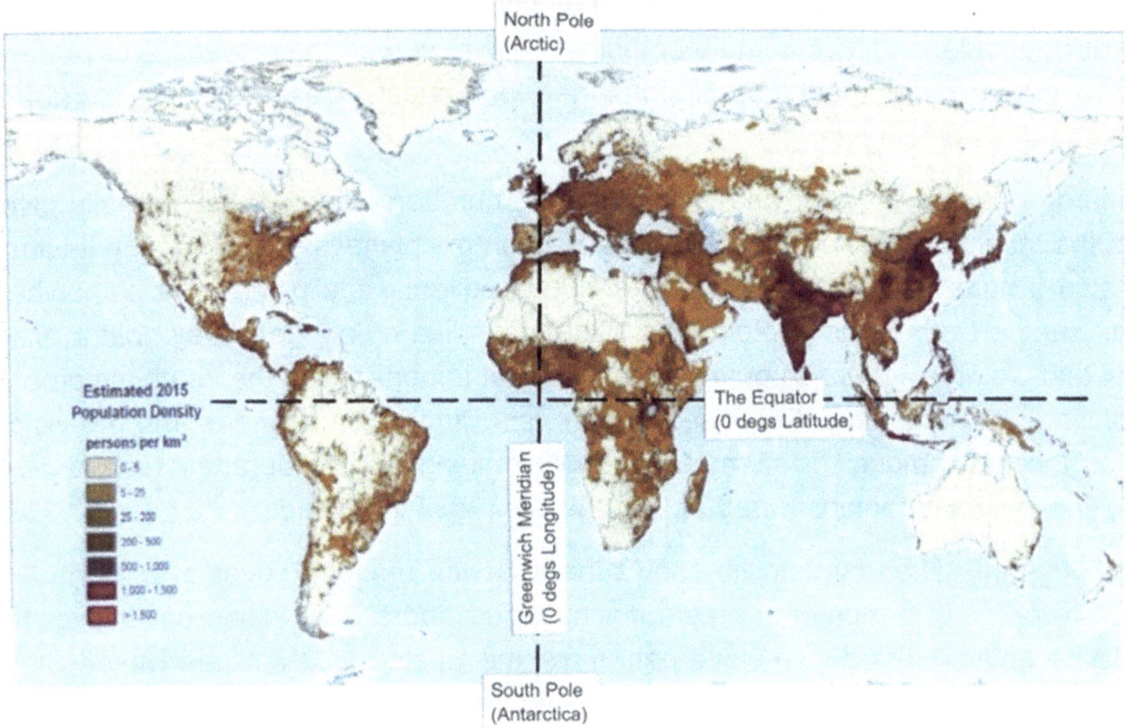

(Source: UN Population Division)

One does not need to be an environmental or climate scientist to appreciate that what the population of the North American Continent slightly does to its atmosphere and climate does not significantly affect the climate of the European and Asian Continents. The reverse of that may not be evident since the population of Europe and Asia is about 7 or 8 times that of the North American Continent. The land mass areas are at about the same ratio of 7 or 8 to 1.

A Harvard student generated an interesting form of the human population distribution of the planet by latitude and longitude as seen below:

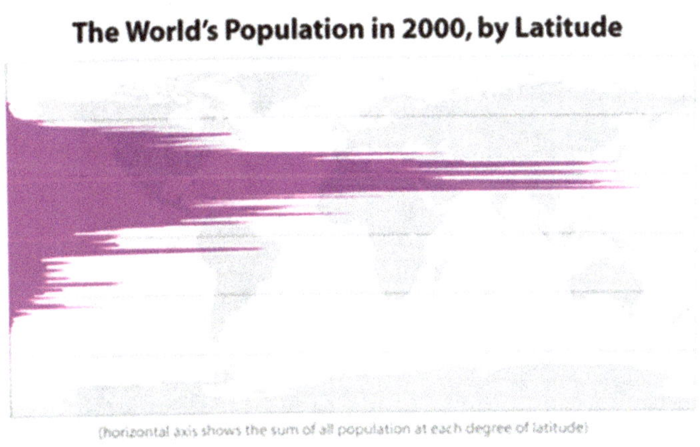

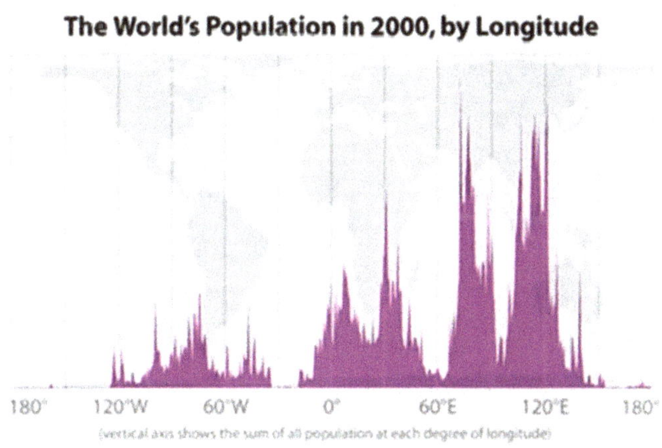

(Source: Harvard University, Unnamed graduate student)

The distribution by latitude (left illustration) is interpreted to show that the human population will choose to reside in a comfortable (temperate) climate – not uncomfortably hot or uncomfortably cold. The distribution by longitude (right illustration) is not surprising in the sense that the population chooses to live on land mass rather than on the ocean waters. The distribution showing the location by density (UN Population Division) is a clear indication of how the human population has chosen to reside at a latitude longitude climate combination that serves their needs and preferences for comfort and for the environment that supports their food supply. Deserts, jungles, and frozen land do not provide that combination.

The very large quantity of water and land area size numbers alone do not visually give the "big picture" of the magnitude of how much total water is on the planet's surface and in its atmosphere. There is also a huge amount of water in the clouds covering the planet in a somewhat random manner as can be seen in the photographs. Clouds act like a sponge – they hold a great deal of water until that water is squeezed out of them. Of great importance is the combination of the huge amount of water on its surface, the unseen water vapor in the atmosphere, and the visible clouds above it all that are among the many factors determining the temperature (seasonal and non-seasonal), the local and regional weather, and the climate of the planet.

Another photograph taken by satellite shows the absence of cloud cover over North America (a relatively rare photo of a moment in meteorology where there is no cloud cover over the United States). To be noted is the dense cloud pattern off the Eastern coast of the United States which influences the weather – especially in the hot summer season - of the East Coast states (from Maine in New England to Florida in the Southeastern region). The dense cloud pattern is above the very warm Gulf Stream current from the Caribbean region and is a huge "bubble" of water vapor in the entire atmosphere from water surface level to the top of the cloud layer.

(Source: NASA)

That frequent meteorological phenomenon seen above is one of the causes for the US East coast to have very high humidity (above 80%) while the US Southwestern states (California, Arizona, and Nevada) to simultaneously have a humidity level below 15 percent and 20 degrees lower temperature. (The Author resided in New York for 23 years and in California for 38 years and can attest to the clothes "sticking" to flesh when the temperature and humidity in New York City are about the same magnitude – 85 degrees Fahrenheit and 85 percent humidity – while in a classroom.) That is a seasonal temperature change phenomenon, not a 10 year climate change phenomenon.

The four seasons in New York make it a most interesting location in which to reside compared to California where there is basically only one season – too warm and very dry. From 1982 to 2020, a local tongue-in-cheek expression by California "dudes" was that "It never rains in California.". The climate factor in recent decades has been changing to cycle between flooding and droughts in California, but there is no factually convincing database that the climate change is due predominantly to Man burning fossil fuels. The heavy Industrial states are in the Eastern half of the United States, but water vapor is clearly a more major "heat-trapper" over the entire "lower 48" region than is carbon dioxide emissions from heavy industries in the eastern half of the United States.

Environmental scientists have chosen to ignore water vapor (despite the concentration ratio of water vapor to all of the "greenhouse gases" in the atmosphere being about 90 to 1) and environmental scientists persistently claiming that it is emissions from the Industrial Revolution that are causing global warming. That "greenhouse gas effect" – is not evidenced in "big picture" satellite photography, but the geophysical factor of latitude is clearly seen in the image on the front cover of this book. It is inconceivable how environmental scientists can use "average global temperature" based on the calculated average of measurements made at on land at over 2,000 land-based meteorological stations (most are at airports) when the planet's surface area is over 70 percent water-covered and at least another 20 percent being covered by deserts and highly mountainous regions.

Large cities are predominantly covered with concrete or asphalt-based black surfaces which retain heat. As the population grows, cities expand, airports are extended, more housing developments arise, and the extent of non-natural areas grow in size. Several book titles and film themes describe the planet well: "The Concrete Jungle". Concrete absorbs heat and radiates that heat when the surrounding air temperature is cool. Since most meteorological measurements were/are made at airports, the illusion that global warming is occurring due to Man burning fossil fuels might better be considered as the effect on "air temperature" from Man covering more of the planet's surface with concrete. As the population grows so does the amount of concrete-covered land mass.

The image on the front cover shows a very powerful effect of latitude on the surface temperature. How can meteorological measurements made on about 10% of the planet's surface area truly represent an "average global temperature" for the entire inhabitable global surface area of the planet? Mathematically it may, but realistically it does not. For environmental scientists' "convenience", a statistical "average global temperature" is a ludicrous unprofessionally deceptive practice that should not justify environmental scientists to set fallacious and unachievable carbon emission reduction goals that are regulatory imposed as US governmental and United Nations international global goals.

Earlier photos in this book (including the front cover photo) reflect the enormity of the bodies of water in liquid state and only partially portray the temperature of those bodies of water. The ice covering both the Arctic region and the Antarctic region are the solid states of water, but since they are in regions where there is little or no population of humans they are usually not considered as being of environmental interest. The tabulations shown earlier are the facts for the largest oceans, seas, gulfs, and bays, and as shown later for large inland seas, bays, and lakes.

For visual completeness of how water is physically distributed – in all three states of liquid, gas (water vapor and clouds), and solid (ice) on the planet, the two diagrams shown next are of the Arctic (North polar region) and Antarctic (South polar region). The two diagrams below of the polar regions should be of assistance to best understand the differences in amount and different type of ice at the two polar regions:

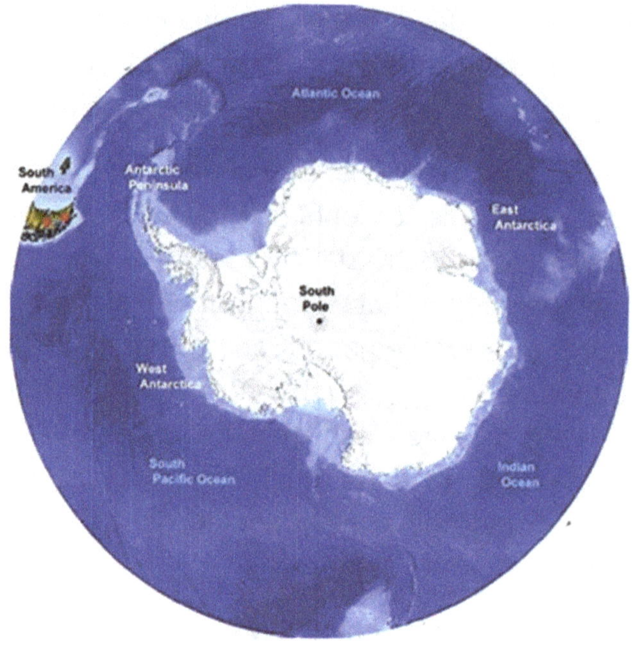

The Arctic Region (North Polar)

Antarctica (South Polar)

(Source: National Geographic Society)

A longer description of the future problem of ocean surface levels rising – and what are the causes for that hazardous eventuality - will appear in a later chapter. That later discussion will also include the three primary causes for ocean water surface levels to be rising and the adverse effects on humans and the need for upgrading of the container ports to sustain human needs.

A major difference between ice in the two polar regions is that the vast majority of ice in the Arctic region is floating "sea ice" (with the ice covering the upper 80 percent of Greenland above the Arctic Circle being on *terra firma*). By contrast, the vast majority of ice in the Antarctic region is on *terra firma* – the Antarctic continent – and only a much smaller percentage being on the ice shelves primarily on the western half of the Antarctic continent. The ratio of the total amount of ice in the Antarctic region to total ice in the Arctic region is about 9:1.

A not-so-mundane problem is that of icebergs, and the epic tragedy of the RMS *Titanic* – the "unsinkable" ship that struck an iceberg and did, most embarrassingly on its maiden voyage from

England to New York. Icebergs do not have a decision-making capability as to where to go after being "calved" from their mother glacier. The winds and ocean currents determine the paths of icebergs, and the surrounding water temperature determines how long they will exist. Most science classes teach that 9/10ths of an iceberg is underwater and unseen and the expression that has become common when addressing only a small part of a bigger problem is: *"That's only the tip of the iceberg."* That is somewhat the same case the research in this book has concluded. Environmental scientists identifying Man burning fossil fuels to be the dominant cause of global warming have only identified the "*tip of the climate change iceberg*".

(Credit to unknown photographer)

A striking example of icebergs moving to where they are least expected and wanted is shown next:

(Credit to unknown photographer) Town on the coast of Greenland

Considering that icebergs are floating on water and have only about one-tenth of their mass above the surrounding water surface level, i.e., nine-tenths of their mass are unseen – the above iceberg was truly a monster. As a positive factor, using the iceberg as a source of unpolluted non-saline water creates a natural resource for the residents to use as needed.

The vast Southern Ocean surrounding the Antarctic continent and the South Polar region is in a very cold but liquid state. As will be discussed in a later chapter, the thawing/melting of floating ice or ice on the ice shelves of West Antarctica does not contribute to ocean surface levels rising as they thaw/melt. It is only the ice on top of *terra firma* East Antarctica that melts and runs off into the surrounding ocean waters that pose a problem as contributory to the ocean surface levels rising in future centuries if/as the atmosphere and ocean waters elevate in temperature.

Two of the three states of water (liquid and gaseous) have a heat-trapping capability which are naturally caused to exist – and far outweigh (by a factor of about 80 or 90 to 1) the heat-trapping effects of the total concentration of natural and Man-made "greenhouse gasses" in the atmosphere. The amount of total carbon dioxide in the atmosphere was recently measured to be 420 parts per million by volume (a concentration of 0.042 percent of the atmosphere), whereas the amount of water vapor in the atmosphere is 1 – 3.1% depending on the temperature of the water below the location where measured). The water vapor over regions that are arid accordingly is lower than 2%, however, 80% of the Earth's surface has a watery surface emitting water vapor that does follow the following relationship:

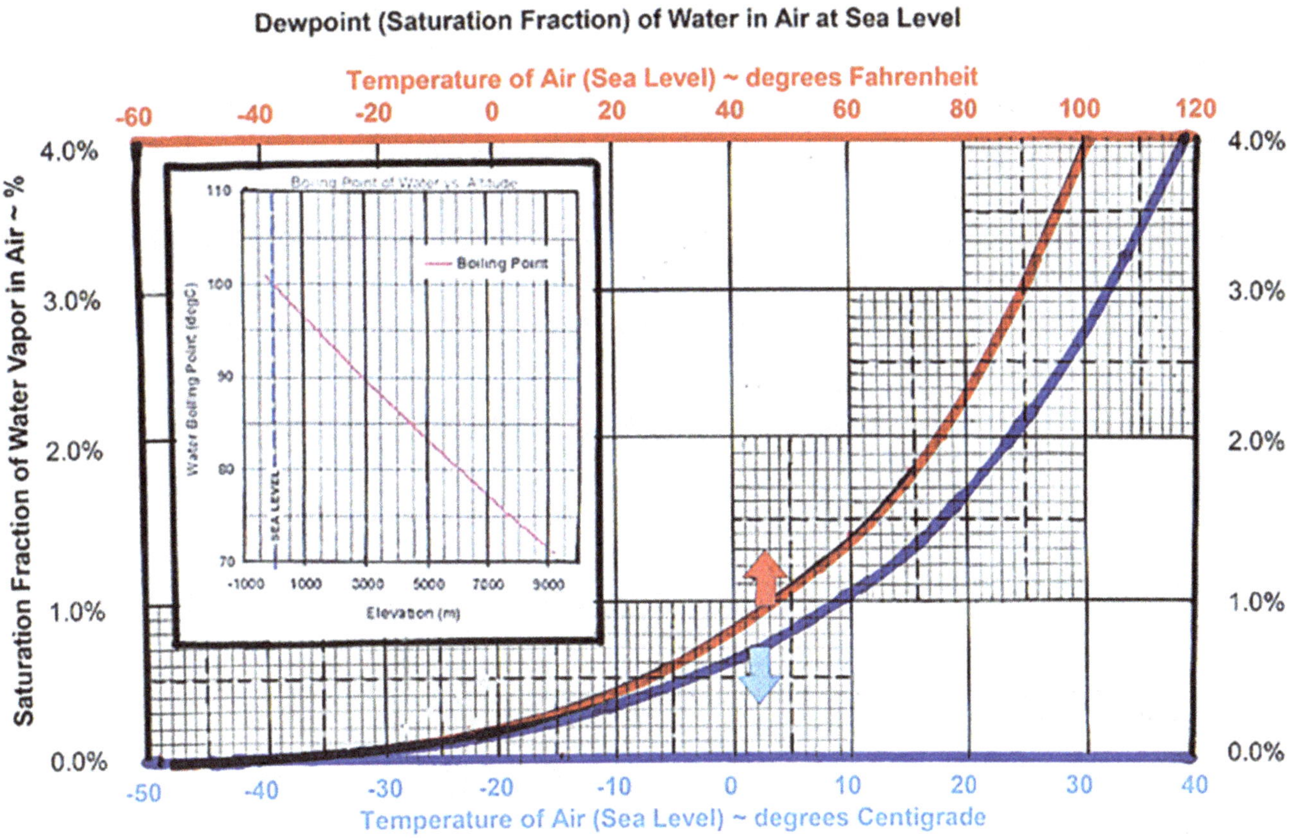

(Source: Reconstructed by Author from NWS/NOAA Database)

Anyone who has boiled a pot of water on a kitchen stove has seen the water bubbling and above the pot a small "cloud" of steam. Steam is hot water vapor with a high temperature. As shown on the small inset on the left of the above diagram, if the kitchen is at sea level elevation, the water boils at 100 degrees Centigrade (Celsius) or 212 degrees Fahrenheit. However, at an elevation higher than sea level, the temperature for boiling water reduces in magnitude such that at an elevation of Denver, Colorado, (the "Mile High City") – 5,280 feet (1,610 meters) – the water will boil at 94 degrees Centigrade (Celsius) or 201 degrees Fahrenheit. In industrial use, steam is used to generate electric power for other uses and that aspect of water will be discussed later in the "sources of energy" chapter of this book.

A simple "test of reasonableness" can be applied to the analysis of the "settled science" segment of environmental scientists, who have "circled their wagons" around their statement that "Man burning fossil fuels and thereby creating "greenhouse gas" emissions is the <u>dominant</u> cause of global warming", have yet to scientifically prove that 0.042 percent of carbon dioxide (and more than half of that 0.042 percent of carbon dioxide being introduced in the atmosphere is attributable to natural causes) that carbon dioxide in the atmosphere is the dominant cause of global warming. <u>With only a maximum of about 0.02% of carbon dioxide in the atmosphere attributable to being Man-made, but with the higher percent (2 – 3%) of water vapor also present in the atmosphere, there is no scientific logic in the statement that Man-made carbon dioxide is the dominant cause of global warming</u>. "Settled science" gets an F grade on the "test of reasonableness" of how such a low concentration of any gas can so hugely affect the entire atmosphere and resulting climate of the planet. The more factual assessment is that water vapor in the atmosphere is the dominant cause for global warming, in locations over 80% of the planet's surface.

For locations in the remaining 20% of the land area of the planet's surface where the highest density of human population exists, especially in cities that are over 50% concrete covered, water vapor in the form of humidity, regardless of the amount of carbon dioxide present in the atmosphere, is of great importance on the comfort of humans residing in that location. That is attested to by the Author who has spent over 20 years living in a high humidity environment in the summer seasons of New York City and of several months in Washington, DC. The previous diagram shows the saturation fraction of the atmosphere, but when there is less than that saturation fraction, the percent of that saturation amount is termed as being humidity. The relationship between humidity and dewpoint with air temperature is indicated in the diagram shown next:

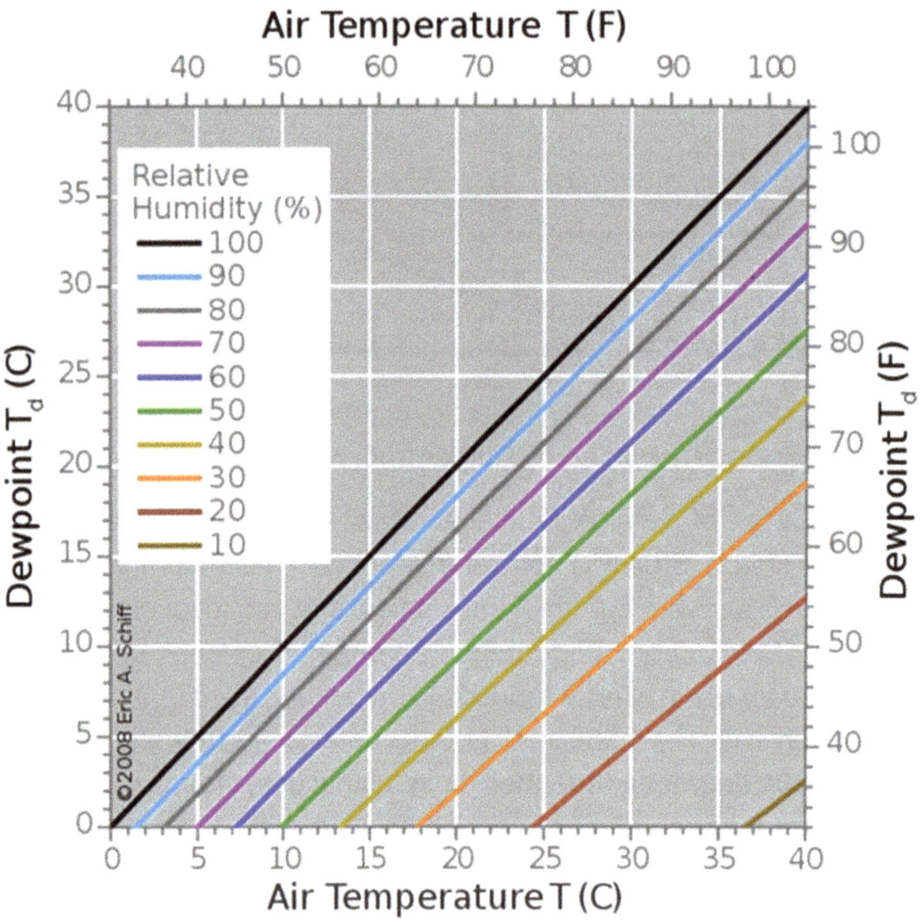

(Source: The National Weather Service, NOAA)

Anyone who has spent much time in both an extremely humid environment ("normal" for the northeast coastal region of the United States in summer months) and an arid environment (extremely low humidity but very high temperatures ("normal" for the Phoenix, Arizona, desert region of the southwest United States in summer months) can describe the difference between "wet heat" and "dry heat". Dry heat is preferred over wet heat by probably 99 humans out of 100. In Chandler (a southern suburb of Phoenix), Arizona, in 2021 and 2022 during the months of July through September, in both years there were periods of about 18 days straight with a temperature of above 100 degrees Fahrenheit but with humidity levels that were consistently lower than 20 percent. With peak temperatures of about 112 to 115 degrees F during those hot weeks, residents did not take pet animals for walks on the hot concrete sidewalks as that would cause burn injuries to the bottoms of their paws. Some residents with pets have "shoes" for their pets. Infrastructure maintenance projects were conducted during the nighttime hours for the dual purpose of least inconvenience to heavy vehicular traffic but also to avoid the excess heat on road maintenance laborers.

High humidity days coupled with high "normal" summer temperatures can create a very dangerous health hazard to humans (especially those with existing respiratory problems or those who engage in some form of exercise or labor that increases their perspiration rate). Spikes in humidity or spikes in a temperature above "normal" that last for less than twelve hours must be considered to be weather-related and not the much longer-term climate related. The National Weather Service (part of the National Oceanic and Atmospheric Administration) has made a "Heat Index" chart available to the public that warns of those combinations of humidity and temperature that are hazardous to human health:

NWS Heat Index — **Temperature (°F)**

Relative Humidity (%)	80	82	84	86	88	90	92	94	96	98	100	102	104	106	108	110
40	80	81	83	85	88	91	94	97	101	105	109	114	119	124	130	136
45	80	82	84	87	89	93	96	100	104	109	114	119	124	130	137	
50	81	83	85	88	91	95	99	103	108	113	118	124	131	137		
55	81	84	86	89	93	97	101	106	112	117	124	130	137			
60	82	84	88	91	95	100	105	110	116	123	129	137				
65	82	85	89	93	98	103	108	114	121	128	136					
70	83	86	90	95	100	105	112	119	126	134						
75	84	88	92	97	103	109	116	124	132							
80	84	89	94	100	106	113	121	129								
85	85	90	96	102	110	117	126	135								
90	86	91	98	105	113	122	131									
95	86	93	100	108	117	127										
100	87	95	103	112	121	132										

Likelihood of Heat Disorders with Prolonged Exposure or Strenuous Activity

Caution · Extreme Caution · Danger · Extreme Danger

(Source: The National Weather Service/NOAA)

If the "settled scientists" modify the claim to be "Total (natural and Man-made) carbon dioxide and water vapor in the atmosphere causes the "greenhouse" effect on global warming to be of urgent concern to the public, then a grade of B+ on the "test of reasonableness" could be assigned. If the "settled scientists" were to modify their claim to include astrophysical, geophysical, total carbon dioxide, and water vapor effects (such as humidity) on global warming, then a grade of A could be assigned for that test. But that IF is not likely. IF a bullfrog had wings, it would not bump its rear end so hard when it lands after a leap.

It is troublesome when a scientist(s), after observing a very small change in a variable in a laboratory experiment, over-exaggerates the observation and resorts to "chartsmanship" (the art of selecting scales on graphic illustrations to exaggerate the visual importance of the factor) to impress his/her peers. The average human temperature sensory system cannot distinguish the difference between two temperatures that are a degree or less apart. However, when a scientist records ocean water temperatures that vary seasonally as they normally do, but the "normal" for that season changes by less than one degree in a period of 30 years and reports that temperature database to show that the ocean water temperatures are "soaring" with time due to climate change – that is "chartsmanship".

The "Grim outlook" version (the red markings are those of the Author) of a chart which was printed in a science magazine:

CLIMATE

Grim results of the ocean's hot streak

Off-the-charts temperatures impact corals, sea ice and more

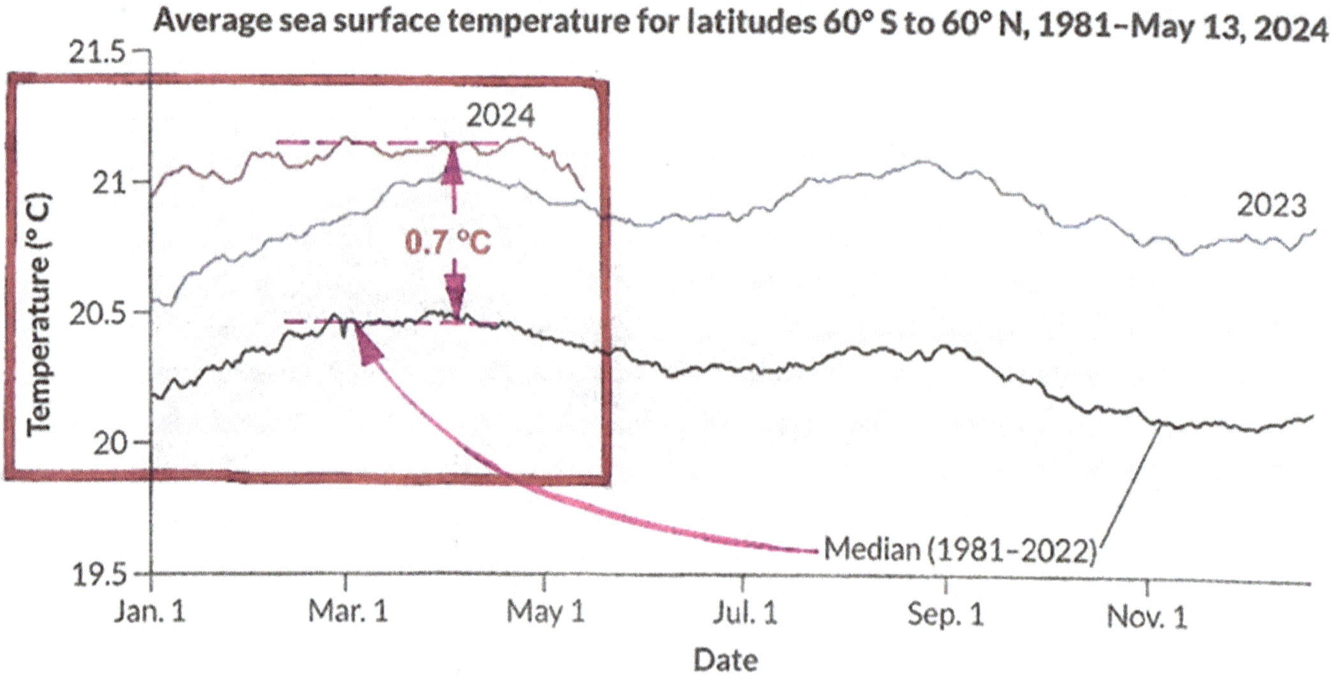

Soaring sea temps Since March 2023, the average sea surface temperature of the extra-polar ocean, at latitudes from 60° S to 60° N, has broken the daily record every single day, NOAA data show. And recent daily temperatures have shattered records that were set last spring.

SOURCE: CLIMATE REANALYZER/UNIV. OF MAINE; NOAA

Shown below is the "white-washed" version (i.e., without emotional descriptors of over-exaggerated significance) used at the UN IPCC Conference of the Parties (COP) deliberations. The COP has been described by pundits as "a group of environmental scientists all drinking their own bath water". The following diagram is shown to address that factual, but less than "angst-provoking" observation in an attempt to deceive the public by scientific "chartsmanship". The exaggerated scale on the left edge of the chart ranges from 19.5 degrees C to 21.5 degrees C which creates no "crisis" level of concern for human safety, health, or comfort.

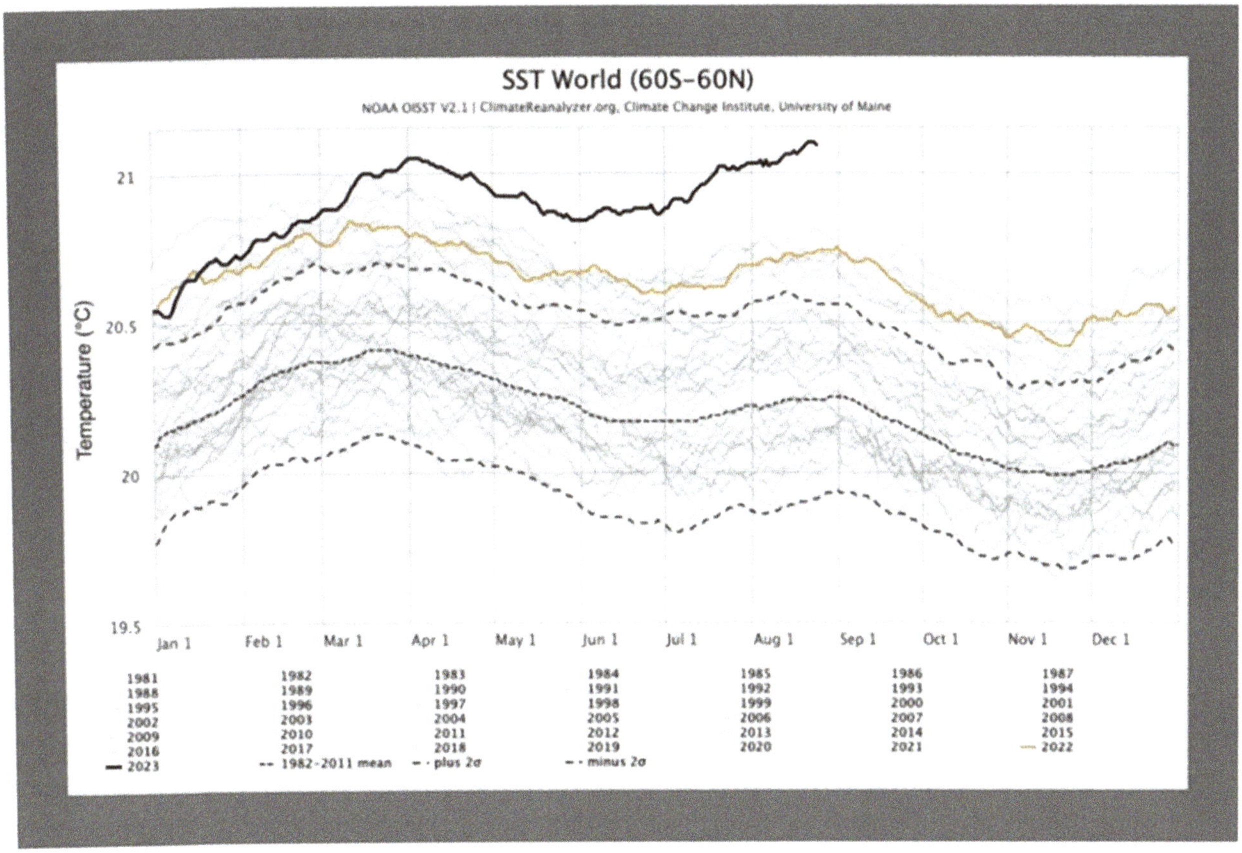

(Source: NOAA, Climate Change Institute, University of Maine)

There may be a "crisis" for some forms of ocean life, especially anemones on the Great Barrier Reef of Australia – which lose their color and "bleach white" – with less than a two degrees of water temperature change.

Readers should note that regardless of the "seasonal" variation in the World's oceans' waters temperature being less than 0.7 degree Centigrade/Celsius from the peak low and peak high temperature for each of the years 1981 to 2022, the water temperature "soaring" for partial year 2023 is apparently 0.2 degrees C above "normal". If there is reportedly an El Nino (warming ocean water temperature flowing from west to east toward the west coast of the USA) condition forming in the Pacific Ocean in 2023, that is not a Man-caused phenomenon caused by the burning of fossil fuels. El Nino originates near the Hawaiian Islands chain, all formed from volcanic eruptions eons ago, but underwater volcanoes are still forming islands that have not emerged above water surface level yet in that chain. That much underwater heating is most likely a contributor to, if not the main cause for, the El Nino warm current.

Of equal and perhaps greater significance as being a heat energy source warming the waters of the ocean floors is the fact there are possibly hundreds of thousands (if not millions) of steam vents continually flowing heated water at steam levels of temperature into the waters of ocean floors. Steam vents have been observed to be in the 15-20% of the ocean floor that has been explored (by non-alarmist oceanographic scientists such as Jacques Cousteau and family). Photographs of steam vents of two different categories (the color and content of their vented steam) are shown next:

(Credit to Unknown Photographers, National Geographic Society)

It has been estimated that over 80% of the ocean floor remains unexplored, and until the underwater steam vents factor is evaluated for its cause of the ocean waters being warmed – as they appear to be regionally – it is very scientifically invalid for environmental protection scientists to proclaim that the ocean levels rising are due to water temperature increases primarily traced to Man burning fossil fuels and creating the heat-trapping "greenhouse gas effect" in the atmosphere many thousands of feet distant from (i.e., far above) the ocean floors.

Supported by later information in this book, environmental scientists may be deluded with "average global temperature" on Earth's land surface area and in the "Average Sea Surface Temperature of the World". The fact remains that the surface temperature on land areas and on expansive ocean areas is dependent on the ocean currents, the wind currents, astrophysical factors, natural weather events locally, and climate change which has been repeatedly pointed out in this book to be regional in latitudinal zones on the planet. The "one-size-fits-all" singular form of "average global temperature" analysis by atmospheric scientists or ocean water scientists so desperate to develop a model for predicting the future of the planet's environment and have lost touch with the factual realities of the problem.

From another perspective of the concern of climate change and ocean water temperatures rising, when water warms it expands its volume and factually does cause rising ocean sea levels to become a human safety concern in the long-term to residences and places of business (especially large shipping ports) or cities presently only a few feet above present sea level. Because the total volume of water of the 70% of Earth's surface is so huge, a few degrees of temperature rise may represent only a very small percent of volume increase, but when the volume is so huge a small percent of temperature increase can expand a huge volume of water which has no elsewhere to expand other than up and over present coastal regions. A later chapter of this book addresses those long-term effects of ocean surface temperatures and rising levels and the need to start planning soon for infrastructure modifications in coastal areas threatened by ocean levels rising.

It is human nature to delay making a decision to relocate elsewhere until the cause for making the move is literally flooding their ankles (as shown in the next photograph):

(Credit to unknown photographer, Miami Dade County August 2023)

The importance of water on our home planet is taught in the grade schools by geography teachers who have possibly never been more than a hundred miles from home and to a beach.

Students in the pre-1950 time period of attendance in public schools were required to read three literary classics – "The Iliad" by Homer, "The Rime of the Ancient Mariner" by Samuel Taylor Coleridge, and "Moby Dick" by Herman Melville. All three literary classics had the theme of ocean adventures and man interacting with his environment. A rhyme from the ancient mariner comes to memory: "Water, water everywhere, but not a drop to drink. Water, water everywhere, and how the boards did shrink." (as the wood planks on the deck of the old sailing ships gradually dried out from the relentless solar rays (known then to the public as "sunrays"). It was not widely taught at that time in students' early development years that ocean water has a high saline content and the "not a drop to drink" lament was a subtle way of teaching that ocean water is too salty to drink. Only fresh water should be drunk – and in large quantity daily – to satisfy the human body needs for water for its internal organs to perform normally and sustain human health.

In some – but not all - high school science classes, it is taught that saline water such as that in the oceans can be made potable (safe to drink) by distilling the water and aerating the distilled water that no longer has a saline content to make it potable – for agricultural uses or for human consumption. But distillation in high school science and chemistry classes was somewhat down-played as to the "big picture" of the oceans evaporating water from their surface into the atmosphere, and then later returning the naturally distilled water vapor from the atmosphere back to Earth's surface as precipitation (rain, snow, hail, graupel, etc.).

Gravity is the ever-present gravitational acceleration pulling every untethered or attached mass towards the center of the planet – such as precipitation in the forms of rain and snow. That precipitation provides the potable water vitally needed for agricultural growth, human health, and less vital as coolants in radiator systems cooling machinery systems for transportation vehicles and for residential and business facilities. The water that is converted into steam (the gaseous form of water vapor) for driving turbines to generate electricity must not have high salt-like impurities as that would corrode the boilers used in the water-to-steam conversion process. In the "big picture" most everything would have a short life if the corrosive salts had not been removed from the water taken into systems for use in human consumption, in transportation machinery, and in industrial processes.

The freshwater available to the human population on the planet is not distributed very usefully and is far less accessible than is the non-potable saline water in the oceans. An estimated 70% of the world's fresh water is located in abundance (26.5 million cubic kilometers of potable water) on the Antarctic continent – but most is in frozen form on the Antarctic Ice Sheet. A smaller portion is on Antarctica's ice shelves and icebergs (which eventually are going to melt into the Southern Ocean and become non-potable) with a portion in glacial form above the Antarctic continent and in liquid state in the several lakes below the ice sheet.

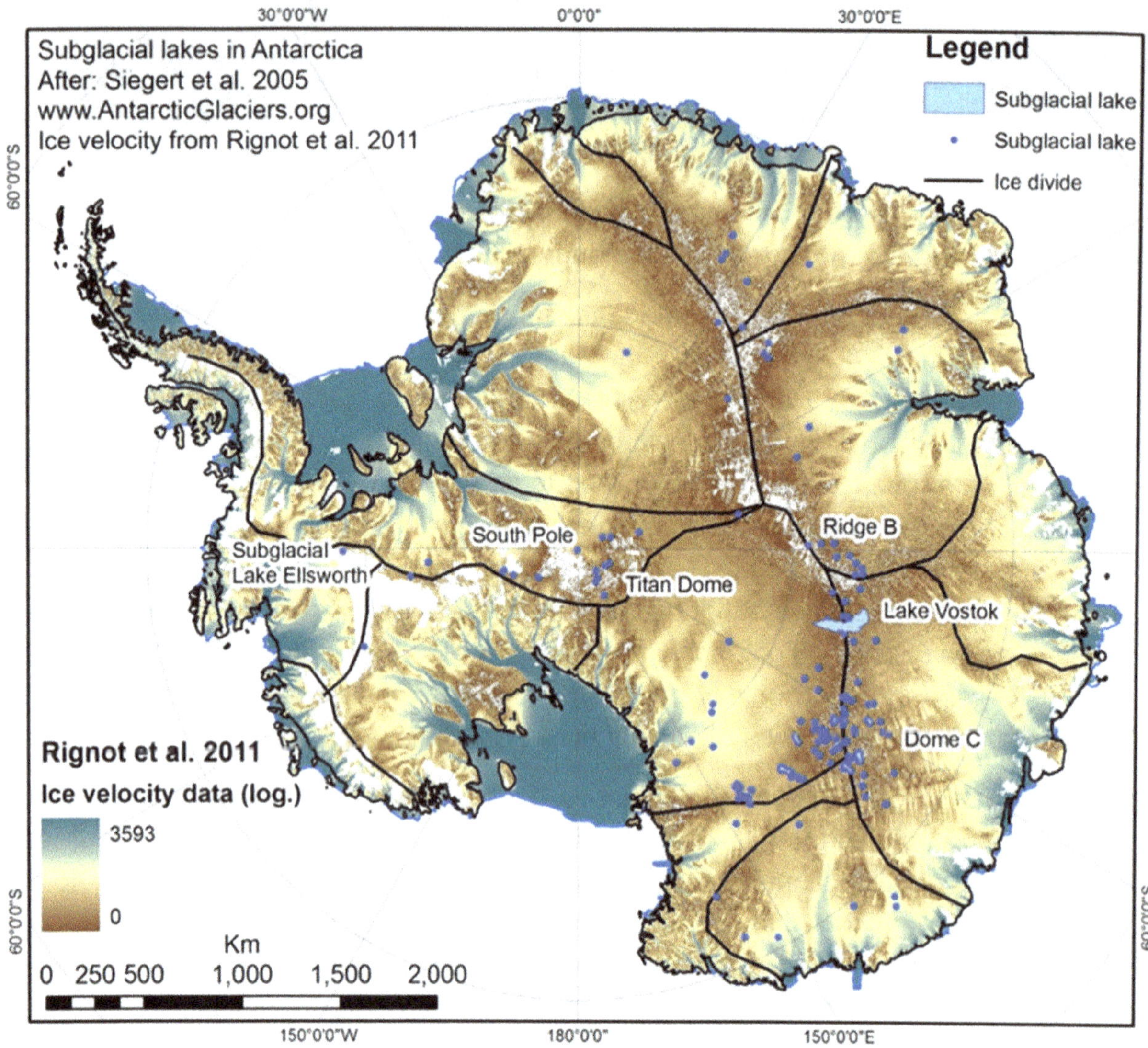

(Source: AntarcticGlaciers.org)

Apparently, the lake water is warmed from geothermal activity beneath the Antarctic plate and the ice sheet acts as an insulating layer to prevent the water from evaporating into the atmosphere.

As this book was being written, there appeared an article in the news media that the world's largest known iceberg A23a – five times the size of New York City and as thick as the Empire State Building in New York had "calved" away from the ice shelf in West Antarctica. Its path is northerly in the Southern Ocean and with a possibility that winds and ocean currents may cause the gigantic iceberg to enter into southern Atlantic Ocean waters. The article shown next provides more information to the public on iceberg (from Germanic "eisberg" meaning mountain of ice) A23a (as it has been labeled for tracking purposes):

World's largest iceberg drifts toward South Atlantic

BY PAUL GODFREY
UPI.com

A gargantuan Antarctic iceberg five times the size of New York City and as thick as the Empire State Building is drifting away from the continental ice shelf toward the South Atlantic Ocean.

Designated A23a, the iceberg had remained in situ - grounded on the floor of the Weddell Sea since breaking from the Filchner Ice Shelf in 1986 - until earlier this year when it began slowly shifting northward.

It is believed A23a finally was able to break free from the seabed after decades of shrinkage.

Driven by wind and ocean currents, its passage recently has picked up pace, and satellite imagery now shows it is about to pass the northern tip of the Antarctic Peninsula.

Posting a map of the iceberg's movements over the past seven months on X, British Antarctic Survey mapping specialist Laura Gerrish said it was the world's largest iceberg.

"This giant berg calved in 1986 and has been grounded in the Weddell Sea since 1991. It started moving again last year, and now that A68 [another giant Antarctic iceberg] has broken up, it reclaims its title as the world's biggest iceberg right now at 1,544 square miles."

In 2020, there were concerns that A68 might run aground on the sub-Antarctic islands of South Georgia and kill marine life on the sea floor and subsequently affect the food supply for penguins and seabirds. Those fears were averted when the monster berg broke apart, which experts said was the likely fate of A23a.

Russia, which had built a research station on the Filchner Ice Shelf, was forced to mount a rescue mission in 1987 after the base found itself adrift on A23a. The base was quickly replaced with a new facility on nearby Cape Norvegia, partly using equipment and prefabricated structures salvaged from the old base.

(Source: United Press International)

Letting the human mind wander, it would be an amazing feat if that monster size iceberg – containing potable water (suitable for human consumption or agricultural irrigation) when thawed – could be tugged by a number of sea-going tugs up to the coast of South America nearest the Amazon basin. The Amazon basin is currently suffering extreme drought conditions, and environmental scientists estimate that the ancient forest stores 123 billion metric tons of carbon; that is more than three times as much carbon as humans emitted in 2022. If scientists could devise an efficient way to transfer that much water in the A23a iceberg to irrigate the Amazon basin, the population of South America could benefit tremendously from that irrigation of the Amazon drought region. If nothing can be done with A23a, it will only melt into the Southern Ocean as it approaches warmer waters to the north and be wasted.

Since the A23a ice is already in a floating state, all melt-off will dissipate into the Southern Ocean and not raise the surface level of the Ocean – floating ice (including that on the opposite polar region in the Arctic Ocean) does not increase ocean surface levels when it melts.

A human-developed scientific process known as "reverse osmosis", basically to filter the unwanted salt from ocean water, has been applied in several large water purification applications throughout the world. However, that process has been found to be of limited use for only local human consumption – and not practical in the large scale for large agricultural needs (such as the Central Valley of California which suffers drought conditions for periods of forty years or more). The irony of that is that less than 80 miles away from the central valley of California lies an almost infinite amount of water – the Pacific Ocean. Granted, there is a mountain range – the Coastal Range – over which water would have to be pumped to make the potable water accessible for distribution onto the agricultural fields and groves of nut and fruit trees and cattle ranches, but solar energy collection (abundant solar energy is being deposited on the planet's surface in that region) could ease the energy needs for that pumping and distribution process. A few reverse osmosis facilities have been built on the California coast but the process is not of a large enough scale to totally satisfy the high population of the state and/or the Central Valley agricultural needs.

Presently, the water for southern California comes from a few sources – a portion from the Colorado River east of the central valley as well as from lakes and reservoirs naturally formed or Man-constructed in southern California. A larger portion – serving the largest city in California, Los Angeles, with a population of 18.64 million per the 2020 Census - comes via the Los Angeles Aqueduct system built by civil engineer William Mulholland (1855 - 1935). The use of aqueducts to supply a city with water is documented in history, and perhaps the most famous is the aqueduct system in ancient Rome.

The Los Angeles Aqueduct system originates in the Owens Valley (above Yosemite National Park) in the northern part of the state and flow (north to south – i.e., "downhill" by gravity) in the Aqueduct relies on the melt-off of the snowpack in the Northern California mountain region. Where there are some long hills preventing water in the Aqueduct to flow over by gravity flow alone, there are pumping stations to lift the water over the hill and back into the downhill portion of the Aqueduct. Again, every region of the United States must develop a unique water system supply for its regional population. It seems that every small town in the United States has a water tank atop a steel tower (much like a gigantic golf ball on a tee – or as likened by the Author's children on driving tours to be "a giant spider with a cap") boasting the name of the town on its side. It is the responsibility of officials for any region where there is a large population to assure the public safety and health from whatever type catastrophe has the moderate to high possibility of happening.

While researching the topic area of desalination of ocean water, it was a pleasant surprise to become better informed on how much and what types of desalination are already in operation around the planet. An interesting illustration is shown next:

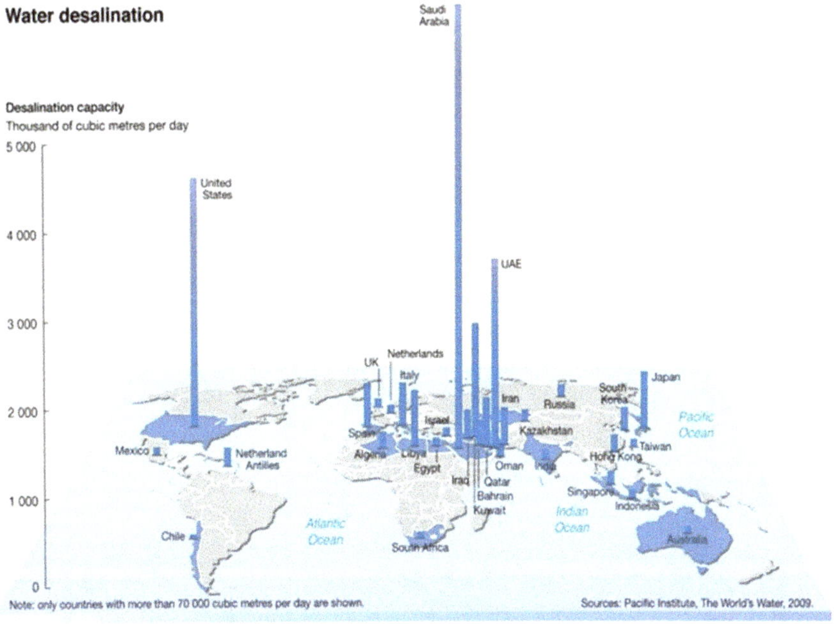

(Source: Pacific Institute, 2009)

The primary users of desalination are the countries in the Middle East countries (Saudi Arabia, United Arab Emirates, Bahrain, Dubai, etc.) using water from the Arabian Sea. The largest desalination facility using the reverse osmosis process of desalination is in Dubai: while the largest desalination facility using the distillation method is in Kuwait.

(Credit to unknown photographer in Dubai)

Desalination facilities in the United States are becoming more popular. Most exist on the west coast and in California:

Existing Seawater Desalination Facilties

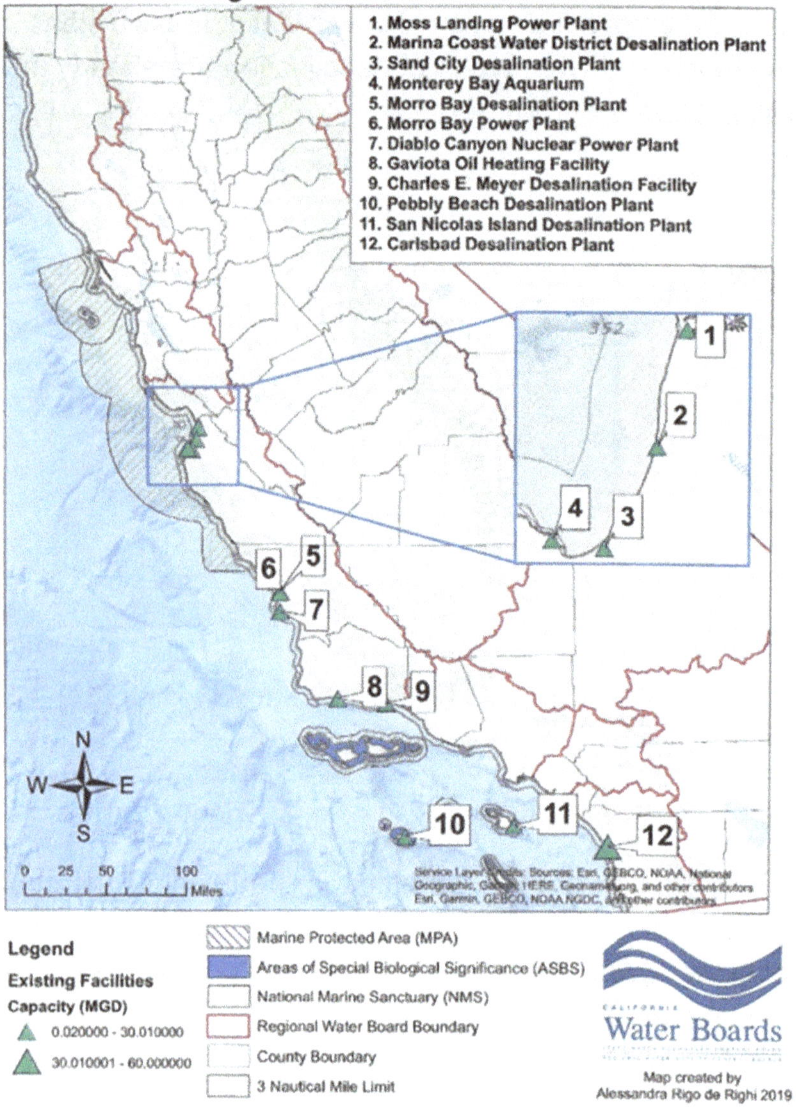

(Source: Water Boards of California)

Almost all known desalination facilities (in the USA and most of the planet) are used for water supply to the local human population and little for the agricultural supply. However, with the world population growing in number exponentially coupled with there being more drought areas, the need for larger volumes of desalinated water than currently developed (and using the slow reverse osmosis process) must increase use of the distillation process of desalination technology powered perhaps by solar energy. Two of the 24 weighted trade studies in the final part of this book assess and prioritize the desalination technologies.

The climate is changing in southern California where it once was boasted: "It never rains in California". In recent years that boast is not heard very often as it has alternated between too much water, causing flooding in areas where no flood control channels were constructed and in turn causing mud slides that devastate residences in the hills for which the entertainment industry has drawn residents from other arid states, and too little water. The central California region which grows the agricultural products for much of the “48 lower states” is at the same time experiencing drought conditions.

Environmental scientists point the finger of blame at the Man-made “greenhouse gasses effect” but the amount of the greenhouse gasses being produced are in a steadily growing manner and fail to confirm that the “scourge of Man” (burning fossil fuels) is the primary cause of the alternating “too much or too little’ water in the southwestern United States.

Short term geophysical events such as earthquakes and landslides are usually experienced with little or no warning and are over in a matter of minutes. However, the residual long-term after-effects may dwell for a generation. Longer term flooding on a level plain is very survivable, but flooding on a sloped land is highly dangerous to human and animal life as well as to any Man-made structure in the path of a flood rampaging down a hillside. It matters little to the victims whether the flooding was caused by too much rainfall in a short period of time or the result of failure of a Man-made structure such as a dam, life is much like the bell that has been rung and cannot be un-rung – death cannot be reversed either, no matter the cause or how quickly it occurs. The news media has little respect for ethical and complete reporting, but environmental scientists are showing disrespect for the honorable science profession by pointing the finger of blame for global warming, climate change, and the ocean levels to rise on the small amount of carbon dioxide that is factually in the atmosphere. There is more of it from natural causes than there is from anthropogenic causes. How can the lower amount of carbon dioxide in the atmosphere be the cause of global warming and the higher amount not be? In the laws of chemistry, is carbon dioxide CO_2 no matter how the CO_2 was created?

Returning to how vast the volume of free-standing waters exists on the planet, the Author was given an “all-expenses-paid” cruise courtesy of the US Army who invited him to a beach party in the quaint harbor town of Inchon, South Korea, in the early-1950s. The cruise was on board a “luxurious” troopship (USMS Gen. William Weigel) for 15 days on the Great Circle route in the Pacific Ocean between Seattle, Washington, to Sasebo, Japan, and then on a smaller landing craft up through the eastern Yellow Sea into Inchon Harbor. On a “buy one, get another one free” cruise (again courtesy of the US Army) a year after the first cruise, the fortunate survivors of the Korean War boarded a “luxury liner” troopship (e.g., USMS Gen. John Pope) and were returned across the center of the Pacific Ocean (passing south of, but not stopping in, Hawaii) back through Puget Sound and into Seattle from whence the round-trip started over a year earlier. The return leg across the Pacific Ocean for the Author on that “Great Circle” route took 16 days.

It was those two cruises (totaling a full month “at sea”) that gave the Author a first-hand sight of just how enormous the Pacific Ocean was/is. Knowing that the water is a few miles deep and exists from horizon to horizon and beyond is mind-boggling. The tabular statistics showing the immense quantities of water in and the surface areas of all the oceans and seas of the planet were shown at the start of this chapter.

Later during the Author’s professional career, the opportunity arose to make 18 round-trip flights in commercial airliners over the Atlantic Ocean between the United States and western Europe – mostly to England, Germany and Switzerland. During their retirement years, the Author and his wife also took three cruises – aboard true luxury liners. The first was from Florida through the Panama Canal and to California; the second from the French Cote d’Azur through the full length of the

Mediterranean Sea, through the Tyrolean Sea and Aegean Sea, and through the Adriatic Sea back to Venice, Italy (which was shocking to see how much of Venice is already largely underwater). The third and last cruise was from Seattle up through the Inland Passage (and through Glacier Bay) of the Alaskan Coast to Skagway, Alaska and return to Seattle in the US Pacific northwest. Not only were those ocean and sea voyages very impressive as to the amount of water covering the planet, but it was the educational and visual experiences ashore in the many different countries and seeing how the local human residents adapted to their environment that differs in each country - especially depending on their latitude. (The Author also won a paper airplane contest aboard the ship in the Aegean Sea, but that experience does not lend much credence to the Author's capability for understanding the rules of aerodynamics.)

The Author and his family have memories and a wealth of interesting and learning experiences of having resided in or traveled by land-vehicle or train to/through every one of the fifty United States (the contiguous 48 and Alaska – fishing for King salmon and halibut - and snorkeling in warm Hawaiian waters). Those travels took the family on visits to all national and many state parks, to historical and natural geology museums and Presidential libraries, to military memorial cemeteries, or to visit with family or close friends. Included in those "bragging rights" is the fact that family homes were bought, or apartments leased in eight states (New York, North Carolina, Georgia, Texas, Ohio, California, Arizona, and Idaho in which the Author's family resided for more than two years (the longest in California for thirty-eight years). Personal observations of the different weather events and climate changes in different regions of the world and in the different states (all 50) of the USA have been found to be highly useful to the Author when interpreting the effects of those events and changes.

To see a photograph of a tornado is much less impressive than being physically in the fringes of one. Living in the different environments (weather, climate, and seasonal changes) in the different states and different countries in the Northern Hemisphere (no travel was made in the Southern hemisphere – although there are several family members in Australia, and New Zealand) as well as learning what the local public throughout the planet regards as being most important to their safety, health, and comfort equals the research done through books and papers and news articles. Those many experiences gave a wealth of background observations made as to the prioritized importance of the many factors that the trifecta environment problem (surface temperature extremes, climatic extremes, and ocean life in different temperature categories) has on the human population of many different regions of the planet (albeit all in the Northern hemisphere).

Understanding the factors in a multiple-factor, complex, database of information is made easier if/when there is close personal connection with most of the factors in that database. A major observation this Author has come to realize (from personal observation and with intensive research in the various databases available for scientific analysis) that the Earth's surface temperature is changing very much but more on a regional basis – severely in some regions and no significant change in fortunate others. Where the regions of elevated surface temperatures above the "normal average temperature - referred to as being temperature anomalies - for that region are located appears to have at least one boundary with a large mass of water.

The following chart shows that the different regions of the planet are experiencing different surface temperature anomalies (an "anomaly" is the difference from "normal" for that region):

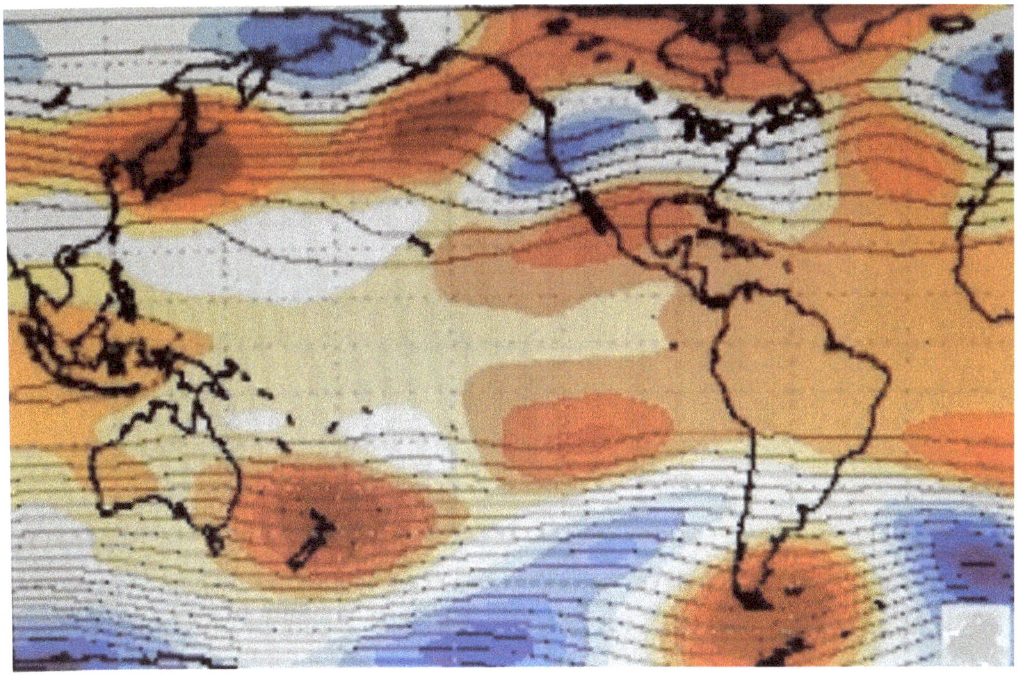

(Source: NASA)

If the environmental scientists can provide a good explanation for public understanding as to how the "greenhouse gas effect" – which purportedly changes average annual global temperature - develops "pockets of anomalies" such as those shown above, it would be most helpful.

This form of temperature observations on Earth's surface (land and water) reflects no scientific reason to believe that Man burning fossil fuels and creating "greenhouse gas effects" in industrial regions but not in non-industrial regions bears any scientific correlation with the measured temperature variations as measured by NASA satellites shown above. In the case of the temperature increasing at the Great Barrier Reef (on the east coast off Australia), it is seen that the reef is on the edge of the "temperature dome" centered almost directly over New Zealand. New Zealand's human population cannot be accused of generating that much "greenhouse gas" to warrant industrial "greenhouse gasses" generated in the upper Midwest of the USA being blamed by NOAA environmental scientists as the dominant cause for the bleaching of anemones on the Australian reefs.

A more plausible explanation, as shown earlier at the beginning of this book, is that the ocean currents are very warm in the southern pacific and affect the temperature of the coral reef which is not very deep below the Equatorial heating of the ocean water:

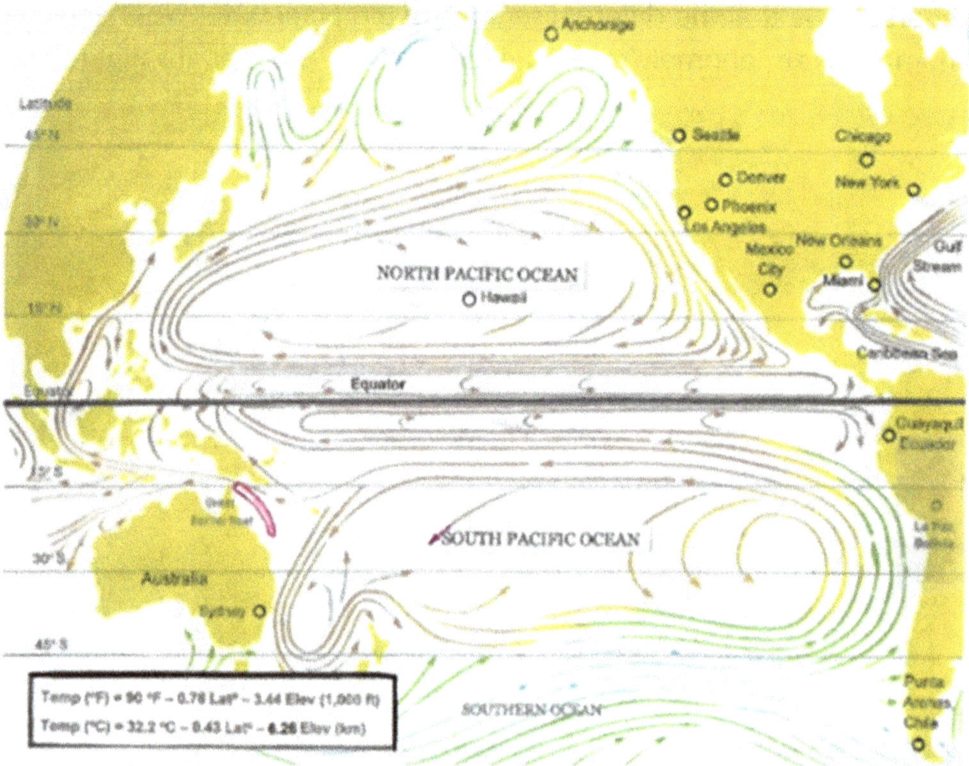

(Source: Rand McNally Atlas of the World, with annotation by Author)

The mathematical expression for temperature in both the Fahrenheit and Celsius scales may be helpful in estimating the average annual temperature of cities at different latitudes due to latitude and elevation. Any effect from warm or cold ocean currents on that temperature estimate are guided by maps of the ocean currents being warmer or colder. The Great Barrier Reef bleaching is due to the very warm Equatorial current which is getting warmer – but not from the "greenhouse gas effect".

Readers are reminded that the illustration on the front cover of this book represents the "average" temperature (collected by satellites not the land-based "average global temperature database from questionable averaging of statistics from a non-uniformly distributed system of meteorological stations located in predominantly the northern hemisphere) against which the temperature anomalies shown above are based. Because of its extreme importance to understanding why this book continues to urge addressing the trifecta problem (global warming, climate change, and ocean levels rising) in a "big picture" perspective, the cover illustration is repeated here for Readers' convenience – and with a color scale shown below the illustration such that numerical temperature variations can be identified for their level of importance to human safety, health, and comfort regionally; i.e., as latitude varies:

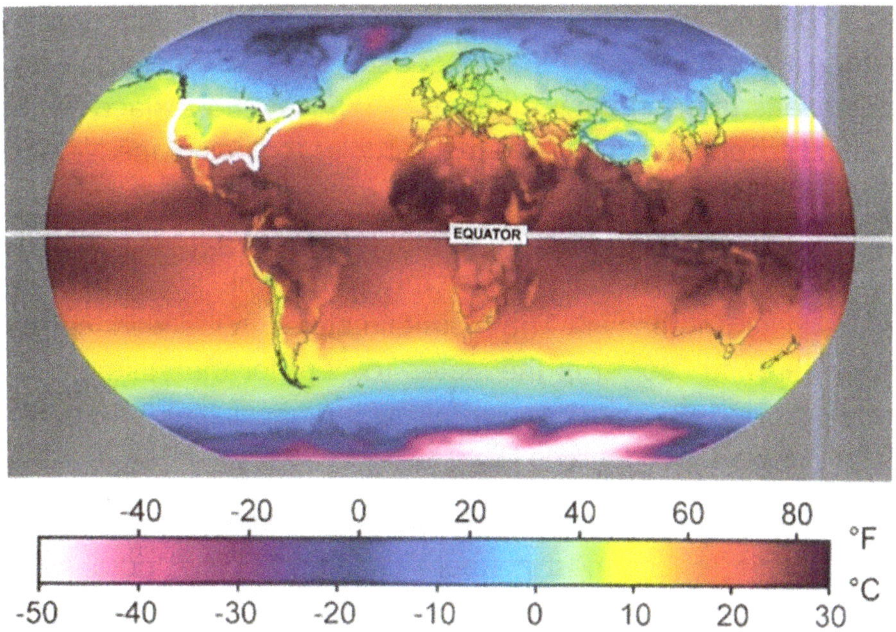

Annual Mean Temperature

(Source: NASA)

Meteorological records for the United States region are consistent with the NASA "big picture" satellite measurements showing the red color (high average temperatures) is predominant in the southeastern and southwestern states comprising the USA. A closer-looking average temperature database for the "lower 48" states of the USA is shown next:

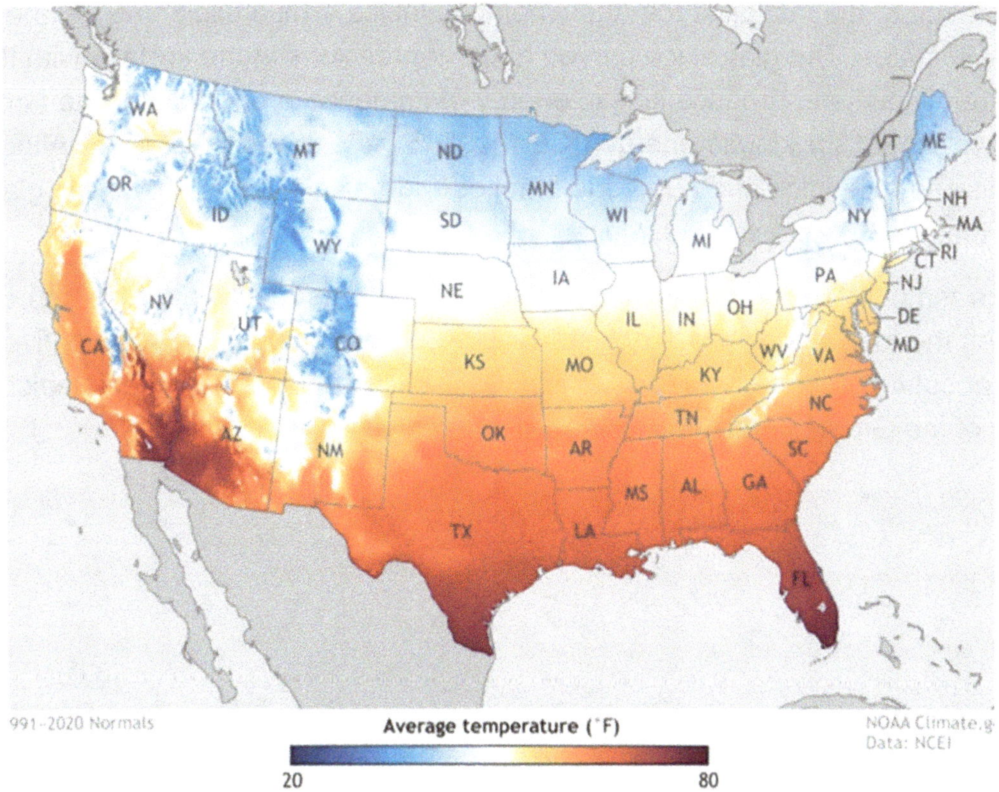

(Source: NOAA, National Weather Service)

The warming effect of the "Gulf Stream" (which originates in the Caribbean Sea) in the Atlantic Ocean on the eastern and southeastern states' coasts, the Gulf of Mexico on the south-central states, and the arid deserts of the southwestern states causes the surface temperatures to be the highest in those regions in deep red compared to the cooler regions of the New England region and the upper central states in light blue. The temperature difference from deep red to light blue on the color scale is seen to be on the order of 50 to 60 degrees Fahrenheit. The temperature in Miami, Florida may be 90 degrees Fahrenheit while at the same time the temperature in Milwaukee, Wisconsin, is at 30 degrees Fahrenheit). The mountain states in the west-central states (because of their high elevation) are the coolest of the states (except for the state of Alaska which is very much northwest of the "lower 48" states and at a much higher latitude between Ketchikan (55.3 degrees N) and Point Barrow (71.4 degrees N).

Within a region the size of the United States may also be characterized by identifying the extremes of its most southern latitude and its most northern latitude. For that purpose, the latitude of the southernmost point of latitude of the contiguous ("lower 48") United States is that of Key West, Florida at 24.5 degrees North (and longitude -81.8 degrees West). The latitude of the northernmost point of latitude (with a significant population density) of the contiguous United States is that of Bellingham, Washington at 48.8 degrees North (and longitude -122.5 degrees West).

Larger regions that have one or more boundaries with a major ocean, sea, gulf, or bay may also have an inland lake or series of lakes that total o be a highly significant influence on surrounding regions. The United States is one such region that is blessed to have three of its boundaries with ocean and gulf interfaces but has the series of lakes known as the Great Lakes. These five large lakes are the result of the last Little Ice Age when glacial ice a mile thick covered the northern half of the "lower 48" states. The glacial ice carved huge depressions in the surface with their immense power when they "flow" and produce kinetic energy of immense proportions. (The familiar scientific formula in Physics 101 for kinetic energy (K.E.) is: K.E. = ½ mass x Velocity^2. Although the land speed V of the one-mile thick and hundreds of miles wide glacial ice is very slow, glacial ice mass m is so huge that the K.E. is extremely huge.)

As the surface temperature of the region increased (due to astrophysical reasons) the glacial ice thawed and left the massive amounts of water in the depressions carved and the five Great Lakes were formed about 4,000 years before the present. The illustration shown next depicts the outlines and locations of the Great Lakes region of the USA:

(Source: U.S. National Geodetic Survey)

The statistics of the five Great Lakes and the amount of surface area in the United States they cover is as impressive when shown tabulated as the earlier tabulation of the major oceans and seas were shown earlier:

	Lake Erie	Lake Huron	Lake Michigan	Lake Ontario	Lake Superior
Surface area	25,700 km^2 (9,910 sq mi)	60,000 km^2 (23,000 sq mi)	58,000 km^2 (22,300 sq mi)	19,000 km^2 (7,340 sq mi)	82,000 km^2 (31,700 sq mi)
Water volume	480 km^3 (116 cu mi)	3,500 km^3 (850 cu mi)	4,900 km^3 (1,180 cu mi)	1,640 km^3 (393 cu mi)	12,000 km^3 (2,900 cu mi)
Elevation	174 m (571 ft)	176 m (577 ft)	176 m (577 ft)	75 m (246 ft)	182.9 m (600.0 ft)
Average depth	19 m (62 ft)	59 m (195 ft)	85 m (279 ft)	86 m (283 ft)	147 m (483 ft)
Maximum depth	64 m (210 ft)	228 m (748 ft)	282 m (925 ft)	245 m (804 ft)	406 m (1,333 ft)

Note: The total water surface area of the five Great Lakes is 244,700 sq kilometers (94,250 sq miles)

(Source: National Geological Survey database; tabulation by Author)

The lakes were created from the melt-off of the glaciers from the last "Little Ice Age" and were formed in shape by the advancing and retreating glaciers grinding depressions in the surface of Earth's surface in which the melting glaciers (fresh water) remained in liquid state. Pollution of the Great Lakes by human and industrial processes and inconsiderate practices is discussed in later chapters as being more of an issue than is the "greenhouse gas" crisis contrived by environmental protection scientists. <u>Far more attention and concern should be given to Man-made pollution than to Man-made global warming.</u>

When the states of Alaska and Hawaii) were entered into the United States as the 49th and 50th states – Alaska in January 1959 and Hawaii in August 1959 – the total United States land and water surface area was increased and became as shown next:

that generate the largest amount of "greenhouse gasses" in the United States - are not the regions that are exhibiting the highest temperature anomalies in the United States. If one also considers that the vast majority of the humans in the United States and their unquenchable consumption of electrical energy consumption are in the most densely populated regions of the USA – i.e., in the northeastern (New England) states and in California – neither region having the extensive amount of "Industrial Revolution" as the Great Lakes and southeastern regions, how do environmental scientists rationalize that it is the carbon emissions in the atmosphere caused by the Industrial Revolution processes and the human demands for energy that are the dominant cause for global warming?

Two other "big picture" questions that arise, if environmental scientists are blaming industry and humanity for causing climate changes that are feeding back to cause the frequency and intensity of both tornadoes and hurricanes to be increasing tremendously;

1) Why is it that tornadoes are formed in the southeastern Gulf states and proceed (on their seasonal "tornado alley" path from that low industrial activity and not extensive population activity and energy demands region northeasterly into the increased industrial activities and very high density populated regions of the USA?

2) Why is it that tropical depressions originate in the western North African desert regions, where there is almost zero industrial activity and no population concentrations with high demands for electrical energy, become of higher intensity as they pass over waters where again there is no industrial activity and population of humans, and then take Interstate A (west of the Florida peninsula) or Interstate B (east of the Florida peninsula) when turning northeasterly to batter the southeastern region of the United States where there again is almost no industrial activity and very low population density with low energy demands?

Neither the Tornado Reporting Center or the National Hurricane Tracking Center have a database being described openly and clearly to the public to explain those illogical circumstances. Instead, there is a continuing stream of news media articles that point the finger of blame on the public for being responsible for the tornadoes and/or hurricanes that reportedly are becoming more frequent and of higher severity due to climate change. As a point of information, hurricanes, typhoons, and cyclones have different names (depending on the latitudinal region where they are spawned) but have the same general adverse characteristics of causing flooding, property damage, and loss of human lives. For Readers' information, the following diagram indicates the regions where hurricanes, typhoons, and cyclones are formed and the typical variation in paths taken after formation:

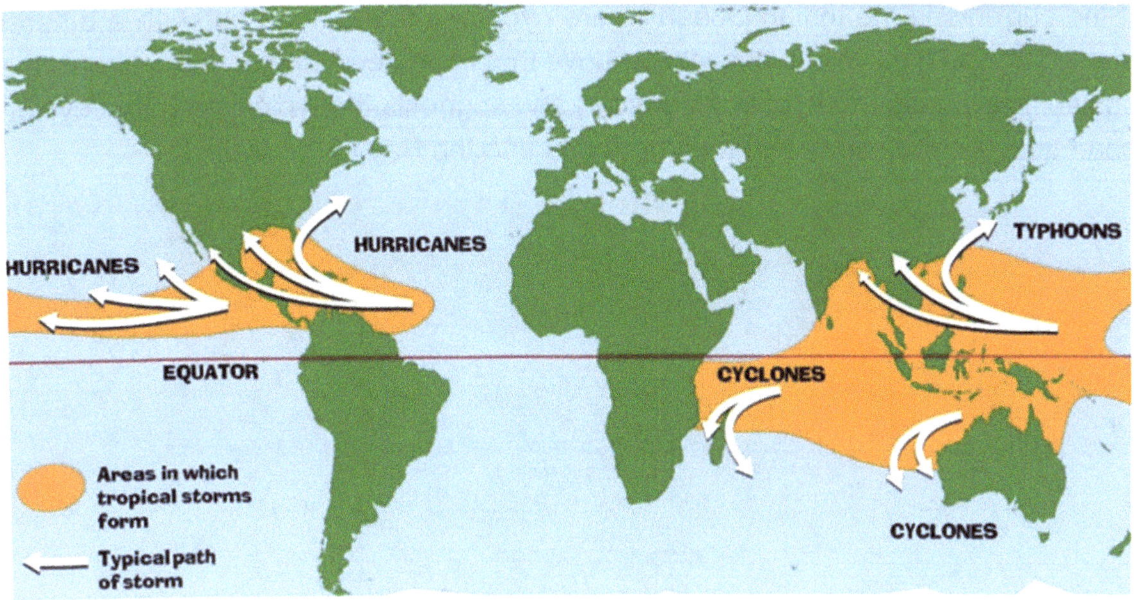

(Source: NOAA)

There has been a recent update to the above chart to add tropical storms that form in the Indian Ocean in the region of East Africa. Those tropical storms that form in that region have an unusual – and unexpected flow as cyclones moving northeasterly and entering the Bay of Bengal. If entering the Bay of Bengal, those cyclones turn northerly and impact the country of Bangladesh which has the highest density of human population on the planet. The similarity of the Bangladesh area (at the delta region of the Ganges River into the Bay of Bengal, and the situation of the city of New Orleans on the delta of the Mississippi River in Louisiana where the delta is formed on the Gulf of Mexico is uncanny in that similarity.

A view of the Category 5 unnamed cyclone that made landfall on Bangladesh on April 29, 1991, was track, size, and intensity-monitored by NASA satellite imagery and provided early warnings to Bangladesh:

(Source: NASA)

A chart of the currents in the Indian Ocean where cyclones (hurricanes but with a different name) are spawned – such as the cyclone described above that impacted Bangladesh – shows the location of those spawning grounds (north of Madagascar) and suggests the typical path that cyclones in the Indian Ocean take (northeasterly and then northerly into the Bay of Bengal).

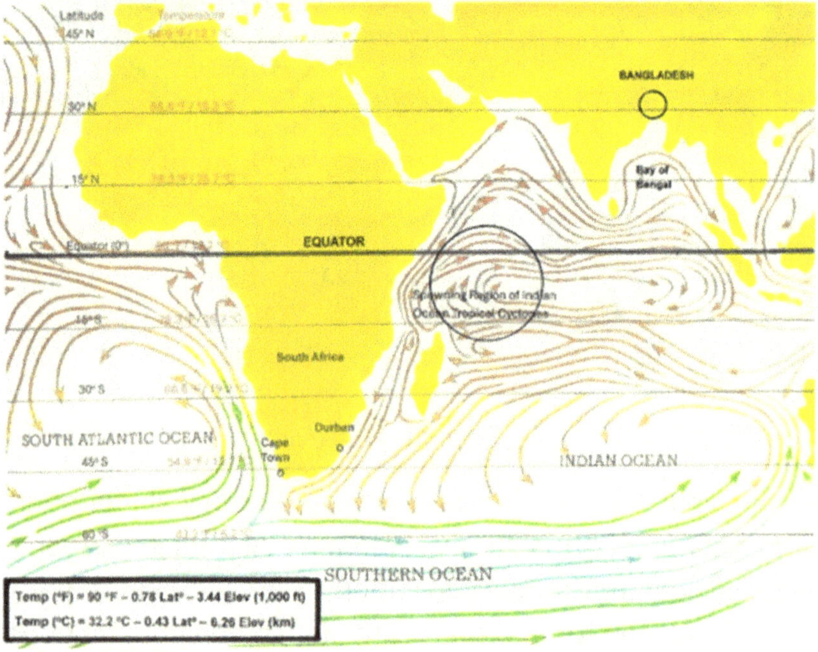

(Source: Rand McNally Atlas of the World, with annotations by the Author)

This event was the largest natural catastrophe in the world in the twentieth century. There were over 135,000 fatalities known to have occurred and over 10,000 people whose remains were never found. Over 10 million people were left homeless, and over 1 million cattle (chiefly Brahmans) were destroyed by drowning. The cyclone destroyed most of the agricultural crops in the region and the many millions of survivors faced a starvation crisis. To understand why the losses were so extensive, two maps of Bangladesh reveal some of the reasons (shown as follows:)

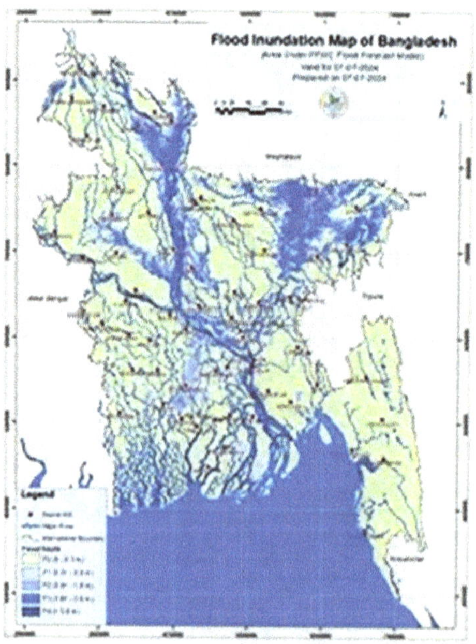

(Source: WorldAtlas.com)

The left map shows the topography (low elevation, hills, mountains) and the location of the major cities comprising Bangladesh. The total population of Bangladesh is 172, 954,319 in a 130,170 square kilometers (50,218 square miles) land area heavily laced with waterways (as seen in the left illustration above). The large Ganges River from India and Pakistan is merged by the Meghna River from Burma and there are literally dozens of smaller rivers that carry the drainage water of the mountains of Nepal to the North into the Bay of Bengal. That topography and high population density (1,329 persons per square kilometer – 3,444 per square mile) the perfect ingredients for cyclone disasters such as the one that struck Bangladesh on April 29, 1991.

Compounding the high population density on a land area that is extremely prone to flooding from heavy rainfall is the low elevation of its major cities. A tabulation of the largest ten cities of Bangladesh having high population density but low elevation is shown next:

Low elevations of Cities in Bangladesh

City	Elevation	Population
	meters (ft)	(As of 2023)
Dhaka (Capital)	4 (13)	23,935,652
Chittagong	11 (36)	5,379,660
Khulna	6 (20)	2,613,385
Rajshahi	21 (69)	2,909,622
Sylhet	35 (115)	999,000
Barisal	3 (30)	533,000
Comilla	13 (43)	1,364,172
Narayanganj	5 (16)	967,951
Mymensingh	14 (46)	576,927
Jessore	9 (30)	298,000

Note: Dhaka has the highest density of World cities at 23,234 people per sqare kilometer. Bangladesh has a total population (2024) of 172,954,319.

(Source: Tabulation by Author from United Nations Population Division database)

The capital city of Bangladesh is Dhaka and is seen to have a population of almost 24 million people residing on land at an elevation of 4 meters or 13 feet. Dhaka has 4 million people greater than are in New York City which is the largest city in the United States with a population of 19,034,000 in 2024. Of all cities that will be hit the hardest in the world by ocean surface level rising, the first largest world city to be impacted the most will be Dhaka if the ocean surface level exceeds a rise of 9-10 feet and the region were to be impacted by a hurricane and surge of 5 feet in water height.

An excellent series of articles by BBC covering the devastation of Hurricane Helene (in October 2024) that took an unusual track from the Gulf of Mexico north through the Pensacola region of Florida as a Category 4 on the Saffir-Simpson scale. The Saffir-Simpson scale is described next:

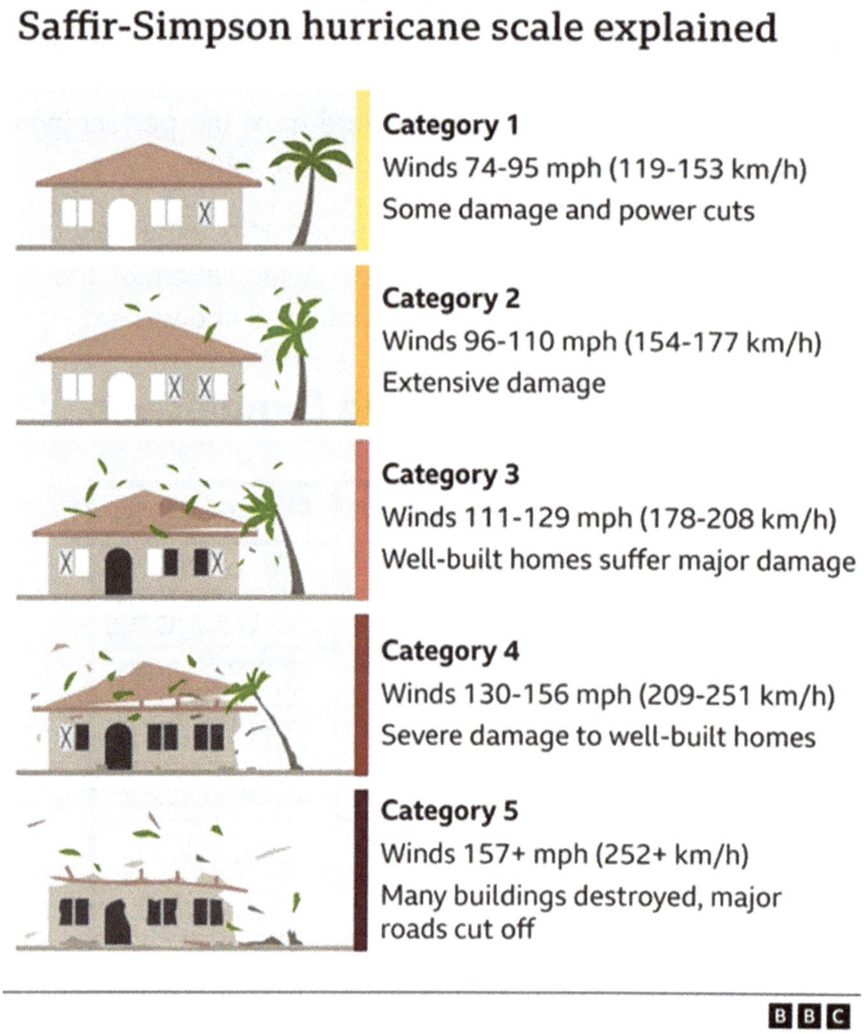

(Source: BBC News Service and NOAA)

From the perspective of human safety, the 2-3 days forewarning given by the National Hurricane Tracking Center of a Category X level (based on its wind velocity) is of immense value in saving human lives. Homeowners who choose to purchase or build a home in a "hurricane-alley" location should know to what extent of hurricane level has their home been built to withstand. Every community and depending on where the community is located – near or in a "hurricane alley", or "tornado alley", or known "flood zone" should have a minimum construction safety code to guide existing or potential residents planning to buy/build their home in that community capable of withstanding the natural wind forces and flooding potential known for that location. It may sound cruel or bureaucratic, but residents living in a trailer home not anchored securely should not be allowed to take the risk of knowingly placing themselves "in harm's way".

An accompanying graphic indicated the major ingredients of hurricanes, cyclones and typhoons), all in the category of tropical storms that have intensified to different levels of the Saffir-Simpson scale, to help the public understand why hurricanes and typhoons form, and that graphic by BBC/NOAA is also repeated as follows:

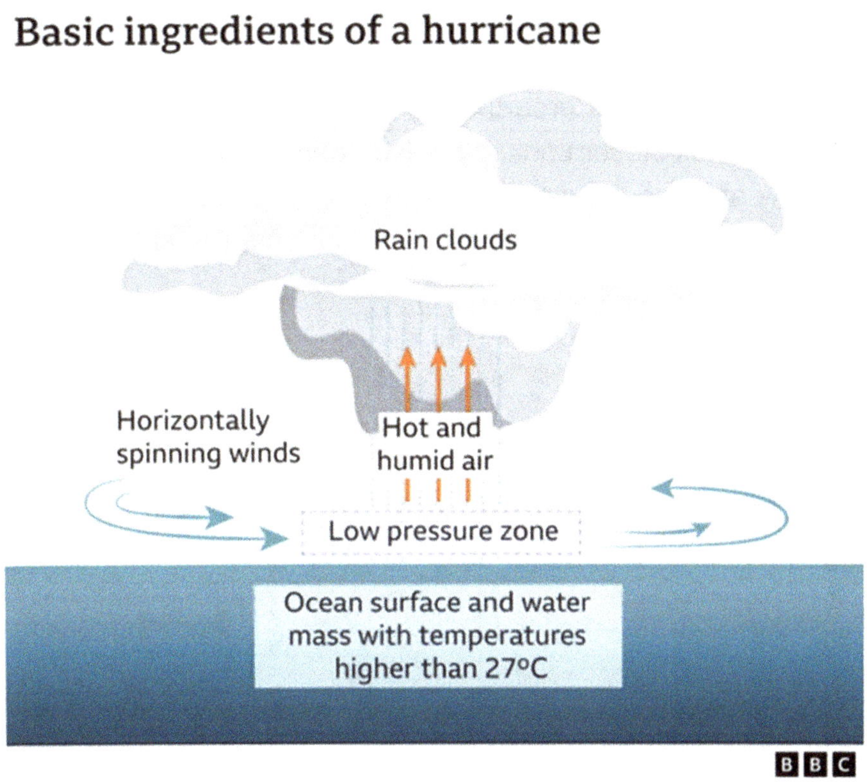

(Source: NOAA)

The temperature of 27 °C or Celsius is the equivalent of 80.6 °F. Those high temperatures are associated with the ocean waters in the Equatorial zone.

The ocean currents have a generally typical circulatory current within land coastal boundaries and move the heated waters as if within a large basin. The more regular prevailing wind currents in the atmosphere move unrestricted above the surface, regardless of being water or land, and in more regularly defined patterns:

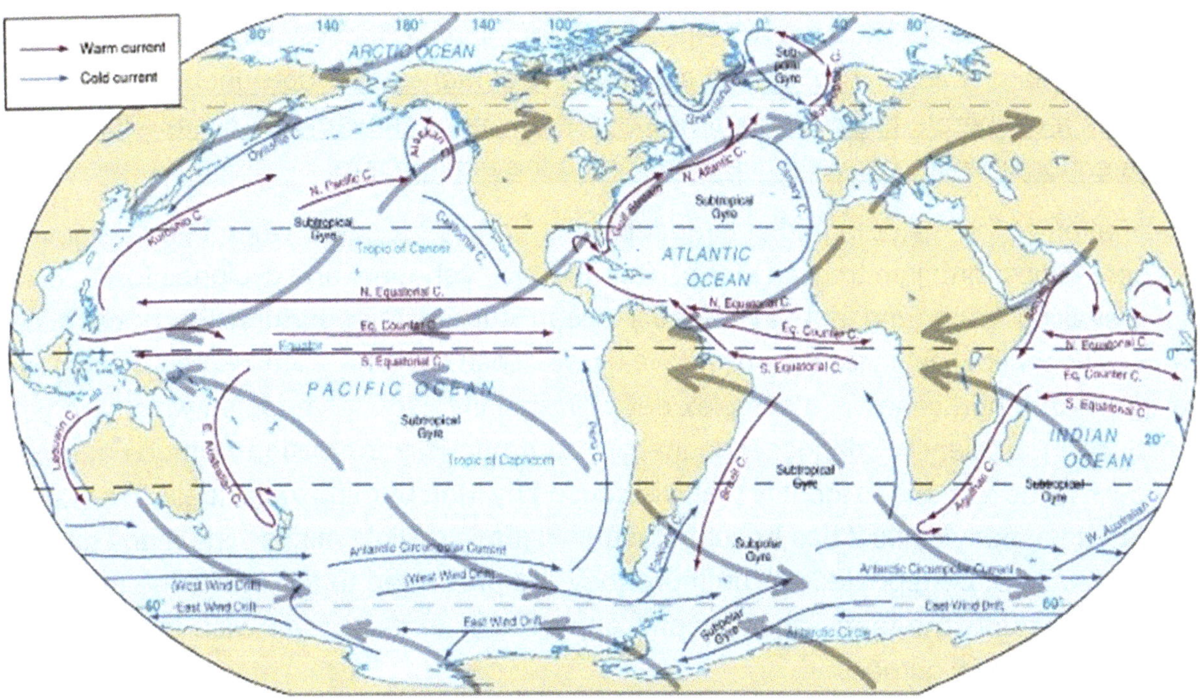

(Source: NOAA)

There is far more (by a ratio of 3% water vapor to 0.04% carbon dioxide = 75:1) heat-trapping content in the atmosphere – causing water vapor to be a much more causative variable on "average global surface temperature" than Man-made greenhouse gasses as Dr. S. Fred Singer revealed in his atmospheric chemistry series "A Closer Look at the Numbers". In that analysis, Dr. Singer compared the "settled science" view of "greenhouse gasses" (which excludes consideration of water vapor) with the ethical scientists' view of all heat-trapping components (by any name) and which includes water vapor:

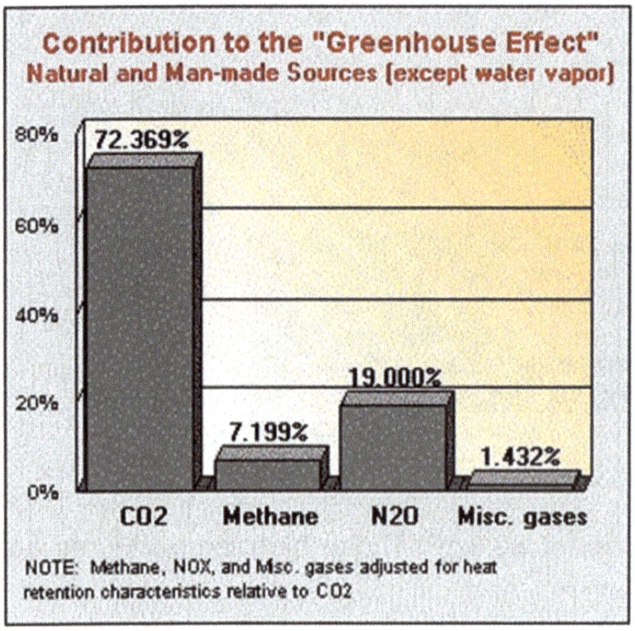

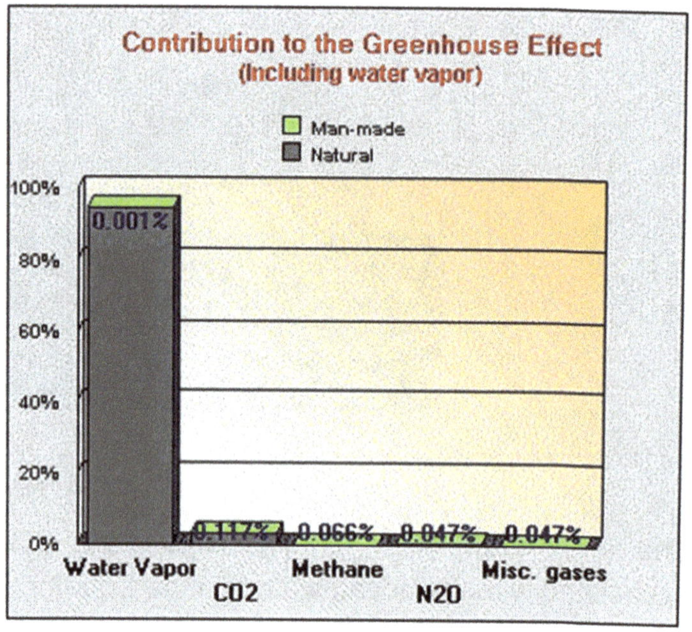

(Source: Dr. S. Fred Singer, University of Virginia, "A Closer Look at the Numbers")

This fact vs. fallacy will be shown more than once throughout this book because of its high significance to being factually accurate and revealing the fallacy of the "settled science" faction in environmental science who disregard water vapor (as a part of their process of "when everything else is eliminated") despite water vapor being a more dominant effect (by a factor of 75:1) on global warming and climate change. The "settled science" set of Man-made "greenhouse gasses" without water vapor is a deliberate fallacy – not a hoax or lie as many "naysayers" claim – it is a scientific fallacy created for scientific convenience in modeling the atmosphere.

The combination of the major ingredients (independent and dependent variables) that govern how tropical depressions form into tropical storms (hurricanes, typhoons, and cyclones form), combined with the prevailing local (and not "averaged") ocean current temperatures and ocean currents irregular curvilinear flow paths, and winds above those ocean surfaces all mixing/combining to cause them to "grow" into hurricanes is a complex set of irregularly varying variables which are yet to be fully explained in a scientific manner. The inability to do so is a frustration to those attempting to "model" how it all occurs and to identify that it is caused by Man burning fossil fuels. That fallacy has been repeated so many times it has become factual in many gullible minds. The many examples of "propaganda" will not be repeated as the more they are presented to the public the more they are believed to be factual. The term "inconvenient" applies only to those who wish to misinterpret the facts for a non-scientific purpose.

Perhaps the most significant "eye-opener" diagram found during the research on truth in climate change for this book dealing with the natural events affecting the weather and climate is that shown next historically recorded for the different levels of severity of tropical storms that have developed into severity sufficient to be categorized on the Saffir-Simpson hurricane scale developed by the National Oceanic and Atmospheric Administration (NOAA):

Historically, including those before and since the Industrial Revolution, tropical storms of all variation in intensity have taken the following tracks:

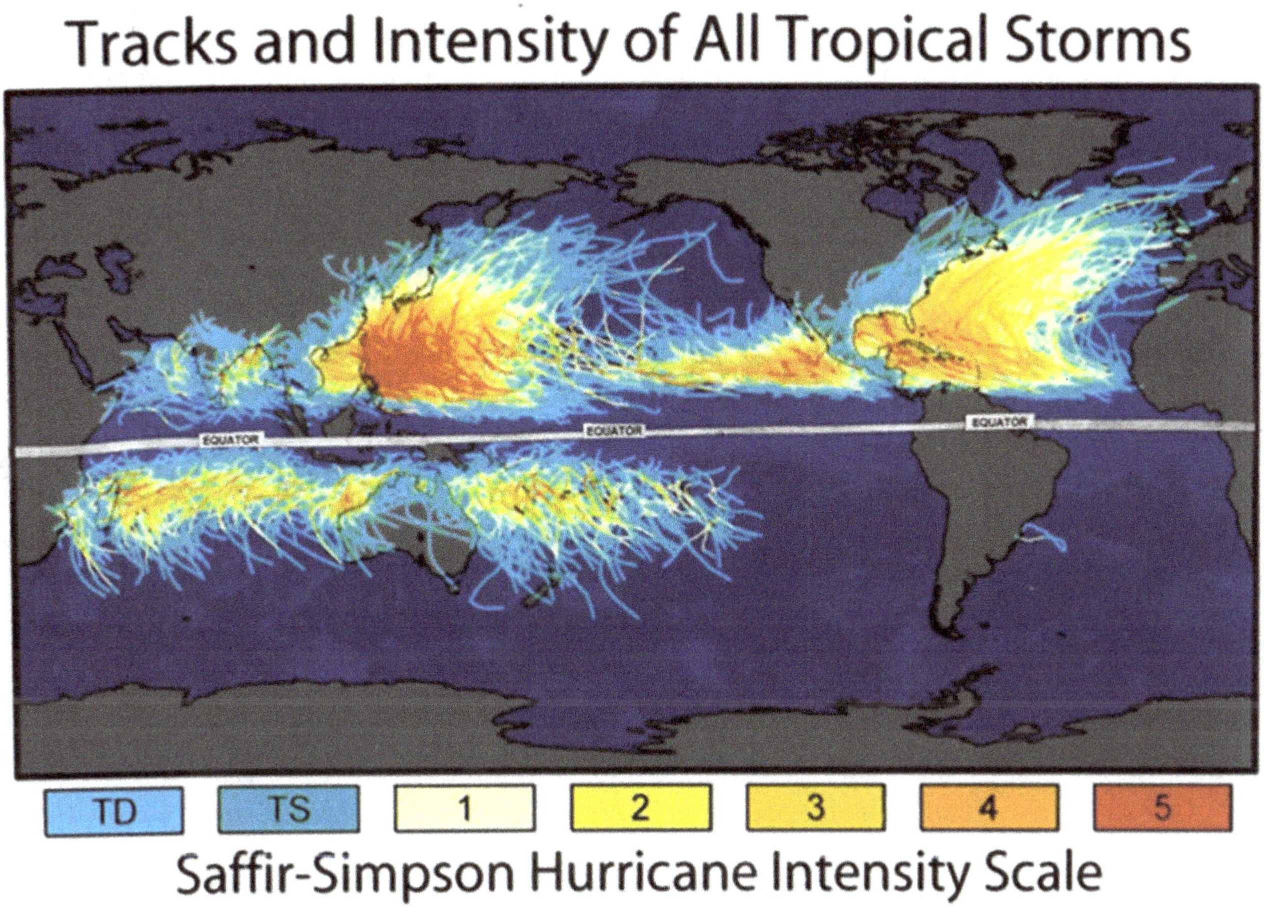

(Source: The National Hurricane Tracking Center, NOAA)

If the causes of those storm tracks were factually due to Man burning fossil fuels and introducing "greenhouse gasses into the atmosphere, it should show that the origins (the spawning conditions) should be over the Industrial regions most responsible for generating the "greenhouse gasses". The historical composite above shows that not to be the case. It is the combination of astrophysical, geophysical, and natural causes that generate larger tropical storms, and primarily on a seasonal basis not all year-long.

There is the possibility that the "greenhouse gas effect" is in small part – but definitely not dominant - due to Man-made emissions causing the waters north of the Equatorial zone such as those above the tropic of Cancer (north of the Equator by 23.5 degrees in latitude) where more than 90% of the heat-trapping "greenhouse gas" generation is produced. Neither natural nor Man-made "greenhouse gasses are being circulated southerly from the heavy industrial areas into the Equatorial zone by ocean currents or by winds. The Gulf Stream which is flowing clockwise in the Atlantic Ocean is northerly along the US East coast. The prevailing winds are also flowing northerly above the US East coast. Before ships were propelled by mechanical power, the sailing ships conducted trade by sailing with the wind current in the Atlantic Ocean.

That became known as “the trade route – from the British Isles southerly along the coasts of France, Portugal, Spain, and north Africa to the Palo Verde Islands where the ships changed course to westerly both with the winds and ocean current. They generally stopped at Barbados and other ports in the Caribbean region to offload manufactured goods from Europe and take on cargoes of sugar, molasses, and rum. From the Caribbean, the sailing ships sailed (again with the wind and ocean currents) to “the Carolines” where they offloaded more European goods and replaced them with cargoes of cotton and tobacco. From the ‘Carolines”, the sailing ships continued on the north moving Gul Stream to ports in the New England area to pick up any passengers returning to the British Isles and took the north Atlantic wind and ocean current back to the British ports (primarily London). That clockwise circular route of the sailing ships is written in the Maritime Museum as being the “Trade Triangle”. Any carbon dioxide concentration in the atmosphere – natural or Man-made – would be airborne above the same “Trade Triangle” route.

The previous diagram by the National Oceanic and Atmospheric Administration and the satellite images from NASA in the image above clearly indicate that the Equatorial zone of the planet contains the spawning grounds for the major tropical storms. Once spawned, the paths of the storms are influenced by the Coriolis Effect, acting on the heavier mass of the moisture-laden atmosphere, to travel in the direction of the ocean currents and prevailing winds. The author has modified the image with a white line (to contrast with the dark blue of the ocean waters background) indicating the Equator. It is readily seen that the major recorded storm tracks (from the 1700s to the 2000s) are typically clockwise north of the Equator and counterclockwise south of the Equator.

The news media, fed with the fallacious information from the environmental “settled science” faction, repeatedly refers to the Man-made “greenhouse gas effect such as in the BBC article (October 2024): *Assessing the precise influence of climate change on individual tropical cyclones is challenging. The storms are relatively localized and short-lived, and can vary significantly in any case. But rising temperatures do affect these storms in several measurable ways:*
Firstly, warmer ocean waters mean storms can pick up more energy, leading to higher wind speeds.
Record high sea surface temperatures were a key reason why US scientists forecast an above-normal Atlantic hurricane season for 2024.
The high temperatures are mainly due to long-term greenhouse gas emissions.
Secondly, a warmer atmosphere can hold more moisture, leading to more intense rainfall.
Climate change made the extreme rainfall from Hurricane Harvey in 2017 around three times more likely, according to one estimate.

The BBC article does show an increase in the Saffir-Simpson ratings varying from 1920 to 2013 as in the next diagram:

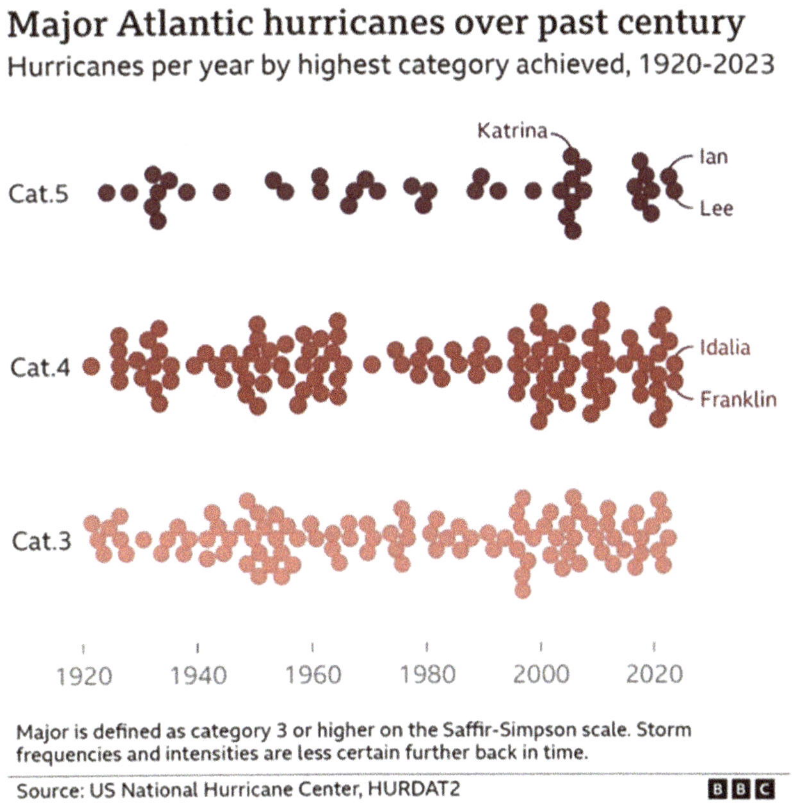

If the "settled science" has "eliminated everything else" except "greenhouse gas effect", the conclusion could be (and with "blinders-on" has been) reached – fallaciously – that it is "greenhouse gas-caused climate change" increasing the frequency and intensity of major Atlantic hurricanes. The spawning grounds of almost all Atlantic tropical depressions which grow into tropical storms and later into hurricanes are off the west coast (near the Palo Verde Islands) of the immense North African desert region where there is absolutely no industrial or human activity that creates an iota of Man-made "greenhouse gasses". The frequent formation of tropical depressions in this region has a well-known basis (from the atmospheric science associated with low-pressure cells and their wind patterns forming the commonly seen small "dust devils" in the southwestern US desert region):

(Photograph from Author's personal photography file)

With the assistance of improved imagery of satellites, photographs (taken in 2022) of the spawning grounds in the atmosphere west of North Africa reveal that the high winds and wind current not unexpectedly lifts enormous amounts of dust from the vast Sahara Desert into the atmosphere. Readers can envision the image below (with the dust appearing as a blurry region in the left half of the image) as being a very large magnification of the "dust devil" patterns of winds in a spiraling motion upward and carrying the dust particles not firmly attached to the ground surface to a higher elevation.

(Source: NASA)

There is not much kinetic energy expended in accelerating the small mass of one dust particle into motion in the plane of the Earth's surface, but to elevate the thousands of tons of dust particles from sea level to an elevation in the range of 1,000 to 5,000 feet is an indicator of the enormous amount of heat energy that is being drawn from the eastern North Atlantic Ocean in the region of the Cape Verdes island group. When the mass of the water vapor molecules (seen in the cloud formations) is added to the phenomenon, and the resulting total mass of dust and water vapor molecules and the mass of all of the gases contained in the atmosphere that is set into a westward motion to possibly become a tropical storm or hurricane, the power of Nature becomes apparent. The addition of any small amount of "greenhouse gasses" into the tropical storm/hurricane phenomenon is virtually of insignificance.

The ocean currents and the prevailing winds in the atmosphere turn due westerly at the Palo Verde islands and cross the southern Atlantic at about 10-12 degrees north latitude. After leaving the hot North African region, the hot winds increase the temperature of the water and continue doing so where they meet the hot waters from the Equatorial current. It is at that juncture that the warm water increases the intensity of tropical storms and they typically begin the climb on the Saffir-Simpson force scale as they approach Barbados. A graphic of the ocean currents in the Atlantic Ocean between the African continent and the North American Continent should be helpful to the public The spawning grounds of the tropical depressions and storm is indicated by a large hexagonal shaped figure and labeled as such:

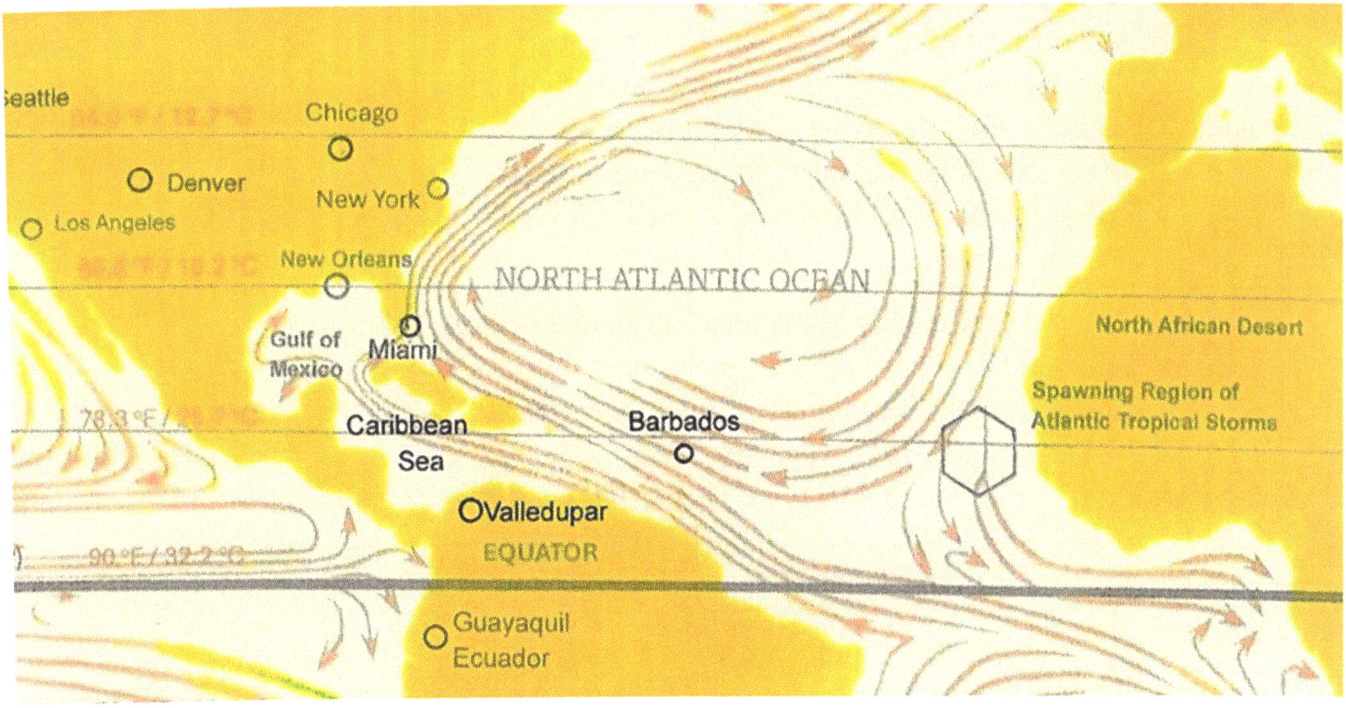

(Source: Rand McNally Atlas of the World, with Author annotations)

One of the largest hurricanes that devastated Barbados (which has a long list of hurricanes in its history since the island's discovery in the early 1600s) was that of the Great Hurricane of October 10, 1780 (before the Industrial Revolution creation of Man-made "greenhouse gasses"). There was no Saffir-Simpson scale to be assigned in 1780 and estimates from the extent of damages – every structure of wooden construction was blown down according to the parish church records on file in the Barbados Archives – the rating would have been most likely a Category 4 or 5. There was no Industrial Revolution in 1780; therefore the question becomes: "What caused that magnitude of hurricane without any Man-made "greenhouse gas emissions" causing or even contributing to that monster. The amount of carbon dioxide in the atmosphere from natural causes would have been below 200 ppmv (0.02%) as there is no ecologic growth creating or absorbing carbon dioxide above the Atlantic Ocean between the Palos Verde Islands and Barbados.

It has been shown in the previous few diagrams that the major hurricanes are at their highest intensity in the Equatorial Zone but lessen intensity as they progress into the regions above the tropic of Cancer. Since it is above the tropic of Cancer where the winds flowing the "greenhouse gasses" from the industrially generated emissions (90% of which are generated well above the tropic of Cancer) away from the Equatorial zone, how can the news media and/or "settled science" maintain the fallacious claim that Man-made greenhouse gasses are increasing the frequency and severity of hurricanes in the Atlantic?

It should be noted also that by far the highest concentration of the Category 5 cyclones have occurred in the west and southwest Pacific and not in the Atlantic.

The effect hurricanes and cyclones have when and where they make landfall is dependent on several factors – the elevation of the land in the region of the landfall, the extent of the tidal ocean level at the time of the landfall, the time of day, and of course the human and animal population "in harm's way". There cannot be enough plaudits and expressions of gratitude given to the organizations providing advance warning given to those in the projected path of major storms – such as the National Weather Service, the National Hurricane Tracking Center, and the NASA Earth Resources Satellites operators – for providing advance warnings and if local safety officials advise evacuations for advising the best evacuation routes to be taken and enough time in advance for safe evacuation. The same plaudits go to those rescue personnel who assist those families who remain to protect their property and animals as best they can during the storm and survive any flooding after the storm passes.

The USA has a Federal Emergency Management Agency (FEMA) that provides shelter and food and humanitarian services to those who lose their homes and "everything else but the shirt on their back" (such as from massive wildfires and floods), but the typical destruction of the infrastructure in the regions hit hardest by wildfires and floods deny access of FEMA – and all other relief organizations - from getting adequate food, water, and health care services to those residents affected the most severely. It takes days and often weeks to restore power losses to the regions affected and that is critical to the totals of fatalities and losses of homes how long the time period without power lasts.

On the preceding chart it is also seen that the largest concentration of red color (Saffir-Simpson Hurricane Scale 5) indicating that the cyclones (as they are named in the Eastern hemisphere) of highest intensity are most frequent in the Bay of Bengal (near India and Burma) off the Indian Ocean and which historically have devastated the country of Bangladesh on repeat occasions.

The statistics are such that the Bangladesh region has the highest number of cyclones than any other region, but the second most hazardous region prone to hurricane damage is the Louisiana Gulf Coast. The Gulf Coast has the highest number of hurricanes of all scales (including some up to scale 5) to make landfall.

Geographically, the northern section of the South American Continent serves as a deflector of the Equatorial current between Africa and North America after passing the island of Barbados, and the chain of islands (Haiti, Dominican Republic, and Cuba) split that deflected hot current in the southern part of the North Atlantic clockwise current with a segment flowing into the Caribbean Sea and Gulf of Mexico. (The Author's ancestral family were/are resident on Barbados, and on family visits, the Author is given the option to body surf in the Atlantic Ocean warm waves rolling onto the Eastern Coast beaches of Barbados or to swim on the placid West Coast of Barbados said to be in the even warmer Caribbean Sea. Some decisions in life are very hard to make.)

The tropical depressions that form (in the region marked with a hexagonal figure on the previous diagram) and develop into tropical storms and hurricanes typically flow toward Barbados. Those that pass north of Barbados generally flow with what is known as the Gulf Stream (with water temperature of about 80 °F (26.7 °C) that approaches the Miami part of Florida before turning northerly or northeasterly. The lower segment that passes south of Barbados mostly flows into the Caribbean Sea and then into the Gulf of Mexico. The Gulf of Mexico acts as a basin and can generate tropical storms within the Gulf that may further develop into named hurricanes with Force

Level 1 to 5 and those are generally the ones that devastate the coast cities of Mississippi, Louisiana, Alabama, and Florida.

There are many in the "settled science" faction of environmental science that attribute the frequency and severity of the Atlantic hurricanes to being contributory to the frequency and severity. There does not appear to be any principle of atmospheric, thermodynamic, or hydrodynamic physics that supports the influence of the "greenhouse gasses" to change either the frequency or severity of ocean hurricanes, cyclones, or tornadoes. Hurricanes, cyclones, and tornados are seasonal which confirms they are primarily influenced more by astrophysical and meteorological factors and very little, if any, by Man-made atmospheric emissions.

A composite of Hurricane Scale Level 3 named and unnamed hurricanes that have made landfall in the New Orleans delta (of the Mississippi River) region from 1852 to 2015 (163 years) is shown next:

(Source: National Hurricane Tracking Center of NOAA)

The first major hurricane to make landfall on the United States and officially recorded was in 1852. Before hurricanes make their landfall on the United States, they seem to randomly enter the Gulf of Mexico and then turn northerly, or (for unknown natural reasons) to turn northeasterly before making landfall. The state of Florida has some influence on that variation in track, but it is unknown scientifically and factually why and how. Shown next is a composite of the Category 4 and Category 5 hurricanes that have crossed southern Florida in the period 1910 to 2018.

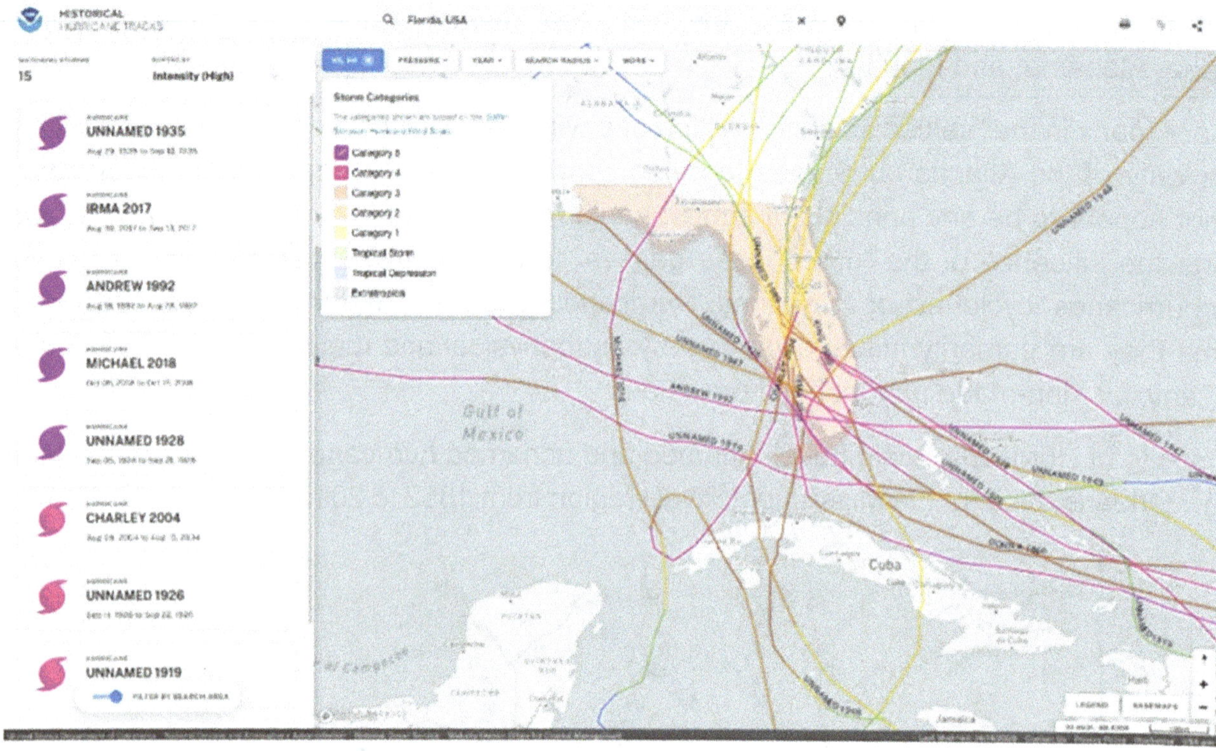

(Source: National Hurricane Tracking Center of NOAA)

One hurricane is seen not to have wanted to leave the Cuba region and made a farewell circular loop before heading toward its Florida target. The favorite city for hurricanes to "visit" in Florida is Naples, on the southwest coast (in the Gulf of Mexico) of the Florida peninsula.

The Atlantic hurricane season is generally in the six months from June through November of each year but there can be hurricanes occurring before June or after November. The bar chart shown next indicates the number of Atlantic hurricanes that have occurred from 1923 to 2023:

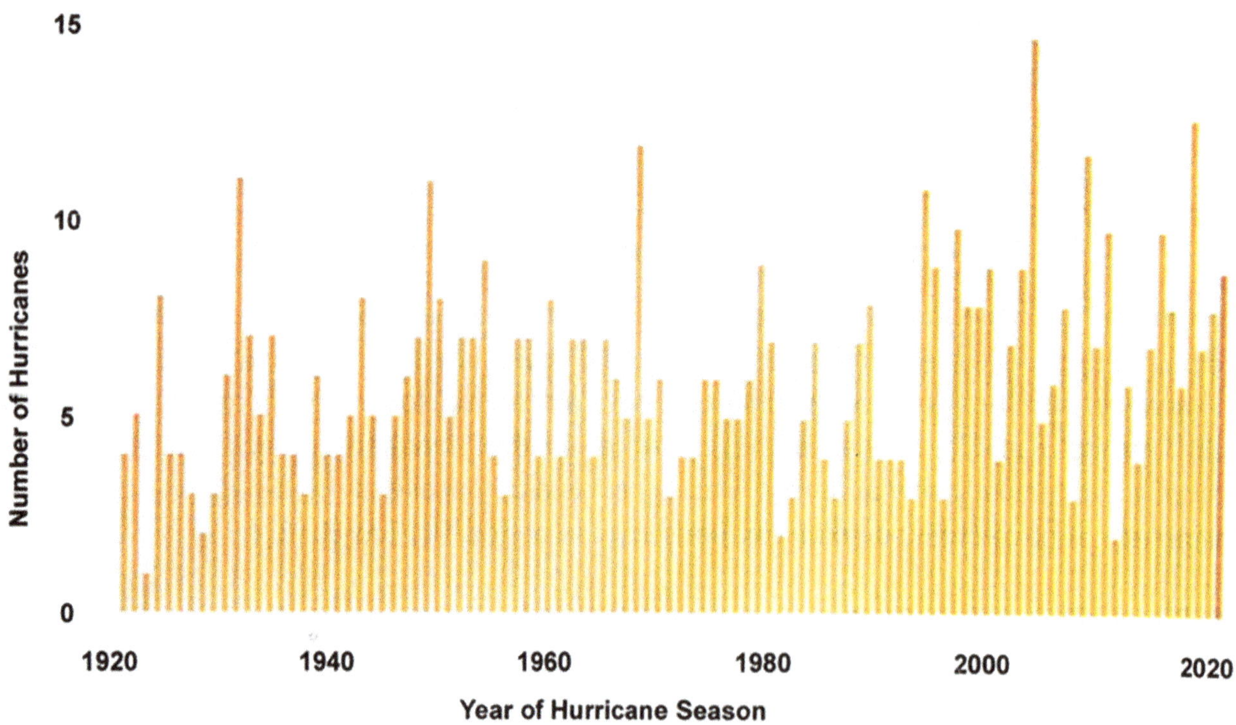

(Source: Colorado State University, Department of Atmospheric Science)

The trends being developed by the environmental science faction to reflect that Man-made "greenhouse gasses" are causing the intensity and frequency of the hurricanes is somewhat a non-scientific achievement to the human population who suffer the loss of lives and property whenever and wherever a major catastrophic event occurs.

The most devastating hurricane that has recently struck the United States was Hurricane Katrina which made landfall as Category 4 on the Louisiana Gulf of Mexico coastline on August 29, 2005. Shown next was a satellite image of Hurricane Katrina as the eye was over the New Orleans region:

(Source: NASA)

To be noted in the photo is that the fringe bands of cloud formations indicate the highest areas of rainfall, and it is these fringe clouds that have the highest winds and contain the most water that causes the wind damage and flooding across a much larger area of the land surrounding the eye. Hurricane Katrina caused 1,577 fatalities, mostly due to severe flooding, and caused about $125 billion in damage.

A map of the state of Louisiana shows the number of Parishes and cities that are in the Louisiana Gulf Coast region prone to hurricane damage:

(Source: Atlas of the World, Rand-McNally)

Although Hurricane Katrina's eye came ashore in lower Plaquemines Parish of Louisiana, the resulting storm surge resulted in the failure of multiple levees of the Mississippi River in the New Orleans delta area, flooding approximately 80% of the city. Some areas of New Orleans were inundated by more than 15 feet (4.6 meters) of water. A photo of a flooded area of New Orleans (near the Superdome) is shown next:

(Credit to Unknown Photographer)

The failures of the levees in Orleans Parish were considered the worst engineering disaster in the history of the United States. The population of Orleans Parish in Thousands of people were stranded inside their homes or on rooftops and required rescue from boats and helicopters. Many buildings and homes were damaged, with 134,000 housing units – approximately 70% of residences in New Orleans – were impacted to some degree. The Superdome, which was sheltering many people who had not evacuated, sustained significant damage. Two sections of the Superdome's roof were compromised, and the dome's waterproof membrane had essentially been peeled off. The famous French Quarter and Garden District escaped flooding because those area are above sea level. A total of 573 deaths occurred in New Orleans alone.

Several other parishes were severely impacted by the storm. Many areas outside of New Orleans also suffered wind damage, especially St. Tammany and Washington parishes. According to the United States Department of Housing and Urban Development, in St. Bernard Parish, 81% (20,229) of the housing units were damaged. In St. Tammany Parish, 70% (48,792) were damaged, and in Plaquemines Parish 80% (7,212) were damaged. Nearly 900,000 Louisianians were left without electricity. Throughout the state, 1,577 fatalities were reported. Overall, it is estimated that Hurricane Katrina caused $125 billion in damage throughout the Southern United States, with much of the damage occurring in Louisiana and neighboring Mississippi.

The effect of natural catastrophes such as Hurricane Katrina on Orleans Parish in 2005 on the human population is not fully depicted by only the number of fatalities and cost of property damage. At the time Katrina impacted Orleans Parish in 2005, its population was 454,863. Most of those people had been given forewarning by the National Weather Service of the approaching monster (Category on the Saffir Simpson Scale) and evacuated low-lying areas further inland. Having lost their residences many decided not to return to Orleans Parish. In the aftermath of the Hurricane Katrina disaster, the Orleans Parish population in 2007 had declined to about 320,000. The population in the Parish stabilized in headcount and in the 2020 Census the population had re-grown to 383,997. The poverty in the Orleans Parish since the 2020 Census has worsened and caused another decrease in the population of Orleans Parish to presently (2024) be 358,355. Louisiana was once a very popular visitor location – especially at Mardis Gras time – but the visitor rate has also dropped considerable further in the past two decades affecting the entire southern Louisiana region economy. The trend of increasing frequency and increasing overall severity rating of hurricanes in the Gulf of Mexico region is having a profound impact of the economy of the Gulf Coast states.

The environmental scientists in the “settled science” fraternity are claiming that climate change is causing those changes in trends, but climate is the statistically dependent variable in the equation of what is happening with the independent variables in the Gulf region changing and it is mathematically a misnomer to assign blame for weather catastrophes to be due to climate change. The “tail does not wag the dog”. The public has been insidiously misled into believing that Man burning fossil fuels is the dominant cause for climate change. The fallacious tangential inference fostered by most environmental scientists (including those in the United Nations Intergovernmental Panel on Climate Change, IPCC) has set a “one-size-fits-all” goal that Man-made emissions must be reduced to net zero by 2050 to bring climate change to a halt.

The analysis done for this book has shown that despite Man-made emissions being reduced (in the USA and Europe, but not in China and India) over the past two decades, the surface temperatures continue to increase, more in some regions than merely on an "average global temperature" basis. A very revealing illustration of that fact is shown in the following illustration (shown earlier but repeated here for emphasis of its importance):

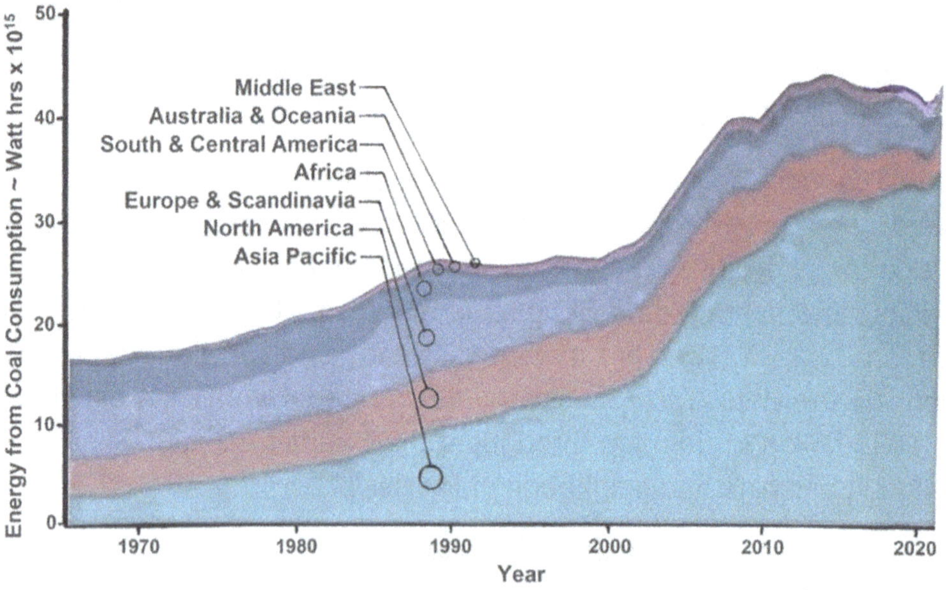

(Source: British Petroleum Corporation, LLC)

Articles in the news media have repeatedly stated that the frequency and the intensity of hurricanes are both in an increasing trend and those increasing trends are due to climate change making the Atlantic Ocean waters "soaring" in temperature. Those claims are not supported by any scientifically valid database found during the research and analysis of the hurricane database in the public domain. With the tangential claim that climate change is due to Man burning fossil fuels being the dominant cause for global temperatures rising, the Man-made emissions labeled as "greenhouse gas emissions" do not have the "spikey" frequency history shown in the above chart for the frequency of the Atlantic hurricanes and therefore Readers should not panic over the fact that even with the reductions in carbon dioxide in the atmosphere that are Man-made and being scientifically measured in a low concentration of 140 parts per million (0.014%) – with 295 parts per million of carbon dioxide (0.0295% being in the atmosphere before the Industrial Revolution – Man should not to be blamed as being the dominant cause for the naturally and seasonally occurring hurricanes that strike the United States southeastern coastal states.

Among the greatest leaps in the ability to monitor the track of hurricanes is attributed to NASA with the development of satellites, originally for communication, navigation, and military purposes, that are now also in the inventory of the National Oceanic and Atmospheric Administration for meteorological tracking of weather-related phenomena. With highly advanced optical imagery coupled with the satellite "big picture" perspective, the "eye" of the hurricane as well as the fringe waves, which typically cause the most flooding damage when passing over a land mass, are measured and tracked. The National Hurricane Center is a vital public safety National organization for providing days of warning to evacuate before an approaching "monster" attacks. An example of that photographic definition of a hurricane, that of Hurricane Dorian (Category 5 with 185 mph winds) which battered Abaco Island in the Bahamas off the east coast of Florida on September 1, 2019, is shown next:

(Source: The National Hurricane Center, NOAA)

Most importantly, the path of the tropical depression which grew into a hurricane being tracked quite accurately on a daily basis for location and strength of the "eye" is shown in the next illustration:

(Source: The National Hurricane Center, NOAA)

With this tracking information for Hurricane Dorian, human populations on Caribbean islands and in the southeastern United States were given as much advance warning as possible and far enough in advance that loss of life and property damage was drastically minimized. Even so, Hurricane Dorian took 84 lives, 245 still missing (and presumed dead) and caused over $5.1 billion in damages – most on Abaco Island in the Bahamas off the Florida coast on September 1, 2019 from Hurricane Dorian (which existed from August 24, 2019, to September 10, 2019). The colored circles reflect the intensity of the storm from its formation as a tropical depression in the warm south Atlantic Ocean waters to maximum Category 5 (in the hotter Caribbean Sea waters) and dissipation in the colder North Atlantic waters.

Still more impressive tracking with satellite imagery is that of an electro-optical thermal image showing the energy concentration of Typhoon Mawar which struck the island of Guam and caused damage to military facilities at Andersen AFB on Guam on May 24, 2023: (the terminology "typhoon" and "cyclone" are used interchangeably)

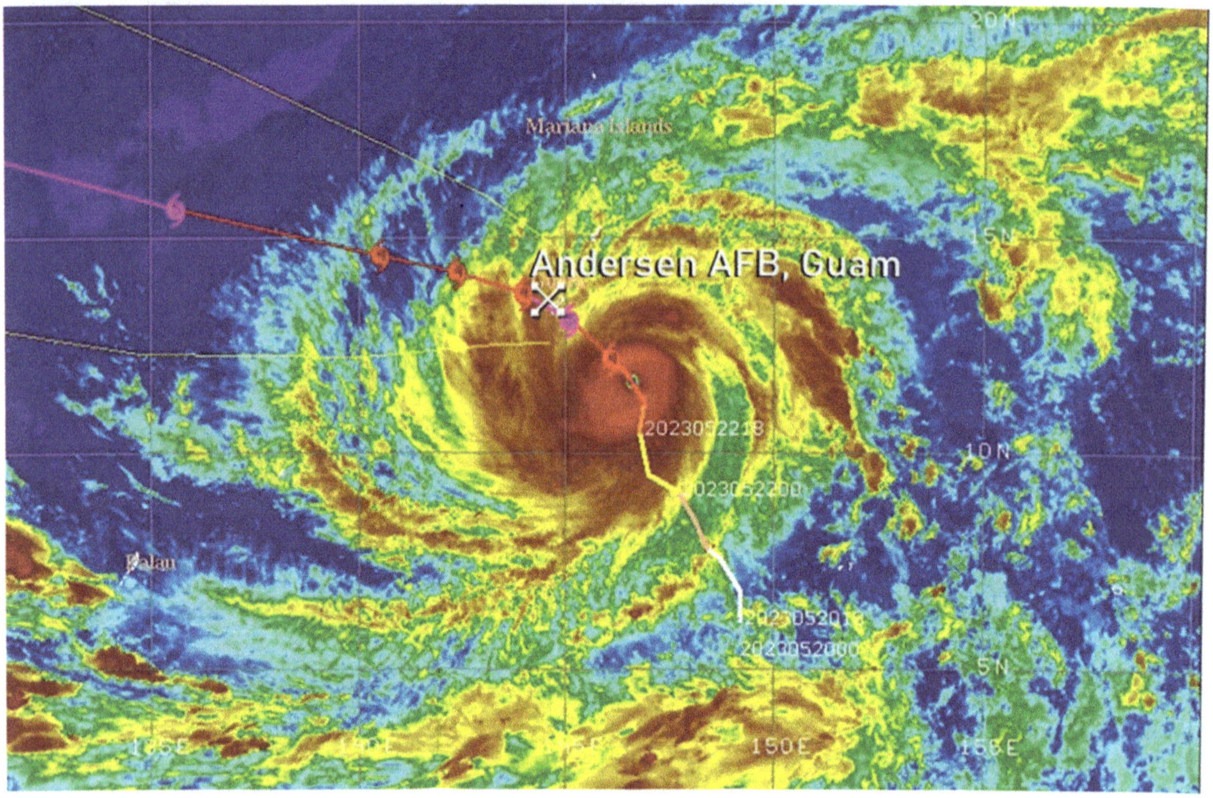

(Source: USAF Military Satellite Imagery, Typhoon Mawar at Guam)

Fortunately, thanks to advanced warning of the intensity (winds up to 175 mph) and accurate tracking of the super typhoon (Mawar), there were no deaths reported on Guam and less than $125 million in damage.

As this book was nearing completion, Hurricane Helene struck the United States Gulf Coast as a Category 4 (on September 25, 2024) and took a surprisingly inland route northerly through Florida, Alabama, Georgia, South Carolina, western North Carolina, Tennessee, Virginia, West Virginia, Kentucky, Ohio, Indiana, and Illinois. The path that Hurricane took - far inland and away from the Atlantic coast – left residents somewhat in disbelief that a hurricane would strike that far inland. Hurricane Helene also was unique in that it spawned near the Yucatan Peninsula and moved almost due northerly with only a few small serpentine-like deviations before deteriorating in the upper midwest. It also was a moderately slow moving hurricane which usually creates worse flooding than the faster-moving hurricanes. A map of the track Helene followed is shown next and how its intensity varied (the color dots representing the Saffir-Simpson scale):

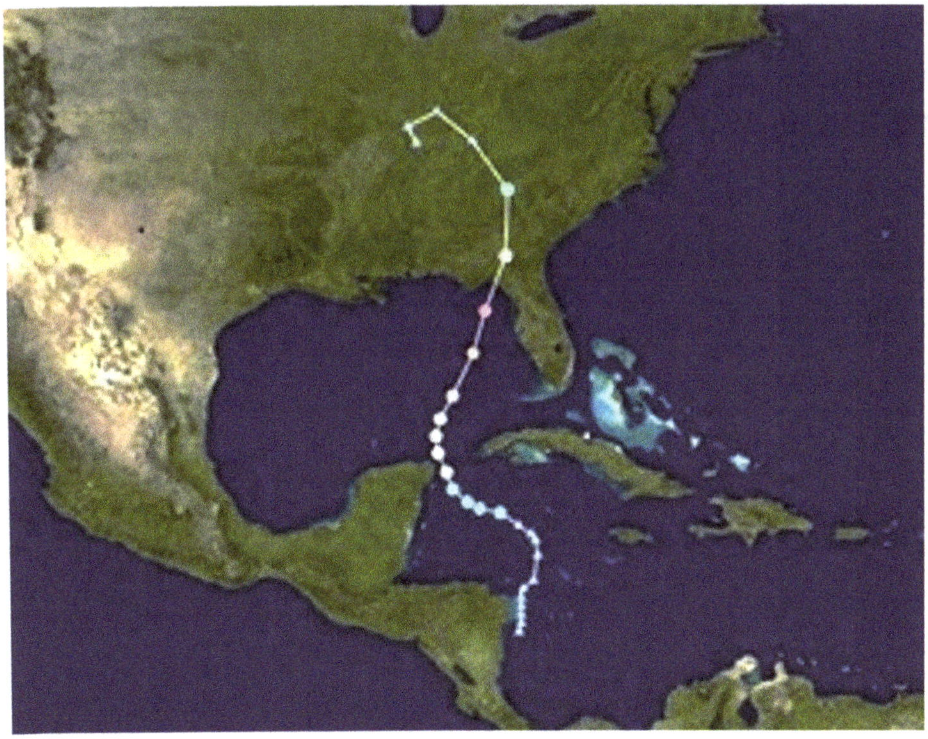

(Source: The Hurricane Tracking Center, NOAA)

The color-coding of the Saffir-Simpson scale is provided for reference:

Saffir–Simpson scale

- Tropical depression (≤38 mph, ≤62 km/h)
- Tropical storm (39–73 mph, 63–118 km/h)
- Category 1 (74–95 mph, 119–153 km/h)
- Category 2 (96–110 mph, 154–177 km/h)
- Category 3 (111–129 mph, 178–208 km/h)
- Category 4 (130–156 mph, 209–251 km/h)
- Category 5 (≥157 mph, ≥252 km/h)
- Unknown

Storm type:

● Tropical cyclone

■ **Subtropical cyclone**

▲ Extratropical cyclone, remnant low, tropical disturbance, or monsoon depression

The impact (fatalities, missing, and direct property damage) by state is as shown next:

Impact of Hurricane Helene by State
(September 24-29, 2024)

State	Deaths (Missing)	Damage (US $)
Florida	26	>$21.1 billion
Georgia	33	>$6.88 billion
South Carolina	51	Unknown
North Carolina	102 (26)	$53.6 billion
Tennessee	18	Unknown
Virginia	2	Unknown
Indiana	1	Unknown
Total	233 (26)	>$87.9 billion

(Source: Wikipedia)

Hurricane Helene was the eighth named tropical storm, fifth hurricane, and second major hurricane of the 2024 Atlantic hurricane season. The hardest hit state was North Carolina (the Author's home state) and the damages and fatalities were primarily in the western mountainous part of North Carolina which received up to 19 inches of rain within a 24 hour period. Power was lost in many areas of the Black Mountains, and the magnificent Blue Ridge Parkway was closed for freely-moving traffic. More than 400 roads were closed (many that will not reopen for at least 18 months), and over 200 people had to be rescued from the floodwaters. The mountainous regions typically cause roads through them to be built following the low and very curvy lying land at the base of the mountains. Nature causes streams and rivers to also follow at that same low-lying elevation and through the same curvatures as the roads. When there is a once-in-a-hundred-years flood – as there was with Hurricane Helene – roads designed next to roadside streams are built for normally placid water flow conditions. The roads in western North Carolina have not been built for the flow of water associated with flooding of the scale of that produced by Hurricane Helene. The section of Interstate I-40 shown below (and the Pigeon River four miles from the North Carolina border with Tennessee) is an example of that Inadequate road construction and also shows the erosive power of the fast moving very turbulent water:

(Credit to unknown photographer for North Carolina Dept. of Transportation)

Because of the high number of road closures into the area most impacted, the only access was by helicopter for food, water, and emergency medical supplies. Many of the homes had emergency generators but the fuel for those quickly was consumed and a large number of solar-powered generators were flown in to partially assist the emergency facilities in treatment of those injured or with pre-existing medical conditions.

The track that Hurricane Helene took in late September 2024 was most untypical of other hurricanes and no factual data or valid scientific explanations to either support or counter the claim that its cause, severity, and/or track could be related to Man burning fossil fuels has been uncovered. More than probably, it is not related since its origin was in a low population area and the earliest 3/4s of its track was through non-industrial "greenhouse gas emitting" regions. It is more likely that the generation of hurricanes (and tornadoes) in the southeastern United States are well known to be seasonal and the origin of Hurricane Helene was due to some unusual astrophysical anomaly in the causes of that seasonal effect.

A more detailed account of the causes of the deaths and the damages in each of the more severely impacted states from Hurricane Helene and from the five tornados that were spawned by the hurricane over land is beyond the scope of this book.

While continuing massive disruption of power caused by Hurricane Helene and with the road system into and from western North Carolina still impassable, another tropical storm that became a hurricane formed on October 4, 2024. It formed in an unusual location in the Gulf of Mexico but took a path of travel that was unlike that of Hurricane Helene less than two weeks earlier. Given the name Milton, the Hurricane Tracking Center of NOAA projected the path of what was to become Hurricane Milton as shown:

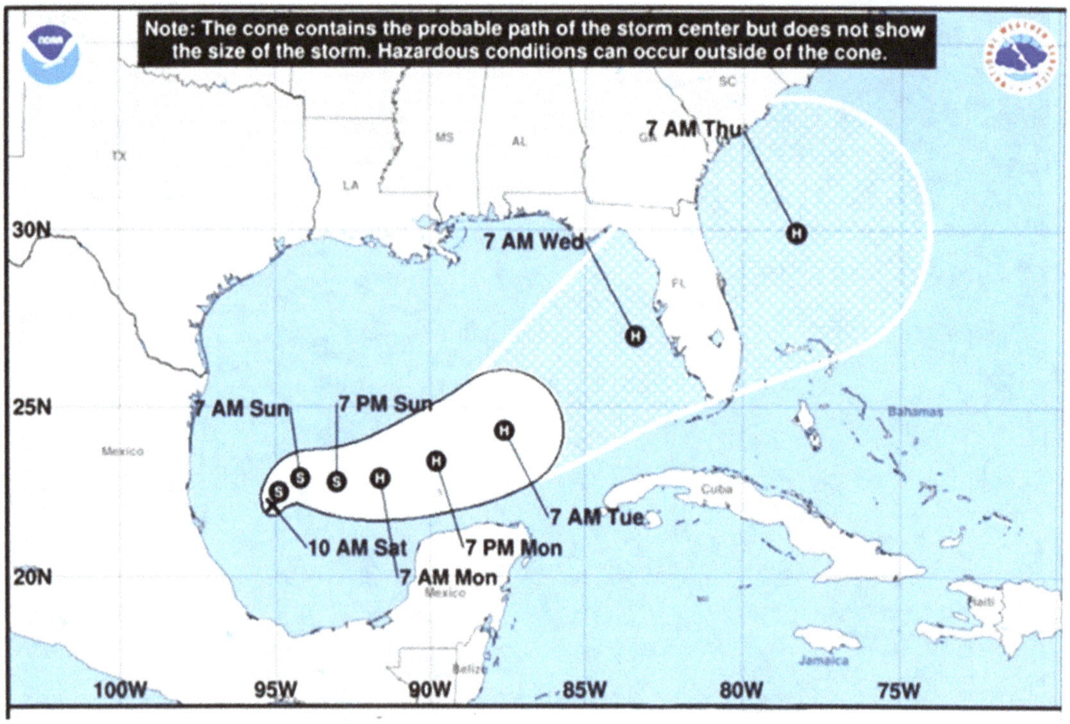

(Source: National Hurricane Tracking Center, NOAA)

The path that Hurricane Milton actually took during its seven day life span from October 5th to October 12th, 2024, is shown as follows:

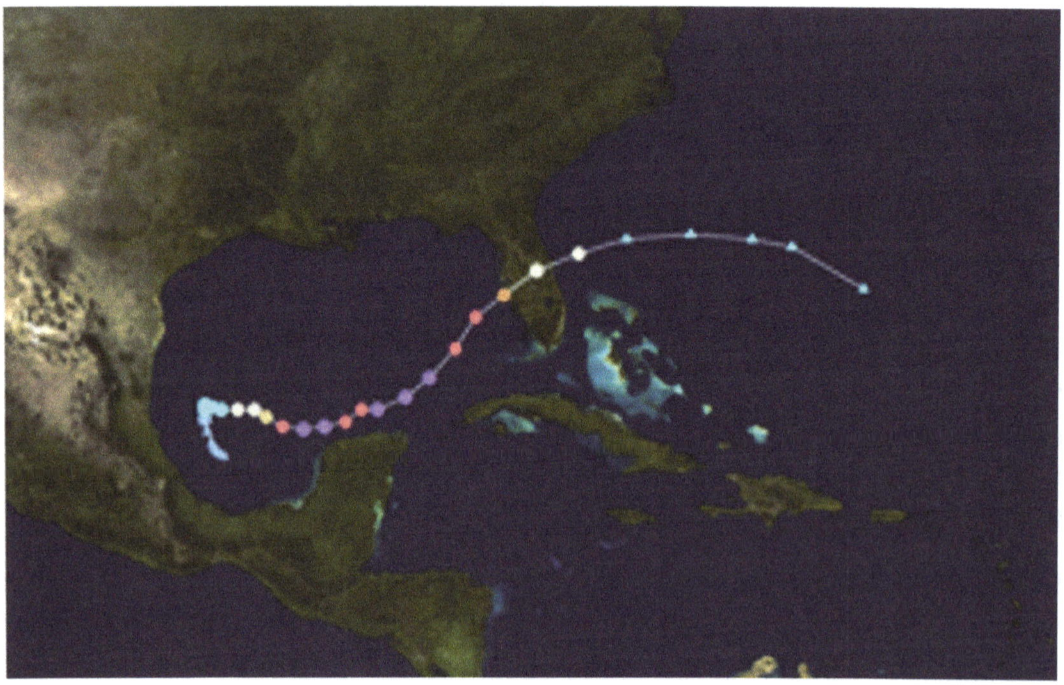

(Source: Hurricane Tracking Center, NOAA)

Hurricane Milton became a hurricane (Category 1) early on October 7th but underwent an "explosive intensification" (NOAA description) and later that same day became a Category 5 with 180 miles per hour winds. Milton became the thirteenth named storm, ninth hurricane, fourth major hurricane, and second Category 5 hurricane of the 2024 Atlantic Hurricane Season. As can be seen, Milton weakened considerably to Category 3 as it made landfall on October 9th near Siesta Key, Florida, and continued weakening as it crossed the state of Florida, being rated as an extratropical storm on October 10th. Unlike Hurricane Helene, fewer deaths were incurred (35) and fewer people were/are listed as missing (6) because of its shorter dwell time over populated land. Unofficial estimates of the damage – mostly on the coastal regions to boats and small beach-front residences – at about $85-90 billion. The damage is coincidentally almost the same as that of Hurricane Helene.

Waterspouts (a close counterpart of tornadoes on land) also have their origin in the warmer waters of offshore ocean currents as shown in the following diagram:

(Credit to unknown photographer)

It must be noted that there is a heavy low cloud cover from which the vortex flow extended down to the water surface and literally sucks the water upward as if in a big straw. The area surrounding the waterspouts is also seen to have much moisture content either as raindrops or a very high concentration of water vapor. Waterspouts, however, do negligible damage to property (only of occasional maritime vessels) or cause losses of human lives as do their land "cousins" tornadoes.

The most common cause for both tornadoes on land and waterspouts on water lies in the "shear" in the wind currents in the upper atmosphere, and by the pressure system between the surface and the upper atmosphere. The public sees only the letters H (denoting a high barometric pressure in a small region) and L (denoting a low barometric pressure) on weather maps in the newspaper - but that is because newspapers are two-dimensional and cannot indicate the more complete three-dimensionality of the phenomenon. A more scientifically described H and L system appears next:

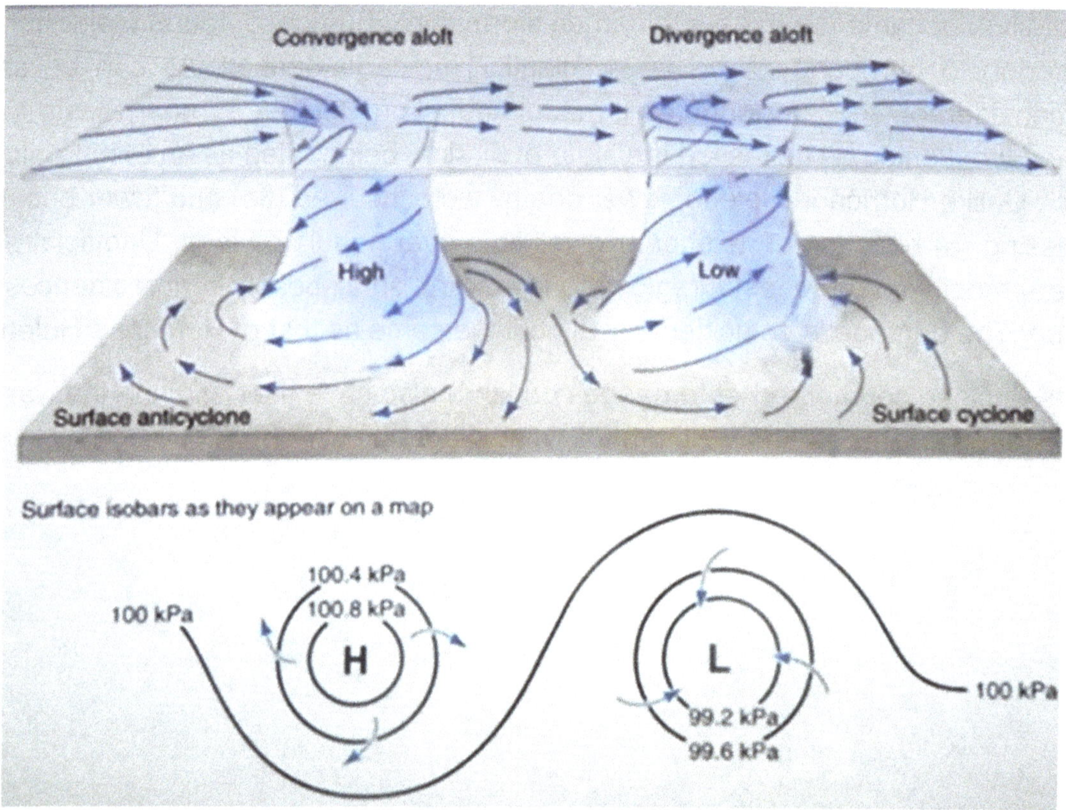

(Source: World Meteorological Organization)

On weather maps are also shown hot and cold "fronts" which in essence are lines of constant barometric pressure – either high barometric pressure or low barometric pressure. The curved line between the H and L systems shown above is a line (or "front") of that constant barometric pressure. In general, the weather associated with an H or along a warm front is an indicator of being fair (sunny with little chance of precipitation). Conversely, the weather associated with an L or along a cold front is an indicator of their being a high probability of adverse weather – cool or cold temperature, high winds, and high probability of precipitation. It is the rotational winds around the L which spawns hurricanes, cyclones, tornadoes, or just tropical depressions that will either increase or decrease in wind velocity and content of water.

It is rather well understood by meteorologists how tornadoes form and is due to the factor of wind shear. When a mass of wind-driven air moves past a different level of winds and direction of wind in an adjacent cell, there is a "shear" causing the mass of air at that juncture to transform from linear motion into rotational motion. The damages and fatalities incurred in the United States due to tornadoes is discussed in more detail in the next chapter, but the point to be made here is that the fallacious assumption apparently made by some environmental scientists is that the "Greenhouse gas emissions" (carbon dioxide, methane, etc.) are introduced into the global atmosphere in a homogenous manner such that emissions in the Northern Hemisphere will have global warming effects the same in the Southern Hemisphere as they do in the Northern Hemisphere.

Not to belabor the point, but environmental scientists using an "average global temperature" and the setting of a "one-size-fits-all" goal of reducing Man-made emissions to limit the rise in global surface temperature to 1.2 degrees Celsius by 2030 and achieve a net-zero in Man-made emissions by 2050 is unprofessionally fallacious. The wind current diagram above indicates just how fallacious the assumption of homogeneity of the atmosphere by environmental scientists is in the meteorological science profession. The medical profession takes severe sanctions against its members who commit malpractice (in which a fallacious act is intentionally perpetrated) against a huma, but fallacious acts against the environment are not subject to such human-protection sanctions.

Of the highest of significance to the weather, weather events, and climate of the United States Atlantic Coast states is the Gulf Stream shown in the next photograph:

Source: National Oceanic and Atmospheric Administration)

The general ocean current of the northern Atlantic Ocean is a clockwise current (being north of the Equator). It is off the West coast of Africa that this current develops the pressure and temperature flowing over the vast desert areas of west central Africa and is the spawning grounds of tropical depressions that are formed by natural causes – without any influence of Man burning fossil fuels for industrial, residential, or transportation fuel combustion engines. When the ocean current and prevailing wind currents (labeled as the “westerlies” by old time mariners) make a turn at about the Cape Verde islands group, they proceed to about the island of Barbados (at the bottom of the Caribbean chain of mostly volcanic origin islands) in the West Indies.

Geographically, the west coast of Barbados is in Caribbean Sea waters and the east coast of Barbados is in Atlantic Ocean waters. The island is somewhat pear-shaped and is 14 miles from north to south tips and 21 miles from west to east beaches. (The Author’s heritage is from this small island and on which the weather was found to be paradisical 99% of the time.) Unlike the other West Indian islands to its north being primarily of volcanic origin, Barbados is the exposed 165 square miles of land on top of a coral reef millions of years old that was thrust upward from the ocean floor.

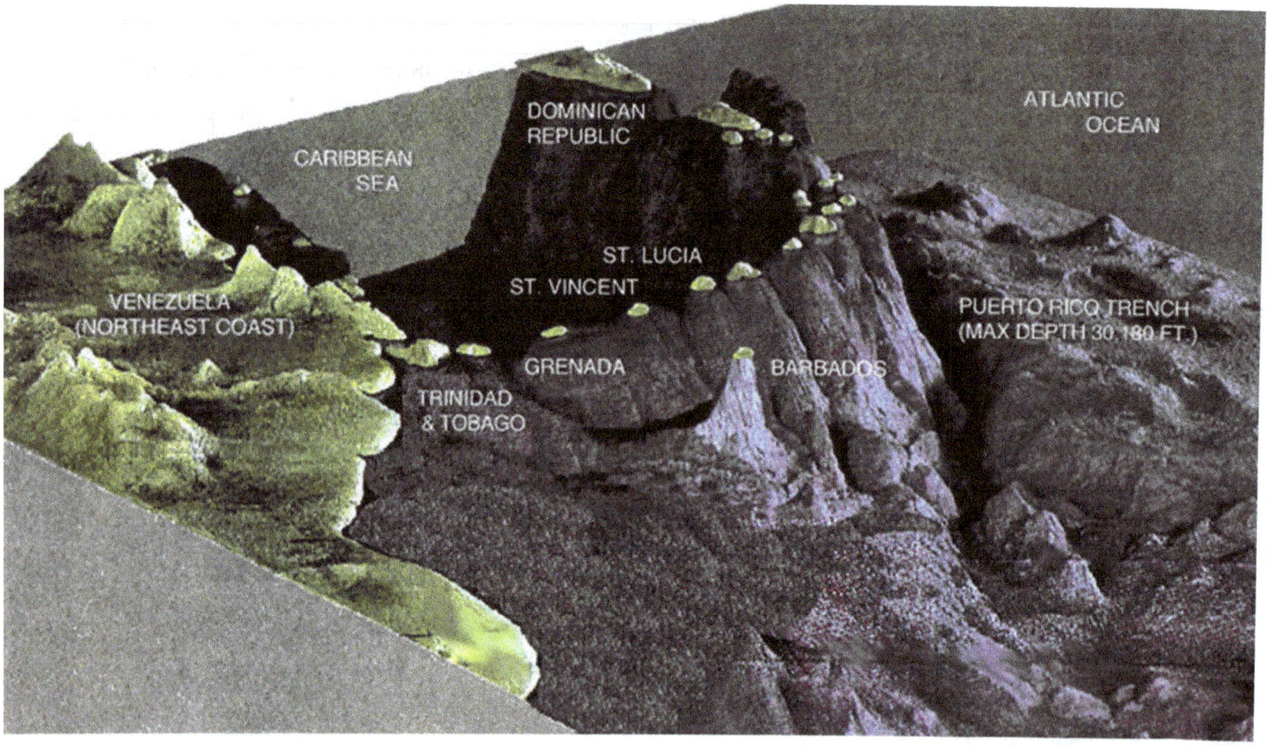

(Source: "Genealogical History of the Hutchinson and Edghill Families of Barbados")

The geological significance of that set of facts is that Barbados serves as the first point of land mass for ships sailing on the ocean and wind currents that are westerly from the Cape Verde islands group off the West coast of Africa. The island of Barbados therefore is the first populated area to be exposed seasonally to the tropical depressions and hurricanes of any Force category that originate in the vicinity of Cape Verde and forewarns the east coast of the southeastern United States of those approaching hurricanes and storms. In general, those storms and hurricanes follow a path such that those north of Barbados generally proceed northerly along the heavily populated southeast and east coast states of the United States. Storm and hurricane paths passing south of Barbados are of highest probability to proceed into the Gulf of Mexico before making the turn to the north and northeast. On that more southern and westerly path, the storms accumulate more heat energy from the very warm waters of the Caribbean Sea and the Gulf of Mexico than their more northern "siblings" and when making their typical turn toward the northeasterly path direction the typically more damaging and injurious hurricanes pass over the Gulf coast states (southeast Texas, Louisiana, Alabama, and Florida Panhandle regions).

Hurricanes are given a name for record-keeping purposes and continuous tracking of each storm/hurricane is done for human safety warnings. (The naming originally was a female first name because their "personality" and direction of movement is as unpredictable as those of a misbehaving teenage female, but more recently are also given male first names to remove any sexist connotation.) Regardless of name, the advance warning of an approaching natural potential disaster is vital to minimize the adverse effects of the approaching monster.

(Author's Note: After retirement from a 40+ years career in aeronautical engineering – which entailed the analysis of large, complex databases of scientific and technical/technology information, the Author applied that analytical experience to generating a genealogical history of his family. That genealogical database revealed that his genealogy (starting with his parents) was from the past 14 generations of personages born on the island of Barbados from 1905 back to 1635. The genealogical database on the island is comprised of the Church Parish Registers (Barbados is composed of eleven parishes named after saints in the Church of England) for births, baptisms, marriages, and deaths.)

In addition to the vital statistics of the parishioners – later used as the genealogical history of the island - the Parish cleric typically wrote accounts of the devastation done to the residences and government and public buildings on the small island when hurricanes made landfall on the island "in harm's way". On October 10, 1780, an extremely powerful (there was no severity rating system at that time) ravaged the small island of Barbados. The parish cleric (John Orderson) of St. Michael parish recorded in St. Michael Parish Register RL1/5 an unimaginable description of the "1780 Great Hurricane".

The devastation done to the wooden construction buildings on the island and the 4,500 deaths on Barbados alone due to that hurricane (now estimated by scientists at the Hurricane Tracking Center of the National Oceanic and Atmospheric Administration) to have been larger than a Force 5), prompted John Orderson to scribe in the Parish Register: *"To give a description of the terrible storm is beyond the abilities of man; but in order that posterity may be put in mind of the dismal scene which was before us, it is here faintly recorded. Adequate description of the accumulated horrors of the night is impossible, but the roaring of the tempest, the noise occasioned by the fall of groans of the wounded, who were hurt by, and buried under the ruins, the insipient flashes of lightning, and unabating torrent of rain, formed together such a scene of terror as has seldom been equaled in any part of the world."* He added (in Latin): *"Crudelis ubique luctus, ubique pauor, eplurima mortis imago"* meaning *"Suddenly everywhere there was mourning, everywhere great terror, and many images of death appeared."*

The Great 1780 Atlantic Hurricane, which formed on October 9, 1790, and dissipated on October 20, 1780, continued on a path over the other islands of the Lesser Antilles and over the larger islands in the Caribbean Sea. The total death toll of that singular hurricane was estimated at a minimum of 22,000 with many humans missing (presumably their bodies taken out into the ocean) to a total of possibly 28,500, making it the deadliest hurricane in the recorded history of the Atlantic hurricane basin. Man was not burning any significant amount of coal in 1780 to have affected that hurricane.

For Reader interest, a tabulation below reflects the Top 10 Deadliest Atlantic Hurricanes recorded starting in 1775 and to the second deadliest Hurricane "Mitch" (Force 5) in the 1998 Atlantic hurricane season (with 11,374+ deaths):

Deadliest Atlantic hurricanes			
Rank	**Hurricane**	**Season**	**Fatalities**
1	? "Great Hurricane"	1780	22,000–27,501
2	5 Mitch	1998	11,374+
3	2 Fifi	1974	8,210–10,000
4	4 "Galveston"	1900	8,000–12,000
5	4 Flora	1963	7,193
6	? "Pointe-à-Pitre"	1776	6,000+
7	5 "Okeechobee"	1928	4,112+
8	? "Newfoundland"	1775	4,000–4,163
9	3 "Monterrey"	1909	4,000
10	4 "San Ciriaco"	1899	3,855

(Source: US National Oceanic and Atmospheric Administration)

It must be said – from personal experience – that from every problem, there is often some good that results. In the engineering profession it is referred to as "Lessons Learned". After the population of Barbados recovered from their massive losses in life and property caused by the "Great Hurricane of 1780, Barbados rebuilt the buildings all over the small island, but they were rebuilt from coral blocks cut from the coral rock quarries on the eastern half of the island and not from the soft wood structure that could not withstand the hurricanes that have impinged on the island almost every Atlantic hurricane season. Except for the Parish church records (Barbados is comprised of 12 Parishes named after saints in the Christian Bible, and each Parish has a church serving that parish), detailed statistics are unknown about the Great Hurricane of 1780.

The second deadliest hurricane shown on the list above was Hurricane Mitch in 1998 and that is one of the best documented hurricanes from when it was created as a tropical depression. Hurricane Mitch was singled out by this Author, whose background includes several aircraft accident investigations for accidents involving multiple fatalities, to draw a parallel if any between the root cause for high number of fatalities accidents and with the objective to identify what can be done to prevent such massive fatalities from recurring. It was felt that understanding hurricanes better might help to understand the largest and most frequent natural event that kills so many humans every year and gain a better appreciation for the hazards that Man encounters living on the planet.

The US National Hurricane Tracking Center of NOAA is one of the best public service groups in the United States and it is from their database of information that enabled a rather fascinating history of the path that Hurricane Mitch took in October and November 1998 in the Central American region before passing over southern Florida and into the "hurricane graveyard" of the north Atlantic Ocean. The track of Hurricane Mitch is shown next:

(Source: Atlas of the World (Rand McNally) and National Hurricane Center database)

Most hurricanes originate as being an unnamed Tropical Depression (TD), and on October 22, 1998, the NHTC detected a tropical depression in the southwestern Caribbean Sea near the cities of Cartagena and Valledupar in Colombia. The TD began moving westerly and turned northwesterly in the same manner most Atlantic hurricanes do, but it gained hurricane Category 5 so quickly that even hurricane trackers were astonished the outer fringes of the eye were extensive and deluged the Panama and eastern Cost Rica with over 24 inches of rain in a two-days slow-moving but very water-laden precipitation. To avoid a lengthy narrative, a table was constructed showing the countries that were impacted, the number of fatalities that were incurred (mostly due to flooding and mudslides) in densely populated regions and losses of many thousands of homes.

Regional Fatalities and Damage from Hurricane Mitch 1998

	Region - Date(s)	Fatalities	Damage Costs (Structures & Infrastructures)	Rainfall Amount ~ inches	
1	Panama - October 22-23	3	$50,000	24+	High Winds; Flooding
2	Costa Rica - October 24-25	7	$92 Million	24+	Flooding; Mudslides
3	Jamaica - October 25	3	Unknown	15+	High Winds; Flooding
4	Nicaragua - October 25-26 (Cat 5)	3,800	$1 Billion	50+	High Winds; Flooding; Mudslides
5	Belize - October 26-27 (Cat 5)	11	$50,000	15+	Flooding
6	Honduras - October 27-29 (Cat 4)	7,000	$3.8 Billion	36.5	High Winds; Flooding; Mudslides
7	Offshore - October 28	31 (Crew)	$1 Million	Unk (50 ft waves)	Sank Windjammer "Fantome"
8	El Salvador - October 30 (Cat 3)	240	$400 Million	24+	Flooding; Crop & Livestock
9	Guatemala - October 30-31 (Cat 2)	268	$748 Million	24+	Flooding; Mudslides
10	Mexico - November 1- 2	9	$1 Million	24+	Flooding; Landslides
11	United States - November 5	2	$40 Million	10	5 Tornadoes F1 (4) - F2 (1)
	Total	11,374	$6.8 Billion		

Note: Tropical Depression formed on October 22, 1998, in SW Caribbean Sea near Cartagena

(Source: Tabulation by Author of multiple databases of countries involved)

It was noted that a common hazardous factor was the slow-moving hurricane and tropical depression inundating each of the ten countries in Central America and in Mexico with water depths varying from one foot to five feet. To ascertain if there was any correlation between the number of fatalities and extent of damages with the depth of precipitation, the Author constructed the diagram shown next (the fatalities shown in red, and the damage costs shown in blue as they increased with increased water depth.

Fatalities and Damages Proportional to Rainfall Amount
(Based on Hurricane Mitch in October 1998 in Central America Region)

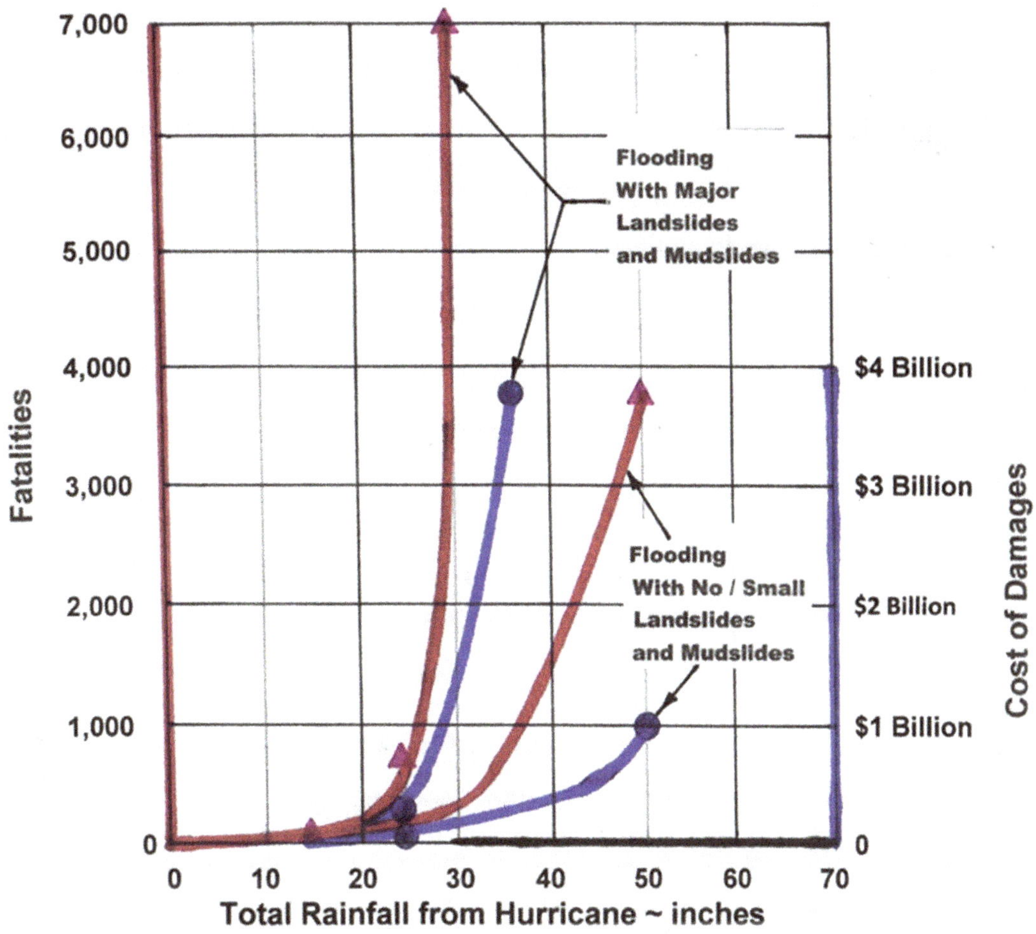

(Source: Author generated from previous tabular database)

The primarily conclusion drawn, not unexpectedly, was that the most useful "Lesson Learned" was that the soil composition on slopes of hills and low mountains can handle up to about fifteen inches of rainwater before causing massive flooding but at about twenty inches of rainwater in less than two days will create landslides and mudslides in hilly regions which are the main cause for high numbers of fatalities and high costs of property damages. That obviously may be different for flatter terrains where flooding becomes more of the hazard to which humans, animals, and properties are exposed. There may be some correlation that the red and blue lines that are at the left are deaths and damages due to flooding and with major landslides and mudslides, whereas the red and blue lines at the right are lower deaths and damages due to large depths of floodwaters but with no or only minor landslides and mudslides accompanying the flooding.

With reference to the deaths and damages in the United States caused by Hurricane Mitch, the database showed that there were 5 tornadoes spawned from Hurricane Mitch – four at Force 1 category and one at Force 2 category. As noted by the National Hurricane Tracking Center, Hurricane Mitch indicates that a slow-moving tropical storm can cause more fatalities and damages than a fast-moving higher Force-rated hurricane because of the lesser depth of water being able to be absorbed more gradually and not causing landslides and/or mudslides.

This book has thus far emphasized the many causes for natural disasters, but the other "side of the coin" is the more important humanitarian side – humans being both the victims of those disasters and perhaps being the cause for some of the collateral factors that exacerbate the losses caused by the disasters.

A primary natural factor largely overlooked by environmental scientists is that of gravity. The flooding causing much damage and a large number of fatalities is due to the rate at which the rain saturates the ground and therefore moves laterally on its surface before it can be absorbed. It is the acceleration of gravity causing a mass of water containing debris (natural and Man-made) and most importantly the mud with the flow. In physics, rather than refer to the weight of something it is more scientifically appropriate to refer to its mass – mathematically, mass is equal to its weight (W) divided by the gravitational acceleration (g) acting on it which is about 32.2 feet per second^2. Astronauts have an equal mass whether standing on the surface of Earth or on the surface of the Moon, but they "weigh" about 1/6th on the Moon than what they "weigh" on Earth. The Moon's gravitational acceleration (5.34 ft/sec^2) is about 1/6th that of Earth's (32.17 ft/sec^2).

The kinetic energy imparted by gravity into the mass of moving water, debris, and mud and moving towards a lower elevation is the factor that causes the number of fatalities and property damages to soar – not merely the amount of precipitation that has fallen and at what rate. The mass of debris/mud adds to the mass of the water and the kinetic energy, driven by gravity, is known (from Physics 101) to be KE = ½ mass x Velocity^2. Recognizing that fact, a major factor in the extent of the fatalities and injuries to humans and the extent of damage to property is what type and how many of them are in the path of the high energy flood movement.

The debris in a "mudflow" or "landslide" adds to its mass as it progresses. Even a slow-moving mudslide can have high kinetic energy from the high mass moving at slow velocity. The mudslide will move in the direction over the local terrain and at a rate dependent on the slope of the terrain in that topography. A mudslide – or avalanche – moving rapidly down a hillside is of no great concern to the public (but is to environmental protectionists) if there is no sizeable population or large number of properties in its path.

A tragic example of those combinations of different factors – geophysical, natural, and human – is seen in the mudslides that occurred in the regions of eastern Spain on the northern Mediterranean Sea in the nine days of late 2024 (October 29 to November 8). A photograph of the city of Valencia (population about 808,000) and inland area above Valencia is shown next:

(Credit to unknown photographer)

This is the hillside above the city of Valencia (seen in the background of the photograph with multi-story buildings), at latitude 29.47° N in the central east coast of Spain and adjacent to the generally placid Mediterranean Sea. There are only warm currents in the northern Mediterranean Sea in the coastal regions of Spain and southern France. There are occasional low-pressure systems throughout the lengthy Mediterranean Sea but they are not seasonal in nature. These depressions are termed as "Medicanes" and typically are not as strong as Atlantic Ocean hurricanes or Indian Ocean cyclones. The photo shows the origin of the mudslides in the region of Valencia – which caused 226 deaths, 14 still missing, and property damage over US $3.8 billion (about 3.5 billion euros).

Rather than attempt to describe the details of the Valencia mudslide, the Author has included an article by the Reuters News Service that is very credible and presents a large number of photographs which describe the physical destruction, and the harm caused to residents in the city regions. The article is included in its near entirety with over 40 emotionally "wrenching" photographs in Appendix 2 of this book. The mudslide flowed rapidly and with so much kinetic energy from the debris and mud in the flowing mass that it caught residents in underground parking facilities and literally buried them alive. Two photographs are extracted to attach a visual example of the problem of debris and mud being mixed in with the excessive flood waters forming a "battering ram" effect more destructive than just the hydrodynamic pressure of the flow of water without any debris:

(Credit for photos above to Reuters News Service photographer)

The debris in the high kinetic energy mudflow not only added to the magnitude of the clean-up of the mud and debris but denied access to the impacted area by rescue/recovery and clean-up crews to the scenes of the mess within the city of Valencia. There is not much preventative action that can be taken to avoid such problems in cities like Valencia, but municipal leaders in cities everywhere should ensure that, when such unavoidable tragedies do and will strike, the resources needed to recover from the tragedies are available on an emergency basis. Debate over what caused the mudslides is of secondary importance – human health and safety are of far higher priority than where to point the finger of blame.

There is also a very humanitarian aspect of tragedies that are caused by rogue natural events in a region. Neighboring regions should assist in the rescue and recovery efforts as a humanitarian act and not bill the impacted region for those already insurmountable loss of lives and properties. Homeless families need and deserve a place to take shelter from merciless natural elements that ensue tragedies such as that in Valencia. Reduction of "greenhouse gasses" in the atmosphere may be of primary importance to environmental scientists to protect the environment, but it is a far higher priority to protect the human population of the planet.

For the purposes of this book, the occurrence of hurricanes being seasonal and/or regional is yet another confirmation that environmental scientists should no longer use the "average global temperature" as the entire basis for setting goals to control Man's emissions of "greenhouse gasses". In a later chapter of this book, the "Lessons Learned" from Hurricane Mitch and other Atlantic hurricanes will be included as the basis for one conclusion and recommendation made to improve the safety of the human population of the United States region. It is strongly recommended that the building codes of future structures (residential and commercial) be changed to withstand those excessive forces of nature unleashed in hurricanes and tornadoes that will most likely get worse than before they get better.

A composite of several NASA satellite photographs tends to verify the wind currents near the Equator with the cloud formations:

(Source: NASA)

From the observation that the clouds are not homogenously distributed in the atmosphere, there should also be a question raised about Man-made "greenhouse gasses being homogenously mixed into the atmosphere at all latitudes. There can be little disagreement by any branch of science that near the surface of the planet – where the humans inhabit the land area and industrial processes and the transportation in land-based vehicles travel - that whatever emissions are created by any process (natural or anthropogenic) in the Northern hemisphere are not transferred into the Southern hemisphere and vice versa. There are very few coal burning facilities south of the Equator as there is almost negligible heavy industry and much less movement of vehicles on roads in the infrastructure below the Equator. This factor alone would cast great doubt on the environmental scientists claim that carbon dioxide emissions that are Man-made in the industries in Europe and the United States causing the coral reefs near Australia to be bleaching white from stress or causing such environmental chaos in Bangladesh. If stating factual data is a form of "naysaying" then there is little hope that environmental scientists will ever admit to the public and to Government environmental regulators the fallacy that Man-made "greenhouse gas emissions" are not the dominant or even highly significant contributors to the planet's "average global temperature" not being a credible scientific indicator of climate change.

The Equator is a most important dividing line in earth and climate science. Accordingly, the Author will repeat again for emphasis that the planet cannot be analyzed for global warming, climate change, and/or ocean levels rising with the "average global temperature" reference from which temperature and climate anomalies or present ocean levels are rising but must be analyzed in logically chosen regions (of maximum size being the continents, and of minimum size being but preferably within latitudinal zones. Logically, there can be no region on one side of the Equator being affected by natural-caused effects and/or anthropogenic lack of consideration of the environment of a region on the other side of the Equator. By the same logic, the consideration of gravity, the principles of the Coriolis effect, and the principles of water vapor in the atmosphere cannot be different between regions on the same side of the Equator.

Historical records and documents show that there are about 1,400 reported tornadoes annually. Every continent except Antarctica has experienced tornadoes. With about 300 tornadoes reported in Europe annually, the United Kingdom having about 33 reported tornadoes a year has the most of any European country. Tornadoes can occur at any latitude on the rotating spherical mass called Earth., and a photograph shown next was taken of a tornado in Australia:

(Credit to unknown photographer in Australia)

A rogue tornado in Bangladesh in 1989 left 1,300 people fatally injured by the tornado and another 80,000 displaced from their homes – built without construction safety standards - which were never generated and still do not exist in too many cities in lesser developed regions. However, it is the United States where tornadoes occur with disturbing regularity and high severity and predominantly in the eastern half of the United States annually that is the focus of this book.

Various scientific interpretations exist for the reason why the eastern half of the United States has the historical reputation for having the most frequent and most severe tornadoes (with the greatest property and human lives losses). The illustration shown next for the tornado history in the United States clearly indicates where the highest annual frequency of tornadoes has been reported:

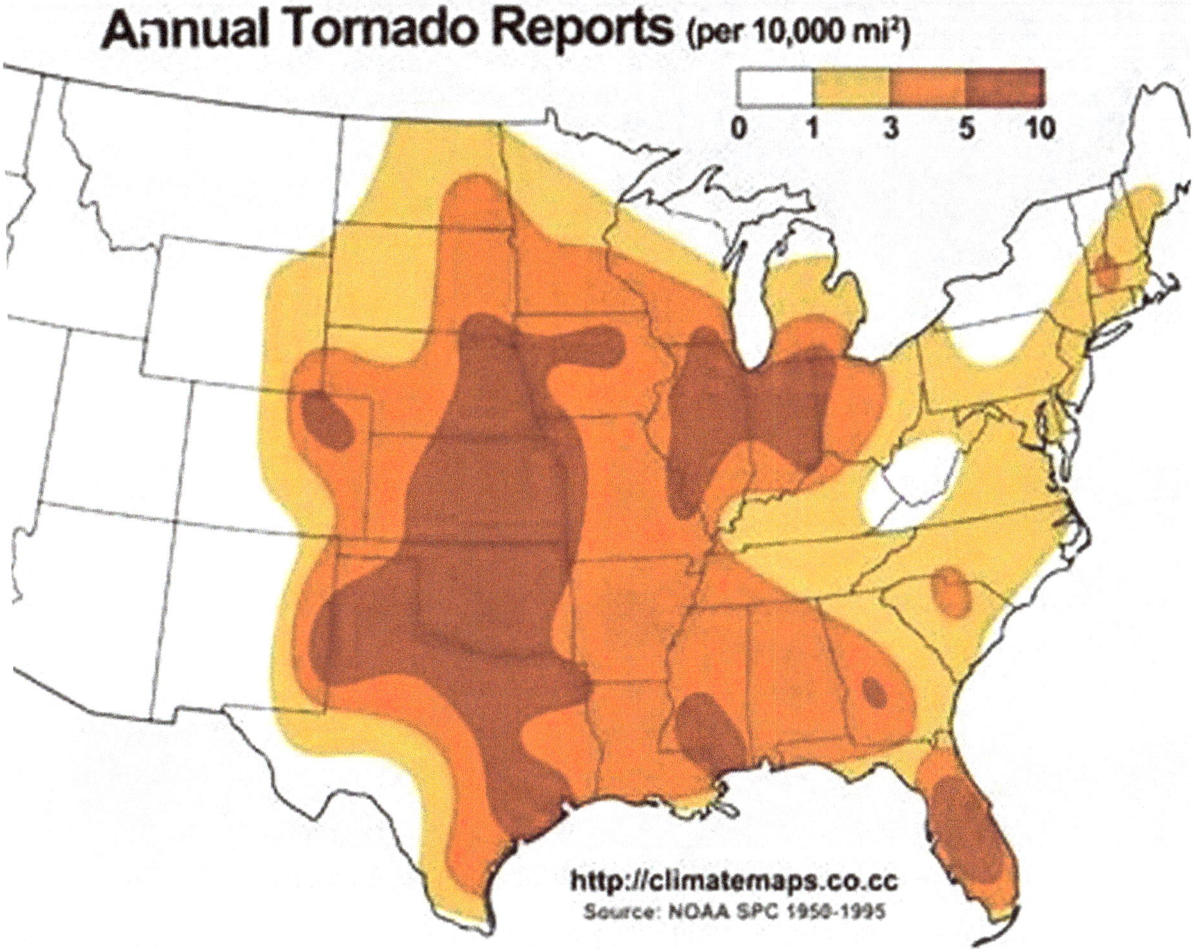

(Source: NOAA)

The rotation of tornadoes and hurricanes are opposite in the Northern and Southern hemispheres, but they have the common ability to cause weakly constructed structures to collapse or be torn apart and the flimsy roofing and wall covering debris (typically wood planks) to be lifted into the atmosphere and deposited up to hundreds of miles elsewhere in the path of the tornado. Generally, but certainly not exclusively, the most severe (i.e., highest energy content) tornadoes are again seen to occur in the states of Texas and Oklahoma. Also, generally most tornado damage is done by the high wind energy and the very low atmospheric pressure contained within the highly vortex shaped core of the tornado. There is not normally any extreme amount of precipitation associated with tornadoes except for the possibility of large hail stones on the hurricane's outer fringes which fall more vertically rather than being driven horizontally. (This Author and his family resided in Kettering, Ohio, for twenty years and were property damage victims from the hail in the periphery of the disastrous Xenia Tornado on April 3, 1974. The town of Xenia is located 15 miles directly east of Kettering in the extreme southwestern corner of Ohio. A photograph (from an Ohio friend's memorabilia) of the Force 5-6 tornado as it was entering Xenia (passing through the Arrowhead subdivision of Xenia) is shown next:

(Source: Co-worker (J. Shoates) of author – Xenia, Ohio, April 3, 1974)

The tornado passed ½ mile south of the author's residence in Kettering and battered the homes in the Author's neighborhood (Oak Creek) with high winds and hail the size of lemons. The hail did not break any glass in the Author's house but did fracture the cold asphalt-based shingles of the roof and caused the entire roof (and of all neighbor's) of the house to need replacement. In addition, it was popular to have corrugated aluminum panels over the outdoor patios of the houses and the Author's patio cover resembled having been struck by a thousand ball peen hammers damaging the patio cover so that it too needed replacement. But that is minor annoyance compared to the 33 persons who lost their lives in Xenia and the losses of over 300 family homes in the Xenia area – the homes were rebuilt but the deaths could not be reversed. By good fortune, the tornado struck in the late afternoon and after the school children had vacated the school and grounds. The school was demolished 1 ½ hours later by the merciless tornado. The latitude of both Kettering, Ohio, and Xenia, Ohio, is approximately 39.7 degrees North.

Tornadoes have the unpredictable variables of width of the swath as well as the length of "dwell" time the swath will exist on the ground. The Xenia tornado is an example of those that have very wide-swaths and "bull-doze" through neighborhoods such as that of the Arrowhead sub-division of Xenia. A photograph, with annotations by the Author to illustrate the swath width and the direction of the Xenia tornado, is shown next:

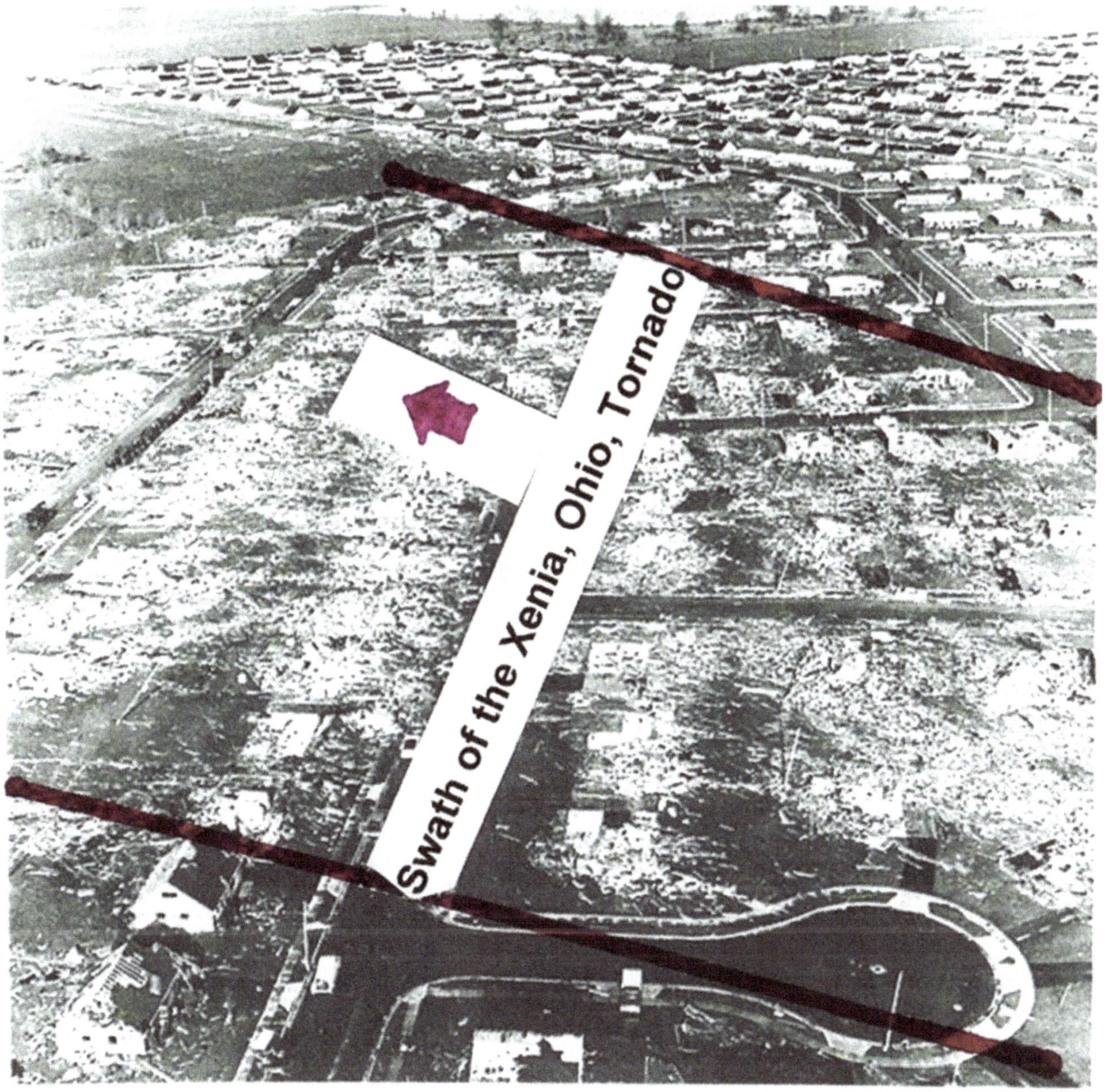

(Source: Author's personal photographs file with annotations)

The "misbehaviors" of water on the planet were accounted in the news media, most being largely factual as to the fatalities caused and the vast amount of damages (and high cost thereof to repair and replace). But many – and increasingly more fallacious - report the misbehaviors and weather-related catastrophes to be due to the Man-made "greenhouse gasses" in the atmosphere as the dominant factor causing a heat energy imbalance that instigated the weather to misbehave.

An example of a typical article – one of the better-written ones - in a reputable newspaper is shown next in its complete form that illustrates that the public is assumed to have accepted the continuously repeated mantra of the environmental science profession that it is "settled science" that humans creating "greenhouse gas emissions are the dominant cause of climate change". Any fallacious claim (to any degree) that is repeated often enough and widely enough for a lengthy period eventually becomes factual in the public interpretation.

Record heat unleashes deadly floods from New York to Libya

BY BRIAN K. SULLIVAN
Bloomberg News

Extreme heat is usually associated with drought and wildfires. But across five continents this year, it's also unleashed deadly flooding.

Cities around the world have seen record rainfall 139 times in 2023. A rare hurricane-like storm inundated Libya last month, killing thousands. More than 100 people died across Asia during an intense monsoon season in July. After fatal floods pummeled the U.S. Northeast over the summer, torrential rains paralyzed New York City last week.

Soaring temperatures were at the root of all these weather calamities. June, July and August were the hottest ever for that period. Higher air temperatures mean the atmosphere can hold more moisture, leading to heavier rainfall, while record-hot oceans provide fuel for storms. And as the climate warms, flooding is poised to get worse.

"We are now seeing record-smashing ocean temperatures, record-smashing global temperatures and record-smashing floods," said Jennifer Francis, a climate scientist at the Woodwell Climate Research Center. "It's all connected."

In the U.S., two major flood events this year – in California from January to March and the Northeast in July – killed 32 people and caused $6.7 billion in damages, according to the U.S. National Centers for Environmental Information. The damages are the highest since 2019, when accounting only for flood incidents totaling more than $1 billion.

Global damages are difficult to tally because not all countries keep accurate records, but flood events have left more than 17,000 people dead or missing in 2023. On average, there were 216 floods or landslides worldwide per year from 2019 to 2022, compared with an average of 160 over the previous four years, according to the International Disaster Database.

In large part, the ever-worsening downpours are caused by what scientists call a "positive feedback loop."

The Earth's temperature rises in response to greenhouse gases like carbon dioxide and methane. Hotter air holds more water vapor, which then absorbs heat and prevents it from escaping the atmosphere – resulting in even more moisture in the air and heavier rainfall.

As the atmosphere gets warmer, so have the world's oceans. Higher water temperatures heat up the air above the sea surface, generating energy for hurricanes and tropical storms to form. The Atlantic saw 10 storms develop from Aug. 20 to Sept. 16, the most on record for that period, according to Phil Klotzbach, a Colorado State University hurricane researcher.

The storm that devastated Libya last month was a "medicane," a name that's an amalgamation of Mediterranean and hurricane. These weather systems are smaller than large Atlantic storms but can strike with Category 1 force on the five-step Saffir-Simpson scale, bringing winds greater than 74 mph and flooding rains.

(Source: Durham, NC "News & Observer" October 9, 2023)

Three factors seldom, if ever, mentioned in news media articles are 1) that it is high heat energy that generates water evaporated from the water surface of the planet into the atmosphere; 2) the ever-present gravitational acceleration (gravity is not a force but an acceleration factor) which causes the massive amount of water in the atmosphere to fall back (i.e., be accelerated down toward the center of the Earth) onto the Earth's surface; and 3) the human-built infrastructure that was designed for "normal" environmental conditions but is not sufficiently strong to withstand anomalistic weather and climatic extreme events (seasonal or random).

Environmental scientists seem to also be "settled" on there being a substantial "feedback" cycle resulting from "climate change", as indicated in the chart shown next:

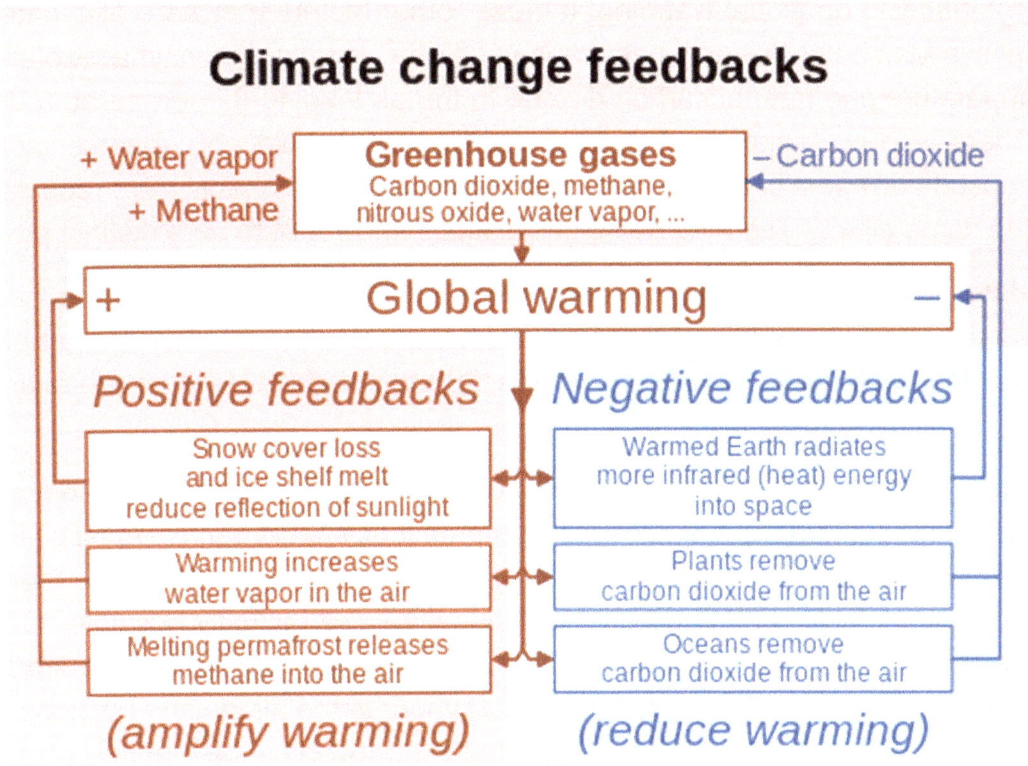

(Source: NOAA)

It should be noted that the chart indicates that the feedback described (both positive and negative) are on "Greenhouse gases" and "Global warming", but the title of the chart is "Climate change feedback". It appears that the conceiver of the chart has mistakenly assumed that "global warming" and "climate change" are one and the same. (In an earlier chapter, the variables of surface temperature and precipitation are defined individually as "weather" in the short term (days) and as a combination over a long term (minimum 25 years) produce a database for calculating "climate". If the chart is mislabeled (should not the title more correctly be "Global warming interactions"?) that would explain part of the mistitling.

But how can the "after the fact" climate average of more than 25 years of weather factors effect (as "feedback" or any form of "effect") those weather factors that occurred as much as 25 years earlier? That doubt of scientific inconsistency in the environmental scientists' hypothesis causes the probability of scientific validity of the hypothesis being low to nil. There can be an "interaction" between the weather and the atmosphere and vice versa but they are contemporary actions, not "feedback" from a future state (of the climate) to a current state (of the weather). The "tail does not wag the dog". Using the term "feedback" in electronics may seem appropriate because of the speed of electrons – contrarily, the time period between "weather" and the "climate" resulting from the weather is measured in years of time. The weather today adds to, but does not define, the climate tomorrow or next month.

It should also be noted on the chart that the chart originator does not show any effect of astrophysical nature on global warming (as discussed in the earlier chapter of this book). The other databases of astrophysical nature (verified as existing in different cyclical periods - 40 years, 90 years, 100,000 years to 400,000 years – by the Milankovic Cycles astrophysical calculations, and evidenced by analysis of the Vostok Ice Core Samples) are not shown in the chart as having an effect (direct or indirectly as a feedback) on global warming. If those "other factors that were eliminated" to bolster the fallacy that it is Man being the dominant cause of global warming, then what explains the cycles in surface temperature and natural carbon dioxide in the air lagging the temperature by about 800 years when there was no human presence on the planet? Would not whatever combination of causes that caused that effect of causes to happen centuries ago probably have recurred years later to generate the same effect? The Milankovic calculations show that to be a distinct probability.

The current database of information from satellites launched into orbit around the Earth by NASA and other countries and observations from the International Space Station are not shown on the above chart as identifying the other factors that affect Earth's surface temperature regionally and by latitude.

As a constructive criticism by the Author of the fallaciously titled 'feedback" chart above, the amount of the individual positive interactions and negative interactions should each at least be identified with an approximate percentage of their individual magnitude on the total interactive loop with global warming. All interacting factors (both positive and negative) are factually possible, but the amount of methane being released by melting permafrost is negligible compared to the amount of water vapor being in the atmosphere. The increase in water vapor in the air caused by global warming far outweighs the amount of carbon dioxide in the air that is removed by the oceans.

It must be noted more importantly though that the interaction chart above gives recognition to the irrefutable fact that water vapor has a heat-trapping effect just as do the Man-made "greenhouse gases" whereas it was previously not considered by environmental scientists who maintained that "by eliminating all other contributors (which included water vapor), settled science concurs that Man burning fossil fuels is the dominant cause for global warming". The charts shown earlier in this chapter - generated by Dr. S. Fred Singer (Professor Emeritus of Climate Science at the University of Virginia) - dispels that original hypothesis that was first stated in the mid-1950s yet fallaciously persists to the present time. The environmental scientists, the "green" activists, the news media, and worst of all the governmental organizations that are spending taxpayers' monies to "combat climate change" have "circled the wagons" and have totally misled and confused the public thinking. It is hoped that this book will cause common-sense to prevail once again in the scientific records of the Earth's turbulent history.

Nothing humans can do will change the effect of gravitational acceleration when it moves a flood of water downhill, other than perhaps to create diversion channels that change the rampart water to flow into reservoirs or catch basins for later use to offset any drought episodes that may and probably will occur. But there is much the human population of a region can do to make their infrastructure able to withstand earthquakes, floods, tornadoes, and hurricanes. It is disturbing when reading that weather-related catastrophes have taken thousands of human lives and caused billions of dollars in property damage, but the news media seldom reports all the factors that contributed to the causes of the catastrophe. In the zeal to tag humans and their dreaded greenhouse gas emissions with the responsibility for the catastrophes reported above, some of the more important contributing factors – far more consequential than the small amount of greenhouse gasses created in Libya for example was the fact that the pressure of the flood waters had caused a poorly constructed dam to burst when excess water behind the dam exceeded the design structural strength of the dam (to withstand the pressure of the water behind the dam). The collapsed dam in Libya sent millions of gallons of water racing downhill towards the coast of the Mediterranean Sea and washed thousands of victims residing below the dam into the Mediterranean waters.

The news media however contends that all the flooding in Libya was triggered by Man burning fossil fuels and causing "greenhouse gasses" to trap all the heat of the Libyan desert and lift that much water into the atmosphere above Libya. Repeating a photo from satellite optical records of the planet shows that the atmosphere above the major deserts of the planet – including Libya in North Africa – is "normally" devoid of cloud cover and significant water vapor:

(Source: NASA)

It is more probable that winds at altitude had blown the water-saturated clouds seen just south of the North African region over onto the location above Libya that experienced the natural catastrophe. The news article inferred that the cause of the natural and Man-made error in dam design was "The Earth's temperature rises in response to greenhouse gases like carbon dioxide and methane." And subliminally causes the public to accept the scientific hypothesis that "Man is to blame for having burned a massive amount of fossil fuels in that region".

Before continuing with other pertinent topics, a very noticeable coloration in the above photographs indicates the extensive areas of the planet that are categorized as desert. A table below indicates where the largest deserts of the world are located and their vast size:

Largest Deserts of the World

(Arranged by Planet Quadrant and Surface Area)

			Desert Area		% of Quadrant
Desert Name	**On Continent**	**In Quadrant on Planet**	~ Sq. Miles	~ Sq. Kilometers	Surface Area
Sahara	Africa (Northern)	Northeast	3,552,140	9,207,502	7.21%
Arabian	Asia (West)	Northeast	1,000,000	2,592,100	2.03%
Gobi	Asia (East)	Northeast	500,002	1,296,055	1.02%
Rub' Al Khali	Asia (Middle East)	Northeast	250,000	648,025	0.51%
Syrian	Asia (Middle East)	Northeast	193,051	500,407	0.39%
Karakum	Asia (Central)	Northeast	135,136	350,286	0.27%
Kyzylkum	Asia (Central)	Northeast	115,831	300,245	0.24%
Taklamakan	Asia (China)	Northeast	104,248	270,221	0.21%
Thar	Asia (Indian Subcontinent)	Northeast	77,220	200,162	0.16%
		Total	5,927,628	15,365,003	
		Quadrant Surface Area	49,240,436	127,636,134	**Total 12.04% (Sand)**
Great Basin	North America	Northwest	189,845	492,097	0.39%
Sonoran	North America	Northwest	119,692	310,253	0.24%
Mojave	North America	Northwest	47,877	124,102	0.10%
		Total	357,414	926,452	
		Quadrant Surface Area	49,240,436	127,636,134	**Total 0.73% (Sand)**
Kalahari	Africa (Southern)	Southeast	347,492	900,734	0.71%
Great Victoria	Australia	Southeast	134,650	349,026	0.27%
Great Sandy	Australia	Southeast	110,036	285,224	0.22%
Namib	Africa (Southern)	Southeast	61,776	160,130	0.13%
		Total	653,954	1,695,114	
		Quadrant Surface Area	49,240,436	127,636,134	**Total 1.33% (Sand)**
Patagonian	South America	Southwest	259,847	673,549	0.53%
Atacama	South America	Southwest	54,054	140,113	0.11%
		Total	313,901	813,662	
		Quadrant Surface Area	49,240,436	127,636,134	**Total 0.64% (Sand)**
Antarctica	Antarctica	South Polar	5,500,024	14,256,612	**Total 100% (Ice)**

Notes:

Planet is assumed to be spherical with average diameter of 7,918 miles (12,742 km). Area is 196,961,745 sq mi (510,065,665 sq km)

Quadrants are defined relative to Equator (0 degs latitude) and Prime (Greenwich) Meridian (0 degs Longitude).

The total area of all major deserts on the planet is 12,752,921 sq miles (33,056,843 sq kilometers).

6.47% of the planet's surface area (196,961,745 sq miles) is covered by desert sand (or desert ice).

(Source: Author tabulation from databases in Encyclopedia Brittanica)

Scientifically, a desert region is classified as being a desert due to lack of annual precipitation. By that definition, the entire continent of Antarctica is categorized as being a desert due to almost zero precipitation in that polar region where the average temperature at that southern latitude is about -76 degrees Fahrenheit in the high elevations and 14 degrees Fahrenheit in the coastal regions (Southern Ocean). The coldest temperature ever recorded on Antarctica was -128.6 degrees Fahrenheit (-89.2 degrees Celsius) at the Vostok Ice Research Station at latitude 75° South in July 1983.

Technically, because of its dry climate (having an average temperature year-round of below -20 degrees Centigrade (- 4 degrees Fahrenheit) and therefore no water vapor in its local atmosphere), the Antarctic Continent is classified as a desert. Much like the numbers in the table of oceans being huge, the total of the desert areas (to be shown in tabular numerical format) are slightly smaller than are the oceans, but with the surface of the planet being over 80% water, and the deserts (sands plus Antarctica's ice) covering another 6.4% of the planet's surface, the world population of 8 billion humans have less than 13.6% (about 1/7th) of the planet's surface area on which to reside.

A perhaps better visual indication of the percentage of area of desert land on the planet's surface than that of the large table of numbers listed earlier is reflected in the following graphic diagram:

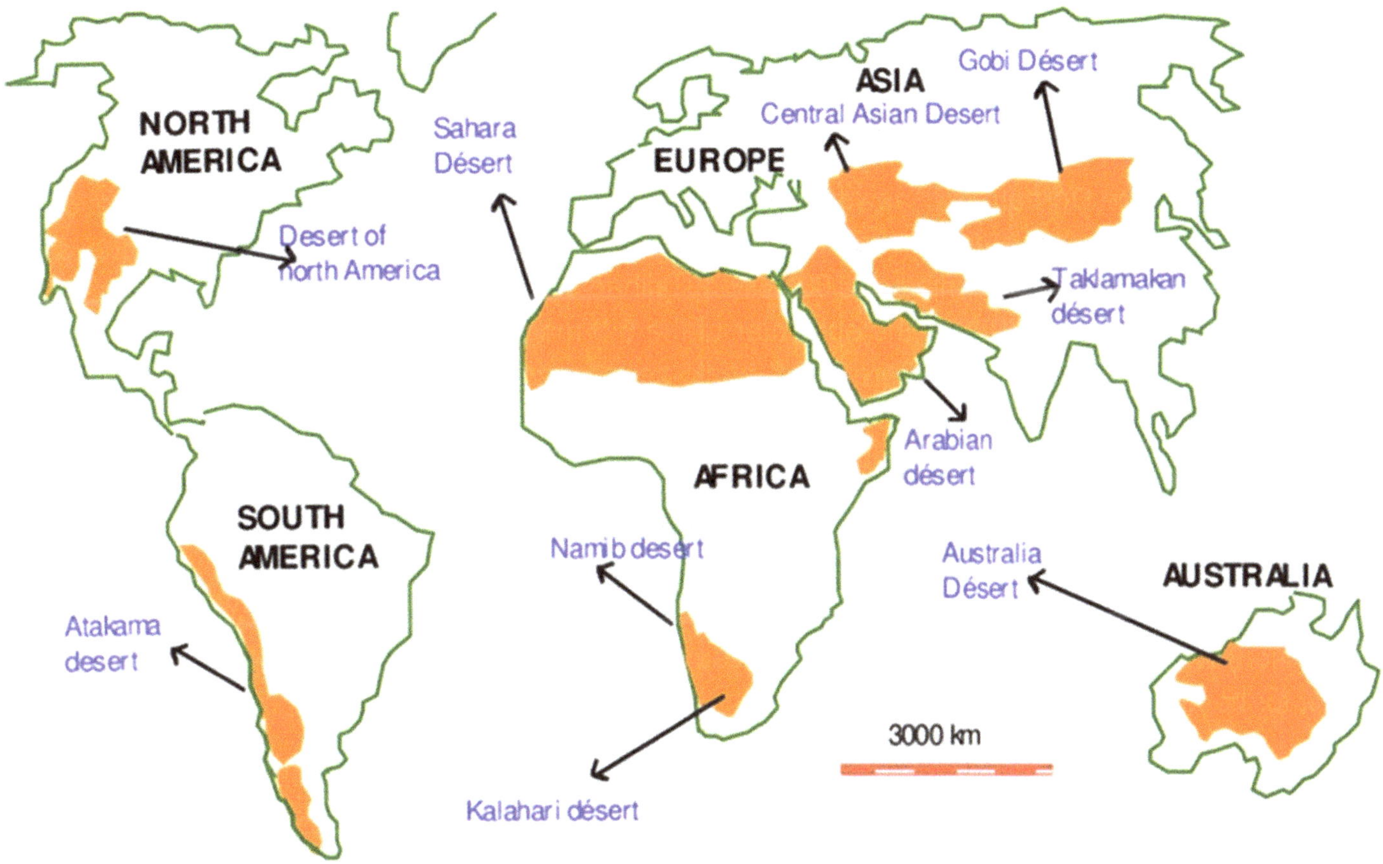

(Source: National Geographic Society)

While there are relatively few population centers in the deserts, there are populated areas on the edges of the deserts, especially those on the shores of large bodies of water. A scene (in the Arabian Desert) from a National Geographic Magazine shows the means of travel across desert sands for trade purposes and for supplying of food from one region to another:

(Source: National Geographic Society)

There is a distinct lack of water in the vicinities of deserts and there seldom are any clouds bringing rains to the deserts. For reference, Antarctica is also considered as a desert as there is no precipitation in Antarctica as the temperature is on average well below freezing.

On page 8A in the Durham, NC News Observer on October 9, 2023, a news media article describing "Record heat unleashes deadly floods from New York to Libya" was overshadowing the article "Quake devastates W. Afghanistan; 2,000 killed, Taliban says". In the Afghanistan earthquake article is stated: "Afghan officials said the epicenter of the 6.3 magnitude earthquake was in two districts, Zinda Jan and Ghurianan, where mud brick houses collapsed within seconds of the initial earthquake, leaving residents with no time to escape."

Poor construction standards of human-occupied buildings and proximity to the origin of the force creating the catastrophe have been and are a contributing factor in natural-caused catastrophes of hurricanes, monsoons, flooding, landslides, tornadoes, earthquakes, and volcanic eruptions throughout the planet's history. Later in this book the choice of residency (type construction and location) will be described that humans choose for economic and lifestyle enjoyment reasons and not for best safety or health considerations.

Few articles attribute any of the climate changes being caused by changing astrophysical and geophysical changes over which Man has no control to abate – most attribute climate change to be caused by Man-made "greenhouse gas (GHG) emissions". In the article above, water vapor in the atmosphere is mentioned in passing, but no mention is made that it is the water vapor that is retaining most of the heat energy in the atmosphere in a manner about 90 times more effectively than does the relatively small concentration (total 0.0435% - of which 0.0295% is of natural origin and 0.0140% is Man-made) of greenhouse gasses (such as carbon dioxide and methane) in the atmosphere. <u>The "settled science" hypothesis that Man burning fossil fuels is the dominant cause for global warming is clearly fallacious – but it must be recognized that the environmental protection</u>

scientists generating that fallacious claim did so without the benefit of many of the updated databases researched by the Author and which came into existence after the fallacious "settled science" conclusion was derived.

Returning to the region of the United States, it is generally known that the geophysical effect of the cool Pacific Ocean current flowing clockwise from Aleutian waters southerly down the West Coast states (Washington, Oregon, and California) generates lower anomalies in temperature in those three western states than in the mountain and central states. The West Coast Pacific Ocean waters flowing clockwise cause the water on "cool" (anything "cool" in California is not all that bad) southern California beaches to range from a cool temperature of 55 degrees Fahrenheit to a hot temperature of 65 degrees Fahrenheit – slightly warmer when an "El Nino" much warmer than the remainder of the Pacific Ocean current flows southerly along the Washington and Oregon coasts and the entire California coast. (It is for that reason primarily why all surfers on the West Coast wear thermally insulated "wet suits" to prevent getting hypothermia during long "hanging-ten" sessions. For unknown reasons (probably for warmth rather than for aesthetics), the wet suits are most often produced in dark color giving the surfers an appearance of resembling an elephant seal or walrus which are high on the menu for great white sharks – but that aquatic hazard to humans is not directly related to climate change.)

The East Coast Atlantic Ocean current is also clockwise but brings the very warm "Gulf Stream" current from the southern Caribbean region and from the Gulf of Mexico currents flowing northerly in parallel with the East coast of the United States. The water temperature of the Gulf Stream ranges from an almost never low of 75 degrees Fahrenheit to sometimes over 85 degrees Fahrenheit and almost all year round. It is little wonder why there is so much travel by human "snowbirds" during the colder months in the northern US and southeastern Canadian provinces to the Florida beach cities and towns. Wind surfers in Florida have no need for wet suits as do the "cool" dudes in California.

The upper central states near the Great Lakes experience more and different problems than do the southeastern states. The winters in the upper central states are normally very cool – due to the "Lake Effect". However, those states just east of the Great Lakes experience blizzards and very heavy snowfalls by the warmer and more moist winds flowing to the east and encountering the warmer Great Lakes waters where they pick up additional moisture (water vapor in the air) and that high water vapor content condensing out as heavy snowfall when the moisture laden winds encounter the very cool surface temperatures of western Pennsylvania and "upstate" New York. Buffalo, New York, is notorious for having multiple blizzards of greater than 36 inches of snow during the Winter and with drifts of blown snow of 6 feet or greater. The high winds cause the snow to drift and bury cars on the highways – presenting life-threatening hazards to motorists stranded for sometimes 12-15 hours before being rescued.

Shown next is what that type of scene is like all too often in the northern states of the USA (those having a "rush hour") during a winter season:

(Source: Automobile Association of America)

Resemblance of the above scene to that of a herd of lemmings headed for a catastrophic end result is coincidental.

The northeastern states (from Maine to Delaware enjoy the "four seasons" but are clearly not paradise in the hot humid summer season. The color of the leaves on the trees change to beautiful "Autumn leaf colors" but snow is in the forecast soon after. The temperatures are often in the 80-85 degrees range during the summer but the high humidity 75-90%) associated with the high temperature can be miserable (as attested by the Author who grew up in New York City and with much time in the "upstate" region.) A scene such as above will never be seen on the Florida Turnpike – unless of course it is far in the future (10,000 years or more) when the Antarctica ice core samples indicate there will likely be another "Little Ice Age". Could it be possible that future Man may need to burn everything that can burn to survive the ice age low temperatures? Once again, the public and the environmental scientists are reminded that it is Man who must survive the extremes of the weather and the climate not the average of those extremes. The Author is not a botanist but is well aware that orchids do not bloom in outside gardens in the colder climates of the United States, but tropical flowers do bloom in "greenhouses" almost anywhere. Many "greenhouse" operators inject carbon dioxide inside the greenhouse to facilitate the chlorophyl process.

Scientists differ from the average member of the public in how to resolve a large problem being encountered such as that of the causes and effects of changes in the natural environment and by/on the activities of humans populating that environment. The average human has not been so fortunate (by choice or because of lack of funds) to possess the level of education needed to do a scientific analysis of the very complex problem that involves many factors. Perhaps this book will serve two purposes: 1) to provide the public with a simplistic "big picture" set of information, and 2) to provide the scientific profession with a different perspective from using a "one-size-fits-all" approach and "average global temperature" variable to "combat climate change". A subset of 2) above is to inform teachers of science to educate the young people in the public-school systems with the more scientifically correct causes and effects of climate change than the fallacious present set of information being taught in schools as "settled science" based on the fallacious hypothesis that "Man burning fossil fuels is the dominant cause for the climate to be changing."

Credit for the concept of analyzing the climate on the planet by regions rather than by "global averaging" should be given to the German Russian climatologist and botanist Wladimir Köppen (1846 – 1940) and joined later by Rudolf Geiger (1894 – 1981). Without the benefit of any "big picture" satellite imagery of the climate being regional in nature, Köppen and Geiger developed a climate classification scheme for the planet but with attention to how the climate affected botanical "victims" not the human population. Their climate zone classification scheme also did not consider any effect of human-generated "greenhouse gasses" on the environment or on vegetation. Their system was much like a 3-column tabular arrangement where the first column designated a climate zone (with five climate zones generated during the "Age of Discovery" 500 years ago). The second column designated the type of precipitation that affects botanical vegetation (with four levels of "wetness"). The third column designated the general warmth of the seasons (with nine types of seasonal temperature variation).

The tabulation shown next illustrates the three-letter "code" system that Köppen and Geiger used in analyzing the five different climate groups of the planet in regions that had a unique three-letter code and color assigned to each region.

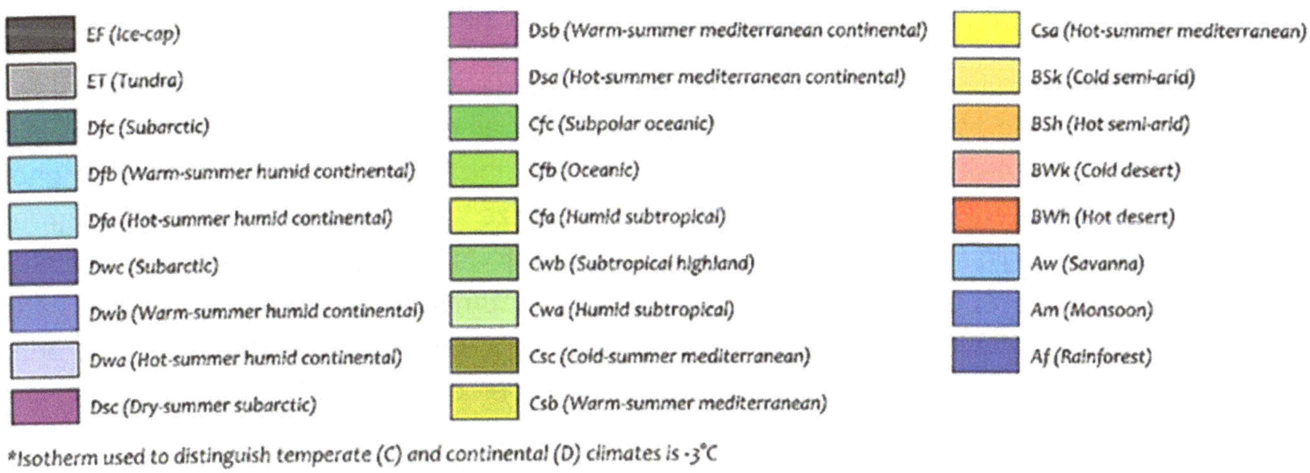

(Source: Britannica database of Köppen - Geiger Climate Classification)

The associated map with the regions color-coded per the Köppen Geiger Climate Classification method is shown next:

Köppen - Geiger Climate Classification (Botanical)

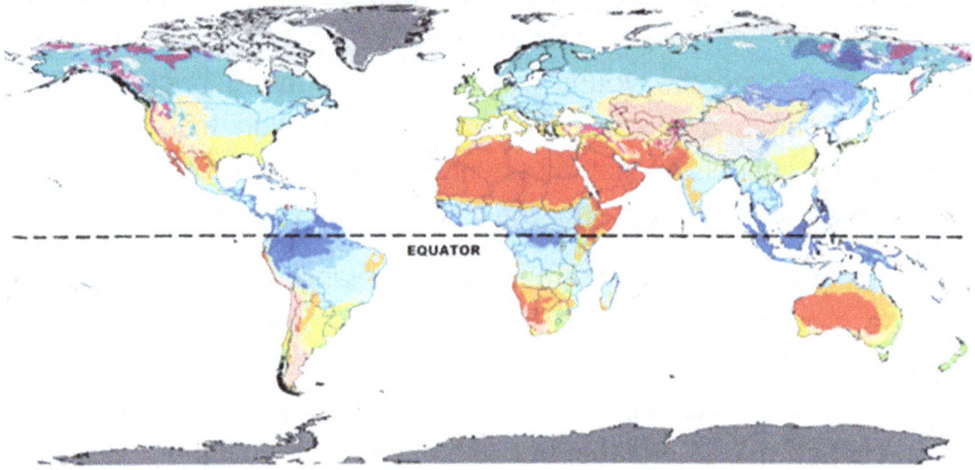

(Source: Encyclopedia Britannica)

Readers are reminded that this climatic- regions chart was generated for botanical purposes and not for any implications of the effect of climate change on the human population of the planet. However, it reveals that reputable scientists recognize the value of classifying the regions of the planet for scientific needs into regions rather than by the "one-size-fits-all" approach by environmental scientists using the "average global temperature" method.

Another observation made from the Köppen - Geiger colored regions map points out the large proportion of the map being in the red. The red area designates the arid – i.e., the lack of water for growing vegetation - conditions of the desert. A later chapter describing why humans populate favorable climatic conditions in many different regions of the planet shows the Köppen – Geiger climate maps for six different regions of the planet in greater detail than that shown above.

It would be wise for the United Nations Intergovernmental Panel on Climate Change (IPCC) to be the "Maestro" for the regional assessments to be done by all of the member countries within the present IPCC purview. A good Maestro does not "direct" the different groups in the symphonic orchestra how to play their instruments, only to "harmonize" their music with the other groups. Similarly, the IPCC would serve to minimize duplication of efforts and waste of resources in pursuing activities that do not apply to each country's unique set of circumstances. Each region has its natural environment, human needs and demands (for safety, health, and comfort), and different priorities, regional alternative energy resources available, absence or presence of coastal regions, being located in different latitudinal climatic regions of the planet, etc.).

The IPCC should not be setting "one-size-fits-all" goals for the control of any human activities needing to be met by an arbitrarily chosen date as a "tipping point" warning sign. The goals for regions can be set by a group of countries in a common region with IPCC acting as the mediator but not as the director. The strategy of "combatting climate change" must be revised to being one of "détente" and adaptation between the human population residing in a region to be in harmony with the environment of that same region.

Humans being in harmony with their environment includes not polluting the air, water, or land of the environment. Environmental scientists have resorted to identifying Man-made carbon dioxide as the dominant cause for global warming to fallaciously capture the minds of the public, but the analysis in this book disproves that claim. Their own measurements of the "greenhouse gasses" in the atmosphere and the measurements by international scientists of carbon dioxide in the atmosphere shown by the Antarctic Ice Core samples shows that Man-made carbon dioxide gas comprises 0.013% of the atmosphere and natural-caused carbon dioxide comprises 0.029% of the atmosphere. That contradicts present day environmental scientists claiming that "greenhouse gas " carbon dioxide generated by Man burning fossil fuels is the dominant cause of global warming.

The ice core sample analysis also shows that the peak in total carbon dioxide in the atmosphere lags the temperature peak by about 800 years. That factual data indicates that it is not the carbon dioxide in the atmosphere at that time (120,000 years ago when there was no Man on Earth burning coal) that drives the average global temperature. Instead, the ice core samples show that it is the average global temperature which influences how much carbon dioxide will be in the atmosphere – a contradiction of the claim by present day environmental scientists.

Environmental scientists may – and probably will very strongly – dispute those contradictory facts that the environment (temperature and atmospheric content) at 40° North latitude (amidst the highest World population density where the global warming is of highest concern to present day environmental scientists) is not the same environment as at 78° South latitude (at the Vostok Antarctica Ice Research Station. The effect of latitude on the temperature on cities at any latitude is factually a strong cause and effect factor as shown in an earlier chapter (describing geophysical factors affecting global warming and climate change.) When the average annual temperature for 40 major cities (as recorded by the World Meteorological Organization) is properly analyzed, it was shown that the latitude and elevation of the cities varies very strongly with latitude. A repeat is made of the graphical illustration for the effect of latitude on eight cities in the USA which differ by only 22° in latitude (Miami, Florida, at about 26° N, and Seattle, Washington, at nearly 48° N):

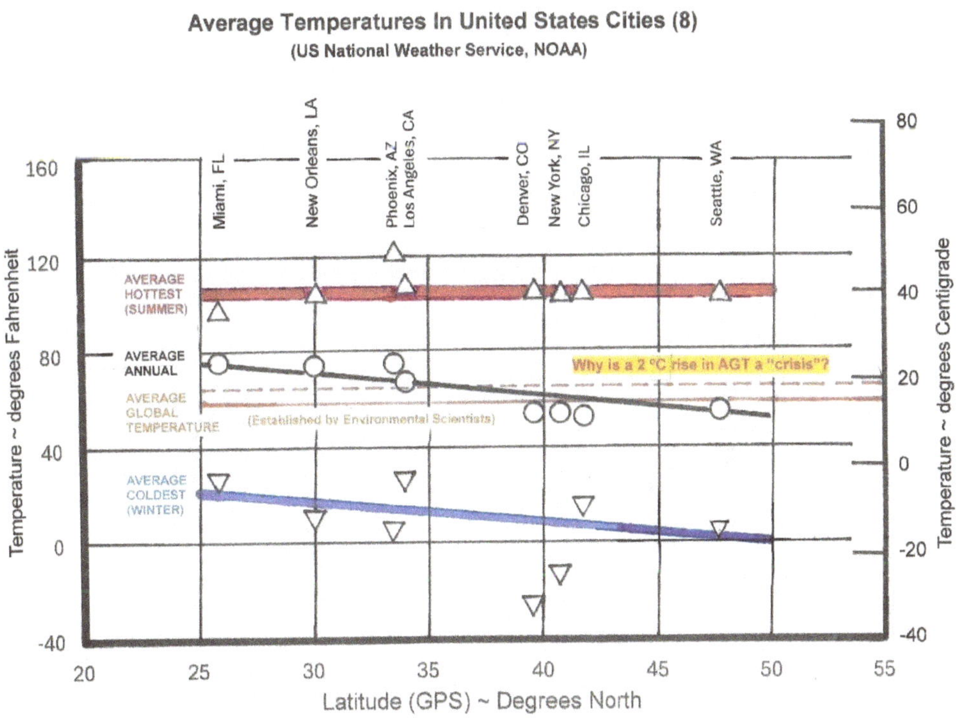

(Source: National Weather Service, NOAA)

As seen above (the data with a solid black line), the human residents in Miami near the lowest latitude of the USA are exposed to an average annual temperature of 76 °F (24.4 °C) but the residents of Seattle near the highest latitude of the USA are exposed to an average annual temperature of 55 °F (12.8 °C). That is a 21 °F (11.6 °C) difference in average annual temperature primarily due to the latitude difference of 22°. The first of three questions that environmental scientist have yet to answer clearly to the public is: **1a)** Why does a 2 °C rise in average global temperature (the dashed red horizontal thin line at the center of the illustration and goal of the UN IPCC to be not exceeded from any cause including Man burning fossil fuels) constitute a near-term global "crisis" when all cities in the USA at or above 38° latitude are already more than 2 °C (3.6 °F) cooler than the environmental scientists' "crisis" threshold average global temperature?

1b) Why has there not been a "crisis" declared for all US cities below 38° latitude and having an average annual temperature already a few degrees above the environmental scientists' "crisis" threshold temperature of 18 °C (64.4 °F)?

The average hottest temperatures (bold red line) for the eight cities are shown in the illustration and surprisingly (to this Author) are seen to be a flat-line variation at about 105 °F (40.6 °C) with no significant effect of latitude. The city of Phoenix, Arizona, is seen to be the furthest difference (hotter) from the solid bold red line, but that is most probably due to Phoenix being in the very hot and arid Southwest Desert region of the USA. A photograph of the Phoenix area as seen from space shows the absence of green color and the predominant tan color of desert sand:

(Source: NASA) Phoenix, Arizona

The Author was residing in a southern suburb (Chandler) of Phoenix in the summer of 2023 when the temperature was at or above 100 °F (37.8 °C) for 23 consecutive days, above 105 °F (40.6 °C for 7 consecutive days, and reached 115 °F (46.1 °C) as the peak maximum temperature in the Phoenix area. Humans do survive extreme temperatures such as those by adapting to them, not "combatting" them. The second question the environmental scientists have yet to answer clearly to the public is:

2) With the average hottest temperature (which occurs in the summer season) of the entire USA being at 105 °F (40.6 °C) and which is 40 °F (22.2 °C) hotter than the environmental scientists' "crisis" temperature threshold of 18 °C (64.4 °F), why has the entire population of the USA not already succumbed to the hyperthermia "crisis?

The use of "average global temperature" and "average annual temperature" for a city masks the seasonal effects on temperature that occurs on all parts of the planet's surface area. The prior graphical illustration shows the two extremes of heat and cold that each city undergoes and within which human residents must adapt to those factual extremes - not their average - to survive. Averaging of any of the meteorological variables (temperature, precipitation, wind speeds, humidity, % cloud cover, H and L cells, etc.) may be convenient for scientists to attempt to be modeled for predictive purposes but that averaging process is not what humans and all living forms of life experience daily in the real world. Fatalities of humans can and do occur both from hyperthermia (excessive heat) and hypothermia (excessive cold) but almost zero fatalities due to heat at the fallacious "average global temperature".

The average coldest temperatures (which occur during the winter season) for the cities considered in the graphical illustration above are shown (in the solid blue line category). The difference in temperature between the hottest and coldest extremes are 85 °F (47.2 °C) at latitude 25° N and 105 °F (58.3 °C) at latitude 50° N. The third question for environmental scientists yet to answer clearly to the public is:

3) Why is a 2 °C (3.6 °F) rise in average global temperature considered a "global crisis" despite there being more deaths of humans from hypothermia than there are from hyperthermia in the World in regions at very different latitudes?

To change the tone of this book from being a report on the many problems that are caused by too much water, or not enough water, or water that is abundant in quantity but unsafe for human or agricultural use because of its high saline or other pollutant content, or water rampaging in a flash flood washing away vehicles that are occupied, there is an aspect of water vapor and rain precipitation in the atmosphere that always brings a pleasurable sight to human eyes. Seeing a rainbow in the sky is one of those pleasures in life that cannot be bought – as shown by a photograph taken in Prescott Valey, Arizona, in September 2023, as a weather front was approaching:

(Credit to Diane Stilson, Author's daughter; Prescott Valley, Arizona)

The occasion of seeing a rainbow is made more enjoyable by seeing a double rainbow as shown in the photograph. There are many stories told about rainbows – leprechauns guarding a pot of gold at the end of the rainbow, etc., but the scientific reason for a rainbow to be formed is much like the theme of this book – there are facts and fallacies when seeing a phenomenon of Nature. The fact is that, unlike some environmental scientists' who choose to disregard water vapor in the atmosphere, there occasionally are small droplets of water in the atmosphere in the air ahead of heavy rainfall. When the rays from the Sun, which contains light having a spectrum of colors having a unique wavelength for each color, encounter a droplet of water, the solar ray is refracted into its spectrum of colors according to each color's wavelength. The phenomenon can be duplicated in a laboratory with a glass prism, but in the atmosphere the prism is replaced by the billions of water droplets. The simple arrangement of the Sun, the rain droplets, the angles, and the position of the human's eye are all factors – and the factors must be in the correct positions for the rainbow to be seen. The arrangement of those factors appears next:

(Source: Sciencetalk.com)

The illustration shown next shows the process with one droplet of water, but the sun's rays are so widespread the rainbow appears to arc across a large portion of the sky.

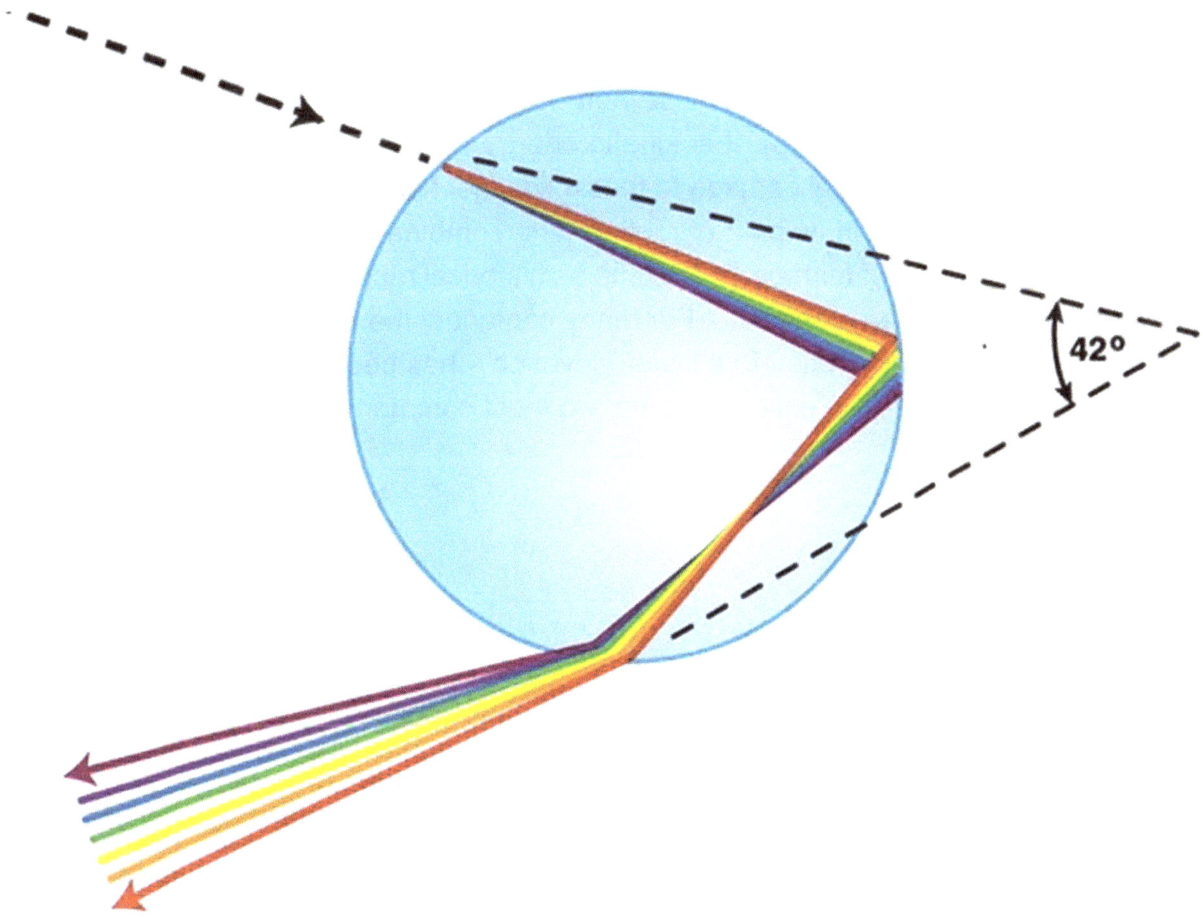

(Source: Sciencetalk.com)

The colors in the spectrum of a rainbow are always in the same variation – which the Author was taught in eighth grade by his science teacher, Mr. Roy G. Biv. (Readers are asked to pardon the Author's attempt at humor but the Author's science teacher's name in eighth grade actually was Mrs. Margaret Thompson and who made it easy for her students to remember the color spectrum using the letters in the name of the mythical character (perhaps a leprechaun's name?) Roy G. Biv – red, orange, yellow, green, blue, indigo, and violet. There are some shades of color before the red, and after the violet but they are in wavelengths shorter than red and longer than violet that are not discernible by the human unaided eye, and of course there is black (no color) and white (all colors). If any Reader doubts that the color spectrum is always in that same sequential order (in the same order of alphabetic characters as Roy G. Biv spells "his name") the Reader is invited to compare the color sequence in the illustration above with the color sequence shown in the actual double rainbow presented in the previous photograph taken in Prescott Valley, Arizona.

It is also important to note that there is no space between each different color as the wavelengths are a continuous variation and not in steps of increasing wavelength to give the separate colors in the diagram above. The rainbow photograph shows the rainbow colors variation without there being any distinct color boundary line between them. From a climate perspective, while the rainbow is a beautiful sight, the more important aspect of sunlight impinging on a water molecule in the atmosphere is that a small amount of solar radiation is also trapped in the molecule as heat energy. The effect of water vapor in the atmosphere is far more important as a heat-trapping phenomenon than is that which is attributed to very small concentrations of "greenhouse gasses" in the atmosphere.

It was mentioned earlier that when gravity accelerates a mass of water downhill, that mass of water possesses kinetic energy (which Sir Isaac Newton developed a set of mathematical laws for "how stuff moves" hundreds of years ago). It is kinetic energy when it strikes a solid object – as flood waters often do - causing physical damage to that solid object and can cause injuries to a human or other properties also that are in its path of motion. The combination of gravity and water in nature can be fearsome at times - but for most of the time as historical records document, gravity and water can do wondrous things when harnessed. For many centuries, the flow of water in a river – the flow being always from a higher elevation to a lower elevation – has been harnessed with water mills of many varieties and purposes. An example of the two most common water mills is shown as follows:

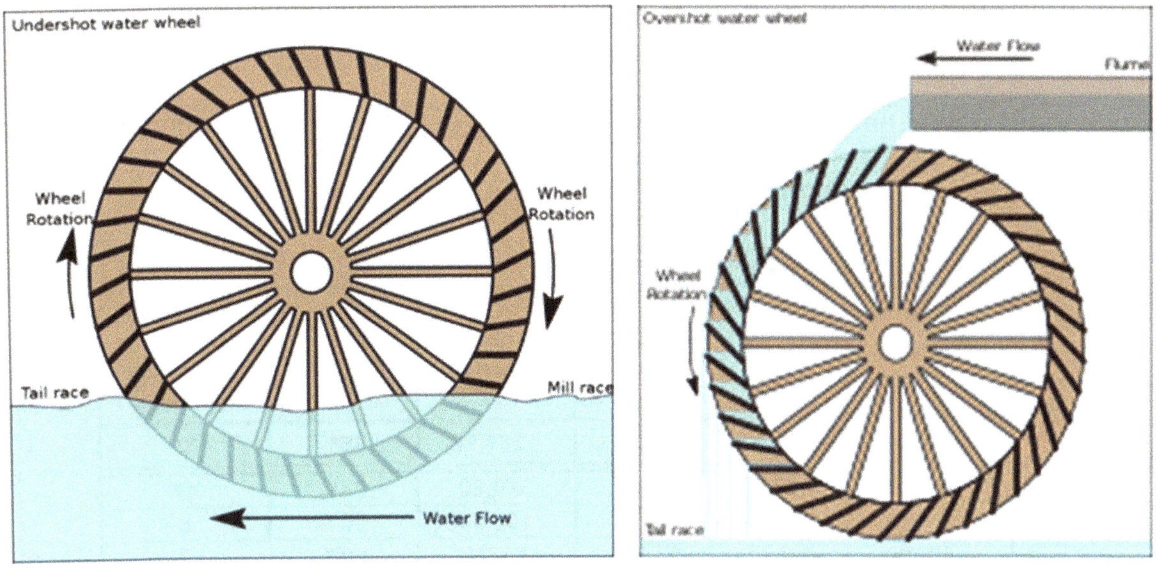

(Credit to Unknown Graphic Illustrator)

The working water mill wheel is mounted on the side of a facility housing the mechanical system (the kinetic water energy being generated by gravity to be converted into mechanical energy to power the grinding system) as shown in the next two illustrations:

(Credit to unknown photographer and unknown graphic illustrator)

Water powered mills were used for grinding grains into flour, sawing tree logs into planks of lumber, drawing hot metal into wire, separating seeds from cotton, and other useful purposes that contributed to the development of the country where used. As electric power became available to drive the mechanisms, the water-powered mills were obsoleted and water-powered mills entered the annals of history. The current generations have little recollection of them, but they exemplified that Man can use the elements of nature (water and gravity) without damaging his environment.

This technology led to the much larger practice of using flowing rivers and the creations of dams near the bottom of which the water pressure was converted into mechanical and hydrodynamic energy for driving turbine machinery used to generate electric power on a huge scale. An unverified estimate is that there are over 800,000 dams of significant size in use on the planet.

How hydrodynamic power has grown in use world-wide is evident in the following table.

Hydrodynamic Power Useage by Top 22 Countries
(As of 2022)

		Power Generated				
	Country	Hydro Power (TWh)	Per Cent of Total Country Power	Capacity (GW)	Capacity Factor (%)	Number of Dams Over 3,000 MW (see Note)
	World	4183.41	14.2%	1267.90	38%	38
1	China	1226.00	13.0%	370.60	38%	17
2	Brazil	428.65	60.4%	109.90	50%	8
3	Canada	364.20	57.5%	83.31	50%	2
4	United States	233.96	5.5%	86.66	31%	2
5	Russia	200.87	17.1%	50.57	45%	4
6	India	149.17	7.6%	47.33	36%	0
7	Norway	135.96	88.5%	34.40	45%	0
8	Vietnam	80.90	29.3%	22.64	41%	0
9	Japan	74.50	7.4%	28.22	30%	0
10	Sweden	66.07	39.7%	16.40	46%	0
11	Venezuela	65.68	77.6%	16.81	45%	1
12	Turkey	63.72	19.9%	31.78	23%	0
13	Colombia	54.24	62.5%	13.21	47%	0
14	France	53.19	10.4%	24.14	25%	0
15	Paraguay	43.87	99.7%	8.81	57%	2
16	Austria	39.79	59.4%	14.71	31%	0
17	Switzerland	39.00	54.8%	15.28	29%	0
18	Italy	37.94	14.5%	18.85	23%	0
19	Pakistan	37.90	23.5%	10.64	41%	1
20	Laos	33.40	72.7%	9.65	40%	0
21	Germany	19.47	3.9%	5.74	39%	0
22	Tajikistan	18.66	89.4%	5.76	37%	1
	58 Other	716.27	1.8% - 100	0.81 - 8.44	15 - 85%	0

International Energy Agency (IEA)

Note: 6 large dams (over 3,500 MW) are under construction: 2 in Myanmar (7,110 MW and 3,600 MW); 2 in Pakistan (4,500 MW and 4,320 MW); 1 in Ethiopia (5,150 MW); 1 in Tajikistan (3,600 MW)

(Source: International Energy Agency (IEA)

Very few accounts exist in the current news media describing how Man has used water and gravity ingenuously in past decades, but the Author's favorite account of a amazingly "green" project that both of his grandfathers worked on constructing in the first decades of the 1900s is hard to describe its immeasurable importance to the World in a few sentences. That "wonder of the world" project was/is the Panama Canal.

The Panama Canal is high on the list of the "greenest" ventures successfully accomplished by humans in modern years, but it was built more for ocean transportation improvement not for improving the environment. A magnificent accounting of the unusual history of the construction of the Panama Canal is available for Readers of this book in "The Path Between the Seas" by the renowned author David McCullogh. (Author's Note: The Author of this book on global warming is a contributor to the historical documents on file at the Barbados Museum and Historical Society on the island of Barbados in the Caribbean Sea and Atlantic Ocean. The Author donated a complete US Postage stamp collection of Panama and Canal Zone postage stamps, a large collection of post cards of scenes while the Panama was being constructed (by the French initially and then completed by the United States), and his large collection of postage stamps for the 400 years old island of Barbados (his family's ancestral home). In addition, the Author has written articles about the Barbadians who were part of the Panama Canal creation team. The article "Commemorating the Barbadians who Excavated the Panama Canal" appeared in the December 2008 *Journal of the Barbados Museum and Historical Society*. Subsequently, in collaboration with historians on the Barbados Museum Staff, the Author was instrumental in causing the Panama Canal to be designated as a UNESCO World Heritage Site.)

The importance of water and gravity on the planet cannot be better captured than describing the example of the Panama Canal. The Panama Canal story – for this book – starts with the Panama Railroad. Before the Panama Canal was constructed, the Panama Railroad was constructed across the Panama Isthmus. Passengers and cargo in the Atlantic Ocean waters wishing to reach a destination on the West Coast of the USA (a popular route inasmuch as there was no Transcontinental Railroad yet between the coasts of the USA) would debark at the port of Colon in Limon Bay, load onto the Panama Railroad to travel across the isthmus past the towns of Gamboa and Miraflores to Panama City. There they off-loaded from the railroad onto a passenger sailing ship in Panama Bay for completion of their trip Northerly in the Pacific Ocean to the West Coast USA – principally San Francisco (where they bought mining equipment in preparation for search of California gold during the Gold Rush years). The eventual Panama Canal shipping route was/is still very much aligned with the original railroad track route. (One point of trivia though that must be stated is that in traveling from the Atlantic Ocean (at Gatun) to the Pacific Ocean (at Panama City), a ship passenger is unexpectedly confused when subjected to traveling from northwest to southeast due to the reflex curvature of the isthmus at that location).

While the original attempt – which was doomed to failure - to construct a canal across the Panama Isthmus goes back to the 1860s and with the International Canal Commission issuing a construction contract to the French entrepreneur Ferdinand de Lesseps (credited with creation of the Suez Canal), this book fast forwards into the years in the first decade of the 20th century (1900) as the United States Canal Zone project under President Theodore Roosevelt. President Roosevelt saw that a canal at Panama was a boon not only to commercial shipping trade and passenger travel efficiency but was a necessity for military purposes as well. The US Navy has maintained large fleets of warships in both the Atlantic Ocean and the Pacific Ocean since before World War I (which started in 1917), and President Roosevelt saw the canal as a strategic war asset.

The route the Panama Canal was selected to take across the Panama isthmus connecting Central America and the North and South American continents, for good reason follows the path of the trans-isthmus Panama Railroad.

(Source: Panama Canal Museum, Miraflores, Rep. of Panama)

Virtually all of the water used in the functioning of the Panama Canal and its system of locks is provided by rainfall in the mountainous region of North Panama – none of that water is pumped anywhere in the system by electric-driven pumps. The flow is gravity accelerated only. The collected rainwater starts in the upper right corner of the above map in the higher elevations of Panama and flows, as the Chagres River, southwesterly through the town of Gamboa and into the immense Gatun Lake. The Chagres River - before the Gatun Dam and Gatun Lake were created - flowed northwesterly toward the Atlantic Ocean end of the Panama Canal at the city of Gatun. The Gatun Dam was created across the Chagres River exit into the Atlantic Ocean and eventually the water level of Gatun Lake would be 85 feet above the sea level of the Atlantic Ocean.

Before showing the scenes of the Panama Canal, a profile is necessary to be shown explaining how ships enter the system of locks at one end of the canal into and up a "water ladder", travel across the isthmus at the upper level of the lakes (Gatun Lake and Miraflores Lake) and then descend down the "water ladder" to exit into the ocean at the other end 50 miles away. That "Cross Section diagram" is shown next:

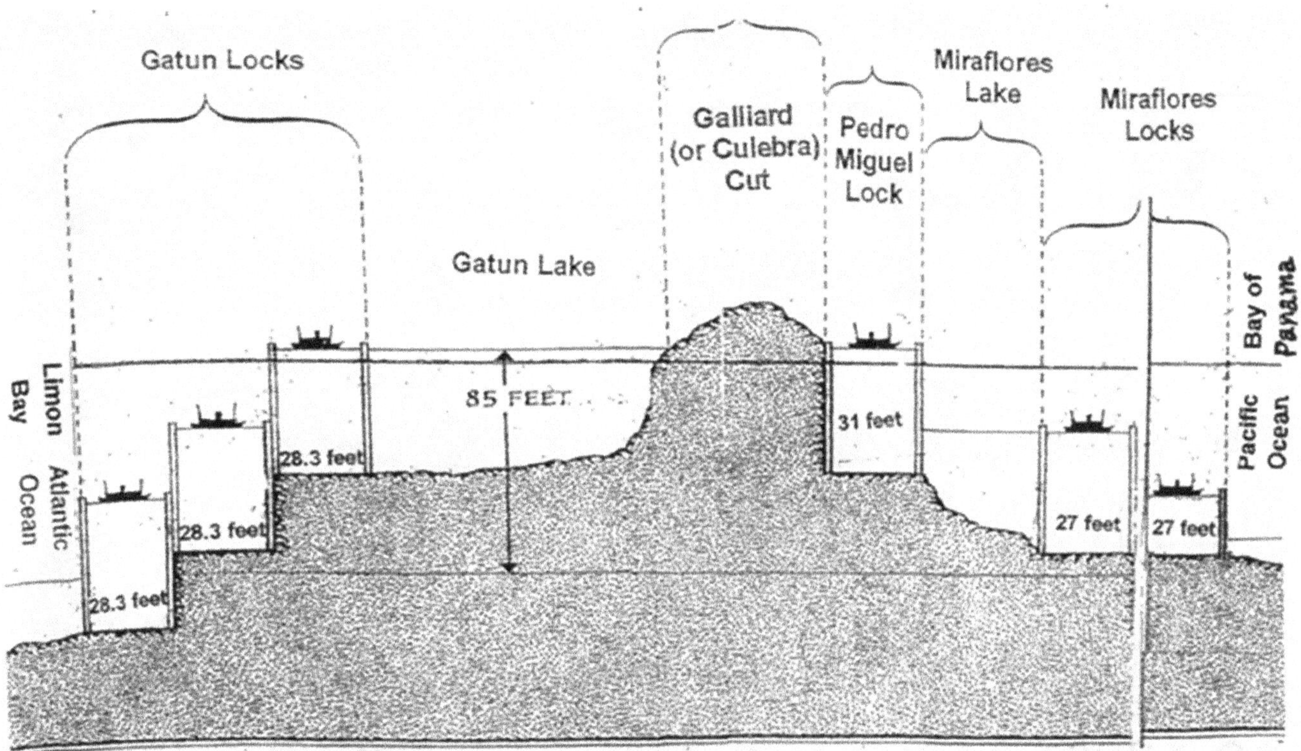

Cross section of the Panama Canal

Each lock is 74 feet tall and 110 feet wide (or 33.5 meters).

(Source: The Panama Canal Museum at Miraflores, Rep. of Panama)

Among the earliest activities at Panama was the construction of the Gatun Dam across the Chagres River at Limon Bay in the Atlantic Ocean – while the six locks were each being constructed as a pair of chambers, 12 chambers in all, to permit ships passing in either direction (Atlantic to Pacific or Pacific to Atlantic). The initial purpose of the Gatun Dam was to create Gatun Lake, and then subsequently to generate all of the electrical power for the functional operation of the canal's locks and lighting system.

As a unique approach to obtaining the construction materials for the construction of the dam and the several lock chamber walls and center island, some of the "spoil" from the excavation blasting of the Culebra Cut through the mountain peaks of the Continental Divide was used (after being crushed and sieved to the needed size). The clay-like "spoil" mixed with rocks was used to fill the cavity between the Gatun Lock face and interior wall, while the enormous boulders that rolled down into the cut were hauled away on railroad flat cars to be used in construction of the breakwaters at both ends of the canal (in Limon Bay and in Panama Bay).

The Miraflores Lake was an existing natural lake and was used in conjunction with the Pedro Miguel single lock as the intermediate path leading to the pair of Miraflores Locks at the Panama Bay end of the Panama Canal.

A photograph of the completed Gatun Dam, Gatun Lake behind it, and the Gatun Electricity Generation Powerhouse below it is shown next.

Gatun Lake, Gatun Dam, and the Gatun Hydroelectric Power Station
(Photo from Postcard Collection of Author on File at Barbados Museum and Historical Society)

The Gatun Lake was/is formed and is maintained full of the heavy rainfall in the Panama Mountains, with the creation of the upper surface of the lake 85 feet above the seal level of the Atlantic Ocean at Limon Bay. The Gatun Dam is a magnificent example of the use of renewable energy to operate a facility as important as the Panama Canal. The electricity generated at the Gatun Hydroelectric Power Station makes the Panama Canal system of locks for transporting ships across the Panama Isthmus between the Atlantic Ocean and Pacific Ocean totally a self-powered operation. By creating the Gatun Lake 85 feet above sea level, the three Gatun Locks (each 28.5 feet high) of the Panama Canal form a 3-step "water ladder up" to lift and 3-step "water ladder down" to descend massive ships using gravity alone to fill a lock chamber from the chamber or lake above it or to empty into a chamber next to or below it.

Gravity flow through the successive three chambers constitute the "water ladder" but the gates at each chamber end are electrically opened and closed to permit the ship to pass from one chamber to the next. There are no electric-driven pumps pumping water up or down in operation of the "water ladder" system of locks. In the center wall of each pair of lock chambers, a large culvert having multiple gates (manually controlled in the gates control house and electrically operated by power from the Gatun Dam hydroelectric power station) can be opened to allow gravity to flow the water either from one chamber laterally to its twin chamber or from a higher chamber to a lower chamber. The center culvert is about 20 feet in diameter and is shown next when the lock was under original construction: (The culvert is not visibly seen by humans as the water level never gets below the keel level of ships).

(Source: The Panama Canal Museum, Miraflores, Rep. of Panama)

The opening and closing of each of the massive lock gates at the ends of each chamber (12 pairs of gates in all) are electric powered using the power generated at the Gatun Hydroelectric Power Station. A photograph of a ship descending the Gatun locks – with the gates in the two chambers in an opening mode to allow the ship to pass from one chamber to the next – is shown next. The gates can be opened only when the water level is the same level to prevent a rush of water into the chamber with the lower surface. The gates were designed to be flush with the inner walls of each chamber so that, when fully open, the gates do not present an obstruction for the side of the passing ship to snag the gate and damage it. Regrettably, the page constraints of this book do not permit a fully detailed description of the ingenuity of the Panama Canal system and down to the individual gate controls in the gate control house (with a single person moving all those large ships, and chamber gates, and multi-millions of gallons of water in a perfectly orchestrated manner).

A scene from a high deck on a ship transiting through the Gatun Locks is shown next:

(Source: The Panama Canal Museum – Gatun Locks in Operation)

The electric power to operate all the gates in the12 chambers of the lock system is generated by the singular hydrodynamic electric generation facility at the base of the Gatun Dam (shown in an earlier photograph and not seen in the photograph above - only a short distance across the original path of the Chagres River to the left of this photograph). That generation facility also provides the electric power for the controls operating the gates, for the extensive lighting system, and to power the electric "mules" that pull the larger ships through each chamber to give the Panama Canal a 24 hours per day full operating capability. Small ships with less draught are self-powered through the series of six chambers in each transit of the Panama Canal – three chambers up and three chambers down.

Since all water to operate the six chambers for each full transit of the Panama Canal by a ship of any size is drawn from the fresh water Gatun Lake or Miraflores Lake, each full transit of the ship flows a total of 53 million gallons (200 million liters) of fresh water (from rainfall alone) into the saline oceans (Limon Bay of the Atlantic Ocean and Panama Bay of the Pacific Ocean). The barnacles that grow on the keels of ships while in saline ocean waters die off when entering the first freshwater chambers and drop to the bottom of the first two chambers – requiring periodic maintenance (cleaning) of the barnacles' debris from those chamber floors. With an average of close to 40 ships per day transiting the Panama Canal (in both directions), if each ship causes 53 million gallons of freshwater to be transferred out into the oceans at its ends, 40 ships per day would cause the loss of about 2.12 billion gallons of water per day. In recent months the rainfall in the upper Panamanian mountains has decreased somewhat (natural causes) and as this book is being written, a limit has been placed on the transit of ships per day to be an average of 32 transits to preserve the level of

Gatun Lake as being 85 feet above the Atlantic Ocean surface level. If the ocean levels rise 20 feet, this would obviate the need to use the lowest of the three Gatun Locks and the two Miraflores Locks.

The most labor-intensive portion of the Panama Canal construction was the excavation through the hill (part of the Continental Divide) known as the Culebra Cut (later renamed to be the Gaillard Cut in honor of Major David Gaillard in the US Army Corps of Engineers who orchestrated the cut). The "cut" was excavated several times in the same place because the heavy rainfall at Panama would cause mudslides down into the cut and bury men and/or their equipment without ample warning. On many days, the excavation crew would complete a section of the excavation cut by the end of the workday, but when they returned the next morning for another day at the "big dig" they found all the previous day's excavation was filled in by an overnight mudslide. Gravity can be a wonderful assistant or a formidable challenge, changing from one event to the next.

Shown in the next photograph is the ill-fated US Navy battleship *Arizona* on a passage (circa 1936) past the Culebra/Gaillard Cut sailing towards its ignominious fate at Pearl Harbor, Hawaii, on December 7, 1941. Culebra is the Spanish word for "snake" as the shape of the canal "snakes" through that cut.

(Source: The Panama Canal postcard collection of the Author) Gaillard Cut

The canopy over the bow section of the Arizona was to allow the crew to witness the historic transit through the Panama Canal without being burned severely by the Sun's rays and the extremely high humidity in the hours The massive amount of coal used to power the steam powered excavation equipment and the steam engine trains to haul away the debris, to construct the Gatun Dam, the Gatun Hydroelectric Power Station, the five pairs of major lock chambers, and to power the "mules" that originally towed the ships through the locks will be described later. The many millions of tons of explosives used in the construction of the Panama Canal would probably be prohibited by today's environmental protection rules. More important, if the burning of several hundreds of thousands of tons of coal powering steam shovels and locomotive engines had not been allowed to be used, the Panama Canal would not exist.

This is perhaps the best example of expending the huge amount of fossil fuels to avoid the many millions of tons of coal to power steamships the shorter route between the two major oceans and avoid the huge amount of coal burned for those steamships by not having to sail over 2,000 miles on a ship route down and around the southern tip of South America (sailing in the hazardous Cape Horn region – "round the Horn" in old-time maritime historical annals). There was no concern of the creation of "greenhouse gas emissions" in the early 1900s and affecting global temperature – only the concern of building the Panama Canal to allow passage of commercial freighters and warships to transit the Panama Isthmus in the minimum time and expenditure of coal fuel.

The details of the achievement of the Panama Canal as a "green" project are unable to be thoroughly described herein but the Panama Canal became a vital part of the Author's family - both of the Author's grandfathers were part of the construction team that created the Panama Canal. With ingenuous planning and design – plus supported with adequate funding resources – the Panama Canal has become the most "green" venture successfully accomplished by humans in modern years. A magnificent accounting of the history of the Panama Canal is available for Readers of this book in "The Path Between the Seas" by the renowned author David McCullough.

Author's Note: The Author of this book on global warming is a Contributor to the holdings on file at the Barbados Museum and Historical Society on the island of Barbados in the Caribbean Sea and Atlantic Ocean, and donated a complete stamp collection of Panama Canal and Canal Zone postage stamps, a large collection of post cards of scenes while the Panama was being constructed (by the French initially and then completed by the United States), and his large collection of postage stamps for the 400 years old island of Barbados. In addition, the Author has written articles about the Barbadians who were part of the Panama Canal creation team. Those articles appear in the *Journal of the Barbados Museum and Historical Society.* In collaboration with historians on the Barbados Museum Staff, the Author was instrumental in causing the Panama Canal to be designated as a UNESCO World Heritage Site.)

The Panama Canal could well serve as a model for present day scientists and engineers and physicists on how to use the climate of a region, the material resources needed for that project being all naturally available on site, the acceleration of gravity endlessly both assisting and hindering the process of constructing the elements of the Panama Canal system, and using the persistence and ingenuity innate to some humans to create a "green" project before "green" was the challenge on every member of the public's mind. The irony however, is that without humans burning fossil fuels to power the machinery to excavate the canal path is of minor concern when compared to the huge amount of fossil fuels that were avoided being burned by thousands of ships not having to travel an extra 8,000 miles (and many days) at sea (via the southern tip of South America) to propel themselves from the East coast Atlantic Ocean ports of the United States and of the West coast of European countries to West coast Pacific Ocean ports and to ports everywhere around the Western coastal regions of the Pacific Ocean (e.g., to Japan, China, and the regions of Southeastern Asia). But a picture is worth a thousand words and shown below is a map illustrating the savings for ships transiting between the port of New York in the Atlantic Ocean and San Francisco in the Pacific Ocean:

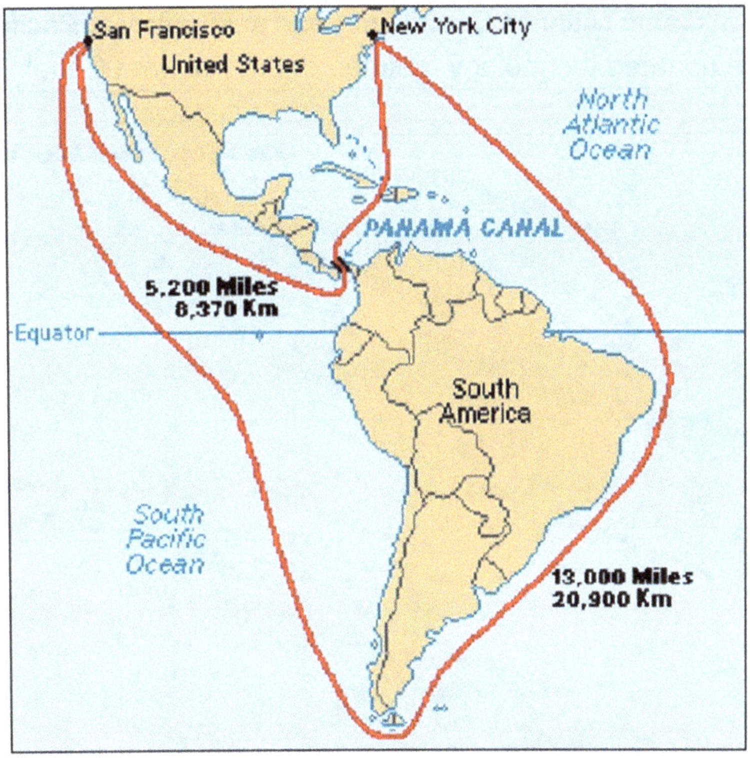

(Source: The Panama Canal Museum at Miraflores, Rep. of Panama)

A seemingly trick question in The Game of Trivia is "What direction does a ship travel when transiting through the Panama Canal from the Atlantic Ocean to the Pacific Ocean?" The Panama Isthmus, a part of the Continental Divide mountain range, has a geographically reflex curvature and the correct answer is "southeasterly" and not "westerly" as most people would answer.

To be factually correct, the Panama Canal was seen needed by President Theodore Roosevelt, a former Secretary of the Navy, to provide the shortest distance and shortest time to move US Naval vessels from one ocean to the other for wartime needs. Perhaps the most historically important event of a single ship passing through the Panama Canal was that of the USS Arizona on her voyage from the Atlantic Ocean (where she was part of the Atlantic Fleet) to her new addition to the US Naval Pacific Fleet headquartered at Pearl Harbor, Hawaii (before Hawaii was a State of the USA). That photograph of the battleship USS Arizona at the Panama Canal was shown earlier.

After President Jimmy Carter granted ownership of the United States Canal Zone (and the Panama Canal in its entirety) to the Panamanian Government in 2000, design of "super-size locks" was begun to handle modern ships that had been designed (some intentionally and some unintentionally) to be too wide and/or too long to fit in the original lock chambers. While the Super locks were under construction, the Panama region coincidentally endured a drought lasting 20 years which began to constrict the number of ships that were permitted to utilize the original locks (which causes over 3 million gallons of rainwater from Gatun Lake per each ship transiting the Panama Canal to be lost into the ocean). Construction of the new Super locks was accelerated to reduce the effects of the drought conditions as well as to accommodate the transport ships being of larger dimensional size. Designated as Super locks size (i.e., wider and longer than the original lock chambers dimensions), the new Super locks were constructed at both ends of the Panama Canal. They opened for ships passage in June 2016. A diagram showing the visual layout of the new Super-locks relative to the original lock chambers at Gatun (Atlantic Ocean end) and at Miraflores (Pacific Ocean end) is shown for brief overview of the updated technology features of the Panama Canal:

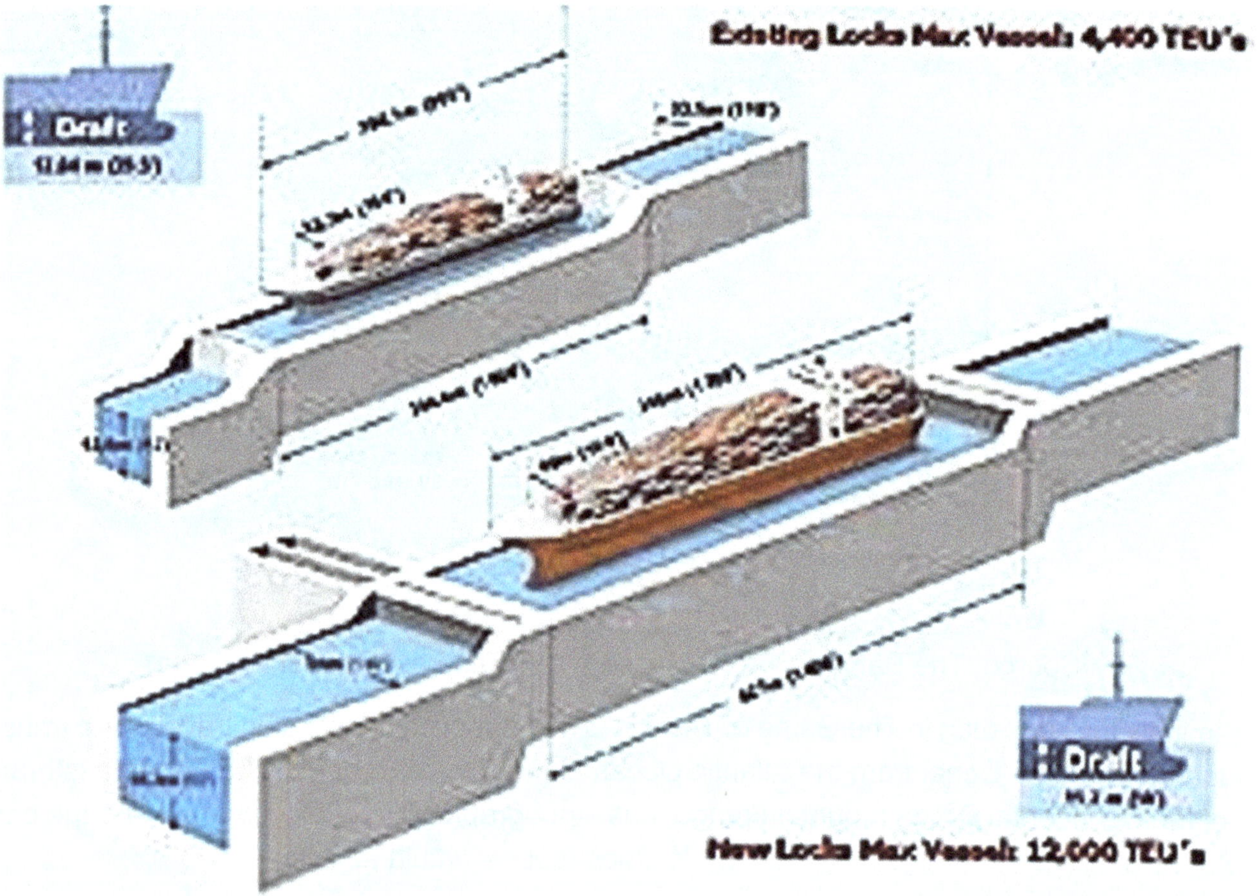

(Source: Panama Canal Historical Museum at Miraflores, Panama)

The new Super locks also feature three lateral basins (water reservoirs) located to the side of the original locks and hugely conserve on the amount of water that was lost (through the last gate) by each transit of a ship through the original locks.

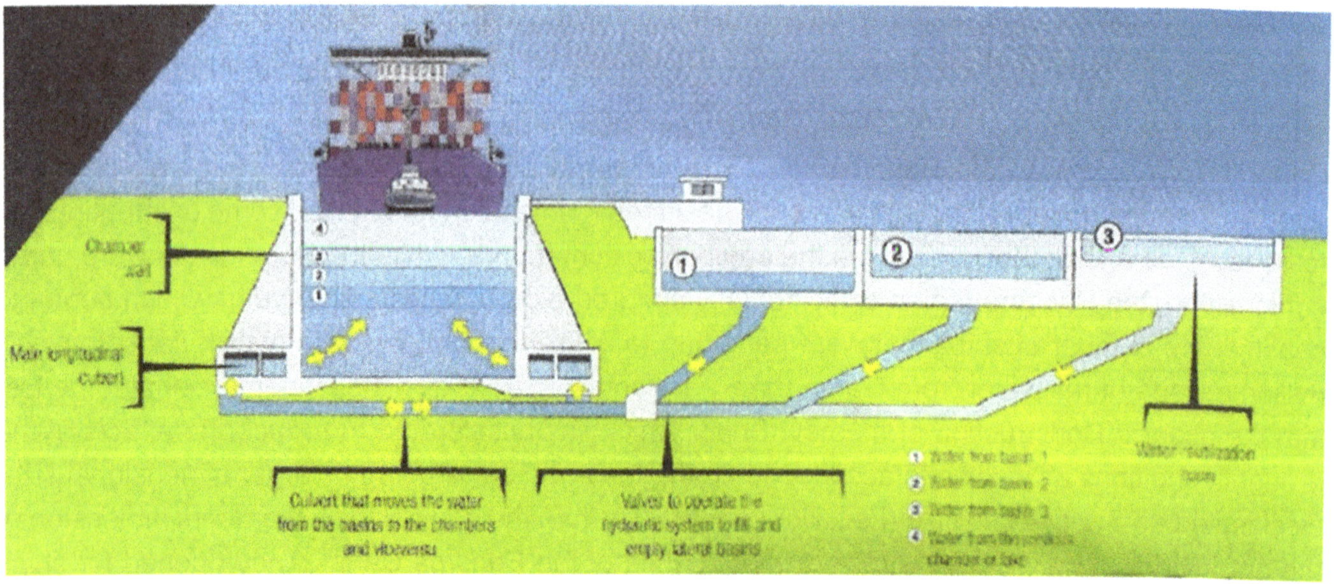

(Source: Panama Canal Museum at Miraflores, Panama)

An overall "big picture" photograph of the Super locks at the Panama Canal is seen as follows:

(Credit to unknown photographer)

This book will not belabor the details of the new Super locks other than to iterate that the new technological capabilities of the Panama Canal use the same principle of no electrically-driven pumps – only gravity – to transfer water into the "water ladders" at both ends of the Panama Canal. The gravity-powered transfer of water into and from the lock chambers to the three side reservoirs is seen in the above diagram. The Panama Canal is regarded by many as the Seventh Wonder of the World and for good reason. The first time a person sees the huge ships squeezing into the lock chambers and magically being lifted without any sounds of a mechanically driven pump system or a hydraulic pump with a heavy lift mechanism, the comment overheard is "I wonder how that is done?" Even seasoned travelers are awe-struck when seeing the new Super locks (officially known as Panamax) in operation. The following photograph shows the ship that set a new record for the amount of cargo moved through the Panama Canal on one ship. Ships this size must be assisted by tugs to position the ship inside the first chamber and then be assisted by electric tow locomotives (known as "mules") to maintain the ship inside the chamber away from the chamber walls as it ascends and descends. The large propellor blades used to power the large ships are not allowed to be used to propel the ship while in a chamber as they damage the bottom of the chambers. The tugboats are used to pull the ship from the last lock chamber (as seen in the above photograph) to send it on its way self-propelled.

(Credit to unknown photographer at the Panama Canal)

Prior to the opening of the Super locks, container ships that were too large for the original lock chambers used the practice of off-loading their containers onto the flat bed cars of the Panama Rail Road trains (still in operation today) that transported the containers through the Panama Canal by the rails laid by the French in the late 18th century (in the first failed attempt by Ferdinand de Lesseps – builder of the Suez Canal). At the far end of the canal, those containers were reloaded onto empty container ships to continue the rest of their shipment. One can only imagine the logistics and planning that must accompany one such unload and reload operation when a fully loaded container ship (of the type shown below) arrives at the first lock chamber:

(Credit to unknown photographer)

It is important to point out how two technologies have synergistic effects on each other and on other peripheral technologies. The canal technologies designed for one vintage shipping are upgraded to handle larger size ships (with a host of different cargoes). Taking advantage of the larger canals, more ships are designed larger to transport more cargo of larger size and/or volume. A limitation on the increased use of natural gas to replace coal has taken place throughout many nations, including the US, as indicated in the US Department of Energy database:

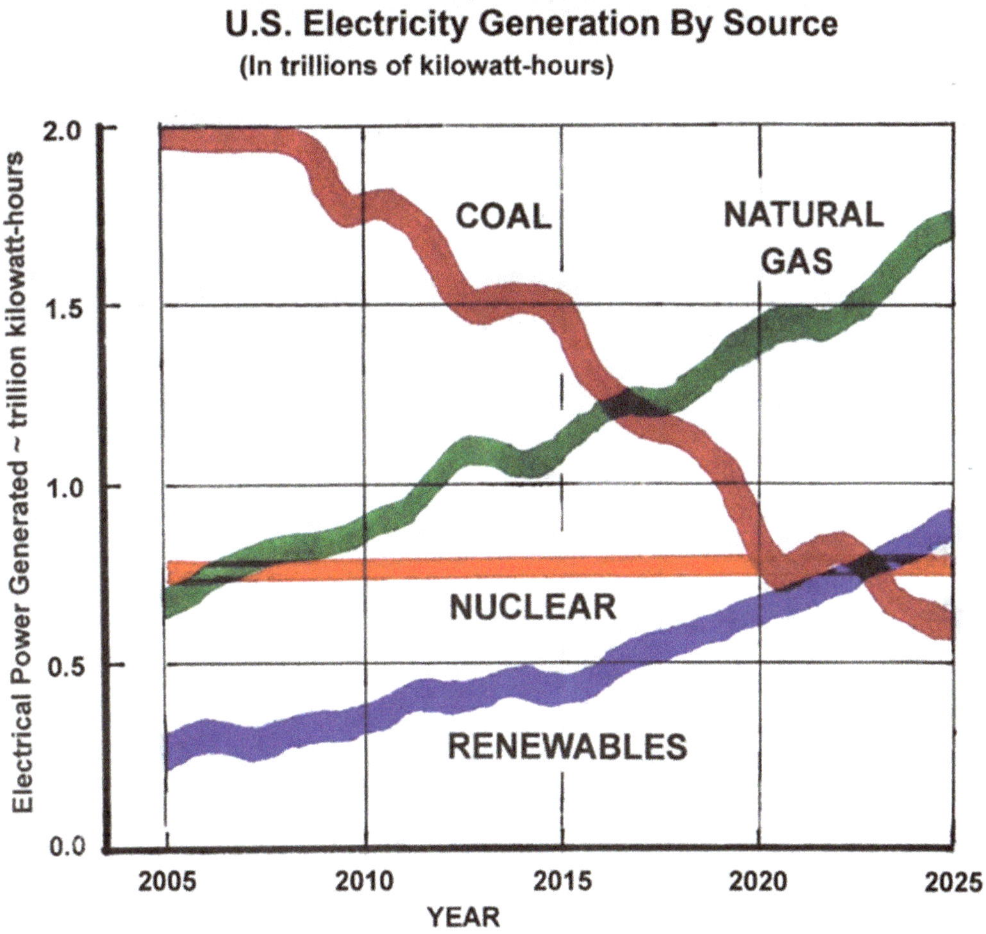

(Source: US Department of Energy DOE)

The technology of natural gas hinges on its availability to the users, and causes the need to transport the natural gas long distances to the users. In the technology domino effect, the means of transporting natural gas developed large tanker ships to transport the natural gas in large volume spherical tanks and with the natural gas in liquid state rather than the voluminous form of gas. Further along in the domino effect, large tanks of liquified natural gas have been installed on rail cars as well as on tanker ships. To be most efficient, the tanks must be of extremely large size, and the most efficient shape of the tank is to be spherical. Combining all the several requirements for the technology of using clean-burning natural gas to replace coal, the transportation technology has evolved large liguefied natural gas (LNG) tankers that were too wide to pass through the original size locks at the Panama Canal which had been designed for the passage of US naval battleship dimensions in the early twentieth century.

One of the most meaningful photographs in the two books describing climate change by this Author is the following, showing a new technology LNG tanker passing through the middle chamber of the new Super locks at the Panama Canal. The new lateral basins three per chamber and three chambers at the two ends of the Panama Canal – Pacific and Atlantic – show once again that water is vitally essential to the shipping transportation technology in partnership with the natural acceleration of gravity as a “benefit for humanity”.

(Credit to unknown photographer in Panama)

As a passing remark, environmental scientists advocated the cessation of the use of fossil fuels of all types – coal, oil, and natural gas. The decision made and goals established to reduce coal were partially valid because of the ash residue pollution (toxins and carcinogens) properties of coal that are known harmful to both the human population and to the ecology. To the contrary, this book has shown that there is no pollution (of major significance) of natural gas and only small side pollution effects of the use of oil. To categorically strive to eliminate all "greenhouse gas emissions" from the use of all types of fossil fuels was/is a major scientific blunder (or was it intentional?) by environmental scientists. (In the engineering profession, it is a basic theorem to "Don't throw out the baby with the bathwater".)

All regions on the planet that have a significant growth of trees are vulnerable to there being wildfires that destroy one of the natural resources that touches many aspects of human life. Some wildfires are started by Man (intentionally and unintentionally), but the vast majority are started by lightning strikes. Lightning strikes have their origin in the clouds and their termination on the surface of the planet, as shown in the next pair of photographs:

(Credit to unknown photographers)

There has been no scientific model yet developed for accurately predicting where and when a lightning strike will occur whether or not a fire will be ignited by that strike. It is most common for lightning to strike from a cloud mass above an area and more common to strike a taller feature within that area. Anyone who has visited the Sequoia National Park in California cannot help but feel saddened when viewing the giant sequoia trees with their top portion decapitated by lightning strikes. The best estimate for predicting a lightning strike is the probability that lightning strikes are far more common in the hotter months than in the colder months, and the location of the strikes being in the southeastern quadrant of the contiguous US (Conus). A distribution of lightning strikes in the United States is shown on the following databases (the frequency distribution in the USA shown first):

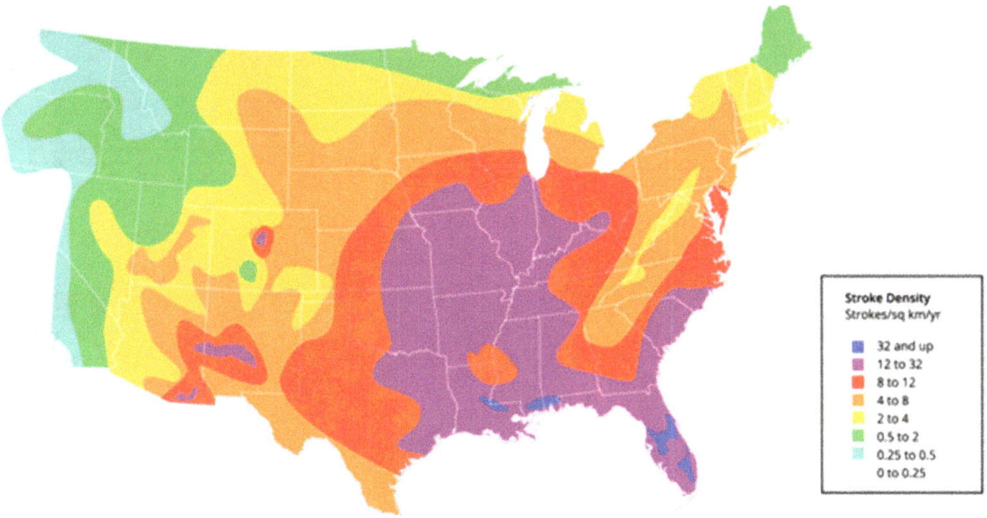

(Source: NOAA)

The time of year – i.e., the seasonal effect - for the highest probability of lightning strikes is shown next as the average monthly temperatures (including 2023) which the travel industry derived from the National Weather Service meteorology. (2023 has been declared to be the hottest year on record for Conus):

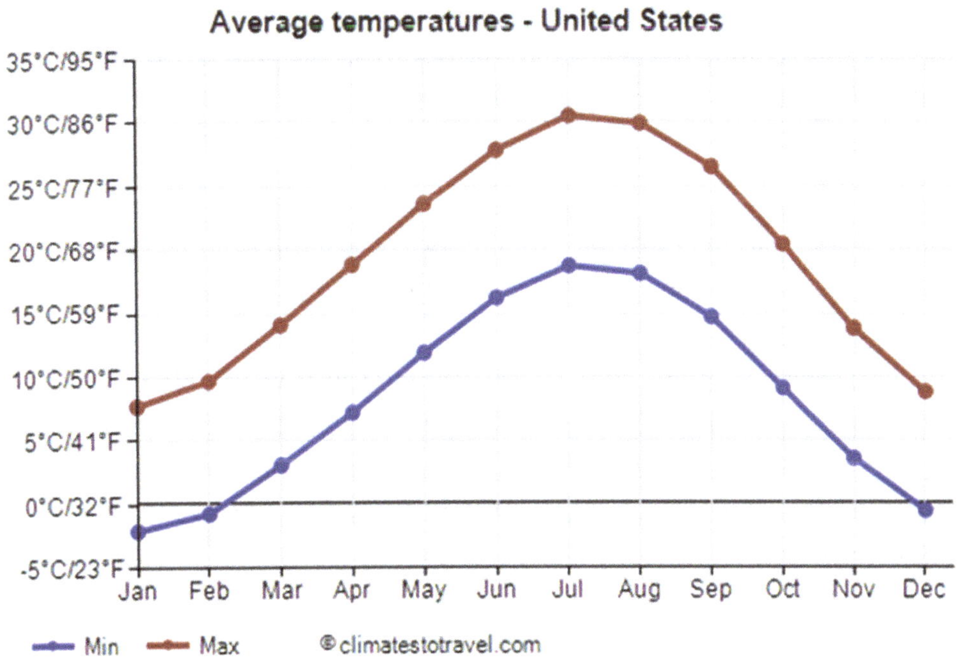

(Source: climatestotravel.com)

The frequency of lightning strikes is proportional to the atmospheric temperature and existence of a high presence of water vapor (typically with cloud formations near ground level) as the primary cause for the changes in lightning strike frequency – not the factor of Man burning fossil fuels and creating "greenhouse gasses" in the atmosphere. The water vapor in the atmosphere is a strong factor that is too variable to be modeled with high confidence by scientists. Nevertheless, the water vapor does exist and cannot simply be eliminated from consideration. In 2023, the year in which the NWS recorded the highest average monthly temperatures for Conus, the US National Lightning Detection Network recorded an astonishing number of lightning strikes:

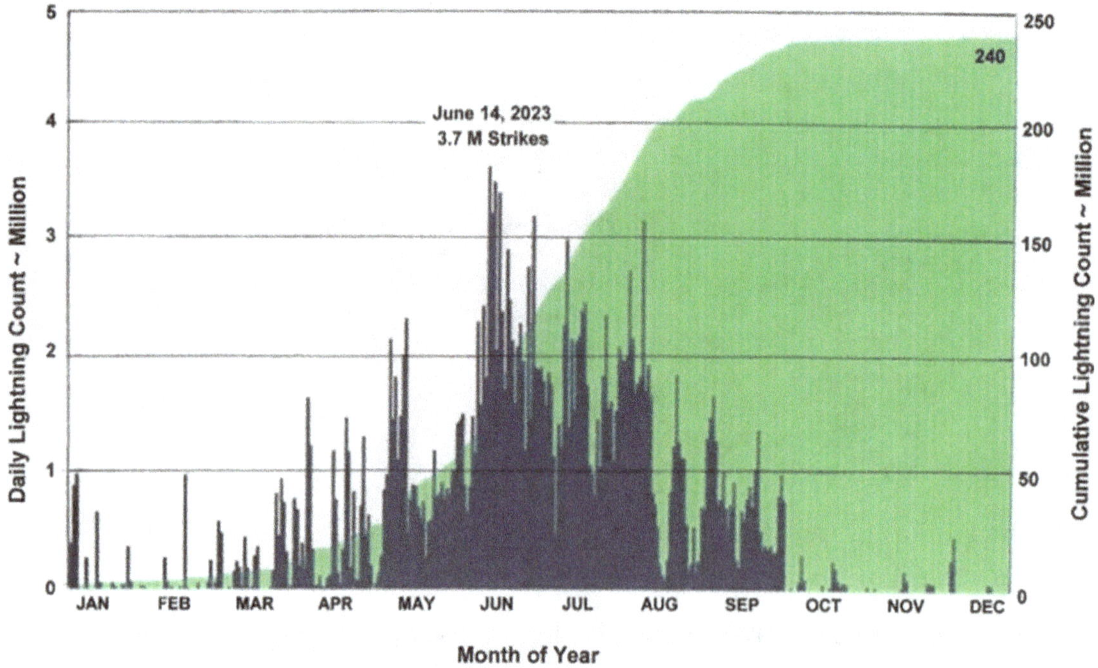

(Source: US National Lightning Detection Network)

It should be noted that the overall shape of the average maximum temperature variation in the CONUS temperature database is much the same overall shape as the frequency of lightning strikes. (An analogy can be stated that the average maximum temperature has smooth skin of a baby porcupine, but later when the porcupine grows bristles it appears more as the shape of the lightning strike frequency database.)

Temperature is a primary factor (with the presence of water vapor as a secondary causative factor) driving the frequency and intensity of lightning strikes. In researching the lightning strike database for other possible causes, a recent issue (June 15, 2024) of the Science News magazine reflects the findings of a scientist 50 years ago in 1974 having correlated the lightning strike phenomenon with solar activity ("sunspots"). In the June 22, 1974 issue of Science News, the link between solar activity and lightning strikes was as follows: *"Using data from 1930 to 1973, a researcher reports (May 24, 1974 issue of Nature) that in spite of year-to-year variations in the incidence of lightning, there is an underlying cyclic variation* (in sunspot activity) *with a period of about 11 years which is in phase with the solar cycle."*

The entire topic area of solar activity and the electromagnetic field surrounding the planet are beyond the scope of this book to be analyzed. The planet is known to have an electromagnetic field existing in its exosphere as indicated next:

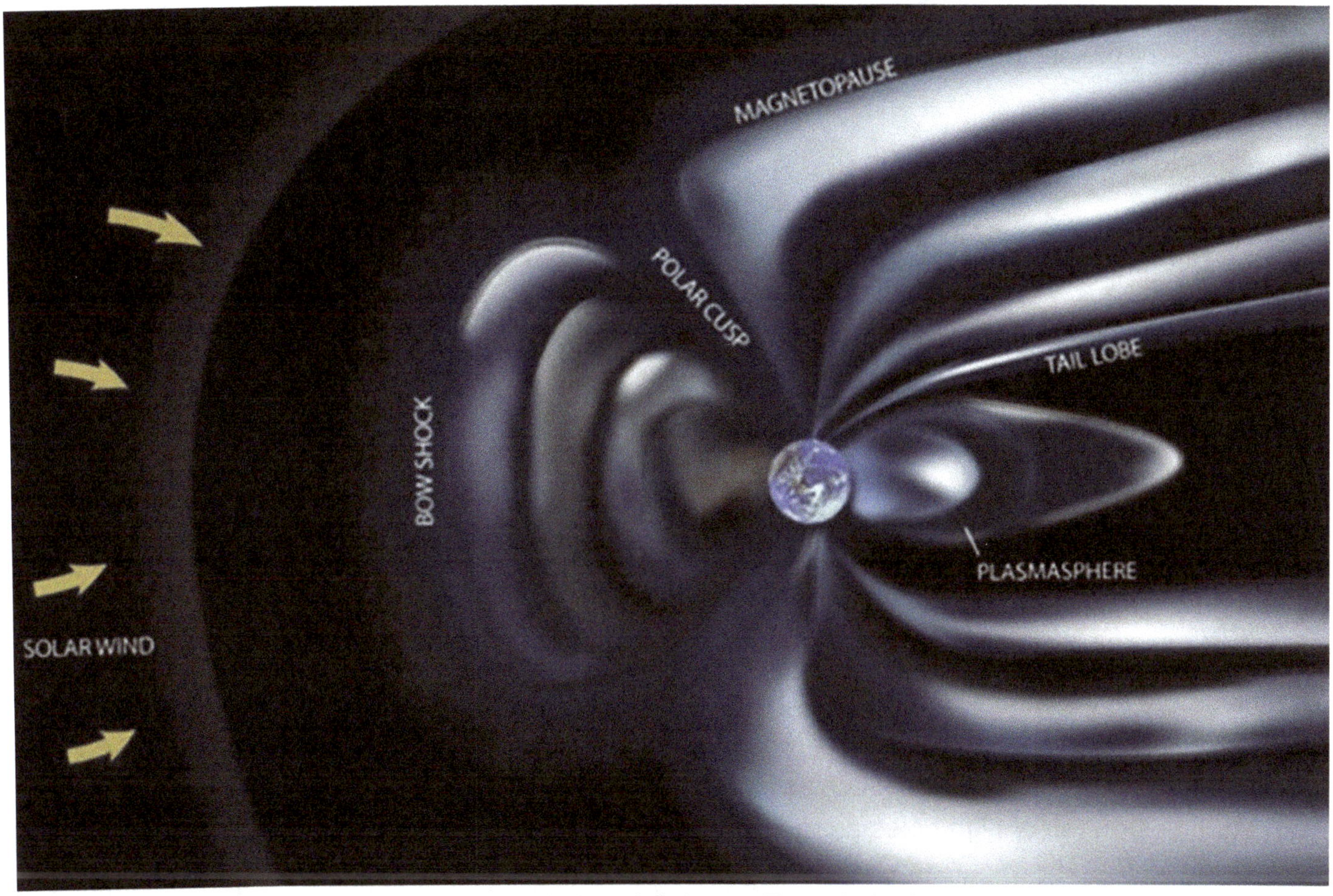

(Source: National Science Foundation)

The effect that the electro-magnetosphere has on the ozone layer and on the climate of the planet's surface is beyond the scope of this book to describe (the Author knows far better how air molecules behave than how electrified electrons behave), but again the physics of electromagnetism is left to those whose careers have specialized in the mushrooming fields of science based on electromagnetism. That admission is necessary to serve as a "Lesson Learned" from several complex accident investigations in which the Author has participated. A primary "lesson learned" is that when confronted with an unknown factor in solving a problem, it is best to consult with others having expertise in that aspect of STEM – science, technology, engineering, and mathematics. By no means is it acceptable for eliminating a factor from further consideration simply because that factor is not fully understood – and what it factually affects elsewhere in the system.

The public is given a glimpse of the artistry of Nature – which results from the electromagnetic field surrounding the planet - when in the northern latitudes of the planet and are treated to a display of aurora borealis known also as the "Northern Lights" and as "the Aurora". In those northern latitudes (Scandinavia and Iceland and Greenland), aurora tourism is one of the largest tourist attractions and source of much welcomed visual entertainment. The awesome sights of volcanic and geothermal activities is commonplace on Iceland. A photograph of the aurora borealis is shown next to illustrate the beauty – and the geologically chaotic environment – Nature causes on the planet.

(Credit to Unknown Photographer in Iceland)

Returning to what is known about the effects lightning strikes have on the environment (and indirectly on the safety of the human population), in areas of flat or nearly flat terrain, lighting tends to strike a high protuberance on the flat terrain – either of natural origin (tall trees being a primary "lightning rod") or Man-made construction (such as a metallic tower or a tall building). It was shown earlier that wind turbines on top of high poles are a frequent lightning attach point.

Shown next is a lightning strike that typifies the high voltage level to cause a large lightning bolt to flash in a matter of a second or two during its short-lived existence:

(Source: The National Interagency Fire Center and of the National Weather Service, NOAA)

Although the above photograph does not reveal any fire having resulted from the lightning strike that struck it, the strike did cause a large limb to be broken from the tree. The limbs of trees that are broken off trees that are struck by lightning may cause damage to adjacent electrical power lines and disrupt the power supplying a large area for a very lengthy period, hampering also rescue efforts to the area. The power outage poses hazards ranging from the inconvenience of not being able to watch television for a few hours to the greater danger as safety and/or health issue to humans tethered to life support equipment that is powered by electricity. Fatalities may, and frequently do, occur in power outages lasting for days.

When lightning-started wildfires start they are fed by any combustible fuel - such as the natural debris on the forest floor or from the wood in structures in the path of the fire. The fire rapidly grows depending on both natural environmental surroundings and meteorological conditions. Every wildfire has a unique "personality" - creating its own wind drivers - and has a pattern of erratic behavior.

While the above photographs do not reveal any fire having resulted from the lightning strikes, when a lightning-started wildfire does get ignited it rapidly grows depending on many variables of the local weather and environmental surroundings.

The numerical statistics indicate the large number of wildfires that are started by lightning strikes in forested areas of the United States as shown in the following table:

Lightning-Started Wildfires Reported by Geographic Area - United States

YEAR	Alaska (partial)	Northwest	Northern California	Southern California	Northern Rockies	Great Basin	Western Great Basin	Southwest	Rocky Mountains	Eastern Area	TOTAL (US)
2001	29	159	2,238	832	1,041	2,405	1,697	2,298	2,114	889	13,702
2002	165	1,797	301	179	1,130	1,602	556	2,469	2,039	372	10,610
2003	72	1,605	966	428	1,921	2,004	569	2,702	1,918	102	12,287
2004	270	2,042	670	323	1,090	1,760	781	2,062	1,340	88	10,426
2005	311	901	186	272	748	1,345	536	1,935	1,398	175	7,807
2006	54	2,170	948	409	1,970	2,259	943	3,220	2,479	256	14,708
2007	201	1,486	574	291	1,363	1,434	463	1,869	1,672	330	9,683
2008	75	1,624	1,400	174	679	835	224	1,027	941	171	7,150
2009	199	1,843	890	179	1,212	1,086	478	1,546	1,090	62	8,585
2010	330	1,110	441	216	633	1,036	273	947	941	169	6,096
2011	138	808	301	258	944	1,168	432	1,678	1,328	161	7,216
2012	141	856	180	266	1,328	1,259	577	1,224	1,992	205	8,028
2013	212	2,282	596	274	1,305	1,441	542	1,238	1,166	54	9,110
2014	55	2,417	477	259	1,019	937	284	840	689	34	7,011
2015	417	1,705	785	397	1,571	1,111	N/A (~300)	986	650	86	7,708
2016	229	437	97	96	916	812	N/A (~300)	1,413	1,247	107	5,354
2017	155	1,254	728	188	1,254	1,139	N/A (~300)	922	1,014	59	6,713
2018	138	977	174	131	604	1,145	N/A (~300)	1,216	1,086	69	5,402
2019	371	1,430	420	76	686	921	N/A (~250)	786	736	38	5,093
2020	168	746	430	124	625	804	N/A (~250)	746	789	171	4,603
2021	122	1,000	341	241	1,071	1,087	N/A (~250)	747	1,170	130	5,709
2022	282	1,225	185	115	1,061	949	N/A (~250)	754	808	107	5,485
TOTAL	4,134	29,874	12,788	4,896	24,546	28,539	8,355 (14)	32,625	28,607	3,835	178,486
AVG	188	1,358	581	223	1,116	1,297	597	1,483	1,300	174	8,317 / YR

Legend:

>AVG	>210	>1,400	>650	>250	>1,200	>1,400	>650	>1,600	>1,400	>200	>8,400
AVG	160-210	1,300-1,400	550-650	200-250	1,000-1,200	1,100-1,400	550-650	1,400-1,600	1,200-1,400	150-250	8,200-8,400
<AVG	<160	<1,300	<550	<200	<1,000	<1,100	<550	<1,400	<1,200	<150	<8,200

Note: The statistics for Alaska are only those reported near inhabited areas. Wildfires in uninhabited areas are occurring but are not reported.

(Source: The National Interagency Fire Center of the US Bureau of Land Management)

The color coding red and green of the table (described in the footnote of the table) reveals an interesting statistic that rebuts news media statements (fed to the news media by over-exaggerated observations of some environmental protection scientists and activists) that Man-generated "greenhouse gasses" have increased the frequency and severity of wildfires with time. By comparing the trend of above (red) and below (green) relative to average (in white) it is seen that there is a trend just the opposite to what the public is being told. There are clearly more wildfires numerically above average in the decade of 2000 to 2010 and fewer wildfires numerically below average in the decade of 2011 to 2022.

A graphical presentation of the wildfire database is shown as follows:

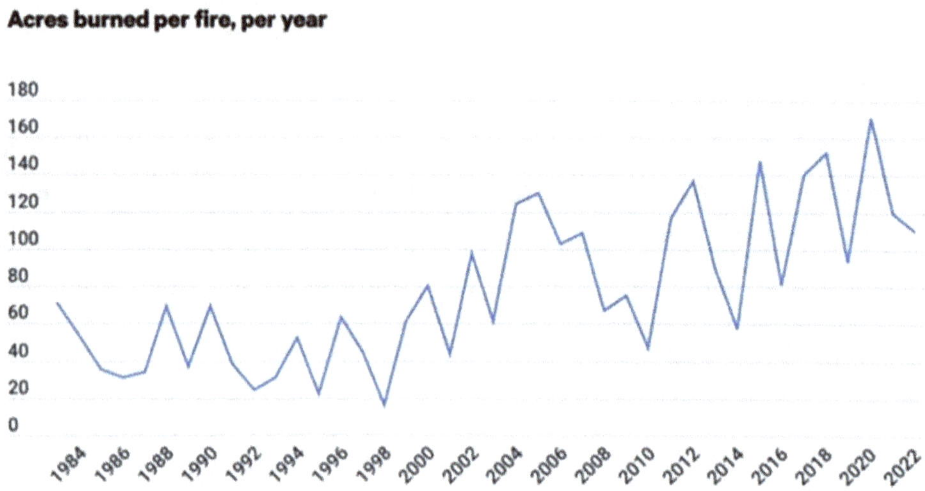

(Source: National Interagency Fire Center (NIFC))

When viewed in graphical form, the trend in wildfires is seen to have been somewhat flat-line from 1984 to 1988, but there is a distinct increase in the average trend of acreage burned per fire from 2000 to 2022 (as the tabulated data also showed by the color-coded data).

Each wildfire consumes a widely varying number of acres (square miles or square kilometers), but when their frequency is as high as shown in the table presented above, the totals of areas burned are incredibly high:

Wildfires Area Burned Cumulative USA 2001 to 2022

Year	Number of Wildfires	Cumulative Number of Wildfires	Number of Acres Burned ~ millions	Square Miles Burned	Cumulative Square Miles Burned	Square Miles Burned per Wildfire
2001	13,702	13,702	7.4	11,840	11,840	864
2002	10,610	24,312	3.6	5,760	17,600	543
2003	12,287	36,599	7.2	11,520	29,120	938
2004	10,426	47,025	4.0	6,400	35,520	614
2005	7,807	54,832	8.1	12,960	48,480	1,660
2006	14,708	69,540	8.7	13,920	62,400	946
2007	9,683	79,223	9.8	15,680	78,080	1,619
2008	7,150	86,373	9.3	14,880	92,960	2,081
2009	8,585	94,958	5.3	8,480	101,440	988
2010	6,096	101,054	5.9	9,440	110,880	1,549
2011	7,216	108,270	3.4	5,440	116,320	754
2012	8,028	116,298	8.7	13,920	130,240	1,774
2013	9,110	125,408	9.3	14,880	145,120	1,633
2014	7,011	132,419	4.3	6,880	152,000	981
2015	7,708	140,127	3.6	5,760	157,760	747
2016	5,354	145,481	10.1	16,160	173,920	3,018
2017	6,713	152,194	5.5	8,800	182,720	1,311
2018	5,402	157,596	10.0	16,000	198,720	2,962
2019	5,093	162,689	8.7	13,920	212,640	2,733
2020	4,603	167,292	4.6	7,360	220,000	1,599
2021	5,709	173,001	10.1	16,160	236,160	2,831
2022	5,485	178,486	7.1	11,360	247,520	2,071
	Avg 8,317		Avg 7.0	Avg 11,251		Avg. 1,528

1 million acres = 1,600 sq miles

(Source: The National Interagency Fire Center of the US Bureau of Land Management)

There is not a conclusive trend in the above table showing a decrease or an increase in acreage burned by wildfires but the acreage per wildfire does show a slight trend toward an increase in acreage burned per wildfire. That slight trend is somewhat surprising and disturbing because of the large improvement in observing wildfires in their infancy (soon after being ignited by the lightning strike(s) and with continuing improvement of forestry management (more intentional burns to thin out fuel on the forests floors) should show a trend toward less acreage burned per wildfire.

A photograph that adds realism to the effect wildfire has on the environment is shown as follows:

(Source: The National Interagency Fire Center and of the National Weather Service, NOAA)

While it is obvious that many trees are destroyed that could have been put to better use – such as the lumber needed to construct new residences for the exponentially growing population – there is the loss of so many trapped animals and other wildlife that are destroyed. Still worse is the loss of human lives and loss of their property. Wildfires have no "pro" – only "con"s.

From a geographical viewpoint, the next illustration (bar chart) indicates the states that have the most damage resulting from wildfires:

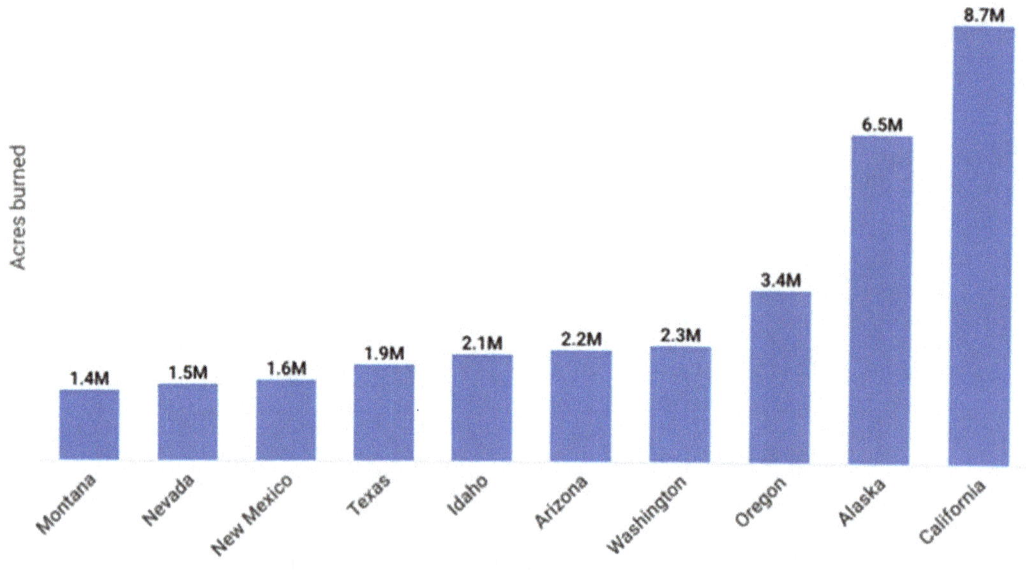

(Source: National Interagency Fire Center (NIFC))

From an impact on the amount of property damage and number of human fatalities perspective, but not to underplay the effect wildfires have on the environment, this entire chapter could be filled with praise for those heroic individuals who volunteer to be on firefighting, rescue, and recovery teams. There can be no labeling of their inefficiency or lack of bravery to be on the front line with attempts to control the wildfire and assisting humans with recovery of their lives during the aftermath.

The magnitude of losses shown on the table as "Cumulative Number of Wildfires" (total of 178,486 during the period 2001 to 2022) and "Cumulative Square Miles Burned" (total 247,520 during the period 2001 to 2022) do not adequately portray the type of damages done by wildfires differently from those caused by tornadoes and hurricanes. The worst example is that of the wildfire which ironically devastated the town named Paradise, in northern California, in 2018. On the morning of November 8, 2018, a strong downslope windstorm in Butte County in Northern California blew down a high voltage (115 kilovolt) electric transmission line which ignited a fire in the brush below the transmission tower. This account will not highlight the reportedly poor maintenance of the suspension system (e.g., C-hooks) holding the electrical transmission lines in place on the tower; suffice it to say that the massive damage and cost of the fire resulted in the power company filing for bankruptcy. But it does point to the need for power distribution companies to assure that power transmission lines are safe for the extreme environmental conditions possible everywhere along the high voltage distribution route. There also is a need to study and enforce the transmission technologies as more demand for electricity increases due to population increases, especially the need to route high voltage lines at ground level (and safely insulated) through heavily forested regions.

The already high winds blew the fire (titled as the “Camp Fire” as it was ignited near Camp Creek Road – not by embers from a human-made campfire) through the dense forestry downwind and toward several small communities. Wildfires create their own local weather and especially add wind velocity from the heat expanding the surrounding air and causing the unusual – but common in wildfires – presence of a fire tornado. A photo of a fire tornado is shown next:

(Source: Photograph by photographer of Butte County Fire Department)

It can be seen by the bending of the tree foliage that the velocity of the winds was high, and in the direction ahead of the advancing fire toward doomed residential construction. Some early warnings were given to residents to evacuate immediately but many of those warnings were either not received or foolishly went unheeded. Compounding the problem was that communities situated in forested areas have but one two-lane route servicing the area – and those evacuating on that single one lane escape route are sometimes jammed by a failure of one vehicle causing a traffic jam. A photograph of the vehicles burned on the escape route of the Camp Fire is shown next:

(Source: Photograph by Photographer of Butte County Fire Department)

The burned vehicles shown (many had their tires burned by the heat as they were jammed in escaping traffic) have been moved from the traveled lane to the shoulder by tow trucks. Occupants of these vehicles tried to flee by foot, and some were rescued in fire engines; others were fatally destroyed in their vehicles by the rapidly moving flames.

Shown is a scene of firefighting equipment unable to get into the towns as the fire had overrun the single lanes into and out from those communities:

(Source: Butte County Fire Department)

Photos of several communities all show the total devastation of the entire community as the fire raced through the communities as in a row of falling dominos.

(Source: Butte County Fire Department)

(Source: Butte County Fire Department)

Damage of a residence by wildfire is more extensive than that caused by a tornado since in a tornado the damage (loss of rooftop or collapse of the structure) is done in a matter of seconds whereas in a wildfire the fire consumes everything for minutes and hours until there is no fuel (anything consumable by flame or heat) left to be burned. Nothing is salvageable or recyclable when burned into ash.

(Source: Butte County Fire Department)

The loss of 85 lives, loss of 18,661 structures (predominantly residential), the loss of 154,000 acres of forest to the Camp Fire were/are incredibly steep; however, the loss of the pet and wildlife animal denizens within the forest as well as the loss of the carbon-dioxide absorbing leafage on the trees and plants in those 154,000 acres burned is unmeasurable. As seen in some of the foregoing photographs the smoke pollution into the atmosphere from wildfires is an unseen hazard – worse than the emission of carbon dioxide - that lasts for weeks and months after the wildfire has been contained or extinguished.

Stating (from personal experience of having resided in California for 38 years) a lesser- known fact to the public that living very close to a wildfire is less hazardous to the respiratory system of humans than it is to those living 3-8 miles downwind of the burned area. The reasoning is based on the larger particles of soot and ash that fall back to the ground close to the fire. The smaller particles of soot and ash – the type that can do serious harm to the alveoli (air sacs) in human lungs – travel for a downwind distance of 3-8 miles from the fire zone depending on the winds aloft.

When humans breathe in, air containing oxygen enters their windpipe, passes through the bronchi and then reaches the air sacs. These air sacs, called alveoli, look a bit like tiny grapes at the end of the bronchial branches. Healthy lungs have about 300 million air sacs in them. Each air sac is surrounded by a network of fine blood vessels (capillaries). The oxygen in inhaled air passes across the thin lining of the air sacs and into the blood vessels. This is known as diffusion. The oxygen in the blood is then carried around the body in the bloodstream, reaching every cell. When oxygen passes into the bloodstream from the air sacs, carbon dioxide leaves the blood and passes into the air sacs. Carbon dioxide (CO_2) is a waste product of reactions in the cells. Humans get rid of it when they exhale. Because these two gases pass between the bloodstream and air sacs in opposite directions, this is referred to as "gas exchange" in the lungs.

A consideration also about wildfires (as well as the other forms of natural catastrophic events – seasonal and anomalistic extreme weather conditions, tornadoes, hurricanes, floods, volcanic eruptions, landslides, etc.) is the survivors of most natural catastrophes losing their places of residence and often a loss of their food supply. This adds to the already disturbing fact of homelessness in urban areas, which is exacerbated if the weather outside is in an extreme state of cold, hot, or precipitating high amounts of rain or snow. In developed countries such as the USA, there have been established various organizations and governmental programs providing emergency services for victims in communities (such as that of Paradise, California), but many undeveloped countries do not assist their residents with loss of residences or loss of food supply.

Where there is a lack of firefighting services to quickly extinguish fires from spreading in the entire region when driven by high wind conditions, the loss of lives and cost of fire damage grow in proportion to the number of facilities involved. The fires that destroyed the beautiful town of Lahaina on the Island of Maui in the state of Hawaii on August 8th through the 11th, 2023, killed at least 98 persons and left 31 persons missing (presumed either destroyed in the fire or from jumping into the nearby ocean and drowning or being taken by sharks). Damage to property in the 2,170 acres burn zone was almost entire. This fire at Lahaina is especially tragic to the Author who, with his wife, enjoyed visiting the small but hugely interesting Whaling Museum section of the Lahaina Heritage Museum in the Lahaina Historic District on Front Street in Lahaina on two occasions. In front of the

Lahaina Heritage Museum was a large old, iron anchor from a sailing ship which was seen on a television news report of the Lahaina Fire and being the only recognizable feature surviving the firestorm in Lahaina.

Shown next is the town of Lahaina with the fire shortly after its outbreak. The fringe winds (30-45 mph with gusts to 60 mph) caused by Hurricane Dora (Category 4) over six hundred miles south of Hawaii drove the fire storm quickly across the entire town seen blowing from the initiation area (where a downed power line – blown down by the high wind gusts - started the fire) virtually horizontally very close to the ground.

(Source: The Federal Emergency Management Agency (FEMA))

Firefighting equipment was unable to prevent the fire from spreading rapidly and trapping humans because of the high winds and extensive area involved. The homes had no built-in water sprinkler systems and were built of almost all wooden construction.

In other more common circumstances, that of fire in a tall commercial structure, putting water on the fire can be a formidable challenge:

(Credit to Unknown Photographer)

The humans least hazarded by wildfire are those living downwind from the main fire and being exposed to the particulate matter in the wildfire smoke – a form of natural pollution of the atmosphere. People living within about 3 miles from the wildfire when active are exposed to large soot particles which are less hazardous to human respiratory systems because they fall to the ground due to gravity and are not inhaled. People beyond that distance are exposed to the fine particulates in the atmosphere that are unhealthy. Two exposures to smoke from wildfire affecting mass populated areas were the wildfires in Northern California that lingered in the Lops Angeles area for a few days (starting June 20, 2020) and caused uncounted health injuries, some undoubtedly fatal to persons in already abnormal respiratory ill health. The Los Angeles skyline resembled that in the early twentieth century before stringent EPA smog-reduction regulations were enforced on motor vehicles.

(Source: Los Angeles Times, Unknown Photographer)

The Author and his family resided near Los Angeles for 34 years and are enormously grateful to the EPA for their smog-reduction activities in southern California. A later section is discussed dealing with the human transportation system (with the transition to all-electric vehicles) expands on the pros and cons of what is being done in the human transportation systems globally.

Another example of wildfire smoke and hazardous particulate matter in the atmosphere downwind was that in New York City in September 2023 caused by wildfires in southeastern Canada. The iconic Empire State Building appeared as shown in the next photograph:

(Source: New York Journal American, Unknown Photographer)

Numerous business offices in the downtown New York City area directed their employees to stay at home in outlying residential areas (in Brooklyn, Queens, and on Long Island). Again, there are no medical statistics for the number of fatalities among the 20 million inhabitants in the Greater New York area. The Canadian wildfires were naturally caused as were the winds that carried the smoke toward the USA, and therefore the fatalities incurred cannot be directly attributed to Man burning fossil fuels changing the climate as having been the cause for them.

A not conducive to health status report caused by air pollution was recently posted (April 25, 2024) by CBS News:

"Almost four out of every 10 people in the United States live in a place where air pollution is considered bad enough to put their health at risk, the American Lung Association warned in its latest "State of the Air" report released on Wednesday. That proportion of people — about 39% of the population — had risen sharply since earlier rounds of pollutant data were analyzed for the annual report last year, and the trends were especially pronounced in certain parts of the country.
This year's air quality report was based on pollution data collected in 2020, 2021 and 2022 by the Environmental Protection Agency.
The American Lung Association, a charity organization focused on improving lung health and addressing lung disease, has released "State of the Air" reports annually since 2000. The new report's findings show that roughly 131 million Americans were breathing unhealthy air during the three-year monitoring period. That number had jumped by almost 10% since the lung association issued its "State of the Air" report in 2023, when data showed 11.7 million fewer people had been regularly exposed to toxic and potentially deadly pollutants. Echoing longstanding concerns from experts and advocates across the board, the newest findings highlight a disproportionate environmental threat to people of color. Since the Clean Air Act was signed into law in 1970, emissions of the six most common air pollutants have fallen by 78%, according to the EPA. But

experts warn that climate change has made air quality harder to manage despite policies designed to protect it, especially as growing sections of the country grapple with soaring temperatures, longstanding drought and unprecedented wildfires."

The report added: *"High ozone days and spikes in particle pollution related to extreme heat, drought and wildfires are putting millions of people at risk and adding challenges to the work that states and cities are doing across the nation to clean up air pollution,"* the report's authors wrote. Two groups of pollutants were examined: fine particles, which are emitted directly into the air from combustion and some chemical reactions, and ozone, which is usually produced when other air pollutants react with sunlight in the atmosphere. For fine particles, the pollution trends were broken down into short-term and long-term patterns. The American Lung Association ranked U.S. cities with the worst air quality based on daily particle pollution, annual particle pollution and ozone between 2020 and 2022. The report assigned many cities a failing grade when repeatedly poor readings on the Air Quality Index pushed an area outside the bounds of accepted general health standards."

Except for those wildfires that are caused by Man (by unintentional carelessness with camping fires not being properly extinguished, or intentionally by arson), the Author found no database that would challenge the annual "State of the Air" report. However, the "State of the Air" report does not associate Man creating "greenhouse gasses" to be a leading cause for the wildfires or for the ozone concentration in the near atmosphere "which is usually produced when other air pollutants react with sunlight in the atmosphere". It is beyond the scope of this book to delve into the ozone increasing or decreasing phenomenon – that is left to atmospheric physicists and chemists to further investigate and to control if possible.

Water as a resource has had a pronounced effect on where a human chooses to establish his/her residency (although most humans remain within 100 miles of the place where they were born for family purposes). The larger port cities of the world were developed by sailing ship trade routes for cargo and/or passengers, and countries that have a large body of water (coastal or inland) suitable for recreational as well as for aquatic food sources also tend to have a higher density of their human population near those oceans, lakes, rivers, etc. Humans are humans and many prefer to prioritize their lifestyles and recreational needs over health, safety, and comfort, regardless of how large bodies of naturally placed water sometimes "misbehave". Homeowners pay extra insurance premium fees to reside in known flood zones taking the gamble that no flooding will occur.

There has been found no database of factually valid (not over-exaggerated, not "cherry-picked", or based on creating fear) information that supports the environmental scientists claims that climate change is having a feed-back effect causing the onset, frequency and/or severity of natural catastrophes (tornadoes, hurricanes, wildfires, volcanic eruptions, earthquakes, floods, droughts) to be increasing – either globally or regionally. Similarly, the meager database of measurements of the ocean water temperatures rising (a few millimeters change which appears to be equally seasonal as it is to water temper volume increase due to all the combined effects causing atmospheric warming) and discounting the cyclic nature of El Nino and La Nina ocean phenomena), or the meager database on ocean levels rising more than should be expected. The probability of the glacial ice sheets on Greenland and in Antarctica suddenly melting to raise ocean levels to a crisis level is very low – but some planning should/must be started to relocate human residences, commercial, and transportation on the infrastructure to higher elevations further inland in some regions of the USA and world-wide should it eventually become necessary.

The inner core regions of the Earth are made of molten iron and nickel, and the Earth is known to be a rotating sphere. Anything having a ferrous (iron) content when rotated in an electric field is known to generate electrical current flow in the mass of iron. If the electric field shown in the aurora borealis is applied to the planet having a large core of iron (ferrite), volcanologists are theorizing there may be a resulting magnetic force causing the molten ferrite to be attracted toward the electromagnetic field – and resulting in a volcanic eruption. Once again, the Author will leave that possible electromagnetic phenomenon to be researched for its possibility of causing ("aiding and abetting") the eruptions that recently occurred in Iceland (as shown next):

(Credit to Unknown Photographer in Iceland)

That flow of magma in Iceland was the fastest ever seen in modern volcanology. The flow of magma into a 15-kilometer-long crack ahead of the recent volcanic eruptions in Iceland occurred at the highest rate observed anywhere in the world for this kind of natural event (bearing no influence on or of Man-made carbon dioxide emissions into the atmosphere from the burning of fossil fuels). This volcanic activity began with magma accumulating several kilometers beneath the Svar sengi region in Iceland and not far from the site of a geothermal power plant that supplies hot water to the capital city of Reykjavik. More discussion of geothermal power appears in a later chapter on the current and potential use of geothermal power as an alternate and clean energy source which is abundant in many regions of the world.

During the research on this topic area, much has been found on the severe problem of <u>pollution</u> (air, water, and land) that is clearly Man-made, and which poses near-term sanitation, health, and safety problems to humans that are more ominous and urgent than whatever the valid scientific concerns about global warming and climate change portend.

To conclude this chapter, it is a general conclusion based on all of the factual scientific information in the preceding chapters (updated to 2024) on factors (astrophysical, geophysical, natural, and anthropogenic) that most significantly affect the human population and the atmospheric environment that the opening hypothesis of environmental scientists: "When all else is eliminated, Man burning fossil fuels is the dominant cause of global warming and climate change" is unprofessionally fallacious. It remains that water vapor in the air is scientifically factual to be a far greater partial cause of the surface temperature of the planet to be increasing. The foregoing chapters have shown that the "science is settled" claimants do not acknowledge that fact and are wastefully expending much effort and an incredibly size number of financial resources – taxpayers monies – to "combat" climate change by eliminating the use of fossil fuels. Furthermore, the miniscule effect of burning fossil fuels on disrupting or displacing the astrophysical, geophysical, natural causes of "global warming" is not of "crisis" level of urgency.

It is hastened to be added that this book does not advocate that the use of fossil fuels to satisfy Man's need for heat energy should continue as it has been for the past century. As with virtually all technologies developed by Man – including the technology of combusting fossil fuels as an energy source – if the technology is generating any form of pollutant to the planet's essential resources (air, water, or land) the technology must be improved or abandoned. The technology of combusting coal is different from the technology of combusting oil, and natural gas in that the combustion of coal (which contains variable amounts of impurities when in solid state) generates both heat and an ash residue which contains toxic and carcinogenic components in its ash residue. Disposal of post-combustion coal ash is known to be polluting the planet's water and land environment, and it is for that reason that the cessation of the coal burning technology be rapidly set in place by United States government anti-pollution regulations. Other regions not under United States jurisdiction should make the same type of assessment that has been made for the United States technology developments to eliminate pollution of their regionally environment.

The following chapters place emphasis on the variety and scope of the many aforementioned challenges that result in not only the need to survive the natural extremes that produce the changes in temperature and in the climate but also all of the natural hazards which dictate the technological and physiological efforts that the human population must generate to survive living on this magnificently beautiful but unpredictable hazardous planet.

Chapter 3
Other Factors; Conclusion Statements; Technology Factors to be Assessed

It is now better appreciated how Homer of ancient Greece must have felt ten centuries ago when he compiled his "Odyssey of Homer". The Odyssey was translated from a combination of facts and myths in ancient Greek literature/mythology and published in 1614 in Great Britain. The Odyssey rapidly became a required book to be read in History classes of most high schools because it described many of the natural forces that existed in the World environment – although embellished quite a bit to emphasize an event or a situation. This Author feels obligated to explain to Readers for having written this book using some phraseology and nomenclature which must "sound like Greek" in places. The only recourse to describing a scientific observation is to use the scientific terminology that defines the observation and then to describe that scientific "Greek" with more common words – "street language".

This chapter first provides information about other factors that must be considered – integrated with those factors already presented and discussed – and the conclusions that have been drawn to this point in this book. Those conclusions serve to identify numerous trade studies that will be useful in guiding the planning and technology development areas for the population of the United States region to adapt to and mitigate the effects of the planet's surface temperatures changes (in regions, not in an average global manner), climate changes (again regionally not average globally), and ocean sea levels rising. Near the end of this chapter, a most important first step in the technology assessment process will be presented – the factors of importance against which the pros and cons of each technology will be tested. Those factors of importance were compiled from the research done over the past several years and having observed and calibrated how important they were, are, and probably will be in the real natural world.

Having collected information into a database about a topic does not automatically increase valid knowledge about the topic – the database must be as complete as it can be and interpreted as scientifically valid as it can be. As this book has shown, assembling the large number of complex facts about global warming and climate change was much like sorting through a landfill for the pertinent facts and not being distracted by the non-pertinent fallacies. Sorting through a pile of trash for the right pieces is a lot different than assembling a jigsaw puzzle (when it is known that all the pieces are color coded and will fit together and make the "big picture" complete).

Just as there are natural disasters causing fatalities, there are also disasters that occur in other fields of technology that cause fatalities and large property losses. Readers who have only glanced at the interior contents of this book and are reading these Concluding Comments probably question the qualifications of the Author of the book to have drawn those "big picture" conclusions. This Author can only show what officials at high levels of Government conducting national interest levels of safety analyses at the President's request state about the Author's qualifications to conduct such "big picture" analyses. The following is an extract of a 1982 letter from a former Secretary of the Air Force (Dr. John McLucas) to the Vice Commander of the US Air Force (General Robert Mathis) in 1981 regarding the Author's qualifications demonstrated in response to then-President Ronald Reagan's request for the Air Force to assist the Federal Aviation Administration in the certification process of the first commercial aircraft to fly with a complement of a two-member flight crew instead of the standard three member flight crew. Two key issues were how that reduction from a three to two-member flight crew affected the safety of transporting more than 200 passengers in the advanced technology aircraft over long distances above vast ocean waters and the enormous saving in cost to not have to train flight engineers being worth the technology risk taken.

> **The Aircraft Crew Complement study centered on the ability of a flight crew composed of two crew members instead of the conventional three crew members on the flight deck safely operating the aircraft, with advanced technology instruments and displays, in heavy airport traffic and under harsh environmental conditions. Carrying out our study and preparing our report would have been difficult and perhaps impossible had it not been for the excellent cooperation of the Air Force which made available to us a wide range of talent and expertise.**
>
> **This is to thank you specifically for making available Mr. Herbert A. Hutchinson, Chief, Flight Technology Division, ASD Deputy for Engineering, Wright Patterson AFB. Mr. Hutchinson served as leader of one of our panels that was responsible for studying the Federal Aviation Administration's flight safety certification process for new commercial aircraft. He provided excellent leadership to the Panel staff, and his outstanding analytical talents and extensive system safety training and military aircraft crash accident investigation experience contributed very substantially to the work of the Task Force.**
>
> **My colleagues, Lt. Gen. Howard Leaf of the Air Force and Mr. Fred Drinkwater of NASA, and I thank you and other senior officials of the Air Force for the strong support given to our Task Force and for providing the invaluable system safety engineering services of Mr. Hutchinson.**

(Source: Author's Personal Correspondence File)

The Author is retired from a career (a personal odyssey) of over forty years designing aircraft and developing technologies for aircraft to fly safely and efficiently within the "flight envelope" which is unique for each different design. Aircraft must be able to fly at a range of speeds and altitudes, through a range of temperatures, and perform safely within the limits of those variables comprising the "flight envelope". Performing that capability (in street language: "Just do it") requires burning an amount of fuel (JP-4, which is a form of fossil fuel used in motor vehicles, and which was held in on-board tanks) that permits the aircraft to safely carry a large payload (which may consist of over two hundred passengers, or cargo or a mix of both) for long distances over hazardous terrain or large bodies of water) within that large envelope of altitude and speed. Almost every system and machine have some form of an operating envelope within which the system or machine must operate efficiently and safely. The human population should be thought of as a human subsystem (the "crew and passengers") of how the planet as a system factually operates. Instead of an aircraft and a safety flight envelope, the human population must thrive and survive an extreme temperature envelope – very hot or very cold - without being injured or succumb to the effects of hyperthermia or hypothermia.

When this research into the causes of atmospheric and planet surface temperatures changing and affecting the climate was accelerated over fourteen years ago, there were no preformed conclusions of what that research might lead to be. From experience in several prior technology developments, it was anticipated that the "When I turned over a stone, I found more stones to be overturned" experience would repeat itself. It is that type of research that is classed as being "heuristic". The most fundamental heuristic is trial and error, which can be used in finding the values of dependent variables in algebra problems – or addressing complex topics like the climate. In applied mathematics, some common heuristics involve the use of visual representations, additional assumptions, forward/backward reasoning and interpretive logic. In this book, the forward/backward forms of heuristic reasoning and common-sense logic - backed up with photographs and illustrations confirming "that's the way it was and is" - were used extensively.

There is also much similarity of flight systems with the human systems performing (eating local foods as fuel, drinking local water as lubricants, and breathing local air – but exhaling carbon dioxide from the digestion of food during the entire process) within a "temperature and climate envelope" on the planet's surface. One human exhalation introduces a small amount of carbon dioxide into the atmosphere; however, the cumulation of several billion humans exhaling carbon dioxide into the atmosphere every few seconds is not insignificant.

There is also the necessity of humans having to wear different types and insulation thickness of clothing to remain healthy and comfortable throughout the geophysical envelope within which they live while performing the everyday tasks required to sustain and enjoy life. Environmental scientists and "Green" activist groups and organizations seem to perceive the safety and health of plant and animal life in the "big system" of Earth to be of more importance than that of human life. In doing "combat with climate change" they "have identified the enemy, and it is us".

An illustration of many of the factors the Author identified that affect global warming and the climate on the planet, some more significantly than others, provides an illustration of many of the factors this Author evaluated in assessing the impact Man is having on the planet:

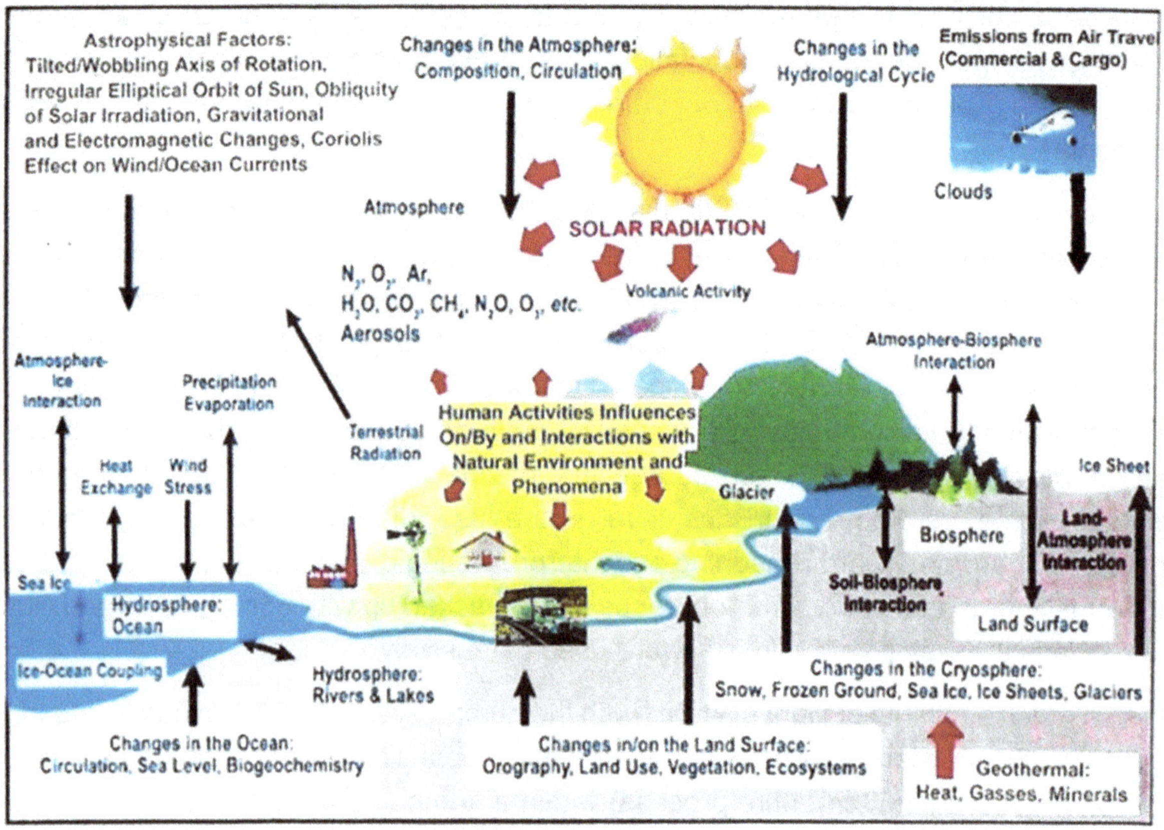

The many variables affecting and affected by the Planet
(Diagram generated by Author from multiple databases)

The above diagram is a simplified "picture" (much like the old photography technology of being a "Polaroid snapshot") and Readers should take notice that the many dynamic factors that are occurring and changing continuously must also be accounted for in the "big picture" of global warming and climate change. As the Earth rotates, wobbles, and changes location in its orbit of the Sun, those astrophysical factors change the angle (obliquity) at which the solar rays of heat energy impact the surface of the Earth. It was painstakingly shown throughout this book that there are daily changes in atmospheric conditions (the Weather Report for the Day showing temperature extremes for the day, projected highs and lows in pressure cells (the Hs and Ls moving across the video screen), the warm and cool fronts, the probability of cloud cover and precipitation, wind conditions, and times for daylight and darkness. Humans move through life like the hands on a clock. The weather changes rapidly as the relative movement of the second hand, the seasons change relatively slower like the minute hand, and the climate changes relatively very slow like the hour hand.

The individual factors in the diagram were in some way observed to have caused an effect on the environment, on the surface temperature of the planet, on the climate of a region on the planet, on the life of every living lifeform and on the most important element on the planet – the human population. The above process is known as a "multi-variable analysis" in the field of mathematics and is used routinely by those in engineering whose specialty is in the category of system engineering.

The intentional omission or ignoring of information that does not support a preconceived notion desired to be confirmed or selecting only a partial piece of information is known as "cherry picking" of the complete database. (The wikipedia website has a definition of cherry picking: "*Cherry picking, suppressing evidence, or the fallacy of incomplete evidence is the act of pointing to individual cases or data that seem to confirm a particular position while ignoring a significant portion of related and similar cases or data that may contradict that position. Cherry picking may be committed intentionally or unintentionally.*") There is a large difference between "cherry-picking" and "separating the wheat from the chaff" as those in the crop-harvesting community well know. The same practice exists in science and technology where the crop is knowledge.

Regrettably, cherry-picking was found to have been practiced by some environmental scientists - as has been shown in the early portions of this book in their *"after eliminating eveything else"* (such as the totally valid Vostok Ice Core database) in deriving their now recognized fallacious hypothetical conclusion that "*Man burning fossil fuels is the dominant cause of global warming.*" The environmental scientists made a poor judgment back then (around 1960) and have "circled the wagons" around it – and are still not yet admitting that new data changes that conclusion. This Author did his best – as a public service - in sorting the valid pieces of information from the fallacious pieces as well as searching for the most recent information. The Vostok Ice Core database and NASA satellite measurements of the variables in environmental science of the surface and atmosphere of the planet, and in a "big picture" perspective to arrive at the factual conclusions summarized in these Concluding Comments, were and are available to environmental scientists but who chose to disregard the full set of database information.

Earlier sections of this book entailed a review of the factual databases of information somewhat chronologically as they became known, but it quickly became evident that unethical suppression of data and/or cherry-picking was being practiced. The incident known as "Climategate" (that name being assigned for its resemblance to the circumstances of the "Watergate" scandal in politics in 1972) occurred in late 1999 and early portion of 2000 with the proceedings of the UN Intergovernmental Panel on Climate Change. This book does not include much of the details of the Climategate incident involving cherry-picking. The important fact is that it did happen and besmirches environmental science history.

In retrospect, the most important database reviewed for this book was that of the Vostok (Antarctica) Ice Core Samples database which spans the period from about 1990 backward to over 4.5 million years before now. Near the start of this book were shown the two statues in front of the National Archives (where the history of the facts for the United States is collected) with the two inscriptions "Study the Past" and "The Past is Prologue". The Russian atmospheric and climatology scientists began the ice core drilling in Antarctica in the years after World War II to address the observation Dr. Svante Arrhenius had made in the late 1890s of the increasing carbon dioxide emissions from the Industrial Revolution in Europe in the atmosphere having a near-term beneficial "hothouse effect" on the planet's surface temperature in the northern latitudes of Scandinavia.

That prospect had been recognized by Dr. Arrhenius who received the Nobel Prize for Chemistry in 1903 for that brilliant scientific observation. But Dr. Arrhenius also extrapolated his findings into a far-term projection that too much carbon dioxide in the atmosphere could lead to excessive heat on the planet's surface and hazard all forms of life on the planet. How the news media in the United

States presented the brilliant interpretations of Dr. Arrhenius' laboratory findings in 1903 illustrates how "cherry-picking" a database can be used as a tool of propaganda to misinform the public:

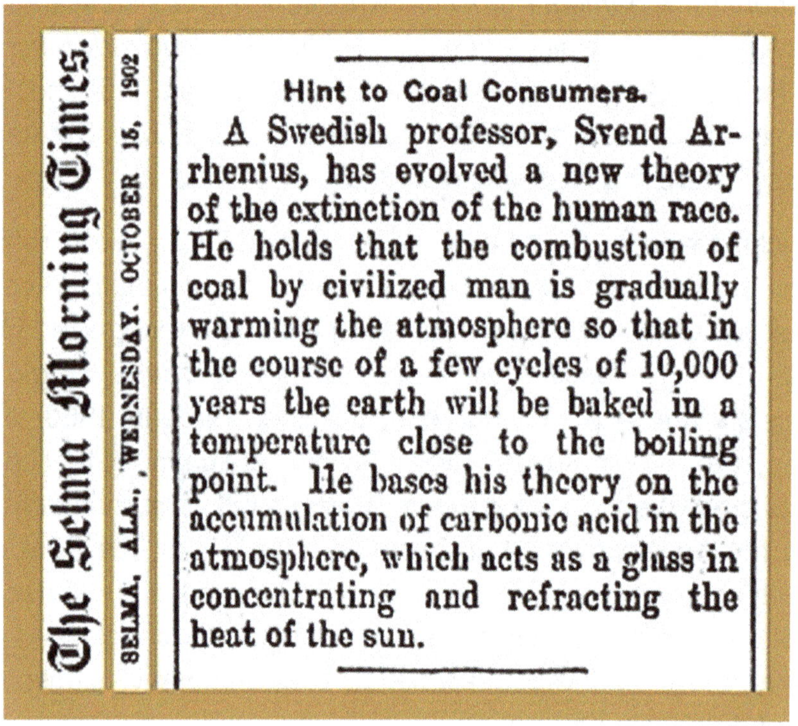

The Selma Morning Times.

SELMA, ALA., WEDNESDAY. OCTOBER 15, 1902

Hint to Coal Consumers.

A Swedish professor, Svend Arrhenius, has evolved a new theory of the extinction of the human race. He holds that the combustion of coal by civilized man is gradually warming the atmosphere so that in the course of a few cycles of 10,000 years the earth will be baked in a temperature close to the boiling point. He bases his theory on the accumulation of carbonic acid in the atmosphere, which acts as a glass in concentrating and refracting the heat of the sun.

(Source: The US National Archives, Washington, DC)

Readers of this book have been provided with extracts of the news media providing "cherry-picked" information about the causes of global warming, and climate change being the result of Man burning fossil fuels. It is professionally embarrassing to report that in the research done for the global warming and climate change "crisis" (per the news media), no news media has mentioned the existence much less the findings of the Russian Vostok Ice Research project (which contradicts the environmental scientists' hypothesis that Man burning fossil fuels is the dominant cause of global warming and climate change).

Analysis of the Vostok Ice Core database was initiated by this Author about 40 years ago (when initial reports of its existence became known in science and engineering literature). At the time, the Author was employed as Chief of the Flight Technology Division of the USAF Aeronautical Systems Division and needed to examine the ice core content about the atmosphere that might affect future military aircraft flight in that better defined atmosphere. That research by the Flight Technology Division for the USAF was initiated with no pre-conceived expectation of what useful data might be found. Research and assessment of the Vostok ice core research in Antarctica had to be suspended because of other more immediate technological developments needed by the USAF. About 15 years ago, this Author resurrected (pursuing a personal interest) the analysis of the Russian Vostok ice core experiments which by then had been more widely published on the internet. That research led to the Author concluding that a factual book (this book in hand) was in need by the public to confirm or reject the claim that Man burning fossil fuels was/is the dominant cause of global warming and climate change. The primary intent was to identify the technologies that are needed to mitigate the effects of climate change depending on the more complete set of factual causes for global warming and climate change being identified.

The Vostok Ice Core samples database was accomplished to extend scientific knowledge and not for any politically motivated purpose. A small photograph of one of the thousands of ice cores analyzed (bisected to expose its inner core for forensic analysis and examination without concern for external contamination by the drilling process itself) is shown as follows:

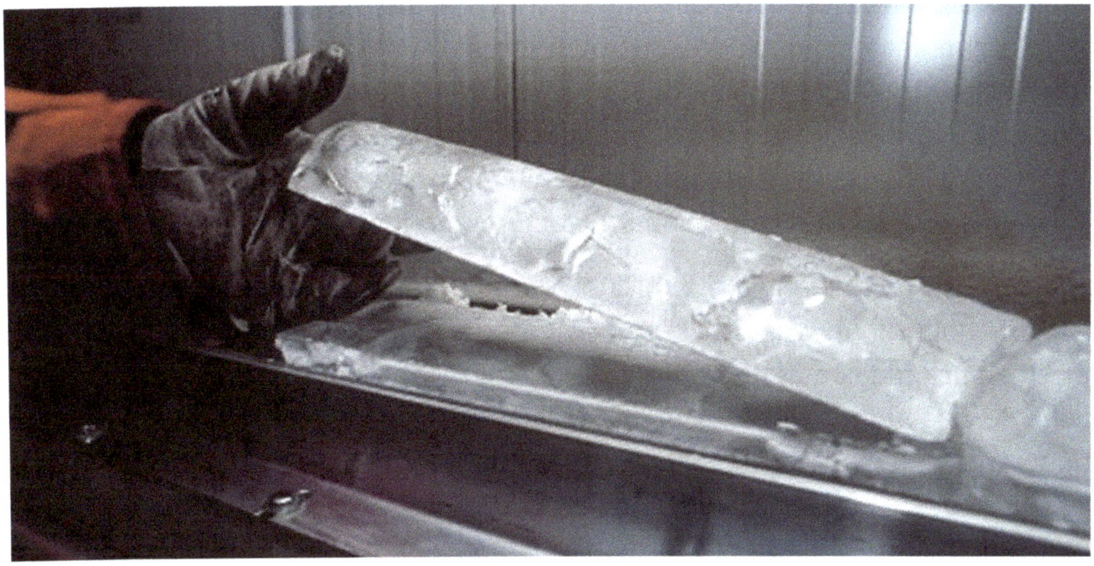

(Source: Smithsonian Institute for Science)

Ice research scientists analyzed the inner core of the ice samples and correlated the findings (deduced from spectrographic analysis and other carbon dating instruments measuring the benthic forams – single-cell forms of life on the ocean floor) to establish the approximate age in years at that depth of drilling through the ice sheet and the temperature and content of the gases trapped in the bubbles in the ice core. The analysis did not identify what force or influencing factor caused the temperature differences from the normal temperature at that latitude (78° South) or chemical contents in the ice cores, only their existence and magnitude.

The resulting Vostok Ice Core database analysis is summarized in a "big picture" form of graphic summarization and is graphically presented in the following illustration:

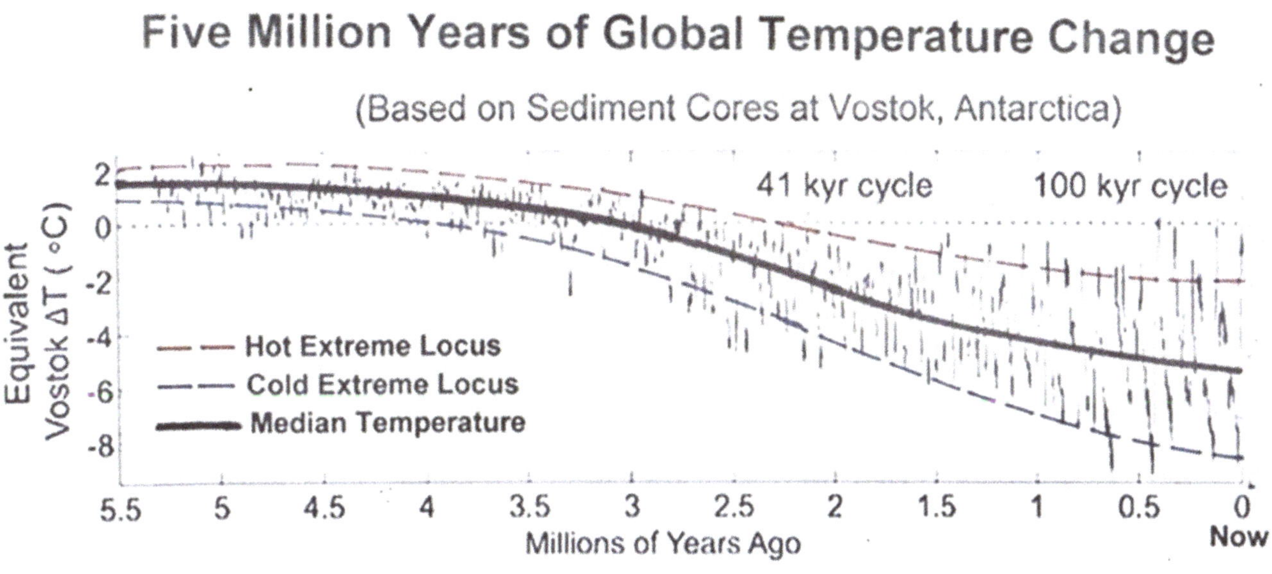

(Source: Smithsonian Institute for Science with Author additions)

The “raw” ice core sample data (seen in the graph above without the added 3 lines -locus lines bounding the extreme hot and cold temperature changes from normal and the median temperature of the extremes drawn in by the Author - was noted to be much like an earthquake seismograph of acoustic variations but instead with a high frequency of surface temperature changes with long periods of time. The immediate observation made was that the multitude of near-vertical fluctuations in temperature at that location (latitude 78° South in East Antarctica) were caused by other factors than those generated by Man since there was no known presence of Man on the planet that long ago. The Industrial Revolution did not begin until approximately 200 years before now (2024) and therefore could have no effect on any but the last vertical line showing the surface temperature to be increasing – due to a combination of natural factors and those of Man.

The interpretation can be made that 5.5 million years ago, the Earth’s average global temperature was about 1.5 °C hotter than what its “normal” average is considered now. If that interpretation is correct, then the black median temperature line indicates the Earth has been cooling with time and is now about 6 °C cooler than what it was 5.5 million years ago. The immediate conclusion is that the term “global warming” very much depends on what time period is being analyzed. The difference between the locus lines of the hot extreme peaks and the cold extreme peaks is also increasing with time and suggests that a complex combination of factors is causing the increase in temperature of the hot extreme is an additional factor contributing to the present global warming.

A question that arises is whether the environmental scientists’ use of an “average global temperature” factually represents the global warming to be the same at all regions of the planet. The early explorers (mariners in sailing ships and land-travelers such as Marco Polo) mapped the perceived temperature and climatic conditions as they varied with location in the ocean and on the land areas and their records were converted into a global “big picture” of the climate to appear as the following shows:

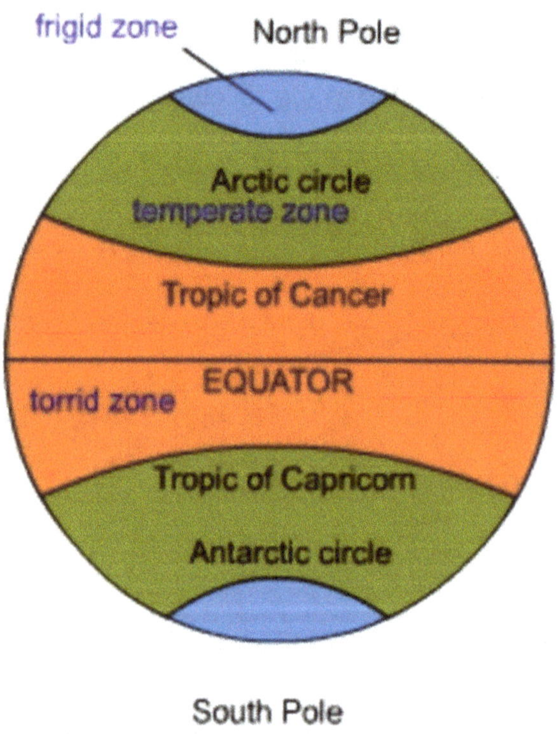

(Source: Encyclopedia Brittanica)

As crude as the above diagram is, it represents the different climates of the planet varying with latitude north and south from the Equator and is more meaningful than the use of "average global temperature". The positional system of latitude and longitude had not been developed at that point in history. An early chapter of this book describes how the latitude and longitude system of positioning was developed and has been updated with satellite technology to have become the Global Positioning System (GPS) currently in use.

A recent illustration appeared in the New York Times (which has been embellished for emphasis by the Author) showing that the use of coal has in fact been reducing toward its elimination from use in the United States, and that users of coal are largely converting their coal-burning processes to the combustion of clean, abundant natural gas. The trends in the use of coal, natural gas, nuclear, and renewables energy sources for generating electricity in the United States for the time span 2005 to 2025 reveal a very interesting and important factor – that the decrease in use of coal is being replaced by the increased use of natural gas. The trends of those four categories reflect the impact that environmental scientists, the Environmental Protection Agency, and the US Department of Energy in their efforts to reduce emissions and in turn improve climate change:

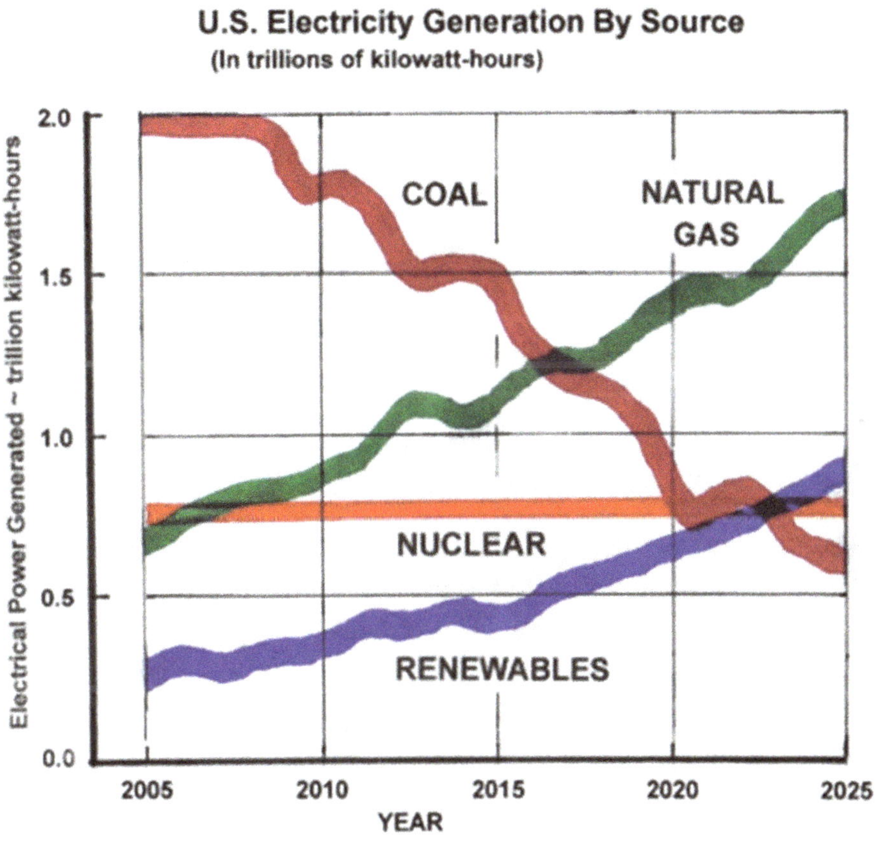

(Source: US Department of Energy, reported in New York Times Feb 2025)

The renewable energy sources to replace fossil fuels have proven to be more useful as sources of energy for small communities and farming lands but have been found to have insufficient reliability for use in large industrial applications. Solar energy is collected only when the solar cells are receiving solar insolation and at an acceptable angle of obliquity and that is dependent on solar illumination which is not more than 17 hours per day in some latitudes and less than 8 hours per day at other latitudes. When there is high precipitation in any form (rain, snow, hail, fog, sleet, etc.)

or smoke or clouds obscuring the Sun's rays, the solar insolation is also reduced to insufficient level. There are technologies for storage of electricity that can improve the reliability of solar energy but long periods of no input recharging electricity is another cause of unreliability. Solar panels were originally developed for use on residential roofs and carport roofs and that is an entirely suitable use for solar energy. It is not well suited for large scale industrial or commercial use. A con of solar panels is that it requires the use of significant land area, as well as requiring cleaning in areas that are subject to dust storms. They are vulnerable also to large hailstones which can damage an entire solar farm with one hailstorm event. Solar energy is a good candidate for use at desalination and water purification plants that utilize the flash distillation process.

Wind generators to produce electric power are even more unreliable than solar energy because there is not very reliable periods predictable for the wind to be blowing. Off-shore wind farms are being developed, but this Author foresees that the corrosive salt atmosphere damaging moving machinery is a maintenance nightmare. Because wind generators are mounted on top of tall posts, they become very much like a lightning arresting rod and attract very damaging lightning bolts at high frequency. They also are not suitable in high wind areas such as "hurricane alley" and "tornado alley".

Nuclear energy is a reliable and clean source but is a highly costly venture to erect a safe, large, efficient nuclear facility near a highly populated area. The emerging technology of the use of thorium instead of uranium has a good set of "pros and cons" that make it a very good candidate for use in some regions.

The use of oil for electricity generation is not shown on the above illustration; however, the oil reserves of the world are best dedicated to the world transportation needs because of its attractive set of pros and cons. The principal "con" with oil (and to a lesser extent natural gas" is the leakages and spillages that may occur and damage the environment. Most regions can obtain oil supply on land surfaces without the hazard of spills and leakages from off-shore facilities fouling beaches and destroying many aquatic life-forms and shore birds.

The numerous (37) conclusions to be shown in the next pages are provided somewhat as a checklist for use as a summary of what technologies were assessed to be potentially developed. It was found in the Author's previous book "The Facts and Fallacies of Climate Change on Our Planet" (summarized as Interim Conclusions IC-1 through IC-12 shown in the Introduction) and as a starting point for identifying the technologies that hold potential promise for mitigating the effects of climate change. Those technologies that have been developed to remove carbon from the atmosphere are of scientific interest but are not of practical use in the real-world environment. If the carbon removed from the atmosphere is used in another potentially useful technology area, then it should be further advanced as a topic of scientific interest but not a technology depended upon to satisfy the high demand for generating clean electrical energy.

Each Conclusion is identified with a **C** prefix and numbered **1** to **37.** Conclusions **C1** through **C23** were formed by this Author after analysis of several databases of factors causing global warming, climate change, and ocean levels rising that were analyzed during the generation of the "Facts and Fallacies of Climate Change on Our Planet" and for the continuation of that original analysis in this book.

Conclusions C24 through C37 were formed after conducting the assessment of the pros and cons (attributes and detriments) of a series of 24 technologies that have some potential for mitigating the effects of climate change caused by astrophysical, geophysical, or natural factors beyond Man's capability to be controlled. Conclusions C24 through C37 will be found at the end of this book in the Concluding Comments section.

The full set of 37 Conclusions are well-founded observations and interpretations of the data found in the different databases, and in numerous items of information for Readers (and hopefully other scientists, analysts of climate change, technology developers, politicians and subcommittees in Government, budget planners, environmental protection regulators, science teachers, etc.) to understand and can put into effective use for mitigating the effects of inevitable global warming, climate change, and ocean levels rising. How the news media will cherry-pick the contents of this book for its sensational observations and conclusions is beyond this Author's ability to control.

C1) Analysis of the factors– astrophysical, geophysical, natural, and Man burning fossil fuels, in some combination – led to the early conclusion that the dominant causes for global temperature change are not Man-made since the presence of Man was only in the most recent 200,000 years of the entire 5.5 million of years of information and data captured in the Antarctica ice cores. The Industrial Revolution began emitting significant amounts of "greenhouse gases" (predominantly carbon dioxide) into the atmosphere less than 200 years ago, and their significance for effect on global warming was not identified scientifically until about 1900 by Dr. Arrhenius in Denmark.

C2) The magnitude of temperature change is shown in the Vostok Ice Core database data to be varying cyclically and with increasing change in hot extreme temperature and cold extreme temperature relative to "average normal temperature" at latitude 78° South, but may exhibit different temperature change trends at other latitudes of other regions on the planet. Based on the scientific research in the Galapagos Islands by Dr. Charles Darwin (documented in 1869 in "On the Origin of the Species"), any life forms that were on the planet during the first 5.4 million years of the planet's environment either adapted to the temperature extremes and climate conditions at their latitude or went extinct.

At the compressed scale for time (the lower horizontal axis labeled as "Millions of Years Ago") to represent 5.5 million years, the selection of a time period ("slice of the data") less than about 10,000 years span at any time between the 5.5 million years (left edge of graph) and Now (right edge of graph) will provide a valid scientifically significant indication of global temperature variation but only during that period. No valid conclusion can be formed as to whether the planet was in a long-term global warming or long-term global cooling trend (or had a peak during the selected period and the temperature trend reversed) when only a small "slice" of the database is considered. The entire known database must be considered, and the median temperature line provides a long-term trend. Environmental scientists using inferential logic could fallaciously interpret that to mean that because the presence of Man burning fossil fuels is in the recent 200 years, the Earth is in a warming trend caused by Man burning fossil fuels. The scientific data measured and analyzed are correct; however, the flawed interpretation "after everything else is eliminated" by environmental scientists is not 100% valid.

C3) The long-term trend is seen to be that the planet has been cooling from the point in time 5.5 million years ago, but the short-term trend (for the ice core depths nearest the top of the ice sheet) is that the planet is in a warming trend. Any effect on cooling or warming of the planet caused by Man is insignificant relative to whatever are the other factors causing changes in the planet's short-term and long-term temperature.

Another interpretation is that humans appeared on Earth within the past 200,000 years but have been burning fossil fuels only in the past 300 years. To examine a recent "slice" of the Vostok ice core database closer, the Vostok database for the past 420,000 years was examined:

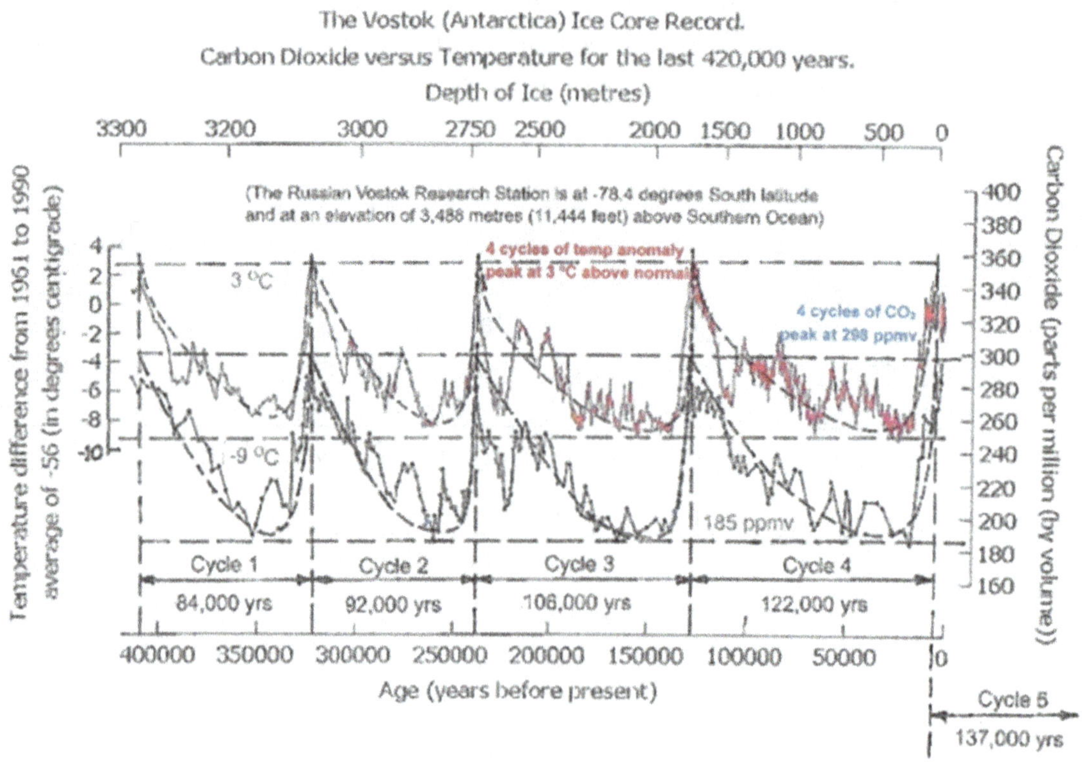

(Source: Smithsonian Institute for Science with Author additions)

The data in red color refer to the temperature anomaly (difference from "normal") and the data in blue refer to the carbon dioxide concentration (measured in parts per million by volume, ppmv) in the atmosphere. For Readers clarification, a concentration of gas in ppmv can also be expressed in percentage form – e.g., 298 ppmv is the equivalent of 0.0298 %. Although not scientifically incorrect, the 298 ppmv "looks" larger to a non-scientist than does 0.0298%. With the above expanded scale on time, the second Vostok Ice Core database reveals other most important aspects of global temperature changes that adds to the "learning curve" about global warming and climate change of the planet. The Author has added some "smoothing of data" u-shaped lines (red for temperature and blue for carbon dioxide in the atmosphere). Also added are the horizontal dashed lines representing the locus of the peak extremes of the temperature (in red) and in the carbon dioxide concentration (in blue). The locus lines for the temperature at the peak hot is 3 °C anomaly and at the peak cold is at or slightly lower than -9 °C anomaly.

With the additional smoothing and locus lines for the temperature cyclical variation during the 420,000 year time span, the data analysis of temperature leads to an important conclusion:

C4) The temperature – shown in red color - at the Vostok Ice Station (at latitude 78° South) unmistakenly indicates that in the prior 420,000 years period before now (0 years on the Age scale is 2024 on the calendar), there have been four (4) distinct cycles from a hot extreme 3 °C (or 5.4 °F) above normal to a cold extreme 9 °C (or 16.2 °F) below normal and return to the hot extreme peak. The second hot extreme of the fourth cycle occurred about 7,000 years ago (before the beginning of the Industrial Revolution) but the rounded peaky nature of the end of the fourth cycle is not as well defined as it is in the prior three cycles. Whether that rounded peaky nature is influenced by a "greenhouse gas effect" was the main subject for the early part of this book.

The carbon dioxide (CO_2) – shown in blue color - in the atmosphere at the Vostok Ice Research Station is seen to also exhibit the same four cycles of u-shaped variation as the temperature in the prior 420,000 years. Similar to that done for the temperature data, the Author added the same type u-shaped "smoothing of the data" lines and the dashed horizontal locus lines through the peak high concentration (298 ppmv) and peak low concentration (185 ppmv) for the carbon dioxide (in blue). Readers are reminded that the 298 ppmv concentration of CO_2 is a percentage equivalent of 0.0298% concentration in the atmosphere, and 185 ppmv is the equivalent of 0.0185%.

The carbon dioxide measured during the prior 420,000 years and to the recent 300 years is also all of natural source (again no presence of Man affecting the ice core measurements), leading to another important conclusion.

C5) There is no Man-made CO_2 "greenhouse gas" – only natural CO_2 - affecting the CO_2 concentration in the atmosphere during those 419,700 years ending 300 years before now (about 1720). The four u-shaped cycles of temperature anomaly and of carbon dioxide were caused by factors other than by any Man-made factors.

It was also observed that the four cycles exhibited hot and cold extremes of the same magnitude but the time span between the starting hot peak and the ending hot peak was growing larger with each cycle. Having shown that Man and Industrial "greenhouse gas emissions" are not significant factors influencing the cycles, it can be reasonably expected that there will be a fifth cycle following the fourth cycle but having a longer period between peak hot temperatures. A quadratic function approximating the time period between peaks of the first four cycles produced an estimate of the fifth probable cycle to be starting soon (possibly 50 – 100 years) and to be about 137,000 years in period from the end of the fourth cycle.

C6) If the current global warming ends in the next 50 to 100 years (about 2100) and transitions into the start of the fifth cycle, there is the high probability that the planet will be trending toward another ice age, probably 90,000 to 100,000 years from now. The probable fifth cycle was added (at the lower right corner) to the Vostok Ice Core analysis chart shown above.

A question that arose was whether carbon dioxide (from any source, natural or Man-made) was the driver of the temperature or was it that temperature was the driver of the carbon dioxide? An international team of scientists (Dr. Nicolas Caillon from France, and others from Sweden, and Russia) examined the Vostok ice core database on a smaller time scale than the previous two Vostok Ice Core samples charts and made the following observational fact:

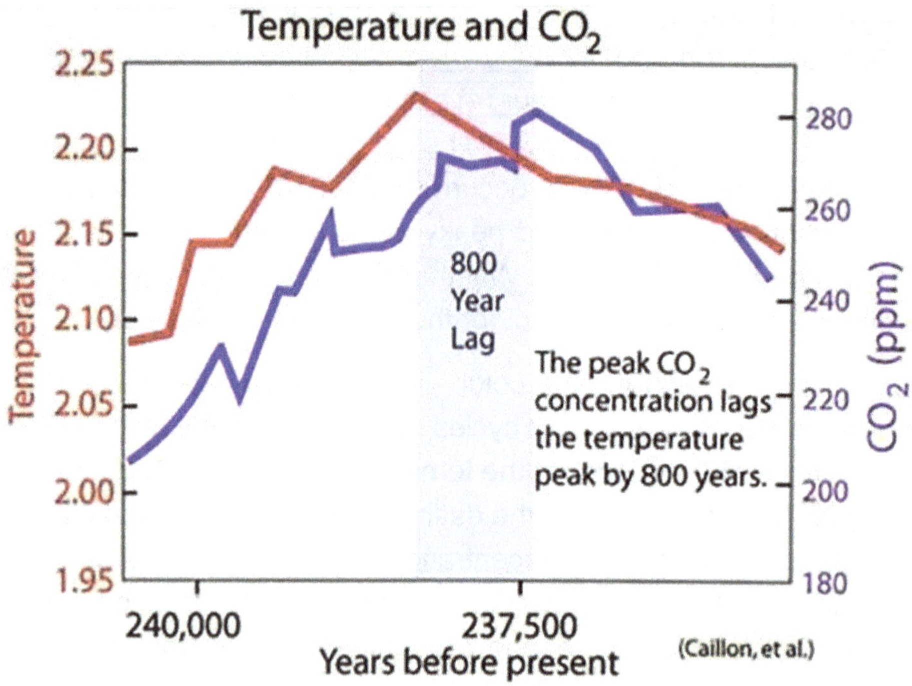

(Source: Smithsonian Institute for Science)

Dr. Caillon, et al. assessed the time relationship of the Vostok ice cores in the time period about 237,500 years before the present (at the end of the second cycle of the temperature and CO_2 database analyzed by the Author in the previous chart) and came to the conclusion (included in this book as being scientifically valid):

C7) The peak CO_2 lags the temperature peak by 800 years. Readers should interpret this conclusion to mean that it is temperature that drives the carbon dioxide in the atmosphere, not the opposite that carbon dioxide (natural or Man-made) drives the atmospheric temperature.

C8) Combining the preceding (**C1**) to **C7**) conclusions above lead the Author to state with scientific and logical confidence that the original environmental scientists' hypothesis: *"It is a settled science conclusion that, after eliminating everything else, Man burning fossil fuels is the dominant cause of global warming."* has been proven by valid scientific database analysis to be fallacious. Whatever technologies Man is developing now (to "combat" climate change will need to be adapted and augmented with other technology developments toward mitigating a climate environment with average global temperature 8 to 10 °C (14 to 16 °F) below the average global normal currently existing.

With the exponentially growing population and the highly likely need for a proportional increase in the demand for energy to heat rather than cool residences and businesses, future generations in the regions with higher latitudes will be faced with a true global cooling long-term "crisis" not a fallacious global warming near-term "crisis" of the entire planet. Increases in Man-made emissions and the additional coverage of large land areas with concrete (which traps heat and reduces drainage leading to occasional flooding) may have a minor effect on regional climate changes, but definitely not to a "global crisis" level of concern. With population growth, the increased areas of land to house the increased number of populists and the increase in infrastructure to support that increased population will cause yet larger areas of land to be covered by concrete. Controlling Man's emissions are worthwhile to do, but only from a pollution avoidance perspective not as a control of climate change perspective.

The daily weather reports are important to the population for inferring what type of clothing should be worn for the day and the probability of what type and amount of precipitation could occur and alter the travel time to work or school or a special event. When planning to travel some distance to visit a relative on an important holiday or family-related event, the weather forecast for both the point of departure and point of destination is provided to the public in useful weather forecasts in the news media:

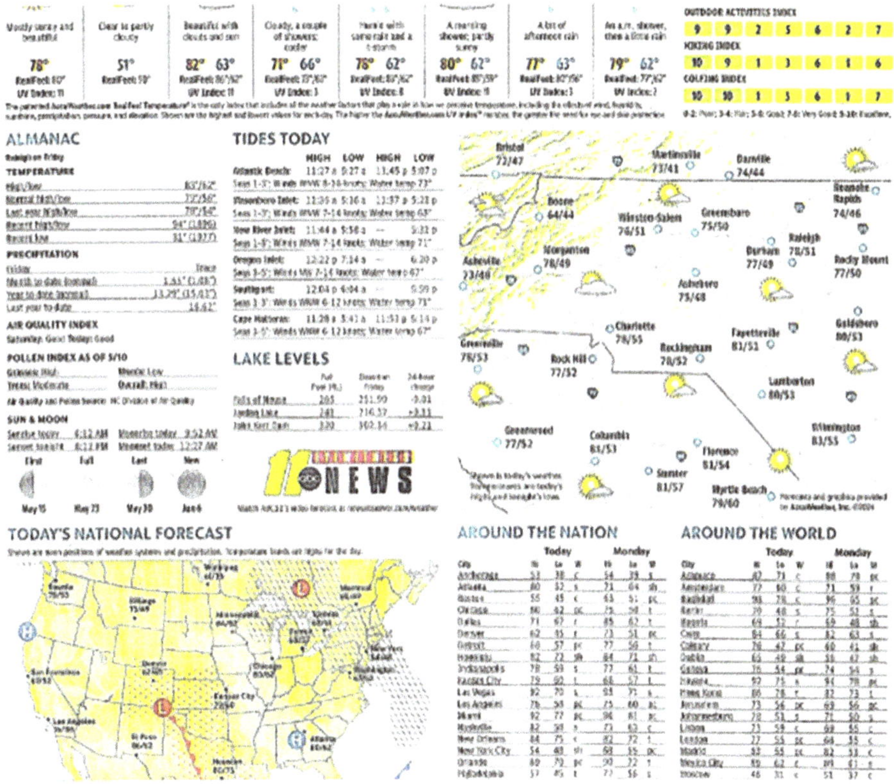

(Source, The Herald-Sun, Durham, North Carolina; May 12, 2024)

As a piece of trivia, the letters N, E, W, and S in "news" refers to the primary points of the compass as well as to new happenings in the four quadrants of the United States– North, East, West, and South. That original acronym has become the common term "news" to designate anything out of the ordinary – i.e., the new events - that are happening anywhere on the planet. Weather is the news; seasons are seen on the calendar pages, and climate is the historical record of the daily and weekly weather and seasonal news.

The claim by some environmental scientists that "climate change" is having a "feedback effect" on the frequency and severity of natural phenomena (hurricanes, tornados, wildfires ignited by lightning strike, typhoons, cyclones, etc.) is a fallacy just as history does not have a feedback effect on current events. The effect of temperature (a common variable in weather and climate) is an effect or an "interaction" not "feedback". The analogy given earlier that the weather is like the second hand on a clock, the seasonal effect like the minute hand, and the climate like the hour hand is more than a "cute" analogy, it is factual and shows that weather and seasonal effects are additive in determining the climate – the climate (hour hand) has no influence on either the weather (second hand) or the seasonal effects (minute hand). Climate is a result, not a cause. The temperature and barometric pressure of the atmosphere (which affect the water vapor in the atmosphere and thereby the precipitation leading to droughts or floods) are interactive factors with the weather and the seasonal effects. Therefore, the climate does not have a "feedback" effect causing droughts or floods, only a possible interactive effect. The term "feedback" "or "static" is used in electronics and acoustics such as at outdoor Taylor Swift performances if a speaker is positioned improperly with a transmitter and produces an acoustic interaction (varying in perception of the frequency of the composite wavelengths as a hum or a screech) more properly termed as "noise" – in engineering, it is not "feedback".

C9) No scientifically valid data was found confirming a feedback effect of climate change on the frequency or severity of natural and seasonal or out-of-season natural phenomena. The effect of changing atmospheric temperature or barometric pressure on the land surface area and the ocean waters.is an interactive effect not a feedback effect. What results from the lack of, or too much, precipitation (drought or flood) may later become a record in climate history (meteorology) but climate does not have a feedback that can change the present weather not yet recorded in climate history.

Probably over 99% of the public are interested in what the weather will be like in the coming week. Less than 1% are interested in a crystal ball forecast of what the climate will be 50 years from now and to which future generations may be exposed. Psychologists are unable to change the public attitude of *"Live today for tomorrow you may die*". From the human genealogy perspective, there are three generations per century in a family group that are of importance – the parents' generation, the children's generation, and the grandchildren's generation, each of which will by necessity adapt to the climate during that generation's life span. The public, however, anticipates the correlation between the date on the calendars and the cyclical seasonal events that are changing insidiously. In short, humans adapt to the weather and climate changes daily, seasonally, and regionally with little regard for what future generations in an adjacent region or in a different hemisphere are forecast to be exposed.

The many photographs shown in this book capture "Polaroid" snapshots of those very dynamic events as they are unfolding and capture the massive damages they cause to the environment and injuries/fatalities to humans and animals. Added to the data and factual information which the contents of this book (and the Author's interpretation of that enormous database) provided to the public is the reality of the world surrounding us. Environmental scientists have ignored those many dynamic factors of astrophysics, geophysics, and natural dynamic phenomena and have been shown in this book to have focused only on Man burning fossil fuels as the dominant factor causing the temperature and climate to change by one or two degrees in a decade. The tongue-in-cheek quip that *"We are choking on the minnows as the whale is swallowing us*" is unfortunately a very somber realism describing how fallacious is the environmental scientists' hypothesis: *"The science is settled. By eliminating everything else, the conclusion is reached that Man burning fossil fuels is the dominant cause for global warming and climate change."* Having read the contents of this book before arriving at this point, Readers should have asked the rhetorical question to themselves: "On what basis and with what logic was "everything else" - except Man burning fossil fuels – so quickly eliminated?" A corollary question should have been "Why was vital information eliminated – or concealed?" The answer is best answered by those who cherry-picked" the entire database and presented only that which was intended to mislead the public and government regulators.

The Author reviewed the differences stated and published by different groups of scientists in research journals of the changing temperature of the atmosphere and noted the large difference between various professional areas of findings about the cause for the temperatures (in the atmosphere and on the ground) and climate on the planet's surface. Those differences triggered this Author to independently research several databases expressing scientific interpretations of atmospheric behavior and changes in the climate which existed in the nineteenth century and early twentieth century and still exist between and among different scientific groups.

What became obvious to be missing from the environmentalists' concept of the climate was the inclusion of what was and continues happening below the surface of the planet that was and continues affecting the surface conditions. This book captured the more significant findings identified "along the way" during several decades of research and technology advancement. For these Concluding Comments, only the highlights of many of the major findings during the research process were summarized. This book became more of a status report than being a final analysis report and should be regarded as such. Readers who are interested in a more in-depth explanation of the significance of the many findings will find the scientific details within one of the previous chapters – in text form and in the form of numerous pertinent photographs and illustrations to attach "reality" to the accompanying narrative explanations.

The text within the main body of this book is the Author's interpretation of the facts – accompanied by the photographs and examples of what the public has been shown and told by the news media and by other biased documents (which are the result of "cherry-picking", "distorting", "over-exaggerating", and/or "masking" the facts.) A quote by Sir Walter Scott that sums up the findings in this book concisely is: "*Oh what a tangled web we weave, when first we practice to deceive.*"

On a more constructive note, challenges bring opportunities. To resolve the more important challenges, numerous technologies were initiated or improved, and some existing technologies were found in need of being updated partially or in total. Findings are of little value if they are not acted upon, and technologies are the actions to be applied in resolving the "big picture" problems as they occur. A somewhat amusing description of humans was heard some years ago that some people have personalities of a common chameleon – changing "colors" to fit in with the surrounding societal environment.

The Author has written each major chapter in this book in a somewhat stand-alone format and content. The most important facts may be found repeated in various chapters but that is the interactive nature of the problem of global warming, climate change, and the ocean levels rising. The causes and effects, the "pros and cons" of technologies, the important and less important facts, and the Author's emphasis and deemphasis are difficult to summarize – they do not sort nicely into the *"I see the cat. The cat sees me."* level of terseness. The Author does not apologize for the facts – only for the hutzpah in daring to reveal them. With a plea "Please do not shoot the messenger", the many conclusions being described came from the many databases that were analyzed.

Following the sequence of steps in order of conduct in accident investigations, the significant global temperature and/or climatic databases were analyzed in chronological order. Starting with the oldest of the databases first, a scientifically important climate model of the planet was formulated by Dr. Wladimir Köppen (1846-1940) of Russia, Germany, and Austria, and published in 1884. Dr. Rudolf Geiger (1894-1981) of Germany updated the original Köppen climate model in 1954 and again in 1964. The color-coding of the Köppen-Geiger climate model they developed and maintained as the climate changed (from all causes) was explained in detail in a previous chapter of this book. A summary illustration of the Köppen-Geiger color code is shown below. Their climate model with color-coded climate categories therein appears as shown next (followed by an application of their color-coded climate model applied to the planet Earth and then applied to the United States region):

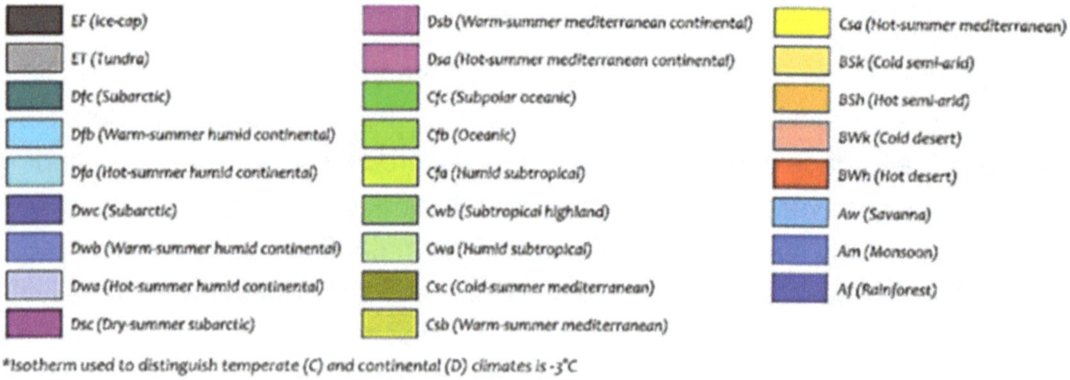

(Source: Encyclopedia Brittanica)

Categorizing the climate of a region was developed from the scientific inquisitiveness and extensive travel of Drs. Köppen and Geiger to almost all regions of the world. The realism of the actual meteorological and climatic environment at full scale in each region was obtained on-site, not estimated by any small-scale experiments in a controlled laboratory environment. In contrast, environmental scientists use the small-scale experiments (observing the heat trapping characteristics of a high concentration of carbon dioxide in a beaker in a controlled laboratory environment) of 1900-1960 vintage with impressive accuracy and extrapolate those experiments to pronounce an "average global temperature" and "average global climate change" due to Man burning fossil fuels and thereby creating "greenhouse gases" in the enormous atmosphere. Readers need not be further lectured on the significant scientific differences between the Köppen-Geiger method and the current environmental scientists' method.

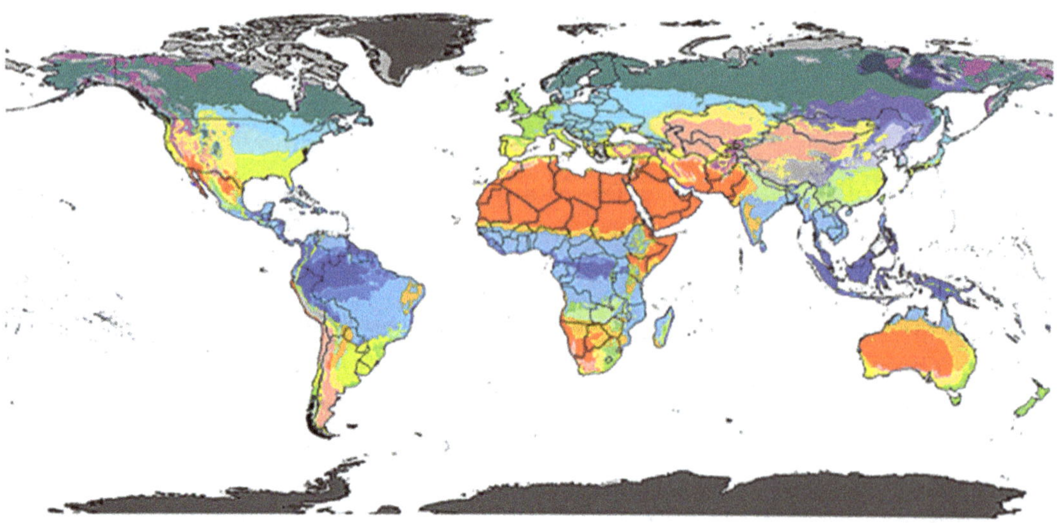

(Sources: Smithsonian Institute for Science and the World Meteorological Organization)

The extensive travel and documentation of Drs. Wladimir Köppen and Rudolf Geiger results in the identification of the climate of the planet varying (as shown above with the Köppen-Geiger color coding) by regions and sub-regions. Left unexplained were the scientific causes for the climatic differences. With primary interest in factually describing the changes which were occurring in the many regions of the World, the Köppen-Geiger climatic model revealed the climate differences within sub-regions of the United States region (shown below) – and with the same color-coding as used for the World climate variation by region.

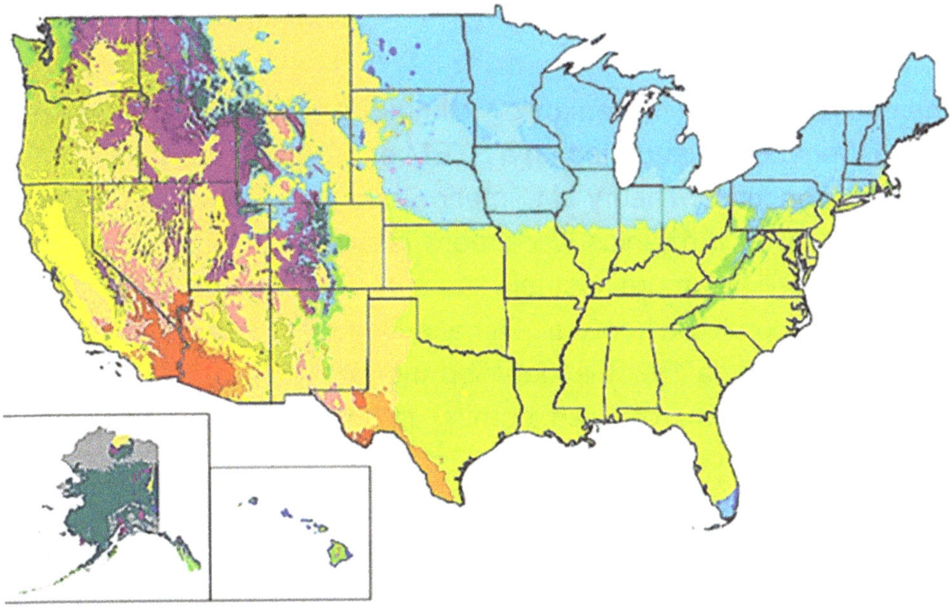

(Source: Smithsonian Institute for Science)

The above illustration was selected to be the front cover of this book because of its importance in showing there may be subregions of climate within a larger region.

There should be little question in Readers' minds as to which of the climate models better represents the climates of the regions and sub-regions of the planet. The question persists in the Author's mind: *"How much longer will it be until present day environmental scientists – and the UN Intergovernmental Panel on Climate Change (IPCC) - revert to the use of the Koppen-Geiger method for the "baseline" regional climates against which to assess ad apply realistic current levels of "temperature anomalies and climate change"?"*

To be observed with the United States climate modelling, the greens and shades of blue coloration represent the temperate climate of the eastern half of the "lower 48". Over 50% of the population of the contiguous United States resides in the eastern half (east of Houston, Texas) of the United States. About 65% live in the temperate West Coast states, Gulf Coast states, or Southeastern Coast states. With the use of inferential logic, the majority of the population of a region will choose to live in the more temperate climates and the more developed sectors of the region. The factors of climate, employment opportunities, and proximity to family are intertwined as to where the population is more dense than another.

C10) Regardless of whether it is a small or large region or a country or a continent, the fact is that the climate may be different at the same latitude because the many factors such as topography, prevailing wind and ocean currents, Hs and Ls on the weather maps, percent of cloud cover, etc.

are different at various points at the same latitude at the same time. It should be evident to Readers that the Köppen-Geiger Climate model is a much more professional description of the many different climate regions of the planet and in the United States than is the "average global temperature" method for describing climate change being championed by environmental scientists since about 1960. The Author questioned (in 1964 when the Keeling carbon dioxide measurements began to be released to the public, and now again in the first quarter of the twenty-first century) why current environmental scientists have not accepted that the climate is different in different regions (as Köppen and Geiger evaluated many decades ago in their color-coded evaluation of the climates in regions).

In 1924, mathematician Milutin Milankovic, climatologist Wladimir Köppen, and German astrophysicist Alfred Wegener collaborated on a STEM effort which was published as "Mathematical Climatology and the Astronomical Theory of Climatic Changes". The scientific community wished to believe that the findings of Dr. Arrhenius were more important than those of Milankovic, Koppen, and Wegener, and the brilliant more factually accurate works published in the "Mathematical Climatology and the Astronomical Theory of Climatic Changes" were scientifically diminished into obscurity. That time period (circa 1928) is identified then as the origin of the current but fallacious environmental science hypothesis of "Man burning fossil fuels is the dominant cause of global warming and climate change".

In 1939, as the political environment in Europe was leading to turmoil and the threat of war on a large scale, the brilliant astronomer and mathematician Milutin Milankovic (1879 – 1958) produced a calculation database describing the astrophysics and dynamics of planet Earth in the solar system. A visual summary of the dynamics of the planet that are happening and in combination and at different periods: Precession – the wobbling of Earth's rotational axis tilt angle; Tilt of the rotational axis of rotation of the planet varying several degrees (between 21.5 degrees and 24.5 degrees, now at 23.5 degrees and decreasing) cyclically; and Eccentricity of the irregular elliptical orbital path the planet follows in each annual orbit of the Sun – are summarized in the illustration below:

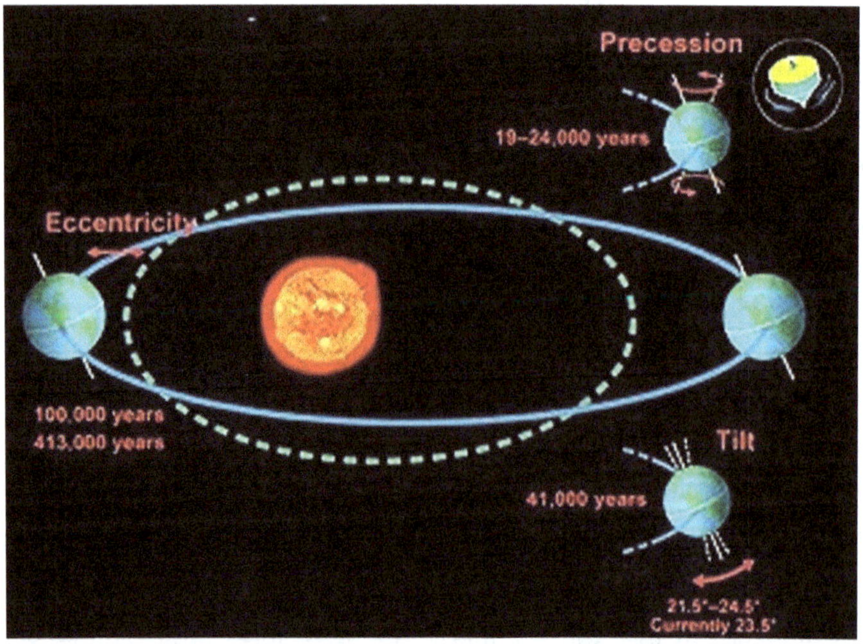

(Source: Smithsonian Institute for Science)

The Milankovic team of mathematicians, employing the mathematician Johannes Kepler's (1571 – 1630) astrophysical laws of motion, calculated the distance changes, timing and synchronization of the astrophysical dynamics the planet exhibited during its irregular shaped orbit about the Sun, and which included explanation for the effects of seasonal temperature changes on Earth's temperature) has on the solar radiation on Earth. Readers should note that there is no "average" movement of Earth in its irregular orbit and precessing ("wobbling", as does a toy top) axis of rotation that results in seasonal and long-term variations of temperature on planet Earth.

C11) Earth's tilt angle variations alters the angle ("Obliquity") at which the Sun's radiation waves impinge on the Earth and is the cause for the Arctic regions to "normally" average about 10 °C (50 °F) while the "normal" temperature of the Antarctic averages about -35 °C (-31 °F). The "normal" average temperature of the Equatorial zone is about 30 °C (86 °F) and changes very little as Earth's tilt angle changes. Global "averaging" of temperature by environmental scientists masks both the latitudinal and cyclical seasonal effects which dominate the temperature and climate differently in different regions of the planet. An identified shortcoming of the Milankovic calculations is that they did not consider that the solar radiation passes through the planet's atmosphere containing a set of gases including carbon dioxide before it impinges on the surface of the planet and alters the solar radiation impingement termed as "insolation" (incoming solar radiation).

The Milankovic Cycles calculations reported in 1939 were done for latitudes of 55°, 60°, and 65° North (corresponding to northern Scandinavia and into the Arctic circle). The calculations were done more to explore warming changes in the cold latitudes north of Stockholm rather than exploring the effect of global warming in the more temperate latitudes of Europe. His classic report published in 1924 was "Canon of Insolation and the Ice-Age Problem", reflecting that Milankovic was addressing the warming of cold climates rather than the global warming effect on the latitudes below 55° North. Environmental scientists in more recent decades seized on that fact and have published rebuttals of the type "Why the Milankovitch Cycles Cannot Explain Climate Change".

The "squiggles" in the serpentine curves shown below are of different amplitude (height and depth of the peaks) at other latitudes but their spacing and repetition on their periodic frequency rate shown are of similar periodicity for latitudes other than 65° N.

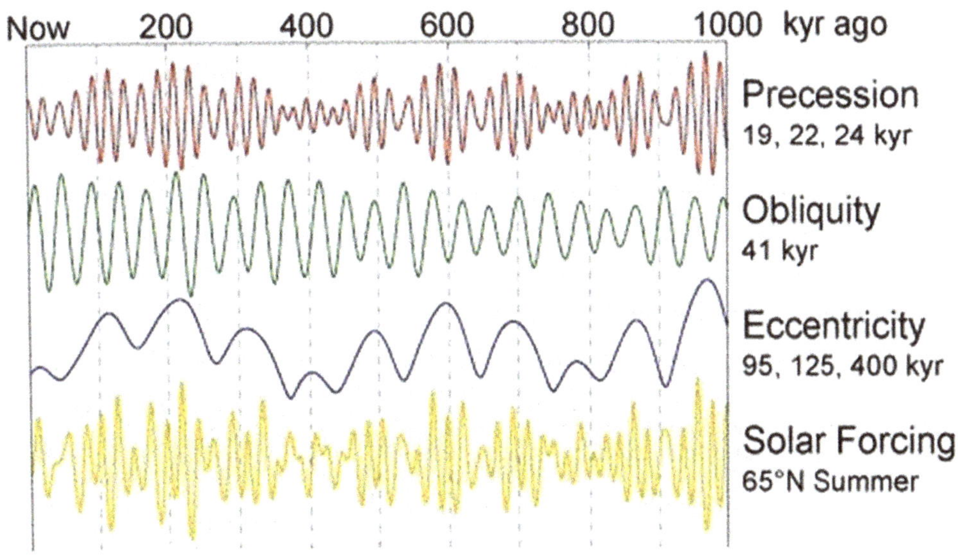

(Source: Smithsonian Institute for Science)

The magnitude of each of the three variations (precession, obliquity, and eccentricity) calculated by the Milankovic-led team of mathematicians is of relative unimportance, but more important is the cyclic variation of each being on periods of different length of time – in thousands of years - that are multiple in number and combine differently to "force" the solar insolation (bottom yellow) to vary in a "chaotic" manner. The astrophysical solar forcing calculations by Milankovic were found to correlate quite well with other databases (especially the Vostok Ice Core samples).

Comparison of Milankovic Cycles with Vostok Ice Station Data

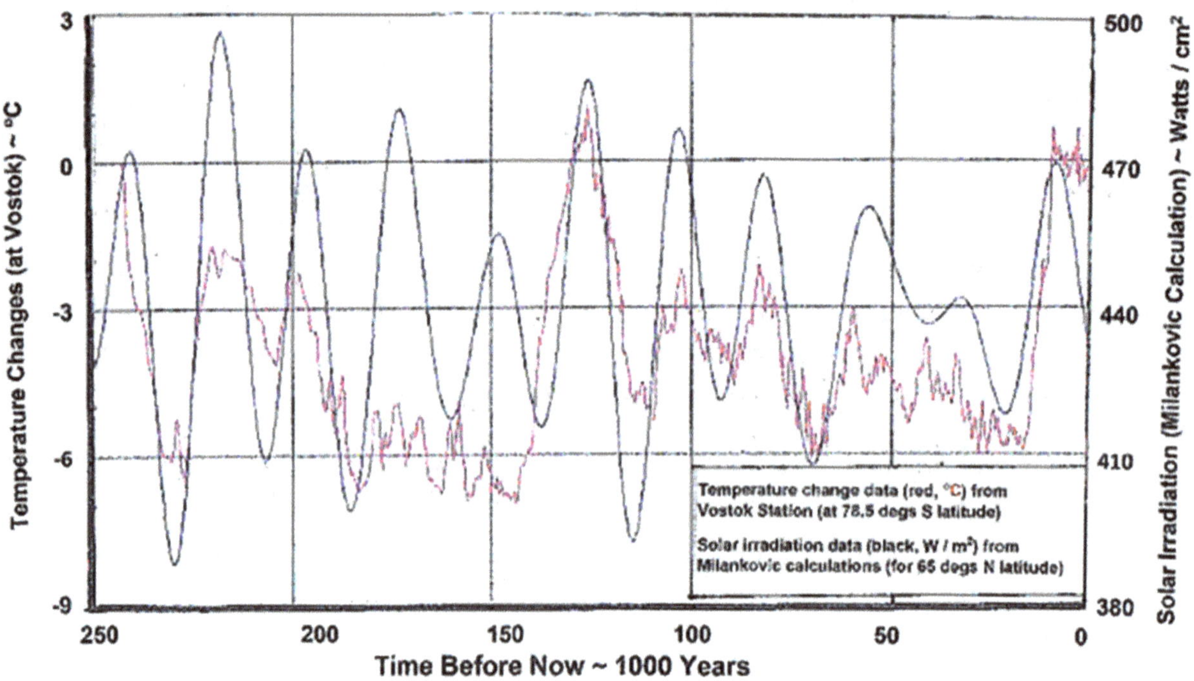

(Source: Smithsonian Institute for Science)

It must be commented that there was no digital computer technology existing at that time (1939) and required the Milankovic team of mathematicians to do manual calculations with great care. It also required them to actually think with their own brain and not rely on a computer to think for them. The computer is a very useful tool but is not a brain that can discern abstract subtleties in a database. The world of art does not give highest credit to the brushes used by the master artists (Rembrandt, Monet or Van Gogh) in painting their artwork. Artificial Intelligence is an emerging technology that holds much promise for assisting Man, not to replace him. Science is much more than learning which apps for averaged and linearized models to call up and which buttons to push on a keyboard.

The atmospheric gases measurements of Dr. Charles Keeling (1928 – 2005), starting in about 1960 at the Mauna Loa Atmospheric Observatory and continue on-going today by a team of NOAA scientists (which includes Dr. Charles David Keeling's son Dr. Ralph Franklin Keeling at the Scripps Institute of Oceanography in California) of the increasing level of "greenhouse gases" in the atmosphere has been discussed in several sections earlier in this book. It is that database that environmental scientists have used to support their hypothesis (which also is echoed repeatedly throughout this book "*When everything else is eliminated, it is a "settled science" conclusion that Man burning fossil fuels is the dominant cause of "greenhouse gas" emissions causing global warming.*")

C12) <u>It was the "Keeling Curve", starting to be published in science journals about 1960, that created the current tsunami (wave would be a gross understatement) of the environmental science segment that has used fallaciously "massaged" and "cherry-picked" data to misinform the public and government leaders around the world.</u>

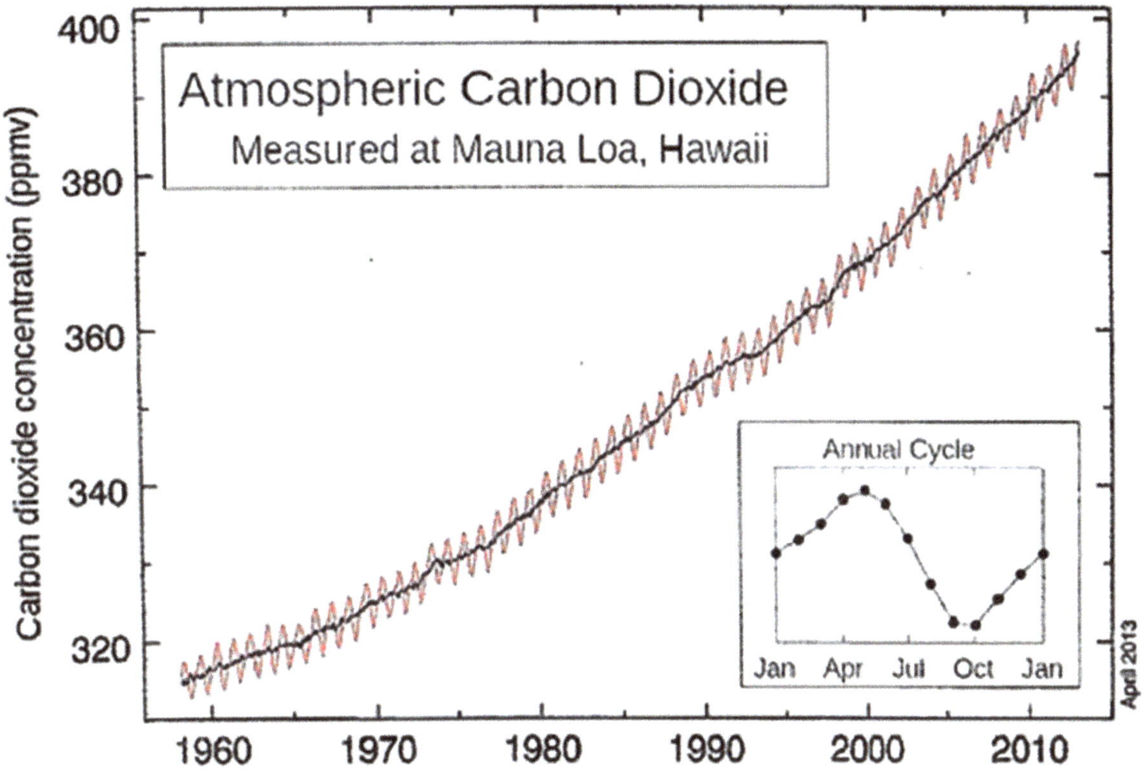

(Source: NOAA) Keeling Curve

Environmental scientists interpreted the "Keeling Curve" above and using inferential logic to mean what scientists wanted it to mean in misleading the public and governmental regulators – that Man was generating carbon dioxide into the atmosphere by industrial activities and those "greenhouse gases" were causing the "average global temperature" to rise at a "crisis rate". The public is not informed that the Keeling Curve is a composite of both natural carbon dioxide and Man-made carbon dioxide and the red line shows the seasonal effect on the measured data. The Keeling data are not misleading to scientists – but the environmental scientists' "spin" on how the data are presented to the public is not in keeping with the ethics of the science profession. When presented in a percentage form, the extrapolation of the carbon dioxide in the atmosphere (total of natural and Man-made) for the near-term is shown to be almost negligible:

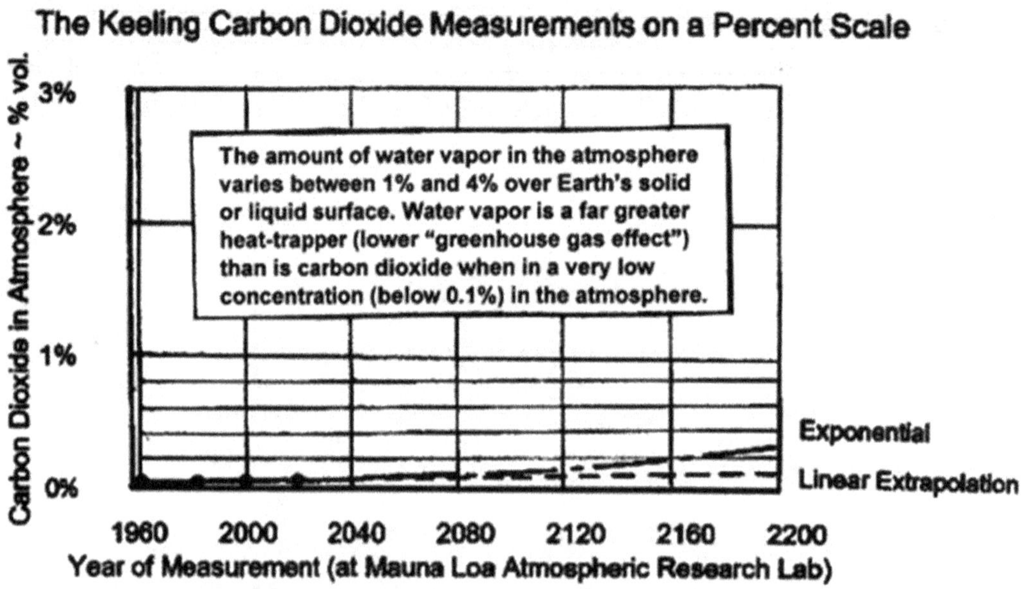

(Source: NOAA Database in alternate percentage format by Author)

The Vostok ice core database shows (being at the end of the fourth cycle) that if the planet is at the near-maximum peak of temperature the amount of carbon dioxide in the atmosphere without any Man-made addition is at 298 ppmv (0.0298%). The Keeling Curve indicates the rise in carbon dioxide to be about 400 ppmv (0.04%) of which 298 ppmv (0.0298%) is of other than Man-caused. With little mathematics involved, the rise in Man-made carbon dioxide from the pre-Industrial Revolution carbon dioxide (298 ppmv) in the atmosphere baseline to the present (ppmv) is 0.0102% of the atmosphere.

What the Keeling Curve also shows to a statistical analyst who researched the databases to determine what factors other than the industrial factor are in the Keeling Curve, the curvilinear (black line) variation of carbon dioxide - which is not the direct measurement of the total carbon dioxide present in the atmosphere – is the result after removing the effect of seasonal temperature changes on water vapor. Also not mentioned is that the effects of water vapor (about 3% at that latitude and elevation) have also been "adjusted" from the data. The direct measurements are those shown in red serpentine form and increasing with time and are the scientifically factual series of measurements made monthly. It should be noted that the measurements (in red) have a sinusoidal cyclic nature referred to by professional data analysts as "serpentine" because of its snake-like shape. The sinusoidal pattern (red), seen to be occurring one per year, is the seasonal effect on temperature varying caused by astrophysical effects – not predominantly caused by Man or by any of Man-related activities.

That misinterpretation was not realized by the Author until he researched another database – that of the Vostok ice core samples (shown earlier) - that shows it is temperature which affects the natural carbon dioxide in the atmosphere. The correct scientific conclusion to be derived from the Keeling Curve is that the red line variation shown is a total of all factors that affect the temperature and carbon dioxide content of the measurements. The red line measurements also include the seasonal effects (seasonal variation of "average surface temperature" shown by NOAA of the oceans waters) on water vapor in the atmosphere. Water vapor at a concentration of about 3% in the atmosphere is a far more efficient "heat-trapper" than is carbon dioxide at a concentration of only 0.04% (the equivalent of 415 parts per million by volume of total carbon dioxide) in the atmosphere. The black Keeling Curve results from the red line variation of carbon dioxide with the seasonal temperature and water vapor effects removed from the measured data. That is not an error in the science of the measurements by Dr. Keeling – but is an unprofessional attempt by environmental scientists to correlate the isolated effect of Man-made carbon dioxide on the "average global temperature".

C13) An important aspect of the Keeling Curve is that it represents the total carbon dioxide in the atmosphere being measured – comprised of that part which is naturally generated and that part which is Man generated. The natural (pre-Industrial Revolution) carbon dioxide in the atmosphere provided from another database (the Vostok Ice Core samples) shows that natural part to be about 290 ppmv (0.029%). Subtracting that natural part from the Keeling Curve measurements results in there being 125 ppmv (0.0125%) comprised of mostly Man-made carbon dioxide but containing the natural amount of carbon dioxide generated by the average global temperature rise since about 1900). Dr. Keeling did not remove the effect of seasonal variations on temperature and of water vapor to mislead the public – but the environmental scientists deceptively suppressed those scientifically-valid adjustments to the measured data under the cloak of the fallacious logic of the hypothesis: "*by eliminating everything else, settled science concludes that Man burning fossil fuels is the dominant cause of global warming*".

In keeping with the earlier caution given by the Author that databases must be updated and re-analyzed, a recent release of the measurements of carbon dioxide in the atmosphere reveals the more current status of that database:

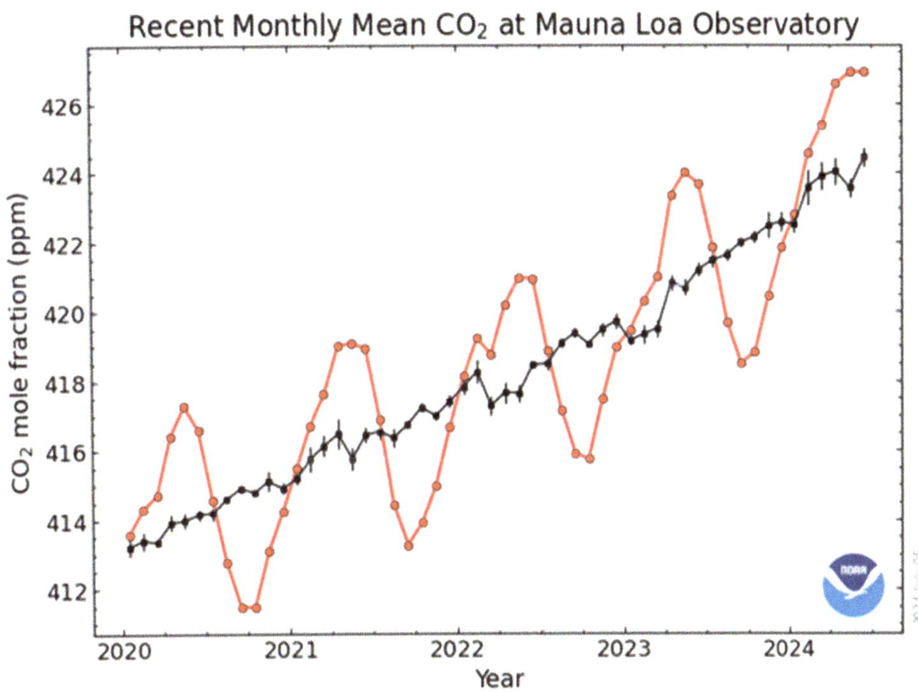

(Source: NOAA)

Readers are reminded that the red color measurements are the "raw" measurements containing the total CO_2 in the measurement and is a result of all causative factors (astrophysics, geophysics, natural, and anthropogenic). The black color line has had the effects of water vapor (which are dependent on seasonally varying temperatures) mathematically removed. In addition, the measurement in 2020 of CO_2 to be 413 parts per million by volume (ppmv) includes the "pre-Industrial Revolution" baseline (reference) which is all-natural. Subtracting the pre-Industrial Revolution baseline of 298 ppmv connotes there being 111 ppmv of Man-made CO_2 in 2020. Similarly, the measurement in 2024 of CO_2 being 424 ppmv translates to Man-made CO_2 being 122 ppmv in 2024. The rise of Man-made CO_2 of 11 ppmv in 4 years is an average of 2.75 ppmv (0.000275%) per year rise in Man-made CO_2.

The above chart with the raw and adjusted measured amounts of total CO_2 rising on a steep rate gives the Reader the illusion that there is a very rapid rise in Man-made CO_2. That is exactly the intent of environmental scientists – the visual perception of a steep slope inferring a "crisis". Environmental scientists have proliferated the misinformation of the very valid scientific measurements of carbon dioxide in the atmosphere and have intentionally caused the public to believe that "settled science" has proven that the Man-made carbon dioxide is on a steady increase with time (in the black line shown on the data record) since the Industrial Revolution. In fact, it is the red line seasonal variation of the measured carbon dioxide which contradicts that portion of the environmental scientists' hypothesis.

The raw CO_2 measurements are the red data points and contain water vapor in the measurement – the temperature effect on water vapor is not Man-made. By mathematically removing the "greenhouse gas" heat trapping effect of the water vapor, the red line with water vapor that is varying in temperature seasonally is smoothed out to the black line (without the effect of water vapor) and reflects only the increase in total carbon dioxide (natural and Man-caused) varying more curvilinearly in a slightly exponential manner.

The variation in temperature causes the water vapor in the atmosphere to vary in the same pattern as the temperature, but it is not without a limit. Heating water increases its water evaporative rate depending on the temperature of the water, The air surrounding the water however has a saturation limit as to how much water can be held in the air since gravity is acting on that mass. Science has developed that relationship between the temperature of water and the percent concentration of water vapor that the air can sustain before precipitation of the water occurs.

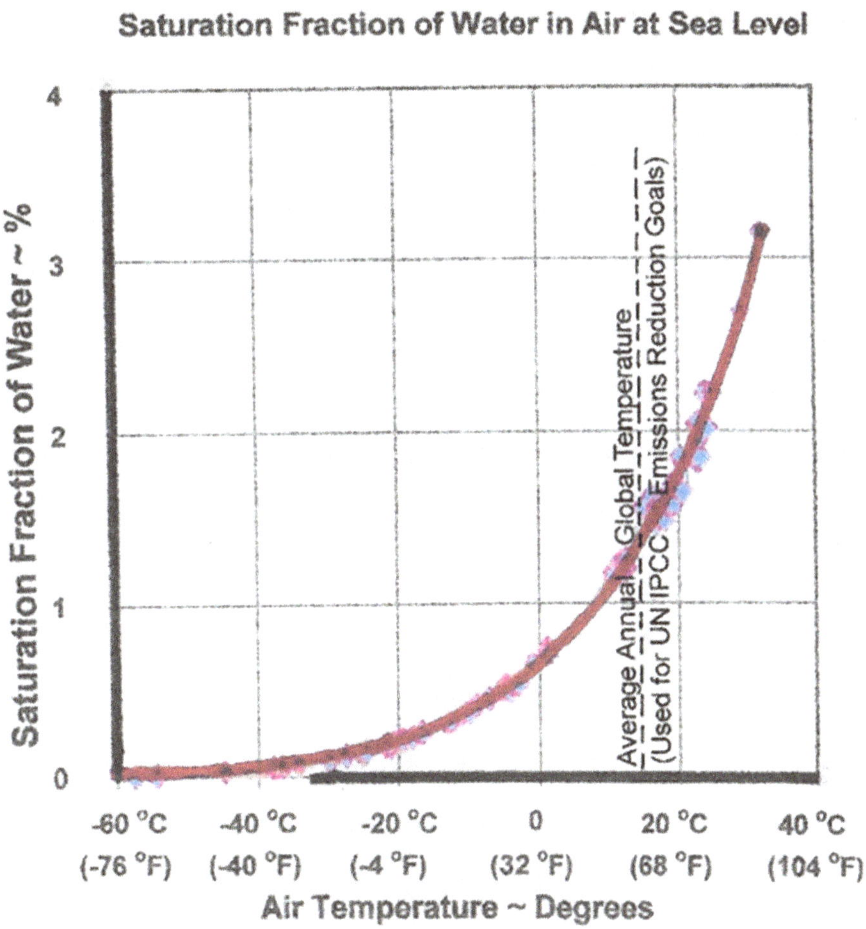

(Source: World Meteorological Organization)

This experimental data indicates that at a water temperature above 8 °C (the temperature at which is almost all ocean water between latitudes 80° above the Equator and 75° below the Equator), there is less than 1% of water vapor in the atmosphere. For colder water at -40 °C (-78 °F) or below, there is essential no water vapor evaporating from the water. Antarctica is categorized as a desert climate because there is almost no precipitation since there is no water vapor to produce precipitation. At a water temperature of about 32 °C (89.6 °F) - the highest point of the data) the maximum percent (saturation) of water vapor that can be held at altitude above a free-standing water surface is about 3.2%. The "average annual global temperature" used by environmental scientists and the United Nations Intergovernmental Panel on Climate Change is added (as a vertical dashed line at 16 °C (about 60 °F) to indicate the discrepancy in their setting of goals to limit the increase in that temperature by reducing anthropogenic "greenhouse gas emissions" - but at the same time neglect to consider water vapor as the far more heat-trapper, at an average concentration of water vapor being naturally in the atmosphere at 1.4%, at that average annual global temperature.

C14) Making valid atmospheric chemistry and physics corrections to the raw (red) data removes that water vapor and after corrections the CO_2 measured data adjusted to "dry air" conditions is made known to the public and entered into the "combat of climate change" theater of operations. In essence, all the many other factors causing the seasonal effect overshadows the singular Man-made Industrial Revolution effect on the measured CO_2 data. The measured CO_2 data are not wrong – the manipulations of the data is scientifically not wrong but the inferential logic does lead to an intentionally wrong interpretation. (In street language, that is "cooking the books" – and using global warming hype about Man burning coal as the fuel.)

With the current multiple technological activities toward reduction of Man-made emissions, the projection of the carbon dioxide concentration in the atmosphere will probably continue to increase but at a lower exponential rate. It is possible the concentration of both natural and Man-made "greenhouse gas" type emissions will cause a reduction in carbon dioxide in the long term (as the Vostok ice core samples project when extrapolated. None of the Readers of this book will be on the planet in 2200 so this Author is relieved to not have to "eat crow" if those extrapolations and projections are or are not as shown earlier.

C15) A rise of 0.000275% of any natural variable (such as water vapor) or Man-made variable (such as CO_2) in a year cannot summarily be eliminated from consideration but definitely does not support the environmental science claim that the increase of Man-made emissions of carbon dioxide in the atmosphere constitutes alarming the public that the rise is at a "crisis" rate. If the gas was a biologic nerve-gas instead of the gas carbon dioxide, that might be a valid concern.

The Milankovic calculations of 1939 (described earlier in this book) and other geophysical database information show that the wobbling, tilted axis of rotation of the planet as it travels in an irregular elliptical orbit (which in itself has varying eccentricity) causes the four seasons on the planet in which the surface temperature varies in a sinusoidal wavelength manner that correlates with each quarterly travel (of approximately 3 months duration) of one orbit of the Sun.

Another misleading act is not showing the fact that in the numerical mole fraction reflected on the left scale (vertical), there is about 295 ppm (shown in the Vostok ice core data at the peak occurring at the end of cycles 1, 2, 3, and 4) which is devoid of any Industrial Revolution effect. The Vostok ice core database has not been "corrected" to dry air conditions to remove water vapor – if that water vapor elimination correction had been made, how would the environmental scientists explain the formation of the ice in which the "impurities" in the core samples are embedded? In addition, the temperature of the atmosphere is driving the carbon dioxide levels in the ice core samples (shown by the 800 years lag of the peak in carbon dioxide after the temperature has peaked), just as the seasonal variation of water vapor in the atmosphere is seen in the Mauna Loa database (in the red measured data points of total CO_2 – i.e., containing water vapor, containing the pre-Industrial Revolution level of CO_2, and whatever has been added by Man since the start of the Industrial Revolution.

Despite satellite images from NASA becoming available and criticism by many fellow climate scientists (some of the books are identified in the References section near the end of this book), the environmental scientists have maintained that the one-size-fits-all "average global temperature" goals set at the 2017 Paris Accord to avert a global "crisis" must be met or a disaster will befall the planet. Left undefined were the critical questions of how critical the "crisis" is that is being alerted by the environmental science fraternity. There are no specific details provided as to who or what would be affected, and when (to the nearest century) would the threat of an apocalypse be experienced.

There is no question that the climate is warming – however, the climate change is different in different regions and to a different extent. The United Nations Intergovernmental Panel on Climate Change is well-advised to spend more effort and resources on updating the Köppen-Geiger method and applying that regionally instead of adhering to the fallacious goals established in the 2017 Paris Accord to limit global warming with the "line in the mud" one-size-fits-all "average global temperature" method presently in use. There is no discredit in having to update a technology or a climate model to account for changes that are continuing to happen in the real world – from both natural and human causes. The Author is appalled (and professionally deeply saddened) to learn that a Department of Environmental Justice has been created in the United States to legally force states to meet the fallacious Paris Accord goals to control atmospheric temperature. Climate change is a scientific topic, not a political or judicial football.

The environmental scientists in the Department of Energy are even more creative with the amount of Man-made carbon dioxide being introduced into the atmosphere. The following table is an extract from a Department of Energy report in October 2000:

TABLE 1.
The Important Greenhouse Gases (except water vapor)
U.S. Department of Energy. (October. 2000) (1)

(all concentrations expressed in parts per billion)	Pre-industrial baseline	Natural additions	Man-made additions	Total (ppb) Concentration	Percent of Total
Carbon Dioxide (CO2)	288,000	68,520	11,880 (2)	368,400	99.438%
Methane (CH4)	848	577	320	1,745	0.471%
Nitrous Oxide (N2O)	285	12	15	312	0.084%
Misc. gases (CFC's, etc.)	25	0	2	27	0.007%
Total	289,158	69,109	12,217	370,484	100.00%

(Source: US Department of Energy, DOE)

(Readers should carefully note that Man-made additions of carbon dioxide are 11,880 ppb (with parts per <u>billion</u> being expressed in font size half of the other lettering. (The "fine print" in insurance contracts is another example of how to "put one over".) The chart portrays the percent (99.438%) to show that carbon dioxide is almost the entire amount of "greenhouse gases" in the atmosphere. Not shown in percentage is that Man-made carbon dioxide is only 0.0322% (11,880 / 368,400 x 100) of the dry atmosphere. The notation in parentheses "(except water vapor)" is left unquantified but in the "real" world, water vapor averages about 3% of the planet's atmosphere – 93 times more water vapor than carbon dioxide. (3% = 3 pphv = 30,000 ppmv = 30,000,000 ppbv). When water vapor – the most powerful "heat-trapper" – is accounted for, the Man-made addition of total "greenhouse gasses" becomes almost negligible. This is yet another rebuttal of *"settled science"* concluding that *"By eliminating everything else* (which includes: astrophysics, geophysics, natural, and water vapor) *Man-made "greenhouse gases" are the dominant cause of global warming and climate change".*

The Vostok ice core sample database was not available to Dr. Svante Arrhenius in Sweden in 1902 who developed the environmental hypothesis that carbonic acid (a chemical compound of carbon dioxide combining with water) in the atmosphere had a heat-trapping characteristic (the "hothouse effect") and might cause the planet's surface temperature to elevate. Had the Vostok ice core database been available to Dr. Arrhenius, there is little doubt (being the professional and ethical scientist he was) he would have modified his conclusions to reduce the exaggerated current hype of "burning coal creates carbon dioxide and other trace amounts of gas having a major "greenhouse" effect". The fact remains, Dr. Arrhenius came to his "hothouse effect" copornclusion in 1902 that was later proven – by heuristic analysis using updated technology databases – to be correct at the laboratory scale and when in high concentration but overexaggerated for its effect at full scale.

C16) <u>When diluted to low concentration in the planet's total atmosphere, the effect of a 0.02% concentration of Man-made carbon dioxide in the atmosphere is nearly insignificant in the near-term, and only of significance at the present growth rate in the long-term. The probability that astrophysical, geophysical, and natural factors driving the temperature, which in turn drive the carbon dioxide in the atmosphere will, based on the four cycles shown in the Vostok Ice Core database reverse any global warming trend and the planet will enter into a fifth cycle beginning with a global cooling trend</u>. The actions the world population will need to take to mitigate global cooling and the possibility of another ice age are beyond the scope of this book to address.

The Author arrived at that same conclusion after researching the astrophysics database. NASA had measured the temperatures of the planets and their mean distances of the planets from the Sun to be as shown in the table below (in vertical format to fit the page):

Surface Temperatures on the Sun and Four Inner Planets & Earth's Moon

Planet	Distances from Sun ~ million km			Solar Radiation at Planet ~ watts / m^2
	Closest	Mean	Farthest	(Planet at Mean distance)
Sun	0	Reference	0	64 x 10^6
Mercury	46.001	57.9	69.817	9116
Venus	107.476	108	108.943	2611
Earth	147.098	149.6	152.098	1366
Earth's Moon	146.714	149.6	152.482	1366
Mars	206.655	227.9	249.232	589

Planet	Surface Temperatures (observed/measured) ~ degrees: Hottest			Mean			Coldest		
	K	°C	°F	K	°C	°F	K	°C	°F
Sun	5873	5600	10112	5780	5507	9941	4273	4000	7232
Mercury	703	430	806	398	125	267	93	-180	-292
Venus	746	473	883	737	464	867	726	453	847
Earth	344	71	160	288	15	59	184	-89	-128
Earth's Moon	407	134	275	263	-10	-14	120	-153	-243
Mars	293	20	68	208	-65	-85	133	-140	-220

Notes: K (Kelvin) = C (Centigrade) + 273 F (Fahrenheit) = 9 / 5 C + 32

1) Kelvin is the absolute temperature scale; coldest temperature in science is liquid hydrogen at/near 0 K

2) Water changes state from liquid to solid ("freezes to ice") at 0° C = 32° F = 273 K

3) Water changes state from liquid to gas (boils to steam") at 100° C = 212° F = 373 K

4) Carbon dioxide (CO_2) changes state from gas to solid ("sublimation" to "dry ice") at -78.5° C = -109.3° F = 194.5 K

(Source: NASA)

NASA had also measured gases in the atmospheres of the planets:

Atmospheres of the Planets (Courtesy of NASA)

Planet	Atmospheric Mass ~ Kilograms	Nitrogen	Hydrogen	Oxygen	Argon	Helium	Carbon Dioxide	Carbon Monoxide	Methane	Sodium	Other	* Water Vapor
Sun	3.0×10^{30}		71%			26%					3%	
Mercury	1,000		22%	42%		6%				22%	8%	
Venus	4.8×10^{20}		4%				96%					
Earth	1.4×10^{21}	77%		21%	1%		0.04%	0.01%			0.95%	0.1 - 4%
Moon	100,000				70%	21%				1%		
Mars	2.5×10^{16}	2.7%			1.6%		95%				0.7%	
Jupiter	1.9×10^{27}		89.8%			10.2%						
Saturn	5.4×10^{26}		96.3%			3.2%					0.5%	
Uranus	8.6×10^{25}		82.5%			15.2%			2.3%			
Neptune	1.0×10^{26}		80%			19%			1%			

Notes

* Earth is the only planet having significant water on its surface or water vapor in its atmosphere.

Earth's surface is covered by more than 75% of its area with water (in its oceans, seas, gulfs, bays, lakes, rivers, etc.).

(Source: NASA)

It was noted by the Author that the NASA database (two tabulations above) indicated Venus had an extremely high surface temperature and had an extremely high (96%) concentration of carbon dioxide (CO_2) in its atmosphere. Earth was seen to have a much lower surface temperature. Using Johannes Kepler's astrophysics principle that the temperature of a planet is inversely proportional to its distance from the Sun, and by referencing the ratio of the temperature (in Kelvin units) of the planets Mercury, Venus, Earth (and its Moon), and Mars to the temperature of Mercury (the closest planet to the Sun), a graphical summary of those facts was generated by the Author which showed Venus to be very much deviant from Kepler's distance line.

Effect of Atmosphere on Inner Planets' Temperatures

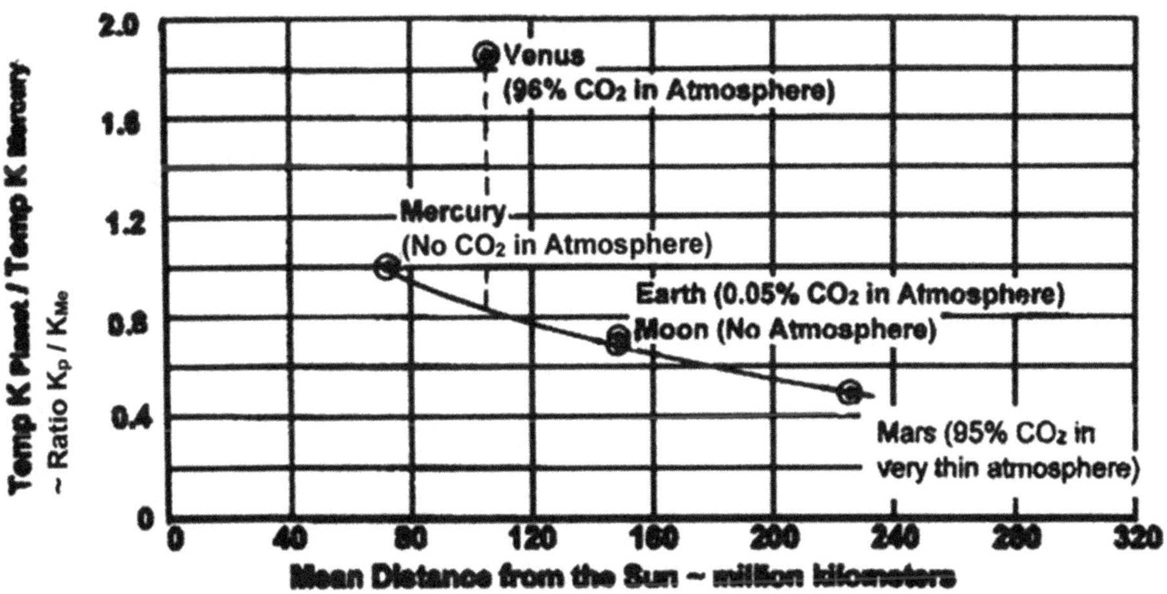

(Source: Author generated chart using NASA database on planets)

Assuming the NASA astrophysical measurements are correct, there is the most scientifically plausible explanatory reason for Venus to have the anomalistic temperature and deviating so greatly from the Kepler distance/temperature variation is due to the high percentage (96%) of carbon dioxide in its atmosphere. Whereas, planet Earth (and its Moon) has about 0.04% carbon dioxide in its atmosphere from all causes (natural and Man-made) and is on the Kepler distance/temperature variation. That conclusion by this Author also brings Dr. Arrhenius' conclusion of near-term low concentration of CO_2 in its atmosphere being potentially beneficial, whereas Dr. Arrhenius' projection that in a few 10,000 years in the future, the CO_2 in the atmosphere would be at a high concentration and cause Earth to not be habitable by humans.

If NASA recants its database on the planets and/or environmental scientists have a different interpretation of the NASA database, this Author would be most interested in learning that alternate explanation. Pending such information, it must be concluded that

C17) a low concentration below 5% (equal to below 50,000 ppmv) of carbon dioxide in the atmosphere of a planet will have minimal effect on the planet's surface temperature and human safety. If the concentration of CO_2 in the atmosphere of a planet is above 5% (above 50,000 ppmv) concentration, that concentration should be of major concern to human safety.

Approximately six decades later (circa 1970), environmental scientists, using more accurate and more complete measured meteorological data obtained from more measurement stations and with improved technology measurement instruments, derived their "settled science" hypothesis: *"When everything else is eliminated, Man burning fossil fuels is the dominant cause for the surface temperature to be rising"* (and the climate to be changing). In that category of "*everything else*" being eliminated were the astrophysical (confirmed by the Milankovic calculations), the geophysical (confirmed by the Vostok ice core database), and the natural (confirmed by satellite imagery and measurements) databases.

An article in the Climate section of a Science magazine gave an indication of what is factually troubling the environmental scientists about global warming – of the planet's land surface and the warming of the oceans which cover more than 70% of the planet's surface. The Author took the liberty of red underlining the emotionally driven statements that were made in "Grim results of the ocean's hot streak" article by a very passionate young scientist regarding the average sea surface temperatures:

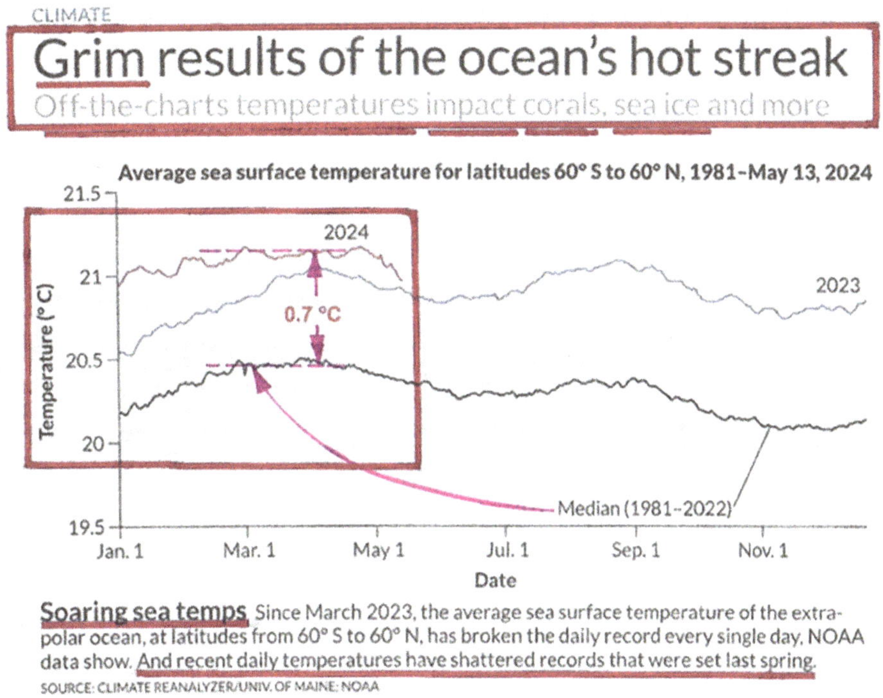
CLIMATE

Grim results of the ocean's hot streak

Off-the-charts temperatures impact corals, sea ice and more

Soaring sea temps Since March 2023, the average sea surface temperature of the extra-polar ocean, at latitudes from 60° S to 60° N, has broken the daily record every single day, NOAA data show. And recent daily temperatures have shattered records that were set last spring.

SOURCE: CLIMATE REANALYZER/UNIV. OF MAINE; NOAA

(Source: NOAA)

Using sophomoric analysis methods of "averaging" and "linearizing" – almost nothing in the natural world and real environment moves linearly or in an average manner – environmental scientists created a fictitious "crisis" of Man burning fossil fuels that were 1) generating "greenhouse gases" which in turn were raising the extremely simplified one-size-fits-all "average global temperature" of the land surface; 2) creating the "devastating" concentration (Man-made CO_2 of 0.02%) of "greenhouse gas" in the atmosphere; and 3) the "soaring", "shattering records", "average" oceans' water surface temperatures" (which NOAA data reveals rose 0.75 °C in the past 40 years) everywhere on the planet.

A recently published NASA satellite database for the temperature anomaly of the planet in 2018 to 2022 compared to the 1951 to 1980 "average global temperature" is shown next:

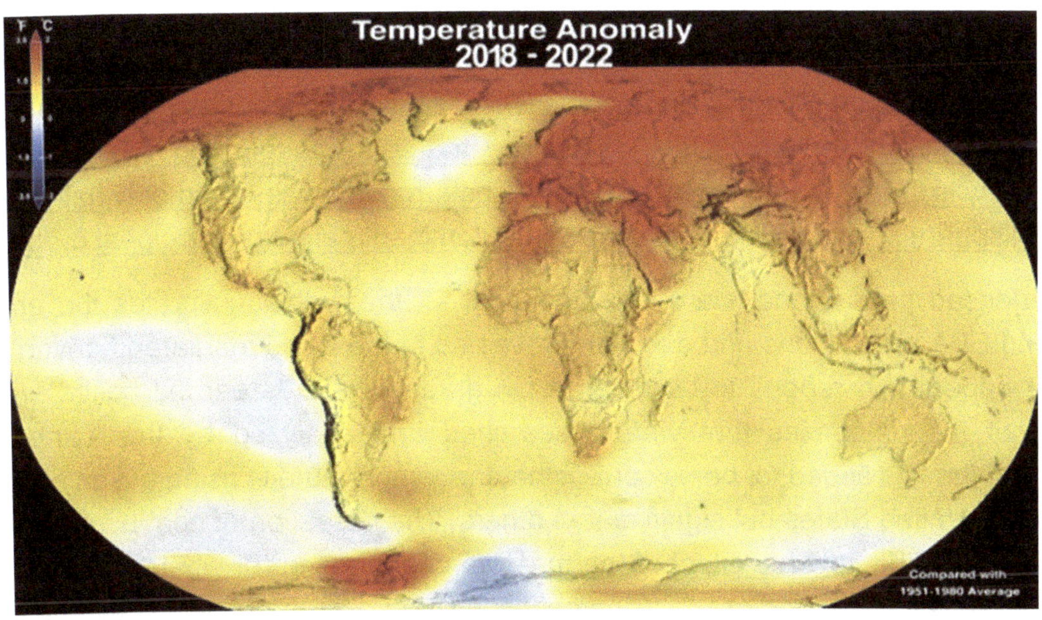

(Source: NASA)

The amount of red indicates a temperature anomaly of 1 °C (1.8 °F) is absent in the United States and most of Canada, but apparently exists for most of Europe, the Mid-East, northern Asia, in the entire Arctic region, and a small portion of western Antarctica. Rather than a conclusion, (but given a conclusion (**C**) number for tracking purposes), there are multiple questions left to be clarified by the environmental scientists comprising "settled science":

C18) Three questions arise: 1) What is the basis for global warming to be declared at "crisis" level of severity for the entire planet? 2) Does the EPA of the State of California realistically expect that mandating the sale of all-electric vehicles by 2035 on California roads to eliminate "greenhouse gas emissions" will alter the distribution and/or level of the temperature anomaly in the regions shown in the chart above?) 3) If a 2 °C (3.6 °F) rise in atmospheric temperature causes oceanographers to declare a corresponding temperature rise of the ocean waters will cause the coral reefs to bleach white, is that the basis for environmental scientists in the UN IPCC to base their claim that a 2 °C (3.6 °F) rise in average global temperature constitutes a "crisis" to the entire planet?

The Author made an observation in the form of a scientific question when analyzing the Vostok Ice Core database if the temperature variations and the carbon dioxide variations with time were applicable to that latitude of 78° South but perhaps not applicable to other latitudes? The above NASA satellite update on the temperature anomalies clearly answers the question the Author posed (and which must be in the mindset of every scientist who considered the Vostok Ice Core database) of whether surface temperature anomalies and carbon dioxide in the atmosphere are latitudinally and regionally sensitive. The analogy of "turning over a stone led to the finding of other stones needing to be overturned", the Author's original question of the latitudinal effect on temperature anomalies in the Vostok Ice Core database spawns five new questions deserving answer to other scientists and perhaps the public by environmental scientists: The five new questions are grouped into one conclusion C number for tracking purposes:

C-19) If the "Man burning fossil fuel causes average global temperature change" hypothesis was/is correct: 1) Why is there a deep red color (1 °C temperature anomaly) in the region (Goreland) above the Arctic Circle where there is almost no coal being burned? 2) Why is there a deep red color in the latitudinal (40° to 70° N) region of Manchuria and Siberia (where there are few industries burning fossil fuels) but no deep red in that same latitudinal zone (40° to 70° but S of the Equator? 3) Why is there almost complete deep red in Europe and western Asia but almost no red in the United States and Canada (at nearly the same latitude) where there is coal-burning? 4) In a "big picture" perspective, why is there so much deep red in the Eastern hemisphere and only half as much in the Western hemisphere if "greenhouse gases" are the dominant cause of the global problem?

As an experienced professional statistical data analyst, the Author calibrates the environmental scientists' and EPA conclusions and decisions revealed in the news media and government (state and federal) papers and reports to be given a D+ grade. The D is earned based on the overall assessment of the databases that were researched and analyzed by the Author and found incomplete and the + is added for being correct that the temperature is rising but on a regional basis – minimal in the United States but significant in others. The cessation of coal-burning in the United States will not undo the adverse effect of the massive coal-burning (or whatever other factors are) causing the highest regional temperature increases in Europe, in the Mid-East (where coal-burning is nil), and in northern Asia. A creative reporter in a newspaper article advised the readers: *"Never forget that the coal burned in Virginia causes temperature rise in Viet Nam and Venezuela."* A nicely presented thought but factually fallacious.

Most daily newspapers show the high and low extremes of the temperature and an estimate of the weather projected for the "Today and Tomorrow" forecast for the Nation (USA) and the World (an example is shown below): The list is presented in alphabetical order of the names of the cities. If the listing of the cities was arranged by latitude (most northern at the top and descending to most southern – but without the latitude numerically listed - readers would perhaps notice that there is a general effect of latitude on the temperature. (At present, the list in alphabetical order is about as interesting as a page in a telephone directory – if the public remembers what a telephone directory was in the "old days".)

AROUND THE NATION

	Today			Monday		
City	Hi	Lo	W	Hi	Lo	W
Anchorage	34	30	i	36	31	i
Atlanta	49	31	s	50	35	s
Boston	22	13	s	27	23	pc
Chicago	35	32	pc	40	30	c
Dallas	60	52	s	64	57	c
Denver	62	36	pc	56	31	c
Detroit	28	24	pc	35	32	sn
Honolulu	83	71	pc	83	71	pc
Indianapolis	35	26	pc	39	33	r
Kansas City	50	38	pc	50	32	c
Las Vegas	64	46	pc	68	46	s
Los Angeles	68	48	pc	65	51	pc
Miami	74	64	pc	76	67	sh
Nashville	45	30	s	54	40	s
New Orleans	56	45	s	67	55	s
New York City	26	17	s	32	27	s
Orlando	70	55	s	73	59	c
Philadelphia	29	16	s	32	28	s
Phoenix	76	51	pc	74	50	pc
St. Louis	46	37	s	52	37	r
Salt Lake City	47	38	c	48	35	c
San Francisco	62	56	c	64	55	c
Seattle	51	46	r	54	47	r
Washington, DC	35	21	s	37	29	s

AROUND THE WORLD

	Today			Monday		
City	Hi	Lo	W	Hi	Lo	W
Acapulco	87	73	pc	86	72	t
Amsterdam	45	42	r	47	39	pc
Baghdad	65	38	s	67	38	pc
Berlin	45	35	sh	40	35	r
Bogota	67	49	r	67	48	r
Cairo	70	55	c	68	52	s
Calgary	43	29	pc	44	26	pc
Dublin	44	36	pc	52	50	c
Geneva	46	33	r	37	28	sn
Havana	72	57	c	79	59	pc
Hong Kong	65	56	pc	65	55	sh
Jerusalem	61	44	c	61	43	s
Johannesburg	74	61	t	74	59	t
Lisbon	62	49	s	61	43	s
London	45	38	pc	45	42	s
Madrid	54	37	pc	53	36	pc
Mexico City	64	46	sh	66	47	pc
Moscow	29	20	s	34	31	c
New Delhi	72	49	pc	64	53	pc
Paris	48	39	sh	45	36	s
Rome	56	42	r	51	36	s
Stockholm	38	31	c	34	23	i
Sydney	80	67	pc	81	60	pc
Tokyo	54	38	pc	52	38	s

Weather(W): s-sunny, pc-partly cloudy, c-cloudy, sh-showers, t-thunderstorms, r-rain, sf-snow flurries, sn-snow, i-ice

The average member of the public (residing at any latitude) does not panic if the weather report in the daily morning newspapers forecasts there will be a difference of 25 - 30 degrees Fahrenheit from a Lo in the morning to a Hi in the early evening, or if the weather (W) is forecast to be snow (s) or rain (r) – the meteorological forecast is used as a guide as to what clothing to wear for the whole day away from his/her domicile. "His" or "Her" clothes closet probably also includes too many outer clothes items to survive the seasonal weather difference of 50 – 60 degrees Fahrenheit average temperature variation between summer and winter.

A 2 °C rise in the average global temperature in the next 10 years is a non-crisis to the human population regardless of season unless the Heat-Index for the extreme hot temperatures of summer in a region is exceeded for a few days. The Heat Index and Wind Chill factor indexes are shown again as a reminder that it is the extremes of temperature and humidity or wind that the human population must adapt to on a regional basis. The Heat Index is of no concern to those humans residing above latitude 70° North or below 70° South. The Wind Chill Index is of no concern to humans residing between 15° North and 15° South of the Equator. The concern for a heat-wave or a cold-wave preempt the concern for whether the "average global temperature" will be exceeded by 2° C or not by 2050.

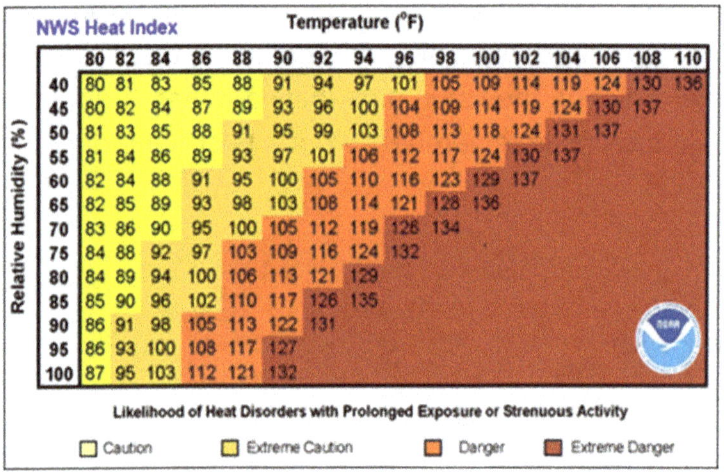

RH (%)	80	82	84	86	88	90	92	94	96	98	100	102	104	106	108	110
40	80	81	83	85	88	91	94	97	101	105	109	114	119	124	130	136
45	80	82	84	87	89	93	96	100	104	109	114	119	124	130	137	
50	81	83	85	88	91	95	99	103	108	113	118	124	131	137		
55	81	84	86	89	93	97	101	106	112	117	124	130	137			
60	82	84	88	91	95	100	105	110	116	123	129	137				
65	82	85	89	93	98	103	108	114	121	128	136					
70	83	86	90	95	100	105	112	119	128	134						
75	84	88	92	97	103	109	116	124	132							
80	84	89	94	100	106	113	121	129								
85	85	90	96	102	110	117	126	135								
90	86	91	98	105	113	122	131									
95	86	93	100	108	117	127										
100	87	95	103	112	121	132										

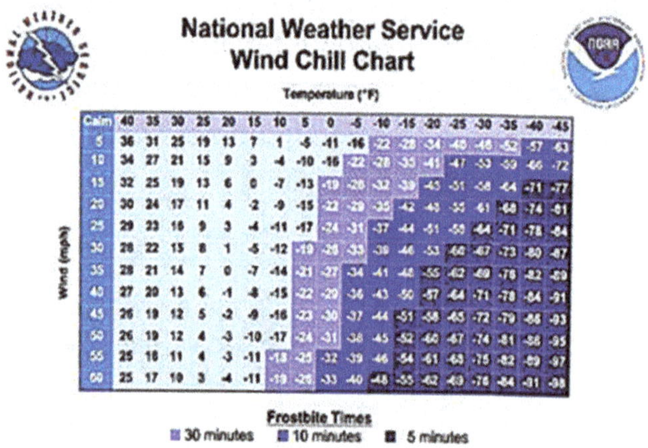

(Source: NOAA National Weather Service)

After noting on the front cover graphic that the bands of color representing numerical temperatures were latitude dependent, the Author was curious to investigate if the effect of latitude was a scientific-based trend or merely coincidental. 20 world and US cities in the Northern hemisphere were selected and their elevations, latitudes, and average annual temperature were listed (in order of descending latitude)

World and US Cities Showing Latitude Effect on Temperature

(Northern Hemisphere)

	City	Elevation	Latitude	Avg. Ann. Temp.
1	Rejkjavik, Iceland	26 ft (8 m)	64° 08'	42.5 °F (5.8 °C)
2	Oslo, Norway	610 ft (186 m)	59° 55'	42.3 °F (5.7 °C)
3	Stockholm, Sweden	85 ft (26 m)	59° 20'	43.9 °F (6.6 °C)
4	Copenhagen, Denmark	30 ft (9 m)	55° 40'	48.4 °F (9.1 °C)
5	London, United Kingdom	138 ft (42 m)	51° 30'	52.3 °F (11.3 °C)
6	Paris, France	114 ft (35 m)	48° 51'	54.1 °F (12.3 °C)
7	Seattle, WA USA	131 ft (40 m)	47° 48'	55 °F (12.8 °C)
8	Chicago, IL USA	579 ft (177 m)	41° 53'	51 °F (10.6 °C)
9	New York, NY, USA	62 ft (19 m)	40° 43'	53 °F (11.7 °C)
10	Denver, CO, USA	5,525 ft (1,684 m)	39° 44'	54 °F (12.2 °C)
11	Los Angeles, CA, USA	791 ft (241 m)	34° 03'	67 °F (19.4 °C)
12	Phoenix, AZ, USA	1,086 ft (331 m)	33° 27'	76 °F (24.4 °C)
13	New Orleans, LA, USA	-7 ft (-2 m)	29° 58'	75 °F (23.9°C)
14	Miami, FL, USA	6 ft (2 m)	25° 47'	76 °F (24.4 °C)
15	Guadalajara, Mexico	5,062 ft (1,543 m)	20° 40'	69.6 °F (20.9 °C)
16	Mexico City, Mexico	7,347 ft (2,240 m)	19° 26'	63.5 °F (17.5 °C)
17	Managua, Nicaragua	636 ft (194 m)	12° 8'	81.1 °F (27.3 °C)
18	Valledupar, Colombia	538 ft (164 m)	10° 28'	85.6 °F (29.8 °C)
19	Panama City, Panama	20 ft (6 m)	8° 59'	82.6 °F (28.1 °C)
20	The Equator	0 ft (0 m)	0° 0'	85.0° F (29.6° C)

Note: The "average global temperature" calculated by WMO in 1990 became 16 °C (60.8 °F)

(Source: temperatures database of the World Meteorological Organization)

Using a technique learned earlier in his career, the Author routinely graphed the meteorological data for the 20 world cities listed above (the independent variable latitude on the horizontal X-axis and the dependent variable on the vertical Y axis) and used their identification number instead of name to identify each data point on the graph shown next:

Effect of Latitude on Average Annual Temperature (20 World and US Cities)

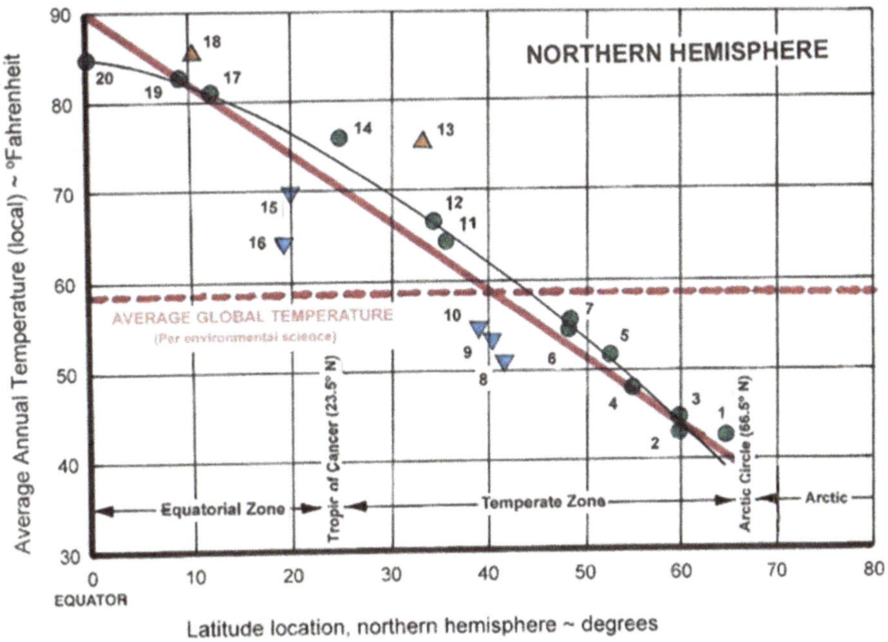

The cities numbered and analyzed in the above illustration were identified in the tabulation:

When a parabolic shaped "fit" was drawn (thin solid black line), it was realized that a simpler linearly sloped line (solid red sloped line) was also a good "fit" of the data showing a very definite trend that might be useful later when analyzing if/how latitude and the "greenhouse gas effect" could be influencing meteorological data. The dashed horizontal red line is the environmental one-size-fits-all "average global temperature". It is left to Readers to make their own interpretation as to whether the sloped solid red line better represents the meteorological database for average annual temperature or whether it is the environmental scientists' "average global temperature" better representing the average annual temperature of US cities at different latitude and having different elevation.

C20) Most important to the world population, to the UN Intergovernmental Panel on Climate Change, and to government leaders in all developed countries, the above illustration reflects that based on factual meteorological data in the World Meteorological Organization database (when graphically shown as above) - the 2017 Paris Accord goal of limiting "average global temperature" to rise to no more than 2 °C by 2035 has little practical credibility. The need to avert a "crisis" that does not exist for the human population by banning the use of all fossil fuels is fallacious and intentionally misleading. The politically and economically motivated "green deals", "carbon credits", "cap and trade", etc., regulations will have no significant impact on global warming or climate change, and substantially cause the cost to the consumer to escalate for all products produced by an energy source aimed at a fallacious goal of net-zero carbon by 2050.

The planet surface temperature with the alias "global warming" and the associated climate change are proven by the several databases analyzed in this book to be dominated by astrophysical, geophysical, and natural factors over which no human or organization has any control.

While analyzing what might be causing the data points far below or slightly above the solid red line, it was noted that the higher the city was in elevation above sea level, the cooler (lower the temperature) the city was based on its meteorological average annual temperature. Having completed a career in the design of aircraft, it was well known that the temperature of the atmosphere decreases nearly linearly from a standard 59 °F (15 °C) at Sea Level (elevation = 0) to -65 °F (-53.9 °C) at 36,089 feet altitude (elevation above sea level) (the equivalent elevation of 11,000 meters).

C21) As a far better approximation of the surface temperature of the planet's surface than that of environmental scientists' "average global annual temperature", mathematically combining the two point-slope lines – temperature linear variation with latitude and temperature linear variation with elevation, a "model" equation for the combined effect of latitude and elevation was derived (in Imperial convention of units, and in Metric convention of units):

(Imperial): Temp (°F) = 90 °F – 0.78 Lat° – 3.44 Elev (1,000 ft)
(Metric): Temp (°C) = 32.2 °C – 0.43 Lat° – 6.26 Elev (km)

It was further noted on the next map of the ocean currents of the world's oceans that coastal cities at low elevation could be cooled or warmed depending on whether the adjacent water temperature was cooler or warmer than the average annual temperature for that latitude and 0 elevation. The Author chose the two cities of Durban and Cape Town in South Africa to illustrate the ocean current effect (if any exists in the region being analyzed):

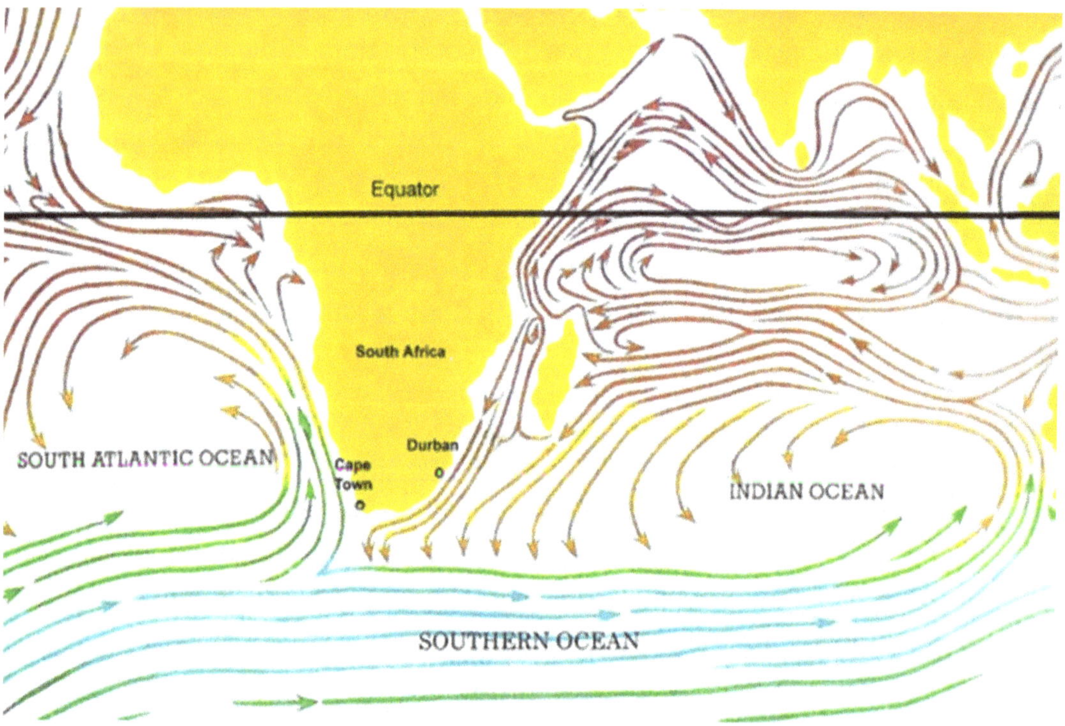

(Source: Rand McNally "Atlas of the World"; notations by Author)

C22) The Author concluded when observing that increments of temperature for the effect of latitude and elevation had a mathematical basis (but ocean proximity did not) that a mathematical model (as shown in Imperial and in Metric units would be improved by visually making a judgmental increment for ocean current temperature using ocean current maps of the three major oceans (Pacific, Atlantic, and Indian)

Imperial) Temp (°F) = 90 °F – 0.78 Lat° – 3.44 Elev (1,000 ft) + dT °F (Ocean)

(Metric) Temp (°C) = 32.2 °C – 0.43 Lat° – 6.26 Elev (km) + dT °C (Ocean)

The Author dubbed the mathematical/map model as LEO (Latitude, Elevation, and Ocean proximity) that has been shown to be much more meaningful than using the present "average" and "linearization" techniques and "models" (which do not consider the Vostok Ice Core and Milankovic factual findings) by the environmental scientists.

A series of charts were developed for the major oceanic regions, and the chart for the United States (with the Atlantic Ocean currents flowing past its East coast, the Pacific Ocean currents flowing its West coast, and the Gulf of Mexico forming its southern coast) is shown as an example (with the LEO Model in the lower left corner):

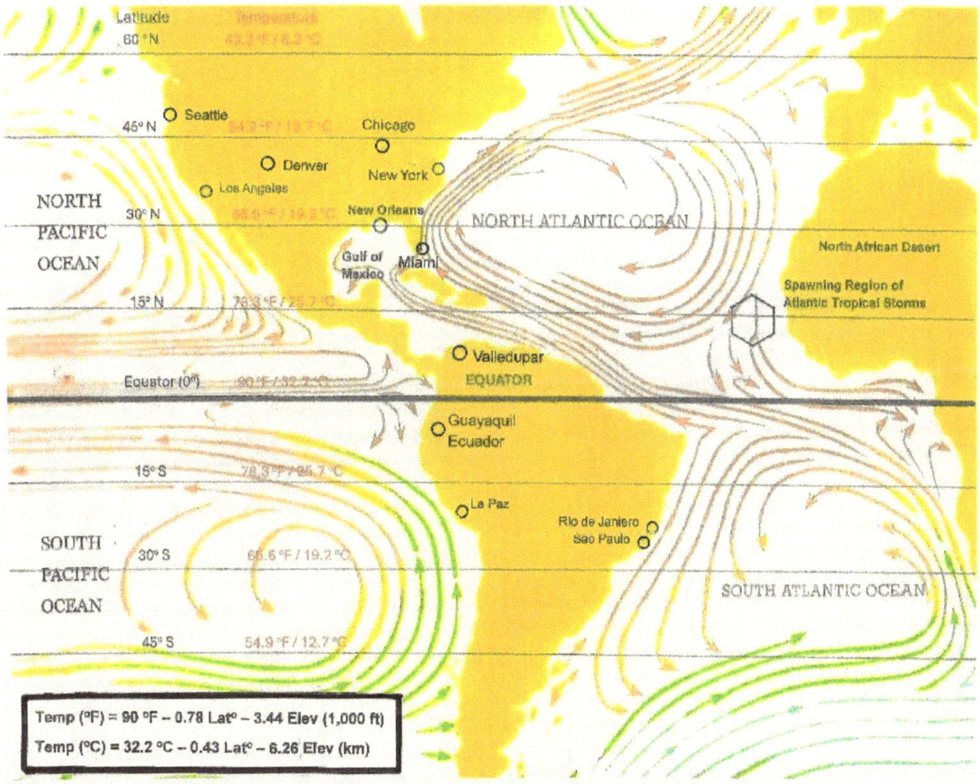

(Source: Author notations on a Rand McNally Map of Ocean Currents)
(The LEO formulation is included in the black block in the lower left corner)

The Author added a hexagonal shaped outline of a region (near right center of diagram) to denote where the spawning ground was for the Tropical Depressions that form off western North Africa (near the Cape Verde Islands) and show how those tropical depressions get warmer and more intense by following the warm Atlantic Ocean current paralleling the Equator. In Hurricane Season, the basic "track" for Atlantic hurricanes is to "ride" the Atlantic Ocean current west (pushed by the prevailing westerly "trade winds"). Once in the Caribbean Sea, the very warm "Gulf Stream" current takes over and the hurricanes proceed northeasterly along the US east coast. Environmental science activists have attempted to make the public believe that the "greenhouse gas effect" is adding to the seasonal occurrence frequency and severity of the hurricanes to further blame Man burning coal as causing the hurricanes. The Man-made "greenhouse gases" in the middle of the atmosphere trapping heat below them have essentially a long-term constant effect not a seasonal effect or on a "track" affecting tropical storms below but above the water surface.

After considering that the LEO model was valid for the range of latitudes between 60° North and South from the Equator, the average annual temperatures at the North Pole (90° N) and at the South Pole (90° South) were entered into the original LEO model. Realizing that the latitude effect on annual average temperature was due to the tilted axis of the Earth causing the obliquity angle effect that Milankovic had recognized in his calculations, the graphical shape of the LEO model with the added polar temperatures became as shown graphically in the following graph:

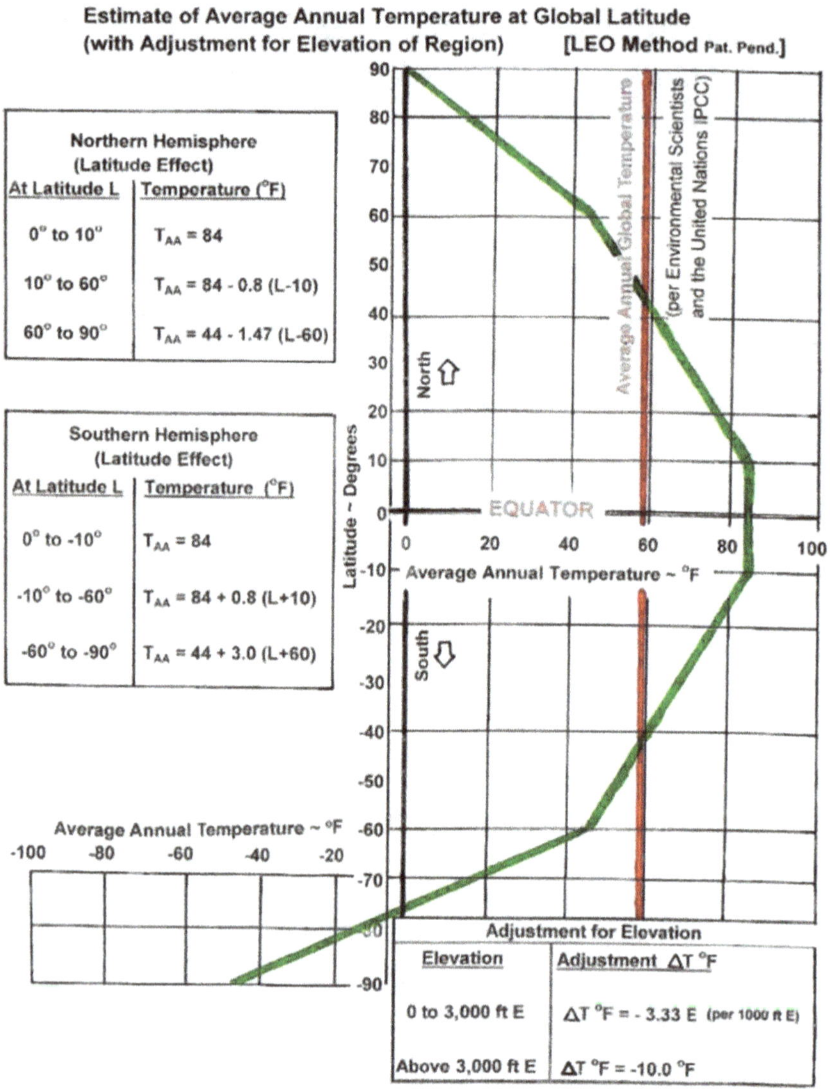

(Source: The World Meteorological Organization database as analyzed by Author)

The five-segmented green line represents the final version of the LEO Model for estimating the average annual temperature of a region at any latitude "between the poles" and accounting for the elevation of the region and allowing an adjustment to the estimate for the effect of proximity to an ocean current on the estimated temperature. The Author conducted an accuracy analysis and found that the final LEO Method estimates the average annual temperature of any region on the planet within 2 °C accuracy. That accuracy is realized by comparison with the vertical red line on the above diagram labeled as the environmental scientists' and UN IPCC one-size-fits-all "average annual global temperature". The LEO Model (for temperature but not precipitation) coupled with the Koppen Geiger climatic model (which considers temperature and precipitation as being climate of a region) is a major advancement in how to estimate the effect of global warming and climate change for any region on the planet.

A very major conclusion derived in this book is that the goal of limiting average global temperature rise to 2 °C by 2035 may be of "crisis" level of importance to environmental scientists (and their EPA and UN IPCC allies) but is relatively meaningless to the human population. Every form of life in the region of the planet where that life is sustained must survive the extremes of temperature – not just the average - in that region. The lifeforms, their food and water supply, their clothing needs, and all the technologies they have developed or need to develop must survive the extremes of heat, extremes of cold, extremes of lack of water, and extremes of too much water. **C-23**) <u>A "temperature envelope" should be defined for each region having a large population. For the United States region (entire 50 states), the human population temperature envelope (in green color) was developed:</u>

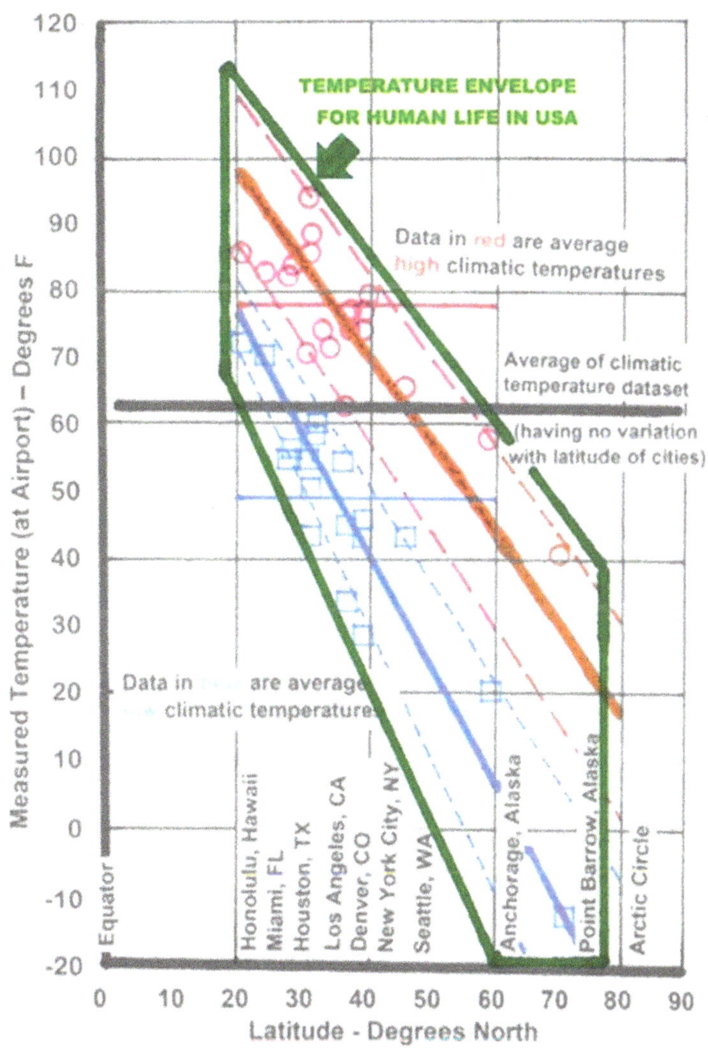

(Source: National Weather Service of NOAA database)

For reference purposes, the environmental scientists' "crisis" level Global Annual Average Temperature (59 o F increased by 2 °C or 3.6 °F) to 62.6 °F) is shown (the horizontal black line) to illustrate why that 2 °C rise in temperature does not constitute a "crisis" to human life in the United States – or a major "crisis" to any region at any latitude having a human population..

In the geophysical chapter, there also was a further refinement for the New York City region to account for the longer winter "Great Lake (Erie) effect" (cooling the average annual temperature of the New York state region east of Lake Erie). It is Buffalo, New York that has epic snow blizzards from the "Lake effect". There also was a temperature increment (warming) refinement for the city of Denver ("the Mile-High" city) to account for the warming effect of the surrounding Southwest Desert Effect on average annual temperature. (The acronym LEO had "stuck" by then, so the LEO formulation was not changed to apply another increment for the desert effect). The LEO effects are far more dominant on the average annual temperature of the region than possibly could an environmental scientists' "settled science" Man-caused one-size-fits-all greenhouse gas effect of 2 °C.

The final topic for which the global warming, climate change, and oceans levels rising concern was researched as a potential future problem is that of the ocean levels rising. Recent information on the ocean levels rising is shown as follows

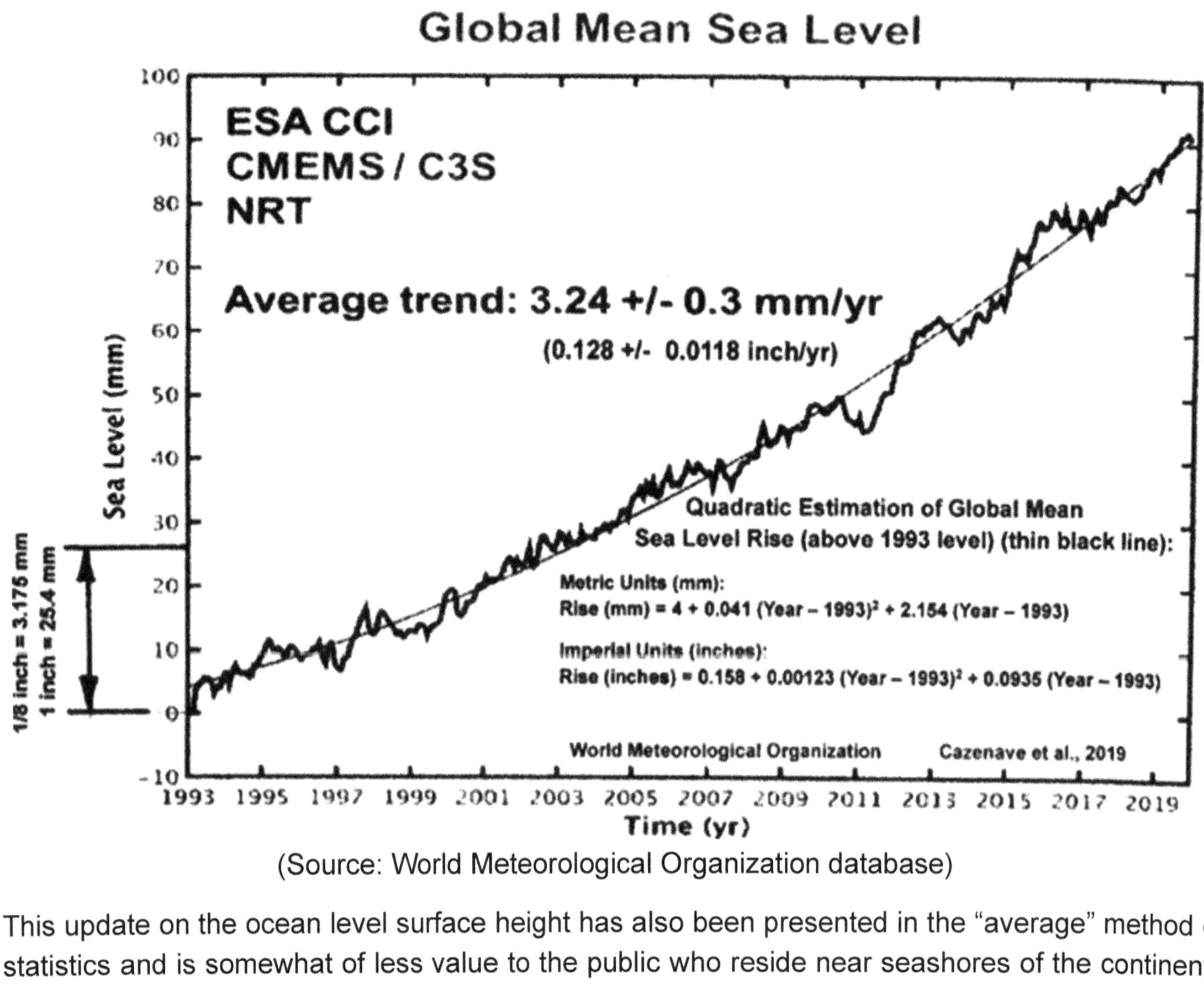

(Source: World Meteorological Organization database)

This update on the ocean level surface height has also been presented in the "average" method of statistics and is somewhat of less value to the public who reside near seashores of the continents or on islands offshore. As shown in the earlier charts on ocean currents (in color showing warm currents and cool currents of the major oceans and seas), there can be no model derived for the temperature of the oceans and seas. The above model for the average trend in Global Mean Sea Level is of scientific interest. The Author has added (lower left corner) a conversion of the measurement being in Metric to Imperial units of measurement. Whereas the scientists have developed the average trend (linearized) to mathematically be expressed as

Average trend in rise = 3.24 mm +/- 0.3 mm/year

The same trend in Metric convention converted to Imperial convention becomes

Average trend in rise = 0.128 inch +/- 0.0118 inches/year

The increase of 1/8th of an inch of rise in the ocean surface level per year in a century (100 years) is then 100/8 inches or 12.5 inches (1 foot ½ inch). Inasmuch as the two factors that affect the ocean levels are the increase in temperature of the ocean and the melting of ice in the two polar regions (Arctic and Antarctica), it is the combined effect of water temperature and ice melt producing the 1 foot of rise in water level per century. Since the ice in the Arctic is primarily floating ice and does not increase water level as it melts, the two effects are then reduced to being the rise in water temperature and melting of the Antarctic ice shelves. The public is apparently less concerned about the ocean levels rising than the environmental science alarmists. One must question the wisdom of building construction companies and Public Works administrators in the erection of so many large buildings on a sandbar, but that is beyond the scope of this book to suggest otherwise. *Caveat emptor.* Any consideration of a safety risk melts when a visitor to south Florida immerses himself/herself in the warm (80 °F) Gulf Stream current.

(Credit to unknown photographer) Southern Florida

Considering the Vostok Ice Core database showing the planet to be nearing the peak hot extreme (3 °C anomaly) in the next 100 years to likely reverse from a warming trend to a cooling trend, the water temperature and the rate of melt of the Antarctic ice shelves should also reverse, resulting in the water level decreasing with time after about a century from now. The conclusion then is that if there is no major flooding of land more than 1 foot above present sea level, ocean level rising is not a major problem for the near term and less for the far term. This conclusion is not given a C designation number as are the other conclusions which reflect that some action be taken to abate the problem. However, if some other event (an "unk-unk") were to happen to increase the water temperature to rise at a greater rate than at present, planning should begin for those now residing less than 1 foot above present sea level to move to higher ground.

With the multitude of conclusions (**C1** through **C23**) formed proving that it is not Man burning fossil fuels that is having a significant effect on the global surface temperature or on the climate, and after identifying the many causes for climate change were identified (astrophysical, geophysical, and natural) the question that logically followed became "What technologies have been or are being developed to mitigate the effects of inevitable global warming, climate change, and ocean levels rising in the next few decades?

The technologies for solving a problem in real life are typically developed from the pros and cons of a previous technology that were developed earlier to solve a similar problem. As the problem changes (either increased or reduced with time), so changes in the remedial technologies for mitigating the problem must be changed accordingly. As an example in the field of environmental science, the development of wind power was vital to the Netherlands centuries ago to control the flooding that ravaged that region – until the dykes and windmills, and system of canals were developed.

Environmental scientists advanced windmill technology into wind turbine technology to harness the energy of wind motion but found, and are finding, that the wind is not as reliable as desired and needed. Solar power technology came into existence more recently but suffers the similar unreliability as wind power. Wind power and solar power technologies are useful but only when the wind is blowing (within tolerable velocity limits – or "envelope") and the sun is shining (at most half of the day) and when skies are not weather-occluded.

A primary conclusion described in this book indicates that technologies which appear to be promising in laboratory experiments and scientific analyses at small scale create knowledge of their pros but fail to identify the cons when placed into use in the full/large scale environment which has many more detractive and/or interacting factors than exist in the laboratory-controlled environment.

As stated near the beginning of this chapter, the Conclusions **C24** through **C37** are better located after the assessment of the technologies to place them in the proper "big picture" of what humans must do to mitigate the effects of global warming, climate change, and oceans' levels rising. They can be found therefore in the Conclusions and Concluding Comments near the end of this book. of this book.

The real naysayers about climate change are the ones in "settled science" who fallaciously hypothesize that the anemones on the coral reefs in the oceans are in crisis of bleaching white because of Man burning any form of fossil fuel (coal, oil, or natural gas).Copernicus and Galileo were branded as being naysayers for challenging their peers who held that the Sun revolved about the Earth, as were others who challenged that the Earth was not flat but was a large sphere. This book may be interpreted by some as "naysaying" by a would-be Galileo admirer but the facts in the numerous databases analyzed completely and in depth presents the "inconvenient" truth. The cartoon on the front cover of this book was chosen as a means to start this book and is repeated as a way to end this book with a lasting impression that scientific theory does not translate to practice and reality. "A picture is worth a thousand words."

Chapter 4
Identifying Candidate Mitigation and Alternative Energy Technologies

The numerous (38) conclusions provided in this book are somewhat useful as a checklist for use as a summary of what was found in the Author's previous book "The Facts and Fallacies of Climate Change on Our Planet" and as a starting point for identifying the technologies that hold promise for mitigating the effects of climate change. Each Conclusion is identified with a **C** prefix and numbered **1** to **37.**

Conclusions **C1** through **C23** were formed by this Author after analysis of several databases of factors causing global warming, climate change, and ocean levels rising that were analyzed for this book. Conclusions **C24** through **C37** were formed after conducting the assessment of the pros and cons (attributes and detriments) of the series of chosen 25 technologies that have some potential for mitigating the effects of climate change caused by astrophysical, geophysical, or natural factors beyond Man's capability to be controlled. Those conclusions will be provided to open the Conclusions and Concluding Comments at the end of this book. A conclusion **C38** is included as the last comment in the Conclusions and Concluding Comments section of this book and is a major suggestion to improve the education of the young people of the United States who will be expected to live and function in safety and health in the future environment of the United States.

Those conclusions **C1** through **C23** were presented in Chapter 3 to mainly provide the basis for the need to reduce the use of coal as a primary source of energy to generate electric power of the level useful to industrial processes. Those same conclusions also form the basis for the new or advanced technologies that are needed by the population of the US region to mitigate the effects that are continually being imposed on the US population by the host of geophysical factors (presented in Chapter 1 and the natural factors (presented in Chapter 2) that befall residents in the United States. Chapter 3 added some other factors that also lead into the need to improve existing technologies and/or re-vectoring them to be less oriented toward reducing emissions and oriented toward more meaningful challenges and opportunities. Readers may wish to review those conclusions C-1 through C-23 before continuing with this chapter.

Just as there are natural disasters causing fatalities, there are also disasters that occur in other fields of technology that cause fatalities and large property losses. Readers who have only glanced at the interior contents of this book and are reading these Concluding Comments probably question the qualifications of the Author of the book to have drawn those "big picture" conclusions that affect the choice of technologies and how those technologies can be assessed for their "pros" and "cons". This Author can only show what officials at high levels of Government conducting national interest levels of safety analysis at the President's request state about the Author's qualifications to conduct such "big picture" analyses. The following is an extract from a former Secretary of the Air Force (Dr.

John McLucas) to the Vice Commander of the US Air Force (General Robert Mathis) in 1981 regarding the Author's qualifications demonstrated in response to President Ronald Reagan's request for the Air Force to assist the Federal Aviation Administration in certifying the first commercial aircraft to fly with a complement of a two-member flight crew instead of the standard three member flight crew. A key issue was how that reduction from three to two pilots affected the safety of transporting more than 200 passengers over long distances above vast ocean waters.

> **The Aircraft Crew Complement study centered on the ability of a flight crew composed of two crew members instead of the conventional three crew members on the flight deck safely operating the aircraft, with advanced technology instruments and displays, in heavy airport traffic and under harsh environmental conditions. Carrying out our study and preparing our report would have been difficult and perhaps impossible had it not been for the excellent cooperation of the Air Force which made available to us a wide range of talent and expertise.**
>
> **This is to thank you specifically for making available Mr. Herbert A. Hutchinson, Chief, Flight Technology Division, ASD Deputy for Engineering, Wright Patterson AFB. Mr. Hutchinson served as leader of one of our panels that was responsible for studying the Federal Aviation Administration's flight safety certification process for new commercial aircraft. He provided excellent leadership to the Panel staff, and his outstanding analytical talents and extensive system safety training and military aircraft crash accident investigation experience contributed very substantially to the work of the Task Force.**
>
> **My colleagues, Lt. Gen. Howard Leaf of the Air Force and Mr. Fred Drinkwater of NASA, and I thank you and other senior officials of the Air Force for the strong support given to our Task Force and for providing the invaluable system safety engineering services of Mr. Hutchinson.**

(Source: Author's Personal Correspondence File)

The Author is retired from a career (a personal odyssey) of over forty years designing aircraft and developing technologies for aircraft to fly safely and efficiently within the "flight envelope" which is unique for each different design. Aircraft must be able to fly at a range of speeds and altitudes, through a range of temperatures, and perform safely within the limits of those variables comprising the "flight envelope". Performing that capability (in street language: "Just do it") requires burning an amount of fuel (JP-4, which is a form of fossil fuel, and which was held in on-board tanks) that permits the aircraft to safely carry a large payload (which may consist of passengers, or cargo or a mix of both) for long distances over hazardous terrain or large bodies of water) within that large envelope of altitude and speed. Almost every system and machine have some form of an operating envelope within which the system or machine must operate efficiently and safely. The human population should be thought of as a human subsystem (the "crew and passengers") of how the planet as a system factually operates.

When this research into the causes of atmospheric and planet surface temperatures changing and affecting the climate was accelerated over fourteen years ago, there were no preformed conclusions of what that research might lead to be. From experience in several prior technology developments, it was anticipated that the "When I turned over a stone, I found more stones to be overturned" experience would repeat itself. It is that type of research that is classed as being "heuristic". The most fundamental heuristic is trial and error, which can be used in finding the values of dependent variables in algebra problems – or addressing complex topics like the climate. In applied mathematics, some common heuristics involve the use of visual representations, additional assumptions, forward/backward reasoning and interpretive logic. In this book, the forward/backward forms of heuristic reasoning and common-sense logic - backed up with photographs and illustrations confirming "that's the way it was and is" - were used extensively.

There is also much similarity of flight systems with the human systems performing (eating local foods as fuel, drinking local water as lubricants, and breathing local air – but exhaling carbon dioxide from the digestion of food during the entire process) within a "temperature and climate envelope" on the planet's surface. One human exhalation introduces a small amount of carbon dioxide into the atmosphere; however, the cumulation of several billion humans exhaling carbon dioxide into the atmosphere every few seconds is not insignificant.

There is also the necessity of humans having to wear different types and insulation thickness of clothing to remain healthy and comfortable throughout the geophysical envelope within which they live while performing the everyday tasks required to sustain and enjoy life. Environmental scientists and "Green" activist groups and organizations seem to perceive the safety and health of plant and animal life in the "big system" of Earth to be of more importance than that of human life. In doing "combat with climate change" they "have identified the enemy, and it is us".

An illustration of many of the factors the Author identified that affect global warming and the climate on the planet, some more significantly than others, provides an illustration of many of the factors this Author evaluated in assessing the impact Man is having on the planet:

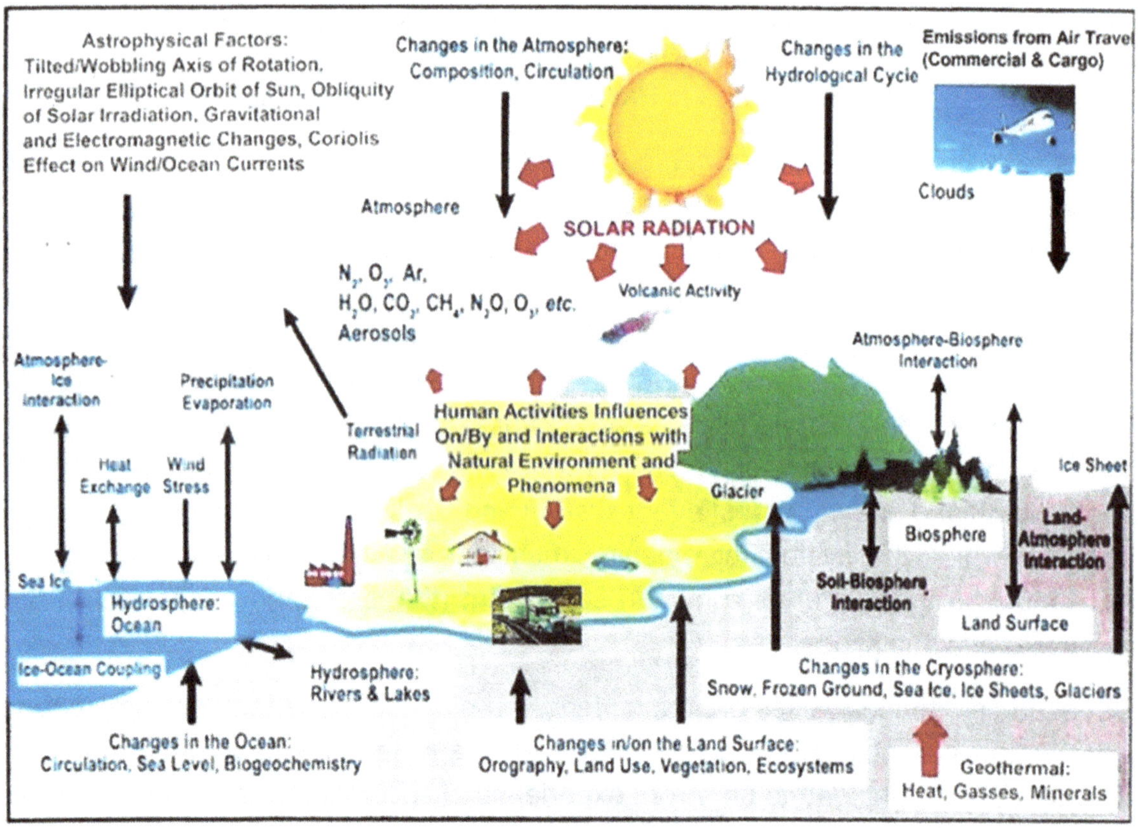

The above diagram is a simplified “picture” (much like the old photography technology of being a “Polaroid snapshot”) and Readers should take notice that the many dynamic factors that are occurring and changing continuously must also be accounted for in the “big picture” of global warming and climate change. The astrophysical, geodynamics, and most of the natural phenomena have been discussed in the early portions of this book and for this chapter can be set aside from the diagram above. There are a number of “man” factors left on the diagram and show as small technology icons – using energy or generating energy.

The diagram is very generic and any region will have only some of the above factors influencing the temperature and climate of that region. Much has been presented to indicate that the regions have an average annual temperature that is unique for the latitude, elevation of the region, and proximity to an ocean where prevailing winds affect that region. The effect is not merely on the environment, but affects virtually everything that must function within the temperature envelope derived for the region. The reality is that the environmentalists’ use of an “average global; temperature” is not scientifically – or technologically – useful for assessing the technology needs for the region.

Using a technique learned earlier in his career, the Author routinely graphed the meteorological data for the 39 world cities and the Equator and created the LEO Model summarized in the graphical representation shown as follows:

Estimate of Average Annual Temperature at Global Latitude
(with Adjustment for Elevation of Region) [LEO Method Patent Pending]

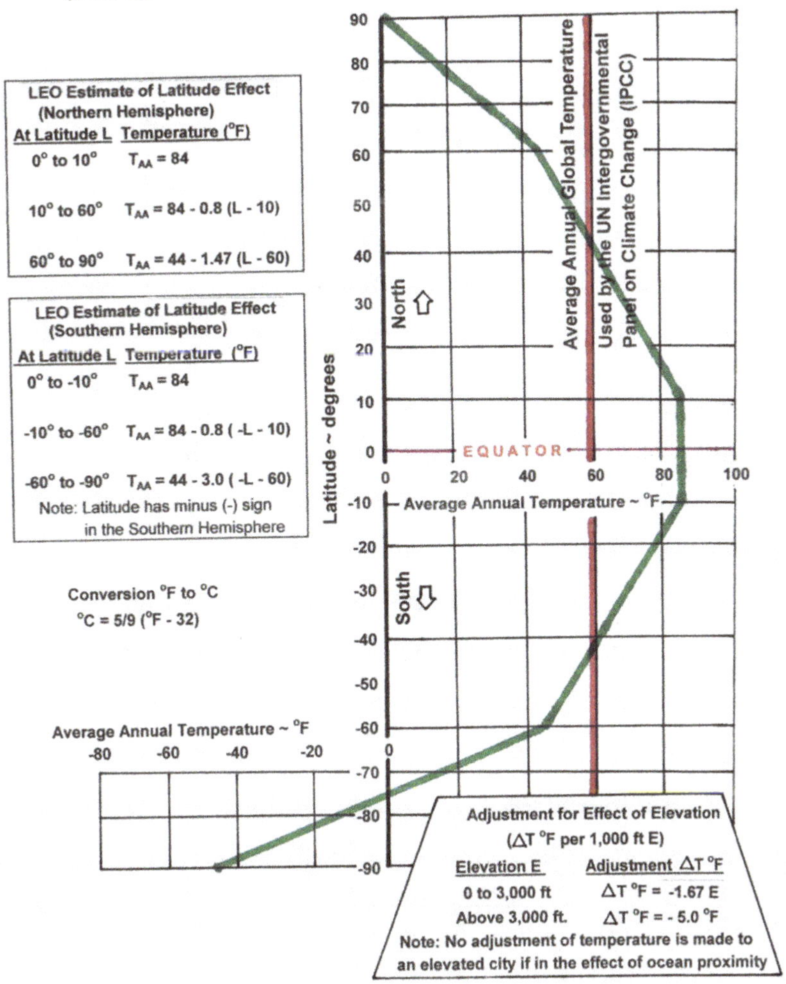

(Source: The World Meteorological Organization database as analyzed by Author)

The five-segmented green line represents the final version of the LEO Model for estimating the average annual temperature of a region at any latitude and accounting for the elevation of the region and allowing an adjustment to the estimate for the effect of proximity to an ocean current on the estimated temperature. The Author conducted an accuracy analysis and found that the final LEO Method estimates the average annual temperature of any region on the planet within 2 °C accuracy. That accuracy is realized by comparison with the vertical red line on the above diagram labeled as the environmental scientists' and UN IPCC one-size-fits-all "average annual global temperature". The LEO Model (for temperature but not precipitation) coupled with the Koppen Geiger climatic model (which considers temperature and precipitation as being climate of a region) is a major advancement in how to estimate the effect of global warming and climate change for any region on the planet. The environmental science and climate science segments of science and technology developers would be well advised to adopt the above LEO model for use in any technology ventures in the future.

The conclusion **C24** will be used as a starting point for the identification of the technology needs for the United States region.

C24) No matter the cause or combination of causes of global warming or of climate change, the conclusion reached by any scientifically valid analysis or interpretation is that the time has come for the use of coal technology as a heat energy generating fuel to be reduced and ultimately to be eliminated in the US region.

The Author evaluated the feasibility and "big picture" value of several types of technology that can help mitigate the effects of planet surface temperatures varying regionally (not just "average global temperature" linearized and purged of the effects of water vapor – which is tantamount to ignoring the effects of gravity on causing flooding). It is fairly obvious that the environmental scientists can serve the public in a far more beneficial manner by advancing the set of technologies analyzed in this book to mitigate the effects of the changes that are inevitable. In addition, while weaning away from the use of coal to reduce pollution, also weaning away from the fallacious goal of reducing carbon emissions to "combat" climate change is highly recommended. Instead, it would benefit the human population if environmental protection scientists would develop a process (or technology) for using the wasteful products discarded by humans as trash (and becoming pollutants) such as plastics which have become a major problem to the environment and to the safety and health of the population in all populated regions.

Before addressing the technological aspects of this book, conclusion **C25** should be presented here to acknowledge the fact that although the reduction of the use of coal was controversial in many respects the outcome is that that effort has spawned the development of a host of technologies in the energy and transportation sectors of life on planet Earth.

C25) Criticism of the environmental scientists' motivation to ease the effect of the burning of fossil fuels on atmospheric "greenhouse gas" emissions, to prevent or at least reduce the extent to and rate at which the "average global surface temperature" is increasing, is moderated to being credited with the cessation of the use of coal in the United States. The cessation of coal burning will reduce the amount of pollution of US land, water, and atmospheric quality by reduction of the disposal of coal ash residue containing toxic and carcinogenic materials in landfills and abandoned pits. The cessation of the use of coal will have little or no effect on improving global warming or climate change for which it was originally targeted as needing elimination.

A tangential outcome from the Author conducting the in-depth research for the global warming, climate change, and ocean levels rising factors was identifying and tracking how the existing energy generating technologies were meeting the energy needs for the United States. The New York Times recently published the following chart corroborating how the various technology groups have been developed and applied in the United States from 2005 to the present:

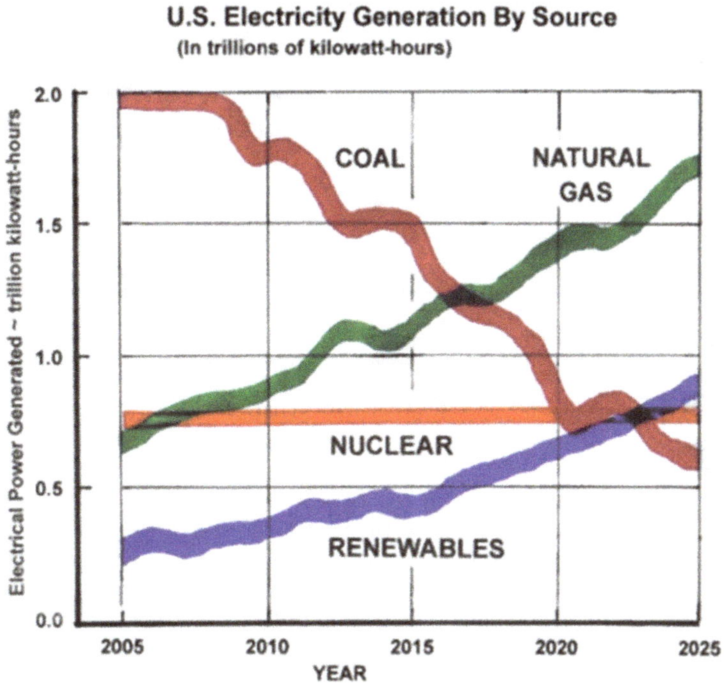

(Source: New York Times, January 2025)

The use of renewables (solar and wind) constitutes about half that of natural gas for. electricity generation in the USA but the trend of increasing use rate of the renewables is low and at a nearly constant gradient with time. Nuclear is seen to be almost a constant trend, most likely attributed to the fact that developing new nuclear facilities are enormously costly and have the ever-present concern for nuclear accidents if located near a large, populated area.

"Pie charts" on how much energy is created by which technology areas are useful but do not reveal the trends with time as meaningfully as the above graphical trend summary. Technologies are developed to solve real problems, not hypothetical scientific ones. The factual data indicating the weaning away from coal use is consistent with improving the environment but from a pollution reduction perspective not because of the reduction of "greenhouse gas" emissions. Despite the environmental scientists theorizing that the use of all fossil fuels must be eliminated to avert a global "crisis" (which is non-existing), the use of natural gas for electricity generation is seen to be on the opposite use rate trend as coal. The use of oil is predominantly in the transportation sector of technology and is understandably not included on the above chart of electricity generation technologies.

Another "eye" was kept on the prospects for using the existing technologies in alternate – not their primary - scenarios, and that resulted in conclusion **C26** shown as follows:

C26) <u>The use of wind turbines should be continued for rural regions (farming, livestock ranching, etc.) but the proposed locations of wind turbines for offshore harvesting of wind energy is not recommended. The increased offshore winds may improve the reliability some, but the reliability will be decreased more by the corrosive effect of saline water on the turbine machinery, the battering of the pylons by wave action, and inevitable storm damage to the entire network</u>.

A few of the alternative energy technologies, however, show promises of being more efficient, safer, collecting the mineral resources most available and accessible in the region, providing opportunities for employment, are affordable, and reliable 24/7/365 etc. while also being clean. The Department of Energy certainly must recognize that the technology of hydrogen has greater potential than does that of solar power and wind power because of its higher reliability and availability of resources involved – as DOE presentation charts and a DOE video by former-Secretary of Energy Granholm reveals.

C27) <u>The technology of geothermal heat energy is seen in the unbiased weighted trade study analysis of the 25 candidate technologies to have the greatest all-around ("big picture") value – even surpassing that of the DOE-preferred hydrogen fuel energy because of the high cost to generate it. Geothermal heat energy lies in and below the Earth's outer crust under all regions of the planet and is the cleanest, most reliable, and most abundant source of heat energy without having to be generated using some other technology.</u>

The most promising geothermal energy technology to meet the growing population demands for heat energy is to use the heat energy that is most abundant at all locations of the planet, independent of latitude or any other factor that causes the increased demand for heat energy. The lowest cost technology to achieve that is with the closed loop pipe-in-a-pipe system described earlier in this book. A diagram of that pipe-in-a-pipe system is shown next:

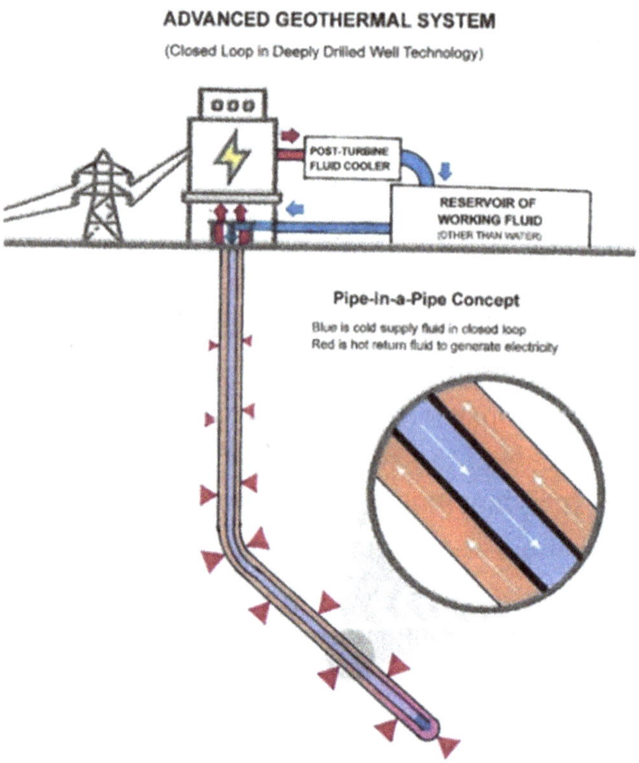

(Source: US Department of Energy database)

The terminology of being a "closed loop" system is that the working fluid (water or other fluids that absorb heat) is inserted under pressure in a cool state through the insulated inner pipe (shown in the insert in blue color) and exits at the bottom of the injection pipe in the cool state into the closed end of the outer pipe. The flow of the working fluid reverses into an outer extraction pipe that is not insulated so that the geothermal heat at the drilled depth is transferred into the return fluid as it ascends (under pressure and in a heated state) where it is used to power the turbines to generate electricity. The "dog-leg" in the well is a developed drilling technology and minimizes the depth of the well while increasing the length of the pipe that will be collecting the surrounding geothermal heat.

The diameter of the well needed to be drilled is the limiting factor on the capacity of the single well pipe-in-a-pipe closed system to generate electricity. However, to increase the capacity of the facility (with the electricity generating turbines and cooling system returning the working fluid in a cooled state back to the injection pumps) can be increased by drilling a well adjacent to the single well system. If more than a single well is used, the hot working fluid from each of the wells could be collected in a plenum chamber before being passed through the turbine(s) generating the electricity. This is the cleanest, most reliable, safest, least-impact on environmental technology for generating electricity meeting the public and industrial needs of a region. The highest cost is in the drilling of the well and the cost of the pipes in the system, therefore the total cost would be multiplied by the number of wells to be drilled to meet the capacity demands of the regional users. The closed loop system can be located as close to or within a populated area as desired without concern for accidents or pollutants entering into the municipal water system.

There are two other very favorable options to the geothermal heat energy sources. On the left is shown a variation of the "closed loop" system where a U-shaped system of wells is drilled deeply into the geothermally hot upper mantel and connected at the bottom of the injection well and the extraction well with a horizontal tunnel. This closed loop system eliminates the need for a pipe-within-a-pipe system but adds the cost of drilling a second extraction well and a horizontal tunnel. The drilling of a second tunnel is not a challenge (but is an additional cost). The challenge (or con) of this type closed loop system is drilling a horizontal tunnel at a large depth and connecting the ends of the injection well and the extraction well. That is beyond the experience level of the Author to evaluate the risks and the costs involved in creating that leg of the closed loop system. The geothermal heat is transferred into the working fluid as it ascends in the extraction well.

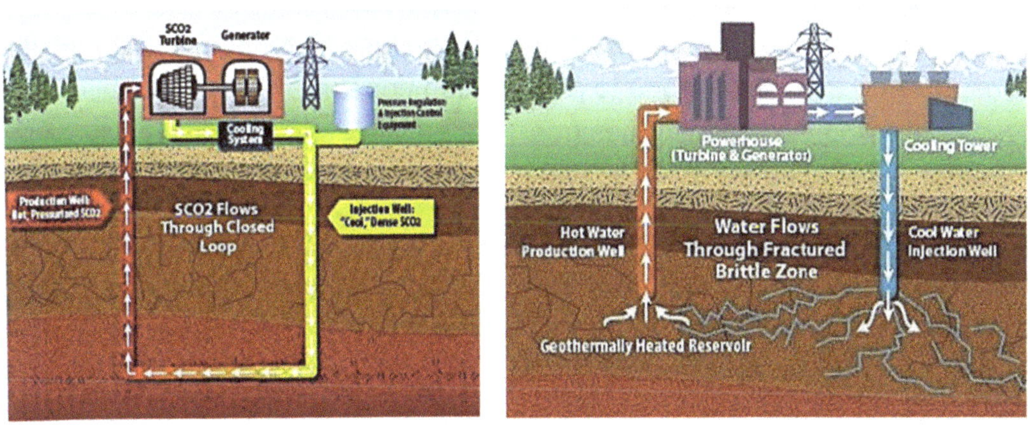

(Source: U.S. Department of Energy database)

On the right side of the diagram is shown the "open loop" version of geothermal energy harvesting at large depths of drilling. The primary difference in the system at the left and the system at the right is that the horizontal tunnel at the bottom of the closed loop system is opened to the working fluid emanating from the bottom of the injection well passing through the shale layer of rock and being geothermally heated as it passes through the hot shale toward the open end of the extraction well. The shale is fractured to open the many passages between the ends of the two vertical wells by the well-developed drilling technique of fracturing the shale known in oil-well technology as "fracking".

The less-positive flow of the working fluid from the injection well through the fractured shale, being heated geothermally by the fractured shale, and entering the bottom end of the extraction well has two cons – one being that there will be losses of the working fluid entering into the fractured shale layer but not entering the extraction well. The amount of the injected fluid that does enter the extraction well will have some small shale debris from the "fracking" process that must be filtered out before the working fluid is passed into the turbine system and generating the electricity from the geothermally heated working fluid. If other than water is used as the working fluid, the valid concern of environmental protectionists would be of the working fluid (possibly containing toxins and/or carcinogens) to not leak when the extraction well is inserted into deep aquafers of the water supply system to any adjacent populated region.

The possibility exists that existing oil and natural gas wells that have been abandoned may potentially be used to serve as injection wells and the extraction wells in geothermal harvesting and avoid much of the additional drilling and "fracking" associated with harvesting geothermal energy.

C28) Because of the looming problem of the population doubling or tripling in the not too distant future, the most efficient use of land and to achieve meeting the demand of the growing population lies in the increased use of nuclear energy – provided its previous problems of failures in its cooling system posing the safety hazard of a nuclear "melt-down" or leakage to the surrounding environment are designed out of those systems. The use of thorium instead of uranium was rated with a slightly better weighted score on its overall set of pros and cons in meeting the needs of the human and industrial electricity demands. The pros of thorium reactors are stated to be:

- Abundance: Thorium is more abundant in the Earth's crust than uranium. Its availability is about the same as for lead.
- Reduced Waste: Thorium reactors produce less long-lived radioactive waste compared to uranium reactors.
- Proliferation Resistance: Thorium reactors produce less plutonium, a material used in nuclear weapons, making them less prone to proliferation concerns.
- Safety: Certain thorium reactor designs, like molten salt reactors, have inherent safety features that reduce the risk of meltdowns.

The cons of thorium reactors are not as well outlined since there are no fully operating thorium nuclear plants in operation in the United States. The only known thorium nuclear facility in operation is one in India, and its economics seem to be its largest con. A few smaller nuclear reactors with thorium have been generated as large scale laboratory research specimens – such as one in the Netherlands:

(Credit to unknown photographer in the Netherlands)

Environmental scientists are quick to comment in the news media: "*Look at all those emissions rising from the smokestacks*". Those "emissions" are mostly steam from the cooling of the nuclear reaction condensing into water vapor. Carbon dioxide is an odorless, invisible gas and is in low concentration in the steam emissions from smokestacks after having passed through environmental protection anti-pollution measures at the facility.

C29) The use of thorium nuclear energy is a proven technology and should be considered for use in either newly constructed nuclear facilities or as a replacement technology in inactive uranium nuclear facilities which were closed for safety reasons (such as those that were deactivated after the Three-Mile Island in Pennsylvania and Indian Point NY nuclear (uranium) facility incidents). Thorium is an abundant element in the United States and in most developed countries of the World. It is "cleaner" to use than uranium. The critical cooling system of the nuclear reactor is much simpler and safer than that for uranium.

The use of the heated cooling water from the nuclear reactors (uranium or thorium) could be used in the flash distillation desalination processes if near a large body of saline water. The heated water could also be used to augment the geothermal energy process using a closed-loop system of harvesting geothermal heat energy. Coupling the four technologies of nuclear energy, solar energy, geothermal energy and desalination technology would be a good way to capitalize on the synergetic effect of combined technologies. The location of combined facilities with two or more technologies must be given careful planning to avoid a natural catastrophic event from damaging/destroying the multi-technology facility (i.e., protect "all the eggs in the one basket".)

When a con in a technology endangers a large number of the population of a region, that con must be eliminated or controlled much more effectively to reduce the fatalities and injuries to near zero. The argument the environmental scientists have used effectively in causing coal technology to be eliminated is to protect the environment. The electric vehicle technology has a con of the same or even more ominous potential for endangerment of human life – resulting from the instability of lithium-ion battery technology. The rise in the number of fatalities per vehicle fire can be correlated with the rise in the number of electric vehicles on the highways of the world and especially in the United States:

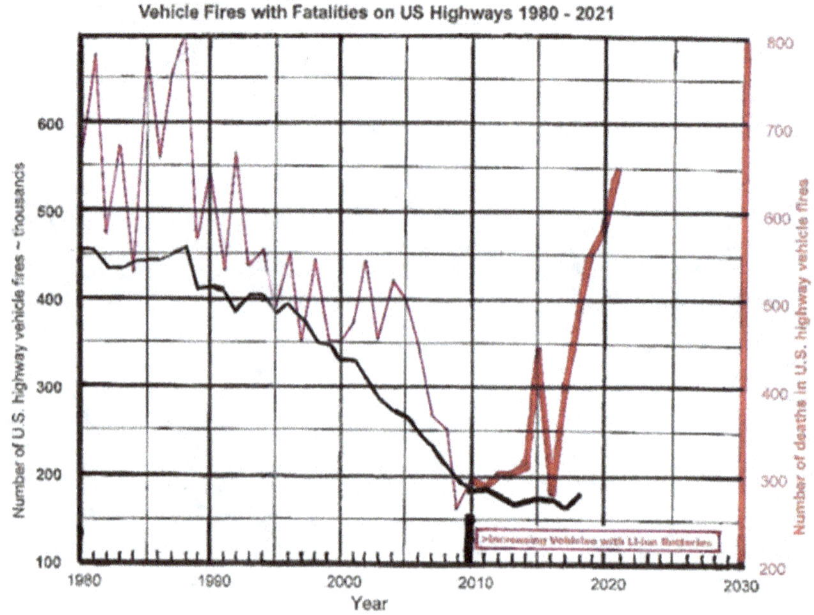

(Source: National Highway Traffic Safety Administration)

Stating that lithium-ion batteries in electric vehicles is an unacceptable hazard is not very attention-getting, but seeing a lithium-ion battery fire as it is exploding and the intensity as it is burning can only be appreciated by “seeing” them (from a safe distance) or to see a photograph of the initial eruption of the high heat-energy lithium ions and the ensuing fire engulfing the passenger compartment of a vehicle. It is important to realize that a fire of the type shown endangers not only the occupants of the affected vehicle but when the vehicle is in the midst of a large number of vehicles (such as in a rush-hour traffic jam on a Los Angeles Freeway twice a day, five days a week, and for 50 weeks of the year are a potential catastrophe that will only get worse as the percentage of the vehicles in that traffic jam are equipped with lithium-ion batteries. The fire departments of many densely populated communities have come to realize that it is too difficult to extinguish lithium-ion fires (water alone does not extinguish them) and have adopted the practice of protecting the surrounding properties of the fire scene and resort to containment of the fire not extinguishing it. Chapter 5 contains additional photographs of other vehicles to reflect that the hazard does not affect just one model of electric vehicle but affects any make or model of electric vehicle equipped with a lithium-ion battery. The hazard is compounded by the hazard being directly beneath the occupants in the vehicle not at a remote location in the front or rear of the vehicle.

(Credit to unknown photographers)

Scenes like the above candid images – not staged - are not uncommon in war scenarios when white-phosphorous munitions are used to destroy a military target. Humans burned so severely and so disfigured express the feeling that it would have been more merciful to not survive the horrible burns.

The lithium-ion battery fire hazard problem has been known to exist with smaller products powered by lithium-ion batteries for over a decade. *"Lithium batteries are generally safe and unlikely to fail, but only so long as there are no defects and the batteries are not damaged or mistreated,"* said Steve Kerber, vice president and executive director of Underwriters Laboratory's (UL) Fire Safety Research Institute (FSRI) in 2015. *"The more batteries that surround us the more incidents we will see."*

In 2016, Samsung issued a global recall of the Galaxy Note 7, citing "battery cell issues" that caused the device to catch fire and at times explode. HP and Sony later recalled lithium computer batteries for fire hazards, and about 500,000 hoverboards were recalled due to a risk of "catching fire and/or exploding," according to the U.S. Consumer Product Safety Commission.

In 2020, the Federal Aviation Administration banned uninstalled lithium-ion metal batteries from being checked in luggage and said they must remain with a passenger in their carry-on baggage, if approved by the airline and between 101-160 watt hours. *"Smoke and fire incidents involving lithium batteries can be mitigated by the cabin crew and passengers inside the aircraft cabin,"* the FAA said.

The safety issue is not the electric vehicle technology *per se*, but with its lithium-ion battery. The repeated instances of fires erupting explosively are from an inherent hazard in lithium-ion battery technology – not caused by Man. Lithium-ion battery fires are cumulatively causing loss of human lives and losses of property that will only increase as the population (and number of vehicle operators) doubles and triples within the next decades. There is no allowable quota of fatalities for a product having an unsafe element, but experience in crash investigations has shown that the sooner a latent hazard is recognized and eliminated, the fewer lives will be lost. The world trend toward battery-powered devices of all sizes and applications should dictate that safe technologies should be incorporated on a higher priority than how many units can be sold quickly before the device is recalled and made safer.

C30) Electric vehicle technology with lithium-ion battery power is unsafe and is recommended to be used with extreme caution by the public in the USA. Electric vehicle technology combined with aluminum-ion or sodium-ion battery technology appears to be an unquestionably better technology for the future human transportation system in the United States as well as in other regions where public safety matters.

Fire containment provisions (which are heavy) have been developed and installed in newer model electric vehicles but there are many hundreds of thousands of electric vehicles without those containment provisions in operation on the US highways system. That number of high-hazard vehicles will grow in the near future to be in the millions. Advertising has been seen that the heavy weight of lithium-ion batteries (weighing about 1,000 pounds per vehicle) adds safety to the vehicles as it reduces roll-over tendencies in a highway crash. Insurance companies increase premiums on lithium-ion equipped vehicles because of their fire propensity but do not decrease the premiums because the electric vehicle has a lower roll-over tendency in a crash.

Better quality control in manufacturing, assembling, and recharging equipment will minimize but will not eliminate the explosive fires igniting. The more electric vehicles, the more lithium-ion battery fires. The better technology solution (as is recommended in this book) is to replace lithium-ion battery technology with less volatile battery technology such as sodium-ion or aluminum-ion have the potential to be. The electric vehicles technology is excellent but the "con" of a spontaneously exploding battery system is too unsafe to continue in use. An improved fire containment system may give the occupants a chance to escape from the vehicle that is afire, but the vehicle with a lithium-ion battery will assuredly be a total loss. In lithium-ion battery fires in open highway scenarios, the remoteness before fire extinguishing equipment or medical rescue services could arrive paints a grim human safety situation. Still more ominous is that there must be a specialized fire extinguishing chemical used – common water does not extinguish most lithium-ion fires.

There are instances of cargo transport ships having sunk (with many hundreds of other new electric vehicles lost) by suspected lithium-ion batteries igniting explosively inside the cargo compartments. Cargo ships have an on-board fire extinguishing capability, but common fire-extinguishing technology is realistically not capable of extinguishing fires of such intense heat. Typically, lithium-ion fires are sprayed with water more as heat containment, but the battery continues to burn even without an oxygen source. The entire vehicle – and perhaps others parked adjacent to the initial vehicle on fire - will most probably be a total loss a few hours later. The toxins and carcinogens inj the debris cannot simply be "swept up".

The potential disaster of one electric vehicle with a lithium-ion battery explosive ignition destroying vehicles surrounding the source (as in a Los Angeles Freeway "rush-hour" traffic circumstances shown next) is too high to accept as a risk: When envisioning a worst-case scenario, if the fire were to occur in an electric vehicle in the middle of a traffic jam when batteries will be overheating, the risk is high that a large number of adjacent vehicles in the traffic jam will be set on fire and burn in a chain-reaction.

(Credit to unknown photographer in Los Angeles)

Without awareness of the potential for catastrophic lithium-ion battery fires in electric vehicles, California Governor Gavin Newsom issued an Executive Order (N-79-20 on September 23,2020) with the following multiple "Whereas"' reasons (noting that the primary is the fallacious "*Whereas the climate change crisis threatening the health and safety of too many Californians is happening now . . .* "):

EXECUTIVE ORDER N-79-20:

WHEREAS the climate change crisis is happening now, impacting California in unprecedented ways, and affecting the health and safety of too many Californians; and WHEREAS we must accelerate our actions to mitigate and adapt to climate change, and more quickly move toward our low-carbon, sustainable and resilient future; and WHEREAS the COVID-19 pandemic has disrupted the entire transportation sector, bringing a sharp decline in demand for fuels and adversely impacting public transportation; and

WHEREAS as our economy recovers, we must accelerate the transition to a carbon neutral future that supports the retention and creation of high-road, high-quality jobs; and

WHEREAS California's long-term economic resilience requires bold action to eliminate emissions from transportation, which is the largest source of emissions in the State; and WHEREAS the State must prioritize clean transportation solutions that are accessible to all Californians, particularly those who are low-income or experience a disproportionate share of pollution; and

WHEREAS zero emissions technologies, especially trucks and equipment, reduce both greenhouse gas emissions and toxic air pollutants that disproportionately burden our disadvantaged communities of color; and

WHEREAS California is a world leader in manufacturing and deploying zero-emission vehicles and chargers and fueling stations for cars, trucks, buses and freight-related equipment; and

WHEREAS passenger rail, transit, bicycle and pedestrian infrastructure, and micro-mobility options are critical components to the State achieving carbon neutrality and connecting communities, requiring coordination of investments and work with all levels of governments including rail and transit agencies to support these mobility options; and WHEREAS California's policies have contributed to an on-going reduction in in-state oil extraction, which has declined by over 60 percent since 1985, but demand for oil has not correspondingly declined over the same period of time; and

WHEREAS California is already working to decarbonize the transportation fuel sector through the Low Carbon Fuel Standard, which recognizes the full life cycle of carbon in transportation emissions including transport into the State; and

WHEREAS clean renewable fuels play a role as California transitions to a decarbonized transportation sector; and

WHEREAS to protect the health and safety of our communities and workers the State must focus on the impacts of oil extraction as it transitions away from fossil fuel, by working to end the issuance of new hydraulic fracturing permits by 2024; and WHEREAS a sustainable and inclusive economic future for California will require retaining and creating high-road, high-quality jobs through sustained engagement with communities, workers and industries in changing and growing industries."

IT IS HEREBY ORDERED THAT:

1. It shall be a goal of the State that 100 percent of in-state sales of new passenger cars and trucks will be zero-emission by 2035. It shall be a further goal of the State that 100 percent of medium- and heavy-duty vehicles in the State be zero-emission by 2045 for all operations where feasible and by

2035 for drayage trucks. It shall be further a goal of the State to transition to 100 percent zero-emission off-road vehicles and equipment by 2035 where feasible.
2. The State Air Resources Board, to the extent consistent with State and federal law, shall develop and propose: a) Passenger vehicle and truck regulations requiring increasing volumes of new zero-emission vehicles sold in the State towards the target of 100 percent of in-state sales by 2035. b) Medium- and heavy-duty vehicle regulations requiring increasing volumes of new zero-emission trucks and buses sold and operated in the State towards the target of 100 percent of the fleet transitioning to zero-emission vehicles by 2045 everywhere feasible and for all drayage trucks to be zero emission by 2035. c) Strategies, in coordination with other State agencies, U.S. Environmental Protection Agency and local air districts, to achieve 100 percent zero-emission from off-road vehicles and equipment operations in the State by 2035.".
It will be interesting for Governor Newsom to explain why the safety hazard of electric vehicles with lithium-ion batteries catching fire explosively in the midst of a traffic jam on Interstate 5 in the Sepulveda Pass area is less of a safety hazard than the emissions from a gasoline powered internal combustion engine vehicle. (This Author moved away from California after 34 years of residency because of the California Air Quality Management gestapo-like mandates made to curtail greenhouse gas emissions.)

C31) Millions of new electric vehicles are equipped with lithium-ion batteries as they presently are in several different manufacturers' designs, and the fatality statistics of the number of fatalities in vehicle fires reflect the increasing number of electric vehicles with unstable lithium-ion batteries are the root cause for the majority if not the total of those car fires. If the use of all-electric vehicles are mandated on California road system, the use of an alternative – and much safer – battery system design should also be mandated, in California and all 50 of the United States for public safety reasoning.

Alternate battery technologies are being developed to eliminate that safety critical con of too many fatalities in electric vehicles equipped with lithium-ion batteries. There has been developed the technology of lithium iron phosphate batteries which are substantially safer than the earlier lithium-ion unsafe batteries. The basic problem appears not to be in the volatility of the metal used but in the electrolyte that is separating the positive and negative ions in the cell. A very thin separator keeps the ions in proper electromagnetic containment and the quality of the separator has lacked the necessary quality to minimize self-ignition of the electrolyte. The basic material lithium, however, is found in useful quantities only in China and Africa but none in the North American continent. The supply of lithium is marginally available to support the current high demand for lithium, and with is low percentage of presence in the Earth's crust, there likely will be a drastic shortage if the exponentially growing population continues at its current growth rate.

Batteries which contain lithium to any degree are costly and extremely heavy, as well as being unsafe when overcharged or overheated. Anyone with experience in magnesium fires can relate to the problem of extinguishing such fires (magnesium, lithium, or phosphorous.) The use of the very light weight magnesium is banned for use in military flight vehicles because of its high volatility and intensive heat when ignited. They cannot be extinguished by water alone.

The basis for a recommendation for replacing all electric vehicle batteries having a lithium content is similar to that made by the environmental scientists who mandate that all fossil fuels be eliminated to reduce emissions in the atmosphere. To eliminate the hazard of fires caused by lithium content all batteries with lithium to any extent should be eliminated. One technology suitable for achieving that In the United States transportation system is summarized in a comparison of aluminum-ion battery technology with lithium-ion technology: (or by any suitable battery such as sodium-ion having similar pros and cons compared to that of lithium-ion)

Feature	Aluminum Ion Batteries	Lithium-Ion Batteries
Cost	Lower	Higher
Abundance	High (aluminum is widely available)	Low (lithium is limited)
Safety	Safer	Less safe (risk of overheating)
Energy Density	Lower	Higher
Charging Speed	Faster	Slower
Environmental Impact	Lower	Higher

(Credit is given to the unknown analyst – not the Author - for the above comparison)

It is hastened to be added that a parallel effort to develop sodium-ion batteries which have many of the same technology pros as does aluminum-ion battery technology and both being much safer technologies than lithium-ion technology. Recognition of the better safety of aluminum-ion and sodium-ion battery technology than that of lithium-ion battery by this Author was observed in the previous decade by a few start-up companies which came to the same conclusion that this Author is revealing to the public in this book as to the increasing frequency and severity of the lithium-ion battery fires in electric vehicles. This Author does not identify which of the two competitive battery technologies is a better alternative to the electric vehicles manufactured with lithium-ion batteries as that opinion could be converted into an advertisement for one technology over another competitor.

C32) The technology of aluminum-ion and sodium-ion batteries has matured sufficiently to require that either or both technologies be developed for all new electric vehicles of any style and size operating on the US highway system and especially in larger cities. Suffice it to say that the motor vehicle manufacturers should observe the need for protecting the safety of the public and advance the technology of the electric vehicles they produce to have sufficient safety being introduced into the millions of their electric vehicles that have been and are being introduced into the public transportation system with an acceptable level of safety risk in the equipment items within their respective vehicle designs.

This Author bases the multiple conclusions in this book using the discipline ingrained in system safety engineers and in crash investigators to identify if, where, and why a safety risk exists in a technology – not to favor or criticize any industrial entity for having applied or excluded that safety risk from their product.

An example of what can – and did – happen when a manufacturer conceals a known safety problem with their product is that experienced six years ago with the international Boeing 737 Max 8 aircraft program.

A professionally-saddening safety situation was exposed after two Boeing aircraft crashed – the first on October 29, 2018, when Lion Air Flight 610, a Boeing 737 Max 8, plunged into the Java Sea and 189 people on board were killed. As the cause of the Indonesian crash was being investigated, a second Boeing 737 Max 8 – Ethiopian Airlines Flight 302 - crashed on March 10, 2019, after take-off from Addis Ababa killing 157 passengers and crew. Crash investigators found a potentially hazardous flaw in the design and safety logic of the flight control system (in the software logic of the angle of attack sensors) that had been experienced on flight simulators on the ground during early development of the Boeing 737 Max 8.

Instead of adequately fixing the safety flaw, the flaw was not made known to the flight crews but was instead buried in the "fine print" of the flight handbook such that if an aircraft being flown on the automatic mode of flight control got into that unsafe flight mode (the nose of the aircraft being commanded down at a dangerous angle), the instructions were to shut off the automatic flight control system and fly the aircraft manually on the basic – and safer - flight control system. However, the airline operators were never made aware that the hazardous safety problem existed because Boeing did not want to ground the entire international 737 Max 8 fleet until the problematic safety flaw could be properly redesigned. To Boeing upper management, time spent fixing a problem was considered as time taken away from making a profit. "Whistle-blowers" simply were dismissed from employment. Those facts were revealed to the public in 2021, and Boeing was subsequently ordered by the courts to recompensate the 346 families of those lost in a $2.5 billion settlement. It would have cost Boeing a very small fraction of that amount to properly fix the problem when it was first identified.

The above example of the Boeing 737 Max 8 safety hazard should serve as a warning to large motor vehicle manufacturers to not sell their all-electric vehicles to the public if there is a known major safety hazard such as having a lithium-ion battery with a very ominous fire safety record. Chapter 7 of this book describes that known safety hazard (erupting in fire for several known reasons in the lithium-ion electricity storage chemistry process or quality of manufacture) in sufficient detail and with photographic evidence to warrant a delivery pause in electric vehicles for a Federal Government level independent fire-safety risk assessment.

C-33) A fire safety assessment of electric vehicles in use by the public and government should be by a team led by a qualified and experienced accident investigator (not by a scientist of the Environmental Protection Agency) and given authority and responsibility equivalent to that which was assigned in the Boeing 737 Max 8 crash investigations. The official assessment should be conducted expeditiously to avoid further loss of human life as well as to minimize disruption to the automotive manufacturing economy and manufacturing schedule. The assessment need not fix blame for the use of the questionably unsafe lithium-ion technology in electric vehicles; only to determine the public safety risk in continued use of the lithium-ion technology and to identify if an alternate battery technology with improved fire safety is required.

Vehicle manufacturers must make their own business assessment of the possibility that a severe technology flaw has been identified and needs to be corrected in a manner so as to least endanger the public until the technology flaw is corrected. Major changes in the design of a company product are considered a senior management and business matter, not a science or engineering call.

If an independent safety assessment of the lithium-ion battery technology in electric vehicles decides that a safer alternate battery technology is needed, the disposal of the replaced lithium-ion batteries creates a pollution concern. Inherently, lithium-ion batteries emit high heat when burning (as well as hazardous gases that would need to be captured and disposed of to meet EPA regulations). A potential technology (in the category of pollution control) would be to sell obsolete lithium-batteries to the US steel industry to be used as fuel for renewing production of virgin steel domestically instead of the present practice of virgin steel being imported from Canada and Mexico and other countries. The practice of importing virgin steel came about by the Environmental Protection Agency imposing overburdening emissions controls (now found to be unjustified as the emissions are proven in this book to not be the dominant cause of global warming or climate change) on the US steel companies burning the fossil fuel coal as the primary high heat energy source needed to produce virgin steel. The US steel companies had little recourse but to instead import the virgin steel needed (and of course to pass the extra import costs on to the user).

The need for steel is another commodity that will grow in proportion to the exponentially growing population, using it productively instead of incurring the cost for safe disposal would avoid the expenses added to the steel being imported and passed along to the user. How the cost to the steel companies to purchase the lithium-ion batteries might be subsidized by the US government is beyond the scope of this Author to be suggested.

A very major conclusion derived in this book is that the goal of limiting average global temperature rise to 2 °C by 2035 may be of "crisis" level of importance to environmental scientists (and their EPA and UN IPCC allies) but is relatively meaningless to the human population. Every form of life in the region of the planet where that life is sustained must survive the extremes of temperature – not just the average - in that region. The lifeforms, their food and water supply, their clothing needs, and all the technologies they have developed or need to develop must survive the extremes of heat, extremes of cold, extremes of lack of water, and extremes of too much water.

C-23) A "temperature envelope" should be defined for each region having a large population. For the United States region (entire 50 states), the human population temperature envelope (in green color) was developed:

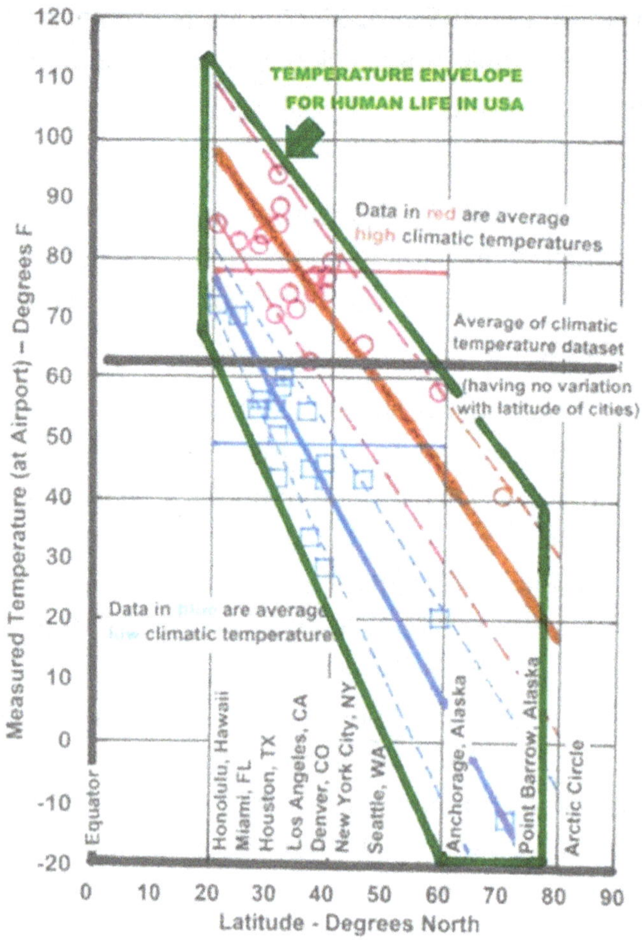

(Source: National Weather Service of NOAA database)

For reference purposes, the environmental scientists' "crisis" level Global Annual Average Temperature (59 o F increased by 2 °C or 3.6 °F) to 62.6 °F) is shown (the horizontal black line) to illustrate why that 2 °C rise in temperature does not constitute a "crisis" to human life in the United States – or a major "crisis" to any region at any latitude having a human population..

In the geophysical chapter, there also was a further refinement for the New York City region to account for the longer winter "Great Lake (Erie) effect" (cooling the average annual temperature of the New York state region east of Lake Erie). It is Buffalo, New York that has epic snow blizzards from the "Lake effect". There also was a temperature increment (warming) refinement for the city of Denver ("the Mile-High" city) to account for the warming effect of the surrounding Southwest Desert Effect on average annual temperature. (The acronym LEO had "stuck" by then, so the LEO formulation was not changed to apply another increment for the desert effect). The LEO effects are far more dominant on the average annual temperature of the region than possibly could an environmental scientists' "settled science" Man-caused one-size-fits-all greenhouse gas effect of 2 °C.

C34) Although not evaluated as a technology area in this book, the numerous photographs after hurricanes, tornados, typhoons, earthquakes, lightning-ignited wildfires, volcanic eruptions, tsunamis, flooding, and mudslides showing the damage done to properties (privately-owned or commercial) highlights the need for all countries to update the building codes used in the structures constructed in their region. In most regions, the probability of one or more of those natural catastrophes occurring and the type of catastrophe is of less importance as to the human and animal population suffering the after-effects of the catastrophe. No region can have a perfect building code, but many regions have disgracefully unsafe building codes that are not enforced. It is not only the climate changing differently in different regions around the World, but people are their likes and dislikes. The attitude that "I have the right to do whatever I want to do" is not a scientific problem but is a societal one. The leaders of a community must establish and enforce building codes, environmental codes protecting humans and all forms of life from pollution of the land, water, and air.

C35) Humans should not choose to locate their homes and places of business in a known flood-zone or a known "hurricane alley" or "tornado alley", on the side of a volcano, over an earthquake-prone fault line, or on a beach less than two or three feet above the "beautiful ocean". Humans enjoy taking risks however (as the croupiers in Las Vegas will confirm), and do not use the common-sense that their parents instilled in them as children before they learned how to spell "hazard". The sport of bungee jumping, skydiving, climbing the sheer vertical side of a cliff, high wire walking with no safety net below, skiing in an avalanche area, etc. are and always will be popular forms of entertainment despite their obvious hazards. No lecture on safety and common-sense will deter humans from placing themselves in "harm's way". This book can only advise, it has no intent of being a "Dear Abby" form of advice for the common problems of merely living on the planet. Those who wish to take risks will do so. But in the basic street language well understood by humans *"If you can't do the time, don't do the crime."*

The last issue surfaced in this book to be given emphasis for resolution is the total problem of human and industrial pollution. In the "big picture", by causing environmental scientists to change their focus from "greenhouse gas emissions" and applying that focus to something else of scientific importance, there would be no greater contribution for the benefit of mankind to resolve the human and industrial pollution problem. That is a problem of far greater magnitude than this book could even make a scratch. Two possibilities proffered in the chapters of this book might be among the solutions investigated.

The most major item of pollution is that of the disposal of plastics that have been recycled into storage warehouses but are going unused because of lack of a suitable reuse. The Author suggested the plastic be heated (by solar energy) and formed into railroad track ties and used to replace the existing wooden ties. Wooden tie technology has reached a state where the cons outweigh the pros of using wood. By reusing excess plastic trash and waste, the use of plastic trash partial solves the disposal of the plastic into landfills and ends up being introduced into ground aquafers – polluting the drinking water of the public or flowing into the oceans and polluting marine life. The second aspect is that by using plastic instead of wood, deforestation to produce wooden ties becomes unnecessary.

C36) The population growth prediction of doubling and perhaps tripling in the next century or two will undoubtedly increase the need for twice or triple the needs for energy, as well as the need to be more efficient at disposal of human-generated waste and trash. The EPA in the United States is an organization that needs to open up its vision about how to handle the threat of pollution on humans far more than the effect of greenhouse gases on the climate. The pollution problem will grow in proportion to human population growth. When scenes like that shown next (in Accra, Ghana) are seen, and it is realized that the world does not know how to handle its present level of waste disposal and pollution, why should there be any expectation that the problem will decrease as the world population doubles or triples? The EPA in the USA and its counterparts in other countries needs to stop "choking on the minnows while the whale is swallowing us". Control of "greenhouse gas emissions" is a much lower hazard to humans than is the hazard of pollution of the land, water, and atmosphere.

(Credit to unknown photographer in Accra, Ghana)

The technologies evaluated in this book may seem unnecessary for the current population's needs and demands for clean energy, but the "big picture" of population growth will require every resource to be used to its maximum availability and every technology to be used to its maximum capacity. The planning for that probability should not be left to "We'll cross that bridge when we come to it." Waiting until water is lapping around our ankles before beginning planning on better flood control and the ocean levels rising is not a risk that should be taken. Plans that go unused can be disposed of easily if later found out not to be necessary but plans that do not exist and must be developed on a crisis-rate to minimize disaster are more difficult to accomplish.

The final topic for which the concern was researched as a potential future problem is that of the ocean levels rising. The most recent information on the ocean levels rising was that shown as follows

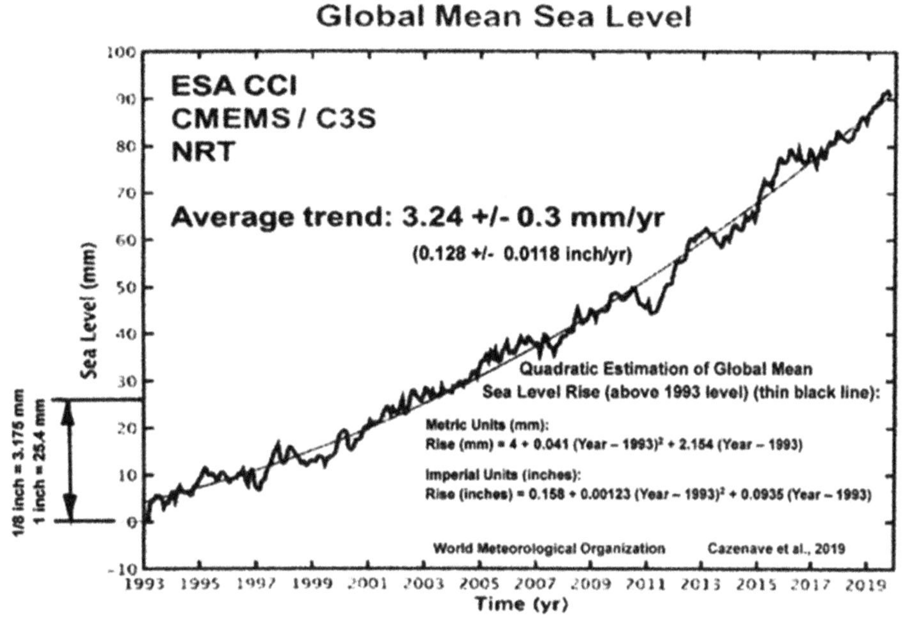

(Source: World Meteorological Organization database)

This update on the ocean level surface height has also been presented in the "average" method of statistics and is somewhat of less value to the public who reside near seashores of the continents or on islands offshore. As shown in the earlier charts on ocean currents (in color showing warm currents and cool currents of the major oceans and seas), there can be no model derived for the temperature of the oceans and seas. The above model for the average trend in Global Mean Sea Level is of scientific interest. The Author has added (lower left corner) a conversion of the measurement being in Metric to Imperial units of measurement. Whereas the scientists have developed the average trend (linearized) to mathematically be expressed as

Average trend in rise = 3.24 mm +/- 0.3 mm/year

The same trend in Metric convention converted to Imperial convention becomes

Average trend in rise = 0.128 inch +/- 0.0118 inches/year

The increase of 1/8th of an inch of rise in the ocean surface level per year in a century (100 years) is then 100/8 inches or 12.5 inches (1 foot ½ inch). Inasmuch as the two factors that affect the ocean levels are the increase in temperature of the ocean and the melting of ice in the two polar regions (Arctic and Antarctica), it is the combined effect of water temperature and ice melt producing the 1 foot of rise in water level per century. Since the ice in the Arctic is primarily floating ice and does not increase water level as it melts, the two effects are then reduced to being the rise in water temperature and melting of the Antarctic ice shelves. The public is apparently less concerned about the ocean levels rising than are the environmental science alarmists. One must question the wisdom of the building construction companies and Public Works administrators in the erection of so many large buildings on a sandbar, but that is beyond the scope of this book to suggest otherwise. *Caveat emptor.* Any consideration of a safety risk melts when a visitor to south Florida immerses himself/herself in the warm (80 °F) Gulf Stream current.

(Credit to unknown photographer) Southern Florida

Considering the Vostok Ice Core database showing the planet to be nearing the peak hot extreme (3 °C anomaly) in the next 100 years to likely reverse from a warming trend to a cooling trend, the water temperature and the rate of melt of the Antarctic ice shelves should also reverse, resulting in the water level decreasing with time after about a century from now. The conclusion then is that if there is no major flooding of land more than 1 foot above present sea level, ocean level rising is not a major problem for the near term and less for the far term. This conclusion is not given a C designation number as are the other conclusions which reflect that some action be taken to abate the problem. However, if some other event (an "unk-unk") were to happen to increase the water temperature to rise at a greater rate than at present, planning should begin for those now residing less than 1 foot above present sea level to move to higher ground.

With the multitude of conclusions (**C1** through **C23**) formed proving that it is not Man burning fossil fuels that is having a significant effect on the global surface temperature or on the climate, and after identifying the many causes for climate change were identified (astrophysical, geophysical, and natural) the question that logically followed became "What technologies have been or are being developed to mitigate the effects of inevitable global warming, climate change, and ocean levels rising in the next few decades?

The technologies for solving a problem in real life are typically developed from the pros and cons of a previous technology that were developed earlier to solve a similar problem. As the problem changes (either increased or reduced with time), so changes in the remedial technologies for mitigating the problem must be changed accordingly. As an example in the field of environmental science, the development of wind power was vital to the Netherlands centuries ago to control the flooding that ravaged that region – until the dykes and windmills, and system of canals were developed.

Environmental scientists advanced windmill technology into wind turbine technology to harness the energy of wind motion but found, and are finding, that the wind is not as reliable as desired and needed. Solar power technology came into existence more recently but suffers the similar unreliability as wind power. Wind power and solar power technologies are useful but only when the wind is blowing (within tolerable velocity limits – or "envelope") and the sun is shining (at most half of the day) and when skies are not weather-occluded.

A primary conclusion described in this book indicates that technologies which appear to be promising in laboratory experiments and scientific analyses at small scale create knowledge of their pros but fail to identify the cons when placed into use in the full/large scale environment which has many more detractive and/or interacting factors than exist in the laboratory-controlled environment.

The weighted trade study analyses conducted for the 24 technologies being developed or strongly considered for development contained technologies mitigating the effects of surface temperatures rising in various regions of the World but more specifically in the United States. Lower importance was given to avoidance of emissions that can be considered in the "greenhouse gas" heat-trapping category. The methodology used for the weighted trade studies was discussed in an earlier chapter. The laboriously detailed worksheets used to "grade the report card" (i.e., assess the net value of the technology area by weighing the importance of each pro and con of the technology) are contained in the last chapter.

The weighting method used for the technology assessments considered how important are the several factors associated with the effects of global warming and climate change and the effect they are having on the human population (safety, health, comfort), the environment, the national economy, the cost and availability of resources, and status of technology development in the region. The effects of the technologies assessed are listed (not in order of importance) below in twelve categories:

1. Status of Technology Development (risk, schedule, cost, potential value)
2. Reliability and Sufficiency (time of day/year, weather dependent, climate dependent)
3. Regional Economy and Resources (mineral, water, labor force to operate/maintain)
4. Cost of Resources (domestic and imported, mode of harvesting / distribution)
5. Importance to Industry, Government, and Human Safety, Health, Comfort
6. Importance and Urgency to Economy and Employment
7. Pros and Cons of By-Products (emissions, pollutants, hazards)
8. a. Effect on Human Population (regional in USA) - Residency.
   b. Effect on Human Population (regional in USA) – Food Supply.
   c. Effect on Human Population (regional in USA) – Safety, Health, Comfort
9. a. Effect on Infrastructure (regional in USA)
   b. Effect on Transportation (regional in USA)
   c. Effect on International Trade
10. a. Effect on Environment (regional in USA) – Atmosphere
    b. Effect on Environment (regional in USA) – Land
    c. Effect on Environment (regional in USA) – Water
11. a. Effect on Animal Life (regional in USA) – Land
    b. Effect on Animal Life (regional in USA) – Aquatic
    c. Effect on Animal Life (regional in USA) – Avian
12. Effect on Forestry and/or Agriculture (regional in USA)

On the scoring worksheet, each of the applicable twelve factors were assigned a numerical importance indicator in a relative manner such that the sum of the numerical importance indicator is 100 points (similar to a percentage of the total importance of the technology being rated to the overall mitigation effort.

The "big picture" summary of the evaluations of the pros and cons for each technology and the prioritization of the weighted scores for the technologies are grouped into the following categories (numbered as follows in the Group column):

1. Fossil fuels (coal, oil, and natural gas) or their derivatives
2. Non-fossil fuels
3. Alternative energy sources – geophysical
4. Alternative energy sources - nuclear
5. Technologies for improving human transportation and the infrastructure
6. Technologies for improving food and water supply for humans

It is important to note that the factors of importance, the technologies considered and the scoring are as they relate to the region in which they will be applied. For this book, the region selected was that of the United States and its unique resources indigenous to the United States. Other regions can use the weighted trade study process and select the factors, the technologies, and assign the weighted ratings/scores judged to be of importance to the region.

The 25 technology areas included in the weighted importance scoring process of the trade studies reflect the pros and cons of each technology in that overall process, and applicable to the United States, are listed below in numerical order with a Trade Study identifier (e.g., TS-1) and within the categories indicated:

Fossil Fuels / Derivatives

TS-1 Use of Fossil Fuel – Coal
TS-2 Use of Fossil Fuel – Oil
TS-3 Use of Fossil Fuel – Natural Gas (Pipeline)
TS-4 Use of Fossil Fuel – Natural Gas (Liquified / Transportable)
TS-5 Use of Fossil Fuel Derivative – Propane

Non-Fossil Fuels

TS-6 Use of Non-fossil Fuel – Hydrogen
TS-7 Use of Non-fossil Fuel – Ethanol

Alternative Energies - Geophysical

TS-8 Use/Development of Hydrodynamic Energy (River Flow)
TS-9 Control of Hydrodynamic Events (Coastal Regions)
TS-10 Use/Development of Solar Energy (Large Industrial / Commercial)
TS-11 Use/Development of Solar Energy (Residential / Rural / Town)
TS-12 Use/Development of Wind Energy (On Land)
TS-13 Use/Development of Wind Energy (Off-Shore)
TS-14 Energy Storage - Stationary/Large Size (Electrical)

TS-15 Use/Development of Geothermal Energy (Surface)
TS-16 Use/Development of Geothermal Energy (Deep-Drilled)
TS-17 Energy Storage - Stationary/Large Size (Thermal)

Alternative Energy - Nuclear

TS-18 Use/Development of Nuclear Energy - Uranium
TS-19 Use/Development of Nuclear Energy - Thorium

Transportation

TS-20 Use of All-Electric Motor Vehicles – Commercial / Government
TS-21 Use of Electric Plug-In / Fuel Hybrid Vehicles – Commercial / Public
TS-22a Energy Storage - Mobile/Small Size (Electrical) Lithium-ion
TS-22b Energy Storage - Mobile/Small Size (Electrical) Aluminum-ion or Sodium-ion

Food/Water Supply – Human, Animal, Agricultural

TS-23 Use/Development of Desalination - Reverse Osmosis
TS-24 Use/Development of Desalination – Distillation (Solar)

Group	Trade Study No.	Technology Area	Total Wghtd Score	Comment (Suggested Use and Development for USA Region)
①	TS-4	Use of Fossil Fuel - Natural Gas (Liquified)	95%	Increase useage; improve transportability
	TS-3	Use of Fossil Fuel - Natural Gas (Pipeline)	93%	Install more pipelines (distribution to ex-coal users)
	TS-5	Use of Fossil Fuel Derivative - Propane	90%	Increase useage; improve storage safety
	TS-2	Use of Fossil Fuel - Oil	89%	Continue but reduce use - to no imported oil
	TS-1	Use of Fossil Fuel - Coal	74%	Phase out coal burning technology (to reduce toxic and carcinogenic land/water pollution) Close deep mining.
②	TS-6	Use of Non-fossil Fuel - Hydrogen	90%	Increase useage; improve storage & distribution safety
	TS-7	Use of Non-fossil Fuel - Ethanol	86%	Continue useage; co-technology with EV Hybrids
③	TS-9	Control of Hydrodynamic Events (Coastal and Seasonal Event Regions) - Levees / Sea Walls	98%	Highest need but with lack of activity/planning to meet safety and energy needs of growing human population
	TS-8	Use of Hydrodynamic Energy (Dams, River Flow)	96%	Increase existing capacity and additional construction
	TS-14	Energy Storage - Stationary Large (Electrical)	94%	Good technology supplementing that of solar and wind.
	TS-15	Use of Geothermal Energy (Surface)	93%	Good technology for the western states in Conus;
	TS-17	Energy Storage - Stationary Large (Thermal)	92%	Most important adjunct technology with geothermal
	TS-11	Use of Solar Energy (Residential / Rural / Town)	89%	Of most value to small users. Requires battery storage.
	TS-16	Use of Geothermal Energy (Deep-Drilled; Closed Cycle and Open Cycle with Fracking)	83%	Increase useage; Closed cycle preferred to Open Cycle; Recycle abandoned oil wells where feasible
	TS-12	Use of Wind Energy (On Land)	82%	More useful for small industrial, farming, and public use.
	TS-10	Use of Solar Energy (Large Industrial / Commercial)	78%	Not reliable for industries, OK for small; co-technology with other uses (distillation desalination; water purification)
	TS-13	Use of Wind Energy (Off-Shore)	63%	Too harsh environment (corrosion, wave force, storm damage). Recommend not installing off-shore system.
④	TS-19	Use of Nuclear Energy (Thorium)	96%	Among best technologies for mitigating climate change.
	TS-18	Use of Nuclear Energy (Uranium)	95%	Among best technologies for mitigating climate change.
⑤	TS-21	Use of Electric Plug-In/Fuel Hybrid Vehicles (Commercial / Public)	82%	Increase mix of All-Electric vehicles with Electric Plug-In / Hybrid Fuel vehicles with longer range for public.
	TS-20	Use of All-Electric Vehicles (Commercial / Govt.)	81%	Technology maturing. Limit initial use until more mature.
	TS-22b	Energy Storage - Mobile Small (Aluminum-ion or Sodium-ion)	97%	Aluminum-ion or sodium-ion battery technology is safer than lithium-ion in electric vehicles
	TS-22a	Energy Storage - Mobile Small (Lithium-ion)	52%	The use of lithium ion battery technology is unsafe. Replace lithium-ion with aluminum-ion or sodium-ion.
⑥	TS-24	Desalination - Distillation (Solar/Geothermal)	95%	A vitally needed process for potable water to population
	TS-23	Desalination - Reverse Osmosis (Solar/Geotherm	92%	A vitally needed process for potable water to population

(Source: Author's analysis of the existing and emerging technology databases)

This Author has written other books on technology development, ("History of the USAF Lightweight Fighters" and "The F-16 Fighting Falcon Multinational Weapon System") and they have the same basic theme as this book: *"As the twig is bent, so shall the tree grow"*. Readers are reminded that to increase knowledge in any topic that time, effort, and funds must be expended. The useful knowledge gained by such expenditures is typically worth the expenditure, however, speaking from experience, during the early testing of a prototype there is confirmation of the pros of the design but there may be surprise cons identified that reflects the state of knowledge was not as well-founded as it was believed before the test. The same exists when doing an analysis rather than a test. At the start of the analysis of global warming, climate change, and oceans' surface levels rising this Author was appalled to observe that the observations made in a laboratory experiment on a beaker of carbon dioxide at high concentration (above 99%) could be interpreted to have such a profound effect on the total atmosphere and resulting climate of the planet when at a concentration of only 0.035% (and of which 0.030% was naturally generated).

Errors in judgment are expected when exploring a new technology or observation a new chemical experiment, but when errors or misjudgments are caught early (by peers independently reviewing the analysis or test data) and corrected there are valid scientific lessons learned. Copernicus and Galileo are not known to have arrogantly flaunted their findings that the Earth is spherical and not flat, or that the Earth orbited the Sun not the Sun that orbited the Earth. The Author found that the peers of Dr. Charles Keeling have intentionally created such professionally embarrassing misinterpretations of his magnificent observations that compelled this Author to integrate an excessive number of photographs and illustrations and tables/figures of factual data not commonly found in a peer review. It is left there for the many ethical scientists in climate science to rectify the intentional over exaggerations of their unethical peers.

The expenditure on non-valuable knowledge is essentially a wasteful expenditure that could be applied to more useful expenditures on other more valuable technologies. There have been several start-up technologies to remove carbon dioxide from the atmosphere and deposit it into the ground in order to reduce the amount of "greenhouse gas" emissions in the atmosphere. Those expenditures may be of scientific value to the science profession but realistically they do little to increase human safety, health, comfort, and improve the world transportation system.

In the lower grades of education, knowledge is not gained rapidly but it is that base of knowledge – learned by human "twigs" - on which all future gains in knowledge will be built. As older members of humanity – and regardless of language spoken or cultural beliefs – there is an obligation to prepare the youngest members of the region humans reside in for the realities of the world the young people are about to learn if they are to survive. This book hopefully will be "one small step for Man – but a giant leap for Mankind" in the learning process of our youth.

The second President of the United States John Adams (1735-1826) made a wise statement about teaching children: *"It should be your care, therefore, and mine, to elevate the minds of our children and exalt their courage; to accelerate and animate their industry and activity; to excite in them an ambition to excel in every capacity, faculty, and virtue."* Deficiencies in the education system are not as hazardous as those described in the foregoing sections of this book, but part of the mitigation of climate change is to educate the public what can be done to mitigate those hazardous effects.

C37) Our children and grandchildren deserve to be taught the correct science of the World they will live in – and to prepare them to survive the extremes of the daily weather, seasonal climate, and natural hazards of their regional environment, not merely the average global temperature. Children are too valuable to risk their young lives being cut short when being transported in an electric vehicle which has a time-bomb battery beneath their little feet. The world population is also too valuable to risk their lives by being exposed to toxins and carcinogens from coal ash residue, environmental pollution (especially that from plastics), and unsanitary waste disposal poisoning the water they drink, tainting the land which produces their food, and fouling the air they breathe.

The real naysayers about climate change are the ones in "settled science" who fallaciously hypothesize that the anemones on the coral reefs in the oceans are in crisis of bleaching white because of Man burning any form of fossil fuel (coal, oil, or natural gas).

Copernicus and Galileo were branded as being naysayers for challenging their peers who held that the Sun revolved about the Earth, as were others who challenged that the Earth was not flat but was a large sphere. This book may be interpreted by some as "naysaying" by a would-be Galileo admirer but the facts in the numerous databases analyzed completely and in depth presents the "inconvenient" truth. The cartoon on the front cover of this book was chosen as a means to start this book and is repeated as a way to end this book with a lasting impression that scientific theory does not translate to practice and reality. "A picture is worth a thousand words."

(Credit to unknown cartoonist)

This Author expresses his gratitude for having been given the honor of analyzing and interpreting the works of those scientists who created the many factual scientific databases portraying the "inconvenient truth".

Chapter 5
Problem Analysis Using a Weighted Trade Study Process

The differences between the way a scientist makes an observation and strives to formulate a model that describes the repeatability of the observation, and the way that an engineer tries to apply the scientists' observation into a useful tool in the real world is far beyond the scope of this book to attempt to explain the complete set of differences. Readers who have reached this point in the book should already have begun to realize that the system engineering process – referred to as the "big picture" process – can only be accomplished by an analyst who has been educated and is well experienced in the application of the STEM elements of science and engineering. The acronym STEM is reminded to be the four major branches of Science Technology Engineering Mathematics, but the Author has gained the experience to add the word Applied into the acronym such that it becomes STEAM.

The elements of STEM (or of STEAM) have been used to some degree in every technological development known to Man, and that is no more applicable than to the multiple technologies that have caused the development of the United States and many other countries to have advanced in the past two centuries. Libraries are replete with books on major technologies such as the invention of the steam engine in industry and transportation, the generation of electricity to power useful machines (from the common light bulb to the development of digital computers), but also for less scientific purposes such as developing the weapon systems used in wars. There is no better example of a technology developed for one purpose to however become useful for a far better purpose by applying the STEM/STEAM process than that of nuclear energy.

History records that the atomic weapons developed by the Manhattan Project in the early 1940s to end World War II (by the bombing of Hiroshima and Nagasaki in Japan) accomplished their purpose for development – to cause an enemy to surrender rather than face the horrific destructive power of such awesome weapons as the atomic bombs developed by brilliant scientists such as Drs. Enrico Fermi, Albert Einstein, and Robert Oppenheimer. Then-President Harry Truman had to make the decision no man should have to make as a trade-off – destroy many thousands of human lives (military enemy) to avoid the loss of perhaps a few million other human lives (populates of Japan not engaged in the combat and of the servicemen of the US military).

However, there was no formal tradeoff of what height above the ground would the burst of the atomic weapon be to contain the damage to military targets concentrated within the circle of destruction. Because there was a pressure altimeter used in the detonation system and in the two drops (from aircraft named "The Enola Gay" and "Bock's Car") of the atomic weapons ("Little Boy" of spherical shape and "Fat Man" more of cylindrical shape) they both were detonated at the wrong (higher) altitude which caused the circle of destruction on the surface level to have a larger diameter, hundreds of thousands of innocent (i.e., non-military) humans were instantaneously destroyed collaterally in both atomic bursts or died shortly after from radiation poisoning. That account of the two botched technology events in 1945 points to the vital need to identify all factors involved when doing major trade studies having global implications. Ignoring or overlooking a critically important factor because it does not "fit" a preconceived hypothesis can have disastrous results. For this book, the most disastrous oversight has been the environmental protectionist cult prioritizing the safety of the environment to be more important than the safety and health of the human population of the planet.

When the Author was conducting the analysis for his Master of Science thesis at Georgia Tech in 1959 – during the "Cold War" years (14 years after the 1945 atomic bomb disasters to end World War II), - federal grant money became available for developing a more reliable arming and detonator device based on timer technology than the pressure altimeter technology and process for detonating air burst of weapons (having different shapes and possessing an aerodynamic "shape factor" or ballistic coefficient) dropped from an aircraft at variable speed and altitude conditions. A complex thesis research and application of the scientific STEM process in a major multi-variable trade study generated for the innocuously titled Master of Science thesis was created by the Author:

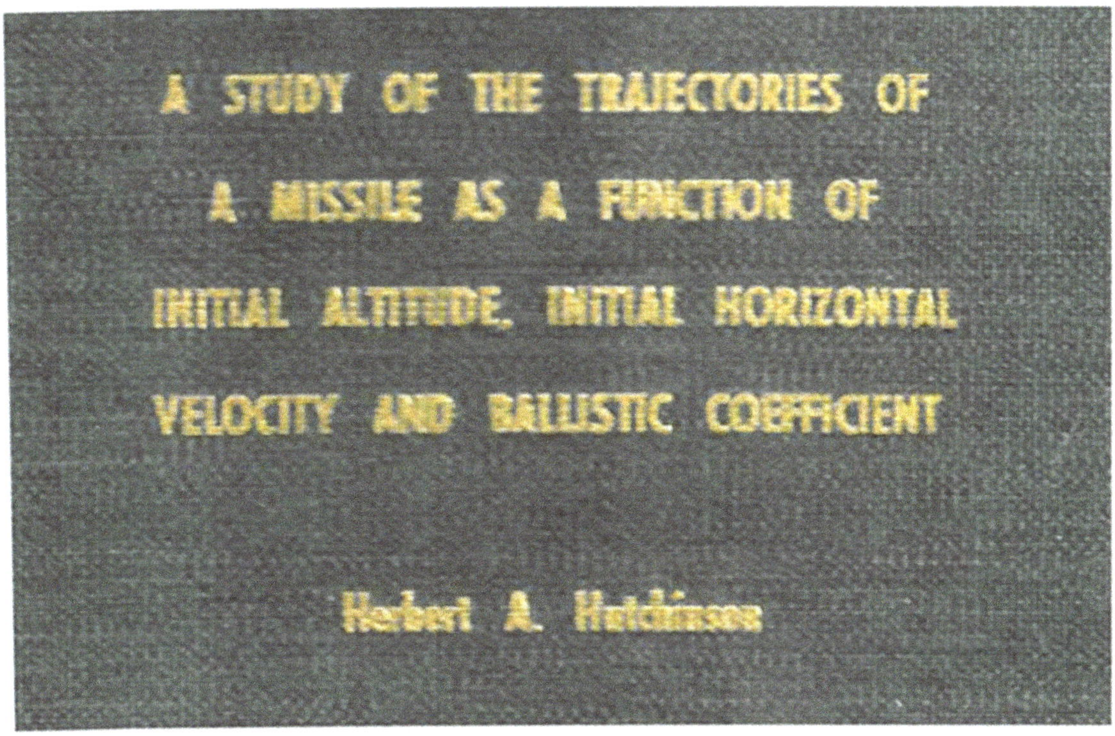

Cover of M.S. in A.E. Thesis at Georgia Tech (1959)

To protect the Georgia Tech Graduate School from any government sanctions on a college developing a weapon of war, the study was conducted strictly for researching the aerodynamics of the trajectories for any shape body (labeled as a "missile"). There were no calculations done involving the amount of warhead material being carried or for estimating the destructive "circle of effect" of the blast (heat and pressure waves) from whatever material might be detonated at altitude. When orally presented to the Thesis Review Committee, the Author presented his thesis research as a means of applying STEM in accurately determining the trajectories of a missile containing highly important papers or products to a point on the surface of the planet having arbitrary GPS coordinates. In other words, it was the experience of learning the multi-variable trade study process and the associated necessary discipline and perseverance of the trade study process that created a package of scientific knowledge. Every trade study has a unique set of variables involved and for a useful purpose and a set of unique "pros and cons" – in simply advancing a potential technological capability, or resolving the adverse effects of global warming, climate change, or oceans' surface levels rising.

Fortunately, the level of "Cold War" decisions made in 1945 (guided by the STEM process of analysis) accomplished the global result desired – detente on the planet but with a terrible impact on humanity and the environment. Following every war, there typically follows a "big picture" period of reconstruction of the material aspects of the war – rebuilding entire cities with better building structures, residential and commercial, and better infrastructure for transportation. It is not common knowledge that President Dwight Eisenhower began the construction of the Interstate Highway system in the United States after seeing that Germany had developed their Autobahn highway system to provide the capability of moving the materials of war from one place to another in a tactical manner. President Eisenhower also built into the Federal Interstate Highway system the capability for military aircraft to use the Interstate Highway as landing sites for military tactical and field support operations. That required a straight segment of the Autobahn every few miles to serve as a runway and with the thickness of the pavement to support the weight of a fully loaded cargo aircraft. The US Interstate highways are designed and paved in a manner mirroring that of the German Autobahn system.

For this book, the lessons learned from the mistakes and misjudgments with previously developed technologies serve to guide the development of other technologies "for the betterment of humanity". The decision to "combat" climate change to avoid damage to the environment by humans and industries has led to the development of technologies for mitigating the effects of global warming, climate change, and oceans' surface levels rising.

The lessons learned from prior technological developments have a history of also having to re-learn (i.e., update) prior lessons learned. In keeping with the purpose and scope of this book, it was shown that development of certain technologies developed for combatting climate change – especially the natural renewable energy technologies of solar and wind – intended to eliminate the "greenhouse gas emissions" from Man and industry burning fossil fuels has not resulted in the technological progress hoped for by environmental scientists and those in the Environmental Protection Agency EPA.

Technology can be developed for one of three purposes: 1) to resolve a "con" with a previously developed technology (with coal technology being a classic example), 2) to resolve a change in one of the many variables that humans encounter in living on the surface of a planet that is an unstable surface that is evolving into a perceived problem (the example being that of the climate warming and posing a "crisis" to some types of life on Earth's surface; and 3) to develop the technology to explore an observation made in a controlled laboratory environment and explore what might become an asset to become of tangible value to humanity.

There is actually a fourth category which those scientists in the field of technology development refer to as "technology for technology sake". The Author was astonished to learn that several start-up companies have been given large amounts of grant money to pursue various means for sucking carbon dioxide from the atmosphere, changing the state of the carbon dioxide from gaseous state to either liquid or solid state and reintroducing that liquid or solid carbon dioxide into the earth at a depth where it can no longer affect the temperature of the planet's surface. The principle is scientifically interesting but is of no practical use toward controlling the climate.

Earlier in this book, the pair of statues in front of the US National Archives in Washington, DC, bore the inscriptions "Study the past" and "What is past is prologue". Geologists studying the layers of sediment of the Earth in the southwestern states of Utah, Arizona, and Montana, have reconstructed much about the past history of conditions on the planet's surface as that surface became buried deeper and deeper under the sediments deposited by water. Other locations such as the Grand Canyon reveal that history as a result of water – the Colorado River – cutting through and ever downward through those sediments. There is a Trail of Time at the Grand Canyon with pieces of the Grand Canyon wall on pedestals for visitors to inspect and touch. The National Park Service has carbon-dated each of the awe-inspiring rocks on display. The following segment of sandstone shows the segment of sandstone that is dated to be 315 million years old. It is seen to have a characteristic red color that comprises much of the Grand Canyon colorful spectacle

(Source: Photograph by Author at Grand Canyon, Arizona)

In another of Nature's historical documentations, the Vostok Ice Core Samples database provided the scientific interpretations of the temperature and the gaseous content (carbon dioxide, methane, etc.) of Earth's atmosphere for the past 420 million years at large depths of the Antarctica ice sheet (at latitude 78° South). It is this database that comprises most of the factual interaction between Earth's surface temperature and the atmosphere surrounding the planet, and which the Author has added additional scientific interpretation.

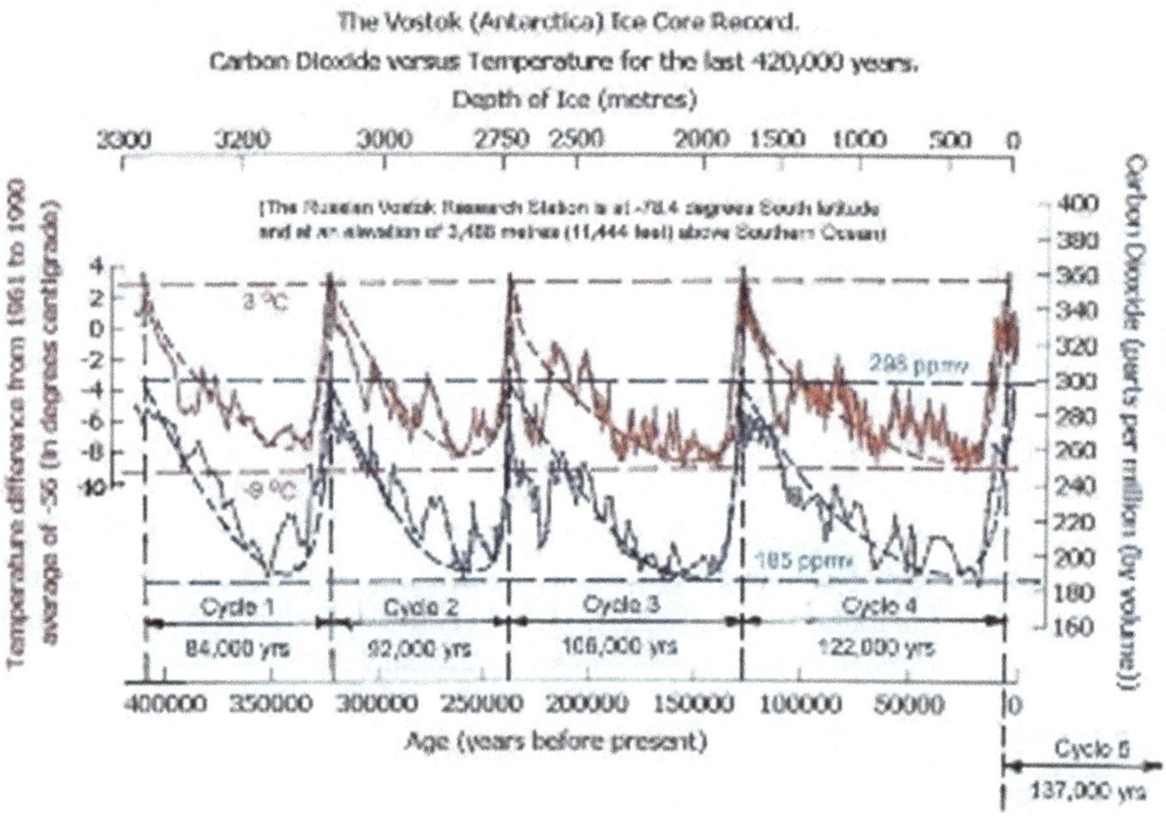

(Source: Smithsonian Institute for Science with Author notations)

"Studying the past" aspects of the Vostok ice core database, it is revealed that there have been four repeatable cycles of temperature and carbon dioxide on Earth's surface in the past 420 million years, but with an apparent increasing length of period from the start of a u-shaped cycle (where temperature (in red) and carbon dioxide (in blue) peak at the high recordings, decrease to low cold recordings and then increase to the high hot recordings. A scientifically valid observation can be made that each of the four cycles were caused by astrophysical, geophysical, and natural factors and without the presence of human and industrial factors such as greenhouse gas emissions of carbon dioxide and methane. The carbon dioxide in the atmosphere of Vostok was therefore not a result of anthropogenic activities such as the burning of fossil fuels.

It should be of interest to notice that about 12,000 years ago, the bottom cold peak of the fourth cycle had a temperature of about 12 ^{0}C colder than it is now. That is referred to as the Last Little Ice Age" and the North American Continent looked very much different than it does 12,000 later.

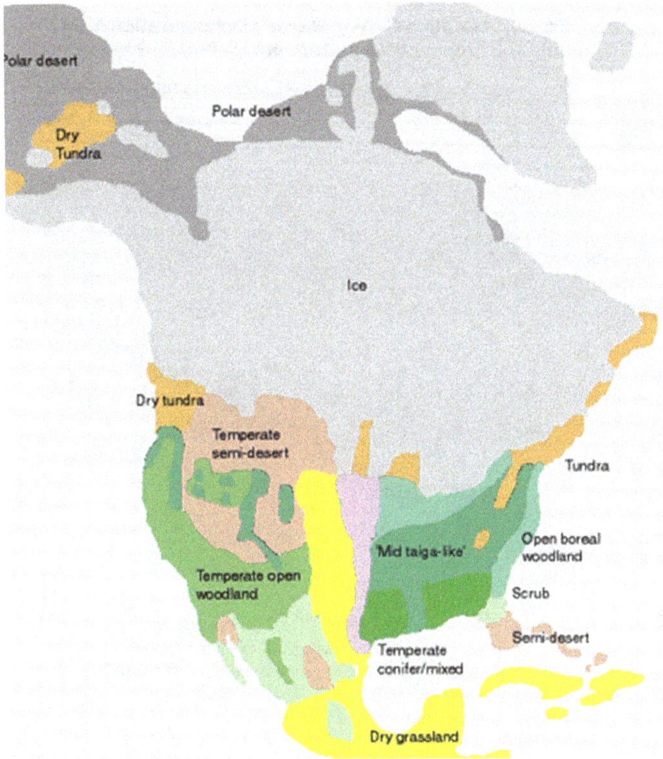

(Source: New York Museum of Natural History)

The Author grew up in the New York City area and during a visit to relatives and to Central Park, a photograph was taken of the glacial grooves etched into a large boulder in Central Park as the ice sheet translated the boulder over a harder subsurface:

(Source: Author's Personal Photography File)

Another important observation is that the end of the fourth cycle is in the general range of time considered as "Now". The high peak of temperature at Vostok based on the scientific interpretation is at 3 °C above the "normal temperature" at Vostok (i.e., at -53 °C). Correspondingly, the high peak of carbon dioxide – all of only natural causes (i.e., "pre-Industrial Revolution") – is seen to peak at about 295 ppmv (0.0295%). In the NASA satellite databases (2000 to 2013), there is shown to be a latitude effect on the temperatures sensed/measured by the satellite at mid-troposphere altitude:

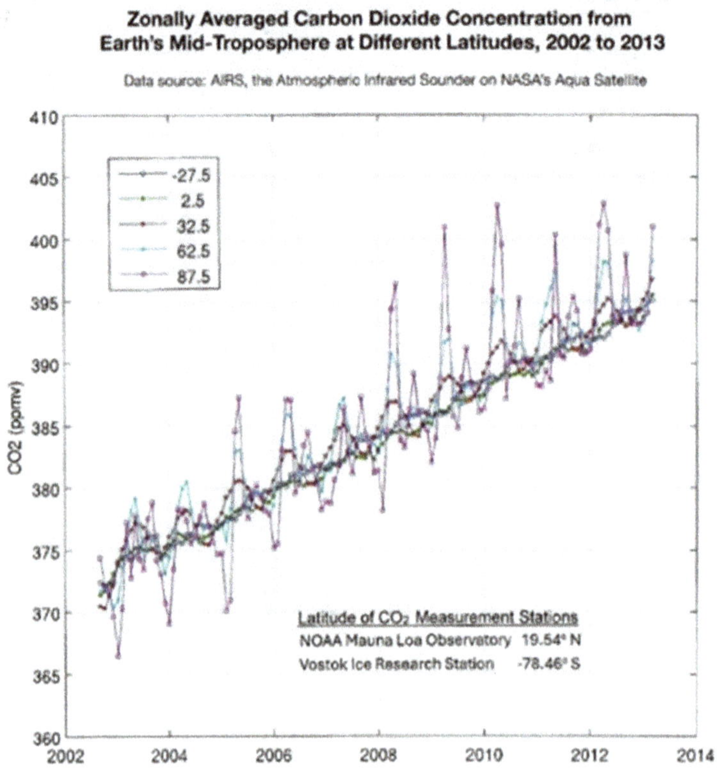

(Source: NASA)

Another important database is that from the NOAA operated Mauna Loa Observatory (where Dr. Charles Keeling managed the measurement operation for many years) (on top of the Mauna Loa extinct volcano in Hawaii) where there are daily recordings of the gases in the atmosphere at that latitude of 20° North, and which contain total contributors (astrophysical, geophysical, meteorological including water vapor, human, and industrial) to that measurement at Mauna Loa. A recent set of the total carbon dioxide measurements is seen as follows:

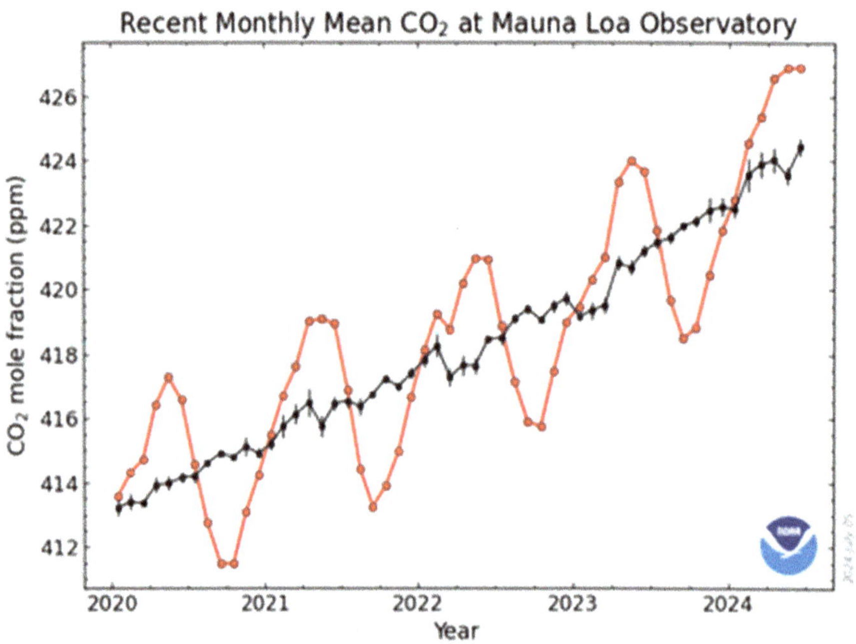

(Source: NOAA)

Readers' attention is called to the variation of the data in red color. That is the total carbon dioxide in the atmosphere measured in parts per million by volume (ppmv). The measurement of 426 ppmv of carbon dioxide seems to be a high amount – 426 units of something has got to be important. When converted to the more commonly understood percentage form, that huge amount of carbon dioxide in the atmosphere is at a concentration of 0.0426% of the atmosphere. NOAA scientists "adjust" the data that is reported to the public by removing the effect of water vapor which gives the red line a cyclical shaped variation of one cycle per year. In other words, the "seasonal" effect – not anthropogenic (human or industrial) related – is removed and the data presented to the public is the black line variation of carbon dioxide "smoothed" to infer that the anthropogenic contribution to heat-trapping "greenhouse gases" is rising at a "crisis" rate. A professional statistical analyst – the Author does not qualify to be named as such - would probably criticize NOAA for "smoothing out" the very important seasonal effects on the measured database. The seasonal effect infers that there was an associated variation in water vapor in the atmosphere, and as Dr. S. Fred Singer has criticized environmental scientists for disregarding the dominant effect water vapor in the atmosphere has compared to the environmental scientists' set of selected "greenhouse gases" the adjustment to the measured CO2 masks the seasonal and water vapor effects that are far more important than are the carbon dioxide and methane emissions from the burning of fossil fuels.

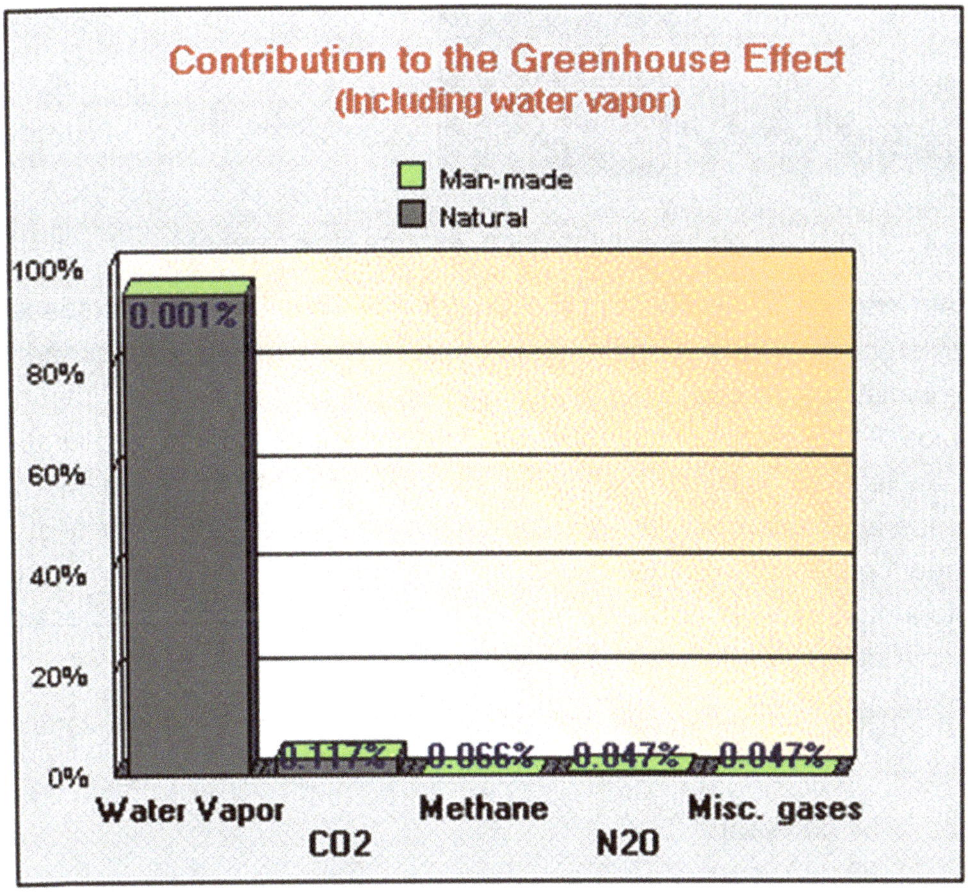

(Source: Dr. S. Fred Singer, Professor Emeritus, University of Virginia)

When converted to percentage, the black line has increased from 413 ppmv (0.0413%) in 2020 to 426 ppmv (0.0426%) in 2024. Does an incremental rise of 0.0013% of anything (except perhaps a nerve gas) in four years constitute a "crisis"? News media would be doing a great service "for the benefit of Mankind" to explain the how/who/why/where/when of that environmental "crisis" looming so ominously to this Author and to Readers. There was nothing found in this comprehensive "big picture" analysis to support the need for "crisis management' of Man burning oil or natural gas to meet the energy needs of the population.

Environmental scientists also have disregarded (intentionally and/or unintentionally) the finding of Dr. Nicolas Caillon – a French scientist conducting ice research at the Vostok Ice Research Station (latitude 78° South) who conducted a closer look at the ice core samples and discovered the fact that the high peak of the carbon dioxide in the air bubbles of the ice core lagged the temperature of the atmosphere by about 800 years.

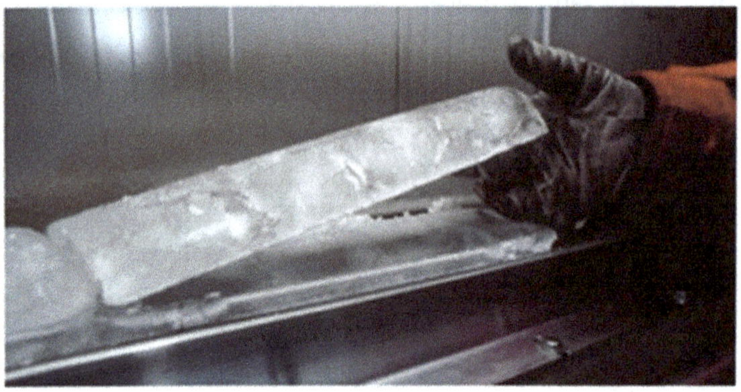

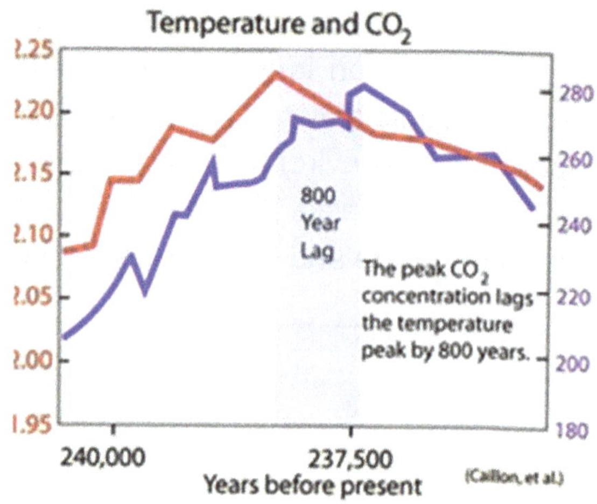

(Source: Smithsonian Institute for Science)

A more important conclusion drawn from the Vostok Ice Core Samples database, originated by Russian scientists shortly after the end of World War II and near the start of the "Cold War" (no inference to the climate conditions of the time), was the "The past is prologue" part of the "Study the past" philosophies. The Author's analysis of the four cycles of varying temperature and carbon dioxide shown earlier was analyzed for a valid scientific estimate to be whether a fifth cycle (of atmospheric temperature and carbon dioxide) is in its early cyclical stage now or in the near future. That is an important projection because the set of technologies to meet the energy demands of the populus and the industrial base of the planet will alter its energy demands because of the large decrease in temperature toward another "little ice age".

Referencing the Vostok ice core sample database showing the temperature variation for the past 420 million years, the four cycles measured periods tabulated below:

Cycle 1	84,000 years
Cycle 2	92,000 years
Cycle 3	106,000 years
Cycle 4	122,000 years (ending approximately now)

Given that pattern of cyclic variation and having found that there is no apparent effect of anthropogenic activities that will affect the planet's surface temperature to any significant amount, the projection can be scientifically made that there will be a Cycle 5 starting in a few centuries from now and lasting for a period of approximately 137,000 years. Within those 137,000 years of Cycle 5, the extremes of temperature (which are dependent upon the latitude, elevation, and ocean currents per the LEO method for estimating the annual average temperature variation) would decrease from what it is now (an average annual global temperature of 16 °C – 61 °F) to about 9 ^{0}C lower than the current average annual global temperature – i.e., to 7 °C – 13 °F).

At that low annual average global temperature, there very well would be the formation of glacial ice covering the upper northern hemisphere to about latitude 35° North. Readers are reminded that there was glacial ice covering Central Park in New York City during the previous "little ice age". Whether that will occur is beyond anyone's capability to predict, but the philosophy of "Study the past. The past is prologue." Suggests that remote possibility. If that should occur, the concern of anthropogenic warming of the planet may be a positive effect on the environment rather than the current negative effect. (A tongue-in-cheek observation can be made that covering the upper half of the northern hemisphere with glacial ice would put an end to anthropogenic pollution.)

Regarding the technological developments that should suffice for the next 40-50 generations (1,500 – 2,000 years), and where the average annual global temperature is + / - 4 ^{0}C (7.2 ^{0}F) from the current temperature of 17 ^{0}C, this book makes an assessment of what set of technological developments are feasible and useful.

It is hastened to be added, the data shown above have been the "post-war against climate change" in the United States but is not representative of that of other countries. Within the United States, the reduction of the use of coal technology to reduce greenhouse gas emissions and developing solar and wind renewable energy technologies to replace the use of fossil fuel (coal, oil, and natural gas) technology did not take place. As shown in the book "Facts and Fallacies of Climate Change on our Planet" and in the previous sections of this book "Dealing with the Real Climate Change" by the same Author, it was realized by the public and by industry that despite the reduction in the use of coal technology – considered to be the worst of the fossil fuel offenders – the surface temperatures, climate, and ocean levels were continuing to change independent from reduction in the use of coal technology. Further as the previous illustration indicates, the almost totally clean burning natural gas technology – despite being a fossil fuel - was being used instead of coal by many coal-burning entities to replace coal. In 2005-2008, it is seen that the coal (red line) usage began its steady rate of decline as the natural gas (green line), renewables (solar and wind and to a small degree geothermal) – purple line – began to rise significantly.

An interesting montage of four photos reflects the modification of coal technology based on its pros and cons and the efforts of the Environmental Protection Agency to spur the reduction of the use of coal because of its offensive greenhouse gas emissions. There was a foregoing discussion in this book of the efforts of the Consolidated Edison Companies to comply with measures for transforming from the burning of coal in the stages shown by the following montage of four photos:

Coal burning facility with no emissions control

Coal burning facility converted to oil burning

Coal burning facility with EPA emissions control

Coal burning facility converted to natural gas

The EPA is commended for imposing the filtration measures (shown in the following diagram) for improving the coal burning technology in the early portion of the twentieth century:

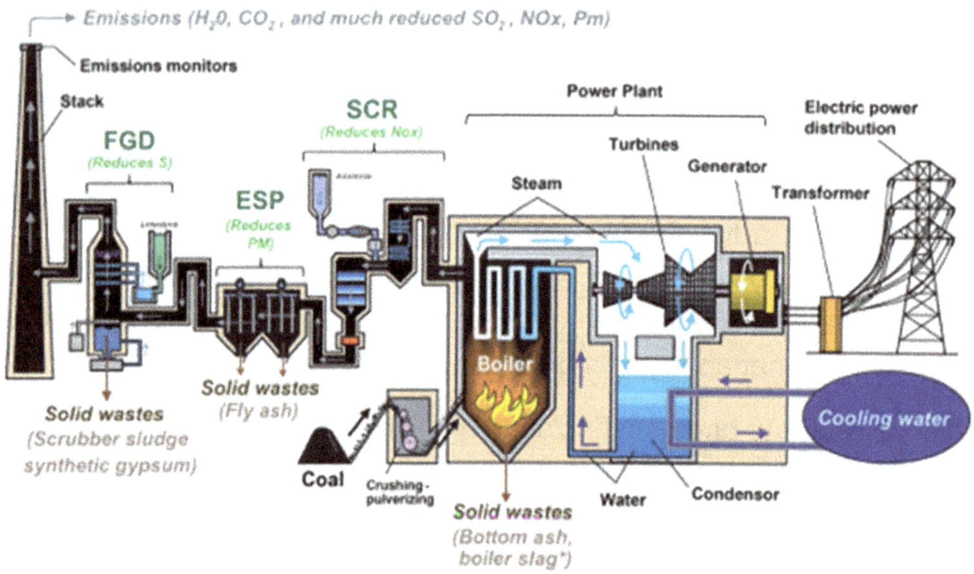

(Source: US Environmental Protection Agency)

Nuclear energy technology (orange line) has remained almost unchanged but the use of thorium instead of uranium as the nuclear fuel promises to cause the nuclear energy line to soon begin an increase also. By 2005, the technologies of natural gas, renewables, and nuclear all have surpassed the use of coal.

Public awareness that coal ash residue contains toxins and carcinogens that are entering into the aquafers below ash landfills and dumps are polluting the water and land used by humans, animals, and agriculture has grown since 2005, but by 2025 that coal ash pollution has been reduced to insignificance. Environmental scientists and the sensational news media have yet to accept what the public now knows – and which the two books by this Author have described as the wrong problem needing to be resolved:

61 SCIENTISTS CALL FOR STATE LEADERSHIP VS. GIANT CORPORATE CLIMATE POLLUTER

(Source: The News Observer, Raleigh, North Carolina, April 20, 2025)

The environmental scientists are perfectly within their rights to address the pollution problems of an industry or any other offender violating the approved anti-pollution laws and regulations enacted to protect the environment. However, they should not have the authority to cause state or federal government leaders to develop or abolish any technology development based on only the scientific suspicion – not backed by scientific evidence "beyond the shadow of doubt" - that the emissions identified as greenhouse gases are polluting the climate. Had the group of 61 scientists making the call above conducted analysis of the pros and cons of wind and solar energies as well as the analysis and tradeoffs of the use of the three fossil fuels sought to reduce greenhouse gas emissions effect on the climate, they would have not found any factual data supporting their fallacious hypothesis that it is the fossil fuel technology that is the root cause for the climate to be changing adversely as claimed. The public has been misinformed for so long by the environmental scientists about the pros of solar and wind energy, but know little about the cons, there are frequent protests against the local energy provider. A photograph of a very recent protest is seen in the following excerpt from the local newspaper:

(Source: The News & Observer, Durham, North Carolina, May 5, 2025)

The above protest essentially is to stop the energy company (the same company referred to in the news media as being a "giant corporate climate polluter") for not using solar energy exclusively as the source of energy for the town. The energy provider is providing energy to the town with the more reliable and cleaner natural gas. The term fossil fuel has been equated in the public mind as being just another fossil fuel creating greenhouse gas emissions.

Before describing the "big picture" system engineering process that was used in the analysis of global warming, climate change, and oceans levels rising, it should be obvious to the Reader that the topics being analyzed are of global nature. Because of that complexity, there can be expected to be many considerations to be included in the analysis process. A partially complete illustration of many of the factors having any degree of influence on global warming and/or climate change is provided as follows to visually remind Readers of the complexity and scope of the analysis to be addressed in this book:

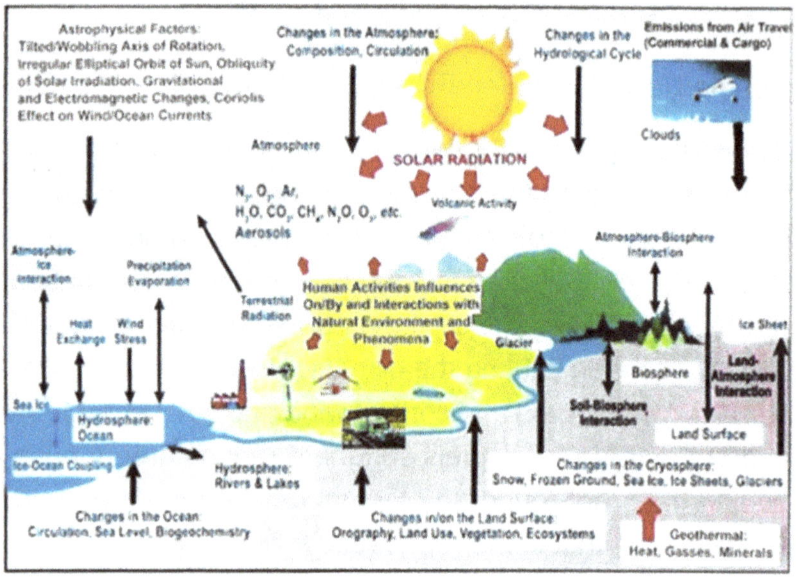

(Source: From the research done by the Author in the past decade)

For this book, the Author (who is retired from a highly successful career in system engineering) has used what has been referred to as a "big picture" approach to the understanding of the complex problem which scientists have titled "global warming and climate change". Environmental scientists have elected to use a "bottom's up" or a "grass roots" approach to describe the same complex problem of global warming and climate change. Not unexpectedly, the conclusions have been found to differ widely between the two approaches, the "big picture" approach being scientifically more accurate.

The purpose of this chapter was stated earlier as being to place into context the actions that should be taken to apply the conclusions that are developed in the "Facts and Fallacies of Climate Change on our Planet" (a previously published book by this Author) to accomplish effective actions "Dealing with the Real Climate Change" (the title of this book by the same Author). For starters, the Readers of either or both of these two books now have been flooded with information mostly of scientific nature to better inform the public that the causes for global warming, climate change, and oceans' surface levels rising are of scientific nature and not of anthropogenic nature. The planet was not created to be a toy for humanity to play with. The planet existed long before it accepted human passengers for an awesome travel at awesome speeds and for awesome distances through space.– not unlike the old carousels at amusement parks featuring horses of different color and bobbing up and down as the carousel rotated to give the urban cowboys on their favorite color wood horse a memorable ride for a short period of time. The reality is that the planet has developed its climate independent of anthropogenic opportunities for amusement or entertainment. The ancient mariners on their amusement rides in sailing ships shared their information in ships logs and created the graphic database for the climate – mainly the temperature – as they experienced it at different latitudes. A "big picture" illustration of the climate on the planet has long existed as shown in the following illustration:

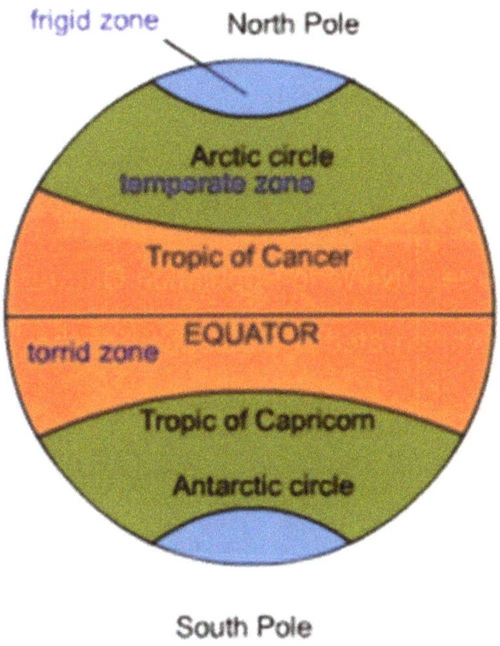

(Source: Encyclopedia Britanica)

What has changed most dramatically on the planet in the past few centuries is the size and distribution of the human population on the planet, and the myriads of technologies that have been developed for "the betterment of Mankind" serving that population. One of those technologies that has developed is the capability of visualizing (qualitatively and quantitatively the plant temperature distribution with latitude as an update to the preceding diagram: (Astute Readers may recognize the next diagram as being the front cover of the "Facts and Fallacies of Climate Change on our Planet" book by this Author.)

The next illustration shows the recent distribution of the population by continent, followed by an illustration of the distribution of the population by latitude

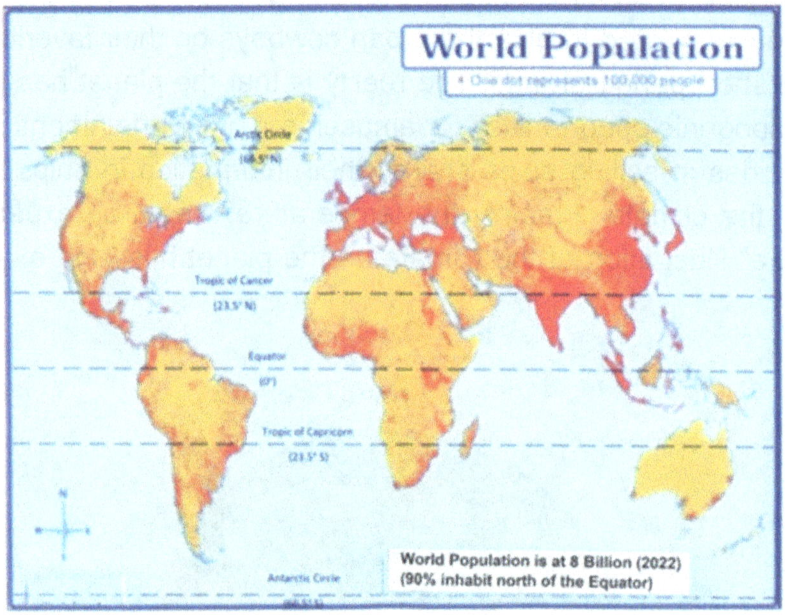

(Source: UN World Population Organization)

Alasdair Rae (affiliation unknown) created a distribution of the World Population as of 2022 by latitude:

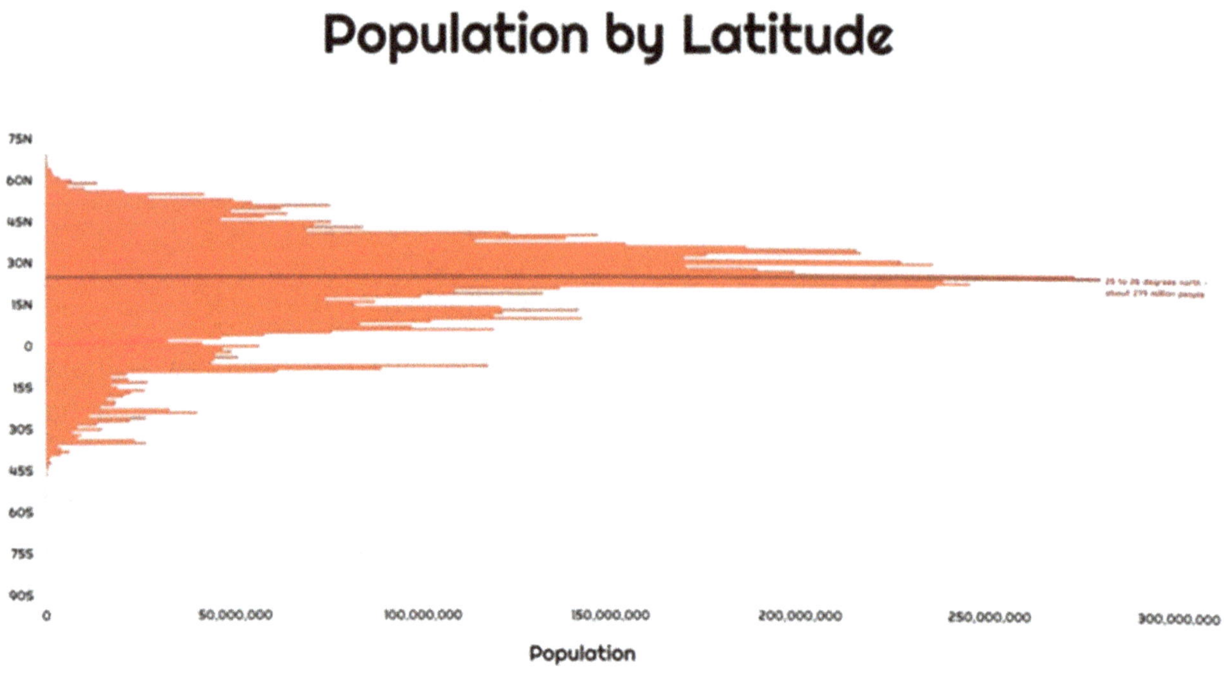

(Source: Alasdair Rae – Affiliation unknown)

It should be evident that there is a powerful effect of latitude on almost everything that touches humans and their likes and dislikes, their choices of where or where not to live, and confirms the findings of this Author that there is also a regional effect on what technologies are in use by the planet's population and in proportion generally to the size of the population in certain regions. Humans do not live year-round in Antarctica where the average annual temperature is -40 °F (which coincidentally is – 40 °C). Over two thirds of the Earth's surface is water and there is practically no human population residing permanently on the water's surface where energy must be provided to that population. The large energy demands are in regions where the population is large. A revealing illustration was found during this research that the demographic aspects of the world's largest energy users are in less than a dozen regions of the planet and are all within the latitudes between 75° North and 50° South where the climate (which includes both temperature and amount of annual precipitation) – as seen in the previous three illustrations – is within a safe and tolerable temperature envelope.

A "Which came first, the chicken or the egg?" quandary exists in the primary regions – did the resources of an energy source in the region draw the population or did the population grow and create the demand for the form of energy most used? In the following illustration, Readers might ask "Did each of the regions draw the population to the region and the technology dominant in use grow to meet the population's demand for energy, or did the availability of an energy source draw the population to the region?" It should become apparent that the question is more complex than an either/or set of conditions, but instead a complex set of factors having an interaction with each other.

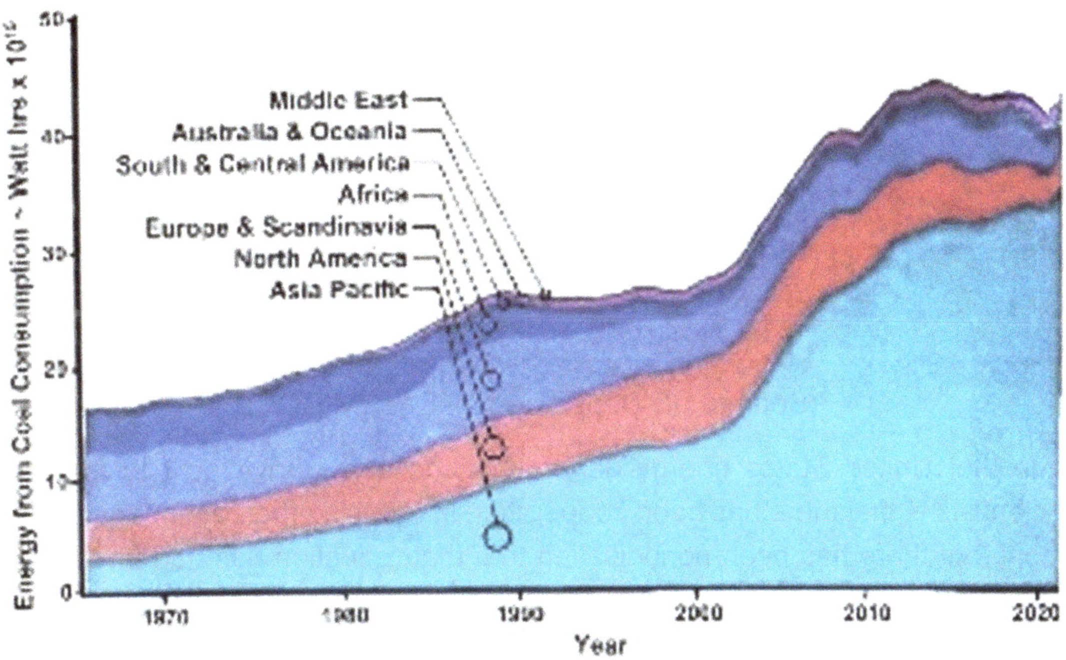

(Source: British Petroleum p.l.c.)

Six of the regions are seen to be reducing their dependency on coal technology but China looks like it is increasing its use of coal. Did China do the research similar to what was done in these two books and conclude that reducing the dependency on fossil fuels to improve the climate wasn't worth the effort? The above illustration reflects why these books by this Author recommends that each region should work the global warming problem as regional warming, global climate change as regional climate change, and oceans' surface levels rising as a regional problem. Noticeably absent above is Russia – one of the biggest regions on the planet. The research found that several regions are reducing their dependency on coal just as the United States has done and is doing but almost no region is reducing their dependency on the other fossil fuels – oil or natural gas.

The conclusions from "Facts and Fallacies of Climate Change on our Planet" leave no doubt that the United States has reduced its dependency on coal as seen in the illustration shown a few times previously in this book and must be shown once again for the important message the trends that technologies have developed in the past twenty years in order to reduce the greenhouse gas effect on global warming and climate change.

The US Department of Energy recently (early 2025) issued the "post-war" status of technology developments for generating electrical energy in the United States:

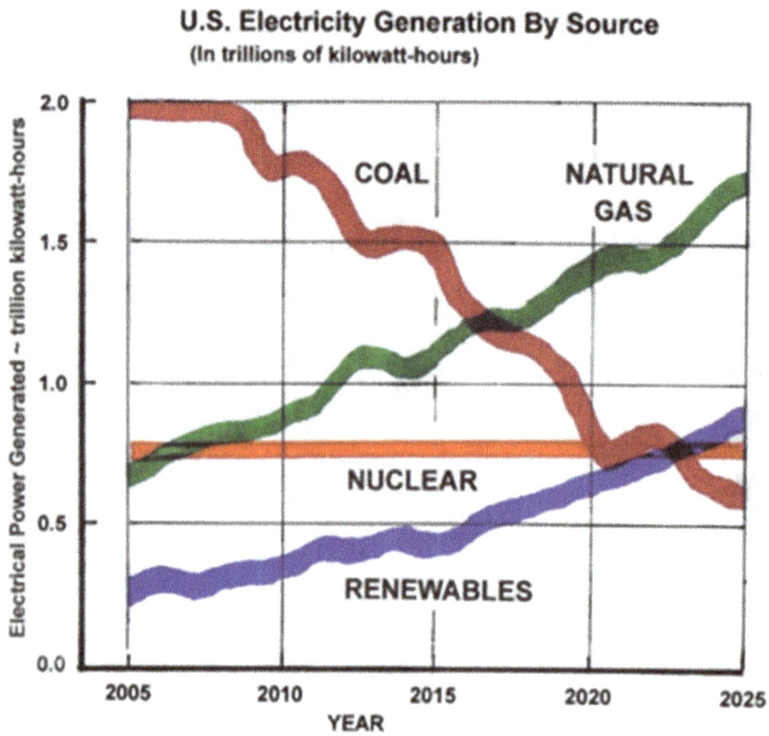

(Source: US Department of Energy, DOE)

The message that United States energy generators and users have sent to the environmental protection scientists is that there has been cooperation in reducing the dependency of the region on the use of coal, but there has been no indication that that reduction has had any significant, if not any effect at all, on altering the changes in average global temperature or on global climate change. The book "Facts and Fallacies of Climate Change on our Planet" has made it more than abundantly clear that the environmental scientists' hypothesis that anthropogenic greenhouse gasses is the dominant cause for global warming and adverse climate change has resulted in failure to change the rising trend in the planet's surface temperature or made any change to either of the two major components of climate – temperature or precipitation. To this Author, the resistance to "see" that the hypothesis is fallacious is beyond scientific understanding and logic.

The time has come to re-vector the entire approach the United States has taken during the past century on the technologies to be developed to mitigate those effects of regional US warming and regional US climate change – and for other regions to neither cooperate or resist the United States from influencing what those other regions should do to mitigate the effects of surface temperature and climate changes unique to their respective regions. It no longer is a problem of global warming or global climate change.

The role that the United Nations has coordinated with the United Nations Intergovernmental Panel on Climate Change should be also re-vectored away from global warming and global climate change. It is beyond the scope and responsibility of this Author to criticize or offer specific constructive criticism to the United Nations. However, in this Author's opinion, the United Nations would do well to serve as a coordinator of the individual regional technology developments that each region pursues to cause a sharing of technical and scientific knowledge being gained during the various regionally unique set of technology developments. Some technologies have pollution (to land, water, and/or atmosphere) that are scientific pollutants and need to be closely controlled for any effect they have on lifeforms and the environment. The United Nations current leadership efforts in controlling global pollution should also be increased, and global pollution will become the "crisis" replacing the now known to be fallacious greenhouse gas emissions crisis. The global population may reach 11 billion humans by 2100 and at the same average rate of pollution (mainly plastics) per person that exists on the planet in 2025 will also increase in proportion to the population growth as it occurs in future years.

The remainder of this book focusses on what the United States could and should do for its unique regional challenges in technological development and within the resources that are unique to the United States – all 50 states. It is a reminder as previously stated that challenges also bring opportunities.

The "big picture" approach used in this research has led to the identification of 25 different technologies to assist the United States public in mitigating the adverse effects that are attendant with increases in the temperature of the planet's surface and the resulting climate. They are not separate problems but are interacting problems, both affecting the natural environment and public safety and health (and several other aspects of life humans enjoy). Readers should also ask: "What gives the Author the impression that he is correct in his assessment of the set of technologies that should be pursued "for the benefit of Mankind in the United States."? Readers wishing to review the Author's education, experience, and personal approach taken in the past sixty years to problem solving are invited to read the About the Author section near the end of this book. A Contact the Author (for Readers with questions, requests for clarifications, and constructive scientific (or any aspect of STEAM) criticism) is also provided.

Midway in the Author's career as a system engineer, he was honored to be chosen to be the Chief System Engineer for a program in the US Air Force (the YF-16 Lightweight Fighter Advanced Technology Demonstrator which evolved into the full-scale development of the F-16 Fighting Falcon Multinational Weapon System). To transition the advanced technology demonstrator into the production program operating in the real-world environment required the use of a series of 28 "weighted trade studies" to "missionize" and "weaponize" the technology demonstrator. The successful use of the "weighted trade study process" was recognized by the USAF for its importance leading to the successful transition program: A visual summary of the 28 advanced technologies that were accepted for application in the F-16 aircraft, including the first all-electric fly by wire flight control system are shown:

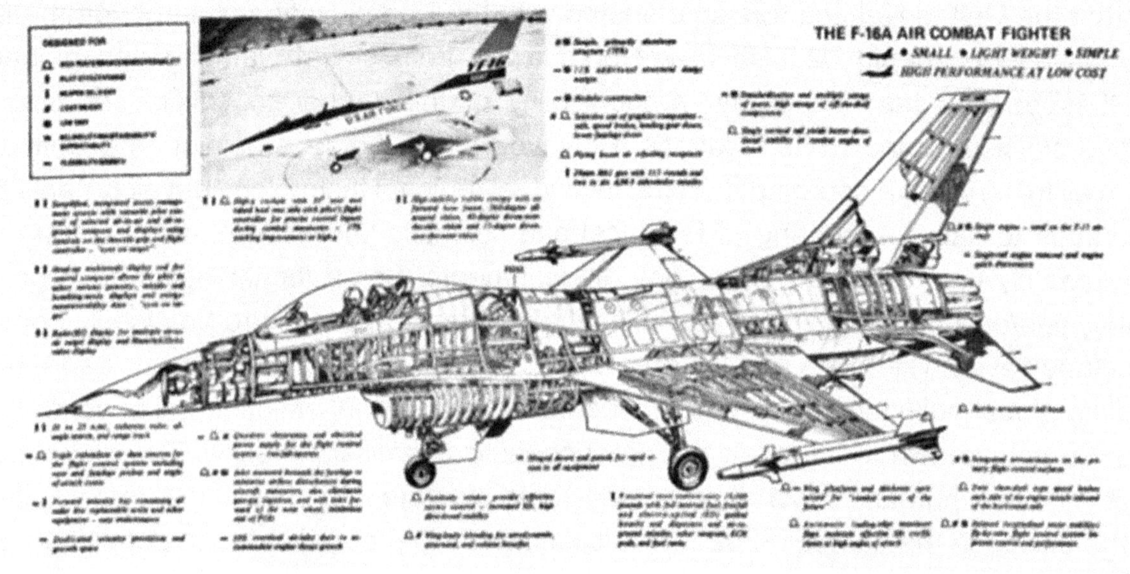

(Source: From Author's Engineering Memorabilia Files)

It is not the physical size of the system that presents challenges as much as it is the functional complexity of the system. Technologies must be integrated to play together harmoniously and not in competition with each other. That same importance of harmony exists with the factors of everything on Earth's surface. The 28 trade studies guided 26 technologies to be integrated into the F-16 weapon system and 2 that were not (either because of insufficient maturity or of having an intolerable "con" that would cause costly and time-consuming operating and maintenance costs The seven factors of importance for the F-16 weapon system are listed in the upper left corner of the diagram and with a small symbol representing the factor. The deciding factors shown in the notations surrounding the F-16 cut-away in the diagram show the symbols of the importance factor that most closely met the functionality of that subsystem technology.

The Author has substituted the many complex variables and technologies comprising the F-16 weapon system program with those comprising the equally complex problem of global warming and climate change and the many technologies that are candidates for the population of the planet to mitigate the adverse effects of global warming and of climate change phenomena.

The conclusions derived in the foregoing analysis (being reported in the two books by this Author) identify the factual scientific causes (derived from the observations and measurements) that indicate a factual rise is occurring (albeit slow and regionally not globally average) in the planet's surface temperatures. In fact, there was a period when measurements of the Earth's surface temperature were debated as to how the data was collected and should properly be interpreted. Those databases – shown as following – illustrates the difference between a biased and an unbiased interpretation:

Observed Global Surface Warming, 1900-1996

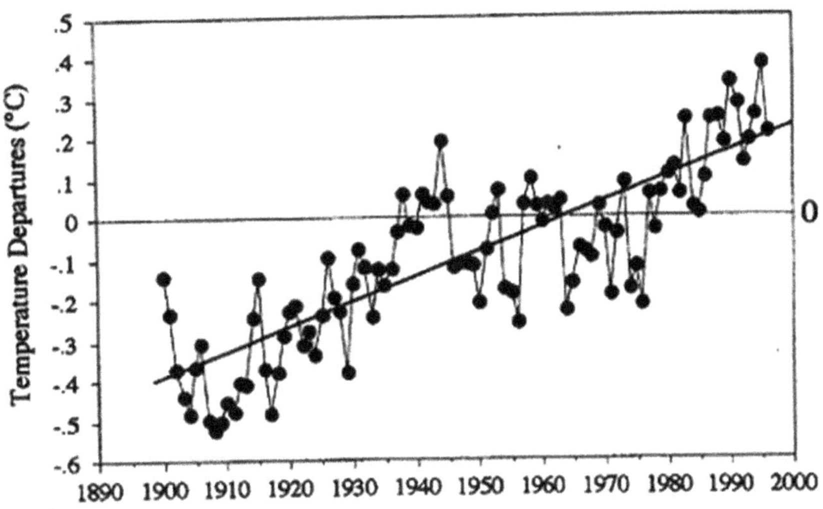

(Source NOAA database – from land-based meteorological stations)

Readers should note that the format in which the data is presented is of critical importance. While the slope of the linearized black line is visually steep, the data factually shows only a change (increase) of 0.6 °C (1.1 °F) in 100 years or 0.006 °C per year.

Satellite-based "big picture" measurements became available starting about 1976 and is the preferred measurement (i.e., does not involve "averaging on a global scale) for measurement of the extent and rate of the trends in global warming and climate change variables.

Global Satellite Temperatures, 1979 to 1999

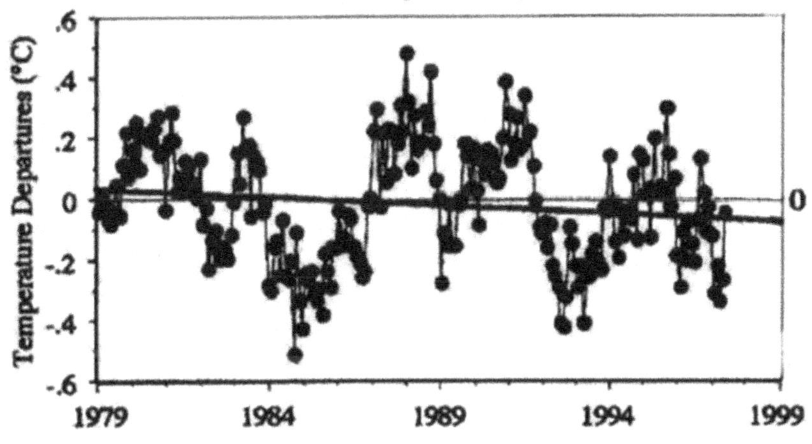

(Source: NASA)

The NASA satellite data shows the same general "variability" in the data with an approximate cyclical period of about 5 years but shows a decline – NOT an increase - in surface temperature from about 0.03 °C departure in 1979 to about -0.05 °C in 1999. Considering the averaging and linearizing of the meteorological data recorded with low technology meteorology deices mostly at airports where the area is surrounded by heat-trapping concrete and asphalt and whereas the satellites have very accurate direct measurement by high technology heat sensing instruments not affected by surround concrete or asphalt, the Author is inclined to trust the NASA satellite database to be more scientifically valid than the NOAA meteorology database.

Factually, and in the "big picture", it may be of low significance as to which is the correct change in global temperature per year, neither rate constitutes a global "crisis" rate. Both methods are relatively short-term measurements, and the long-term astrophysical and geophysical drivers of the planet's surface temperatures are better indicators of long-term climate changes. Of significance though is that the NASA satellite data is also more scientifically valid than the Vostok Ice Core Sample database for the period less than 5,000 years ago as it is real-time data and not that inferred from extrapolation of past data. In that regard the Vostok Ice Core data confirms the very historical astrophysical and geophysical long-term measurements. The Author's analysis showing the probability of a fifth long-term cycle of warming and cooling remains valid since it is based on direct measurement with high technology instrumentation and not the end result of mathematical gymnastics with the inherent containment of human recorded data being masked by "averaging" and "linearizing" – as well as scientifically unethical "cherry-picking" highly probable to have been practiced (the scientifically shameful "Climategate" incident at the University of East Anglia in 2009). The renowned cartoonist Scott Adams portrayed the "naysayer" persona in a clever "Dilbert" cartoon:

(Credit to the talented – and very introspective - cartoonist Scott Adams)

Errors in judgment or interpretation are expected when exploring a new technology or pursuing an observation made during a new chemical experiment, but when errors or misjudgments are caught early (by peers independently reviewing the analysis or test data) and corrected there are valid scientific lessons learned. Copernicus and Galileo are not known to have arrogantly flaunted their findings that the Earth is spherical and not flat, or that the Earth orbited the Sun not the Sun that orbited the Earth. The Author found that the peers of Dr. Charles Keeling have intentionally created such professionally embarrassing misinterpretations of his magnificent observations that compelled this Author to integrate an large number of photographs and illustrations and tables/figures of factual data not commonly found in a peer review. It is left there for the many ethical scientists in climate science to rectify the intentional over exaggerations of their unethical peers.

Given the different interpretations of the databases of meteorological data generated by NOAA with land-based methods and by NASA with satellite measurements, the Author was motivated into developing the LEO model – emulating the skepticism of the "science denier" shown by Scott Adams in his above Dilbert cartoon. The LEO model – which considers the combined effects of latitude, elevation, and proximity of ocean currents – is a far superior model for the average annual temperature of a region at any latitudinal position on our planet.

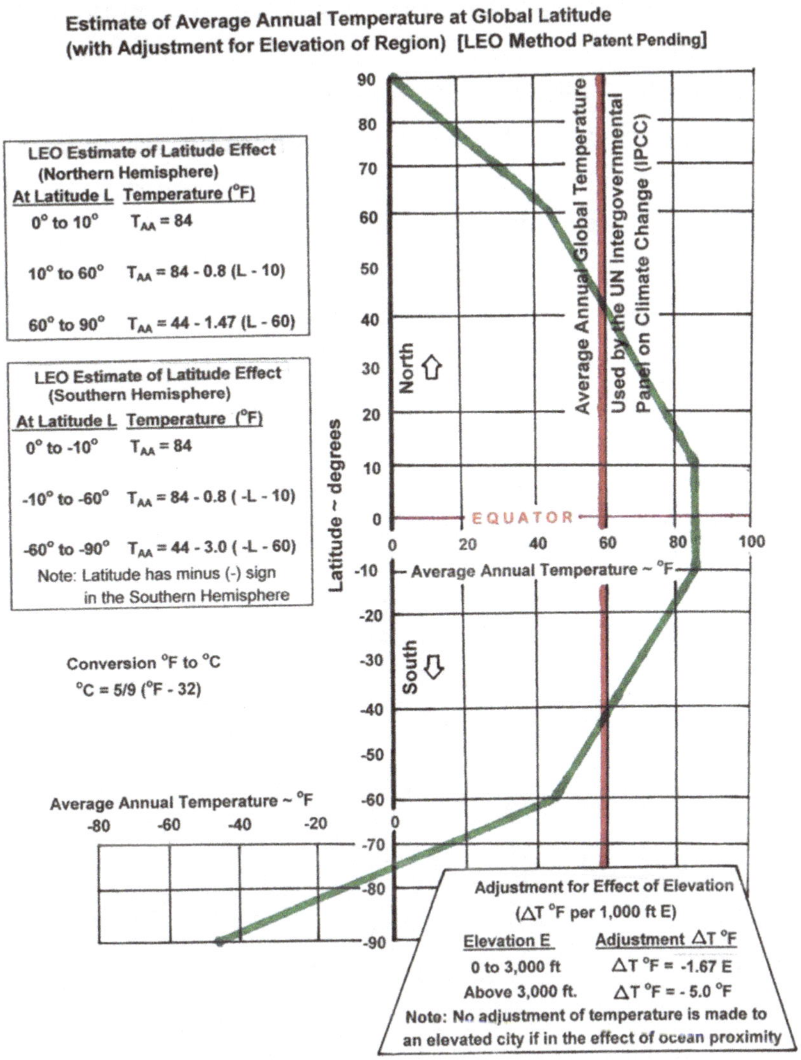

The LEO Method developed by Author

The environmental scientists' "model" for the surface temperature of the planet is seen as the red line in the LEO diagram above – at a constant "average annual global temperature" of near 60 o F (16 o C). The LEO method (green line) obviously reflects the meteorological database varying the average annual temperature as it varies in the real environment at regional locations at any location on the planet between the poles.

The dual benefits of the LEO model are that it not only improves the understanding of the temperatures the humans at a latitude must adapt to, it also provides a temperature level at which the infrastructure and the technologies developed for use at that latitude must also be designed to function within.

The human body cannot distinguish a change in temperature of 1 °C in 12 hours and therefore humans will not change the thermostat setting on their air conditioner or heater unit to use more or less energy for their comfort. From the perspective of large energy providers for a region, less than 1 degree in temperature (any scale) in a year does not alter the technology that should be developed and installed to meet the energy demands of the region. (A tongue-in-cheek comment when long-time data analysts consider databases such as that of NOAA and NASA is "We are choking on the minnows while the whale is swallowing us.")

At the risk of sounding extremely redundant and repetitive, the technologies selected, and the Factors of Importance formulated are those that apply to the United States region. A temperature envelope shown as follows identifies the extremes of hot and cold temperatures to which the population living within the extremes of the latitudes defining the boundaries of the United States must adapt (i.e., wear the appropriate clothing for their health, safety, and comfort. Technologies must also function within those same extremes:

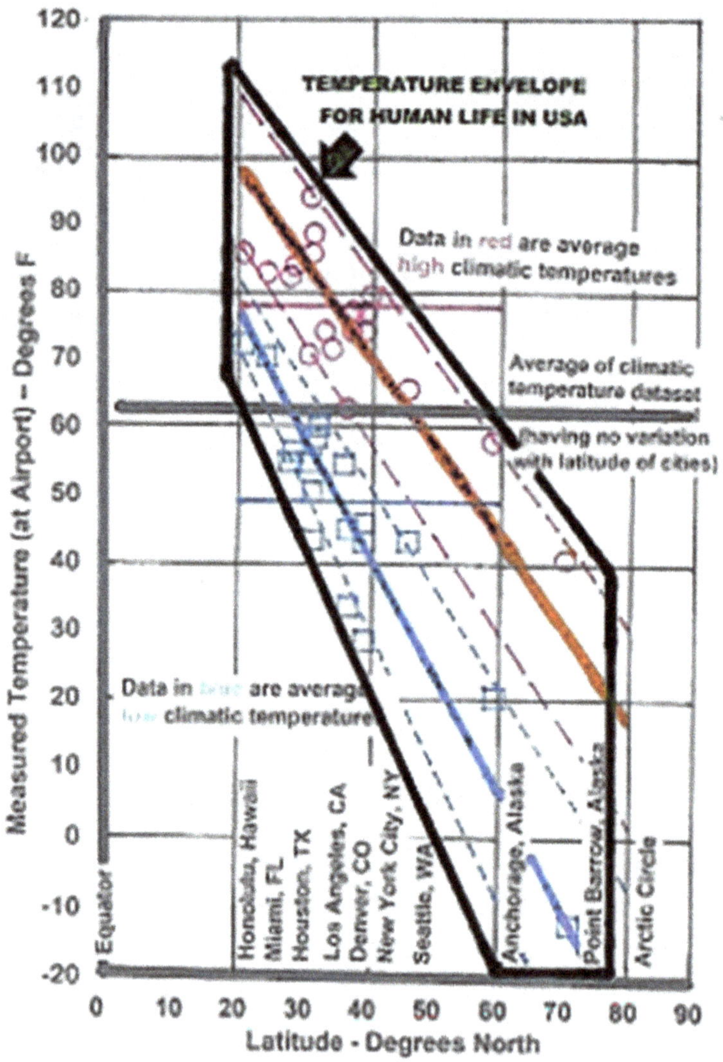

(Source: World Meteorological Organization (WMO) database)

The front cover of this book provides visual indication that the climate varies in a sub-regional manner as determined by Wladimir Köppen and Rudolf Geiger several decades ago and which also reflects the climate – not just the temperature – for the subregions within the United States. The color code information assigner by Köppen and Geiger is provided to clarify the meaning of the color variations shown:

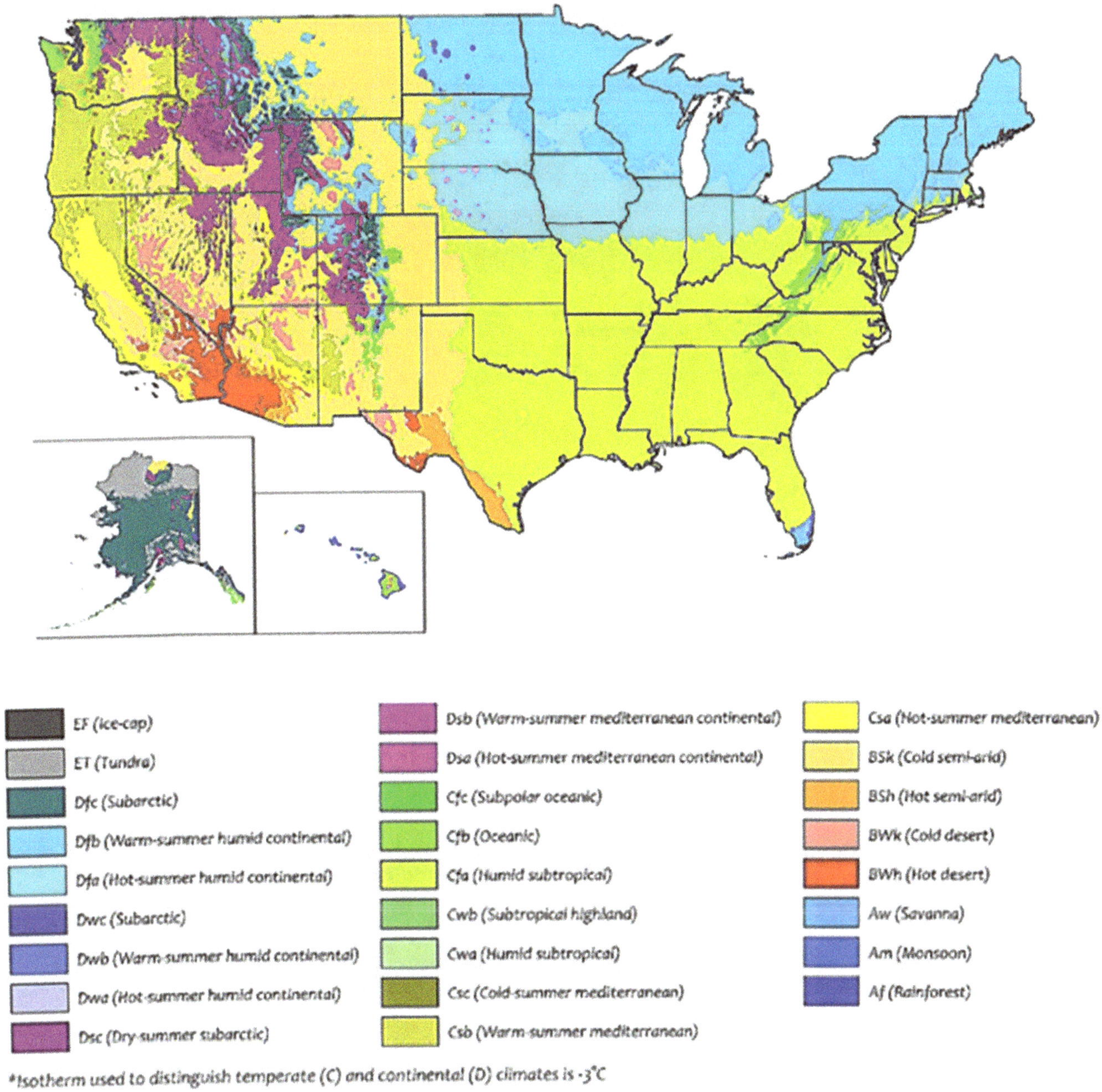

(Source: Smithsonian Institute for Science)

(Author's Note: It was professionally perturbing to find that environmental scientists who established the hypothesis that Man and industry burning fossil fuels was the dominant cause of climate change have not informed the public of even the existence of the Köppen Geiger climate model shown above for the United States region. The UN IPCC is similarly remiss in not identifying the existence of the Köppen Geiger climate model that was done for the major regions of the planet having a significant human population.)

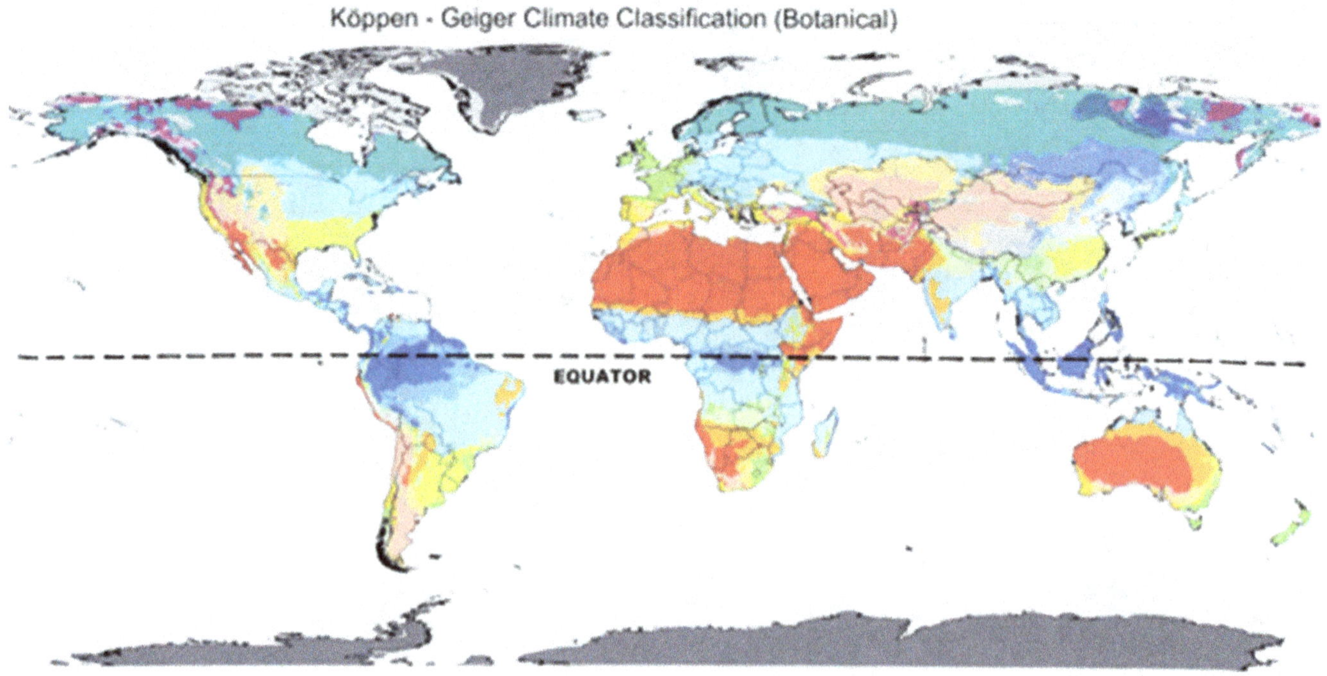

(Source: Smithsonian Institute for Science)

While climate scientists may find it satisfying to categorize the climate of a region with diagrams such as the Koppen Geiger Climate model or the temperature envelope extremes of hot and cold temperature, the human occupant of any region must endure the variation in temperature that occurs daily, and seasonally.

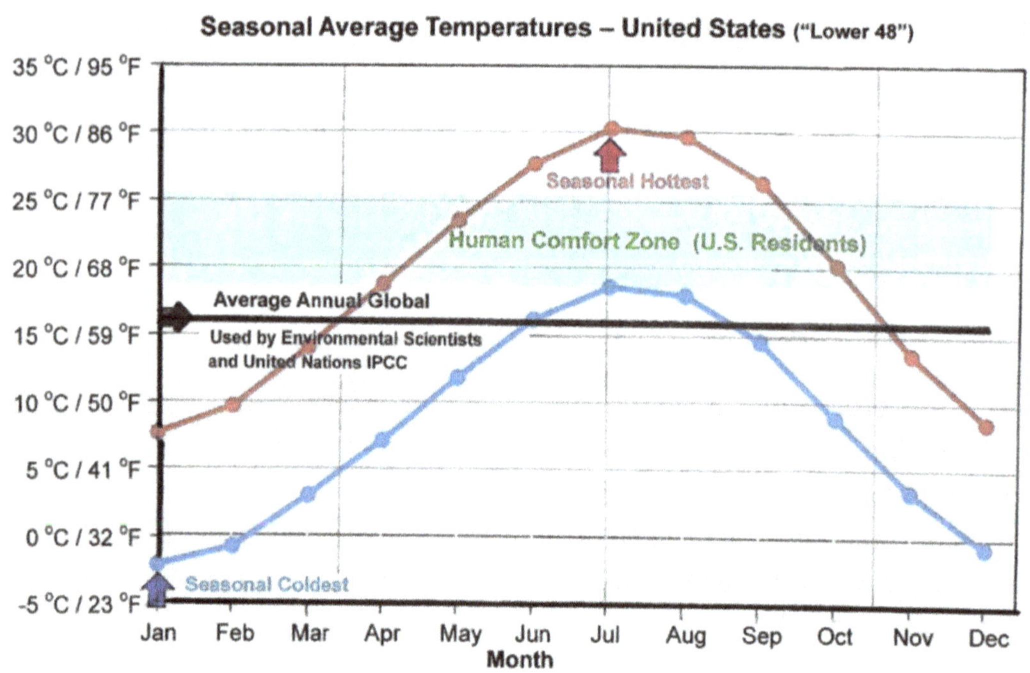

(Source: National Weather Service, NOAA with Author notations)

The seasonal effect (of astrophysical and geophysical influence and of latitude dependency – shown in the following LEO diagram – are major factors in the energy demands for which energy generating and distributing facilities base their energy capacity planning and construction.

In addition, the number of residents and business facilities and whether there are large energy users in the industrial sector of the service area for which the energy provider must consider as well as the periods of peaks in demand for weather normal and anomalies combine to make it very challenging for energy providers to satisfy the many facets of the energy demand. Accordingly, the energy technology each energy provider opts to use in meeting the demand must also consider the availability of the resources required by the technology selected. Typically, there are billions of dollars that hinge on what technology and what capacity must be constructed to meet the energy demands and be within the cost affordability of each customer.

The weighted trade study analyses conducted for the 25 technologies being developed or strongly considered for development contained technologies mitigating the effects of surface temperatures rising in various regions of the World but more specifically in the United States. Lesser importance was given to avoidance of emissions that can be considered in the "greenhouse gas" heat-trapping category. The methodology used for the weighted trade studies was discussed in an earlier chapter. The laboriously detailed worksheets used to "grade the report card" (i.e., assess the net value of the technology area by weighing the importance of each pro and con of the technology) are contained in the Appendix. Although most Readers will find the trade study worksheets to be somewhat pedantic, students of system engineering or in any field involving STEM would benefit from reading at least a few of the worksheets to understand the weighted trade study process as an important tool in their knowledge tool kit.

The weighting method used for the technology assessments considers how important are the several factors associated with the effects of global warming and climate change and the effect they are having on the human population (safety, health, comfort), the environment, the national economy, the cost and availability of resources, and status of technology development in the region. The effects of the pros and the cons of each technology assessed on the set of pertinent Factors of Importance (listed below) provides a comparative "big picture" value for the overall technological capability

1. Status of Technology Development (risk, schedule, cost, potential value)
2. Reliability and Sufficiency (time of day/year, weather dependent, climate dependent)
3. Regional Economy and Resources (mineral, water, labor force to operate/maintain)
4. Cost of Resources (domestic and imported, mode of harvesting / distribution)
5. Importance to Industry, Government, and Human Safety, Health, Comfort
6. Importance and Urgency to Economy and Employment
7. Pros and Cons of By-Products (emissions, pollutants, hazards)
8. a. Effect on Human Population (regional in USA) - Residency.
   b. Effect on Human Population (regional in USA) – Food Supply.
   c. Effect on Human Population (regional in USA) – Safety, Health, Comfort
9. a. Effect on Infrastructure (regional in USA)
   b. Effect on Transportation (regional in USA)
   c. Effect on International Trade

10. a. Effect on Environment (regional in USA) – Atmosphere
    b. Effect on Environment (regional in USA) – Land
    c. Effect on Environment (regional in USA) – Water
11. a. Effect on Animal Life (regional in USA) – Land
    b. Effect on Animal Life (regional in USA) – Aquatic
    c. Effect on Animal Life (regional in USA) – Avian
12. Effect on Forestry and/or Agriculture (regional in USA)

The Factor of Importance "1. (Status of Technology Development)" was listed as the starting Factor of Importance for a special reason. Most of the technologies underway or nearing initiation of development were initiated to reduce greenhouse gas emissions effects on the environment. In a general sense, those technologies need to be re-vectored from the goal of reducing emissions to a more important goal of minimizing adverse effects of the warming of the planet's surface from astrophysical, geophysical, and natural factors beyond human and scientific control.

The "big picture" summary of the evaluations of the pros and cons for each technology and the prioritization of the weighted scores for the technologies are grouped into the following technological categories (numbered as follows in the Grouping of similar candidate technologies):

1. Fossil fuels (coal, oil, and natural gas) or their derivatives
2. Non-fossil fuels
3. Alternative energy sources – geophysical
4. Alternative energy sources - nuclear
5. Technologies for improving human transportation and the infrastructure
6. Technologies for improving food and water supply for humans

It is important to note that the factors of importance, the technologies considered, and the weighted scoring are as they pertain to the region in which they will be applied. For this book, the region selected was that of the United States and its unique resources indigenous to the United States. Other regions can use the weighted trade study process and select the factors, the technologies, and assign the weighted ratings/scores judged to be of importance to the region.

This Author has written other books on technology development, and they have the same basic theme as this book: "*As the twig is bent, so shall the tree grow*". Readers are reminded that to increase knowledge in any topic that time, effort, and funds must be expended. The useful knowledge gained by such expenditures is typically worth the expenditure, however, speaking from experience, during the early testing of a prototype there is confirmation of the pros of the design but there may be surprise cons identified that reflects the state of knowledge was not as well-founded as it was believed to be before the test. The same possibility exists when doing an analysis rather than a test.

The expenditure on non-valuable knowledge is essentially a wasteful expenditure that could be applied to more useful expenditures on other more valuable technologies. There have been several start-up technologies to remove carbon dioxide from the atmosphere and deposit it into the ground to reduce the amount of "greenhouse gas" emissions in the atmosphere. Those expenditures may be of scientific value to the science profession but realistically they do little to increase human safety, health, comfort, and improve the world transportation system.

In the lower grades of education, knowledge is not gained rapidly but it is that base of knowledge – learned by human "twigs" - on which all future gains in knowledge will be built. As older members of humanity – and regardless of language spoken or cultural beliefs – there is an obligation to prepare the youngest members of the region humans reside in for the realities of the world the young people are about to learn if they are to survive. This book hopefully will be "one small step for Man – but a giant leap for Mankind" in the learning process of the youths in the United States.

The second President of the United States John Adams (1735-1826) made a wise statement about teaching children: *"It should be your care, therefore, and mine, to elevate the minds of our children and exalt their courage; to accelerate and animate their industry and activity; to excite in them an ambition to excel in every capacity, faculty, and virtue*." Deficiencies in the education system are not as hazardous as those described in the foregoing sections of this book, but part of the mitigation of climate change is to educate the public about what they can and should do to mitigate the hazardous effects of whatever the climate changes are and will be.

The next chapter is the final chapter describing the "odyssey the Author has taken in the research of the many diverse and complex factors in the "big picture" of the forces and processes on the planet. The many technologies existing and undergoing advancement are added to the 25 technologies that have been identified by this Author as being of significant potential value "for the benefit of Mankind". The 25 technologies are identified, and a summary of their assessment is provided.

Chapter 6
Technologies Needed and Weighted Trade Studies

There are no doubts that the climate factually is changing. The information (data obtained with scientific measurements and professionally statistically analyzed) summarized earlier in this book clarifies the complex set of facts causing the climate changes that are being experienced by humans in their unique regional environments (topographical, meteorological, latitudinal position, proximity to water, etc.). The many factual databases of pertinent information have been comprehensibly researched. Logical interpretation has been derived from the systematic evaluation of several different scientific "databases" that compile factual data and information from different origins, with different states of completeness, and measured with instruments of different technological "vintage".

It has been proven earlier in this book that Man is adding a very small (less than 0.02% concentration) amount of "greenhouse gasses" into the atmosphere concurrently with the larger portions of the temperature rises due to astrophysics (causing seasonal changes), geophysics (changes depending on latitude and elevation), and natural "drivers" (water vapor in the atmosphere which environmental scientists ignore). It is fallacious for environmental scientists to continue with their claim that Man burning fossil fuels is the <u>dominant</u> cause for climate change. Global warming is a scientific misnomer as warming and cooling has been shown (with data collected by satellite-based highly accurate thermal sensors and mathematical analysis of meteorological records of World cities) to be varying differently in different regions at different latitudes on the planet. The thermal image shown on the front cover of this book and similar thermal images and tabular databases within the book define the regional differences and variations.

This book has disclosed – with numbers, photographs, and careful statistical data analysis where possible – that environmentalists have not considered (regardless of whether intentionally or unintentionally) other far more complex but scientifically factual causes that are truly the dominant causes of the global warming and climate change. Those causes not considered far overshadow the effects of Man burning fossil fuels and creating the dreaded "greenhouse gas effect". Naturally generated water vapor (at a concentration of 3%) in the atmosphere must be considered as a heat-trapping "greenhouse gas". Carbon dioxide (at a concentration of 0.043% - of which only 0.02% is Man-generated in the atmosphere - is relegated to being a concern but clearly not a crisis to the human population.

The "soaring" rises in average ocean surface temperature – less than one degree Celsius in recent decades – may explain to environmental scientists why the coral and anemones of the Great Barrier Reef are bleaching white, but the conclusion by environmental scientists that it is emissions from Man and industrial activities (burning fossil fuels) being the dominant (or even a significant) cause for that bleaching pales in validity. Humans (and most animal life according to environmental scientist Charles Darwin) have adapted to their environment and survive 20-30 °F (11 – 17 °C) variation (between low cold temperature and high hot temperature) every day and 40-50 °F (22 - 28 °C) temperature variation in seasonal changes. To human skin, a "soaring" (per environmental scientists) 1 °C (1.8 °F) change – in either direction – is almost imperceptible much less being a "crisis".

There has been, are, and always will be a variety of non-Man-caused catastrophes causing human fatalities and trillions of dollars of property damage. The best goal that Man can substitute for the fallacious "net zero emissions by 2050" is to take action to improve the several candidate technologies existing that are relatively clean but perhaps not so reliable as the best near-term compromise for Man's effect on the environment and the environment's effect on Man. As a long-term goal of leaving a "healthy" planet for the next few generations, there should be development of numerous technologies that are potentially and simultaneously "squeaky clean" and totally reliable.

No single technology presently known can achieve the "ultimate technology" status, but in combination with other net beneficial technologies will make vast improvements over the dedicated-purpose technologies currently in use. This chapter identifies and prioritizes several of the more potentially beneficial (in the net of pros and cons) technologies pertaining to the United States region. The set of unique technologies that are, in the net, beneficial to be used in other regions of the planet can be selected from the technologies set hopefully to be adopted by the enlightened "settled science" scientists and US government legislators after considering the analysis and data presented in this book. Every region has its unique resources available, unique energy demands of its population, unique topography, and unique seasonal effects of climate. There is no individual or set of energy sources and technologies that can be considered as being the "one-size-fits-all" optimum set for the entire planet.

There is a simple "learning curve" process in technology development that humans follow (similar to stepping-stones being used in a sequential order when crossing a stream of water and not wanting to get wet feet). Humans learn by doing, but doing takes time. The science profession initiates the learning curve for a new technology based on observations and exploratory experiments and analysis to determine trends when the different known variables in a technology area are changed one at a time and the effect carefully measured. That process has been labeled as being the "scientific process", and applying the pertinent principles of the four primary elements of the acronym STEM (science, technology, engineering, and mathematics) guide humans through the process.

The primary issues that remain in explaining global warming and climate change are multi-faceted and are perceived by different scientific groups to have different degrees of importance and state of urgency for being resolved. The most common error in any scientific research endeavor is to ignore something that cannot be scientifically explained. An equally deficient step in technology development is to develop something that is not able to withstand its environment when entered into practical use in its intended actual operating environment. That was not carefully thought out when the environmental scientists of the past four or five decades rushed to develop solar power and wind power technology as the replacements for fossil fuels.

The technology development process is by nature often a frustrating one because the process itself depends on the uniqueness and complexity of the technology to be developed and its purpose for being developed. The most common driving need for a technology improvement is to improve an existing technology that has deficiencies in its application – moving up on the "learning curve". Very few technologies are developed to provide a capability which has low practical value. Environmental protection scientists ignore the many attributes that using fossil fuels have brought into practical and affordable use for the betterment of humans and focus only on the detrimental effects combusting coal has on some non-human elements of the ecology.

No technology is perfect – just as there is no "perpetual motion machine". Developers of any technology must account for all attributes ("pros") and detriments ("cons") inherent in that technology and all laws of physics and chemistry and nature that are imposed during that technology's development. Not to be overlooked is that when multiple technologies are being developed in parallel, there can be interactive effects that are either additive or negative in the net use of the technologies. (Additive interactions between technologies are commonly referred to in system engineering as being "synergistic", whereas conflicting interactions between technologies are referred to as being "incompatible").

Very few technologies are developed if there is no demand or if there is no urgency for the capability provided with the new technology. This book by necessity has focused on the technologies that were developed to meet the increasing demand for sources of energy to power the non-military technologies developed for, by, and after the Industrial Revolution – i.e., after 1900. There is sufficient information in the preceding sections and chapters of this book that the technologies identified by environmental protection scientists as causing excessive "greenhouse gas emissions" – those being powered by fossil fuels – does not distinguish the difference between coal and oil/natural gas which produce greatly different amounts of carbon dioxide when combusted, but which also result in greatly different amount of ash residue. Coal ash residue is very detrimental as a pollutant of land, water, and air because that ash residue contains high amounts of toxins (especially mercury and lead) and carcinogens detrimental to human health and safety. There is no physical ash residue from combusting unleaded oil-based fuels and/or natural gas and therefore there can be no claim for oil or natural gas as being serious pollutants.

Small-scale scientific experiments under controlled laboratory conditions produce different outcomes when the scale of the experiment is magnified to that of being planetary ("humongous" in street language) in scope rather than that learned in a finite-sized laboratory. Not only is there a difference in scale of the experiment there is a vast difference between laboratory and actual natural environments that causes results that are vastly dissimilar. The solar power technology works very well in benign environments - but the natural environment can be and often is far from benign. The solar panels not only are dormant during the hours of darkness and when there are heavily overcast skies but also need high maintenance and cleaning especially in arid desert environments because of the blowing sand and dust settling on the almost flat panels. Installing a solar farm in "tornado alley" or "hurricane alley" or any high-risk adverse weather location is a high-risk venture as shown in the next photograph:

(Credit to unknown photographer) Unknown location in “tornado alley”

Locating fragile technology elsewhere than in historically known severe weather locations is a clear preference when installing the solar panels, but there also is no escaping the fact that if there is cloud cover between the Sun and the panels the panel solar radiation capture efficiency will be degraded, During the hours of darkness, there can be zero efficiency of solar radiation collection. Solar energy is a useful technology, but in the right applications in the environment. They are useful as rooftops on residences and small businesses but not so useful in industrial large-manufacturing applications. The solar panels being dormant for many hours each day (during the 8-10 hours of darkness) causes the need for vast improvement of augmenting energy storage technologies that store excess energy during efficient operating periods. Drawing energy from electrical energy storage technologies (batteries or whatever) may be called upon during periods when the solar panels are “resting”.

A similar risk of damage or period of inefficiency - due to natural typically irregularly occurring forces or events - exists with wind turbines technology. When there is no wind or wind below minimum energy generating velocity, there is no wind turbine electricity generated. Too high winds can cause wind turbine damage. Since the wind turbines are elevated above the ground act like a lightning rod, in severe weather conditions, lightning strikes can be a frequent and expensive problem driving high repair and maintenance costs:

(Credit to Unknown Photographer)

Wind energy therefore is of low reliability for industrial applications, but can be useful for small communities if again there is energy storage capabilities to provide the energy when the wind turbines are "resting" or are locked in a non-rotating mode to withstand excessively high winds. There are several communities that reside near a hill with a natural lake or reservoir at higher elevation than the community below. Hydrodynamic power is used to rotate electric generators that supplies electric power when the wind turbines are dormant and electric demand is high. When the wind turbines are rotating in moderate winds, a portion of that electricity being generated is used to pump the water back into the reservoir/lake at higher elevation. Communities that are on flat terrain erect the "big spiders" (water tanks on tall support "legs" – with the obligatory "Welcome to Pumpkin Center" signs painted on the tanks).

Returning to the topic of global warming and the "greenhouse gas" heat-trapping effect, carbon dioxide in a high concentration in a laboratory beaker when heated with an external heat lamp will absorb heat energy. But when a lower concentration (e.g., 0.02% instead of 96%) of carbon dioxide is mixed into the nearly infinite volume of the atmosphere (in with a non-homogenous mixture of other gasses) blanketing the planet and the planet surface heated by a variable heat source, the heat-trapping effect is diluted to being almost imperceptible. That overexaggeration by environmental protection scientists has insidiously been identified as being a global climate change "crisis" without defining to whom, to what, where, and how the crisis is affecting human existence and/or ecological forms. The most common explanation given is that anemones on the Great Barrier Reef of Australia are bleaching white because of the heat stress from Man-generated "greenhouse gasses".

A set of Interim Conclusions (ICs) were presented earlier in this book resulting from an independent and unbiased analysis of the various different databases related to climate change. The set of those twelve Interim Conclusions (in abstract form) are repeated as follows for scientific emphasis and for Readers' convenience:

IC-1) Carbon dioxide in the atmosphere is increasing, but the total carbon dioxide measured (at the NOAA Mauna Loa Observatory on Mauna Loa volcano in Hawaii to be 435 parts per million) contains both natural-caused carbon dioxide and Man-made carbon dioxide in an approximate ratio of 2:1 (295 ppmv natural to 140 ppmv Man-made).

IC-2) Environmental protection scientists have overexaggerated and oversimplified an outdated but otherwise valid database of scientific measurements (principally warping the interpretation of the scientifically valid "Keeling Carbon Dioxide Curve"), and eliminating/ignoring other pertinent databases that also have not been fully updated with known more scientifically accurate and more "big picture" data collection techniques and scientific observations and interpretations.

IC-3) The combined effects of astrophysics, geophysics (tectonic plate activity), and natural factors (long-term cyclical, seasonal and random) on the destruction of human-made structures (for housing needs, food supplies, and the far-worse large numbers of human fatalities, are shown to be collectively far greater than are the man-made adverse effects on atmospheric factors).

IC-4) The scientific facts contained in the databases analyses reflect that it is water vapor in the atmosphere - not carbon dioxide - that is the largest greenhouse gas "driver" of warming surface and oceans temperatures.

IC-5) The need for mitigating and adapting to the astrophysical, geophysical, natural and other non-Man caused factors that are uncontrollable must be better balanced with increased building code regulations and penalties for failure to adhere to the essential improved codes.

IC-6) The combined effects of global warming, climate change, and the oceans levels rising are manifested differently in different regions of the planet as shown by satellite imagery having greatly improved sensors measuring a far larger area simultaneously adding to the "big picture" form of analysis

IC-7) The arbitrary setting of unrealistic goals for controlling global temperatures rising, climate changing, and ocean levels rising are based on misleading "averaging", "linearization", "overexaggerations", and "selective considerations" using databases that have known important updates, but which have been ignored "for scientific convenience in modeling of trends". Humans must survive the extremes of their surrounding regional environment not just the "average global temperature".

IC-8) The presumption that the use of all fossil fuels (coal, oil, and natural gas) must be ceased by 2050 if the climate is to be stopped from reaching an undefined "tipping point" is not supported by the scientific databases that have been correctly analyzed and interpreted with a professionally credible "scientific process".

IC-9) The goal to reduce and eliminate the use of oil and natural gas is not appropriate in many countries, and that "destroy all the eggs in the basket to get rid of one bad egg" decision by devious US environmental protection scientists is deservedly meeting opposition (as shown at the COP28 Conference of Parties in Dubai in late 2023).

IC-10) The pollution aspect of toxic and carcinogenic ash residue (residual from burning coal) causes the need for transitioning away from the use of coal entirely, and to convert coal-burning facilities to oil or natural gas burning in the United States as soon as feasible. That necessity and urgency may not apply to other regions of the planet.

IC-11) The decisions being mandated to shift quickly to the use of all-electric forms of propulsion for commercial and public transportation has proven to be premature as the infrastructure apparently cannot realistically support such an idealistic rapid transition. The more useful transition from fossil fuel-powered vehicles to all-electric vehicles should include a hybrid (electric/fuel) powered vehicle for public use.

IC-12) The primary "one-size-fits-all" goal of reducing Man-made "greenhouse gas emissions" for controlling the climate has been proven to be a wasteful, unrealistic plan with the incorrect goal, and that fact is beginning to be realized. Pursuing the development of alternate technologies to replace the proven but outdated technology of burning coal to satisfy global heat energy demand is timely.

The Interim Conclusions serve to identify what the causes for global warming and climate changes are and are not, and have an indirect impact on the choice of technologies that are developed in a region. For example, developing a technology to remove carbon from the atmosphere may have been a potential solution to the "greenhouse gas effect" before these books were written and published. Those technologies can do no harm to remove carbon from the atmosphere; however, they may also have no significant value for doing so if the measure of merit is to improve the climate.

As the several databases and other scientific documents were being researched and analyzed on the topics of global warming, climate change, and oceans' surface levels rising, there were many descriptions of candidate technologies that were conceived to harness the energy from natural sources. The technologies identified were found to be in different stages of development and were seen to fall into six different categories of applications. The first category and of highest interest for this book was the candidate technologies of the fossil fuels (coal, oil, and natural gas) or derivatives that have been developed by industrialized countries for the purpose of being combusted to produce heat energy to power some form of machinery or to generate electric power for industrial, commercial, and public use.

The process for developing any technology depends on many factors, just as climate change is caused by several uncontrollable causes (one of the lesser important being the "greenhouse gas emissions that are Man-made). The early stages in the development of any technology also depend on many factors of what is needed and what, by when, and how is the technology to be applied as a capability improvement. Each technology has its "personality", its importance level, and its risks. Some technologies have by-products and side-effects separate from the primary output that is being sought, and those side-effects and by-products add to or subtract from the value of developing the technology. For lack of a more familiar terminology, those attributes and detriments are referred to in this analysis as being "pros" and "cons". The "Multi-Variable Analysis" technique is adaptable for use in many areas of technology and is particularly useful when analyzing the many factors that must be considered when accomplishing crash investigations.

For this book, 25 technology items independently have been forensically analyzed, and prioritized for use. One of those 25 trade studies (TS-22) addressing the use of electric vehicles was divided into two parts – part a of TS-22 evolved into the evaluation of electric vehicles equipped with the technology of lithium-ion batteries as their power source. As the evaluation of that pair of technologies was progressing, the need to evaluate the technology of electric vehicles using alternate battery technology to the hazardous lithium-ion technology became apparent. Part b of TS-22 introduced the need to evaluate the electric vehicle being alternatively powered with the safer aluminum-ion or sodium-ion battery technologies which are currently in an adolescent stage of development.

Each technology was evaluated considering the following 12 factors of importance: They are listed somewhat in order of importance, but in the analysis of each technology there is allocated a numerical ranking by order of importance to the applicability and extent (direct or indirect) of the technological development on global warming, climate change, and ocean surface levels rising.

Factors of Importance in Technology Prioritizations Weighted Trade Studies (The prioritization methodology is provided later in this chapter):

13. Status of Technology Development (risk, schedule, cost, potential value)
14. Reliability and Sufficiency (time of day/year, weather dependent, climate dependent)
15. Regional Economy and Resources (mineral, water, labor force to operate/maintain)
16. Cost of Resources (domestic and imported, mode of harvesting / distribution)
17. Importance to Industry, Government, and Human Safety, Health, Comfort

18. Importance and Urgency to Economy and Employment

19. Pros and Cons of By-Products (emissions, pollutants, hazards)

20. a. Effect on Human Population (regional in USA) - Residency.
    b. Effect on Human Population (regional in USA) – Food Supply.
    c. Effect on Human Population (regional in USA) – Safety, Health, Comfort

21. a. Effect on Infrastructure (regional in USA)
    b. Effect on Transportation (regional in USA)
    c. Effect on International Trade

22. a. Effect on Environment (regional in USA) – Atmosphere
    b. Effect on Environment (regional in USA) – Land
    c. Effect on Environment (regional in USA) – Water

23. a. Effect on Animal Life (regional in USA) – Land
    b. Effect on Animal Life (regional in USA) – Aquatic
    c. Effect on Animal Life (regional in USA) – Avian

24. Effect on Forestry and/or Agriculture (regional in USA)

The above set of Factors of Importance for application in the technology prioritization weighted trade study process were derived after having completed the preceding several chapters of this book in which the unbiased and independent analysis of the existing databases existing (astrophysical, geophysical, natural, and anthropogenic (Man-made)) was conducted by this Author. Readers should observe that the single criteria used by environmental protection scientists selecting the "only factor of importance" criteria ((Importance Factor 10 shown in the list above) of the effect on the environment the amount of the singular variable "greenhouse gas emissions" from combusting fossil fuels for producing heat energy has on global warming, climate change, and ocean surface levels rising. In doing so, environmental protection scientists have biased the results of their "blinders-on" sophomoric analysis of all fossil fuels (coal, oil, and natural gas) that considers their continued use being disastrous to the planet as an energy-producing technology.

The attributes ("pros") of using fossil fuels by and for humans and industry apparently were intentionally disregarded by environmental protection scientists, who focused only on the detriments("cons"). That form of "blinders-on" analysis is prohibited when doing a crash investigation. In crash investigations, the entire "big picture" form of analysis investigates all of the significant factors that may have combined (sequentially or collectively) to cause the accident or crash and identifies all the victims of the crash (mechanisms, life forms, properties, and peripheral damage to the environment). It is that system engineering manner of analysis (i.e., using weighted trade studies considering all pros and cons) that is reflected in the Technology Prioritization Weighted Trade Study process in this book.

During the comprehensive system engineering research phase of the complex set of multiple causes for global warming, climate change, and ocean levels rising (considered by the Author as being "natural-caused accidents" affecting safety of both the environment and of humans), there also were seen to be many candidate technologies in various stages of development and extent of use affecting the mitigation of the adverse effects of global warming, climate change, and ocean levels rising. There were six grouping categories for the technologies selected to be analyzed in this book as listed below:

Group 1. Fossil fuels (coal, oil, and natural gas) or their derivatives

Group 2. Non-fossil fuels

Group 3. Alternative energy sources – geophysical

Group 4. Alternative energy sources - nuclear

Group 5. Technologies for improving human transportation and infrastructure

Group 6. Technologies for improving food and water supply for humans

During the comprehensive ("big picture") research phase of the complex set of multiple causes for global warming, climate change, and ocean levels rising (considered by the Author as being "natural-caused accidents" affecting safety of both the environment and of humans), there also were revealed to be many candidate technologies in various stages of development and extent of use affecting the mitigation of the adverse effects of global warming, climate change, and ocean levels rising. The technologies identified were found to be in different stages of development and were seen to fall into the six different groups (shown above) of applications. The first category and of highest interest for this book were the candidate technologies of the fossil fuels (coal, oil, and natural gas) or derivatives that have been developed by industrialized countries for the purpose of being combusted to produce heat energy to power some form of machinery or to generate electric power for industrial, commercial, and public use.

A 7th category dealing with pollution was considered as that is a large problem Man is imposing on the environment and will become larger as the population of the world increases and the use of plastic materials increases proportionately. This book has become extremely large covering the above listed six categories treating the more critical topics affecting global warming, climate change, and oceans' warming causing surface levels to rise, such that the pollution problem can only be identified; however, thoughts for specific resolution actions cannot be discussed herein. One thought for easing the problem of how to recycle plastics rather than simply sending them to landfills or dumping them into the oceans will be mentioned as it has multiple potential advantages to resolution of the plastics pollution problem.

A potential technology for pollution abatement is to recycle the unwanted plastics (much of which is now stored in plastics-recycling warehouses) and which are non-biodegradable into rectangular "logs" and to be used in replacement of the billions of wooden ties in use by the railroad system of the nation. The process could serve other countries (with a railroad system) well also in resolving the world-wide plastics pollution problem. If done, the replacement of the wooden ties that must be replaced because of degradation would avoid the harvesting of timber to produce the needed wooden tie replacements. That lessened deforestation can assist in the natural removal of carbon dioxide from the atmosphere, although that has been shown to not be a very significant factor in climate change. Further, the present process of impregnating new wooden ties with a creosote-based coating to deter biodegradation would become unnecessary with the already non-biodegradable plastic ties. Plastic ties, if/when installed, would require less maintenance and lower replacement frequency than their wooden counterparts.

In addition, heat to warm the plastics (not melt it) so the plastic tie can be compression-formed could be obtained with solar heat collection. Mechanical energy and any electrical energy for the process could be from solar or wind – which suffer from unreliability if the sun is not shining, or the wind is not blowing – as the plastic tie-making process need not be a 24/7/365 process. To serve as a fastening system for "tying" the rails to the ties and at the proper spacing, there can be 4 threaded "L" bolts (as used in concrete construction) with clamp-down plates and attachment nuts per bolt embedded in the molded/compressed ties at four places (two on each end) for attaching the rails and with the correct spacing for the gauge of the railroad system. The metal components could also be made from recycled metals (aluminum, tin, scrap automobile structure, etc.). Those technology details are left for others to develop. A prototype facility should be constructed to work out the logistics and the mechanization/installation of the recycled plastic ties, to ensure that the plastic ties can be used for long-term use in the railroading system environment. Disposal of the replaced wooden ties (with creosote) can be as presently done (or perhaps filled in to obsolete coal mines).

For this book, there have been 25 technology items (selected as potential candidates from the above six categories) independently and forensically analyzed and prioritized for use. Each technology will be the subject of the analysis considering the above-listed 12 Factors of Importance. It is to be noted that the factors of importance, the categories of technologies, and the resources available are those pertaining to the United States region of the planet. The 25 technology items are identified (with a Trade Study identification number TS-XX) and within the six technology groups (bold letters):

Group 1 - Fossil Fuels / Derivatives

TS-1 Use of Fossil Fuel – Coal
TS-2 Use of Fossil Fuel – Oil
TS-3 Use of Fossil Fuel – Natural Gas (Pipeline)
TS-4 Use of Fossil Fuel – Natural Gas (Liquified / Transportable)
TS-5 Use of Fossil Fuel Derivative – Propane

Group 2 - Non-Fossil Fuels

TS-6 Use of Non-fossil Fuel – Hydrogen
TS-7 Use of Non-fossil Fuel – Ethanol

Group 3 - Alternative Energies - Geophysical

TS-8 Use/Development of Hydrodynamic Energy (River Flow)
TS-9 Control of Hydrodynamic Events (Coastal Regions)
TS-10 Use/Development of Solar Energy (Large Industrial / Commercial)
TS-11 Use/Development of Solar Energy (Residential / Rural / Town)
TS-12 Use/Development of Wind Energy (On Land)
TS-13 Use/Development of Wind Energy (Off-Shore)
TS-14 Energy Storage - Stationary/Large Size (Electrical)
TS-15 Use/Development of Geothermal Energy (Surface)
TS-16 Use/Development of Geothermal Energy (Deep-Drilled)
TS-17 Energy Storage - Stationary/Large Size (Thermal)

Group 4 - Alternative Energy - Nuclear

TS-18 Use/Development of Nuclear Energy – Uranium
TS-19 Use/Development of Nuclear Energy - Thorium

Group 5 - Transportation

TS-20 Use of All-Electric Motor Vehicles – Commercial / Government
TS-21 Use of Electric Plug-In / Fuel Hybrid Vehicles – Commercial / Public
TS-22a Energy Storage - Mobile/Small Size (Electrical) Lithium-ion
TS-22b Energy Storage - Mobile/Small Size (Electrical) Aluminum-ion or Sodium-ion

Group 6 - Food/Water Supply – Human, Animal, Agricultural

TS-23 Use/Development of Desalination - Reverse Osmosis
TS-24 Use/Development of Desalination – Distillation (Solar)

Other regions may have very different importance factors, very different technological development goals, and very different resources available.

The interim conclusions (ICs) shown earlier have served to establish the effect of Man-made greenhouse gas emissions to not be the dominant cause of, and crisis level of concern for, global warming, climate change, and ocean levels rising. The set of 25 weighted trade studies came into great use for evaluating and prioritizing the technological activities needed to protect humans (safety, health, comfort, modes of transportation, agricultural food needs, clean water supply for drinking/cooking as well as for mitigating the lack of water during droughts, and/or the contamination of water supplies when floods occur). The trade studies do not focus only on reducing "greenhouse gas emissions" to protect the environment but place greater emphasis on the "big picture" of human needs for future habitation of the planet.

The human digestive system and respiratory systems produce carbon dioxide gas in every exhalation – as well as water vapor. Human exhalation (containing both water vapor and carbon dioxide) is often seen in cold weather as a white "fog" flowing from the human mouth or nose region. Should the 8 billion humans presently exhaling an average of 10 exhalations per minute be prohibited from exhaling and thereby polluting the atmosphere? Carbon dioxide gas cannot be regarded as a pollutant and the court system has denied the environmental protection scientists from imposing anti-pollution penalties for those users of oil and natural gas. Pollution can result from spillages and leaks of oil and natural gas vapor into the surrounding environment (land, water, and/or air) and it is that form of pollution which should be penalized for physical damage done.

It is noted in passing that the environmental scientists and the Environmental Protection Agency scientists and management have created organizations with various titles but are to carry out "environmental justice". There should be environmental polluters brought to responsibility for any form and extent of pollution that endangers the safety, health, and efficiency of humans, animals, or the environment. These books have proven that "greenhouse gas emissions" are not pollutants. It is recommended that any judicial action taken against organizations or industries that emit an insignificant of "greenhouse gas emissions" not be penalized for that "con of fossil fuel burning technologies. The use of the fossil fuel natural gas is shown in this book to be ideal replacement for the burning of coal. If there is any leakage of natural gas that does pollute the surrounding environment, that is a matter of pollution control not a matter of climate change.

It may seem to be a daunting challenge to evaluate the many candidate Trade Study technologies each having many Factors of Importance as variables to distinguish why one technology may be better than another from the system engineering perspective of "all things considered". In the School of Mathematics at colleges emphasizing the STEM aspects of science and technology, there is a process taught for how to use the "Multi-Variable Analysis" technique. That may seem too "nerdy" for Readers – especially those who are challenged in mathematically assessing how many pieces of fruit can you buy if you only have a total of 90 cents to spend and apples are 6 cents apiece and oranges are 5 cents apiece? Multi-variable analysis can be used for the simplest applications of buying fruit or for analyzing the complex, multi-faceted "chaotic" climate of the planet.

Before mathematicians developed the "multi-variable analysis" technique, the same form of analysis logic was used by the renowned fictitious detective Sherlock Holmes of Great Britain's elite Scotland Yard Police Department. At every crime scene, Sherlock meticulously examined all the evidence before identifying the prime suspect who committed the crime. He constantly reminded his associate Dr. Watson: *"When all the impossibilities are removed, all that remains is truth."*

It is hastened to be added that when eliminating the impossibilities, caution must be exercised to not confuse uncertainty with impossibility. The earlier portions of this book have shown that environmental protection scientists have eliminated (some intentionally and some unintentionally) multi-faceted pros and cons of natural forces -, such as gravity and water vapor in the atmosphere), seasonal, and weather-related events that have caused variations in regional surface temperature. By eliminating (or "averaging" or "linearizing") lesser known but large factors, environmental scientists have overexaggerated the small impacts that Man has made on the atmosphere and on the surface temperature and climate of the planet by the burning of fossil fuels of different types. (An associate of the Author described: *"We are choking on the minnows while the whale is swallowing us."*). Using faulty logic: (Combusting coal produces hazardous by-products. Coal is a fossil fuel. Oil and natural gas are fossil fuels. Therefore, environmental scientists have used faulty environmental science logic in declaring the combusting of <u>all</u> fossil fuels must cease simply because one fossil fuel has a very severe "con".)

A similar faulty logic is used by environmental scientists in attributing the effects of "greenhouse gases" on global warming: (Man-made greenhouse gases trap heat and causes global warming. Carbon dioxide is a greenhouse gas. Therefore, carbon dioxide must be eliminated to stop global warming of the planet.) However; water vapor has the most heat-trapping characteristic (and there is over 85 times more water vapor in the atmosphere than there is Man-made carbon dioxide: to wit, there is an average concentration of 3% water vapor randomly located in the atmosphere and in the most recent measurements of carbon dioxide in the atmosphere (420 parts per million or 0.042%) there is only 0.013% of that total that is Man-made. Using the biased logic that water vapor is not Man-made, it is eliminated by environmental scientists. That water vapor elimination allows "settled science" scientists to conclude – fallaciously - that Man-made greenhouse gasses are the dominant factor causing global warming.

Dr. S. Fred Singer at the University of Virginia revealed that faulty environmental science logic in a series of papers and lectures entitled "A Closer Look at the Numbers". The environmental scientists' fraternity branded Dr. Singer as a "contrary scientist" not to be believed – and apparently the public, the news media, and the United Nations have believed the environmental scientists' fallacious conclusions not those of any "contrary" scientist or a several hundred "nay-sayers". (This Author does not anticipate being invited to join the environmental scientists' fraternity after this book is "on the street", and quite honestly is not disappointed to not be invited). Several other well-qualified authors have rebutted the environmental science "conclusions" and have published books like this book – seen in the list of References at the end of the previous book "Facts and Fallacies of Climate Change on our Planet".

A sample worksheet of TS-1 (assessing the pros and cons and relative importance of the technology of coal burning) is presented as follows to show the organization and method how each technology has been scored (prioritized) for its overall value to humans and to the environment. The worksheet is an example of the weighted trade study prioritization process that was used in prioritizing the technology of fossil fuel coal (TS-1). The actual worksheet is larger and it (as are the other several trade studies) is presented in near full scale in the Appendix so that Readers can see the words and numbers that are unique to each worksheet and technology.

TRADE STUDY SCORE WORKSHEET (Applicable to United States of America Region)

(Ratings and scores based on factual information in pertinent databases in Public Domain)

	TRADE STUDY: TS-1 Use of Fossil Fuel - Coal				
	FACTOR CONSIDERED (Being rated for Importance and primary qualitative magnitude)	**ELEMENTS CONSIDERED IN RATING OF FACTOR (Attributes and detriments if/as applicable)**	**RELATIVE IMPORTANCE RI (% or n/a)**	**POINTS SCORED (Rated)**	**WEIGHTED LINE SCORE (WS) (RI x Rated Score)**
	(Minor interactions with/between Elements/Factors are not directly in rating)		RI	XX (of RI)	XX x RI (of RI²)
1	Status of Technology Development	Use of coal (with emission controls) for over 120 years in US requires no further development. Transition to a safer and a cleaner technology is needed/overdue.	12	10 (of 12)	120 (of 144) = 83%
2	Reliability / Sufficiency of Energy	Reliability is 100%; sufficient to meet present demands	12	12 (of 12)	144 (of 144) = 100%
3	Resource(s) Exist Domestically or Approx % Needed to be Imported	There is an abundance of coal in both underground mining (phasing out) and open-pit mining (preferred).	10	10 (of 10)	100 (of 100) = 100%
4	Cost of Resource(s) (Dom. + Imp.)	Cost to consumer is presently low but will increase as coal production is phased out. Law of supply and demand.	10	9 (of 10)	90 (of 100) = 90%
5	Importance to Industry of Region	Reliable, low cost energy vital in industrial areas.	9	8 (of 9)	72 (of 81) = 89%
6	Importance and Urgency to Gov't. Economy and Employment	Coal combustion in industrial use loses importance as cleaner, more reliable energy sources are developed.	9	6 (of 9)	54 (of 81) = 67%
7	By-Products/Emissions / Pollution	Coal ash contains toxic and carcinogenic materials.	8	1 (of 8)	8 (of 64) = 13%
8	Effect on Human Population	High incidence of mine accidents; "Black-lung" deaths; Pollutants are much more hazardous than emissions.	8	1 (of 8)	8 (of 64) = 13%
9	Effect on Regional Infrastructure / Transportation / Commercial Trade	Extent of effect depends on producibility of steel for infrastructure and transportation vehicles/vessels	7	6 (of 7)	42 (of 49) = 86%
10	Effect on Environment / Weather / Climate (Frequency & Severity)	"Greenhouse gas" effects of emissions on severe storm frequency/severity have been exaggerated	5	4 (of 5)	20 (of 25) = 80%
11	Effect on Animal Life (Wild or Livestock) - Land, Aquatic, Avian	Toxins and carcinogens from coal ash disposal are polluting underground aquafers (animal water).	5	1 (of 5)	5 (of 25) = 20%
12	Effect on Forestry and Agriculture	Coal ash pollution hazards affects food supply safety.	5	1 (of 5)	5 (of 25) = 20%
	RECOMMENDATIONS FOR CURRENT AND FUTURE DEVELOPMENT(S): **The reduction of "greenhouse gas" emissions to reduce/eliminate the adverse effects of climate change will not achieve the unrealistic present goal set by the UN IPCC.**		Total A Sum of RI 100	Total B Sum of XX 77 (of 100)	Total C XX x RI (of RI²) 668 (of 902)
			WEIGHTED SCORE =		74%

Because of the serious <u>polluting</u> effects of its ash residue, coal as an energy source should be gradually phased down to elimination. Coal-burning facilities in operation should convert to burn cleaner natural gas as the fuel. Imposing economic penalties on industries for failing to meet unrealistic emissions goals should be terminated. **Every effort, including heavy penalization, should be made to stop the disposal of polluting coal-ash** or chemical by-product of industrial processes with toxic or carcinogenic content.

(Source: Author's Analysis, Trade Study Worksheet) Example for TS-1 Use of Coal

The color coding of the line items is optional but it is useful to quickly visualize and identify if/how a con is affecting the value of the technology candidate. Reading the words without color emphasis does not reveal how serious a con – or a pro – is considered.

The methodology used in the scoring of the technology prioritization weighted trade study worksheets is somewhat like the teacher basing a final grade for a student's report card. Each of multiple factors in a weighted trade study (used to evaluate and derive a prioritization score for a candidate technology) is assigned (based on the education, experience, and judgmental capability of the person scoring the importance) a relative importance factor (RI, expressed as a percentage portion of the entire technology), as shown on the example weighted trade study (TS-1) worksheet. The importance factor is then allocated an estimate (a percentage score, XX%, similar to that on report cards in the school system) of its relative importance rating (of its attributes and detriments – more commonly known as "pros" and "cons"). Multiplying the importance factor by its importance score provides a weighted score. It is that final weighted score (the number in % at the bottom right corner of each worksheet) which is used to guide the planning for that technology and the relative priority. There usually are limited available resources in a region which affect the rate (and risk and cost) at which the technology is to be further developed.

A useful visual aid in interpreting the weighted factors (i.e., to see the "whole picture" – as in assembling a jig-saw puzzle) is to color code the Elements Considered column which otherwise is a lot of words that do not show the highlights of the pros (or attributes) and cons (or detriments). The red, yellow, and green color spectrum is the same general advisory as are the colors to vehicle operators while travelling in a motor vehicle on a city street. The red signifies that forward progress is to slow down and stop, the yellow to proceed but cautiously, and the green to "go for it" at reasonable speed for the conditions.) The Weighted Score (WS, in the lower right corner) is shown in both a percent (%) form and an overall color form.

As a "rule of thumb", a weighted score above 90% indicates a technology that is in the highest priority of eligibility for being developed (or the technology to be improved. A weighted score between 80% and 90% indicates a technology that has much potential for application if the con factors in yellow or red (if any) can be converted to green with available resources for development of that technology.

A weighted score below 80% indicates there are several cons (yellows and reds) advising that the technology should be removed from use/application and supplanted by a more efficient and less risky/hazardous alternate technology. The urgency for replacing a technology is proportional to how much numerically below the 80% rating the score sheet indicates the technology to be. (As on the report card in most school systems, a final grade below 65% is a most serious matter.) Removing a major source of employment (association with a specific technology) will affect many families of the employees losing their primary income, and any such reduction or elimination should be accomplished over a period of a few years not an immediate execution. It is suggested that the removal of a technology for replacement with another technology be accompanied with employment offers to those humans and their families affected to be retrained to transfer their experience and talent in advancing the replacing technology field.

The colors of the individual (12) weighted importance factors are typically scattered as are the pros and cons being different for each technology, but those technologies with a higher Total Weighted Score will be seen to have more green color at the top and less yellow color at the bottom. Those having an average Total Weighted Score will have a few reds in the middle or bottom, and those having a below average Total Weighted Score will be mostly yellow all over and with few reds nearer the top. The worksheets for each of the 25 Technology Prioritization Weighted Trade Studies TS-1 through TS-24 (TS-22 has a TS-22a and a TS-22b variant) , analyzed with a practical form of the multi-variable analysis process, are provided. There is one worksheet per technology - and in large size to be read without a magnifying glass – in the Appendix of this book.

A tabulation showing the Total Weighted Score for each of the 25 trade studies listed within technology groupings, in prioritized TS order within the 6 technology groups, is provided next. The scoring has been transferred from each corresponding Worksheet in the Appendix and contains the highlights of the pros and cons multiplied by the importance "weight" assigned to the importance factor to which it is rated. It is reminded that these scores and pros and cons and weighted importance factors are applicable to the United States region. Other regions can modify the factors, the weight of their importance, and the score assigned to the line of interest that better fits the interests and policies and availability of resources unique to their region. Just as there is no "one-size-fits-all" goal for "greenhouse gas emissions" control that will alter the climate, there is no "one-size-fits-all" worksheet content and set of candidate technologies that fits all regions of the planet.

Depending on the rating score of the pros and cons, government leaders can develop the plans and regulations that need to be developed as necessary. It is reminded that the scoring below was done for the United States region and need to be evaluated individual by region and by country or sub-regions by larger developed countries.

The scoring column – applicable to the USA region - has also been highlighted in color – red, yellow, and green. Red color denotes a technology that should be given low priority or not be pursued in any further development for the region. The red color for a technology is assigned when the Total Weighted Score on its worksheet is below 80%. Yellow is assigned to those technologies having a Total Weighted Score between 80% and 85%. Green is assigned to those technologies having a Total Weighted Score above 85%. Every technology has pros and cons; therefore, no Weighted Score can/will be rated as 100% or 0%. Not surprisingly, the technologies evaluated in this book have without exception been initiated and developed for serving a useful purpose and in most cases are well-thought out. (In a very humbled manner, the Author conducted the evaluation of each of the technologies as representing the product of many hours of research and were interpreted professionally without bias.

The umpire at a baseball game or referee at a football game suppresses any personal bias toward one team or the opponent and applies the set of approved rules for the conduct of the game in an equal manner. (An old-time umpire at a professional baseball game after a controversial "call" was made once stated: *"I calls 'em as I sees 'em"*.) That is not an expression of apology but more to assure the Readers of this book that the scoring of the report card on each technology is a most unbiased and fair one based on the information that was found in years of research and database review, and based on the Author's long time experience about how the planet works and sometimes doesn't work as desired.

After the 25 trade studies of the technology candidates were completed, a "big picture" summary was tabulated for planning purposes. Within each group – which has a varying number of technology candidates – the worksheets are listed in descending total weighted score. It must be emphasized that this prioritization removes the bias – if any – is in the person or group of persons who selected the importance factor and scoring factor attributed to the significance of the net pros and cons for that importance factor.

The best manner of appreciation is for anyone interested in applying the weighted trade study to another region – or repeat the scoring for the same ones presented herein "through a different set of eyeballs" and "through a different logic center of a different human brain" – is for the new evaluator to grind through the process of scoring and evaluating the importance of the "pros and cons". The Author would like to take credit for the weighted trade study assessment and prioritization process but cannot. It is a formalized process in many systems engineering curricula.

The Author will take credit (on behalf of the technical team) for the assessment of the 25 candidate technologies as having been a repeat accomplishment. As the Chief Systems Engineer for development of the USAF Lightweight Fighters (YF-16 and YF-17) in the early 1970s, and for the 28 "missionization" and "weaponization" trade studies that resulted in the YF-16 being developed as the USAF F-16 Fighting Falcon Weapon System (4,578 F-16s produced) and the YF-17 being developed as the USN/USMC F/A-18 Hornet and Super Hornet Weapon System (over 3,000 F/A-18s produced). The following extract of a memorable piece of correspondence "grades the Author's report card":

DEPARTMENT OF THE AIR FORCE
HEADQUARTERS AERONAUTICAL SYSTEMS DIVISION (AFSC)
WRIGHT-PATTERSON AIR FORCE BASE, OHIO 45433

Office of the Commander

28 OCT 1975

Mr. Herbert A. Hutchinson
Deputy for Engineering
Aeronautical Systems Division
Wright-Patterson AFB OH 45433

Dear Mr. Hutchinson

It is a pleasure to award you the Air Force Systems Command Certificate of Merit. Your performance as Chief Systems Engineer, Deputy for the Air Combat Fighter/F-16 System Program Office from July 1974 to July 1975 is deserving of this recognition.

As a result of your dedicated efforts, the Air Combat Fighter made a successful transition from the prototype program, through source selection, into a rigorous full scale development program. The F-16 Air Combat Fighter has introduced many unique technological innovations into weapon systems development.

I commend you for this outstanding performance and wish you continued success.

Sincerely

JAMES T. STEWART
Lt General, USAF
Commander

The "proof of the pudding" of the weighted trade studies done for the 25 technology assessments of the "pros" and "cons" of each in adding value to the technological development of the United States lies in the following summary "big picture" scoresheet:

Group	Trade Study No.	Technology Area	Total Wghtd Score	Comment (Suggested Use and Development for USA Region)
①	TS-4	Use of Fossil Fuel - Natural Gas (Liquified)	95%	Increase useage; improve transportability
	TS-3	Use of Fossil Fuel - Natural Gas (Pipeline)	93%	Install more pipelines (distribution to ex-coal users)
	TS-5	Use of Fossil Fuel Derivative - Propane	90%	Increase useage; improve storage safety
	TS-2	Use of Fossil Fuel - Oil	89%	Continue but reduce use - to no imported oil
	TS-1	Use of Fossil Fuel - Coal	74%	Phase out coal burning technology (to reduce toxic and carcinogenic land/water pollution) Close deep mining.
②	TS-6	Use of Non-fossil Fuel - Hydrogen	90%	Increase useage; improve storage & distribution safety
	TS-7	Use of Non-fossil Fuel - Ethanol	86%	Continue useage; co-technology with EV Hybrids
③	TS-9	Control of Hydrodynamic Events (Coastal and Seasonal Event Regions) - Levees / Sea Walls	98%	Highest need but with lack of activity/planning to meet safety and energy needs of growing human population
	TS-8	Use of Hydrodynamic Energy (Dams, River Flow)	96%	Increase existing capacity and additional construction
	TS-14	Energy Storage - Stationary Large (Electrical)	94%	Good technology supplementing that of solar and wind.
	TS-15	Use of Geothermal Energy (Surface)	93%	Good technology for the western states in Conus;
	TS-17	Energy Storage - Stationary Large (Thermal)	92%	Most important adjunct technology with geothermal
	TS-11	Use of Solar Energy (Residential / Rural / Town)	89%	Of most value to small users. Requires battery storage.
	TS-16	Use of Geothermal Energy (Deep-Drilled; Closed Cycle and Open Cycle with Fracking)	83%	Increase useage; Closed cycle preferred to Open Cycle; Recycle abandoned oil wells where feasible
	TS-12	Use of Wind Energy (On Land)	82%	More useful for small industrial, farming, and public use.
	TS-10	Use of Solar Energy (Large Industrial / Commercial)	78%	Not reliable for industries, OK for small; co-technology with other uses (distillation desalination; water purification)
	TS-13	Use of Wind Energy (Off-Shore)	63%	Too harsh environment (corrosion, wave force, storm damage). Recommend not installing off-shore system.
④	TS-19	Use of Nuclear Energy (Thorium)	96%	Among best technologies for mitigating climate change.
	TS-18	Use of Nuclear Energy (Uranium)	95%	Among best technologies for mitigating climate change.
⑤	TS-21	Use of Electric Plug-In/Fuel Hybrid Vehicles (Commercial / Public)	82%	Increase mix of All-Electric vehicles with Electric Plug-In / Hybrid Fuel vehicles with longer range for public.
	TS-20	Use of All-Electric Vehicles (Commercial / Govt.)	81%	Technology maturing. Limit initial use until more mature.
	TS-22b	Energy Storage - Mobile Small (Aluminum-ion or Sodium-ion)	97%	Aluminum-ion or sodium-ion battery technology is safer than lithium-ion in electric vehicles
	TS-22a	Energy Storage - Mobile Small (Lithium-ion)	52%	The use of lithium ion battery technology is unsafe. Replace lithium-ion with aluminum-ion or sodium-ion.
⑥	TS-24	Desalination - Distillation (Solar/Geothermal)	95%	A vitally needed process for potable water to population
	TS-23	Desalination - Reverse Osmosis (Solar/Geotherm	92%	A vitally needed process for potable water to population

(Source: Author's Analysis of Technology Databases)

The green-scored technologies are those that offer the best set of pros and cons that, in the net, are most valuable for vigorous development in the "lower 48" states of the USA. The TS-10 Trade Study reflects that solar energy has been demonstrated as not being sufficiently reliable by itself for large industrial use unless there is a very large electrical energy storage facility (as TS-14 indicates) located nearby that is sufficiently large to provide sufficient energy to operate the industrial user during the darkness hours or when the weather is such that solar energy has not been collected due to occluded skies. Solar energy is useful for powering other technologies (desalination or water purification) that can be operated intermittently as the need and demand requires regardless of sun insolation conditions (TS-10 and TS-11).

The pink-colored comments are associated with 1) a suggestion the technology be phased out of use (the fossil fuel coal (TS-1) for toxic and carcinogenic ash residue disposal pollution reasons not for "greenhouse gas" effect on the climate considerations); 2) recommended not to be developed for the hazardous environment at the proposed installation location (wind energy in off-shore locations (TS-13), or 3) are of high importance for control of the population choosing housing location on land areas that are prone to flooding or eventually to ocean levels rising to inundate large coastal areas (TS-8).

The red-colored comment for ceasing the production of electric vehicles with lithium-ion technology battery power (TS-22a) is made for the safety of the public and not for any climate change considerations. This recommendation is the most important and far-reaching conclusion from the "big picture" (system safety engineering) analysis of all pertinent databases affecting the human population of the planet. It is hastened to be added that the Trade Study TS-22b (which analyzed the pros and cons of aluminum-ion or sodium-ion battery technology in electric vehicles) makes a very-clear case for human safety considerations to drastically curtail the losses of vehicles and their human occupants to fires on the US highway system when the lithium-ion battery installed in the electric vehicle explodes (caused by a number of highly possible flawed actions). There is nothing recommended about the public drastically changing their choice and use of electric vehicle technology per se but only to protect their safety.

The Concluding Comments section of this book addresses the "big picture" perspective of the technological findings in the research and analysis done in this book.

The USA is large enough (and populated differently between its coasts on the east, west, and south) to warrant sub-dividing it into different climate regions. The Importance Factors and pros and cons scoring of the 25 Weighted Trade Study worksheets (contained in the Appendix) should be modified to account for the uniqueness of that sub-region in a dedicated worksheet.

When considering a technology for a region (such as a group of states within the United States), it is suggested by the Author that the USA region be divided into ten sub-regions (consisting of states grouped according to the USPS Zip code designation as suggested below.

For this discussion in this chapter, suffice it to say the regions following the suggested Zip code grouping are much the same as the professional botanists and climatologists Drs. Köppen and

Geiger described for the climate as it affected the botany in the environment of a region – growing well in favorite climates and not growing well in unfavorable environments.

Region A. Zip codes 0, 1, and 2 (Northeast, Mid-South, Atlantic Coast)

Region B. Zip code 3 (Southeast, Atlantic Coast, East Gulf Coast)

Region C. Zip code 4 (Upper Midwest, Great Lakes)

Region D. Zip code 5 (North Central Plains)

Region E. Zip code 6 (Central Plains)

Region F. Zip code 7 (South Central, West Gulf Coast)

Region G. Zip code 8 (Mountain, Desert Southwest)

Region H. Zip Code 9 So. Cal. (Southwest, Pacific Coast)

Region I. Zip Code 9 No. Cal. (Much Agriculture, Pacific Coast)

Region J. Zip Code 9 OR/WA/AK (Much Forestation, Pacific Coast

Note: Alaskan Aleutians and Hawaii have island climates and ocean rise effects with low population affected that are uniquely different from coastal regions of the "lower 48" contiguous states.

The groupings of the states in the subregions were made in consideration of the surface topography (plains, deserts, mountains), agricultural/forestry, proximity to large body of water, and total population/distribution of the subregion. The next map of the USA shows the ten subregional groupings (outlined in black, labeled with a capital letter of the alphabet, and showing the first two or three digits of the USPS ZIP code system) that are matched in general with the Köppen-Geiger climate designations:

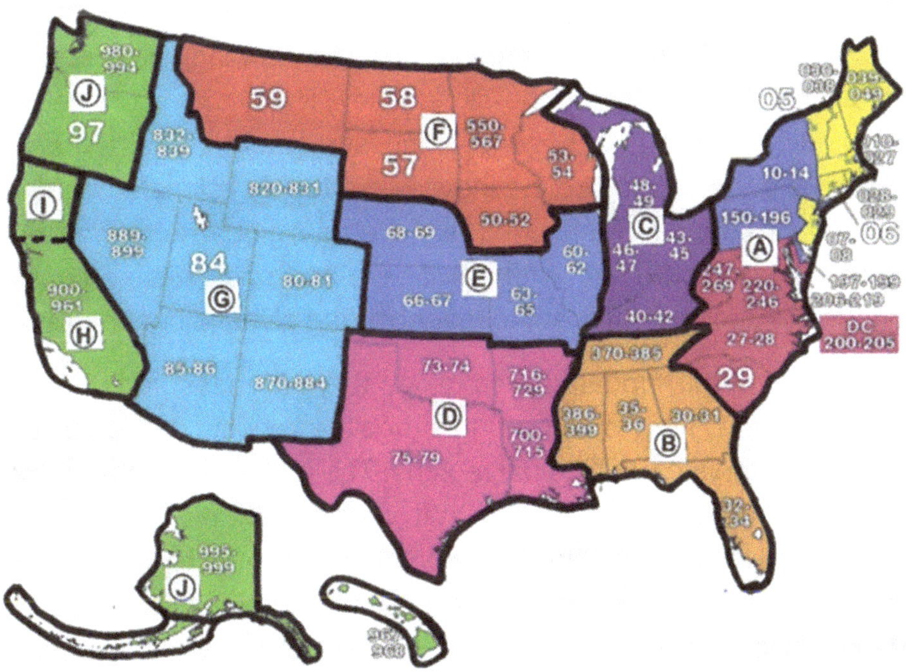

(Sources: US Postal Service Map of ZIP Codes, modified by Author)

There was a consideration of the population distribution and industrial technology needs in the different subregions in the US. Concluding remarks for this book are contained in the next section of the book. Readers are asked only to use common sense as they form their own conclusions as to the validity of the present emissions control goals established from the fallacious environmental "settled science" hypothesis that Man burning fossil fuels is the dominant cause of global warming, climate change, and the ocean levels rising.

If Readers, after digesting 1) all the many facts (and "Facts do matter") revealed in huge (but not exaggerated) numbers in databases not "averaged", "linearized", and "cherry-picked" by environmental scientists; 2) all the images in photographs of actual non-Man caused disastrous events that have killed many hundreds of thousands of humans and destroyed many trillions of dollars of property that have been presented in the previous chapters of this book; 3) fail to recognize the environmental scientists are using unnatural "averaged" and "linearized" mathematical models describing the planet's atmosphere to be filled with overexaggerated concentrations of Man-made carbon dioxide; and 4) do not agree that humans and virtually all ecological life-forms must survive at all temperatures within the Human Temperature Envelope presented in an earlier chapter and not merely a rise of 1.6 degrees Celsius by 2050, reading this book will have been a waste of their time.

Until and unless environmental scientists (who have "circled their wagons" to defend the fallacious hypothesis that Man is the dominant cause for climate change), modify their "combat climate change" strategy to a "mitigate the effects of climate change" strategy, the public will see their tax monies wasted on futile attempts to combat and control climate change.

To their great credit, environmental scientists have awakened the public and government leaders and regulators to the fact that the time has come for the USA (and some other well-developed) countries to improve their sources and uses of energy as technology improvements. The time has come to reduce the technology of using coal as a fuel, because its ash residue is a toxic, carcinogenic pollutant. The reduction of Man-made "greenhouse gas emissions" as the single variable that will cause the climate to behave itself according to environmental scientists' hypothetical desires is an idealistic "pie-in-the-sky" fallacy. The climate change problem is complex beyond full understanding, much less beyond being accurately modelled, and beyond human-control, but much can be done to mitigate the inevitable effects of climate change. It is not a global problem but is a regional problem. Measuring its "average global temperature" with two single-variable thermometers – one in its northern hemisphere and one in its southern hemisphere and taking the average - is the wrong instrument to diagnose the climatic health of the planet.

The NASA satellite image – with updated technology thermal measurement sensors of the type illustrated that produced the front cover of this book - reflects more accurate "big picture" temperatures of the planet varying in latitudinal regions. The Author developed a model entitled LEO (for Latitude, Elevation, and Ocean proximity) that very closely models how those geophysical properties at any populated location at any latitude on the planet are the dominant variables affecting the temperature of a region. Why do environmental scientists not "see" that they are "choking on the minnows while the whale is swallowing them"? Is the health of the anemones on a coral reef more important than the health of the world population?

What also is beyond full understanding is that the environmental scientists "see" – but fail to "understand" the reason why - the goal of limiting the average global temperature to 1.5 degrees Celsius by 2050 (the target agreed to - by the environmental scientists representing many countries

- in Paris in 2015) is unrealistic. In "New Scientist" magazine (24 August 2024 edition), in the article *"1.6° C is now best-case scenario"*, it is stated by an unquestionably sincere environmental scientist: *"It has become essentially impossible to hit the totemic climate goal of keeping warming to no more than 1.5° C above pre-industrial levels".* Based on the Author's experience in combat, efforts and munitions should not be wasted on hitting an impossible to hit target – especially after being advised by reconnaissance it is a decoy target – and would be a better expenditure to hit an alternate target of higher tactical value. The technologies considered as high-value "targets" have been prioritized in this book – but will the science profession use them or continue to hit the fallacious decoy?

From the "big picture" system perspective, it is Man-made (industrial and residential) pollutants needing to be reduced – elimination is preferred but is not likely – and as the more immediate "crisis".

Development of the several technologies chosen to be pursued in a region is a major challenge in manpower and financial resources, but the outcome is well-worth the effort and expenditures "for the benefit of mankind". The improvement of the infrastructure is a parallel effort, and much depends on whether the region is subject to the effects of oceans levels rising both in terms of the magnitude of the rise and the rate of the rise. Much also depends on the population growth of the region. Some regions may have no growth, while others may experience a doubling or tripling of their population of humans. Whatever technologies are chosen to be advanced or developed from its inception will undoubtedly be determined by the population size, by their food and water needs, housing needs, transportation needs, and a myriad of other functions proportional to the population size, its culture, and its state of development.

Historically, developed nations have risen to the needs and wants of its population whatever they are manifested to be – evidenced by this period in history becoming known as the "Age of Technology". A slogan well-remembered from the nations in the world unifying eight decades ago to defeat the formidable enemies in World War II – in two regions of the world (in Europe and in Asia) - was: *"We did it before, and we can do it again".* Combating the enemy is a terrible waste to humanity – much the same as combating gravity. The winner of a war is the one who has lost the least total amount of its resources – human and materiel.

Conclusions (Overall) and Concluding Comments

The scientific research of the environmental scientists' hypothesis (posed over 40 years ago): "The science is settled. After eliminating everything else, it is clear than humans and industry creating "greenhouse gas emissions from the burning of fossil fuels are the dominant causes for global warming and climate change." has been completed. That hypothesis was based on the observation on the increases in carbon dioxide in the atmosphere measured by Dr. Charles Keeling starting in the mid-twentieth century.

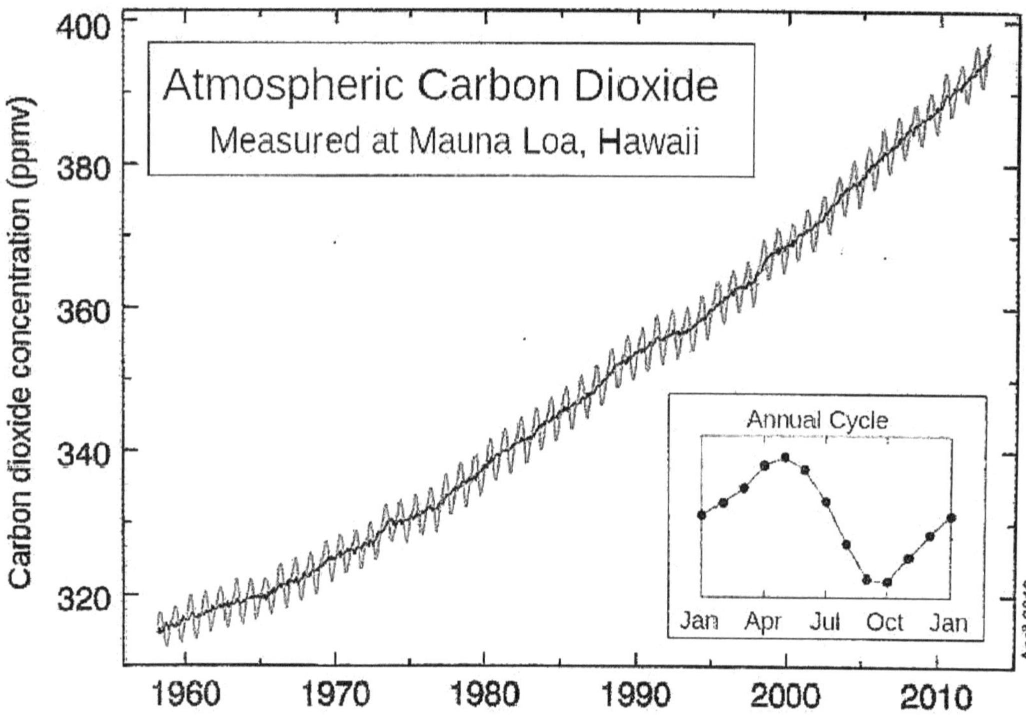

(Source: NOAA)

An independent review of Dr. Keeling's carbon dioxide measurements by his atmospheric science peers earned Dr. Keeling several prestigious scientific awards – all well-deserved – such as the American Meteorological Society (1981) and National Medal of Science (2002). His brilliant attention to detail and insistence on scientific accuracy of his database on the measurement of carbon dioxide in the atmosphere confirmed the earlier proposition (by Dr. Svante Arrhenius in Stockholm, Sweden, in 1896) that there was an anthropogenic contribution of the greenhouse gas effect on global warming. When the numerical analysis of the extent of that effect was examined by the Author, it showed that when the concentration of carbon dioxide gas in the atmosphere was separated into that which is attributed to natural causes (shown in the Vostok Ice Core Samples database to be 295 ppmv at the time (1850) when the Industrial Revolution was in its early development stage), the Keeling Curve database attributed to anthropogenic contributions was only about one third of the total CO_2 measured. For example, the total concentration measurement of 380 ppmv (0.0380%) is separated into the natural - pre-Industrial Revolution - level of 295 ppmv (0.0295%), only 85 ppmv (0.0085%) could be attributed to anthropogenic causes. Another factor found that causes diminishment of the anthropogenic contribution was shown in the original Keeling Curve measurements shown in red on the graph. The red measurements contain the seasonal effect – temperature and water vapor content in the captured atmospheric sample – which were removed scientifically and very accurately from the Keeling curve database. When the Author presented the concentration of carbon dioxide and water vapor in the atmosphere in a percentage presentation, the scientific interpretation (shown graphically below) is that the anthropogenic contribution to the "greenhouse" heat-trapping effect is far more caused by natural water vapor in the atmosphere and little (almost negligible) by the anthropogenic contribution accurately measured by Dr. Keeling.

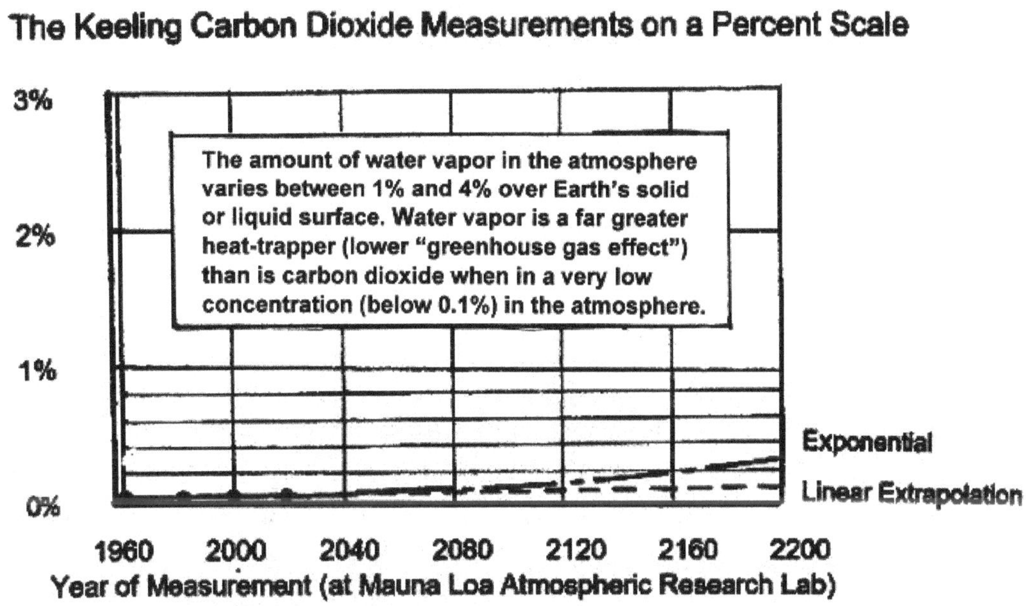

(Source: NOAA Database with Graphics by Author)

This form of presentation is not intended to denigrate Dr. Keeling's measurement database but only to provide a "big picture" perspective on its significance in the real atmosphere of the planet. The public understands the significance of concentration in percentage form rather than its equivalent in scientific parts per million by volume (ppmv) form. The extrapolation of the Keeling database by either a linear or an exponential relationship also indicates there being no cause to believe there is an ominous "climate catastrophe" in the near future if the burning of fossil fuels is not eliminated or reduced to "net zero". Dr. Keeling never claimed there to be an atmospheric crisis in existence, only the possibility of slight global warming from anthropogenic cause – the same as proposed by Dr. Svante Arrhenius in Sweden sixty years earlier. Dr. Arrhenius saw the potential of slight warming in colder latitudes being a benefit by opening up permafrost areas in northern Scandinavia to the growing of farm crops assisting in the food supply increases meeting the demand of a growing population. The primary difference between Dr. Arrhenius' database and that of Dr. Keeling is the measurement of carbon dioxide in a controlled laboratory experiment versus that of the measurement of the carbon dioxide concentration in the real atmosphere but in a closely controlled scientific collection process.

Numerous conclusions are provided herein as a summary of what was found in the Author's previous book "The Facts and Fallacies of Climate Change on Our Planet" – referred to as Book 1. It is demonstrated in the book that the investigation into the causes for global warming, climate change, and ocean levels rising would be conducted with a "big picture" system engineering type approach known to system engineering analysts as a heuristic approach. In lay terms, heuristic approach is one of "turning over a stone to see what is under it and finding more stones to be overturned."

There is no question that the planet is experiencing a warming trend in some regions, but it is not the same trend in all regions where it has been shown to be a mixed combination of warming temperature, cooling temperature, or no significant changes in either trend toward warmer or cooler. Correspondingly, there is no question that the climate of the planet is also changing but again it is changing more in some regions than in others in keeping with the pattern of the temperature changes in regions.

The primary causes for the surface temperature and climate changes have scientifically been proven to be from many factors – astrophysical, geophysical, and natural - over which humans have no control. However, the finger of blame has realistically begun shifting from accusing humans and industry of creating "greenhouse gas emissions", which environmental scientists consider as the dominant cause for changes in temperature and climate, to humans being solely to blame for the true crisis of Man polluting his environment and hazarding his own safety and health.

Before transitioning to the conclusions drawn for the main topic of this book "Dealing with the Real Climate Change", it will be useful to identify and rate in a priority manner the technologies that comprise the challenge for mankind to be able to enjoy a safe, healthy, and efficient life on the planet in the region of their choice - one area of improvement identified and developed in the Author's Book 1 ("Facts and Fallacies of Climate Change on Our Planet") that will assist in choosing the technologies needed in the near future is the improved method for estimating the temperature envelope of a region using what was described in Book 1 as the "LEO Method". The derivation of the LEO Method was described extensively in Book 1, but in short summary for this book, it is a far closer estimation of the average temperature of a region at any latitude and at any elevation above sea level than that used by the environmental scientists – the one-size-fits-all "average global temperature" (approximately 60 °F (26 °C). The chart that best summarizes the LEO Method is shown as follows:

Estimate of Average Annual Temperature at Global Latitude (with Adjustment for Elevation of Region) [LEO Method Patent Pending]

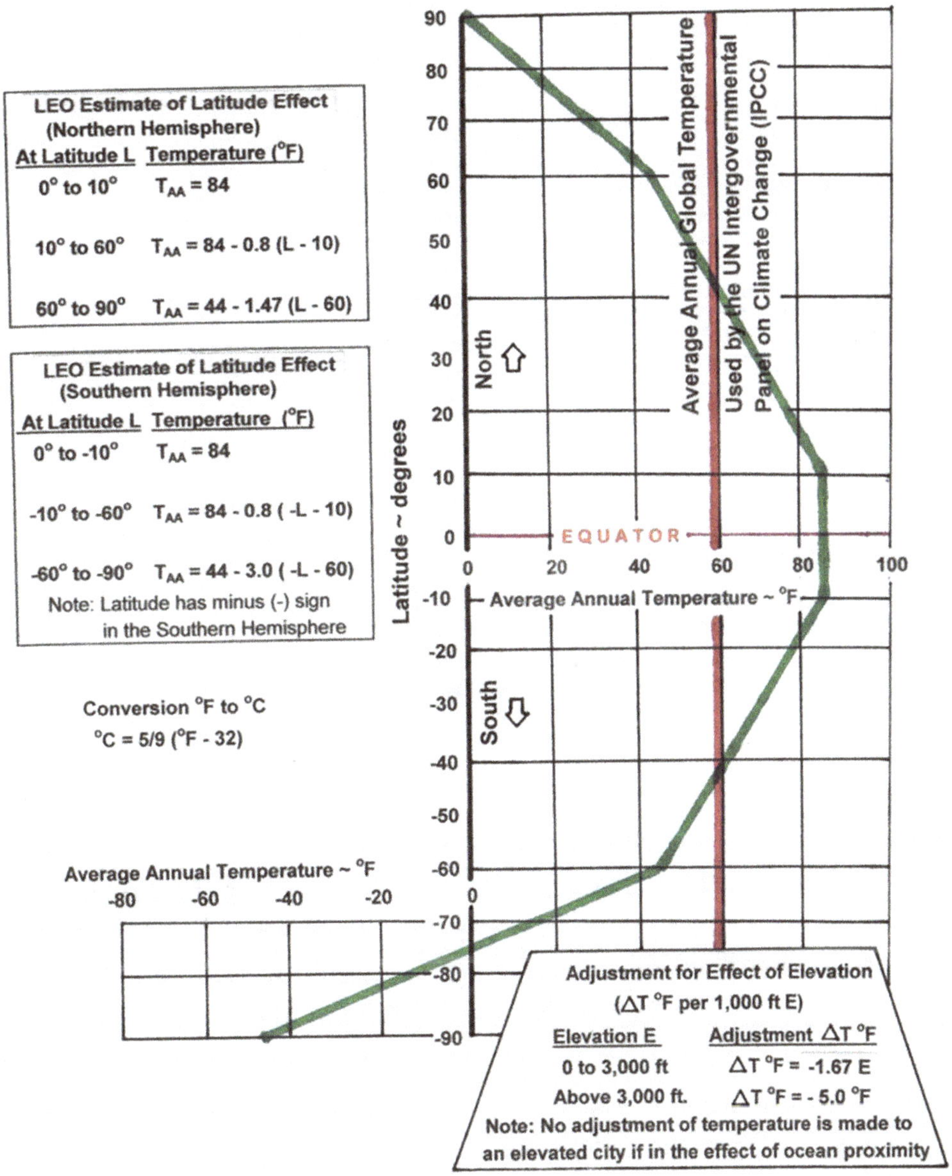

The primary benefit of the LEO Method (green line and mathematical modelling equations) is that it represents the basic average annual temperature as it varies with the latitude of the region anywhere between the North Pole and the South Pole. Because regions are not all at sea level a correction factor for elevation is added to the latitude effect (the elevation factor is included in the lower right of the diagram). The environmental scientists' red line average annual global temperature is seen to represent only the average annual temperature for the entire planet independent of latitude or elevation. Little if anything varies in Nature in an average, linear pattern of variation. It borders on ludicrous to believe that a 2 °C rise in the average annual temperature (the red line) constitutes a "crisis" when the variation from the actual annual temperature (green line) varies as much as 25 to 40 °F (14 to 22 °C) from the LEO temperature estimate. For each region being analyzed, the LEO Method only provides the annual average temperature for the region. Using the "average" masks the extremes of temperature that realistically occur seasonally (i.e., in four approximately 3 months periods) shown as following for the USA region:

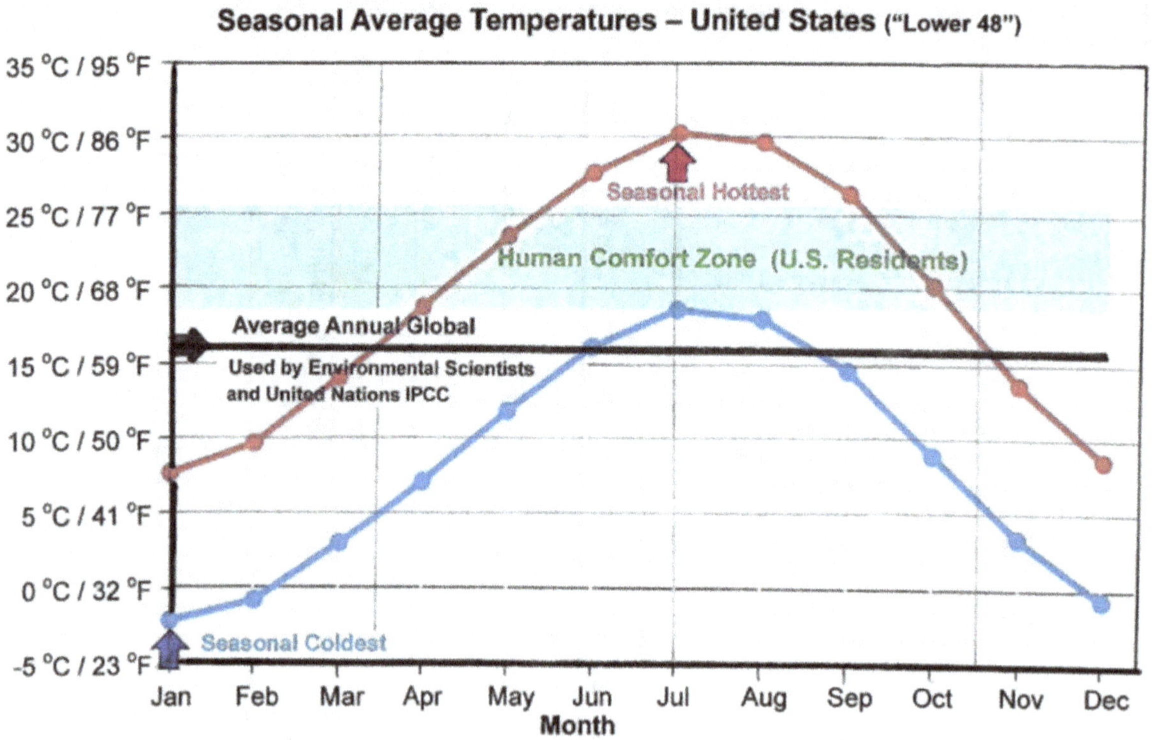

(Source: NOAA)

There are occasions when a combination of natural meteorological evnts cause a region to have an extremely hot or extremely cold temperature lasting perhaps only a few days; nevertheless, all life forms in the region must survive those extremes well above or below the seasonal extremes shown above and in the LEO method for the annual average temperature. The resulting reality is that every region has a temperature envelope within which everything in that region – animate or inanimate – must survive or perish. The extremes were also found in this analysis to vary with latitude. When it is considered that the safety, health, and efficiency of the humans on the planet is a much more urgent need to be protected than the safety, health, and efficiency of life forms on the ocean floor, the concern of environmental scientists must defer to the safety, health, and efficiency of humans and their technological machinery and "toys".

For the United States region, the following temperature envelope contains the record extremes of hot and cold temperatures that were recorded in meteorological records (for the cities shown). The bold black horizontal line (at 62 °F) is the average of the temperatures within the temperature envelope and reflects that whether environmental scientists calculate an average annual temperature for a region or for the entire planet it is misleading and has caused much wasted effort and resources in trying to justify their hypothesis that a "crisis" exists because humans and industry are combusting fossil fuels to meet their electrical and heat energy demands (which will continue to increase as the world and the USA population continues to grow exponentially.

The same temperature envelope shown following should be developed for the other regions of the planet using the meteorological databases in the World Meteorological Organization WMO.

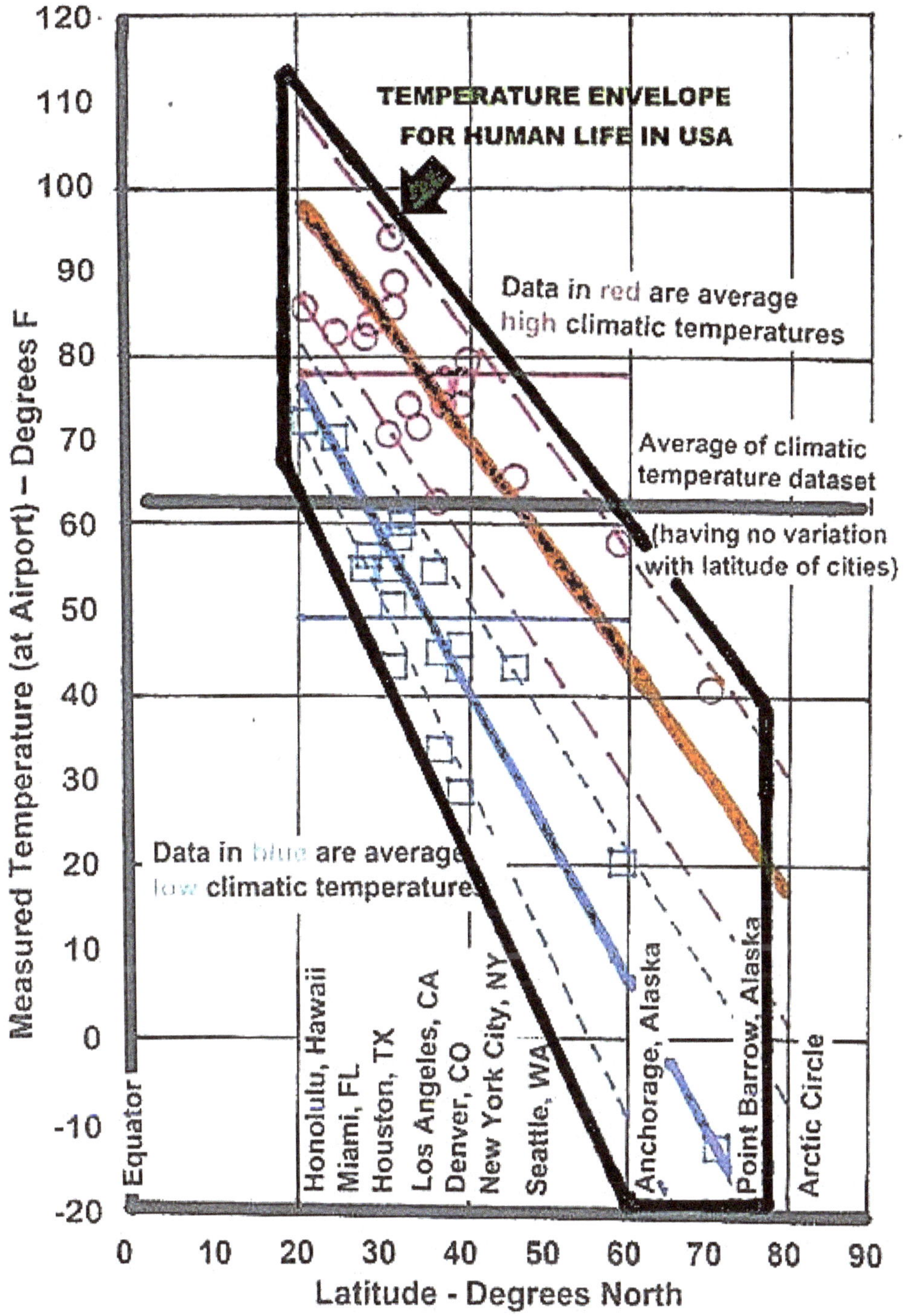

(Source: The meteorological database of the NOAA NWS, compiled by Author)

In addition to the temperature extremes a region experiences, the climate of the region also has a precipitation factor that affects its total climate. For that purpose, the Koppen-Geiger Climate Model is the better model of climate than is the single "average global temperature" that environmental scientists use to determine their version of "climate change". The Köppen-Geiger model climate for the United States is formally contained in climate science as follows:

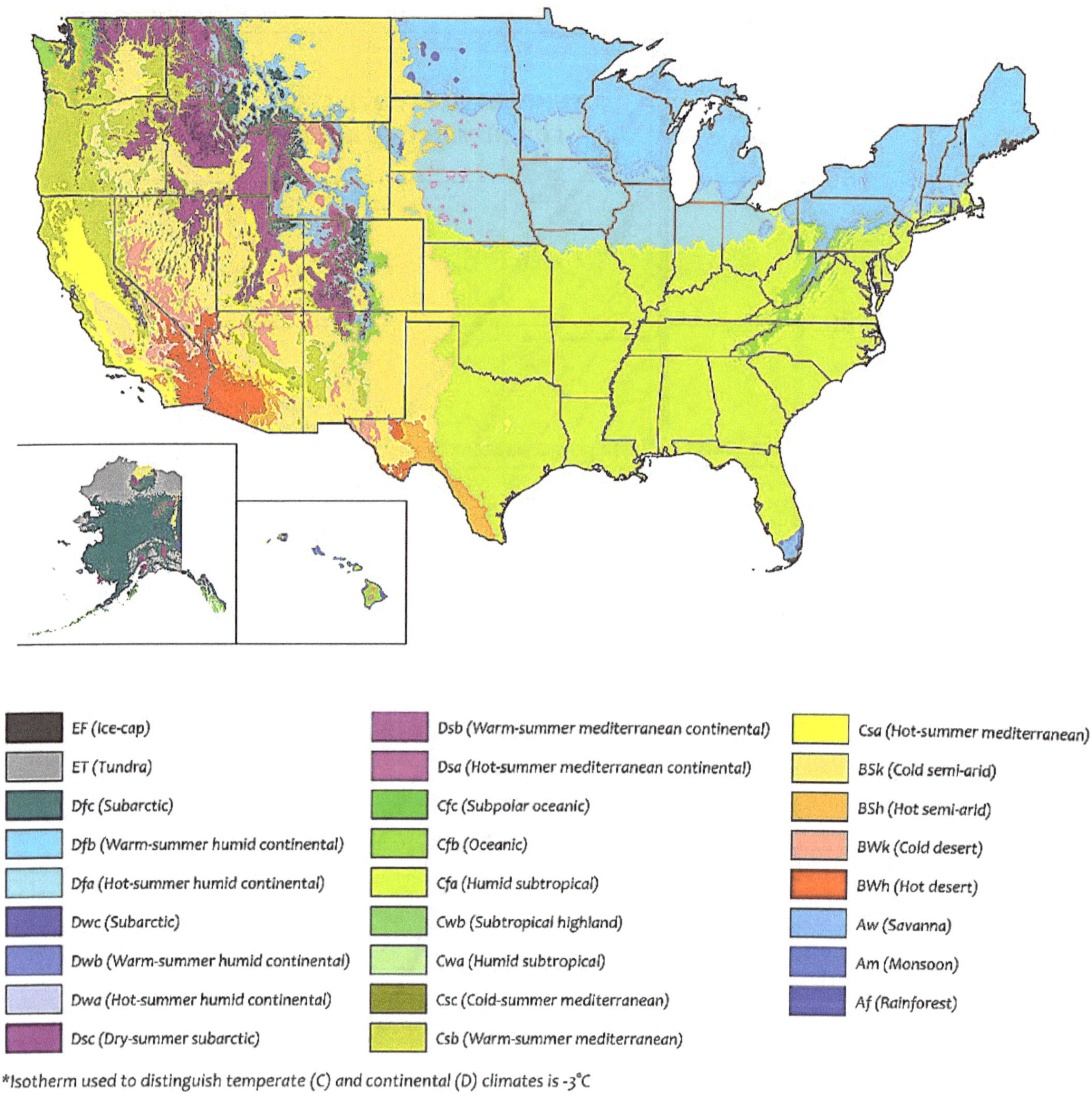

(Source: Smithsonian Institute for Science)

A logical combination of the previously described LEO Temperature Model (developed by the Author for this book) to closely estimate the annual average temperature for a region (considering its latitude and elevation and estimate of prevailing adjacent ocean current and prevailing wind current) with the Koppen-Geiger Model would be most meaningful to the agricultural industry of the region. It should be evident that within the climate of a region may be different sub-climates in subregions, as is well-shown above for the United States. The Köppen-Geiger Climate Model has been constructed for all regions of the world and is shown for the entire planet as follows:

Köppen - Geiger Climate Classification (Botanical)

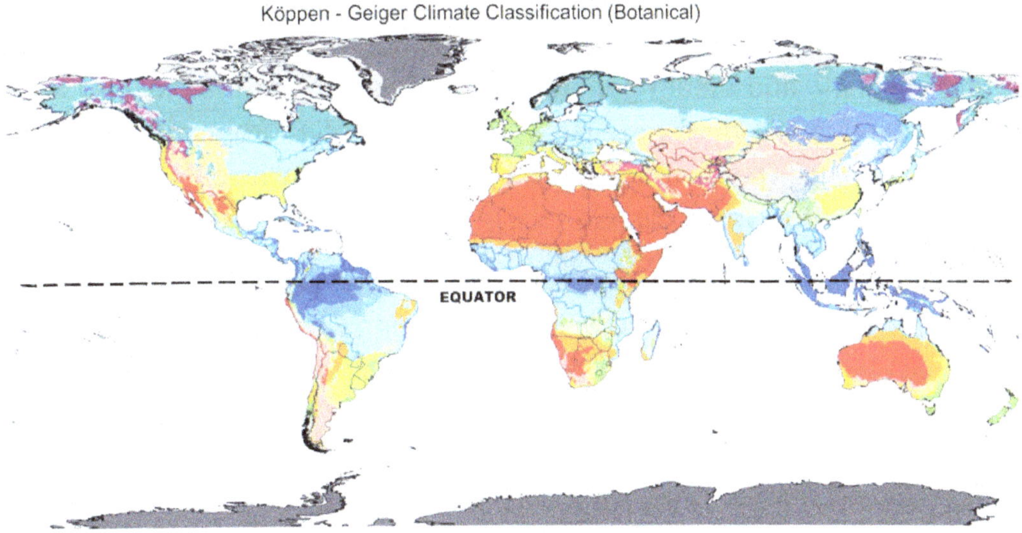

(Source: Smithsonian Institute for Science)

Needless to say, there is no mathematical model that could be developed to describe the Koppen-Geiger Climate Model (although the LEO Method for the temperature consideration of the climate of a region has beem modelled for the latitude and elevation factors).

This Author would be remiss not to give high credit for the Environmental Protection Agency for their diligent efforts to reduce the amount of factual pollutants from the burning of coal being emitted into the atmosphere. A montage of four scenes is shown as follows which visually reflects the accomplishments of the EPA in reducing the pollution of coal burning – by adding capture devices of the carbon-particles from the emissions and by causing the transition from burning coal to that of oil and natural gas.

Coal burning facility with no emissions control

Coal burning facility converted to oil burning

Coal burning facility with EPA emissions control

Coal burning facility converted to natural gas

(Credit to unknown photographers)

Readers should note that the visible emissions (the dreaded "greenhouse gas emissions") in the left two photographs are not seen in the right two photographs. The upper right photograph is of the Consolidated Edison electric and steam generating facilities serving the New York City area, and the lower right photograph is of a facility in Alabama that converted from coal burning to the much-preferred natural gas burning. The environmental scientists who fallaciously claim that the burning of all fossil fuels must cease to avert a climate "crisis" are clearly mistaken.

(Author's note: The Author's father was in the Electric Production Department at the Brooklyn Edison part of the Consolidated Edison Corporation when the decision was made for the Edison Companies to transition from coal-burning to oil-burning. The Author can attest that the decision to make that transition was based on the elimination of the problem of disposal of the coal-ash residue (containing toxins and carcinogens) from the boilers and was not influenced by the coal burning process having any degree of influence on changing the climate. The elimination of the smoke from the four stacks blowing into the Manhattan business district when the winds blew in that direction was also an atmospheric quality improvement consideration but climate change in the 1940s was not a factor entering into the transition from coal to oil burning decision.)

An additional 15 conclusions **C24** to **C38** were derived as to what the human population of the planet can and should do to mitigate the effects of the temperature, climate, and oceanic changes. This Conclusions section provides those 13 Conclusions **C24** to **C36** before presenting the overall Concluding Comments to end this book. Conclusion **C24** is presented first followed by the additional conclusions **C25** to **C36.** Conclusion **C37** is added near the end of the chapter in which 25 technologies were assessed for their net value toward mitigating the effects of global warming and climate change.

The final conclusion **C38**, to be found at the end of this Conclusions and Concluding Comments section, is a recommendation that the young people of the United States be educated with the factual causes and effects of global warming and climate change (as revealed in these two books by the Author). Inasmuch as the purpose of the educational system is to educate (with the factual database of knowledge) the people in how to conduct their lives in the real environment and prepare the students to contend with all the many challenges – and opportunities - that the real world will bring to them. That is the largest reward the Author can hope for with the publication of these two books.

The entire set of 38 conclusions in the two books on climate change by this Author are incorporated in somewhat chronological order of their being made, not in order of importance, and are begun with conclusion **C24** as follows: (only the key sentences of each numbered conclusion are underlined)

C24) No matter the cause or combination of the multiple factors causing global warming and/or climate change, the conclusion reached by any scientifically valid analysis or interpretation is that the time has come for the use of coal technology as a heat energy generating fuel to be reduced and ultimately to be eliminated in the US region. That finding may not apply to other regions of the planet.

At the risk of being redundant, that more urgent action needed is to the reduction of Man-made pollution (with disposal of the toxic and carcinogenic ash residue from the impure coal combustion process) of water, land, and air – not for the purpose of retarding the "greenhouse gas effect" and attempting to control the climate. As coal burning dependent technologies are being phased out, there is an unquestionable need to explore "cleaner" alternative energy sources and other technologies that are reliable and are capable of meeting the demands of a world population that is expected to double or triple in the next century. Reducing negative side-effects (including pollution) from any technology that has many positive attributes is considered a normal function in technological advancement. The environmental science profession is being accused of "gambling with taxpayers' money" to better their reputation as being scientists in the same category as the exemplary Galileo class of scientists, and as a parallel benefit to continue receiving huge grants of money from the US Government to continue the research effort. The findings in the analysis described in this book will regrettably cause the grant support of emissions reduction efforts to be reduced but there should be grant funds supporting efforts to control Man-made pollution which is at far more a critical need than is "greenhouse gas" emissions control.

The Author evaluated the feasibility and "big picture" value of 25 types of technology that can help mitigate the effects of planet surface temperatures varying regionally (not just "average global temperature" linearized and purged of the effects of water vapor – which is tantamount to ignoring the effects of gravity on causing flooding). It is fairly obvious that the environmental scientists can serve the public in a far more beneficial manner by advancing the set of technologies analyzed in this book to mitigate the effects of the changes that are inevitable. In addition, while weaning away from the use of coal to reduce pollution, also weaning away from the fallacious goal of reducing carbon emissions to "combat" climate change is highly recommended. Instead, it would benefit the human population if environmental protection scientists would develop a process (or technology) for using the wasteful products discarded by humans as trash (and becoming pollutants) such as plastics which have become a major problem to the environment and to human safety and health.

The amount of effort and amounts of money being expended (inefficiently and wastefully in reducing the "greenhouse gas emissions" as a means of altering the climate is appalling when considering that climate change is an effect not a cause. Environmental scientists are juggling the books in the laws of science to prove the fallacious hypothesis that Man burning fossil fuels is the dominant effect on global warming. A basic principle in physics is that matter having mass cannot be created or destroyed – matter or mass can only be transformed in state. (Matter matters). The thermodynamics of combusting 1 unit mass of coal requires 2.88 mass units of oxygen. If air is taken from the atmosphere to obtain 2.88 mass units of oxygen that requires 25 mass units of air. The 25 units of air are "borrowed" from the atmosphere, such that when environmental scientists proclaim that burning 1 gigaton of coal produces 26 gigatons of carbon dioxide in the atmosphere, the laws of conservation of matter and mass are being juggled fallaciously. Burning 1 gigaton of coal in solid state may increase the total mass of the atmosphere by however many units of gas volume one mass unit of solid-state coal transforms into gas, but the volume of carbon in gaseous state cannot have a mass of carbon greater than the original mass of coal (at near 100% carbon in solid-state). A loan of air borrowed from the atmosphere is much like a financial loan from a bank. The loan amount of air is paid back with some interest added, the coal representing the payback interest.

C25) Criticism of the environmental scientists' motivation to ease the effect of the burning of fossil fuels on atmospheric "greenhouse gas" emissions, to prevent or at least reduce the extent to and rate at which the "average global surface temperature" is increasing, is moderated to being credited with the cessation of the use of coal in the United States. The cessation of coal burning will reduce the amount of pollution of US land, water, and atmospheric quality by reduction of the disposal of coal ash residue containing toxic and carcinogenic materials in landfills and abandoned pits. The cessation of the use of coal will have little or no effect on improving global warming or climate change for which it was originally targeted as needing elimination.

"Pie charts" on how much energy is created by which technology areas are useful but do not reveal the trends with time as meaningfully as the above graphical trend summary. Technologies are developed to solve real problems, not hypothetical scientific ones. The factual data indicating the weaning away from coal use is consistent with improving the environment but from a pollution reduction perspective not because of the reduction of "greenhouse gas" emissions. Despite the environmental scientists theorizing that the use of all fossil fuels must be eliminated to avert a global "crisis" (which is non-existing), the use of natural gas for electricity generation is seen to be on the opposite use rate trend as coal. The use of oil is predominantly in the transportation sector of technology and is understandably not included on the above chart of electricity generation technologies.

The use of renewables (solar and wind) constitutes about half that of natural gas in U.S. electricity generation, but the trend of increasing use rate of the renewables is low and at a nearly constant gradient with time.

C26) The use of wind turbines should be continued for rural regions (farming, livestock ranching, etc.) but the proposed locations of wind turbines for offshore harvesting of wind energy is not recommended. The increased offshore winds may improve the reliability some, but the reliability will be decreased more by the corrosive effect of saline water on the turbine machinery, the battering of the pylons by wave action, and inevitable storm damage to the entire network.

A few of the alternative energy technologies, however, show promises of being more efficient, safer, collecting the mineral resources most available and accessible in the region, providing opportunities for employment, are affordable, and reliable 24/7/365 etc. while also being clean. The Department of Energy certainly must recognize that the technology of hydrogen has greater potential than does that of solar power and wind power because of its higher reliability and availability of resources involved – as DOE presentation charts and a DOE video by former-Secretary of Energy Granholm reveals.

C27) The technology of geothermal heat energy was found in the unbiased weighted trade study analysis of the 25 candidate technologies to have the greatest all-around ("big picture") value – even surpassing that of the DOE-preferred hydrogen fuel energy because of the high cost to generate it. Geothermal heat energy lies in and below the Earth's outer crust under all regions of the planet and is the cleanest, most reliable, and most abundant source of heat energy without having to be generated using some other technology.

The most promising geothermal energy technology to meet the growing population demands for heat energy is to use the heat energy that is most abundant at all locations of the planet, independent of latitude or any other factor that causes the increased demand for heat energy. The lowest cost technology to achieve that is with the closed loop pipe-in-a-pipe system described earlier in this book. A graphical schematic showing the functional elements of that pipe-in-a-pipe closed loop system concept is summarized in the following diagram:

ADVANCED GEOTHERMAL SYSTEM

(Closed Loop in Deeply Drilled Well Technology)

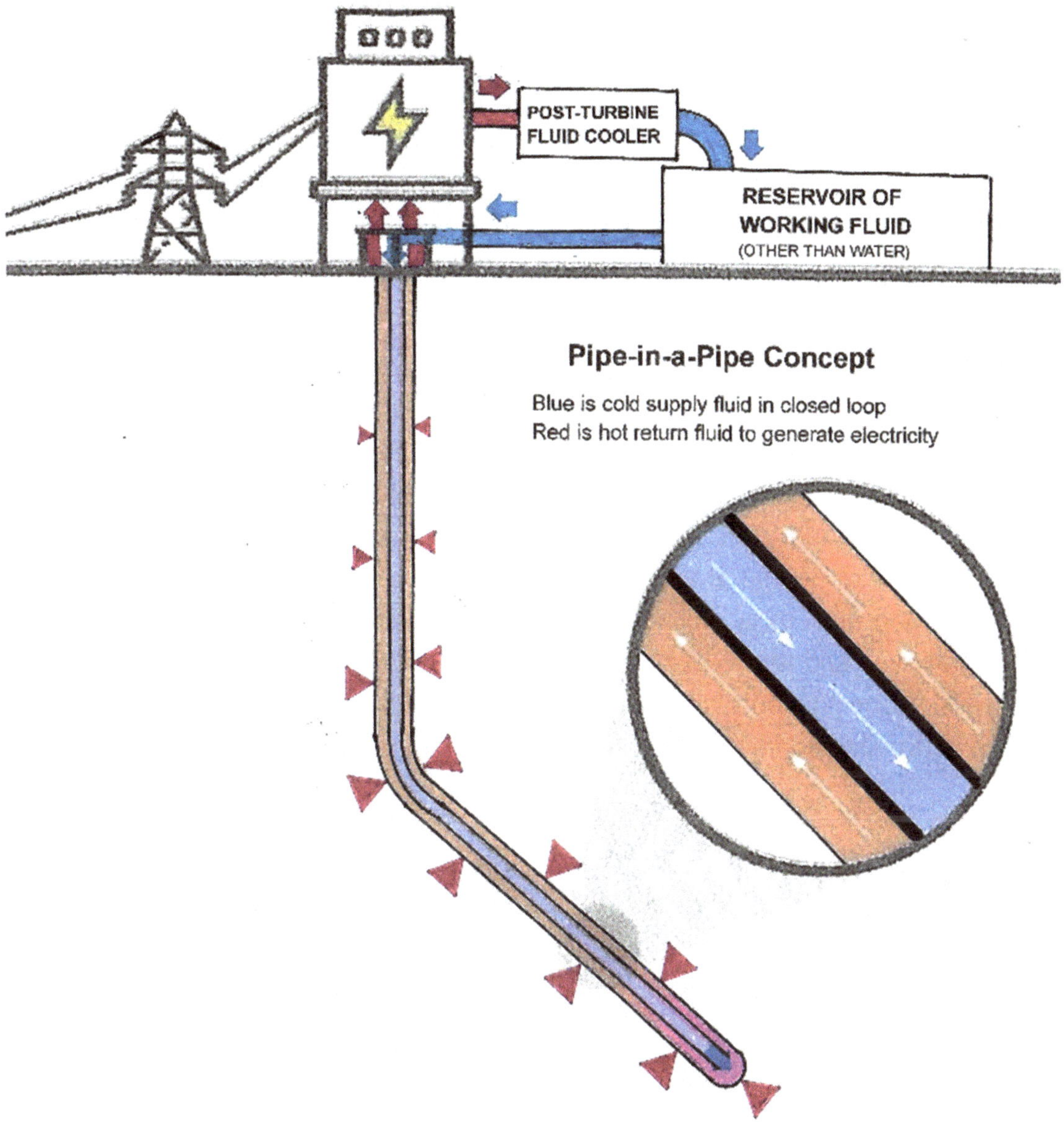

(Source: US Department of Energy database)

The terminology of being a "closed loop" system is that the working fluid (water or other fluids that more efficiently absorb heat) is inserted under pressure in a cool state through the insulated inner pipe (shown in the insert in blue color) and exits at the bottom of the injection pipe in the cool state into the closed end of the outer pipe. The flow of the working fluid reverses into an outer extraction pipe that is not insulated so that the geothermal heat at the drilled depth is transferred into the return fluid as it ascends (under pressure and in a heated state) where it is used to power the turbines to generate electricity. The "dogleg" in the well is a developed drilling technology and minimizes the depth of the well while increasing the length of the pipe that will be collecting the surrounding geothermal heat.

The diameter of the well needed to be drilled is the limiting factor on the capacity of the single well pipe-in-a-pipe closed system to generate electricity. However, to increase the capacity of the facility (with the electricity generating turbines and cooling system returning the working fluid in a cooled state back to the injection pumps) can be increased by drilling a well adjacent to the single well system. If more than a single well is used, the hot working fluid from each of the wells could be collected in a plenum chamber before being passed through the turbine(s) generating the electricity. This is the cleanest, most reliable, safest, least-impact on environmental technology for generating electricity meeting the public and industrial needs of a region. The highest cost is in the drilling of the well and the cost of the pipes in the system, therefore the total cost would be multiplied by the number of wells to be drilled to meet the capacity demands of the regional users. The closed loop system can be located as close to or within a populated area as desired without concern for accidents or pollutants entering the municipal water system.

There are two other very favorable options to the geothermal heat energy sources. On the left is shown a variation of the "closed loop" system where a U-shaped system of wells is drilled deeply into the geothermally hot upper mantel and connected at the bottom of the injection well and the extraction well with a horizontal tunnel. This closed loop system eliminates the need for a pipe-within-a-pipe system but adds the cost of drilling a second extraction well and a horizontal tunnel. The drilling of a second tunnel is not a challenge (but is an additional cost). The challenge (or con) of this type closed loop system is drilling a horizontal tunnel at a large depth and connecting the ends of the injection well and the extraction well. That is beyond the experience level of the Author to evaluate the risks and the costs involved in creating that leg of the closed loop system. The geothermal heat is transferred into the working fluid as it ascends in the extraction well.

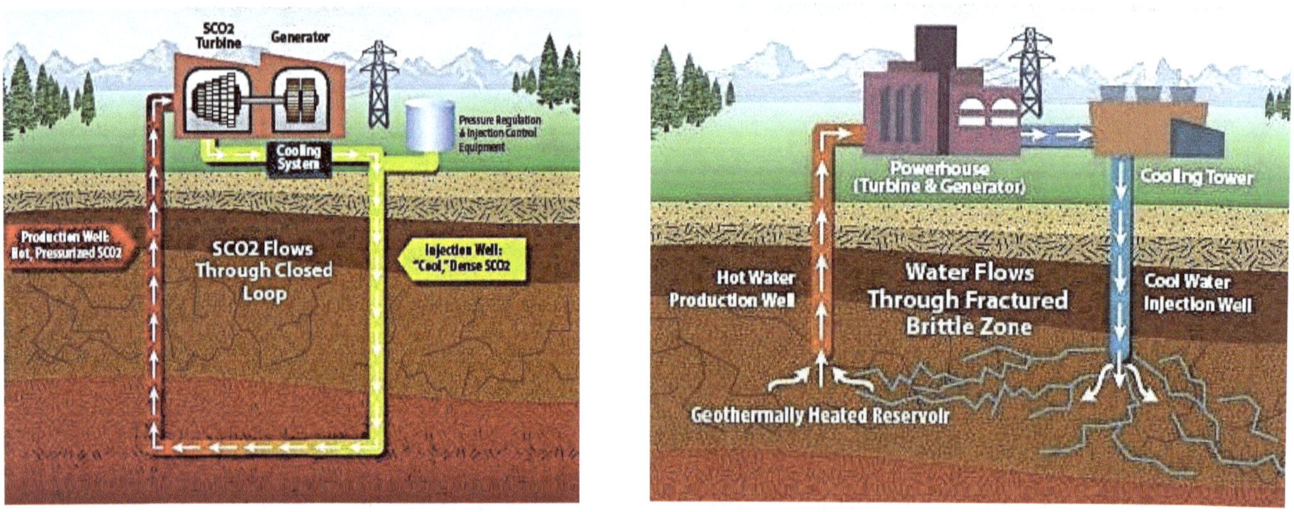

(Source: U.S. Department of Energy database)

On the right side of the diagram is shown the "open loop" version of geothermal energy harvesting at large depths of drilling. The primary difference in the system at the left and the system at the right is that the horizontal tunnel at the bottom of the closed loop system is opened to the working fluid emanating from the bottom of the injection well passing through the shale layer of rock and being geothermally heated as it passes through the hot shale toward the open end of the extraction well. The shale is fractured to open the many passages between the ends of the two vertical wells by the well-developed drilling technique of fracturing the shale known in oil-well technology as "fracking".

The less-positive flow of the working fluid from the injection well through the fractured shale, being heated geothermally by the fractured shale, and entering the bottom end of the extraction well has two cons – one being that there will be losses of the working fluid entering into the fractured shale layer but not entering the extraction well. The amount of the injected fluid that does enter the extraction well will have some small shale debris from the "fracking" process that must be filtered out before the working fluid is passed into the turbine system and generating the electricity from the geothermally heated working fluid. If other than water is used as the working fluid, the valid concern of environmental protectionists would be of the working fluid (possibly containing toxins and/or carcinogens) that does not enter the extraction well inserted into deep aquafers of the water supply system to any adjacent populated region.

The possibility exists that existing oil and natural gas wells that have been abandoned may potentially be used to serve as injection wells and the extraction wells in geothermal harvesting and avoid much of the additional drilling and "fracking" associated with harvesting geothermal energy.

C28) Because of the looming problem of the population doubling or tripling in the not too distant future, the most efficient use of land near heavily populated regions and to achieve meeting the demand of the growing population lies in the increased use of nuclear energy – provided its previous problems of failures in its cooling system posing the safety hazard of a nuclear "melt-down" or leakage to the surrounding environment are designed out of those systems. The use of thorium instead of uranium was rated with a slightly better weighted score on its overall set of pros and cons in meeting the needs of the human and industrial electricity demands. The pros of thorium reactors are stated to be:

- Abundance: Thorium is more abundant in the Earth's crust than uranium. Its availability is about the same as for lead.
- Reduced Waste: Thorium reactors produce less long-lived radioactive waste compared to uranium reactors.
- Proliferation Resistance: Thorium reactors produce less plutonium, a material used in nuclear weapons, making them less prone to proliferation concerns.
- Safety: Certain thorium reactor designs, like molten salt reactors, have inherent safety features that reduce the risk of meltdowns.

The cons of thorium reactors are not as well outlined since there are no fully operating thorium nuclear plants in operation in the United States. The only known thorium nuclear facility in operation is one in India, and its economics seem to be its largest con. A few smaller nuclear reactors with thorium have been generated as large scale laboratory research specimens – such as one in the Netherlands:

(Credit to unknown photographer in the Netherlands)

Environmental scientists are quick to comment in the news media: "*Look at all those emissions rising from the smokestacks*". Those "emissions" are mostly steam from the cooling of the nuclear reaction condensing into water vapor. Carbon dioxide is an odorless, invisible gas and is in low concentration in the steam emissions from smokestacks after having passed through environmental protection anti-pollution measures at the facility.

C29) <u>The use of thorium nuclear energy is a proven technology and should be considered for use in either newly constructed nuclear facilities or as a replacement technology in inactive uranium nuclear facilities which were closed for safety reasons</u> (such as those that were deactivated after the Three-Mile Island in Pennsylvania and Indian Point NY nuclear (uranium) facility incidents). Thorium is an abundant element in the United States and in most developed countries of the World. It is "cleaner" to use than uranium. The critical cooling system of the nuclear reactor is much simpler and safer than that for uranium.

China has recently announced that it has developed a most-promising thorium reactor which burns trash combustibles as part of the heat generation needed in the thorium nuclear process.

The use of the heated cooling water from the nuclear reactors (uranium or thorium) could be used in the flash distillation desalination processes if near a large body of saline water. The heated water could also be used to augment the geothermal energy process using a closed-loop system of harvesting geothermal heat energy. Coupling the four technologies of nuclear energy, solar energy, geothermal energy and desalination technology would be a good way to capitalize on the synergetic effect of combined technologies. The location of combined facilities with two or more technologies must be given careful planning to avoid a natural catastrophic event from damaging/destroying the multi-technology facility (i.e., protect "all the eggs in the one basket".)

When a con in a technology endangers a large number of the population of a region, that con must be eliminated or controlled much more effectively to reduce the fatalities and injuries to near zero. The argument the environmental scientists have used effectively in causing coal technology to be eliminated is to protect the atmospheric environment from greenhouse gas emissions, but the Author has yet been able to find any factual statistics that directly apply to there being a "crisis" affecting the human population.

The electric vehicle technology has a con of the same or even more ominous potential for endangerment of human life – resulting from the instability of lithium-ion battery technology. The rise in the number of fatalities per vehicle fire can be correlated with the rise in the number of electric vehicles on the highways of the world and especially in the United States:

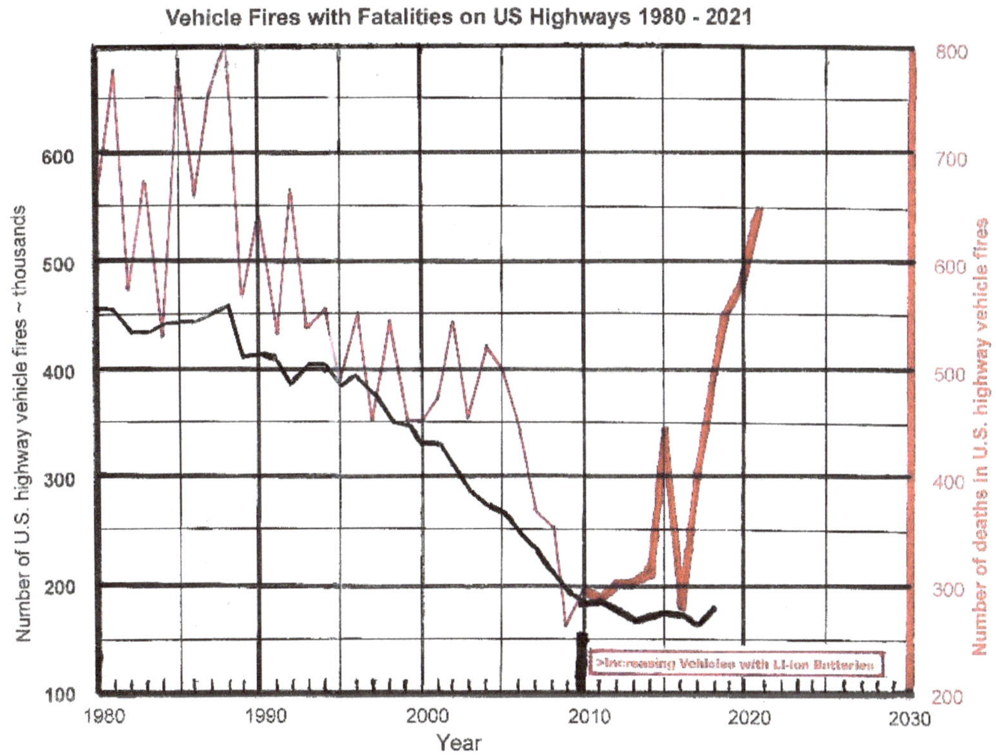

(Source: National Highway Traffic Safety Administration)

Stating that lithium-ion batteries in electric vehicles is an unacceptable hazard is not very attention-getting, but seeing a lithium-ion battery fire as it is exploding and the intensity as it is burning can only be appreciated by "seeing" them (from a safe distance) or to see a photograph of the initial eruption of the high heat-energy lithium ions and the ensuing fire engulfing the passenger compartment of a vehicle.

(Credit to unknown photographers)

It is important to realize that a fire of the type shown endangers not only the occupants of the affected vehicle but when the vehicle is in the midst of a large number of vehicles (such as in a rush-hour traffic jam on a Los Angeles Freeway twice a day, five days a week, and for 50 weeks of the year are a potential catastrophe that will only get worse as the percentage of the vehicles in that traffic jam are equipped with lithium-ion batteries.

The fire departments of many densely populated communities have come to realize that it is too difficult to extinguish lithium-ion fires (water alone does not extinguish them) and have adopted the practice of protecting the surrounding properties of the fire scene and resort to containment of the fire not extinguishing it. Chapter 5 contains additional photographs of other vehicles to reflect that the hazard does not affect just one model of electric vehicle but affects any make or model of electric vehicle equipped with a lithium-ion battery. The hazard is compounded by the hazard being directly beneath the occupants in the vehicle not at a remote location in the front or rear of the vehicle.

Scenes like the above candid images – not staged - are not uncommon in war scenarios when white-phosphorous munitions are used to destroy a military target. Humans burned so severely and so disfigured express the feeling that it would have been more merciful to not survive the horrible burns.

The lithium-ion battery fire hazard problem has been known to exist with smaller products powered by lithium-ion batteries for over a decade. *"Lithium batteries are generally safe and unlikely to fail, but only so long as there are no defects and the batteries are not damaged or mistreated,"* said Steve Kerber, vice president and executive director of Underwriters Laboratory's (UL) Fire Safety Research Institute (FSRI) in 2015. *"The more batteries that surround us the more incidents we will see."*

In 2016, Samsung issued a global recall of the Galaxy Note 7, citing "battery cell issues" that caused the device to catch fire and at times explode. HP and Sony later recalled lithium computer batteries for fire hazards, and about 500,000 hoverboards were recalled due to a risk of "catching fire and/or exploding," according to the U.S. Consumer Product Safety Commission.

In 2020, the Federal Aviation Administration banned uninstalled lithium-ion metal batteries from being checked in luggage and said they must remain with a passenger in their carry-on baggage, if approved by the airline and between 101-160 watt hours. *"Smoke and fire incidents involving lithium batteries can be mitigated by the cabin crew and passengers inside the aircraft cabin,"* the FAA said.

The safety issue is not the electric vehicle technology *per se*, but with its lithium-ion battery. The repeated instances of fires erupting explosively are from an inherent hazard in lithium-ion battery technology – not caused by Man. Lithium-ion battery fires are cumulatively causing loss of human lives and losses of property that will only increase as the population (and number of vehicle operators) doubles and triples within the next decades. There is no allowable quota of fatalities for a product having an unsafe element, but experience in crash investigations has shown that the sooner a latent hazard is recognized and eliminated, the fewer lives will be lost. The world trend toward battery-powered devices of all sizes and applications should dictate that safe technologies should be incorporated on a higher priority than how many units can be sold quickly before the device is recalled and made safer.

C30) Of the 25 technologies evaluated in this book, and, as the prioritized weighted importance scores reveal, electric vehicle technology with lithium-ion battery power is unsafe and is recommended not to be used by the public in the USA. Electric vehicle technology combined with aluminum-ion or sodium-ion battery technology is the unquestionably better technology for the future human transportation system in the United States as well as in other regions where public safety matters..

Fire containment provisions (which are heavy) have been developed and installed in newer model Tesla electric vehicles but there are many hundreds of thousands of electric vehicles without those containment provisions in operation on the US highways system. That number of high-hazard vehicles will grow in the near future to be in the millions. Advertising has been seen that the heavy weight of lithium-ion batteries (weighing about 1,000 pounds per vehicle) adds safety to the vehicles as it reduces roll-over tendencies in a highway crash. Insurance companies increase premiums on lithium-ion equipped vehicles because of their fire propensity but do not decrease the premiums because the electric vehicle has a lower roll-over tendency in a crash.

Better quality control in manufacturing, assembling, and recharging equipment will minimize but will not eliminate the explosive fires igniting. The more electric vehicles, the more lithium-ion battery fires. The better technology solution (as is recommended in this book) is to replace lithium-ion battery technology with a less volatile battery technology such as sodium-ion or aluminum-ion have the potential to be. The electric vehicles technology is excellent but the “con” of a spontaneously exploding battery system is too unsafe to continue in use. An improved fire containment system may give the occupants a chance to escape from the vehicle that is afire, but the vehicle with a lithium-ion battery will assuredly be a total loss. In lithium-ion battery fires in open highway scenarios, the remoteness before fire extinguishing equipment or medical rescue services could arrive paints a grim human safety situation. Still more ominous is that there must be a specialized fire extinguishing chemical used – common water does not extinguish most lithium-ion fires.

There are instances of cargo transport ships having sunk (with many hundreds of other new electric vehicles lost) by suspected lithium-ion batteries igniting explosively inside the cargo compartments. Cargo ships have an on-board fire extinguishing capability, but common fire-extinguishing technology is realistically not capable of extinguishing fires of such intense heat. Typically, lithium-ion fires are sprayed with water more as heat containment, but the battery continues to burn even without an oxygen source. The entire vehicle – and perhaps others parked adjacent to the initial vehicle on fire - will most probably be a total loss a few hours later. The toxins and carcinogens inj the debris cannot simply be “swept up”.

The potential disaster of one electric vehicle with a lithium-ion battery explosive ignition destroying vehicles surrounding the source (as in a Los Angeles Freeway “rush-hour” traffic circumstances shown next) is too high to accept as a risk: When envisioning a worst-case scenario, if the fire were to occur in an electric vehicle in the middle of a traffic jam when batteries will be overheating, the risk is high that a large number of adjacent vehicles in the traffic jam will be set on fire and burn in a chain-reaction.

(Credit to unknown photographer in Los Angeles)

Without awareness of the potential for catastrophic lithium-ion battery fires in electric vehicles, California Governor Gavin Newsom issued an Executive Order (N-79-20 on September 23,2020) with the following multiple "Whereas"' reasons (noting that the primary is the fallacious "*Whereas the climate change crisis threatening the health and safety of too many Californians is happening now . . .* "):

EXECUTIVE ORDER N-79-20:

WHEREAS the climate change crisis is happening now, impacting California in unprecedented ways, and affecting the health and safety of too many Californians; and WHEREAS we must accelerate our actions to mitigate and adapt to climate change, and more quickly move toward our low-carbon, sustainable and resilient future; and WHEREAS the COVID-19 pandemic has disrupted the entire transportation sector, bringing a sharp decline in demand for fuels and adversely impacting public transportation; and

WHEREAS as our economy recovers, we must accelerate the transition to a carbon neutral future that supports the retention and creation of high-road, high-quality jobs; and

WHEREAS California's long-term economic resilience requires bold action to eliminate emissions from transportation, which is the largest source of emissions in the State; and WHEREAS the State must prioritize clean transportation solutions that are accessible to all Californians, particularly those who are low-income or experience a disproportionate share of pollution; and

WHEREAS zero emissions technologies, especially trucks and equipment, reduce both greenhouse gas emissions and toxic air pollutants that disproportionately burden our disadvantaged communities of color; and

WHEREAS California is a world leader in manufacturing and deploying zero-emission vehicles and chargers and fueling stations for cars, trucks, buses and freight-related equipment; and WHEREAS passenger rail, transit, bicycle and pedestrian infrastructure, and micro-mobility options are critical components to the State achieving carbon neutrality and connecting communities, requiring coordination of investments and work with all levels of governments including rail and transit agencies to support these mobility options; and WHEREAS California's policies have contributed to an on-going reduction in in-state oil extraction, which has declined by over 60 percent since 1985, but demand for oil has not correspondingly declined over the same period of time; and WHEREAS California is already working to decarbonize the transportation fuel sector through the Low Carbon Fuel Standard, which recognizes the full life cycle of carbon in transportation emissions including transport into the State; and

WHEREAS clean renewable fuels play a role as California transitions to a decarbonized transportation sector; and

WHEREAS to protect the health and safety of our communities and workers the State must focus on the impacts of oil extraction as it transitions away from fossil fuel, by working to end the issuance of new hydraulic fracturing permits by 2024; and WHEREAS a sustainable and inclusive economic future for California will require retaining and creating high-road, high-quality jobs through sustained engagement with communities, workers and industries in changing and growing industries."

IT IS HEREBY ORDERED THAT:

1. It shall be a goal of the State that 100 percent of in-state sales of new passenger cars and trucks will be zero-emission by 2035. It shall be a further goal of the State that 100 percent of medium- and heavy-duty vehicles in the State be zero-emission by 2045 for all operations where feasible and by 2035 for drayage trucks. It shall be further a goal of the State to transition to 100 percent zero-emission off-road vehicles and equipment by 2035 where feasible.

2. The State Air Resources Board, to the extent consistent with State and federal law, shall develop and propose: a) Passenger vehicle and truck regulations requiring increasing volumes of new zero-emission vehicles sold in the State towards the target of 100 percent of in-state sales by 2035. b) Medium- and heavy-duty vehicle regulations requiring increasing volumes of new zero-emission trucks and buses sold and operated in the State towards the target of 100 percent of the fleet transitioning to zero-emission vehicles by 2045 everywhere feasible and for all drayage trucks to be zero emission by 2035. c) Strategies, in coordination with other State agencies, U.S. Environmental Protection Agency and local air districts, to achieve 100 percent zero-emission from off-road vehicles and equipment operations in the State by 2035.".

It would be interesting if Governor Newsom could explain why the safety hazard of electric vehicles with lithium-ion batteries catching fire explosively in the midst of a traffic jam on Interstate 5 in the Sepulveda Pass area is less of a safety hazard than the emissions from a gasoline powered internal combustion engine vehicle. (This Author moved away from California after 38 years of residency because of the California Air Quality Management gestapo-like mandates made to curtail greenhouse gas emissions.)

C31) <u>Millions of new electric vehicles are equipped with lithium-ion batteries as they presently are in several different manufacturers' designs, and the fatality statistics of the number of fatalities in vehicle fires reflect the increasing number of electric vehicles with unstable lithium-ion batteries are</u>

the root cause for the majority if not the total of those car fires. If the use of all-electric vehicles are mandated on California road system, the use of an alternative – and much safer – battery system design must also be mandated, in California and all 50 of the United States for public safety reasoning.

Alternate battery technologies are being developed to eliminate that safety critical con of too many fatalities in electric vehicles equipped with lithium-ion batteries. There has been developed the technology of lithium iron phosphate batteries which are substantially safer than the earlier lithium-ion unsafe batteries. The basic material lithium however is found in useful quantities only in China and Africa but none in the North American continent. Batteries which contain lithium to any degree are costly and extremely heavy, as well as being unsafe when overcharged or overheated. Anyone with experience in magnesium fires can relate to the problem of extinguishing such fires (magnesium, lithium, or phosphorous.) The use of the very light weight magnesium is banned for use in military flight vehicles because of its high volatility and intensive heat when ignited. They cannot be extinguished by water alone.

The basis for a recommendation for replacing all electric vehicle batteries having a lithium content is similar to that made by the environmental scientists who mandate that all fossil fuels be eliminated to reduce emissions in the atmosphere. To eliminate the hazard of fires caused by lithium content all batteries with lithium to any extent should be eliminated. One technology suitable for achieving that In the United States transportation system is summarized in a comparison of aluminum-ion battery technology with lithium-ion technology: (or by any suitable battery such as sodium-ion having similar pros and cons compared to that of lithium-ion)

Feature	**Aluminum Ion Batteries**	**Lithium-Ion Batteries**
Cost	Lower	Higher
Abundance	High (aluminum is widely available)	Low (lithium is limited)
Safety	Safer	Less safe (risk of overheating)
Energy Density	Lower	Higher
Charging Speed	Faster	Slower
Environmental Impact	Lower	Higher

(Credit is given to the unknown analyst – not the Author - for the comparison)

It is hastened to be added that a parallel effort to develop sodium-ion batteries which have many of the same technology pros as does aluminum-ion battery technology and both being much safer technologies than lithium-ion technology. Recognition of the better safety of aluminum-ion and sodium-ion battery technology than that of lithium-ion battery by this Author was observed in the previous decade by a few start-up companies which came to the same conclusion that this Author is revealing to the public in this book as to the increasing frequency and severity of the lithium-ion battery fires in electric vehicles. This Author does not identify which of the two competitive battery technologies is a better alternative to the electric vehicles manufactured with lithium-ion batteries as that opinion could be converted into an advertisement for one technology over another competitor.

C32) The technology of aluminum-ion and sodium-ion batteries has matured sufficiently to require that either or both technologies be developed for all new electric vehicles of any style and size operating on the US highway system and especially in larger cities. Suffice it to say that the motor vehicle manufacturers should observe the need for protecting the safety of the public and advance the technology of the electric vehicles they produce to have sufficient safety being introduced into the millions of their electric vehicles that have been and are being introduced into the public transportation system with an acceptable level of safety risk in the equipment items within their respective vehicle designs.

This Author bases the multiple conclusions in this book using the discipline ingrained in system safety engineers and in crash investigators to identify if, where, and why a safety risk exists in a technology – not to favor or criticize any industrial entity for having applied or excluded that safety risk from their product.

An example of what can – and did – happen when a manufacturer conceals a known safety problem with their product is that experienced six years ago with the international Boeing 737 Max 8 aircraft program.

A professionally-saddening safety situation was exposed after two Boeing aircraft crashed – the first on October 29, 2018, when Lion Air Flight 610, a Boeing 737 Max 8, plunged into the Java Sea and 189 people on board were killed. As the cause of the Indonesian crash was being investigated, a second Boeing 737 Max 8 – Ethiopian Airlines Flight 302 - crashed on March 10, 2019, after take-off from Addis Ababa killing 157 passengers and crew. Crash investigators found a potentially hazardous flaw in the design and safety logic of the flight control system (in the software logic of the angle of attack sensors) that had been experienced on flight simulators on the ground during early development of the Boeing 737 Max 8.

Instead of adequately fixing the safety flaw, the flaw was not made known to the flight crews but was instead buried in the "fine print" of the flight handbook such that if an aircraft being flown on the automatic mode of flight control got into that unsafe flight mode (the nose of the aircraft being commanded down at a dangerous angle), the instructions were to shut off the automatic flight control system and fly the aircraft manually on the basic – and safer - flight control system. However, the airline operators were never made aware that the hazardous safety problem existed because Boeing did not want to ground the entire international 737 Max 8 fleet until the problematic safety flaw could be properly redesigned. To Boeing upper management, time spent fixing a problem was considered as time taken away from making a profit. "Whistle-blowers" simply were dismissed from employment.

Those facts were revealed to the public in 2021, and Boeing was subsequently ordered by the courts to recompensate the 346 families of those lost in a $2.5 billion settlement. It would have cost Boeing a very small fraction of that amount to properly fix the problem when it was first identified.

The above example of the Boeing 737 Max 8 safety hazard should serve as a warning to large motor vehicle manufacturers to not sell their all-electric vehicles to the public if there is a known major safety hazard such as having a lithium-ion battery with a very ominous fire safety record. Chapter 7 of this book describes that known safety hazard (erupting in fire for several known reasons in the lithium-ion electricity storage chemistry process or quality of manufacture) in sufficient detail and with photographic evidence to warrant a delivery pause in electric vehicles for a Federal Government level independent fire-safety risk assessment.

C-33) <u>A fire safety assessment of electric vehicles in use by the public and government should be by a team led by a qualified and experienced accident investigator (not by a scientist of the Environmental Protection Agency) and given authority and responsibility equivalent to that which was assigned in the Boeing 737 Max 8 crash investigations.</u> The official assessment should be conducted expeditiously to avoid further loss of human life as well as to minimize disruption to the automotive manufacturing economy and manufacturing schedule. The assessment need not fix blame for the use of the questionably unsafe lithium-ion technology in electric vehicles; only to determine the public safety risk in continued use of the lithium-ion technology and to identify if an alternate battery technology with improved fire safety is required.

Vehicle manufacturers must make their own business assessment of the possibility that a severe technology flaw has been identified and needs to be corrected in a manner so as to least endanger the public until the technology flaw is corrected. Major changes in the design of a company product is considered a senior management and business matter, not a science or engineering call.

If an independent safety assessment of the lithium-ion battery technology in electric vehicles decides that a safer alternate battery technology is needed, the disposal of the replaced lithium-ion batteries creates a pollution concern. Inherently, lithium-ion batteries emit high heat when burning (as well as hazardous gases that would need to be captured and disposed of to meet EPA regulations). A potential technology (in the category of pollution control) would be to sell obsolete lithium-batteries to the US steel industry to be used as fuel for renewing production of virgin steel domestically instead of the present practice of virgin steel being imported from Canada and Mexico and other countries. The practice of importing virgin steel came about by the Environmental Protection Agency imposing overburdening emissions controls (now found to be unjustified as the emissions are proven in this book to not be the dominant cause of global warming or climate change) on the US steel companies burning the fossil fuel coal as the primary high heat energy source needed to produce virgin steel. The US steel companies had little recourse but to instead import the virgin steel needed (and of course to pass the extra import costs on to the user).

The need for steel is another commodity that will grow in proportion to the exponentially growing population, using it productively instead of incurring the cost for safe disposal would avoid the expenses added to the steel being imported and passed along to the user. How the cost to the steel companies to purchase the lithium-ion batteries might be subsidized by the US government is beyond the scope of this Author to be suggested.

C34) Although not evaluated as a technology area in this book, the numerous photographs after hurricanes, tornados, typhoons, earthquakes, lightning-ignited wildfires, volcanic eruptions, tsunamis, flooding, and mudslides showing the damage done to properties (privately-owned or commercial) highlights the need for all countries to update the building codes used in the structures constructed in their region. In most regions, the probability of one or more of those natural catastrophes occurring and the type of catastrophe is of less importance as to the human and animal population suffering the after-effects of the catastrophe. No region can have a perfect building code, but many regions have disgracefully unsafe building codes that are not enforced. It is not only the climate changing differently in different regions around the World, but people are their likes and dislikes. The attitude that "I have the right to do whatever I want to do" is not a scientific problem but is a societal one. The leaders of a community must establish and enforce building codes, environmental codes protecting humans and all forms of life from pollution of the land, water, and air.

C35) Humans should not choose to locate their homes and places of business in a known flood-zone or a known "hurricane alley" or "tornado alley", on the side of a volcano, over an earthquake-prone fault line, or on a beach less than two or three feet above the "beautiful ocean". Humans enjoy taking risks however (as the croupiers in Las Vegas will confirm), and do not use the common-sense that their parents instilled in them as children before they learned how to spell "hazard". The sport of bungee jumping, skydiving, climbing the sheer vertical side of a cliff, high wire walking with no safety net below, skiing in an avalanche area, etc. are and always will be popular forms of entertainment despite their obvious hazards. No lecture on safety and common-sense will deter humans from placing themselves in "harm's way". This book can only advise, it has no intent of being a "Dear Abby" form of advice for the common problems of merely living on the planet. Those who wish to take risks will do so. But in the basic street language well understood by humans *"If you can't do the time, don't do the crime."*

The last scientific and environmental issue surfaced in this book to be given emphasis for resolution is the total problem of human and industrial pollution. In the "big picture", by causing environmental scientists to change their focus from "greenhouse gas emissions" and applying that focus to something else of scientific importance, there would be no greater contribution for the benefit of mankind to resolve the human and industrial pollution problem. That is a problem of far greater magnitude than this book could even make a scratch. Two possibilities proffered in the chapters of this book might be among the solutions investigated.

The most major item of pollution is that of the disposal of plastics that have been recycled into storage warehouses but are going unused because of lack of a suitable reuse. The Author suggested the plastic be heated (by solar energy) and formed into railroad track ties and used to replace the existing wooden ties. Wooden tie technology has reached a state where the cons outweigh the pros of using wood. By reusing excess plastic trash and waste, the use of plastic trash partial solves the disposal of the plastic into landfills and ends up being introduced into ground aquafers – polluting the drinking water of the public or flowing into the oceans and polluting marine life. The second aspect is that by using plastic instead of wood, deforestation to produce wooden ties becomes unnecessary.

C36) The population growth prediction of doubling and perhaps tripling in the next century or two will undoubtedly increase the need for twice or triple the needs for energy, as well as the need to be more efficient at disposal of human-generated waste and trash. The EPA in the United States is an organization that needs to open up its vision about how to handle the threat of pollution on humans far more than the effect of greenhouse gases on the climate. The pollution problem will grow in proportion to human population growth. When scenes like that shown next (in Accra, Ghana) are seen, and it is realized that the world does not know how to handle its present level of waste disposal and pollution, why should there be any expectation that the problem will decrease as the world population doubles or triples? The EPA in the USA and its counterparts in other countries needs to stop "choking on the minnows while the whale is swallowing us". Control of "greenhouse gas emissions" is a much lower hazard to humans than is the hazard of pollution of the land, water, and atmosphere.

(Credit to unknown photographer in Accra, Ghana)

The technologies evaluated in this book may seem unnecessary for the current population's needs and demands for clean energy, but the "big picture" of population growth will require every resource to be used to its maximum availability and every technology to be used to its maximum capacity. The planning for that probability should not be left to "We'll cross that bridge when we come to it." Waiting until water is lapping around our ankles before beginning planning on better flood control and the ocean levels rising is not a risk that should be taken. Plans that go unused can be disposed of easily if later found out not to be necessary. Plans that do not exist and are developed at a crisis rate to minimize a fear-provoking but fallacious "catastrophic disaster" are more difficult to accomplish, and typically are found to have lower prospect of being successful.

Concluding Comments

It cannot go without comment that as this book was being prepared for publication in the public domain, the newly elected officials in the White House (in Washington, D.C.) announced plans to reduce the huge expenditures of taxpayers' money allocated to reducing "greenhouse gas" emissions as the approach needed to control surface temperature and climate changes hypothetically approaching "critical" consequences on the environment. Taxpayers should have a major voice in how their taxes are being expended and also a major voice in their governmental leaders to enact regulations and laws "for the benefit of Mankind".

Trump floats ending clean energy agency

BY ARI NATTER
Bloomberg News

WASHINGTON

The U.S. Energy Department is proposing to shut down its Office of Clean Energy Demonstrations and cut roughly $9 billion in awards for programs regarding carbon capture, direct air capture, solar and hydrogen, according to documents seen by Bloomberg.

Under the plan, which isn't final, the $27 billion agency's staff would be reduced to 35 employees, and about $10 billion in projects, including $3 billion for so-called hydrogen hubs, would be kept "as is" and transferred to other parts of the Energy Department.

The office employed about 250 people before President Donald Trump, a climate change skeptic, returned to office.

A decision on the fate of the agency, which was created in 2021 in bipartisan infrastructure legislation, could come as soon as this week, according to Axios, which earlier reported on the proposal.

The Energy Department didn't immediately respond to a request for comment on Friday evening.

Money for hydrogen hubs, in which the gas can be produced and used, is intended to highlight its potential as carbon-free fuel to run factories and power plants. Direct air capture involves technology that sucks planet-warming carbon dioxide out of the air.

Among the funding identified for potential elimination are billions of dollars for hubs in California, the Pacific Northwest, the Mid-Atlantic and the Midwest while hubs in Texas, Appalachia, and the Minnesota, North and South Dakota would be spared, according to the documents.

The plan also recommends keeping $3.5 billion for advanced nuclear reactor projects and $1.9 billion in industrial demonstration program projects.

Also slated for termination are carbon capture projects, battery storage projects, and a direct air capture project by a subsidiary of Occidental Petroleum Corp. that was selected for an award of as much as $1.2 billion by the Biden administration in 2023.

The plan comes as the Energy Department, which has a broad mission ranging from overseeing the nation's emergency reserves of oil to protecting the power grid from cyber threats, has identified 8,500 jobs as "non-essential" as it prepares to cut employees in response to a mandate from Elon Musk's government efficiency team.

(Source: Bloomberg News, "News Observer", Durham, North Carolina, April 9, 2025)

With regard to the statement in the top of the second column that President Donald Trump is a "climate change skeptic", the public should be made aware that this Author had corresponded to President Trump while in office in his first term in 2017 and revealed (in a paper entitled "The Facts About Climate Change") that he had good cause to not accept the environmental scientists' hypothesis that Man and industry burning fossil fuels was the dominant cause of global warming, climate change, and the oceans levels rising. There are numerous scientifically-proven reasons for President Trump to view the "energy and climate crises" with skepticism – in the same manner that Galileo viewed the scientific claims of a "flat earth" and the 'Sun orbits the Earth" with skepticism.

Shortly after submittal of the Author's "The Facts About Climate Change" paper to President Trump, he removed the United States as a signatory and adherent to the United Nations Intergovernmental Panel on Climate Change (UN IPCC) emissions control goals set forth in the document known as the "Paris Agreement" in 2019. In early 2019, the Author received the following response from President Trump (although it appears to have been written by one of his staff personnel):

THE WHITE HOUSE

WASHINGTON

June 11, 2019

Mr. Herbert A. Hutchinson
Paso Robles, California

Dear Mr. Hutchinson,

Thank you for sharing your views on energy and environmental policy.

My Administration is strongly committed to environmental protection, including keeping our air and water clean and being good stewards of our environment. This requires all of us to do our part, and I want to thank you for your commitment to preserving our environment for future generations.

As we continue to protect our environment, we must also ensure that American workers have greater opportunities to provide for their families. That is why my Administration is also committed to protecting American workers and American companies from unnecessary regulatory burdens. Going forward, regulations must balance our stewardship of the environment with the needs of a growing economy.

Thank you again for sharing your views. As President, I am committed to unleashing America's economic potential, while also protecting the environment. I am confident that together we can preserve America's natural blessings for future generations.

Sincerely,

Shortly before the Author received the above letter from President Trump in 2019, he issued an Executive Order that the United States was being removed as a signatory to the UN IPCC Paris Agreement. Somewhat as a déjà vu, in late December 2024, the Author sent the Background portion of this book ("Dealing with the Real Climate Change") to the Mar-a-Lago Resort in Florida where the newly-elected President was selecting the new heads of the Federal Departments (including Department of Energy, and the Environmental Protection Agency) to pursue the many changes that this book was recommending. It is not known to what extent the submittal of that Background information had on the President's planning, but the following Bloomberg News article appeared (in a mid-January 2025 edition of the Durham, North Carolina, Herald-Sun newspaper):

Trump creates uncertainty for major climate report

BY ERIC ROSTON AND SHERYL TIAN TONG LEE
Bloomberg News

Climate diplomats beginning work on the next report from the Intergovernmental Panel on Climate Change - the crucial assessment on global warming that helps shape policy for governments to companies - are meeting in China this week without U.S. officials.

U.S. government scientists participating in the IPCC's global assessments were issued a stop-work order from the Trump administration, according to media reports late last week, and NASA's chief scientist Kate Calvin, who holds a leadership role in the new report cycle, is no longer attending as a result, CNN said, citing a spokesperson from the space agency.

ANDREW CABALLERO-REYNOLDS AFP via Getty Images/TNS

U.S. government scientists participating in the Intergovernmental Panel on Climate Change's global assessments were issued a stop-work order from the Trump administration, according to media reports late last week.

NASA, Calvin and the IPCC did not immediately respond to requests for comment.

The U.S. absence comes amid broader cuts to research funding and a retreat from climate diplomacy under the Trump administration, raising new questions on what the IPCC's future might look like without U.S. leadership. The group's assessments are widely viewed as the world's most trusted source of information on climate change.

"Without the U.S., the IPCC fails," said Benjamin Horton, director of the Earth Observatory of Singapore, who has contributed to past assessment reports. "The U.S. puts more money, more personnel, collects more data and runs more models for climate science than the rest of the world combined."

About 18% of IPCC authors have been from the U.S., more than twice the next biggest national

(Source: Bloomberg News, January 2025, Durham Herald-Sun newspaper)

With challenges comes opportunities. It is the fervent wish of the Author that anyone – scientific professional or non-STEM non-professional employee - who is faced with a career path change not founded on reducing carbon-based emissions will pursue the opportunities being generated by the boom to generate cleaner, more reliable, and more accessible alternative sources of energy. The need and demand for larger amounts of energy are already becoming a challenge and will continue to become more of a challenge as the world population exponentially grows in number. The specific amount of energy needed and demanded has been shown to be determined by the regional population and state of development, not determined by an "average global" level of population or annual "average global temperature".

The technologies that were assessed in this book are those that were identified as being appropriate candidates for consideration in the United States based on the research that was done for the effects of temperature and climate changes that were identified in databases applicable to the latitude of the "Temperature Envelope" that this Author generated for the United States region. Other nations – in other latitudes and on other continents – should conduct similar "big picture" type research for the set of energy technologies that are most appropriate for supporting the unique energy needs and demands, and accessibility of natural geological resources inherent to that region.

The expenditure on non-valuable knowledge is essentially a wasteful expenditure that could be applied to more useful expenditures on other more valuable technologies. There have been several start-up technologies to remove carbon dioxide from the atmosphere and deposit it into the ground in order to reduce the amount of "greenhouse gas" emissions in the atmosphere. Those expenditures may be of scientific value to the science profession but realistically they do little to increase human safety, health, comfort, and improve the world transportation system.

In the lower grades of education, knowledge is not gained rapidly but it is that base of knowledge – learned by human "twigs" - on which all future gains in knowledge will be built. As older members of humanity – and regardless of language spoken or cultural beliefs – there is an obligation to prepare the youngest members of the region humans reside in for the realities of the world the young people are about to learn if they are to survive. This book hopefully will be "one small step for Man – but a giant leap for Mankind" in the learning process of our youth.

The second President of the United States John Adams (1735-1826) made a wise statement about teaching children: *"It should be your care, therefore, and mine, to elevate the minds of our children and exalt their courage; to accelerate and animate their industry and activity; to excite in them an ambition to excel in every capacity, faculty, and virtue*." Deficiencies in the education system are not as hazardous as those described in the foregoing sections of this book, but part of the mitigation of climate change is to educate the public what can be done to mitigate those hazardous effects.

C37) <u>The resultant summary of the 25 technologies assessed in this book is best presented in a tabular form on one page. The weighted scores for the technologies are listed in descending score sequence – i.e., in descending order of importance and of usefulness – within each of the six different technology areas given a Group number:</u>

Group	Trade Study No.	Technology Area	Total Wghtd Score	Comment (Suggested Use and Development for USA Region)
①	TS-4	Use of Fossil Fuel - Natural Gas (Liquified)	95%	Increase useage; improve transportability
	TS-3	Use of Fossil Fuel - Natural Gas (Pipeline)	93%	Install more pipelines (distribution to ex-coal users)
	TS-5	Use of Fossil Fuel Derivative - Propane	90%	Increase useage; improve storage safety
	TS-2	Use of Fossil Fuel - Oil	89%	Continue but reduce use - to no imported oil
	TS-1	Use of Fossil Fuel - Coal	74%	Phase out coal burning technology (to reduce toxic and carcinogenic land/water pollution) Close deep mining.
②	TS-6	Use of Non-fossil Fuel - Hydrogen	90%	Increase useage; improve storage & distribution safety
	TS-7	Use of Non-fossil Fuel - Ethanol	86%	Continue useage; co-technology with EV Hybrids
③	TS-9	Control of Hydrodynamic Events (Coastal and Seasonal Event Regions) - Levees / Sea Walls	98%	Highest need but with lack of activity/planning to meet safety and energy needs of growing human population
	TS-8	Use of Hydrodynamic Energy (Dams, River Flow)	96%	Increase existing capacity and additional construction
	TS-14	Energy Storage - Stationary Large (Electrical)	94%	Good technology supplementing that of solar and wind.
	TS-15	Use of Geothermal Energy (Surface)	93%	Good technology for the western states in Conus;
	TS-17	Energy Storage - Stationary Large (Thermal)	92%	Most important adjunct technology with geothermal
	TS-11	Use of Solar Energy (Residential / Rural / Town)	89%	Of most value to small users. Requires battery storage.
	TS-16	Use of Geothermal Energy (Deep-Drilled; Closed Cycle and Open Cycle with Fracking)	83%	Increase useage; Closed cycle preferred to Open Cycle; Recycle abandoned oil wells where feasible
	TS-12	Use of Wind Energy (On Land)	82%	More useful for small industrial, farming, and public use.
	TS-10	Use of Solar Energy (Large Industrial / Commercial)	78%	Not reliable for industries, OK for small; co-technology with other uses (distillation desalination; water purification)
	TS-13	Use of Wind Energy (Off-Shore)	63%	Too harsh environment (corrosion, wave force, storm damage). Recommend not installing off-shore system.
④	TS-19	Use of Nuclear Energy (Thorium)	96%	Among best technologies for mitigating climate change.
	TS-18	Use of Nuclear Energy (Uranium)	95%	Among best technologies for mitigating climate change.
⑤	TS-21	Use of Electric Plug-In/Fuel Hybrid Vehicles (Commercial / Public)	82%	Increase mix of All-Electric vehicles with Electric Plug-In / Hybrid Fuel vehicles with longer range for public.
	TS-20	Use of All-Electric Vehicles (Commercial / Govt.)	81%	Technology maturing. Limit initial use until more mature.
	TS-22b	Energy Storage - Mobile Small (Aluminum-ion or Sodium-ion)	97%	Aluminum-ion or sodium-ion battery technology is safer than lithium-ion in electric vehicles
	TS-22a	Energy Storage - Mobile Small (Lithium-ion)	52%	The use of lithium ion battery technology is unsafe. Replace lithium-ion with aluminum-ion or sodium-ion.
⑥	TS-24	Desalination - Distillation (Solar/Geothermal)	95%	A vitally needed process for potable water to population
	TS-23	Desalination - Reverse Osmosis (Solar/Geotherm	92%	A vitally needed process for potable water to population

Prioritized Summary of the Assessment of Candidate Climate Mitigation Technologies

As shown in the Weighted Trade Study summary above – applicable to the United States region - the continued use of natural gas (either in liquified or gaseous state) and oil should be continued but to the extent that no oil will need to be imported from the OPEC countries.

Shortly after completing the above chart summarizing all the 25 trade studies into their groupings and in a priority order for pursuit of individual technologies, the New York Times published the following chart (corroborating how the technology groups have been put into application in the real world from 2005 to the present:

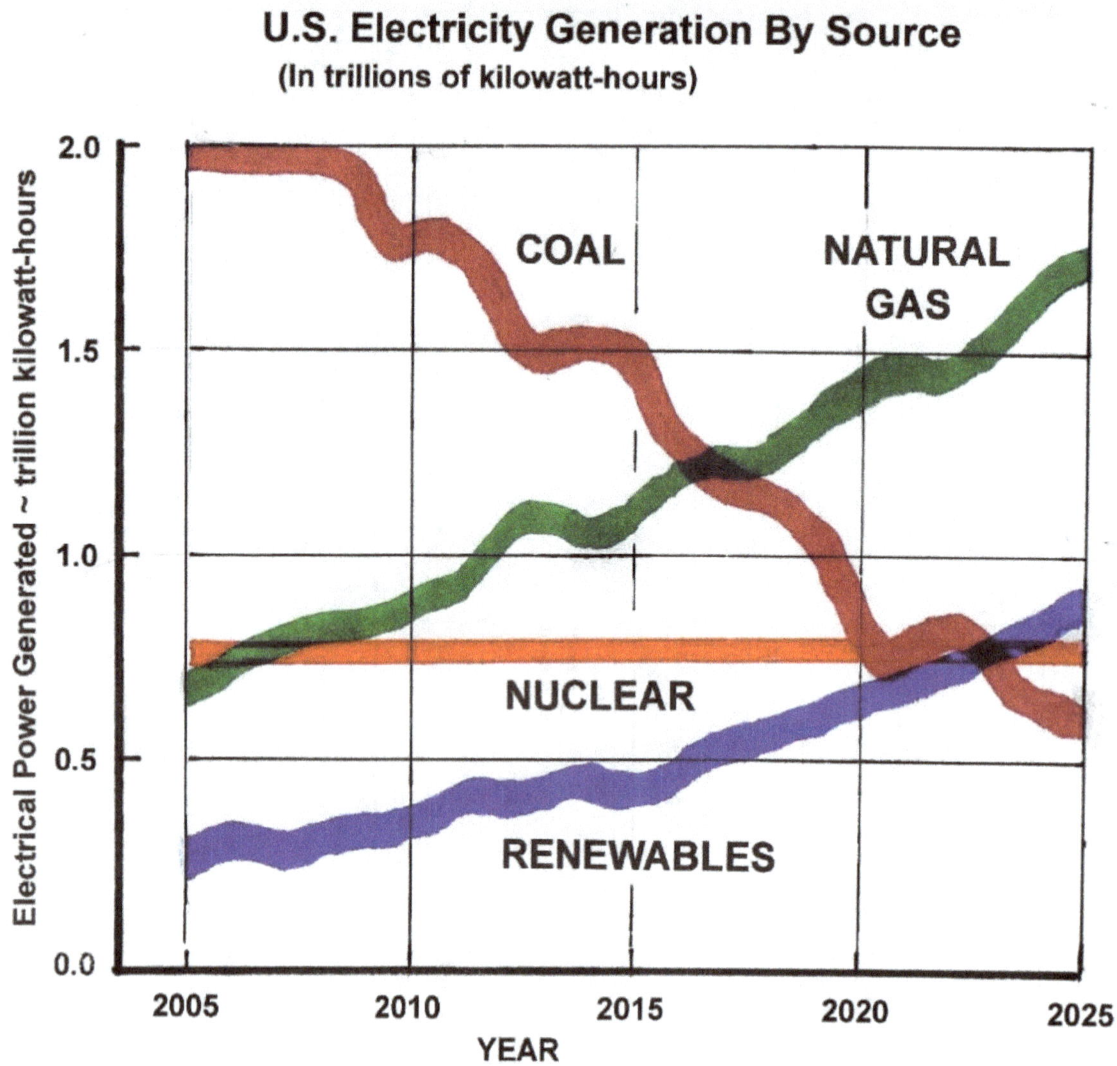

(Source: New York Times, January 2025, colorized by Author)

It should be obvious that the energy-generating corporations have begun the very difficult but necessary trend to decrease the use of coal burning in the USA, regardless of whether it was or was not an action to improve the climate. The technology of coal has reached the point of being obsolete because of the development of other energy generating technologies with lesser cons than that of coal. Equally obvious is that the trend of both natural gas and renewables (solar and wind) technologies have the increasing trend to satisfy the national demand for energy as the coal energy wanes. Natural gas has the very important factor of reliability – it is available regardless of whether the sun is shining on the Earth's surface and regardless of whether the wind is blowing at the speed to properly operate the wind turbines. Too little or no wind or too high a wind (potentially causing blade overspeed damage) and the wind turbine shuts itself down for self-protection. The same unreliability consideration exists with the other primary renewable energy source of solar power. Both agricultural energy demand and low community energy demands can be met with wind and solar energies and their unreliability characteristics. However, industrial demand cannot tolerate the low reliability of solar and/or wind even when supplemented by battery storage. For large scale generation of energy satisfying industrial levels of demand and with near 100% reliability, the two options realistically available are the other fossil fuels (oil and natural gas) or nuclear energy. The public has an innate NIMBY (not in my back yard) phobia about the safety of nuclear energy use but is more than willing for the nearby operation of a large energy producer such as the other fossil fuel alternatives to coal – oil and natural gas. The two photographs shown earlier in this book of the Brooklyn Edison (adjacent to the Greater New York City population of over 3 million people) that converted from coal-burning to oil burning to generate high levels of electric power and steam is shown as follows and also shown is a facility in Alabama that converted from coal burning to natural gas burning to satisfy the high energy demands of the steel industry in the region:

Brooklyn Edison and Navy Yard, NYC

Natural Gas Burning Facility in Alabama

(Credit to unknown photographers)

Note that there is no visible "greenhouse gas emissions" from the smokestacks at either facility oil burning on the left and natural gas burning on the right – but both burning fossil fuels.

There are indications that some environmental scientists are realizing the use of fossil fuels of oil (mostly for the transportation technology sectors) and natural gas (for the industrial users needing a high reliability energy source) are not in the same hazardous category as coal. Unfortunately, there are many environmental scientists who have a vendetta for all fossil fuels to be eliminated – no exceptions. Their fear-mongering rhetoric is apparently etched into their memory. An example of that "inconvenient truth" is reflected in an extract of a full-page advertisement in a North Carolina newspaper that reflects there are yet many environmental scientists who refuse to accept the "big picture" (seen in NASA satellite photographs (such as that shown on the front cover of this book). Their advertisement is a plea to North Carolina Governor Josh Stein to stop Duke Energy using any form of fossil fuel to save the environment – regardless of the unreliability of solar energy with battery storage adjuncts:

61 SCIENTISTS CALL FOR STATE LEADERSHIP VS. GIANT CORPORATE CLIMATE POLLUTER

The science is clear: fossil fuel emissions are heating the planet at an unprecedented rate and making it increasingly unlivable. Duke Energy's leaders had a chance to end their role in the five-alarm fire we find ourselves in today. But they did not.

Now is the moment for Duke Energy to pivot to real climate and energy resilient solutions like rooftop and community-based solar with battery storage, energy efficiency, demand response, and microgrids. These alternatives are readily available and capable of withstanding extreme weather and keeping the lights on for families. Local solar-plus-storage is also the fastest, cheapest and most equitable tool we have to speed the phase-out of fossil fuel-burning power plants, lower power bills, create good jobs and improve public health.

You have the power to confront Duke Energy for all they have done and continue to do to block climate action and entrench the fossil fuel status quo. We implore you to lead in the transition away from fossil fuels and to the renewable, resilient, equitable, affordable, and sustainable energy future that humanity desperately needs.

Sincerely,

Paul A. Baker
Professor Emeritus, Division of Earth and Climate Sciences
Duke University

Drew Shindell
Nicholas Professor of Earth Sciences
Duke University

Dale Evarts
Former Leader, Climate and International Group
US Environmental Protection Agency

Michael E. Mann
Presidential Distinguished Professor
Director, Center for Science, Sustainability and the Media | University of Pennsylvania

(Source: The News & Observer, Raleigh-Durham, North Carolina, April 20, 2025)

An old philosophical observation states: "*There are none so blind as those that will not "see"*. The statement that "The science is clear: fossil fuel emissions are heating the planet at an unprecedented rate and making it increasingly unlivable" is a wish not a fact, as this book abundantly has shown. The claim of there being a "five alarm fire we find ourselves in today." Is the same fear-instilling rhetoric used by Henny Penny to Chicken Little in the children's fable: *"The sky is falling. The sky is falling."* after a heavy rainfall. Oil and natural gas fossil fuels do not have the same polluting ash residue (containing toxins and carcinogens) as that of coal when combusted. It is hastened to be added the Duke Energy Corporation can use many improvements to the installation of the power grid (obviating falling trees cutting power lines or lightning strike igniting wildfires) and the North Carolina Department of Transportation installation of major arteries used by motor vehicles being redesigned and constructed to be more tolerant of flooding wash-outs such as that incurred from Hurricane Helene) isolating many thousands of people in the western North Carolina mountainous region. Those are the "real climate changes" that need to be dealt with – as the title of this book states on the front cover.

There were other technologies considered that were initiated in many countries of the planet intended to extract carbon dioxide from the atmosphere as an attempt to subdue the effects of "greenhouse gas" emissions on global warming and/or climate change. Based on the findings in the two books by this Author revealing that the greenhouse gas heat-trapping characteristic of naturally-generated water vapor in the atmosphere dwarfed the effect of human and industrial created greenhouse gases by a large ratio (an average of 1.4% water vapor in the atmosphere and only 0.0128% of anthropogenic carbon dioxide – a ratio of over 100 to 1), those carbon extraction technologies were not assessed as a potentially useful candidate technology.

In the European countries, the contrails in the high atmosphere left by commercial aircraft have caused environmental scientists to claim commercial flight operations are a contributor to global warming and climate change. A photograph of a heavy set of contrails over Germany illustrates the concern:

(Credit to Unknown Photographers)

The contrails (shown in both photographs above) result from the hot gasses in the exhaust of turbomachinery in commercial and military aircraft (shown at the left above) encountering the colder air of the atmosphere – typically above 25,000 feet. The temperature in the exhaust plume will be well above 800 °F whereas the surrounding air temperature above 25,000 feet altitude is of approximately -60 °F. Any water vapor in the atmosphere is condensed from gaseous state to liquid state (very small droplets; hence the name "contrail" is a contraction of "condensation trail". The white color "smoke" is condensation, not a product of combustion of the fuel that is combusted in the turbomachinery (note in the photo at the left that the contrail forms several feet after the hot exhaust exits the engine exhaust nozzle).

The environmental scientists apparently do not claim that there are "greenhouse gas" emissions (e.g., carbon dioxide or methane) at play but instead claim there is a heat-trapping effect of the contrails playing a role in global warming. Others claim there is a "shadowing effect" of the contrails. Both the heat-trapping and the shadowing effect are valid observations but are of insignificance since they occur at above 25,000 feet AGL – above ground level. If they both were of significance, the shadowing effect is a cooling effect whereas the heat trapping effect is a warming effect, and those two effects would cancel out.

As seen in the above photograph at the right, the white appearance near the Earth's surface is that of water vapor in the lower 5,000 feet of the atmosphere where the temperature is 10 to 15 °F cooler than the environmental scientists" "average global temperature" of 60 °F.. Dr. S. Fred Singer is very critical (in his lectures derived from "A Closer Look at the Numbers" series of papers generated at the University of Virginia) of the environmental scientists for misleading the public that water vapor can be ignored as an effect on global warming. Environmental scientists are rife in showing the left diagram (ignoring water vapor) below but do not reveal the right diagram (showing the dominant water vapor) to the public.

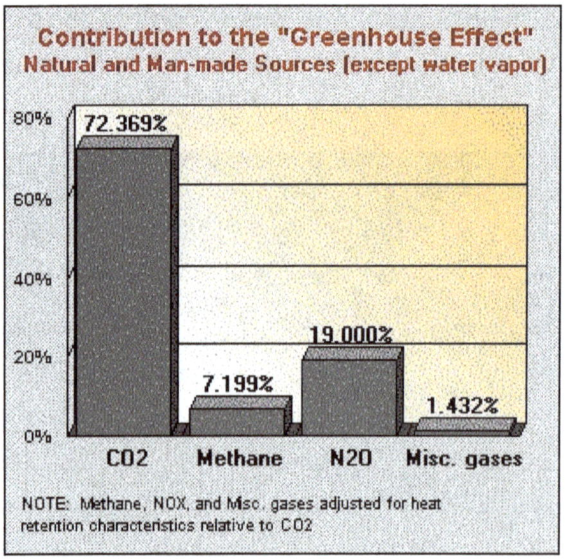

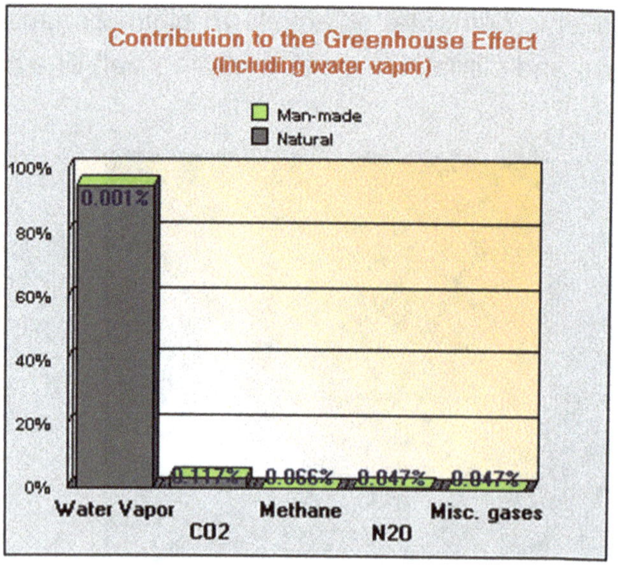

(Source: Dr. S. Fred Singer – Professor Emeritus at the University of Virginia)

Environmental scientists are of course free to explore any scientific observations they wish to, but it is unethical to intentionally withhold from the public an important factor such as that of water vapor being of comparative much greater significance than anthropogenic generated "greenhouse gases".

C38) Our children and grandchildren deserve to be taught the correct science of the World they will live in – and to prepare them to survive the extremes of the daily weather, seasonal climate, and natural hazards of their regional environment, not merely the average global temperature. Children are too valuable to risk their young lives being cut short when being transported in an electric vehicle which has a time-bomb battery beneath their little feet. The world population is also too valuable to risk their lives by being exposed to toxins and carcinogens from coal ash residue, environmental pollution (especially that from plastics), and unsanitary waste disposal poisoning the water they drink, tainting the land which produces their food, and fouling the air they breathe.

The real naysayers about climate change are the ones in "settled science" who fallaciously hypothesize that the anemones on the coral reefs in the oceans are in crisis of bleaching white because of Man burning any form of fossil fuel (coal, oil, or natural gas).

Copernicus and Galileo were branded as being naysayers for challenging their peers who held that the Sun revolved about the Earth, as were others who challenged that the Earth was not flat but was a large sphere. This book may be interpreted by some as "naysaying" by a would-be Galileo admirer but the facts in the numerous databases analyzed completely and in depth in the two books by this Author presents the "inconvenient" truth.

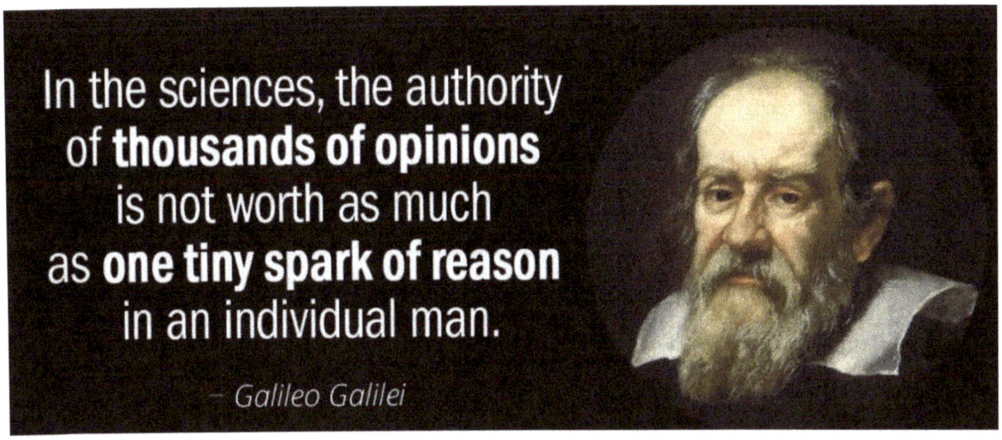

(Credit to unknown graphic artist, Encyclopedia Britannica)

The analogy given at the start of this analysis of the "problems" was that of the resemblance to assembling a jigsaw puzzle with many pieces. This analysis has identified 38 conclusions (numbered **C-1** through **C-38)** but the "inconvenient truth" is that there are many pieces of the global warming and climate change jigsaw puzzle yet to be discovered – and not until all the pieces are found and analyzed correctly will the "jigsaw puzzle" be able to be declared as completed. The many tables of numbers and the many illustrations, diagrams, and photographs shown in these two books provide the clues to what environmental scientists and climatologists should continue searching for – using the 37 conclusions in this book as clues to what paths should be taken in search for the yet missing pieces.

The father of all "naysayers" – Galileo de Gallilei – has been quoted for having stated over fifty philosophical scientifically-based observations about science (especially in the field of astronomy). There are fifty of them shown on websites on the internet. One that is very meaningful and highly relative to the environmental scientists" hypothesis that 'humans burning fossil fuels is the dominant cause for climate change': "*The hypothesis is pretty; its only fault is that it is neither demonstrated nor demonstrable. Who does not see that this is purely arbitrary fiction that puts nothingness as existing and proposes nothing more than simple noncontradiction?"*

That quote serves as an appropriate conclusion of this book and as an invitation for scientists to modify their fallacious hypothesis to be aligned more with the harsh realities of life on the planet as a system engineer has seen those realities in the "big picture". This book has revealed the realism of numbers to quantify the actual problem far better than the fallaciousness of a scientific hypothesis that has no numbers to factually support the hypothesis.

The analysis shown in this book concludes with the excellent editorial by Susan Bass of the EARTHDAY.ORG, confirming most – but not all - of the message she portrays coinciding with the conclusions drawn in this book. Her summation in the last column is well summarized: *"Transitioning to clean energy, reducing plastic waste, and increasing resilience to extreme weather are among the most fiscally responsible actions governments can take. This Earth Day, we must recognize that efficiency isn't just cutting – it's about investing in solutions that protect people and our infrastructure."*

Her statement: *"The link between greenhouse gas emissions and climate change is now scientifically indisputable."* is a fallacious statement which the two books by the Author "Facts and

Fallacies of Climate Change on Our Planet" and this book "Dealing with the Real Climate Change" have the objective of correcting the public and Ms. Bass' understanding of the factual causes (astrophysical, geophysical, and natural, but not anthropogenic) for the inevitable climate change being a regional problem, not simply an "average" global problem. The pollution of the land, water, and air by plastics waste is entirely an anthropogenic-caused problem – and is the true "crisis" (not climate change) for the near future of the exponentially growing population.

We have the power to tackle climate change, protect Earth

BY SUSAN BASS

Progressive Perspectives

Amid the historic and sweeping cuts to federal agencies and programs being carried out by the Trump administration, one truth has been overlooked: If we're serious about cutting waste and protecting public funds, we must confront climate change head-on.

2024 was a disaster for the planet and its people. According to NASA, it was the warmest year since temperatures began being recorded in 1880. In the U.S. alone there were 27 climate and weather events that resulted in at least a billion dollars in damages - second only to 2023, with 28 such events.

These events - wildfires, floods, tornadoes and hurricanes - are becoming the norm, and they're financially devastating.

In January, tens of thousands of acres and more than 16,000 structures burned in southern California. Last month, more than 150 tornadoes tore across the central and southeastern U.S. and, this month, historic floods submerged parts of the Midwest and South.

In the parts of the United States at higher risk for climate-related extreme weather events, insurance claims are increasing in cost and frequency. Not surprisingly, these high-risk areas are also now seeing the highest increase in cancellations for failure to pay premiums and nonrenewals by the insurance companies. Without insurance, homeowners may not be able to rebuild when disaster strikes.

And climate change isn't the only escalating crisis. The world is also drowning in plastic. On Sept. 5, 2024 - Plastic Overshoot Day - the world exceeded its capacity to manage plastic waste. An estimated 220 million tons of plastic are projected to be produced this year alone, and 66% of people live in areas where plastic waste exceeds waste management capacity.

Meanwhile, new research shows just how dangerous plastics are to human health. Microplastics have been found in human brains, and the World Wildlife Fund estimates that we may be ingesting up to five grams of plastic each week - the weight of a credit card. Plastics are now linked to cancer, endocrine disruption, impaired fertility and cognitive development issues. Wildlife, too, is suffering from entanglement, starvation and habitat loss.

Here's the hopeful part: We already have the tools and the power to change this. EARTHDAY.ORG, the network created by the original organizers of the first Earth Day in 1970, is still leading the charge with our campaign, "Our Power, Our Planet." The goal is to help individuals, cities and communities act on the environmental challenges of today.

The economic upside of environmental action is massive. New solutions to our current environmental crisis rest in the hands of the people. On Earth Day, which is today, you can exercise your power. Demonstrate to our leaders in government and business that we are still here, we are a witness to their actions and we will hold them accountable to do right by our planet and its people.

As consumers, we can choose plastic-free products and demand a reduction and transition in the use of plastics from businesses, while at the same time pressuring government leaders to reduce plastic production globally to end the use of toxic ingredients and to improve waste management systems.

We have the collective power not only to protect our planet but also to improve lives and livelihoods. The link between greenhouse gas emissions and climate change is now scientifically indisputable. According to the International Renewable Energy Agency, 90% of global electricity can and should come from renewable sources by 2050. The transition also promises cleaner air, up to 30 million new jobs and stronger energy independence.

Transitioning to clean energy, reducing plastic waste and increasing resilience to extreme weather are among the most fiscally responsible actions governments can take. This Earth Day, we must recognize that efficiency isn't just about cutting - it's about investing in solutions that protect people and our infrastructure. True government efficiency means reducing risk in order to cut costs - not paying billions each year to clean up after preventable disasters.

Susan Bass is senior vice president of programs and operations for EARTHDAY.ORG.

(Source: The News Observer, Durham, North Carolina, April 22, 2025)

Having the power to do something and the actual "doing" something meaningful in applying that power remains to be seen. This Author expresses his gratitude for having been given the honor of analyzing and interpreting the works of those scientists who created the many factual scientific databases portraying the "inconvenient truth". The two books written by this Author had the initial goal of resolving the global warming and climate change "problem" but as the analysis of those "problems" progressed it became abundantly clear that they are not the problem first thought to exist. The effort expended in analyzing the most appropriate databases and applying the multitude of conclusions into the assessment of the many technologies needing to be developed will not have been for naught if these many conclusions and assessments are applied realistically to the problem resolution.

The foregoing accounts (with shocking photographic images) of the fire hazards of lithium-ion batteries self-igniting certainly was not "swept under the carpet" by electric vehicle developers. China has scientists and engineers that are the most ingenuous in the world, and the following news article is shown as an example of how the cons of one technology serve as the seeds for the development of another branch of that same technological development to convert the con of one into the pro of another:

Major advances in EV batteries announced by Chinese company

BY KEITH BRADSHER
NYT News Service/Syndicate Stories

SHANGHAI

The Chinese manufacturing giant CATL, the biggest supplier of batteries for the world's electric cars, said Monday that it had made technological advances that would allow it to produce batteries that are cheaper, lighter, faster to recharge and more resistant to cold, while providing greater driving range.

Most of the changes, which are a couple of years away from being widely available in new cars, could make electric cars more competitive in price and performance with gasoline-powered models.

CATL — its full name is the Contemporary Amperex Technology Company Ltd. — produces a third of the world's electric car batteries and supplies 16 of the world's biggest carmakers, including General Motors and the Shanghai factory of Tesla. Its main rivals for the global market are BYD in Shenzhen, China, which makes about one-sixth of the world's EV batteries, almost entirely for its own cars, and Korean and Japanese battery manufacturers.

CATL executives spoke at a news conference before the Shanghai auto show, which starts Wednesday. The choreographed event evoked the launch of a new car model.

Batteries represent at least a third of the cost of an electric car, making CATL a critical player in the EV supply chain in China and beyond. Many automakers have been watching nervously whether CATL will someday try to establish its own car brand that could overshadow their own models.

The biggest surprise by CATL was an announcement about auxiliary batteries for electric cars. The batteries would share space in the underbody of cars, where there is currently only one large battery.

The auxiliary battery would be the first commercially available electric vehicle battery that would not use graphite as one of its poles, CATL said.

Removing costly graphite will eventually make the batteries cheaper, after some initial costs, and will allow 60% more electricity to be squeezed in each cubic inch of the battery, said Gao Huan, CATL's chief technology officer for electric cars in China. The extra energy density means that the car's driving range can be greater, or the overall size of the battery can be reduced, leaving more room for the car's passenger compartment.

The second battery also would provide backup in case one has trouble. That has become more important as self-driving features, which require uninterrupted electricity, become more common.

Ouyang Chuying, co-president for research and development at CATL, said that auxiliary batteries without graphite would be available in cars in two to three years and possibly sooner. He declined to say which automakers might be the first to use them.

But taking out the graphite has a downside, which is why CATL will only remove it for the auxiliary batteries. Batteries without graphite recharge more slowly, and cannot be recharged as many times as conventional EV batteries before they need to be replaced.

The auxiliary batteries are meant to be used less frequently, on longer drives after the main battery is exhausted.

CATL, which is based in Ningde, China, also said it had made more progress in the speed of charging main batteries. The company said its new system would allow an electric vehicle to be charged enough in five minutes to drive 520 kilometers, or 320 miles.

BYD and Huawei, a Chinese electronics giant that plays an ever-larger role in auto parts manufacturing, have also announced five-minute charging systems, known as supercharging.

CATL also said it would start selling sodium-ion batteries, which can retain more than 90% of their charge even at a temperature of 40 degrees below zero, for use in cars and trucks with internal combustion engines. The sodium batteries could be used by automakers to replace conventional lead-acid batteries, which go dead in very cold weather, and in some electric cars.

Ouyang said that the electricity of these sodium batteries would be compatible with the electrical systems of existing gasoline-powered cars, but the new batteries might not fit in the same space.

CATL said that its first customer for sodium-ion batteries would be freight trucks from First Auto Works, an automaker in Changchun, in China's far northeast, where temperatures frequently fall well below zero. Developing sodium-ion batteries has been a priority for the Chinese electric car industry because the country's northern provinces, bordering Mongolia and Russia's Siberia, have bitterly cold temperatures in winter.

In interviews last autumn, car owners in Urumqi, in the far northwest of China, drivers said that cold weather was why they would not consider buying electric cars.

Battery makers have been working on sodium-ion batteries for many years, but the United States may have a long-term advantage in the technology. Almost all of the world's naturally occurring geological deposits of soda ash, the raw material for sodium-ion batteries, are in southwestern Wyoming.

CATL showed a video of its sodium-ion batteries undergoing stress tests, such as being punctured with a nail or power drill or even cut in half with a power saw, without catching fire or exploding. Just five years ago, CATL had argued that nail tests were unrealistic and batteries should not be expected to withstand them.

(Source: The News Observer, Durham, North Carolina, April 22, 2025)

It must be mentioned that as the research was being conducted for this book, it was a challenge to keep up with the technological developments that are happening around the world. That is the very nature of technological development as the cons of a specific technology often do not appear until that technology is placed into the real-world environment - which is typically different than the design environment known at the start of the technology development process.

Inasmuch as the approach taken at the start of the research into the global warming and climate change "problem" was in the "big picture" mode, it is appropriate to end that research mode with a "big picture" conclusion. The rates of technological developments are faster than the rates at which the temperature and climate of the planet is changing, and it is clear that the problems of surface temperature changes, climate change, and ocean levels rising can be, and are being, mitigated to eliminate the fear of a catastrophic future for the present and near future generations. The technologies being developed are toward more reliable, more efficient, and less costly actions to meet the energy demands of the burgeoning world population and only tangentially toward improving the climate and the future of the ocean floor life. The remaining stain on technological development is that of world pollution caused by humans and industries and the technologies for controlling and improving the world pollution is severely lacking. That is a good sign of net progress in the development of technologies "for the benefit of humanity.".

It is the Author's most professional and sincere hope that the assessment of future technologies provided in this book will assist the technology developers in all regions of the planet to develop - and share the knowledge gained with international counterparts utilizing their unique resources in their respective regions – so that the current and near-future population of the planet will safely, efficiently, and economically benefit from the developments. In parallel, the human population must cease polluting the environment of their regions. The focus of scientists on man polluting the atmosphere and environment with fossil fuels emissions has been shown to be a misinformed approach to the adaptation and mitigation humanity must achieve in harmony with the environment – not destroying it. Even the uneducated common pig knows not to urinate in its own water trough or in that of its neighbors.

APPENDIX
(Pertinent to United States Region)

Included herein are:

1) List of Factors of Importance by Group and Factor
2) List of Trade Studies by Technology Group
3) Worksheets for the 25 Technologies Prioritization Weighted Trade Studies
   TS-1 through TS-24 (TS-22 is in two parts TS-22a and TS-22b)
4) The summary page with the 25 Trade Studies Total Weighted Scores (arranged in order of descending priority within technology group)

Grouping of Factors of Importance for Technology Prioritization Weighted Trade Studies (rated by pros and cons within group)

Group	**Factor**

1. Status of Technology Development (Risk, Schedule, Cost, Potential)
2. Reliability and Sufficiency (Time of Day/Year, Weather Dependent, Climate Dependent)
3. Regional Economy and Resources (Mineral, Water, Labor Force to Operate/Maintain)
4. Cost of Resources (Domestic and Imported)
5. Importance to Industry, Government, and Human Safety, Health, Comfort
6. Importance and Urgency to Economy and Employment
7. Pros and Cons By-Products (Emissions, Pollutants, Hazards)
8. a. Effect on Human Population (Regional in USA) – Residency
   b. Effect on Human Population (Regional in USA) – Food Supply
   c. Effect on Human Population (Regional in USA) – Safety, Health, Comfort
9. a. Effect on Infrastructure (Regional in USA)
   b. Effect on Transportation (Regional in USA)
   c. Effect on International Trade
10. a. Effect on Environment (Regional in USA) – Atmosphere
    b. Effect on Environment (Regional in USA) – Land
    c. Effect on Environment (Regional in USA) – Water
11. a. Effect on Animal Life (Regional in USA) – Land
    b. Effect on Animal Life (Regional in USA) – Aquatic
    c. Effect on Animal Life (Regional in USA) – Avian
12. Effect on Forestry and/or Agriculture (Regional in USA)

List of Technology Groups with Trade Studies Identifiers

1. Fossil Fuels / Derivatives

TS-1 Use of Fossil Fuel – Coal
TS-2 Use of Fossil Fuel – Oil
TS-3 Use of Fossil Fuel – Natural Gas (Pipeline)
TS-4 Use of Fossil Fuel – Natural Gas (Liquified / Transportable)
TS-5 Use of Fossil Fuel Derivative – Propane

2. Non-Fossil Fuels

TS-6 Use of Non-fossil Fuel – Hydrogen
TS-7 Use of Non-fossil Fuel – Ethanol

3. Alternative Energies - Geophysical

TS-8 Use/Development of Hydrodynamic Energy (River Flow)
TS-9 Control of Hydrodynamic Events (Coastal Regions)
TS-10 Use/Development of Solar Energy (Large Industrial / Commercial)
TS-11 Use/Development of Solar Energy (Residential / Rural / Town)
TS-12 Use/Development of Wind Energy (On Land)
TS-13 Use/Development of Wind Energy (Off-Shore)
TS-14 Energy Storage - Stationary/Large Size (Electrical)
TS-15 Use/Development of Geothermal Energy (Surface)
TS-16 Use/Development of Geothermal Energy (Deep-drilled)
TS-17 Energy Storage - Stationary/Large Size (Thermal)

4. Alternative Energy - Nuclear

TS-18 Use/Development of Nuclear Energy – Uranium
TS-19 Use/Development of Nuclear Energy - Thorium

5. Transportation

TS-20 Use of All-Electric Motor Vehicles – Commercial / Government
TS-21 Use of Electric Plug-In / Fuel Hybrid Vehicles – Commercial / Public
TS-22a Energy Storage - Mobile/Small Size (Electrical); Lithium-ion
TS-22b Energy Storage - Mobile/Small Size (Electrical); Aluminum-ion or Sodium-ion

6. Food/Water Supply – Human, Animal, Agricultural

TS-23 Use/Development of Desalination - Reverse Osmosis
TS-24 Use/Development of Desalination – Distillation (Solar)

TRADE STUDY SCORE WORKSHEET (Applicable to United States of America Region)

(Ratings and scores based on factual information in pertinent databases in Public Domain)

#	TRADE STUDY: TS-1 Use of Fossil Fuel - Coal				
	FACTOR CONSIDERED (Being rated for Importance and primary qualitative magnitude)	ELEMENTS CONSIDERED IN RATING FACTOR (Attributes and detriments if/as applicable)	RELATIVE IMPORTANCE RI (% or n/a)	POINTS SCORED (Rated)	WEIGHTED LINE SCORE (WS) (RI x Rated Score)
	(Minor interactions with/between Elements/Factors are not directly in rating)		RI	XX (of RI)	XX x RI (of RI^2)
1	Status of Technology Development	Use of coal (with emission controls) for over 120 years in US requires no further development. Transition to a safer and a cleaner technology is needed/overdue.	12	10 (of 12)	120 (of 144) = 83%
2	Reliability / Sufficiency of Energy	Reliability is 100%; sufficient to meet present demands	12	12 (of 12)	144 (of 144) = 100%
3	Resource(s) Exist Domestically or Approx % Needed to be Imported	There is an abundance of coal in both underground mining (phasing out) and open-pit mining (preferred).	10	10 (of 10)	100 (of 100) = 100%
4	Cost of Resource(s) (Domestic + Import if any)	Cost to consumer is presently low but will increase as coal production is phased out. Law of supply and demand.	10	9 (of 10)	90 (of 100) = 90%
5	Importance to Industry of Region	Reliable, low cost energy vital in industrial areas.	9	8 (of 9)	72 (of 81) = 89%
6	Importance and Urgency to Gov't. Economy and Employment	Coal combustion in industrial use loses importance as cleaner, more reliable energy sources are developed.	9	6 (of 9)	54 (of 81) = 67%
7	By-Products/Emissions / Pollution	Coal ash contains toxic and carcinogenic materials.	8	1 (of 8)	8 (of 64) = 13%
8	Effect on Human Population	High incidence of mine accidents; "Black-lung" deaths; Pollutants are much more hazardous than emissions.	8	1 (of 8)	8 (of 64) = 13%
9	Effect on Regional Infrastructure / Transportation / Commercial Trade	Extent of effect depends on producibility of steel for infrastructure and transportation vehicles/vessels	7	6 (of 7)	42 (of 49) = 86%
10	Effect on Environment / Weather / Climate (Frequency & Severity)	"Greenhouse gas" effects of emissions on severe storm frequency/severity are overexaggerated	5	4 (of 5)	20 (of 25) = 80%
11	Effect on Animal Life (Wild or Livestock) - Land, Aquatic, Avian	Toxins and carcinogens from coal ash disposal are polluting underground aquafers (water	5	1 (of 5)	5 (of 25) = 20%
12	Effect on Forestry and Agriculture	Coal ash pollution hazards affects food supply safety.	5	1 (of 5)	5 (of 25) = 20%
			Total A Sum of RI 100	Total B Sum of XX 77 (of 100)	Total C XX x RI (of RI^2) 668 (of 902)
			WEIGHTED SCORE =		74%

RECOMMENDATIONS FOR CURRENT AND FUTURE DEVELOPMENT(S):

The reduction of "greenhouse gas" emissions to reduce/eliminate the adverse effects of climate change will not achieve the unrealistic present goals set by the UN IPCC. Because of the serious polluting effects of its ash residue, coal as an energy source should be gradually phased down to elimination. Coal-burning facilities in operation should transition to burn cleaner natural gas as the fuel. Imposing economic penalties on industries for failing to meet unrealistic emissions goals should be terminated. Every effort, including heavy penalization, should be made to stop the disposal in landfills of polluting coal-ash or chemical by-product of industrial processes with toxic or carcinogenic content. Technology advancement urgent, not as crisis need.

TS-1 Use of Fossil Fuel (Coal)

TRADE STUDY SCORE WORKSHEET (Applicable to United States of America Region)

(Ratings and scores based on factual information in pertinent databases in Public Domain)

TRADE STUDY: TS-2 Use of Fossil Fuel - Oil

#	FACTOR CONSIDERED (Being rated for Importance and primary qualitative magnitude)	ELEMENTS CONSIDERED IN RATING FACTOR (Attributes and detriments if/as applicable)	RELATIVE IMPORTANCE RI (% or n/a)	POINTS SCORED (Rated)	WEIGHTED LINE SCORE (WS) (RI x Rated Score)
	(Minor interactions with/between Elements/Factors are not directly in rating)		RI	XX (of RI)	XX x RI (of RI^2)
1	Status of Technology Development	There is a need to drill off-shore for oil but with better anti-leak/spillage provisions causing pollution	12	11 (of 12)	132 (of 144) = 92%
2	Reliability / Sufficiency of Energy	Reliability is 100%; sufficient to meet present demands	12	12 (of 12)	144 (of 144) = 100%
3	Resource(s) Exist Domestically or Approx % Needed to be Imported	Government restrictions on pipeline routings reduces the availability of crude oil to many refineries	10	8 (of 10)	80 (of 100) = 80%
4	Cost of Resource(s) (Domestic + Import if any)	Frequent shortage of on-line oil refineries keeps cost of oil/gasoline very high. Transition to all Electric Vehicles will increase cost of oil/gasoline.	10	7 (of 10)	70 (of 100) = 70 %
5	Importance to Industry of Region	Reliable, low cost energy vital in industrial areas.	9	9 (of 9)	81 (of 81) = 100 %
6	Importance and Urgency to Gov't. Economy and Employment	Use of oil in facilities producing energy critical during transition from dependency on coal	9	9 (of 9)	81 (of 81) = 100 %
7	By-Products and Emissions	Much lower emissions than coal. Spillage accidents too common causing local pollution of land and water	8	6 (of 8)	48 (of 64) = 75 %
8	Effect on Human Population	Pollutants are much more hazardous than emissions.	8	6 (of 8)	48 (of 64) = 75 %
9	Effect on Regional Infrastructure / Transportation / Commercial Trade	Current dependence of private and commercial vehicles on oil/gasoline/Diesel for transportation and trade being reduced by electric vehicle technology.	7	7 (of 7)	49 (of 49) = 100 %
10	Effect on Environment / Weather / Climate (Frequency & Severity)	"Greenhouse gas" effects of emissions on frequency/severity have been over-exaggerated	5	5 (of 5)	25 (of 25) = 100 %
11	Effect on Animal Life (Wild or Livestock) - Land, Aquatic, Avian	Frequent spills during land transportation accidents had/has effect on local life forms	5	4 (of 5)	20 (of 25) = 80 %
12	Effect on Forestry and Agriculture	No significant effect of oil used in energy production	5	5 (of 5)	25 (of 25) = 100 %
			Total A Sum of RI 100	Total B Sum of XX 89 (of 100)	Total C XX x RI (of RI^2) 803 (of 902)
			WEIGHTED SCORE =		89 %

RECOMMENDATIONS FOR CURRENT AND FUTURE DEVELOPMENT(S):

The reduction of "greenhouse gas" emissions to reduce/eliminate the adverse effects of climate change will not achieve the unrealistic present goal set by the UN IPCC. Except for the instances of oil spillages on water, beaches, and land from oil drilling leakages and spillages after transportation accidents and pipeline leakages, oil is a very efficient, safe, economic, and mature technology for creating energy both for increasing industrial energy and human requirements as the world population increases. Oil (and natural gas) are well suited as alternatives for coal. Technology advancement routine, no crisis.

TS-2 Use of Fossil Fuel – Oil

TRADE STUDY SCORE WORKSHEET (Applicable to United States of America Region)

(Ratings and scores based on factual information in pertinent databases in Public Domain)

#	TRADE STUDY: TS-3 Use of Fossil Fuel - Natural Gas (Pipeline)				
	FACTOR CONSIDERED (Being rated for Importance and primary qualitative magnitude)	**ELEMENTS CONSIDERED IN RATING FACTOR (Attributes and detriments if/as applicable)**	**RELATIVE IMPORTANCE RI (% or n/a)**	**POINTS SCORED (Rated)**	**WEIGHTED LINE SCORE (WS) (RI x Rated Score)**
	(Minor interactions with/between Elements/Factors are not directly in rating)		**RI**	**XX (of RI)**	**XX x RI (of RI^2)**
1	Status of Technology Development	The growing technology of "fracking" is opening and producing natural gas in abundant supply	12	12 (of 12)	144 (of 144) = 100 %
2	Reliability / Sufficiency of Energy	Reliability is 100%; sufficient to meet present demands	12	12 (of 12)	144 (of 144) = 100%
3	Resource(s) Exist Domestically or Approx % Needed to be Imported	Government restrictions on gas pipeline routings reduces the availability of gas to several regions	10	8 (of 10)	80 (of 100) = 80%
4	Cost of Resource(s) (Domestic + Import any)	Of the 3 fossil fuel types, natural gas is the lowest cost to distribute and per volume transformed into heat energy	10	10 (of 10)	100 (of 100) = 100 %
5	Importance to Industry of Region	Reliable, low cost energy vital in industrial areas.	9	9 (of 9)	81 (of 81) = 100 %
6	Importance and Urgency to Gov't. Economy and Employment	Use of gas in facilities producing efficient energy critical during transition from dependency on coal	9	9 (of 9)	81 (of 81) = 100 %
7	By-Products and Emissions	Almost totally "clean" fuel when burned - no solid residue with toxic/carcinogenic content as pollutant	8	7 (of 8)	56 (of 64) = 88 %
8	Effect on Human Population	Safety hazard of storage of explosive gas in tankage	8	4 (of 8)	32 (of 64) = 50 %
9	Effect on Regional Infrastructure / Transportation / Commercial Trade	Most coal burning facilities in Eastern US have - or plan to - convert from coal to natural gas. Midwest and Western US will convert from coal to oil or natural gas	7	7 (of 7)	49 (of 49) = 100 %
10	Effect on Environment / Weather / Climate (Frequency & Severity)	"Greenhouse gas" effects of emissions on weather events frequency/severity have been exaggerated	5	5 (of 5)	25 (of 25) = 100 %
11	Effect on Animal Life (Wild or Livestock) - Land, Aquatic, Avian	No significant effect of gas used in energy production on almost all life-forms.	5	5 (of 5)	25 (of 25) = 100 %
12	Effect on Forestry and Agriculture	No significant effect of gas used in energy production	5	5 (of 5)	25 (of 25) = 100 %
			Total A Sum of RI 100	Total B Sum of RI 93 (of 100)	Total C XX x RI (of RI 2) 842 (of 902)
			WEIGHTED SCORE =		**93 %**

RECOMMENDATIONS FOR CURRENT AND FUTURE DEVELOPMENT(S):

The reduction of "greenhouse gas" emissions to reduce/eliminate the adverse effects of climate change will not achieve the unrealistic present goals set by the UN IPCC. Leakage of gas into atmosphere after gas pipeline accidents and tank storage leakages may occur increasing fire hazard. Natural gas in either gaseous or liquified form is a highly useful fuel as a complement to alternative (non-fossil) fuels in creating energy for meeting industrial and residential energy needs as the world human population increases. Natural gas (and oil) are well suited as alternatives for coal. Technology advancement routine, no crisis.

TS-3 Use of Fossil Fuel – Natural Gas (Pipeline)

TRADE STUDY SCORE WORKSHEET (Applicable to United States of America Region)

(Ratings and scores based on factual information in pertinent databases in Public Domain)

TRADE STUDY: TS-4 Use of Fossil Fuel - Natural Gas (Liquified)

	FACTOR CONSIDERED (Being rated for Importance and primary qualitative magnitude)	ELEMENTS CONSIDERED IN RATING OF FACTOR (Attributes and detriments if/as applicable)	RELATIVE IMPORTANCE RI (% or n/a)	POINTS SCORED (Rated)	WEIGHTED LINE SCORE (WS) (RI x Rated Score)
	(Minor interactions with/between Elements/Factors are not directly in rating)		RI	XX (of RI)	XX x RI (of RI^2)
1	Status of Technology Development	The growing technology of "fracking" is opening and producing natural gas in abundant supply	12	12 (of 12)	144 (of 144) = 100 %
2	Reliability / Sufficiency of Energy	Reliability is 100%; sufficient to meet demands	12	12 (of 12)	144 (of 144) = 100%
3	Resource(s) Exist Domestically or Approx % Needed to be Imported	Government restrictions on liquid/gas pipeline routings reduces the availability of gas to several regions	10	8 (of 10)	80 (of 100) = 80%
4	Cost of Resource(s) (Dom. + Imp.)	Of the 3 fossil fuel types, natural gas is the lowest cost to distribute and per volume transformed into heat energy	10	10 (of 10)	100 (of 100) = 100 %
5	Importance to Industry of Region	Reliable, low cost energy vital in industrial areas.	9	9 (of 9)	81 (of 81) = 100 %
6	Importance and Urgency to Gov't. Economy and Employment	Use of gas in facilities producing efficient energy critical during transition from dependency on coal	9	9 (of 9)	81 (of 81) = 100 %
7	By-Products and Emissions	Almost totally "clean" fuel when burned - no solid residue with toxic/carcinogenic content as pollutant.	8	7 (of 8)	56 (of 64) = 88 %
8	Effect on Human Population	No physical long term effect. Minor respiratory irritant	8	6 (of 8)	48 (of 64) = 75 %
9	Effect on Regional Infrastructure / Transportation / Commercial Trade	Most coal burning facilities in Eastern US have - or plan to - convert from coal to natural gas. Midwest and Western US will convert from coal to oil or natural gas	7	7 (of 7)	49 (of 49) = 100 %
10	Effect on Environment / Weather / Climate (Frequency & Severity)	"Greenhouse gas" effects of emissions on weather events frequency/severity have been exaggerated	5	5 (of 5)	25 (of 25) = 100 %
11	Effect on Animal Life (Wild or Livestock) - Land, Aquatic, Avian	No significant effect of gas used in energy production on almost all life-forms.	5	5 (of 5)	25 (of 25) = 100 %
12	Effect on Forestry and Agriculture	No significant effect of gas used in energy production	5	5 (of 5)	25 (of 25) = 100 %
			Total A Sum of RI	Total B Sum of RI	Total C XX x RI (of RI^2)
			100	95 (of 100)	858 (of 902)
			WEIGHTED SCORE =		95 %

RECOMMENDATIONS FOR CURRENT AND FUTURE DEVELOPMENT(S):

the reduction of "greenhouse gas" emissions to reduce/eliminate the adverse effects of climate change will not achieve the unrealistic present goal set by the UN IPCC.

Leakage of gas into atmosphere after transportation accidents and pipeline leakages may occur increasing fire hazard. Natural gas in either gaseous or liquified form is a highly useful fuel as a complement to alternative (non-fossil) fuels in creating energy for meeting industrial and residential energy needs as the human population increases.

TS-4 Use of Fossil Fuel – Natural Gas (Liquified / Transportable)

WEIGHTED TRADE STUDY SCORE WORKSHEET (Applicable to United States of America Region)

(Ratings and scores based on factual information in pertinent databases in Public Domain)

#	FACTOR CONSIDERED (Being rated for Importance and primary qualitative magnitude)	ELEMENTS CONSIDERED IN RATING OF FACTOR (Attributes and detriments if/as applicable)	RELATIVE IMPORTANCE RI (% or n/a)	POINTS SCORED (Rated)	WEIGHTED LINE SCORE (WS) (RI x Rated Score)
	TRADE STUDY: TS-5 Use of Fossil Fuel Derivative - Propane (Transportable Tank)				
	(Minor interactions with/between Elements/Factors are not directly in scoring)		**RI**	**XX (of RI)**	**XX x RI (of RI^2)**
1	Status of Technology Development	Mature technology in widespread residential use	**12**	**12** (of 12)	**144** (of 144)
2	Reliability / Sufficiency of Energy	Highly reliable. Wide range of uses in USA	**12**	**12** (of 12)	**144** (of 144)
3	Resource(s) Exist Domestically or Approx % Needed to be Imported	By-product of crude oil refining. None needs being imported.	**10**	**10** (of 10)	**100** (of 100)
4	Cost of Resource(s) (Dom.+Imp.)	None imported. Lower cost than gasoline and Diesel fuels.	**10**	**9** (of 10)	**81** (of 100)
5	Importance to Industry of Region	Not in large use for heavy industry	**9**	**6** (of 9)	**36** (of 81)
6	Importance and Urgency to Gov't. Economy and Employment	Increased use (with decrease in use of other fossil fuels) will increase employment; beneficial to national economy	**9**	**8** (of 9)	**64** (of 81)
7	By-Products and Emissions	Lower emissions than from natural gas	**8**	**8** (of 8)	**64** (of 64)
8	Effect on Human Population	No significant effect when handled properly. Used in hot air ballooning, with some fires.	**8**	**6** (of 8)	**36** (of 64)
9	Effect on Regional Infrastructure / Transportation / Commercial Trade	Transportation by truck and rail in tanks; stored at residential properties in 500 gal tanks. Low accident rate (fire/explosion).	**7**	**7** (of 7)	**49** (of 49)
10	Effect on Environment / Weather / Climate (Frequency & Severity)	Negligible effect on environment, weather, and climate	**5**	**5** (of 5)	**25** (of 25)
11	Effect on Animal Life (Wild or Livestock) - Land, Aquatic, Avian	Negligible effect on all forms of animal life	**5**	**5** (of 5)	**25** (of 25)
12	Effect on Forestry and Agriculture	Negligible effect on forestry and agriculture	**5**	**5** (of 5)	**25** (of 25)
			Total A Sum of RI **100**	**Total B** Sum of XX **93** (of 100)	**Total C** XX x RI (of RI^2) **793** (of 902)
			WEIGHTED SCORE =		**88%**

RECOMMENDATIONS FOR CURRENT AND FUTURE DEVELOPMENTS:

The widespread use of propane in residences, small businesses, and sporting applications has demonstrated credibility and safety of use of propane.

The storage and transport of propane in tanks of various volumetric size causes leaks that are unsafe to humans if inhaled, and accumulated gas can explode or be ignited by ignition source causing property damage. Building safety codes must be updated when propane is used in areas more densely populated than that of rural areas.

TS-5 Use of Fossil Fuel Derivative – Propane

WEIGHTED TRADE STUDY SCORE WORKSHEET (Applicable to United States of America Region)

(Ratings and scores based on factual information in pertinent databases in Public Domain)

TRADE STUDY: TS-6 Use of Non-Fossil Fuel - Hydrogen					
	FACTOR CONSIDERED (Being rated for Importance and primary qualitative magnitude)	ELEMENTS CONSIDERED IN RATING FACTOR (Attributes and detriments if/as applicable)	RELATIVE IMPORTANCE RI (% or n/a)	POINTS SCORED (Rated)	WEIGHTED LINE SCORE (WS) (RI x Rated Score)
	(Minor interactions with/between Elements/Factors are not directly in scoring)		RI	XX (of RI)	XX x RI (of RI^2)
1	Status of Technology Development	Mature technology in widespread use as fuel and as a generator of electricity	12	11 (of 12)	132 (of 144) = 92%
2	Reliability / Sufficiency of Energy	Highly reliable. More useful for electricity geberation than as a common propulsive fuel	12	11 (of 12)	132 (of 144) = 92%
3	Resource(s) Exist Domestically or Approx % Needed to be Imported	Abundance as gas in atmosphere and as a component of water in all regions on planet.	10	10 (of 10)	100 (of 100) = 100%
4	Cost of Resource(s) (Domestic + Import if any)	None imported. Lower cost to produce than gasoline and Diesel fuels.	10	10 (of 10)	100 (of 100) = 100%
5	Importance to Industry of Region	Potential for large use for steel production industry replacing coal.	9	9 (of 9)	81 (of 81) = 100%
6	Importance and Urgency to Gov't. Economy and Employment	Increased use (with decrease in use of other fossil fuels) will increase employment; beneficial to national economy	9	8 (of 9)	72 (of 81) = 89%
7	By-Products and Emissions	No heat-trapping emissions. Unhealthful to human respiratory system.	8	6 (of 8)	48 (of 64) = 75%
8	Effect on Human Population	Has a toxic effect as well as a frostbite threat to humans if in direct contact (liquid state)..	8	6 (of 8)	48 (of 64) = 75%
9	Effect on Regional Infrastructure / Transportation / Commercial Trade	Transported in liquid state by truck and rail in expensive tanks. Exploaive hazard transporting in pipeline through populated areas Handled in slush form on aircraft to minimize volume.	7	3 (of 7)	21 (of 49) = 43%
10	Effect on Environment / Weather / Climate (Frequency & Severity)	Negligible effect on environment, weather, and climate	5	5 (of 5)	25 (of 25) = 100%
11	Effect on Animal Life	Negligible effect on all forms of animal life	5	5 (of 5)	25 (of 25) = 100%
12	Effect on Forestry and Agriculture	Negligible effect on forestry and agriculture	5	5 (of 5)	25 (of 25) = 100%
	RECOMMENDATIONS FOR CURRENT AND FUTURE DEVELOPMENTS: **High abundance in atmosphere and in water makes hydrogen a very clean source of energy, more useful for producing electricity than as fuel to propel a vehicle (on land, water, or in air). Explosivity is hydrogen's major detriment. Technology advancement is routine, no crisis need.**		Total A Sum of RI 100	Total B Sum of XX 89 (of 100)	Total C XX x RI (of RI^2) 809 (of 902)
			WEIGHTED SCORE =		**90%**

TS-6 Use of Non-fossil Fuel – Hydrogen

WEIGHTED TRADE STUDY SCORE WORKSHEET (Applicable to United States of America Region)

(Ratings and scores based on factual information in pertinent databases in Public Domain)

#	FACTOR CONSIDERED (Being rated for Importance and primary qualitative magnitude)	ELEMENTS CONSIDERED IN RATING FACTOR (Attributes and detriments if/as applicable)	RELATIVE IMPORTANCE RI (% or n/a)	POINTS SCORED (Rated)	WEIGHTED LINE SCORE (WS) (RI x Rated Score)
	TRADE STUDY: TS-7 Use of Non-Fossil Fuel - Ethanol				
	(Minor interactions with/between Elements/Factors are not directly in scoring)		RI	XX (of RI)	XX x RI (of RI^2)
1	Status of Technology Development	Mature technology in widespread use as fuel and fuel additive. Damages engine seals in long use.	12	8 (of 12)	96 (of 144) = 67%
2	Reliability / Sufficiency of Energy	Highly reliable. More useful as a common propulsive fuel than for electricity generation.	12	10 (of 12)	120 (of 144) = 83%
3	Resource(s) Exist Domestically or Approx % Needed to be Imported	In abundance as a product of corn growth in middle corn-belt states of US. .	10	9 (of 10)	90 (of 100) = 90%
4	Cost of Resource(s) (Domestic + Import if any)	None imported. Lower cost to produce than gasoline and Diesel fuels.	10	10 (of 10)	100 (of 100) = 100%
5	Importance to Industry of Region	Potential for use as alternative fuel in hybrid vehicles (electric + fuel)	9	8 (of 9)	72 (of 81) = 89%
6	Importance and Urgency to Gov't. Economy and Employment	Increased use (with decrease in use of fossil and other alternative fuels); will increase farm employment; beneficial to national economy	9	8 (of 9)	72 (of 81) = 89%
7	By-Products and Emissions	No heat-trapping emissions. Residue from corn after processing useful as a fertilizer.	8	7 (of 8)	56 (of 64) = 88%
8	Effect on Human Population	Is in wide use in Brazil, Europe, and US with no effects on human health or safety reported.	8	8 (of 8)	64 (of 64) = 100%
9	Effect on Regional Infrastructure / Transportation / Commercial Trade	Transported in liquid state by truck and rail in inexpensive tanks. No hazardous exposure if leak or spill occurs, Good co-technology EVs	7	7 (of 7)	49 (of 49) = 100%
10	Effect on Environment / Weather / Climate (Frequency & Severity)	No effect on environment, weather, or climate. Not useful in cold weather regions.	5	3 (of 5)	15 (of 25) = 60%
11	Effect on Animal Life	No known effect on all forms of animal life	5	5 (of 5)	25 (of 25) = 100%
12	Effect on Forestry and Agriculture	No known effect on forestry; uses about 40% of US corn crop near-term; probable impact on US food supply if US population doubles or triples.	5	3 (of 5)	15 (of 25) = 60%
	RECOMMENDATIONS FOR CURRENT AND FUTURE DEVELOPMENTS: Less costly to produce as fuel than gasoline but use ratio is 1.5 higher in volume used. Good for use in local municipal government and public transportation vehicles and as fuel in private hybrid vehicles (not cold temps).		Total A Sum of RI 100	Total B Sum of XX 86 (of 100)	Total C XX x RI (of RI^2) 774 (of 902)
			WEIGHTED SCORE =		86%

TS-7 Use of Non-fossil Fuel – Ethanol

TRADE STUDY SCORE WORKSHEET (Applicable to United States of America Region)

(Ratings and scores based on factual information in pertinent databases in Public Domain)

TRADE STUDY: TS- 8 Use/Development of Hydrodynamic Energy (Dams, River Flow)

#	FACTOR CONSIDERED (Being rated for Importance and primary qualitative magnitude)	ELEMENTS CONSIDERED IN RATING FACTOR (Attributes and detriments if/as applicable)	RELATIVE IMPORTANCE RI (% or n/a)	POINTS SCORED (Rated)	WEIGHTED LINE SCORE (WS) (RI x Rated Score)
	(Minor interactions with/between Elements/Factors are not directly in rating)		RI	XX (of RI)	XX x RI (of RI²)
1	Status of Technology Development	Technology of Dams for electric generation, flood control, and irrigation is well established. Growing population will require expansion of existing dams and greater number of newer dams where feasible.	12	12 (of 12)	144 (of 144) = 100%
2	Reliability / Sufficiency of Energy	High reliability except in prolonged drought regions.	12	11 (of 12)	132 (of 144) = 92%
3	Resource(s) Exist Domestically or Approx % imported	Resources for dam improvements and new construction for electric power generation and flood control are abundant in US	10	10 (of 10)	100 (of 100) = 100%
4	Cost of Resource(s) (Domestic + Import if any)	Growing population requires maximum dam electric power generation and for irrigation in near future. Augments other energy technology developments.	10	10 (of 10)	100 (of 100) = 100%
5	Importance to Industry of Region	Of utmost importance for power, irrigation, flood control	9	9 (of 9)	81 (of 81) = 100%
6	Importance and Urgency to Gov't. Economy and Employment	Important for economy and more employment. Good augmentation with other energy collection technologies.	9	8 (of 9)	72 (of 81) = 89%
7	By-Products, Emissions, and Pollutants	No emissions or pollutants into surrounding environment Typically large network of power distribution towers/lines	8	7 (of 8)	56 (of 64) = 88%
8	Effect on Human Population	Not aesthetic but not any harmful effect to safety/health..	8	8 (of 8)	64 (of 64) = 100%
9	Effect on Regional Infrastructure / Transportation / Commercial Trade	Has small impact on regional infrastructure (network of power distribution lines) but otherwise no harmful effect.	7	6 (of 7)	42 (of 49) = 86%
10	Effect on Environment / Weather / Climate (Frequency & Severity)	No effect on weather or climate.	5	5 (of 5)	25 (of 25) = 100%
11	Effect on Animal Life (Wild or Livestock) - Land, Aquatic, Avian	Negligible harmful effect on animal life of any type.	5	5 (of 5)	25 (of 25) = 100%
12	Effect on Forestry and/or Agriculture	No effect on forestry. Provides irrigation water that could support agriculture of growing population.	5	5 (of 5)	25 (of 25) = 100%
	RECOMMENDATIONS FOR CURRENT AND FUTURE DEVELOPMENT(S):		**Total A Sum of RI 100**	**Total B Sum of XX 96 (of 100)**	**Total C XX x RI (of RI²) 866 (of 902)**
			WEIGHTED SCORE =		**96%**

The prospect of an exponentially growing population dictates the maximum use of dams in rivers for additional electric power generation, flood control, and irrigation for increased agricultural food supply. Existing dams need upgrading to larger capacity and additional dams to be constructed where feasible. Planning should start for sea walls in coastal areas hazarded by ocean rise.

TS-8 Use/Development of Hydrodynamic Energy (River Flow)

TRADE STUDY SCORE WORKSHEET (Applicable to United States of America Region)

(Ratings and scores based on factual information in pertinent databases in Public Domain)

#	TRADE STUDY: TS- 9 Control of Hydrodynamic Events (Coastal Regions)				
	FACTOR CONSIDERED (Being rated for Importance and primary qualitative magnitude)	ELEMENTS CONSIDERED IN RATING FACTOR (Attributes and detriments if/as applicable)	RELATIVE IMPORTANCE RI (% or n/a)	POINTS SCORED (Rated)	WEIGHTED LINE SCORE (WS) (RI x Rated Score)
	(Minor interactions with/between Elements/Factors are not directly in rating)		RI	XX (of RI)	XX x RI (of RI^2)
1	Status of Technology Development	Growing population and possibility of oceans levels rising creates need for many technology improvements	12	12 (of 12)	144 (of 144) = 100%
2	Reliability / Sufficiency of Energy	Need much improved flood control in flood hazard zones in southeastern states in US.	12	12 (of 12)	144 (of 144) = 100%
3	Resource(s) Exist Domestically or Approx % imported	Resources for mitigating effects of climate change are not sufficient at present. Great need to reprioritize efforts.	10	10 (of 10)	100 (of 100) = 100%
4	Cost of Resource(s) (Domestic + Import if any)	Present and growing population requires whatever is necessary for safety, health, comfort, transportation.	10	10 (of 10)	100 (of 100) = 100%
5	Importance to Industry of Region	Of utmost importance for power, irrigation, flood control	9	9 (of 9)	81 (of 81) = 100%
6	Importance and Urgency to Gov't. Economy and Employment	Important for economy and more employment. Need higher priority of attention to mitigating effects of climate change and natural hazards on US population.	9	9 (of 9)	81 (of 81) = 100%
7	By-Products, Emissions, and Pollutants	Lack of controls to minimize flooding and mudslides causes human and animal safety and health hazards, and infrastructure inadequacies.	8	7 (of 8)	56 (of 64) = 88%
8	Effect on Human Population	Planning needed to mitigate climate change effects on humans and livestock animals	8	8 (of 8)	64 (of 64) = 100%
9	Effect on Regional Infrastructure/ Transportation/Commerce/Trade	Flooding and mudslides causes power outages, road closures and shortage of food/homes for lengthy periods	7	6 (of 7)	42 (of 49) = 86%
10	Effect on Environment / Weather / Climate (Frequency & Severity)	No effect on weather or climate. Reverse of effect is on humans and animals	5	5 (of 5)	25 (of 25) = 100%
11	Effect on Animal Life (Wild or Livestock) - Land, Aquatic, Avian	Flooding and mudslides have harmful effect on animal life of all types.	5	5 (of 5)	25 (of 25) = 100%
12	Effect on Forestry and/or Agriculture	Flooding, mudslides, and lightning strikes have major impact on agriculture and on forestry.	5	5 (of 5)	25 (of 25) = 100%
			Total A Sum of RI 100	Total B Sum of XX 98 (of 100)	Total C XX x RI (of RI^2) 887 (of 902)
			WEIGHTED SCORE =		98%

RECOMMENDATIONS FOR CURRENT AND FUTURE DEVELOPMENT(S):

The prospect of an exponentially growing population dictates all possible actions be taken for mitigating the adverse effects of inevitable climate change, tragic consequences of natural catastrophic events (flooding, mudslides, wildfires, tornadoes, hurricanes, earthquakes, volcanic eruptions, etc.). Greater action and planning is needed on a prioritized basis for near and far term.

TS-9 Control of Hydrodynamic Events (Coastal Regions)

TRADE STUDY SCORE WORKSHEET (Applicable to United States of America Region)

(Ratings and scores based on factual information in pertinent databases in Public Domain)

TRADE STUDY: TS-10 Use/Development of Solar Energy (Large Industrial/Commercial)

#	FACTOR CONSIDERED (Being rated for Importance and primary qualitative magnitude)	ELEMENTS CONSIDERED IN RATING FACTOR (Attributes and detriments if/as applicable)	RELATIVE IMPORTANCE RI (% or n/a)	POINTS SCORED (Rated)	WEIGHTED LINE SCORE (WS) (RI x Rated Score)
	(Minor interactions with/between Elements/Factors are not directly in rating)		RI	XX (of RI)	XX x RI (of RI^2)
1	Status of Technology Development	Solar energy technology is well-developed and in use in all regions of the planet. Technology not well-suited for large industrial or commercial energy use.	12	9 (of 12)	108 (of 144) = 75%
2	Reliability / Sufficiency of Energy	Reliability is limited to solar insolation <50%/day being received. Requires large battery storage	12	6 (of 12)	72 (of 144) = 50%
3	Resource(s) Exist Domestically or Approx % Needed to be Imported	Most materials needed for panel production are abundant in US. Requires import of rare metals	10	9 (of 10)	90 (of 100) = 90%
4	Cost of Resource(s) (Domestic + Import if any)	Cost to industrial/commercial users is presently low but will increase as coal production is phased out. Law of supply and demand.	10	9 (of 10)	90 (of 100) = 90%
5	Importance to Industry of Region	Not reliable for large industrial use. Unimportant.	9	6 (of 9)	54 (of 81) = 67%
6	Importance and Urgency to Gov't. Economy and Employment	Of low importance to large Industrial/Commercial businesses; is of major importance to small users.	9	6 (of 9)	54 (of 81) = 67%
7	By-Products/Emissions / Pollution	Clean technology to operate but needs maintenance.	8	7 (of 8)	56 (of 64) = 88%
8	Effect on Human Population	No direct effect. Large arrays use high acreage.	8	7 (of 8)	56 (of 64) = 88%
9	Effect on Regional Infrastructure / Transportation / Commercial Trade	Not a factor in transportation or commercial trade	7	7 (of 7)	49 (of 49) = 100%
10	Effect on Environment / Weather / Climate (Frequency & Severity)	No effect on weather/climate	5	5 (of 5)	25 (of 25) = 100%
11	Effect on Animal Life (Wild or Livestock) - Land, Aquatic, Avian	No significant effect on animal life. Larger arrays in desert regions disturb desert animal life.	5	5 (of 5)	25 (of 25) = 100%
12	Effect on Forestry and Agriculture	No effect on forestry and/or agriculture	5	5 (of 5)	25 (of 25) = 100%
			Total A Sum of RI 100	Total B Sum of XX 81 (of 100)	Total C XX x RI (of RI^2) 704 (of 902)
			WEIGHTED SCORE =		78%

RECOMMENDATIONS FOR CURRENT AND FUTURE DEVELOPMENT(S):

The reduction of "greenhouse gas" emissions to reduce/eliminate the adverse effects of climate change will not achieve the unrealistic present goals set by the UN IPCC. The technology of solar energy to replace the reduction/elimination of use of coal is highly recommended in small industrial, small commercial, residential, and rural uses, but due to its low reliability 24/7 is not efficient for use in large industrial and large commercial uses. In any use, there must exist the co-technology of a storage capability that is intermittently supplied with electrical energy from soalr energy collection facilities. Solar energy is also well-suited when coupled with wind energy for the operation of large desalination facilities and for water purification facilities for municipalities.

TS-10 Use/Development of Solar Energy (Large Industrial / Commercial)

TRADE STUDY SCORE WORKSHEET (Applicable to United States of America Region)

(Ratings and scores based on factual information in pertinent databases in Public Domain)

TRADE STUDY: TS-11 Use/Development of Solar Energy (Residential / Rural / Town)

#	FACTOR CONSIDERED (Being rated for Importance and primary qualitative magnitude)	ELEMENTS CONSIDERED IN RATING FACTOR (Attributes and detriments if/as applicable)	RELATIVE IMPORTANCE RI (% or n/a)	POINTS SCORED (Rated)	WEIGHTED LINE SCORE (WS) (RI x Rated Score)
	(Minor interactions with/between Elements/Factors are not directly in rating)		RI	XX (of RI)	XX x RI (of RI^2)
1	Status of Technology Development	Solar energy technology is well-developed and in use in all regions of the planet. Technology best suitable for residential, rural, farm, and town uses.	12	12 (of 12)	144 (of 144) = 100%
2	Reliability / Sufficiency of Energy	Reliability is limited to solar insolation <50%/day being received. Requires large battery storage	12	8 (of 12)	96 (of 144) = 67%
3	Resource(s) Exist Domestically or Approx % Needed to be Imported	Most materials needed for panel production are abundant in US. Requires import of rare metals	10	9 (of 10)	90 (of 100) = 90%
4	Cost of Resource(s) (Domestic + Import if any)	Initial installation cost moderately high to residential users but is offset by long-term benefits. Rare metals are imported.	10	9 (of 10)	90 (of 100) = 90%
5	Importance to Industry of Region	Is of major importance to small users.	9	8 (of 9)	72 (of 81) = 67%
6	Importance and Urgency to Gov't. Economy and Employment	Is of major importance to small users.Of low importance to large Industrial/Commercial users.	9	8 (of 9)	72 (of 81) = 89%
7	By-Products/Emissions / Pollution	Clean technology to operate but needs maintenance.	8	7 (of 8)	56 (of 64) = 88%
8	Effect on Human Population	No direct effect. Large arrays use high acreage.	8	7 (of 8)	56 (of 64) = 88%
9	Effect on Regional Infrastructure / Transportation / Commercial Trade	Not a factor in transportation or commercial trade	7	7 (of 7)	49 (of 49) = 100%
10	Effect on Environment / Weather / Climate (Frequency & Severity)	No effect on weather/climate	5	5 (of 5)	25 (of 25) = 100%
11	Effect on Animal Life (Wild or Livestock) - Land, Aquatic, Avian	No significant effect on animal life. Larger arrays in desert regions disturb desert animal life.	5	5 (of 5)	25 (of 25) = 100%
12	Effect on Forestry and Agriculture	No effect on forestry and/or agriculture	5	5 (of 5)	25 (of 25) = 100%
			Total A Sum of RI 100	Total B Sum of XX 90 (of 100)	Total C XX x RI (of RI^2) 800 (of 902)
			WEIGHTED SCORE =		89%

RECOMMENDATIONS FOR CURRENT AND FUTURE DEVELOPMENT(S):

The reduction of "greenhouse gas" emissions to reduce/eliminate the adverse effects of climate change will not achieve the unrealistic present goals set by the UN IPCC. The technology of solar energy to replace the reduction/elimination of use of coal is highly recommended in small industrial, small commercial, residential, and rural uses, but due to its low reliability 24/7 is not efficient for use in large industrial and large commercial uses. In any use, there must exist the co-technology of a storage capability that is intermittently supplied with electrical energy from alternative energy collection facilities. Solar energy is of value when coupled with wind energy for the operation of large desalination facilities and for water purification for coastal municipalities.

TS-11 Use/Development of Solar Energy (Residential / Rural / Town)

TRADE STUDY SCORE WORKSHEET (Applicable to United States of America Region)

(Ratings and scores based on factual information in pertinent databases in Public Domain)

#	TRADE STUDY: TS-12 Use/Development of Wind Energy (On Land)				
	FACTOR CONSIDERED (Being rated for Importance and primary qualitative magnitude)	ELEMENTS CONSIDERED IN RATING OF FACTOR (Attributes and detriments if/as applicable)	RELATIVE IMPORTANCE RI (% or n/a)	POINTS SCORED (Rated)	WEIGHTED LINE SCORE (WS) (RI x Rated Score)
	(Minor interactions with/between Elements/Factors are not directly in rating)		RI	XX (of RI)	XX x RI (of RI^2)
1	Status of Technology Development	Wind energy generation technology is mature but low reliability is a large factor. Requires a co-technology(ies).	12	11 (of 12)	132 (of 144) = 92%
2	Reliability / Sufficiency of Energy	Reliable if wind is blowing within limits. Not when no wind.	12	7 (of 12)	84 (of 144) = 58%
3	Resource(s) Exist Domestically or Approx % imported	Resources mostly in US - small amount of rare earth is imported. Maintenance cost is high.	10	8 (of 10)	80 (of 100) = 80%
4	Cost of Resource(s) (Domestic + Import if any)	Construction cost for large base, tall tower, and large turbine are moderate; turbine maintenance cost high.	10	8 (of 10)	80 (of 100) = 80%
5	Importance to Industry of Region	More useful to mechanical industry and agriculture/farming	9	7 (of 9)	63 (of 81) = 78%
6	Importance and Urgency to Gov't. Economy and Employment	Of low importance to economy and employment. Good augmentation with other energy collection technologies.	9	7 (of 9)	63 (of 81) = 78%
7	By-Products, Emissions, and Pollutants	Low amount of electromagnetic wave energy into surrounding atmosphere, no "greenhouse emissions.	8	7 (of 8)	56 (of 64) = 88%
8	Effect on Human Population	Not aesthetic but not any harmful effect to safety/health..	8	8 (of 8)	64 (of 64) = 100%
9	Effect on Regional Infrastructure / Transportation / Commercial Trade	Has large footprint area but otherwise no harmful effect.	7	7 (of 7)	49 (of 49) = 100%
10	Effect on Environment / Weather / Climate (Frequency & Severity)	No effect on weather or climate.	5	5 (of 5)	25 (of 25) = 100%
11	Effect on Animal Life (Wild or Livestock) - Land, Aquatic, Avian	Hazard to local bird population; not to land animals..	5	4 (of 5)	20 (of 25) = 80%
12	Effect on Forestry and/or Agriculture	No effect on forestry. Uses land area that could support agriculture of growing population.	5	4 (of 5)	20 (of 25) = 80%
			Total A Sum of RI 100	Total B Sum of XX 83 (of 100)	Total C XX x RI (of RI^2) 736 (of 902)
			WEIGHTED SCORE =		82%

RECOMMENDATIONS FOR CURRENT AND FUTURE DEVELOPMENT(S):

Wind energy is one of the older technologies that have allowed Man to develop their potential for doing useful work - typically transforming wind energy into mechanical source of power for a mechanical machine (windmill in the Netherlands, a sugar cane grindi grinding mill on Barbados or in South America, or a lumber mill in Oregon) - that limitation being that it produces power but only when the wind is blowing between the limits of not enough or too much. Its prime value is with small industrial operations, agricultural growing (for pumping irrigation water) or in combination with other emerging technologies dependent on a more reliable source of energy. Wind power technology could be highly useful in combination with the associate technology of generating heat energy for desalination.

TS-12 Use/Development of Wind Energy (On Land)

TRADE STUDY SCORE WORKSHEET (Applicable to United States of America Region)

(Ratings and scores based on factual information in pertinent databases in Public Domain)

#	TRADE STUDY: TS-13 Use/Development of Wind Energy (Off-Shore)				
	FACTOR CONSIDERED (Being rated for Importance and primary qualitative magnitude)	ELEMENTS CONSIDERED IN RATING OF FACTOR (Attributes and detriments if/as applicable)	RELATIVE IMPORTANCE RI (% or n/a)	POINTS SCORED (Rated)	WEIGHTED LINE SCORE (WS) (RI x Rated Score)
	(Minor interactions with/between Elements/Factors are not directly in rating)		RI	XX (of RI)	XX x RI (of RI^2)
1	Status of Technology Development	Wind energy generation technology is mature but low reliability iin off-shore environment s a prohibitive risk.	12	6 (of 12)	72 (of 144) = 50%
2	Reliability / Sufficiency of Energy	Higher winds and salt corrosion will damage electromechanical components. Pedestals must be stronger to endure high wave activity, building underwater pedestals is risky. Blades highly susceptible to ocean storms and corrosive damage. Low reliability even if wind is blowing within safe/efficient limits which is not often.	12	4 (of 12)	48 (of 144) = 33%
3	Resource(s) Exist Domestically or Approx % imported	Resources mostly in US - small amount of rare earth is imported. Maintenance cost is high due to salt corrosion..	10	8 (of 10)	80 (of 100) = 80%
4	Cost of Resource(s)	Construction and maintenance/repair cost prohibitive	10	4 (of 10)	40 (of 100) = 40%
5	Importance to Industry of Region	More useful to coastal community populations; Most useful to desalination technologies on coast.	9	7 (of 9)	63 (of 81) = 78%
6	Importance and Urgency to Gov't. Economy and Employment	Of low importance to economy and employment. Useful augmentation with other energy collection technologies.	9	7 (of 9)	63 (of 81) = 78%
7	By-Products, Emissions, and Pollutants	Low amoiunt of electromagnetic wave energy into surrounding atmosphere, no "greenhouse emissions.	8	7 (of 8)	56 (of 64) = 88%
8	Effect on Human Population	Not aesthetic, but not any harmful effect to safety/health. Oil leakage from turbine could cause toxic pollution..	8	6 (of 8)	48 (of 64) = 75%
9	Effect on Local Infrastructure	Small effect if any on coastal water recreation vehicles	7	6 (of 7)	42 (of 49) = 86%
10	Effect on Environment / Weather / Climate (Frequency &	No effect on weather or climate.	5	5 (of 5)	25 (of 25) = 100%
11	Effect on Animal Life (Wild or Livestock) - Land, Aquatic, Avian	Rotating blades hazardous to coastal bird population; Grid of multiple pedestals hazardous to whales et al..	5	3 (of 5)	15 (of 25) = 60%
12	Effect on Food Supply	Possible effect on local aquatic lifeforms consumed.	5	4 (of 5)	20 (of 25) = 80%
			Total A Sum of RI 100	Total B Sum of XX 67 (of 100)	Total C XX x RI (of RI^2) 572 (of 902)
			WEIGHTED SCORE =		63%

RECOMMENDATIONS FOR CURRENT AND FUTURE DEVELOPMENT(S):

The several hazards associated with constructing strong underwater pedestals, maintaining large but sensitive to salt water corrosion machinery and structures, and the high potential to pedestal and large diameter blade damage by high wave and wind loads from ocean storms and hurricanes are prohibitively formidable. This form of off-shore wind technology is strongly discouraged.

TS-13 Use/Development of Wind Energy (Off-Shore)

TRADE STUDY SCORE WORKSHEET (Applicable to United States of America Region)

(Ratings and scores based on factual information in pertinent databases in Public Domain)

#	TRADE STUDY: TS-14 Energy Storage - Stationary Large (Electrical)		RELATIVE IMPORTANCE RI (% or n/a)	POINTS SCORED (Rated)	WEIGHTED LINE SCORE (WS) (RI x Rated Score)
	FACTOR CONSIDERED (Being rated for Importance and primary qualitative magnitude)	ELEMENTS CONSIDERED IN RATING FACTOR (Attributes and detriments if/as applicable)			
	(Minor interactions with/between Elements/Factors are not directly in rating)		RI	XX (of RI)	XX x RI (of RI^2)
1	Status of Technology Development	The electrical storage technology of large stationary units is nearing maturity; need is for much larger storage capacity for rapidly growing population	12	11 (of 12)	132 (of 144) = 92%
2	Reliability / Sufficiency of Energy	Storage of electricity and retrieval from storage is reliable. Safety of distribution routing of electricity from process dependent on regional weather events	12	11 (of 12)	132 (of 144) = 92%
3	Resource(s) Exist Domestically or % imported	All components of large electrical storage are domestic; energy storage technology being updated	10	9 (of 10)	90 (of 100) = 90%
4	Cost of Resource(s) (Domestic + Import if any)	Cost of resources in/for storage components is reasonable and affordable.	10	10 (of 10)	100 (of 100) = 100%
5	Importance to Industry of Region	Use with solar and wind energy technologies unreliability for large industrial demends increases importance for mitigating that unreliability factor	9	7 (of 9)	63 (of 81) = 78%
6	Importance and Urgency to Gov't. Economy / Employment	Not a factor affecting economy. Has beneficial effect on employment to manufacture, install and maintain the growing capability and need for electricity storage.	9	9 (of 9)	81 (of 81) = 100%
7	By-Products, Emissions, and Pollutants	No by-products, emissions, or pollutants	8	8 (of 8)	64 (of 64) = 100%
8	Effect on Human Population	No effect on human population	8	8 (of 8)	64 (of 64) = 100%
9	Effect on Commerce / Infrastructure/Transportation	No effect on regional Commerce, infrastructure, transportation or trade.	7	7 (of 7)	49 (of 49) = 100%
10	Effect on Environment / Weather / Climate	No effect on weather or climate. Storage yard occupies moderate percent of footprint of facility.	5	4 (of 5)	20 (of 25) = 80%
11	Effect on Animal Life (Wild or Livestock) - Land, Water, Avian	No effect on animal life of any type (low possibility of disturbing desert life (snakes, spiders, tortoises).	5	5 (of 5)	25 (of 25) = 100%
12	Effect on Forestry and/or Agriculture	No effect on forestry or agriculture	5	5 (of 5)	25 (of 25) = 100%
	RECOMMENDATIONS FOR CURRENT AND FUTURE DEVELOPMENT(S): The storage technology for large capacity but variable electrical energy associated with solar and wind energy technologies (each having unique unreliability patterns) is not a limiting factor on solar and/or wind energy harvesting technologies.		Total A Sum of RI 100	Total B Sum of XX 94 (of 100)	Total C XX x RI (of RI^2) 845 (of 902)
			WEIGHTED SCORE =		94%

TS-14 Energy Storage - Stationary/Large Size (Electrical)

TRADE STUDY SCORE WORKSHEET (Applicable to United States of America Region)

(Ratings and scores based on factual information in pertinent databases in Public Domain)

#	TRADE STUDY: TS-15 Geothermal Energy (Surface or Near Surface)				
	FACTOR CONSIDERED (Being rated for Importance and primary qualitative magnitude)	ELEMENTS CONSIDERED IN RATING OF FACTOR (Attributes and detriments if/as applicable)	RELATIVE IMPORTANCE RI (% or n/a)	POINTS SCORED (Rated)	WEIGHTED LINE SCORE (WS) (RI x Rated Score)
	(Minor interactions with/between Elements/Factors are not directly in rating)		RI	XX (of RI)	XX x RI (of RI^2)
1	Status of Technology Development	Technology has been in use long term; harvesting heat energy from surface or near-surface is well-proven.	12	11 (of 12)	132 (of 144) = 92%
2	Reliability / Sufficiency of Energy	Reliability is 100%; Potential for high sufficiency west of Mississippi R. (N/A east of Mississippi R.)	12	11 (of 12)	132 (of 144) = 92%
3	Resource(s) Exist Domestically	Limited resources for surface only geothermal; however, need to revisit use of heat energy from existing depleted oil wells. No further drilling needed..	10	7 (of 10)	70 (of 100) = 70%
4	Cost of Resource(s) (Domestic + Import if any)	Cost only for surface facility to harvest no-cost geothermal energy at/near surface.	10	9 (of 10)	90 (of 100) = 90%
5	Importance to Industry of Region	Unimportant - resources not near industrial areas.	9	4 (of 9)	36 (of 81) = 44%
6	Importance and Urgency to Gov't. Economy and Employment	Population exponentially growing; clean safe geothermal energy could augment other technologies to generate energy for larger population. Low employment opportunities.	9	6 (of 9)	54 (of 81) = 67%
7	By-Products, Emissions, and Pollutants	Zero emissions after any clean-out drilling of existing bore holes. Pollution free during operation.	8	8 (of 8)	64 (of 64) = 100%
8	Effect on Human Population	No emissions or pollutants to affect humans	8	8 (of 8)	64 (of 64) = 100%
9	Effect on Regional Infrastructure / Transportation/Commercial Trade	No emissions or pollutants to affect infrastructure	7	7 (of 7)	49 (of 49) = 100%
10	Effect on Environment / Weather / Climate (Frequency & Severity)	No emissions or pollutants to affect weather or the climate	5	5 (of 5)	25 (of 25) = 100%
11	Effect on Animal Life (Wild or Livestock) - Land, Aquatic, Avian	No emissions or pollutants to affect animal life (land, aquatic, or avian)	5	5 (of 5)	25 (of 25) = 100%
12	Effect on Forestry and Agriculture	No emissions or pollutants to affect the ecology	5	5 (of 5)	25 (of 25) = 100%
			Total A Sum of RI 100	Total B Sum of XX 86 (of 100)	Total C XX x RI (of RI^2) 766 (of 902)
			WEIGHTED SCORE =		85%

RECOMMENDATIONS FOR CURRENT AND FUTURE DEVELOPMENT(S):

The surface and near-surface geothermal source of energy has the potential for also utilizing currently inactive oil wells in the western half of the United States. Its primary detriment is the lack of surface and near-surface geothermal sources near heavily populated regions (e.g. those east of the Mississippi River). It could augment solar, wind, and nuclear energy generation facilities in the western United States for some types of natural catastrophes (earthquakes, lightning-ignited wildfires, flooding, mudslides, etc.) causing lengthy power outages and access road blockages. Technology advancement routine, no crisis.

TS-15 Use/Development of Geothermal Energy (Surface)

TRADE STUDY SCORE WORKSHEET (Applicable to United States of America Region)

(Ratings and scores based on factual information in pertinent databases in Public Domain)

#	TRADE STUDY: TS-16 Geothermal Energy (Deep Drilled - Open and Closed Cycles)				
	FACTOR CONSIDERED (Being rated for Importance and primary qualitative magnitude)	ELEMENTS CONSIDERED IN RATING FACTOR (Attributes and detriments if/as applicable)	RELATIVE IMPORTANCE RI (% or n/a)	POINTS SCORED (Rated)	WEIGHTED LINE SCORE (WS) (RI x Rated Score)
	(Minor interactions with/between Elements/Factors are not directly in rating)		RI	XX (of RI)	XX x RI (of RI^2)
1	Status of Technology Development	Rapidly growing in recognition of being clean, reliable, available with proven deep drilling technologies (open cycle and closed cycle)	12	11 (of 12)	132 (of 144) = 92%
2	Reliability / Sufficiency of Energy	Reliability is 100%; Potential for full sufficiency	12	11 (of 12)	132 (of 144) = 92%
3	Resource(s) Exist Domestically or Approx % Needed to be Imported	Abundance of heat energy exists at attainable drilling depths. Less complexity with closed system than with open system which involves "fracking".	10	9 (of 10)	90 (of 100) = 90%
4	Cost of Resource(s) (Domestic + Import if any)	Cost of initial drilling is significant (lower for closed cycle than with open cycle). No import.	10	9 (of 10)	90 (of 100) = 90%
5	Importance to Industry of Region	Reliable, low cost energy vital in industrial areas.	9	9 (of 9)	81 (of 81) = 100%
6	Importance and Urgency to Gov't. Economy and Employment	Population exponentially growing; clean safe drilling allows facilities near high-density populated areas.	9	8 (of 9)	72 (of 81) = 89%
7	By-Products, Emissions, and Pollutants	Zero emissions after drilling complete; Pollution free during operation (closed cycle no pollution; open cycle negligible pollution)	8	7 (of 8)	56 (of 64) = 88%
8	Effect on Human Population	No emissions or pollutants to affect humans	8	8 (of 8)	64 (of 64) = 100%
9	Effect on Regional Infrastructure / Transportation/Commercial Trade	No emissions or pollutants to affect infrastructure	7	7 (of 7)	49 (of 49) = 100%
10	Effect on Environment / Weather / Climate (Frequency & Severity)	No emissions or pollutants to affect weather or the climate	5	5 (of 5)	25 (of 25) = 100%
11	Effect on Animal Life (Wild or Livestock) - Land, Aquatic, Avian	No emissions or pollutants to affect animal life (land, aquatic, or avian). Might kill local worms.	5	5 (of 5)	25 (of 25) = 100%
12	Effect on Forestry and Agriculture	No emissions or pollutants to affect the ecology	5	5 (of 5)	25 (of 25) = 100%
			Total A Sum of RI 100	Total B Sum of XX 94 (of 100)	Total C XX x RI (of RI^2) 841 (of 902)
			WEIGHTED SCORE =		93%

RECOMMENDATIONS FOR CURRENT AND FUTURE DEVELOPMENT(S):

The deep-drilled (both open cycle and closed cycle) geothermal source of energy is the near-optimum technology for meeting the needs and demands of the present and near-term population of the contiguous United States, Hawaii, and the Anchorage region of Alaska. When augmented by hydrodynamic and by solar, wind, and nuclear energy generation facilities in regions prone to seasonal storms and naturalcatastrophes (hurricanes, tornados, earthquakes, lightning-ignited wildfires, flooding, mudslides, etc.) causing power outages, geothermal power is of constant availability independent of those catastrophic events on or above surface level of the region involved. Technology advancement routine, not as crisis need.

TS-16 Use/Development of Geothermal Energy (Deep-drilled)

TRADE STUDY SCORE WORKSHEET (Applicable to United States of America Region)

(Ratings and scores based on factual information in pertinent databases in Public Domain)

#	TRADE STUDY: TS-17 Energy Storage - Stationary Large (Thermal)				
	FACTOR CONSIDERED (Being rated for Importance and primary qualitative magnitude)	ELEMENTS CONSIDERED IN RATING FACTOR (Attributes and detriments if/as applicable)	RELATIVE IMPORTANCE RI (% or n/a)	POINTS SCORED (Rated)	WEIGHTED LINE SCORE (WS) (RI x Rated Score)
	(Minor interactions with/between Elements/Factors are not directly in rating)		RI	XX (of RI)	XX x RI (of RI^2)
1	Status of Technology Development	The thermal storage technology of large stationary units is nearing maturity; need is for much larger storage capacity for rapidly growing population	12	11 (of 12)	132 (of 144) = 92%
2	Reliability / Sufficiency of Energy	Storage of thermal energy and retrieval from storage is reliable. Safety of distribution routing of electricity from process dependent on regional weather events	12	11 (of 12)	132 (of 144) = 92%
3	Resource(s) Exist Domestically or % imported	All components of large thermal storage are domestic; energy storage technology being updated	10	9 (of 10)	90 (of 100) = 90%
4	Cost of Resource(s) (Domestic + Import if any)	Cost of resources in/for storage components is reasonable and affordable	10	10 (of 10)	100 (of 100) = 100%
5	Importance to Industry of Region	Use with geothermal energy and desalination technologies for human needs and demands increases in importance as population grows	9	7 (of 9)	63 (of 81) = 78%
6	Importance and Urgency to Gov't. Economy / Employment	Not a factor affecting economy. Has beneficial effect on employment to manufacture, install and maintain the growing capability and need for thermal energy storage.	9	9 (of 9)	81 (of 81) = 100%
7	By-Products, Emissions, and Pollutants	No by-products, emissions, or pollutants	8	8 (of 8)	64 (of 64) = 100%
8	Effect on Human Population	No effect on human population unless need is ignored.	8	8 (of 8)	64 (of 64) = 100%
9	Effect on Commerce / Infrastructure/Transportation	No effect on regional Commerce, infrastructure, transportation or trade.	7	7 (of 7)	49 (of 49) = 100%
10	Effect on Environment / Weather / Climate	No effect on weather or climate. Thermal storage yard occupies moderate % of footprint of facility.	5	4 (of 5)	20 (of 25) = 80%
11	Effect on Animal Life (Wild or Livestock) - Land, Water, Avian	No effect on animal life of any type (low possibility of disturbing desert life (snakes, spiders, tortoises).	5	5 (of 5)	25 (of 25) = 100%
12	Effect on Forestry and/or Agriculture	No effect on forestry or agriculture unless process diverts water from agricultural uses and needs.	5	4 (of 5)	20 (of 25) = 80%
	RECOMMENDATIONS FOR CURRENT AND FUTURE DEVELOPMENT(S): **The storage technology for large capacity but variable electrical energy associated with solar and wind energy technologies (each having unique unreliability patterns) is not a limiting factor on solar and/or wind energy harvesting technologies.**		Total A Sum of RI 100	Total B Sum of XX 86 (of 100)	Total C XX x RI (of RI^2) 830 (of 902)
			WEIGHTED SCORE =		**92%**

TS-17 Energy Storage - Stationary/Large Size (Thermal)

TRADE STUDY SCORE WORKSHEET (Applicable to United States of America Region)

(Ratings and scores based on factual information in pertinent databases in Public Domain)

#	TRADE STUDY: TS-18 Use/Development of Nuclear Energy - Uranium				
	FACTOR CONSIDERED (Being rated for Importance and primary qualitative magnitude)	ELEMENTS CONSIDERED IN RATING FACTOR (Attributes and detriments if/as applicable)	RELATIVE IMPORTANCE RI (% or n/a)	POINTS SCORED (Rated)	WEIGHTED LINE SCORE (WS) (RI x Rated Score)
	(Minor interactions with/between Elements/Factors are not directly in rating)		RI	XX (of RI)	XX x RI (of RI^2)
1	Status of Technology Development	Nuclear energy technology is well-developed, and many nuclear facilities using depleted U-235 exist	12	12 (of 12)	144 (of 144) = 100%
2	Reliability / Sufficiency of Energy	Reliability is 100%; sufficient to meet present demands	12	12 (of 12)	144 (of 144) = 100%
3	Resource(s) Exist Domestically or Approx % Needed to be Imported	Numerous facilities have been constructed but need additional to meet needs of growing population	10	10 (of 10)	100 (of 100) = 100%
4	Cost of Resource(s) (Domestic + Import if any)	Some cost to upgrade numerous existing nuclear energy facilities - higest cost is constructing new ones. In US, use of depleted uranium to be used	10	9 (of 10)	90 (of 100) = 90 %
5	Importance to Industry of Region	Reliable, clean abundant energy vital in large ndustrial and highly populated areas.	9	9 (of 9)	81 (of 81) = 100 %
6	Importance and Urgency to Gov't. Economy and Employment	One of best technologies to be increased in use proportional to population growth.	9	9 (of 9)	81 (of 81) = 100 %
7	By-Products and Emissions	No detrimental emission; exhausted uranium needs special disposal. Heated water should be used in conjunction with other technologies harvesting heat.	8	6 (of 8)	48 (of 64) = 75 %
8	Effect on Human Population	None if safety provisions are correctly installed and maintained. Previous incidents were avoidable.	8	6 (of 8)	48 (of 64) = 75 %
9	Effect on Regional Infrastructure / Transportation/Commerce/Trade	No significant effect. High power transmission lines need safe installation from natural adverse events.	7	7 (of 7)	49 (of 49) = 100 %
10	Effect on Environment / Weather / Climate (Frequency & Severity)	No effect known.	5	5 (of 5)	25 (of 25) = 100 %
11	Effect on Animal Life (Wild or Livestock) - Land, Aquatic, Avian	No effect known.	5	5 (of 5)	25 (of 25) = 100 %
12	Effect on Forestry and Agriculture	No effect known.	5	5 (of 5)	25 (of 25) = 100 %
	RECOMMENDATIONS FOR CURRENT AND FUTURE DEVELOPMENT(S):		Total A Sum of RI 100	Total B Sum of XX 95 (of 100)	Total C XX x RI (of RI^2) 860 (of 902)
			WEIGHTED SCORE =		**95 %**

The reduction of "greenhouse gas" emissions to reduce/eliminate the adverse effects of climate change will not be achieved with the unrealistic present goal set by the UN IPCC. The use of nuclear energy technology is a near ideal fuel for the most energy per unit mass, and for avoiding any of the emissions that are heat-trapping or pollutants that are toxic or carcinogenic.

TS-18 Use/Development of Nuclear Energy - Uranium

TRADE STUDY SCORE WORKSHEET (Applicable to United States of America Region)

(Ratings and scores based on factual information in pertinent databases in Public Domain)

TRADE STUDY: TS-19 Use/Development of Nuclear Energy - Thorium

#	FACTOR CONSIDERED (Being rated for Importance and primary qualitative magnitude)	ELEMENTS CONSIDERED IN RATING FACTOR (Attributes and detriments if/as applicable)	RELATIVE IMPORTANCE RI (% or n/a)	POINTS SCORED (Rated)	WEIGHTED LINE SCORE (WS) (RI x Rated Score)
	(Minor interactions with/between Elements/Factors are not directly in rating)		RI	XX (of RI)	XX x RI (of RI^2)
1	Status of Technology Development	Nuclear energy technology is well-developed, and a few nuclear facilities using thorium already exist	12	12 (of 12)	144 (of 144) = 100%
2	Reliability / Sufficiency of Energy	Reliability is 100%; sufficient to meet present demands	12	12 (of 12)	144 (of 144) = 100%
3	Resource(s) Exist Domestically or Approx % Needed to be Imported	A few facilities have been constructed but need many more to meet needs of growing population	10	9 (of 10)	90 (of 100) = 90%
4	Cost of Resource(s) (Domestic + Import if any)	Some cost to upgrade the few existing nuclear energy facilities - highest cost is constructing new ones. Using thorium vs uranium is desirable.	10	10 (of 10)	100 (of 100) = 100 %
5	Importance to Industry of Region	Reliable, clean abundant energy vital in large ndustrial and highly populated areas.	9	9 (of 9)	81 (of 81) = 100 %
6	Importance and Urgency to Gov't. Economy and Employment	One of best technologies to be increased in use proportional to population growth.	9	9 (of 9)	81 (of 81) = 100 %
7	By-Products and Emissions	No detrimental emission; exhausted thorium needs less special disposal. Heated water could be used in conjunction with other technologies harvesting heat.	8	7 (of 8)	56 (of 64) = 88 %
8	Effect on Human Population	None if safety provisions are correctly installed and maintained. Previous incidents were avoidable.	8	6 (of 8)	48 (of 64) = 75 %
9	Effect on Regional Infrastructure / Transportation/Commerce/Trade	No significant effect. High power transmission lines need safe installation from natural adverse events.	7	7 (of 7)	49 (of 49) = 100 %
10	Effect on Environment / Weather / Climate (Frequency & Severity)	No effect known.	5	5 (of 5)	25 (of 25) = 100 %
11	Effect on Animal Life (Wild or Livestock) - Land, Aquatic, Avian	No effect known.	5	5 (of 5)	25 (of 25) = 100 %
12	Effect on Forestry and Agriculture	No effect known.	5	5 (of 5)	25 (of 25) = 100 %
			Total A Sum of RI 100	Total B Sum of XX 96 (of 100)	Total C XX x RI (of RI^2) 868 (of 902)
			WEIGHTED SCORE =		96 %

RECOMMENDATIONS FOR CURRENT AND FUTURE DEVELOPMENT(S):

The reduction of "greenhouse gas" emissions to reduce/eliminate the adverse effects of climate change will not be achieved with the unrealistic present goal set by the UN IPCC. The use of nuclear technology with thorium is an ideal fuel for the most energy per unit mass, and for avoiding any of the emissions that are heat-trapping or pollutants that are toxic or carcinogenic.

TS-19 Use/Development of Nuclear Energy - Thorium

TRADE STUDY SCORE WORKSHEET (Applicable to United States of America Region)

(Ratings and scores based on factual information in pertinent databases in Public Domain)

#	TRADE STUDY: TS-20 Use of All-Electric Motor Vehicles - Commercial / Government				
	FACTOR CONSIDERED (Being rated for Importance and primary qualitative magnitude)	**ELEMENTS CONSIDERED IN RATING FACTOR (Attributes and detriments if/as applicable)**	**RELATIVE IMPORTANCE RI (% or n/a)**	**POINTS SCORED (Rated)**	**WEIGHTED LINE SCORE (WS) (RI x Rated Score)**
	(Minor interactions with/between Elements/Factors are not directly in rating)		**RI**	**XX (of RI)**	**XX x RI (of RI^2)**
1	Status of Technology Development	The all-electric motor vehicles technology is in an emerging status without a database to support it.	12	9 (of 12)	108 (of 144) = 75%
2	Reliability / Sufficiency of Energy	The electric power source for the All-Electric Vehicles is rechargeable electric storage batteries (lithium ion)	12	9 (of 12)	108 (of 144) = 75%
3	Resource(s) Exist Domestically or Approx % imported	The resources for the motor vehicle are all a domestic resource. The critical material in the batteries - lithium-ion must largely be imported (from Nigeria and Russia).	10	8 (of 10)	80 (of 100) = 80%
4	Cost of Resource(s) (Domestic + Import if any)	The cost of large, all-electric vehicles is high but becoming lower as the technology matures.	10	8 (of 10)	80 (of 100) = 80%
5	Importance to Industry of Region	Applies only to the transportation factors in industry	9	7 (of 9)	63 (of 81) = 78%
6	Importance and Urgency to Gov't. Economy and Employment	Important if All-Electric Vehicles are used as the primary vehicles in Government motor pools at all levels.	9	9 (of 9)	81 (of 81) = 100%
7	By-Products, Emissions, and Pollutants	The main purpose of EV technology is to eliminate emissions and pollutants into the environment.	8	8 (of 8)	64 (of 64) = 100%
8	Effect on Human Population	Lesser effect on the human respiratory system but induces physical hazards from the battery technology.	8	6 (of 8)	48 (of 64) = 100%
9	Effect on Regional Infrastructure / Transportation / Commercial Trade	**Heavy Electric Vehicles are damaging the thinly paved surfaces of roads causing increased road maintenance. Lithium ion batteries cause fires. Using less gasoline will increase gasoline prices.**	7	3 (of 7)	21 (of 49) = 43%
10	Effect on Environment / Weather / Climate (Frequency & Severity)	No significant effect on the adverse weather events, climate change, or on ocean levels rising.	5	5 (of 5)	25 (of 25) = 100%
11	Effect on Animal Life (Wild or Livestock) - Land, Aquatic, Avian	No significant effect on the adverse weather events, climate change, or on ocean levels rising.	5	5 (of 5)	25 (of 25) = 100%
12	Effect on Forestry and/or Agriculture	No significant effect on the adverse weather events, climate change, or on ocean levels rising.	5	5 (of 5)	25 (of 25) = 100%
	RECOMMENDATIONS FOR CURRENT AND FUTURE DEVELOPMENT(S): **The strong recommendation is made that EVs with lithium ion batteries be used only by government and public transportation vehicles and short-haul commercial vehicles until the database of pros and cons is made known to mature the technologies involved.**		Total A Sum of RI 100	Total B Sum of XX 82 (of 100)	Total C XX x RI (of RI^2) 728 (of 902)
			WEIGHTED SCORE =		**81%**

TS-20 Use of All-Electric Motor Vehicles – Commercial / Government

TRADE STUDY SCORE WORKSHEET (Applicable to United States of America Region)

(Ratings and scores based on factual information in pertinent databases in Public Domain)

TRADE STUDY: TS-21 Use of Electric Plug-In / Fuel Hybrid Vehicles - Commercial / Public

#	FACTOR CONSIDERED (Being rated for Importance and primary qualitative magnitude)	ELEMENTS CONSIDERED IN RATING FACTOR (Attributes and detriments if/as applicable)	RELATIVE IMPORTANCE RI (% or n/a)	POINTS SCORED (Rated)	WEIGHTED LINE SCORE (WS) (RI x Rated Score)
	(Minor interactions with/between Elements/Factors are not directly in rating)		RI	XX (of RI)	XX x RI (of RI^2)
1	Status of Technology Development	The all-electric motor vehicles technology is in an emerging status without a database to support it.	12	10 (of 12)	120 (of 144) = 83%
2	Reliability / Sufficiency of Energy	The electric power sources for the Electric Plug-In/Hybrid Vehicles are rechargeable lithium-ion batteries	12	9 (of 12)	108 (of 144) = 75%
3	Resource(s) Exist Domestically or Approx % imported	The resources for the motor vehicle are all a domestic resource. The critical material in the batteries - lithium-ion must largely be imported (from Nigeria and Russia).	10	8 (of 10)	80 (of 100) = 80%
4	Cost of Resource(s) (Domestic + Import if any)	The cost of large, Electric Plug-In / Hybrid vehicles is high but becoming lower as the technology matures.	10	8 (of 10)	80 (of 100) = 80%
5	Importance to Industry of Region	Applies only to the transportation factors in industry	9	7 (of 9)	63 (of 81) = 78%
6	Importance and Urgency to Gov't. Economy and Employment	Important if Electric Plug-In/Hybrid vehicles are used in Government and commercial motor pools at all levels.	9	9 (of 9)	81 (of 81) = 100%
7	By-Products, Emissions, and Pollutants	The main purpose of EV and EV Plug-In/Hybrid technology is to reduce emissions into the environment.	8	8 (of 8)	64 (of 64) = 100%
8	Effect on Human Population	Lesser effect on the human respiratory system but induces physical hazards from the battery technology.	8	8 (of 8)	48 (of 64) = 100%
9	Effect on Regional Infrastructure / Transportation / Commercial Trade	**Heavy Electric Vehicles are damaging the thinly paved surfaces of roads causing increased road maintenance. Lithium ion batteries cause fires. Using less gasoline will increase gasoline prices.**	7	3 (of 7)	21 (of 49) = 43%
10	Effect on Environment / Weather / Climate (Frequency & Severity)	No significant effect on the adverse weather events, climate change, or on ocean levels rising.	5	5 (of 5)	25 (of 25) = 100%
11	Effect on Animal Life (Wild or Livestock) - Land, Aquatic, Avian	No significant effect on the adverse weather events, climate change, or on ocean levels rising.	5	5 (of 5)	25 (of 25) = 100%
12	Effect on Forestry and/or Agriculture	No significant effect on the adverse weather events, climate change, or on ocean levels rising.	5	5 (of 5)	25 (of 25) = 100%
	RECOMMENDATIONS FOR CURRENT AND FUTURE DEVELOPMENT(S): **The strong recommendation is that a high mix of All-Electric Vehicles and Electric Plug-In/Hybrid Fuel vehicles be developed to allow the maturity of the All-Electric vehicles but retain the longer range capability of Electric Plug-In/Hybrid Fuel vehicles for public use.**		Total A Sum of RI 100	Total B Sum of XX 85 (of 100)	Total C XX x RI (of RI^2) 740 (of 902)
			WEIGHTED SCORE =		**82%**

TS-21 Use of Electric Plug-In / Fuel Hybrid Vehicles – Commercial / Public

TRADE STUDY SCORE WORKSHEET (Applicable to United States of America Region)

(Ratings and scores based on factual information in pertinent databases in Public Domain)

	TRADE STUDY: TS-22a Energy Storage - Mobile Small (Electrical) - Lithium-Ion				
	FACTOR CONSIDERED (Being rated for Importance and primary qualitative magnitude)	**ELEMENTS CONSIDERED IN RATING FACTOR** (Attributes and detriments if/as applicable)	**RELATIVE IMPORTANCE** RI (% or n/a)	**POINTS SCORED** (Rated)	**WEIGHTED LINE SCORE (WS)** (RI x Rated Score)
	(Minor interactions with/between Elements/Factors are not directly in rating)		RI	XX (of RI)	XX x RI (of RI^2)
1	Status of Technology Development	The electrical storage technology in mobile vehicles is in crisis. The use of lithium ion battery technology in large vehicles is a major safety & technology error.	12	6 (of 12)	72 (of 144) = 50%
2	Reliability / Sufficiency of Energy	The explositivity of lithium ion batteries is causing thousands of injuries to humans, is causing airplane crashes, and causes container ships to be sunk.	12	5 (of 12)	60 (of 144) = 42%
3	Resource(s) Exist Domestically or % imported	Most components are domestic; Lithium in lithium ion batteries is imported from Nigeria and Russia.	10	5 (of 10)	50 (of 100) = 50%
4	Cost of Resource(s) (Domestic + Import if any)	The cost of imported lithium and the associated cost of human injury combine to be excessively costly.	10	7 (of 10)	70 (of 100) = 70%
5	Importance to Industry of Region	Using lithium ion batteries to transport and/or distribute humans and industrial products should be eliminated	9	4 (of 9)	36 (of 81) = 44%
6	Importance and Urgency to Gov't. Economy / Employment	Not a factor affecting economy. Is a "crisis" affecting human life and property losses globally in developed countries.	9	4 (of 9)	36 (of 81) = 44%
7	By-Products, Emissions, and Pollutants	Lithium ion batteries cause extensive injuries and property losses	8	4 (of 8)	32 (of 64) = 50%
8	Effect on Human Population	Is at crisis level to human safety in wide use	8	3 (of 8)	24 (of 64) = 38%
9	Effect on Commerce / Infrastructure/Transportation	Shipment of lithium ion batteries is banned on aircraft and on some shipping carriers for safety reasons.	7	2 (of 7)	14 (of 49) = 29%
10	Effect on Environment / Weather / Climate	No effect on weather or climate.	5	5 (of 5)	25 (of 25) = 100%
11	Effect on Animal Life (Wild or Livestock) - Land, Water, Avian	No effect on animal life of any type.	5	5 (of 5)	25 (of 25) = 100%
12	Effect on Forestry and/or Agriculture	No effect on forestry or agriculture	5	5 (of 5)	25 (of 25) = 100%
	RECOMMENDATIONS FOR CURRENT AND FUTURE DEVELOPMENT(S): **The environmental scientists may be content with the use of lithium ion battery use in electric vehicles and the elimination of fossil fuels to protect the environment but the human and property loss aspects of lithium ion technology should ban its use.**		Total A Sum of RI 100	Total B Sum of XX 55 (of 100)	Total C XX x RI (of RI^2) 469 (of 902)
			WEIGHTED SCORE =		**52%**

TS-22a Energy Storage - Mobile/Small Size (Electrical); Lithium-ion

TRADE STUDY SCORE WORKSHEET (Applicable to United States of America Region)

(Ratings and scores based on factual information in pertinent databases in Public Domain)

	TRADE STUDY: TS-22b Energy Storage - Mobile Small (Electrical) Aluminum-ion or Sodium-ion				
	FACTOR CONSIDERED (Being rated for Importance and primary qualitative magnitude)	**ELEMENTS CONSIDERED IN RATING FACTOR (Attributes and detriments if/as applicable)**	**RELATIVE IMPORTANCE RI (% or n/a)**	**POINTS SCORED (Rated)**	**WEIGHTED LINE SCORE (WS) (RI x Rated Score)**
	(Minor interactions with/between Elements/Factors are not directly in rating)		**RI**	**XX (of RI)**	**XX x RI (of RI²)**
1	Status of Technology Development	The use of safer aluminum-ion or sodium-ion batteries replacing lithium-ion batteries in electric vehicles is an essential need for safer use by general public	12	10 (of 12)	120 (of 144) = 83%
2	Reliability / Sufficiency of Energy	Far safer and more reliable aluminum-ion and sodium-ion technology in batteries for EVs in USA	12	12 (of 12)	144 (of 144) = 100%
3	Resource(s) Exist Domestically or % imported	Aluminum and sodium are in abundance in the geophysical resources of the USA. No importing.	10	10 (of 10)	100 (of 100) = 100%
4	Cost of Resource(s) (Domestic + Import if any)	Aluminum and sodium are low cost materials to extract, produce, and apply domestically in USA	10	10 (of 10)	100 (of 100) = 100%
5	Importance to Industry of Region	There is extreme urgency to create full-rate production capability of aluminum-ion or sodium-ion batteries to fulfill production and retrofit of all EVs	9	9 (of 9)	81 (of 81) = 100%
6	Importance and Urgency to Gov't. Economy / Employment	Aluminum-ion and sodium-ion battery technology in millions of electric vehicles produced is essential to public safety, to the economy, and to employment	9	9 (of 9)	81 (of 81) = 100%
7	By-Products, Emissions, and Pollutants	There is no known toxic or carcinogenic by-product, emission, or pollutant in production or in operation of aluminum-ion or sodium-ion batteries for EVs	8	8 (of 8)	64 (of 64) = 100%
8	Effect on Human Population	Far safer and quicker recharge capability than lithium-ion. Technology is resistant to highway FOD hazards	8	8 (of 8)	64 (of 64) = 100%
9	Effect on Commerce / Infrastructure/Transportation	Aluminum-ion and sodium-ion battery technologies are near ideal forms of battery power in transportation safety	7	7 (of 7)	49 (of 49) = 100%
10	Effect on Environment / Climate	No adverse effect on environment or climate. Disposal of lithium-ion batteries has a non-polluting option.	5	5 (of 5)	25 (of 25) = 100%
11	Effect on Animal Life	No adverse effect on animal life of any type	5	5 (of 5)	25 (of 25) = 100%
12	Effect on Forestry/Agriculture	No adverse effect on forestry or agriculture	5	5 (of 5)	25 (of 25) = 100%
	RECOMMENDATIONS FOR CURRENT AND FUTURE DEVELOPMENT(S): **There is great urgency to accelerate production of aluminum-ion and/or sodium-ion batteries for public safety and use in recent and new produced electric vehicles currently in operation or planned to meet transportation needs of the US population.**		Total A Sum of RI 100	Total B Sum of XX 99 (of 100)	Total C XX x RI (of RI²) 890 (of 902)
			WEIGHTED SCORE =		**97%**

TS-22b Energy Storage - Mobile/Small Size (Electrical); Aluminum- or Sodium-ion

TRADE STUDY SCORE WORKSHEET (Applicable to United States of America Region)

(Ratings and scores based on factual information in pertinent databases in Public Domain)

TRADE STUDY: TS-23 Use/Development of Desalination - Reverse Osmosis

#	FACTOR CONSIDERED (Being rated for Importance and primary qualitative magnitude)	ELEMENTS CONSIDERED IN RATING FACTOR (Attributes and detriments if/as applicable)	RELATIVE IMPORTANCE RI (% or n/a)	POINTS SCORED (Rated)	WEIGHTED LINE SCORE (WS) (RI x Rated Score)
	(Minor interactions with/between Elements/Factors are not directly in rating)		RI	XX (of RI)	XX x RI (of RI^2)
1	Status of Technology Development	There are numerous desalination facilities performing well in many countries. The technology is proven.	12	12 (of 12)	144 (of 144) = 100%
2	Reliability / Sufficiency of Energy	The desalination process is reliable but the membrane used for reverse osmosis limits efficiency	12	10 (of 12)	120 (of 144) = 83%
3	Resource(s) Exist Domestically or Approx % imported	All resources for the reverse osmosis form of desalination are abundantly available in the USA.	10	10 (of 10)	100 (of 100) = 100%
4	Cost of Resource(s) (Domestic + Import if any)	Cost to construct the desalination facility is typical of other industries, but cost of saline water used in process is negligible. Membrane servicing cost high.	10	8 (of 10)	80 (of 100) = 80%
5	Importance to Industry of Region	Of low importance to industry of region. Highly important to the human population.	9	7 (of 9)	63 (of 81) = 78%
6	Importance and Urgency to Gov't. Economy and Employment	Of low importance to economy and employment. Good augmentation of water for agriculture (human foods).	9	8 (of 9)	72 (of 81) = 89%
7	By-Products, Emissions, and Pollutants	No harmful by-products, emissions, or pollutants generated by by process.	8	8 (of 8)	64 (of 64) = 100%
8	Effect on Human Population	Desalinated water is generated for human benefit.	8	8 (of 8)	64 (of 64) = 100%
9	Effect on Regional Infrastructure /Transportation/Commerce/Trade	Process has no impact on infrastructure, transportation, commerce, or trade.	7	7 (of 7)	49 (of 49) = 100%
10	Effect on Environment / Weather / Climate (Frequency & Severity)	Process has small effect on local environment and no effect on weather or climate.	5	5 (of 5)	25 (of 25) = 100%
11	Effect on Animal Life (Wild or Livestock) - Land, Aquatic, Avian	No hazard to local bird population; or to land animals. Disposal of residual saline water back into ocean may affect local marine life.	5	4 (of 5)	20 (of 25) = 80%
12	Effect on Forestry and/or Agriculture	No effect on forestry. Desalinated water can be used to support agricultural food of large US population.	5	5 (of 5)	25 (of 25) = 100%
			Total A Sum of RI 100	Total B Sum of XX 92 (of 100)	Total C XX x RI (of RI^2) 826 (of 902)
			WEIGHTED SCORE =		92%

RECOMMENDATIONS FOR CURRENT AND FUTURE DEVELOPMENT(S):

The desalination of saline water (in oceans, gulfs, seas) is important to the coastal regions of the United States and is a vital technology for the arid countries of the Mid-East. The reverse osmosis process is well-proven, but does require the use of a membrane and which lowers the overall efficiency of the process while increasing its cost for frequent maintenance of that membrane. The other form of desalination - by a distillation process - avoids the detriment of the use of the high-maintenance membrane.

TS-23 Use/Development of Desalination - Reverse Osmosis

TRADE STUDY SCORE WORKSHEET (Applicable to United States of America Region)

(Ratings and scores based on factual information in pertinent databases in Public Domain)

#	TRADE STUDY: TS-24 Use/Development of Desalination - Reverse Osmosis				
	FACTOR CONSIDERED (Being rated for Importance and primary qualitative magnitude)	ELEMENTS CONSIDERED IN RATING FACTOR (Attributes and detriments if/as applicable)	RELATIVE IMPORTANCE RI (% or n/a)	POINTS SCORED (Rated)	WEIGHTED LINE SCORE (WS) (RI x Rated Score)
	(Minor interactions with/between Elements/Factors are not directly in rating)		RI	XX (of RI)	XX x RI (of RI^2)
1	Status of Technology Development	There are numerous desalination facilities performing well in many countries. The technology is proven.	12	12 (of 12)	144 (of 144) = 100%
2	Reliability / Sufficiency of Energy	The desalination process is reliable and the distillation process achieves high reliability with low maintenance.	12	12 (of 12)	144 (of 144) = 100%
3	Resource(s) Exist Domestically or Approx % imported	All resources for the reverse osmosis form of desalination are abundantly available in the USA.	10	10 (of 10)	100 (of 100) = 100%
4	Cost of Resource(s) (Domestic + Import if any)	Cost to construct the desalination facility is typical of other industries, but cost of saline water used in process is negligible. Low maintenance costs.	10	9 (of 10)	90 (of 100) = 90%
5	Importance to Industry of Region	Of low importance to industry of region. Highly important to the human population water and agricultural foods.	9	7 (of 9)	63 (of 81) = 78%
6	Importance and Urgency to Gov't. Economy and Employment	Of low importance to economy and employment. Good augmentation of water for agriculture (human foods).	9	8 (of 9)	72 (of 81) = 89%
7	By-Products, Emissions, and Pollutants	No harmful by-products, emissions, or pollutants generated by the distillation process.	8	8 (of 8)	64 (of 64) = 100%
8	Effect on Human Population	Desalinated water is generated for human benefit.	8	8 (of 8)	64 (of 64) = 100%
9	Effect on Regional Infrastructure /Transportation/Commerce/Trade	Process has no impact on infrastructure, transportation, commerce, or trade.	7	7 (of 7)	49 (of 49) = 100%
10	Effect on Environment / Weather / Climate (Frequency & Severity)	Process has small effect on local environment and no effect on weather or climate.	5	5 (of 5)	25 (of 25) = 100%
11	Effect on Animal Life (Wild or Livestock) - Land, Aquatic, Avian	No hazard to local bird population or land animals. Disposal of residual salt and/or saline water into the environment or into ocean may affect local ecology.	5	4 (of 5)	20 (of 25) = 80%
12	Effect on Forestry and/or Agriculture	No effect on forestry. Desalinated water can be used to support agricultural food of growing population.	5	5 (of 5)	25 (of 25) = 100%
	RECOMMENDATIONS FOR CURRENT AND FUTURE DEVELOPMENT(S): **The desalination of saline water (in oceans, gulfs, seas) is important to the coastal regions of the United States and is a vital technology for the arid countries of the Mid-East. The distillation process is well-proven, and does not require the use of a membrane. The desalination by distillation avoids the detriment of the use and cost of the high-maintenance reverse osmosis membrane.**		Total A Sum of RI 100	Total B Sum of XX 95 (of 100)	Total C XX x RI (of RI^2) 860 (of 902)
			WEIGHTED SCORE =		**95%**

TS-24 Use/Development of Desalination – Distillation (Solar)

Group	Trade Study No.	Technology Area	Total Wghtd Score	Comment (Suggested Use and Development for USA Region)
①	TS-4	Use of Fossil Fuel - Natural Gas (Liquified)	95%	Increase useage; improve transportability
	TS-3	Use of Fossil Fuel - Natural Gas (Pipeline)	93%	Install more pipelines (distribution to ex-coal users)
	TS-5	Use of Fossil Fuel Derivative - Propane	90%	Increase useage; improve storage safety
	TS-2	Use of Fossil Fuel - Oil	89%	Continue but reduce use - to no imported oil
	TS-1	Use of Fossil Fuel - Coal	74%	Phase out coal burning technology (to reduce toxic and carcinogenic land/water pollution) Close deep mining.
②	TS-6	Use of Non-fossil Fuel - Hydrogen	90%	Increase useage; improve storage & distribution safety
	TS-7	Use of Non-fossil Fuel - Ethanol	86%	Continue useage; co-technology with EV Hybrids
③	TS-9	Control of Hydrodynamic Events (Coastal and Seasonal Event Regions) - Levees / Sea Walls	98%	Highest need but with lack of activity/planning to meet safety and energy needs of growing human population
	TS-8	Use of Hydrodynamic Energy (Dams, River Flow)	96%	Increase existing capacity and additional construction
	TS-14	Energy Storage - Stationary Large (Electrical)	94%	Good technology supplementing that of solar and wind.
	TS-15	Use of Geothermal Energy (Surface)	93%	Good technology for the western states in Conus;
	TS-17	Energy Storage - Stationary Large (Thermal)	92%	Most important adjunct technology with geothermal
	TS-11	Use of Solar Energy (Residential / Rural / Town)	89%	Of most value to small users. Requires battery storage.
	TS-16	Use of Geothermal Energy (Deep-Drilled; Closed Cycle and Open Cycle with Fracking)	83%	Increase useage; Closed cycle preferred to Open Cycle; Recycle abandoned oil wells where feasible
	TS-12	Use of Wind Energy (On Land)	82%	More useful for small industrial, farming, and public use.
	TS-10	Use of Solar Energy (Large Industrial / Commercial)	78%	Not reliable for industries, OK for small; co-technology with other uses (distillation desalination; water purification)
	TS-13	Use of Wind Energy (Off-Shore)	63%	Too harsh environment (corrosion, wave force, storm damage). Recommend not installing off-shore system.
④	TS-19	Use of Nuclear Energy (Thorium)	96%	Among best technologies for mitigating climate change.
	TS-18	Use of Nuclear Energy (Uranium)	95%	Among best technologies for mitigating climate change.
⑤	TS-21	Use of Electric Plug-In/Fuel Hybrid Vehicles (Commercial / Public)	82%	Increase mix of All-Electric vehicles with Electric Plug-In / Hybrid Fuel vehicles with longer range for public.
	TS-20	Use of All-Electric Vehicles (Commercial / Govt.)	81%	Technology maturing. Limit initial use until more mature.
	TS-22b	Energy Storage - Mobile Small (Aluminum-ion or Sodium-ion)	97%	Aluminum-ion or sodium-ion battery technology is safer than lithium-ion in electric vehicles
	TS-22a	Energy Storage - Mobile Small (Lithium-ion)	52%	The use of lithium ion battery technology is unsafe. Replace lithium-ion with aluminum-ion or sodium-ion.
⑥	TS-24	Desalination - Distillation (Solar/Geothermal)	95%	A vitally needed process for potable water to population
	TS-23	Desalination - Reverse Osmosis (Solar/Geotherm	92%	A vitally needed process for potable water to population

Summary of Trade Studies (Prioritized within Technology Group)

About the Author

Herbert A. Hutchinson graduated from the nationally renowned Brooklyn Technical High School (Arista) in New York, and appears as one of a hundred Notables in the BTHS "100 Years of Excellence" who graduated in the first fifty years of Brooklyn Tech. He received a Bachelor of Science degree in Aeronautical Engineering at Georgia Tech in Atlanta, Georgia, in 1953. After serving as a Combat Engineers platoon leader in the U.S. Army Corps of Engineers in the Korean War, he returned to Georgia Tech to earn a Master of Science in Aeronautical Engineering degree (1959).

The initial nine years (1952-1961) of his professional engineering career were as a wind tunnel test engineer (Lockheed C-130) at the Georgia Tech Research Institute, as an aerodynamicist (F-84, F-105B and XF-103) at Republic Aviation in New York, and as a flight test performance engineer (B-58) at General Dynamics in Texas.

The middle half of his career (1961 – 1982) was as the Chief of the Aeromechanics Branch of the Air Force Flight Dynamics Laboratory, and as the Chief System Engineer (GS-15) for several U.S. Air Force weapons systems developments, including the U.S. Air Force Lightweight Fighter program and the U.S. Air Force F-16 Fighting Falcon Multi-National Weapon System, developed by the U.S. Air Force at Wright-Patterson Air Force Base in Dayton, Ohio. He was on the team conducting exploitation testing of foreign flight system technologies at the formerly very highly classified clandestine flight test site now known as Area 51. With extensive applied system engineering experience, he was selected to serve on three Presidential Advisory Committees, participated in numerous aircraft crash investigations (with numerous fatalities), and served three years as Chief of the USAF Flight Technologies Division (Aeronautical Systems Division).

Based on his having been the Chief System Engineer for the USAF YF-16 Lightweight Fighter Prototype Program and the ensuing F-16 Fighting Falcon multinational fighter for the USAF and twenty-eight other countries, former pilots of the advanced technology F-16, used in the several Gulf Wars such as Desert Storm, have awarded him the tactical call name "Stuffer". He has authored two books: "The Inside History of the USAF Lightweight Fighters, 1900 to 1975" and "The F-16 Fighting Falcon Multinational Weapon System, 1972 to 2019", both published by the Xlibris Publishing Company, and which reflect his understanding of the complexities and interactions that exist in applying emerging technologies in "big picture" development of large, complex systems.

The final twelve years (1982 – 1994) of Hutchinson's career were as a Manager of System Engineering at the Northrop Aircraft Division in California managing the system engineering of a new fighter aircraft design for NATO forces in Europe and of the design of stealthy flight vehicles (the Tri-Service Standoff Attack Missile (TSSAM) and the YF-23 Advanced Tactical Fighter (YF-23). He retired from his 41 years professional career at the end of 1994.

After retirement, he applied his system analytical and database research experiences to becoming a world-recognized genealogist and created the extensive family tree histories for his family, of English and British West Indian descendancy (fourteen generations 1595 to 1995), and that of his wife's family, of European descendancy (thirteen generations 1620 to 2000).

In collaboration with the Barbados Museum and Historical Society, he capitalized on the fact his two Barbadian grandfathers were part of the labor force in creating the Panama Canal (in this book as an example of a highly successful early "green" international environmentally-friendly project) to support the Barbados Government and UNESCO in naming the Panama Canal Zone as a Historical World Heritage Site. He invented "An Optronic System for the Measurement of Vehicle Traffic" and holds US Patent 6,750,787, measuring vehicles speeding on neighborhood streets to improve pedestrian safety in neighborhoods. His extensive experience in genealogy and in patenting an invention for improving human safety provided a wealth of knowledge about the human factors involved in everyday living in different regions and migratory habits of humans in different climates.

He is the author of this book: "Dealing with the Real Climate Change", and also the contemporary book "Facts and Fallacies of Climate Change on our Planet", both written as a public service and intended to inform the public worldwide that inevitable climate change is happening but at different rates and with different effects and different state of urgency to be mitigated depending on the latitude of the region on the planet. In his books on global warming and climate change, Hutchinson provides data and a system-level, "big picture" analysis concluding that there is no "one-size-fits-all" solution or goal for mitigating the environmental and human problems brought about by global climate changes occurring in different regions of the planet. His fourth book assesses a set of twenty-five energy and transportation technologies being developed to mitigate climate change effects.

Hutchinson is an Associate Fellow Emeritus (seventy-four years continuous membership) in the American Institute of Aeronautics and Astronautics (AIAA), is a life member of the Veterans of Foreign Wars, is a member of the Air Force Association, and is a member of Mensa. His name is on the Smithsonian Institute Wall of Honor. He and his wife Arline reside in Durham, North Carolina. Herb continues to be active in the aeronautical engineering profession, passing along his extensive experiences to any group (educational, industrial, or Governmental) pursuing the application of new technologies. A visual indicator of his dedication and passion for describing STEM activities applied in the "real world" is shown as follows:

(From Author's Memorabilia – at Lockheed Martin, Greenville, SC)

Contact the Author:

The Author of this book "Dealing with the Real Climate Change", and the prior Xlibris books: : "Facts and Fallacies of Climate Change on our Planet", "The Inside History of the USAF Lightweight Fighters" and "The F-16 Fighting Falcon Multinational Weapon System", may be contacted with questions or be provided constructive criticism comments at: herbineer@aol.com

Fact Sheet

Based on an independent, unbiased analysis of valid databases of scientific information conducted "for the benefit of humanity", the facts and fallacies about global warming and climate change are summarized in this Fact Sheet as follows:

1) There were no discernible anthropogenic greenhouse gas effects found in the several scientific databases when assessed in a professionally ethical "big picture" system engineering analysis approach by the unbiased Author. Environmental scientists disregarding the complex factors causing climate change and "cherry-picking" the total scientific databases and, in collusion with the sensationalistic news media, have intentionally misinformed and alarmed the public to believe there is an imminent catastrophic environmental crisis.

2) The surface temperatures of the planet (land and water) are rising but are found to be caused by combined astrophysical, geophysical, and natural factors beyond human control, NOT by "anthropogenic greenhouse gas" emissions. Mathematical calculations by Dr. Milutin Milankovic confirm the cyclical frequency at which geophysical parametric variables appear in meteorological databases.

3) The combined heat-trapping natural (geophysical) greenhouse gas emissions (measured in Vostok Antarctica pre-Industrial Revolution ice core sample analysis to be 295 ppmv parts per million by volume (0.0295% concentration) and water vapor (at a global average of 1.4%) total to approximately 1.43% average concentration of natural heat-trapping gases in Earth's atmosphere affecting different regions to different extent depending on their latitude, and not in a "one-size-fits-all" average global manner.

4) Concentrated analysis of the Vostok ice core sample database (by ice research scientist Dr. Nicolas Caillon of France) found that it is temperature that drives natural carbon dioxide in the atmosphere and not the converse. The lag of carbon dioxide peak behind the temperature peak (and with the absence of human influence) was determined to be approximately 800 years.

5) The amount of anthropogenic (human and industrial generated) greenhouse gases measured by NOAA at Mauna Loa are currently (April 28, 2025) at 135 ppmv (430 ppmv total – 295 ppmv natural) or 0.0135% concentration in the atmosphere. The small concentration of anthropogenic emissions is an insignificant factor to global warming or climate change.

6) The ratio of natural (atmospheric chemistry + water vapor) greenhouse gas concentration to that of anthropogenic (human and industrial) greenhouse gas concentration is 1.43% to 0.0135% or 106 to 1 (as shown in climate scientist Dr. S. Fred Singer' "A Closer Look at the Numbers" analysis). Environmental scientists disregard the effect of water vapor which contradicts their fallacious hypothesis that anthropogenic greenhouse gases dominate the global warming effect on climate change.

7) The statistical approach used by environmental scientists of "averaging" and "linearizing" meteorological data masks the factual seasonal temperature extremes of hot and cold at regional topographical elevations to which all life forms in a latitudinal region (not merely globally) must adapt. A better method (considering latitude, elevation, and ocean proximity effects) for estimating average annual temperature was developed using World Meteorological Organization database statistics. (LEO)

8) The technology of coal burning in the USA must be reduced because coal ash residue contains toxins and carcinogens which, when disposed in landfills factually pollute underground aquafers endangering life (human, animate and agricultural). Combustion of fossil fuels – oil, petroleum fuels, and natural gas - generates no emissions affecting regional or global surface temperatures, climate, or ocean levels.

It is scientifically unrealistic – not "naysaying" - to expect that reducing anthropogenic greenhouse gas emissions by eliminating combustion of all fossil fuels (coal, oil, and natural gas) will have a significant effect on retarding or avoiding global warming, climate changes, or ocean levels rising. Technologies for mitigating the effects of the inevitable surface temperature rises, climate changes, and ocean surface levels rising and the urgent need to better control anthropogenic pollution (especially plastic materials and leakages/spillages of oil and natural gas) must be escalated to a high priority – national and international - as the projected growing population will exacerbate human safety, health, sanitation, and/or environmental pollution hazards.

www.ingramcontent.com/pod-product-compliance
Lightning Source LLC
LaVergne TN
LVHW082022200926
848481LV00010B/2903
* 9 7 9 8 8 9 3 9 5 1 0 1 1 *